PROGRESS IN BRAIN RESEARCH

VOLUME 171

USING EYE MOVEMENTS AS AN EXPERIMENTAL PROBE OF BRAIN FUNCTION

A SYMPOSIUM IN HONOR OF JEAN BÜTTNER-ENNEVER

Other volumes in PROGRESS IN BRAIN RESEARCH

Volume 135: Do Seizures Damage the Brain, by A. Pitkänen and T. Sutula (Eds.) – 2002, ISBN 0-444-50814-7.
Volume 136: Changing Views of Cajal's Neuron, by E.C. Azmitia, J. DeFelipe, E.G. Jones, P. Rakic and C.E. Ribak (Eds.) – 2002, ISBN 0-444-50815-5.
Volume 137: Spinal Cord Trauma: Regeneration, Neural Repair and Functional Recovery, by L. McKerracher, G. Doucet and S. Rossignol (Eds.) – 2002, ISBN 0-444-50817-1.
Volume 138: Plasticity in the Adult Brain: From Genes to Neurotherapy, by M.A. Hofman, G.J. Boer, A.J.G.D. Holtmaat, E.J.W. Van Someren, J. Verhaagen and D.F. Swaab (Eds.) – 2002, ISBN 0-444-50981-X.
Volume 139: Vasopressin and Oxytocin: From Genes to Clinical Applications, by D. Poulain, S. Oliet and D. Theodosis (Eds.) – 2002, ISBN 0-444-50982-8.
Volume 140: The Brain's Eye, by J. Hyönä, D.P. Munoz, W. Heide and R. Radach (Eds.) – 2002, ISBN 0-444-51097-4.
Volume 141: Gonadotropin-Releasing Hormone: Molecules and Receptors, by I.S. Parhar (Ed.) – 2002, ISBN 0-444-50979-8.
Volume 142: Neural Control of Space Coding, and Action Production, by C. Prablanc, D. Pélisson and Y. Rossetti (Eds.) – 2003, ISBN 0-444-509771.
Volume 143: Brain Mechanisms for the Integration of Posture and Movement, by S. Mori, D.G. Stuart and M. Wiesendanger (Eds.) – 2004, ISBN 0-444-513892.
Volume 144: The Roots of Visual Awareness, by C.A. Heywood, A.D. Milner and C. Blakemore (Eds.) – 2004, ISBN 0-444-50978-X.
Volume 145: Acetylcholine in the Cerebral Cortex, by L. Descarries, K. Krnjević and M. Steriade (Eds.) – 2004, ISBN 0-444-51125-3.
Volume 146: NGF and Related Molecules in Health and Disease, by L. Aloe and L. Calzà (Eds.) – 2004, ISBN 0-444-51472-4.
Volume 147: Development, Dynamics and Pathology of Neuronal Networks: From Molecules to Functional Circuits, by J. Van Pelt, M. Kamermans, C.N. Levelt, A. Van Ooyen, G.J.A. Ramakers and P.R. Roelfsema (Eds.) – 2005, ISBN 0-444-51663-8.
Volume 148: Creating Coordination in the Cerebellum, by C.I. De Zeeuw and F. Cicirata (Eds.) – 2005, ISBN 0-444-51754-5.
Volume 149: Cortical Function: A View from the Thalamus, by V.A. Casagrande, R.W. Guillery and S.M. Sherman (Eds.) – 2005, ISBN 0-444-51679-4.
Volume 150: The Boundaries of Consciousness: Neurobiology and Neuropathology, by Steven Laureys (Ed.) – 2005, ISBN 0-444-51851-7.
Volume 151: Neuroanatomy of the Oculomotor System, by J.A. Büttner-Ennever (Ed.) – 2006, ISBN 0-444-51696-4.
Volume 152: Autonomic Dysfunction after Spinal Cord Injury, by L.C. Weaver and C. Polosa (Eds.) – 2006, ISBN 0-444-51925-4.
Volume 153: Hypothalamic Integration of Energy Metabolism, by A. Kalsbeek, E. Fliers, M.A. Hofman, D.F. Swaab, E.J.W. Van Someren and R.M. Buijs (Eds.) – 2006, ISBN 978-0-444-52261-0.
Volume 154: Visual Perception, Part 1, Fundamentals of Vision: Low and Mid-Level Processes in Perception, by S. Martinez-Conde, S.L. Macknik, L.M. Martinez, J.M. Alonso and P.U. Tse (Eds.) – 2006, ISBN 978-0-444-52966-4.
Volume 155: Visual Perception, Part 2, Fundamentals of Awareness, Multi-Sensory Integration and High-Order Perception, by S. Martinez-Conde, S.L. Macknik, L.M. Martinez, J.M. Alonso and P.U. Tse (Eds.) – 2006, ISBN 978-0-444-51927-6.
Volume 156: Understanding Emotions, by S. Anders, G. Ende, M. Junghofer, J. Kissler and D. Wildgruber (Eds.) – 2006, ISBN 978-0-444-52182-8.
Volume 157: Reprogramming of the Brain, by A.R. Møller (Ed.) – 2006, ISBN 978-0-444-51602-2.
Volume 158: Functional Genomics and Proteomics in the Clinical Neurosciences, by S.E. Hemby and S. Bahn (Eds.) – 2006, ISBN 978-0-444-51853-8.
Volume 159: Event-Related Dynamics of Brain Oscillations, by C. Neuper and W. Klimesch (Eds.) – 2006, ISBN 978-0-444-52183-5.
Volume 160: GABA and the Basal Ganglia: From Molecules to Systems, by J.M. Tepper, E.D. Abercrombie and J.P. Bolam (Eds.) – 2007, ISBN 978-0-444-52184-2.
Volume 161: Neurotrauma: New Insights into Pathology and Treatment, by J.T. Weber and A.I.R. Maas (Eds.) – 2007, ISBN 978-0-444-53017-2.
Volume 162: Neurobiology of Hyperthermia, by H.S. Sharma (Ed.) – 2007, ISBN 978-0-444-51926-9.
Volume 163: The Dentate Gyrus: A Comprehensive Guide to Structure, Function, and Clinical Implications, by H.E. Scharfman (Ed.) – 2007, ISBN 978-0-444-53015-8.
Volume 164: From Action to Cognition, by C. von Hofsten and K. Rosander (Eds.) – 2007, ISBN 978-0-444-53016-5.
Volume 165: Computational Neuroscience: Theoretical Insights into Brain Function, by P. Cisek, T. Drew and J.F. Kalaska (Eds.) – 2007, ISBN 978-0-444-52823-0.
Volume 166: Tinnitus: Pathophysiology and Treatment, by B. Langguth, G. Hajak, T. Kleinjung, A. Cacace and A.R. Møller (Eds.) – 2007, ISBN 978-0-444-53167-4.
Volume 167: Stress Hormones and Post Traumatic Stress Disorder: Basic Studies and Clinical Perspectives, by E.R. de Kloet, M.S. Oitzl and E. Vermetten (Eds.) – 2007, ISBN 978-0-444-53140-7.
Volume 168: Models of Brain and Mind: Physical, Computational and Psychological Approaches, by R. Banerjee and B.K. Chakrabarti (Eds.) – 2008, ISBN 978-0-444-53050-9.
Volume 169: Essence of Memory, by W.S. Sossin, J.-C. Lacaille, V.F. Castellucci and S. Belleville (Eds.) – 2008, ISBN 978-0-444-53164-3.
Volume 170: Advances in Vasopressin and Oxytocin — From Genes to Behaviour to Disease, by I.D. Neumann and R. Landgraf (Eds.) – 2008, ISBN 978-0-444-53201-5.

PROGRESS IN BRAIN RESEARCH

VOLUME 171

USING EYE MOVEMENTS AS AN EXPERIMENTAL PROBE OF BRAIN FUNCTION

A SYMPOSIUM IN HONOR OF JEAN BÜTTNER-ENNEVER

EDITED BY

CHRISTOPHER KENNARD
Division of Neurosciences and Mental Health, Imperial College London, London SW7 2AZ, UK

R. JOHN LEIGH
Neurology Service, Veterans Affairs Medical Center, and Case Medical Center, 11100 Euclid Avenue, Cleveland, OH 44106-5040, USA

ELSEVIER

AMSTERDAM – BOSTON – HEIDELBERG – LONDON – NEW YORK – OXFORD
PARIS – SAN DIEGO – SAN FRANCISCO – SINGAPORE – SYDNEY – TOKYO

Elsevier
Radarweg 29, PO Box 211, 1000 AE Amsterdam, The Netherlands
Linacre House, Jordan Hill, Oxford OX2 8DP, UK

First edition, 2008

British Library Cataloguing in Publication Data
A catalog record for this book is available from the British Library

Library of Congress Cataloging-in-Publication Data
A catalogue record for this book is available from the Library of Congress

ISBN: 978-0-444-53163-6
ISSN: 0079-6123

Printed and bound in Hungary

08 09 10 11 12 10 9 8 7 6 5 4 3 2 1

List of contributors

R. Abadi, Faculty of Life Sciences, University of Manchester, Manchester, UK
T. Akao, Department of Physiology, Hokkaido University School of Medicine, Sapporo, Japan
O.E. Akman, Centre for Systems Biology, The University of Edinburgh, Kings Buildings, Mayfield Road, Edinburgh, UK
D. Anastasopoulos, Department of Physiology, School of Nursing, University of Athens, Goudi, Athens, Greece
Y. Arai, Department of Otorhinolaryngology, Tokyo Women's Medical College, Tokyo, Japan
I. Arend, Wolfson Center for Clinical and Cognitive Neuroscience, Bangor University, Bangor, UK
T. Asaka, Department of Health Sciences, Hokkaido University School of Medicine, Sapporo, Japan
S.T. Aw, Department of Neurology, Royal Prince Alfred Hospital, Camperdown, Sydney, NSW, Australia
L. Bachus, Department of Neurobiology and Anatomy, University of Rochester School of Medicine and Dentistry, Rochester, NY, USA
G.R. Barnes, Faculty of Life Sciences, University of Manchester, Manchester, UK
P. Bays, Institute of Neurology and Institute of Cognitive Neuroscience, UCL, London, UK
G. Beltrami, Department of Biomedical Engineering, University of Pavia, Pavia, Italy
R.A. Berman, Laboratory of Sensorimotor Research, National Eye Institute, National Institutes of Health, Bethesda, MD, USA
G. Bertolini, Dipartimento di Informatica e Sistemistica, Università di Pavia, Italy
R.A. Black, Department of Neurology, Royal Prince Alfred Hospital, Camperdown, Sydney, NSW, Australia
C.J. Bockisch, Department of Neurology, Department of ENT, Department of Ophthalmology, Zurich University Hospital, Zurich, Switzerland
T. Brandt, Department of Neurology, University of Munich, Klinikum Grosshadern, Munich, Germany
A.M. Bronstein, Department of Clinical Neuroscience, Charing Cross Hospital, Imperial College London, London, UK; Academic Department of Neuro-otology, Charing Cross Hospital, Imperial College London, London, UK
D.S. Broomhead, Department of Mathematics, University of Manchester, Manchester, UK
J.A. Büttner-Ennever, Institute of Anatomy, Ludwig-Maximilian University, Munich, Germany
U. Büttner, Department of Neurology and Bernstein Center for Computational Neuroscience, Ludwig-Maximilians-University, Munich, Germany; Yerkes National Primate Research Center and Department of Neurology, Emory University, Atlanta, GA, USA
A. Caspi, Gonda Multidisciplinary Brain Research Center, Bar Ilan University, Ramat Gan, Israel
A.L. Cecala, Department of Neurobiology and Anatomy, University of Rochester, Rochester, NY, USA
R. Chang, University of Michigan School of Literature, Science and the Arts, Ann Arbor, MI, USA
R.A. Clark, Jules Stein Eye Institute, David Geffen Medical School at University of California, Los Angeles, CA, USA
A.H. Clarke, Vestibular Research Laboratory, ENT Department, Campus Benjamin Franklin, Charité Universitaetsmedizin, Berlin, Germany
R.A. Clement, Visual Sciences Unit, Institute of Child Health, University College London, London, UK
B. Cohen, Department of Neurology, Mount Sinai School of Medicine, New York, USA
C.J.S. Collins, Faculty of Life Sciences, University of Manchester, Manchester, UK

S. Colnaghi, Neuro-otology and Neuro-ophthalmology Laboratory, IRCCS "Neurological Institute C. Mondino" Foundation; Department of Neurology, University of Pavia, Pavia, Italy
A. Cortese, Neuro-otology and Neuro-ophthalmology Laboratory, IRCCS "Neurological Institute C. Mondino" Foundation; Department of Neurology, University of Pavia, Pavia, Italy; Department of Neurology, UKE, University of Hamburg, Germany
V. Cosi, Neuro-otology and Neuro-ophthalmology Laboratory, IRCCS "Neurological Institute C. Mondino" Foundation; Department of Neurology, University of Pavia, Pavia, Italy
B.T. Crane, Department of Otolaryngology/Head-Neck Surgery, The Johns Hopkins University, Baltimore, MD, USA
T.B. Crapse, Department of Neuroscience and the Center for the Neural Basis of Cognition, University of Pittsburgh, Pittsburgh, PA, USA
Q.N. Cui, Department of Neurobiology and Anatomy, University of Rochester School of Medicine and Dentistry, Rochester, NY, USA
K.E. Cullen, Department of Physiology, McGill University, McIntyre Medical Sciences Building, Montréal, Québec, Canada
I.S. Curthoys, Department of Neurology, Royal Prince Alfred Hospital, Camperdown, Sydney, NSW, Australia; Vestibular Research Laboratory, School of Psychology, University of Sydney, Sydney, NSW, Australia
M. Dai, Department of Neurology, Mount Sinai School of Medicine, New York, NY, USA
P. Dean, Department of Psychology, University of Sheffield, Sheffield, UK
L.F. Dell'Osso, Daroff-Dell'Osso Ocular Motility Laboratory, Louis Stokes Cleveland Department of Veterans Affairs Medical Center and Case Medical School; Department of Neurology, Case Western Reserve University and University Hospitals Case Medical Center; Department of Biomedical Engineering, Case Western Reserve University and University Hospitals Case Medical Center, Cleveland, OH, USA
J.L. Demer, Jules Stein Eye Institute; Department of Neurology; Neuroscience Interdepartmental Program; Bioengineering Interdepartmental Program, David Geffen Medical School at University of California, Los Angeles, CA, USA
A. Deutschländer, Department of Neurology, University of Munich, Klinikum Grosshadern, Munich, Germany
C. Distler, Allgemeine Zoologie and Neurobiologie, Ruhr-Universität Bochum, Bochum, Germany
A.H. Duc, Institute of Neurology and Institute of Cognitive Neuroscience, UCL, London, UK
M.R. Dürsteler, Department of Neurology, University Hospital Zurich, Zurich, Switzerland
T. Eggert, Department of Neurology, Klinikum Grosshadern, Ludwig-Maximilians University, Munich, Germany
J.T. Erichsen, School of Optometry and Vision Sciences, Cardiff University, Cardiff, Wales, UK
J.N. Eron, Department of Neurology, Mount Sinai School of Medicine, New York, NY, USA
K. Fackelmann, Institute of Anatomy, Ludwig-Maximilian University, Munich, Germany
E.J. FitzGibbon, Laboratory of Sensorimotor Research, National Eye Institute, National Institutes of Health, Bethesda, MD, USA
M. Fouladvand, Division of Neurology and Department of Ophthalmology and Vision Sciences, University Health Network, University of Toronto, Toronto, Ontario, Canada
E.G. Freedman, Department of Neurobiology and Anatomy, University of Rochester, Rochester, NY, USA
P. Fu, University of Michigan School of Literature, Science and the Arts, Ann Arbor, MI, USA
A. Fuchs, Department of Physiology and Biophysics and Washington National Primate Research Center, University of Washington, Seattle, WA, USA
J. Fukushima, Department of Health Sciences; Department of Physiology, Hokkaido University School of Medicine, Sapporo, Japan

K. Fukushima, Department of Physiology, Hokkaido University School of Medicine, Sapporo, Japan
S. Geiner, Institute of Anatomy, Ludwig-Maximilian University, Munich, Germany
St. Glasauer, Department of Neurology and Bernstein Center for Computational Neuroscience, Ludwig-Maximilians-University, Munich, Germany
M.E. Goldberg, Mahoney Center for Brain and Behavior, Department of Neuroscience, Columbia University College of Physicians and Surgeons, and the New York State Psychiatric Institute, New York, NY, USA; Departments of Neuroscience, Psychiatry and Ophthalmology, Columbia University College of Physicians and Surgeons, New York, NY, USA
C.V.P. Golding, Department of Clinical Neurosciences, Imperial College, London, UK
C.R. Gordon, Department of Neurology, Meir Medical Center, Kfar Saba, Israel; Sackler Faculty of Medicine, Tel Aviv University, Tel Aviv, Israel
M.A. Gresty, Department of Clinical Neuroscience, Charing Cross Hospital, Imperial College London, London, UK
G.M. Halmagyi, Department of Neurology, Royal Prince Alfred Hospital, Camperdown, Sydney, NSW, Australia; Department of Medicine, University of Sydney, Sydney, NSW, Australia
M. Hartmann, Perception and Eye Movement Laboratory, Departments of Neurology and Clinical Research, Inselspital, Bern University Hospital, University of Bern, Bern, Switzerland
C. Helmchen, Department of Neurology, University of Lübeck, Lübeck, Germany
B.J.M. Hess, Department of Neurology, University Hospital Zurich, Zurich, Switzerland; Vestibulo-Oculomotor Laboratory, Zürich University Hospital, Zürich, Switzerland
C.W. Hess, Perception and Eye Movement Laboratory, Departments of Neurology and Clinical Research, Inselspital, Bern University Hospital, University of Bern, Bern, Switzerland
S.L. Hicks, Department of Clinical Neurosciences, Imperial College, London, UK
O. Hikosaka, Laboratory of Sensorimotor Research, National Eye Institute, National Institutes of Health, Bethesda, MD, USA
T.L. Hodgson, Department of Psychology, Exeter University, UK
K.-P. Hoffmann, Allgemeine Zoologie and Neurobiologie, Ruhr-Universität Bochum, Bochum, Germany
M.A. Hollands, School of Exercise and Sport Sciences, University of Birmingham, Birmingham, UK
S. Hong, Laboratory of Sensorimotor Research, National Eye Institute, NIH, DHHS, Bethesda, MD, USA
A.K.E. Horn, Institute of Anatomy, Ludwig-Maximilians University, Munich, Germany
K. Hüfner, Department of Neurology, University of Munich, Klinikum Grosshadern, Munich, Germany
M. Husain, Department of Clinical Neuroscience, Imperial College London, and Institute of Cognitive Neuroscience, UCL, London, UK
M. Isoda, Laboratory for Symbolic Cognitive Development, RIKEN Brain Science Institute, Wako, Saitamo 351–0198, Japan
Y. Izawa, Department of Systems Neurophysiology, Graduate School of Medicine, Tokyo Medical and Dental University, Yushima, Bunkyo-ku, Tokyo, Japan
K. Jahn, Department of Neurology, University of Munich, Klinikum Grosshadern, Munich, Germany
J.C. Jen, Department of Neurology, UCLA School of Medicine, Los Angeles, CA, USA
A. Joshi, Department of Biomedical Engineering, Daroff-Dell'Osso Laboratory, Veterans Affairs Medical Center and University Hospitals, Case Western Reserve University, Cleveland, OH, USA
R. Kalla, Department of Neurology, University of Munich, Klinikum Grosshadern, Munich, Germany
S. Karim, Neuroscience Interdepartmental Program, David Geffen Medical School at University of California, Los Angeles, CA, USA
E.L. Keller, Smith-Kettlewell Eye Research Institute, San Francisco, CA, USA
C. Kennard, Department of Clinical Neuroscience, Imperial College London, London, UK
W.M. King, Department of Otolaryngology and Kresge Hearing Research Institute, Ann Arbor, MI, USA

E. Knoth, Department of Neurobiology and Anatomy, University of Rochester School of Medicine and Dentistry, Rochester, NY, USA
P.C. Knox, Division of Orthoptics, University of Liverpool, Thompson Yates Building, Liverpool, UK
Y. Kojima, Department of Physiology and Biophysics and Washington National Primate Research Center, University of Washington, Seattle, WA, USA
O. Kremmyda, Department of Neurology, Klinikum Grosshadern, Ludwig-Maximilians University, Munich, Germany
M. Kunin, Department of Computer and Information Science, Brooklyn College of the City University of New York, Brooklyn, NY, USA
S. Kurkin, Department of Physiology, Hokkaido University School of Medicine, Sapporo, Japan
A.G. Lasker, The Johns Hopkins University, School of Medicine, Baltimore, MD, USA
J. Laurens, Vestibulo-Oculomotor Laboratory, Zürich University Hospital, Zürich, Switzerland
B.-T. Lee, Smith-Kettlewell Eye Research Institute, San Francisco, CA, USA; Institute of Psychological Science, Seoul National University, Seoul, Korea
K.-M. Lee, Smith-Kettlewell Eye Research Institute, San Francisco, CA, USA; Department of Neurology, Seoul National University, Seoul, Korea
N. Lehnen, Department of Neurology, Bernstein Center for Computational Neuroscience, Ludwig-Maximilian-University, Munich
R.J. Leigh, Department of Neurology; Department of Biomedical Engineering, Daroff-Dell'Osso Laboratory, Veterans Affairs Medical Center and University Hospitals, Case Western Reserve University, Cleveland, OH, USA
R. Levite, Department of Neurology, Meir Medical Center, Kfar Saba, Israel
K. Liao, Daroff-Dell'Osso Laboratory, Veterans Affairs Medical Center and Case Western Reserve University, Department of Neurology, University Hospitals, Cleveland, OH, USA
L. Machado, University of Otago, Dunedin, New Zealand
S.K. Mannan, Department of Clinical Neuroscience, Imperial College, London, Charing Cross Campus, London, UK
S. Marti, Department of Neurology, University Hospital Zurich, Zurich, Switzerland
M. Matsumoto, Laboratory of Sensorimotor Research, National Eye Institute, National Institutes of Health, Bethesda, MD, USA
P.J. May, Departments of Anatomy, Ophthalmology and Neurology, University of Mississippi Medical Center, Jackson, MS, USA
L.A. McGarvie, Department of Neurology, Royal Prince Alfred Hospital, Camperdown, Sydney, NSW, Australia
M. McGrath, Bangor University, Wolfson Center for Clinical and Cognitive Neuroscience, Bangor, UK
A. Messoudi, Institute of Anatomy, Ludwig-Maximilian University, Munich, Germany
F.A. Miles, Laboratory of Sensorimotor Research, National Eye Institute, National Institutes of Health, Bethesda, MD, USA
R.M. Müri, Perception and Eye Movement Laboratory, Departments of Neurology and Clinical Research, Inselspital, Bern University Hospital, and University of Bern, Bern, Switzerland
M.J. Mustari, Department of Neurology and Bernstein Center for Computational Neuroscience, Ludwig-Maximilians-University, Munich, Germany; Division of Sensory-Motor Systems, Yerkes National Primate Research Center and Department of Neurology, Emory University, Atlanta, GA, USA
P. Nachev, Department of Clinical Neuroscience, Imperial College London, and Institute of Cognitive Neuroscience, UCL, London, UK
A. Narasimhan, Bioengineering Interdepartmental Program, David Geffen Medical School at University of California, Los Angeles, CA, USA
T. Nitta, Department of Physiology; Department of Ophthalmology, Hokkaido University School of Medicine, Sapporo, Japan

A. Nouriani, Institute of Anatomy, Ludwig-Maximilian University of Munich, Munich, Germany
U. Nuding, Department of Neurology and Bernstein Center for Computational Neuroscience, Ludwig-Maximilians-University, Munich, Germany; Yerkes National Primate Research Center and Department of Neurology, Emory University, Atlanta, GA, USA
T. Nyffeler, Perception and Eye Movement Laboratory, Departments of Neurology and Clinical Research, Inselspital, Bern University Hospital, and University of Bern, Bern, Switzerland
K. O'Beirne, University of Michigan School of Literature, Science, Ann Arbor, MI, USA
W.E. O'Neill, Department of Neurobiology and Anatomy, University of Rochester School of Medicine and Dentistry, Rochester, NY, USA; Center for Navigation and Communication Sciences, University of Rochester School of Medicine and Dentistry, Rochester, NY, USA
D.A. Ogorodnokov, Department of Neurology, Mount Sinai School of Medicine, New York, USA
I. Olasagasti, Department of Neurology, Zurich University Hospital, Zurich, Switzerland
S. Ono, Department of Neurology and Bernstein Center for Computational Neuroscience, Ludwig-Maximilians-University, Munich, Germany; Division of Sensory-Motor Systems, Yerkes National Primate Research Center and Department of Neurology, Emory University, Atlanta, GA, USA
B.S. Oommen, Department of Neurology, Case Western Reserve University School of Medicine, OH, USA; Dell'Osso-Daroff Ocular Motility Laboratory, Veterans Affairs Medical Center, Cleveland, OH, USA
L.M. Optican, Laboratory of Sensorimotor Research, National Eye Institute, NIH, DHHS, Bethesda, MD, USA
G.D. Paige, Department of Neurobiology and Anatomy; Center for Navigation and Communication Sciences; Center for Visual Science, University of Rochester School of Medicine and Dentistry, Rochester, NY, USA
A. Palla, Department of Neurology, University Hospital Zurich, Zurich, Switzerland
C. Pierrot-Deseilligny, Service de Neurologie 1, Hôpital de la Salpêtrière, Paris, France
J. Porrill, Department of Psychology, University of Sheffield, Sheffield, UK
S. Pratap, Department of Neurology, Royal Prince Alfred Hospital, Camperdown, Sydney, NSW, Australia
C. Quaia, Laboratory of Sensorimotor Research, National Eye, Institute, Bethesda, MD, USA
R.D. Rafal, Wolfson Center for Clinical and Cognitive Neuroscience, Bangor University, Bangor, UK
S. Ramat, Dipartimento di Informatica e Sistemistica, Università degli Studi di Pavia, Italy
H.A. Rambold, Department of Neurology, University of Lübeck, Lübeck, Germany; Laboratory of Sensorimotor Research, National Eye Institute, National Institutes of Health, Bethesda, MD, USA
T. Raphan, Department of Neurology, Mount Sinai School of Medicine, New York, NY, USA; Department of Computer and Information Science, Brooklyn College of the City University of New York, Brooklyn, NY, USA
M. Reschke, Neurosciences Laboratories, Johnson Space Center, Houston, TX, USA
V. Rizzo, Sobell Department of Motor Neuroscience and Movement Disorders, Institute of Neurology, UCL, London, UK
T. Ro, Rice University, Huston, TX, USA
M.P.A. Robert, Department of Clinical Neuroscience, Imperial College, London, UK; Institute of Neurology and National Hospital for Neurology and Neurosurgery, London, UK
C.R. Rosenthal, Department of Clinical Neuroscience Imperial College London, Charing Cross Campus, London, UK
C. Rössert, Department of Neurology, Ludwig-Maximilian University, Munich, Germany
J.C. Rothwell, Sobell Department of Motor Neuroscience and Movement Disorders, Institute of Neurology, UCL, London, UK
J.C. Rucker, Rush Medical Center, Chicago, IL, USA
H. Sakai, Department of Anatomy, University of Nagoya, Nagoya, Japan
T. Sander, Department of Neurology, University of Lübeck, Lübeck, Germany

C. Scallan, University of Manchester, Faculty of Life Sciences, Manchester, UK
B.M. Seemungal, Department of Clinical Neuroscience, Charing Cross Hospital, Imperial College London, London, UK
A. Serra, Daroff-Dell'Osso Laboratory, Veterans Affairs Medical Center and Case Western Reserve University, Cleveland, OH, USA; Case Western Reserve University, Cleveland, OH, USA; Institute of Clinical Neurology, University of Sassari, Sassari, Italy; OASI Institute for Research and Care on Mental Retardation and Brain Aging, Troina, Italy
A.G. Shaikh, Department of Neurology, The Johns Hopkins University, Baltimore MD, USA
X. Shan, Department of Neurology, The Johns Hopkins Hospital, Baltimore, MD, USA
J.A. Sharpe, Division of Neurology and Department of Ophthalmology and Vision Sciences, University Health Network, University of Toronto, Toronto, Ontario, Canada
B.M. Sheliga, Laboratory of Sensorimotor Research, National Eye Institute, National Institutes of Health, Bethesda, MD, USA
N. Shichinohe, Department of Physiology, Hokkaido University School of Medicine, Sapporo, Japan
Y. Shinoda, Department of Systems Neurophysiology, Graduate School of Medicine, Tokyo Medical and Dental University, Yushima, Bunkyo-ku, Tokyo, Japan
N. Sinha, The Physiological Laboratory, University of Cambridge, Cambridge, England
S. Sklavos, Department of Physiology, School of Nursing, University of Athens, Goudi, Athens, Greece
R. Soetedjo, Department of Physiology and Biophysics and Washington National Primate Research Center, University of Washington, Seattle, WA, USA
D. Solomon, The Johns Hopkins University, Baltimore, MD, USA
M.A. Sommer, Department of Neuroscience and the Center for the Neural Basis of Cognition, University of Pittsburgh, Pittsburgh, PA, USA
A. Sprenger, Department of Neurology, University of Lübeck, Lübeck, Germany
J.S. Stahl, Department of Neurology, Case Western Reserve University School of Medicine, OH, USA; Dell'Osso-Daroff Ocular Motility Laboratory, Veterans Affairs Medical Center, Cleveland, OH, USA
T. Stephan, Department of Neurology, University of Munich, Klinikum Grosshadern, Munich, Germany
D. Straumann, Department of Neurology, University Hospital Zurich, Switzerland; Vestibulo-Oculomotor Laboratory, Zürich University Hospital, Zürich, Switzerland
M. Strupp, Department of Neurology, University of Munich, Klinikum Grosshadern, Munich, Germany
Y. Sugiuchi, Department of Systems Neurophysiology, Graduate School of Medicine, Tokyo Medical and Dental University, Yushima, Bunkyo-ku, Tokyo, Japan
W. Sun, Department of Ophthalmology, Northwest Hospital and Medical Center, Seattle, WA, USA
S.J. Tabrizi, Institute of Neurology and National Hospital for Neurology and Neurosurgery, London, UK
M. Takahashi, Department of Systems Neurophysiology, Graduate School of Medicine, Tokyo Medical and Dental University, Yushima, Bunkyo-ku, Tokyo, Japan
R.J. Tamargo, Department of Neurology, The Johns Hopkins Hospital, Baltimore, MD, USA; Department of Neurosurgery, The Johns Hopkins University, Baltimore, MD, USA
X.-f. Tang, Institute of Anatomy, Ludwig-Maximilian University, Munich, Germany
A.A. Tarnutzer, Department of Neurology, University Hospital Zurich, Switzerland; The Johns Hopkins University, School of Medicine, Baltimore, MD, USA
M. Tatalias, Neurology Department, Zurich University Hospital, Zurich, Switzerland
J. Tateishi, Department of Neuropathology, Kyushu University, Fukuoka, Japan
T. Theodorou, Hallamshire Hospital, Sheffield, UK
M.J. Thurtell, Department of Neurology, Royal Prince Alfred Hospital, Camperdown, Sydney, NSW, Australia; Department of Medicine, University of Sydney, Sydney, NSW, Australia; Department of Neurology, University Hospitals Case Medical Center, Cleveland, OH, USA
J. Tian, Department of Neurology, The Johns Hopkins Hospital, Baltimore, MD, USA

J.-R. Tian, Jules Stein Eye Institute, David Geffen Medical School at University of California, Los Angeles, CA, USA
C. Tilikete, Unité de Neuro-ophtalmologie, Hôpital Neurologique, Hospices Civils de Lyon, Université de Lyon-I, Lyon, France
M.J. Todd, Department of Neurology, Royal Prince Alfred Hospital, Camperdown, Sydney, NSW, Australia
T. Uemura, Department of Otorhinolaryngology, Kyushu University, Fukuoka, Japan
M. Van Horn, Department of Physiology, McGill University, McIntyre Medical Sciences Building, Montréal, Québec, Canada
M. Versino, Neuro-otology and Neuro-ophthalmology Laboratory, IRCCS "Neurological Institute C. Mondino" Foundation; Department of Neurology, University of Pavia, Pavia, Italy
K.C. Vitorello, Division of Sensory-Motor Systems, Yerkes National Primate Research Center, Emory University, Atlanta, GA, USA
N.H. Wadia, Department of Neurology, Jaslok Hospital and Research Centre, Mumbai, India
J. Wagner, Department of Neurology, University of Munich, Klinikum Grosshadern, Munich, Germany
D.M. Waitzman, Department of Neurology, University of Connecticut Health Center, Farmington, CT, USA
M.F. Walker, Department of Neurology, The Johns Hopkins University, Baltimore, MD, USA; Department of Ophthalmology, The Johns Hopkins University, Baltimore, MD, USA; Department of Neurology, Daroff-Dell'Osso Laboratory, Veterans Affairs Medical Center and University Hospitals, Case Western Reserve University, Cleveland, OH, USA
X. Wang, Mahoney Center for Brain and Behavior, Department of Neuroscience, Columbia University College of Physicians and Surgeons, and the New York State Psychiatric Institute, New York, NY, USA
Z.I. Wang, Daroff-Dell'Osso Ocular Motility Laboratory, Louis Stokes Cleveland Department of Veterans Affairs Medical Center and Case Medical School, Cleveland, OH, USA; Department of Biomedical Engineering, Case Western Reserve University and University Hospitals Case Medical Center; Cleveland, OH, USA
R. Ward, Wolfson Center for Clinical and Cognitive Neuroscience, Bangor University, Bangor, UK
K.P. Weber, Department of Neurology, Zurich University Hospital, Zurich, Switzerland; Department of Neurology, Royal Prince Alfred Hospital, Camperdown, Sydney, NSW, Australia
A.M.F. Wong, Division of Neurology and Department of Ophthalmology and Vision Sciences, University Health Network, University of Toronto, Toronto, Ontario, Canada
R.H. Wurtz, Laboratory of Sensorimotor Research, National Eye Institute, National Institutes of Health, Bethesda, MD, USA
S.B. Yakushin, Department of Neurology, Mount Sinai School of Medicine, New York, NY, USA
H.S. Ying, Department of Neurology, The Johns Hopkins Hospital, Baltimore, MD, USA; Department of Ophthalmology, The Johns Hopkins University, Baltimore, MD, USA
S.H. Ying, The Johns Hopkins University School of Medicine, Baltimore, MD, USA
N. Zaher, Tishreen University, Lattakia, Syria
W.H. Zangemeister, Department of Neurology, UKE, University of Hamburg, Hamburg, Germany
D.S. Zee, Department of Neurology, The Johns Hopkins University, Baltimore, MD, USA; Department of Ophthalmology, The Johns Hopkins University, Baltimore, MD, USA
M. Zhang, Mahoney Center for Brain and Behavior, Department of Neuroscience, Columbia University College of Physicians and Surgeons, and the New York State Psychiatric Institute, New York, NY, USA
N. Ziavra, Academic Department of Neuro-otology, Charing Cross Hospital, Imperial College London, London, UK
A.Z. Zivotofsky, Gonda Multidisciplinary Brain Research Center, Bar Ilan University, Ramat Gan, Israel

Foreword

This volume of *Progress in Brain Research* is the fruit of a conference *Using Eye Movements as an Experimental Probe of Brain Function*, held at the Charing Cross Hospital Campus of Imperial College London, UK on 5th–6th December, 2007. About a hundred basic and clinical scientists contributed to this conference to honour Professor Jean Büttner-Ennever. Jean earned her Ph.D. at University College, London and pursued the itinerant career of a scientist, in America, Switzerland, and Germany, most recently Munich. As the "Principal Anatomist" of the ocular motor system, she has made a number of seminal contributions to the field. Additionally, Jean has collaborated with research laboratories around the world, and promoted the career of young investigators; many of these individuals have contributed to the conference and to this volume. She is widely respected for her many attributes that are summed up in the following poem, *Jean Büttner-Ennever, A Life in Verse*, composed for this celebration by Dr. Michael E. Goldberg:

As oculomotor scientists we're all aware that whenever
We think about anatomy we think Jean Büttner-Ennever.
And if you do a Medline search your work will surely yield
Full evidence that Jean must be the mother of our field.
Good scientists deserve respect; moms have a rank above
Which is why we look at Jean with both respect and love.
Jean was born in Braintree. It's a quiet Essex place,
I looked for fame historical but could not find a trace.
No battles there nor burnings, evil sheriffs, Robin Hoods
Just some sturdy yeomen who wove silk and woollen goods.
Around the fourteenth century the Flemings wandered in.
Is Ennever related to the Flemish word for gin?
Jean grew up in Liverpool, there first heard science call —
If not there might have been a band — George, Ringo, *Jean*, and Paul
Her PhD's from UCL, a single authored paper
On strychnine, rats, how far it takes the EEG to taper.
She also had an abstract about Olaf Lippold's hustle,
That human alpha rhythm comes from extraocular muscle.
And next, job hunting. In the best of Carnaby Street styles,
In bright white boots and miniskirt she interviewed Fred Miles.
But Sussex, charming though it is, clearly lost the battle
Jean went off to Arnie Towe in far away Seattle.
From rats to cats, an upward step. There was nothing like 'er
Insulating spinal tracts by inserting mica.
Four postdoc desks were crowded in a rather tiny room
Seattle's rainy, but a German chased away the gloom.

Although her work with Arnie Towe was not exactly blinding
It is clear that Ulrich Büttner was her major finding.
And then to Göttingen. Ulrich tells a tale to make one weep.
So many trodes, the cat would die or Jean would fall asleep.
And then to Zurich. No more physiology. Don't knock it.
For there she found her metier, Volker Henn and Konrad Akert.
Oculomotor anatomy. The brainstem in its glory.
Omnipausers, motor neurons, each a wondrous story.
Significance illuminated like a master chef
Of the rostral interstitial nucleus of the MLF
A brief sojourn in Dusseldorf and then appointment great
To Ludwig Maximilians Universität.
It isn't just anatomy that makes Jean Alice bonny.
In Geisenbrunn she raised three children Alex, Tom, and Johnny.
Ran Ahornstrasse 14 whose guestbook I love to sign
And talking music, dogs, and science, drinking good red wine.
And long walks in the countryside, a stroke of wondrous luck,
With Ulrich, Jean, and Robin around Furstenfeldenbruck.
And now a new phase in her life, anatomist reborn
The microscope will be for Jean, the grants for Anja Horn.
She will revise Olzewski's tome, a book among the greats
At least she will if Karger Publishers can find the plates.
There are unanswered questions. I hope Jean will make the journey
Across six time zones so she can inject some beasts with Bernie.
Jean now has a bionic knee. And is a pro at prancing.
Ulrich tells me that she wants to take him tango dancing.
She still will play the flute with her extraordinary brio
Bach once in a while with me, more often with her trio.
And Sunday crossword puzzles which she does not do alone
But rather with her sprightly mother on the telephone.
Jean Büttner Ennever! We honour with a love immense
Which is why we came to London at our own expense.

We are grateful to many individuals for making this project a success. For organizing and coordinating the meeting over many months we are very grateful to Margaret Collett. For the smooth running of the conference, we are also grateful to Sharon Baker, Rebecca Graham, Stephen Hicks, Siobhan Keenan, Sabira Mannan, Steve Rawlinson, and Clive Rosenthal. For the timely publication of this volume, we gratefully acknowledge the assistance of Johannes Menzel and Maureen Twaig of Elsevier. Finally, the conference and this volume would not have been possible without support from the Evenor Armington Fund of University Hospitals of Cleveland and the Guarantors of "Brain". We are also grateful for support from Advanced Clinical Instrumentation, Bunneyfoot, Cambridge Research Systems, Chronos Vision, Ober-Consulting Poland, SR Research, and Tracksys LTD.

Christopher Kennard
R. John Leigh

Contents

See Color Plate Section at the end of this book

SECTION 1

Using Novel Techniques to Define the Neural Control of Extraocular Muscles

C. Kennard & R.J. Leigh (Eds.)
Progress in Brain Research, Vol. 171
ISSN 0079-6123

CHAPTER 1.1

Mapping the oculomotor system

Jean A. Büttner-Ennever*

Institute of Anatomy, Ludwig-Maximilian University of Munich, 80336 Munich, Germany

Abstract: Over the last three decades and together with Bernard Cohen, Volker Henn, Ulrich Büttner, and Anja Horn, it has been possible to morphologically identify several functional cell groups in the oculomotor system: the medium-sized horizontal excitatory and inhibitory burst neurons (EBNs, IBNs) in the paramedian pontine reticular formation (PPRF), the more sparsely scattered vertical EBNs in the rostral interstitial nucleus of the MLF (RIMLF), and the typically elongated omnipause neurons (OPNs) in nucleus raphé interpositus — all essential for the generation of saccades. In contrast, the role of the central mesencephalic reticular formation (cMRF) in saccades is more complex, as is the morphological outlining of its borders. A detailed study of the extraocular motoneurons showed that they can be divided into two separate types: those for singly innervated (twitch) muscle fibres (SIFs) and those for multiply innervated (non-twitch) muscle fibres (MIFs). The two motoneuron types receive different premotor afferents, proving that MIF and SIF motoneurons have different functions. The cell groups were outlined by different tract tracing methods including rabies virus. The localization and histochemical characterization of all these functional cell groups in monkey formed the basis for the identification of the homologous groups in the human brainstem. Taken together these studies provide a neuroanatomical background for understanding clinical eye movement disorders.

Keywords: horizontal burst neurons; vertical burst neurons; omnipause neurons; rostral interstitial nucleus of the MLF; interstitial nucleus of Cajal; twitch motoneurons; central mesencephalic reticular formation; rabies virus; non-twitch motoneurons

Introduction

In the early 1970s the introduction of two new techniques made a great impact on the understanding of the central nervous system. First, was the development of stable single-unit recordings in awake mammals, a technique pioneered by K.-P. Schaefer many years ago. Second, was the development of sensitive and reliable tract tracing techniques, based on retrograde and anterograde axonal transport of substances like horseradish peroxidase and radioactive leucine, that replaced the inaccurate degeneration techniques. At this time eye movements were generally considered to be a subfeature of the vestibular system rather than a field of their own. Clinical observations had shown that the paramedian pontine reticular formation (PPRF) was associated with the generation of horizontal conjugate eye movements but the reason for this, the functional cells groups or anatomical pathways involved, were all unknown. At Mount Sinai Hospital New York, Morris Bender and later Bernie Cohen started stimulation experiments in monkeys to locate the horizontal

*Corresponding author. Tel.: +49 89 5160 4851; Fax: +49 89 5160 4802; E-mail: jean.buettner-ennever@med.uni-muenchen.de

DOI: 10.1016/S0079-6123(08)00601-8

eye movement area more exactly (Bender and Shanzer, 1964; Goebel et al., 1971; Cohen and Komatsuzaki, 1972). With the advent of chronic unit recordings it became clear from several parallel studies of the PPRF that the pontine neurons encoded precisely the parameters of the subsequent eye movement, and from their activity one could predict the subsequent saccade (Cohen and Henn, 1972; Luschei and Fuchs, 1972; Keller, 1974). From this point on the analysis of the oculomotor system exploded into one of the most popular fields of investigation, in which physiologists, like Bernard Cohen, and system-modellers like David Robinson, worked together with clinicians and neuroanatomists to understand how the brain moved the eye. In this article I will describe some of the functional cell groups of the oculomotor system, which we have outlined over the last 30 years, in both monkey and man. These studies were only possible because of the long-standing support of Bernard Cohen, Volker Henn, Ulrich Buettner, and Anja Horn.

Neuroanatomical methods

The recent development of highly specific and sensitive immunochemical stains, and new tract tracing techniques offer unique possibilities to study the functional connectivity of neuronal networks (Horn et al. 2008; Wickersham et al., 2007). Neurotropic viruses are particularly effective due to their ability to function as self-amplifying markers, and they produce exceptionally intense labelling. In collaboration with Gabriella Ugolini and Werner Graf, we have injected rabies virus (CVS fixed strain 11) into the lateral rectus muscle of monkey and have been able to visualize many of the premotor cell groups of the oculomotor system which we originally discovered using very different techniques (Ugolini et al., 2006). Rabies virus is only taken up at motor endplates and not by sensory or sympathetic endings, furthermore the virus remains in neuronal systems and is not accompanied by spurious uptake, e.g. in glial systems. After survival times of 3–3.5 days the monkeys were perfused with 4% paraformaldehyde. This period is long enough for transsynaptic transport back to the extraocular motoneuron and further retrograde into premotor networks over at least 2–3 synapses; this was not long enough to produce any rabies symptoms. In this review cases LR2 and LR4 are used to illustrate premotor neuronal populations and not demonstrate their connectivity, which has to be worked out using other techniques.

Premotor cell groups of the oculomotor system

Already in 1982 the combined effort of scientists had worked out the basic scaffolding of oculomotor pathways essential for the generation of horizontal and vertical saccades (Fig. 1A, B). A discrete group of medium-sized neurons in PPRF, called excitatory burst neurons (EBNs), lie below the medial longitudinal fasciculus (MLF) rostral to the abducens nucleus (VI) in part of nucleus reticularis pontis caudalis (NRPC) (Fig. 2A). The EBNs relay a premotor saccadic burst signal, from areas such as the superior colliculus (SC), to the lateral rectus motoneurons and to the internuclear neurons (INT) in the ipsilateral VI. The medial rectus motoneurons receive their saccadic burst signal via the crossed axons of INT in the MLF (Fig. 1B). However the EBNs are under a continual inhibition from omnipause neurons (OPNs), whose activity pauses only before and during horizontal or vertical saccades (Optican, Chapter 2.5 this volume). Just before a saccade, the OPNs are inhibited from ‘higher centres,’ which releases the EBN activity, activates motoneurons and INTs, and generates a coordinated horizontal saccade. There is a second group of burst neurons in the reticular formation ventromedial to VI in a region of the pontomedullary reticular formation called nucleus paragigantocellularis dorsalis (PGD) (Figs. 1A, B, and 2C). These are the inhibitory burst neurons (IBNs), which project to the contralateral VI as well as to the OPNs (Rucker, this volume).

The vertical and torsional components of saccades are elaborated in burst neurons near to the vertical moving motoneurons in the mesencephalon. They lie rostral to the oculomotor and trochlear nuclei in the most rostral tip of the

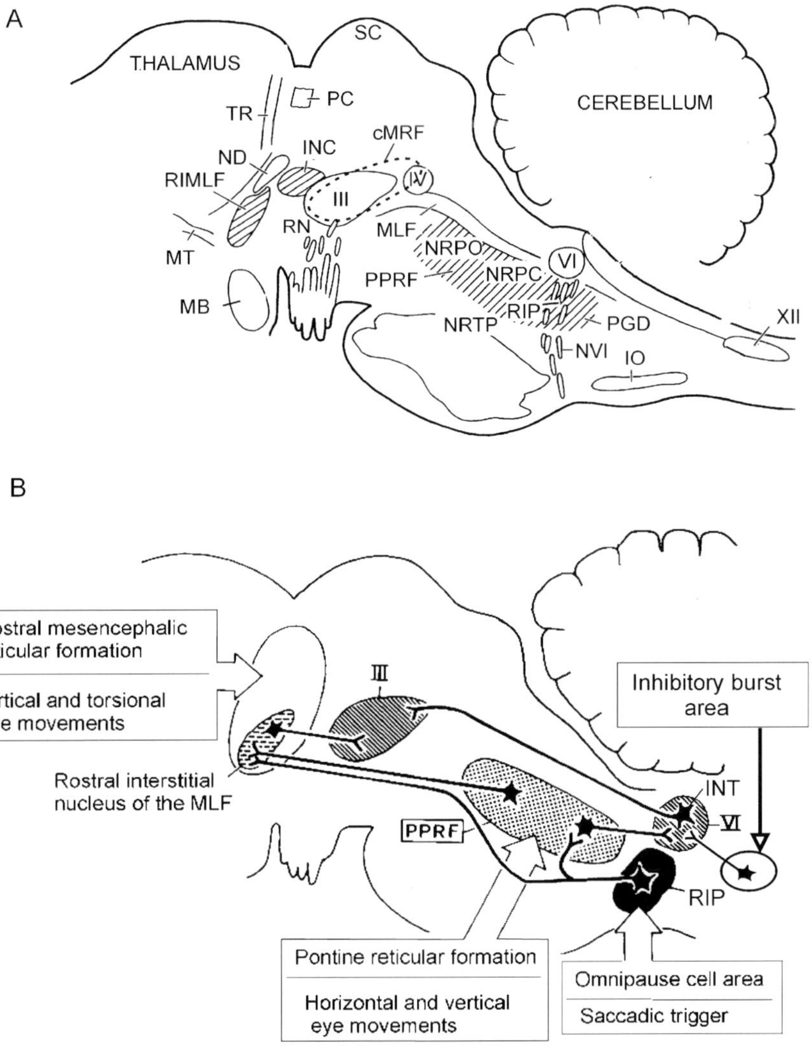

Fig. 1. (A) Drawing of a sagittal view of the brain to show some brainstem areas involved in the generation of eye movements. The dotted line indicates the lateral position of cMRF with relation to III. (B) Drawing of a sagittal view of the brain with some interconnections of premotor cell groups in the brainstem essential for the generation of saccades.

reticular formation, rostral interstitial nucleus of the MLF (RIMLF) (Figs. 1 and 3B–D) (Büttner-Ennever and Büttner, 1978). The long name of RIMLF arose from cumulative attempts to distinguish it from the interstitial nucleus of Cajal (INC), which lies immediately caudal to RIMLF (Fig.3B), and which over time has been given many different names. Bender emphasized the principle that bilateral lesions of the mesencephalon were necessary to produce vertical gaze paralysis. While this is generally correct, some unilateral lesions around the posterior commissure (PC) can give rise to upward or up and downgaze paralysis. Although the medium-sized vertical burst neurons are unobtrusive the RIMLF is clearly outlined by the posterior thalamo-subthalamic paramedian artery (also called the intermediate interpeduncular artery) (Fig. 3D asterisk). The artery arises from the posterior cerebral artery from a single origin: it divides and supplies each side of the brain, thus an

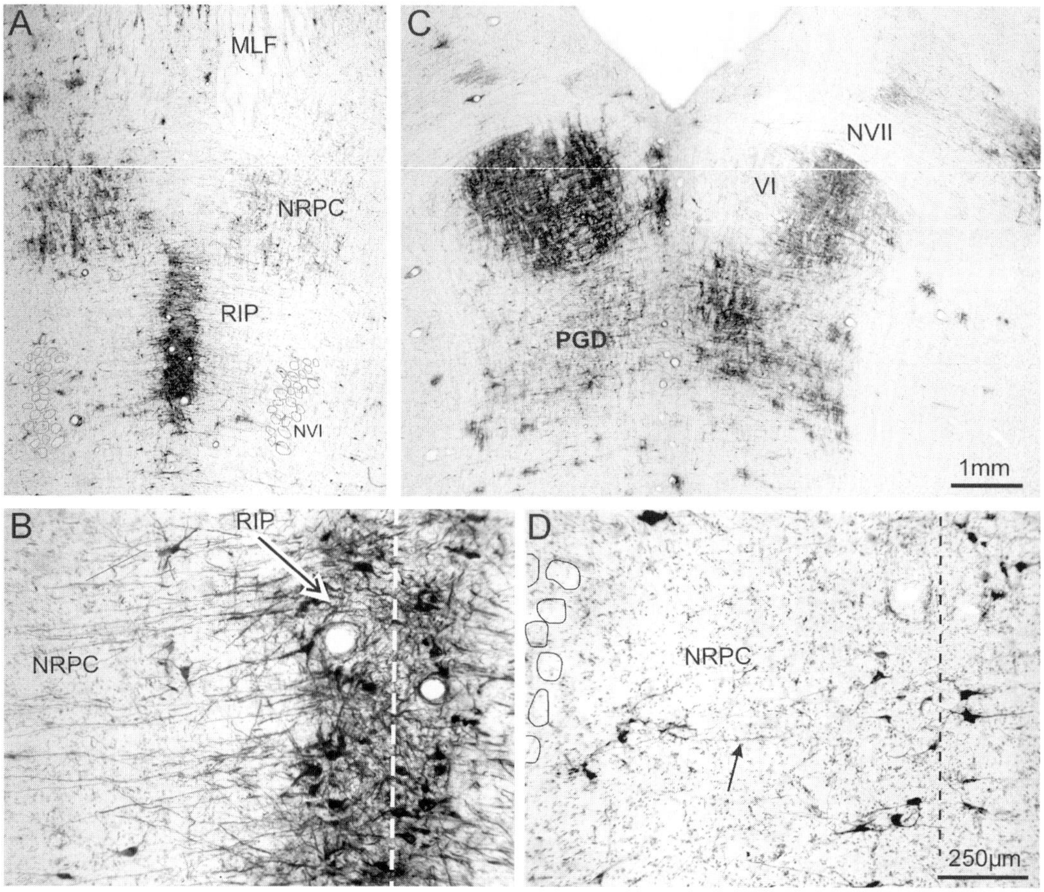

Fig. 2. (A) Labelling in the PPRF with rabies virus after injection into lateral rectus and a survival time which allows retrograde transsynaptic transport over 2–3 synapses (LR4). Note the strong labelling of the OPNs in RIP, and the relatively compact EBN regions in NRPC where the labelled is stronger ipsilaterally (left side). The abducens rootlets serve as a reliable landmark. (B) An enlargement of RIP in (A) to show the morphology of OPNs. Note the labelled neurons scattered laterally. (C) Abducens nucleus of the same experiment is completely filled on the ipsilateral (left) side, but contralaterally only the ABI area is labelled in VI. Ventrally the IBN areas in PGD are labelled, and the contralateral side being stronger. In (C) and (D) the midline is indicated by a dashed line. (D) OPN area of experiment LR2 with a shorter survival time than LR4. It shows the first OPN cells labelled in RIP and the scattered labelled neurons lateral to them with fine projections (arrow) onto OPNs. Note the NVI rootlets as landmark.

infarct of the artery could lead to *bilateral* lesions of RIMLF. Soon after the anatomical and physiological identification of RIMLF (Büttner et al., 1977; Büttner-Ennever and Büttner, 1978) a clinical case of vertical gaze paralysis with increasing drowsiness presented. Later the autopsy verified a bilateral RIMLF lesion which accounted for the vertical gaze paralysis (Büttner-Ennever et al., 1982). The concomitant bilateral loss of ascending brainstem pathways feeding into the medial forebrain bundle, the H-fields and ansa lenticularis accounted for the changes in wakefulness (Saper, 2006).

Before the discovery of these premotor cell groups one lively argument involving Ed Keller, Albert Fuchs, Mike King, and Craig Evinger at the Royaumont Meeting in 1977 stands out in my memory (Baker and Berthoz, 1977). It concerned how stimulation of the OPN area could arrest vertical and horizontal saccades. In independent

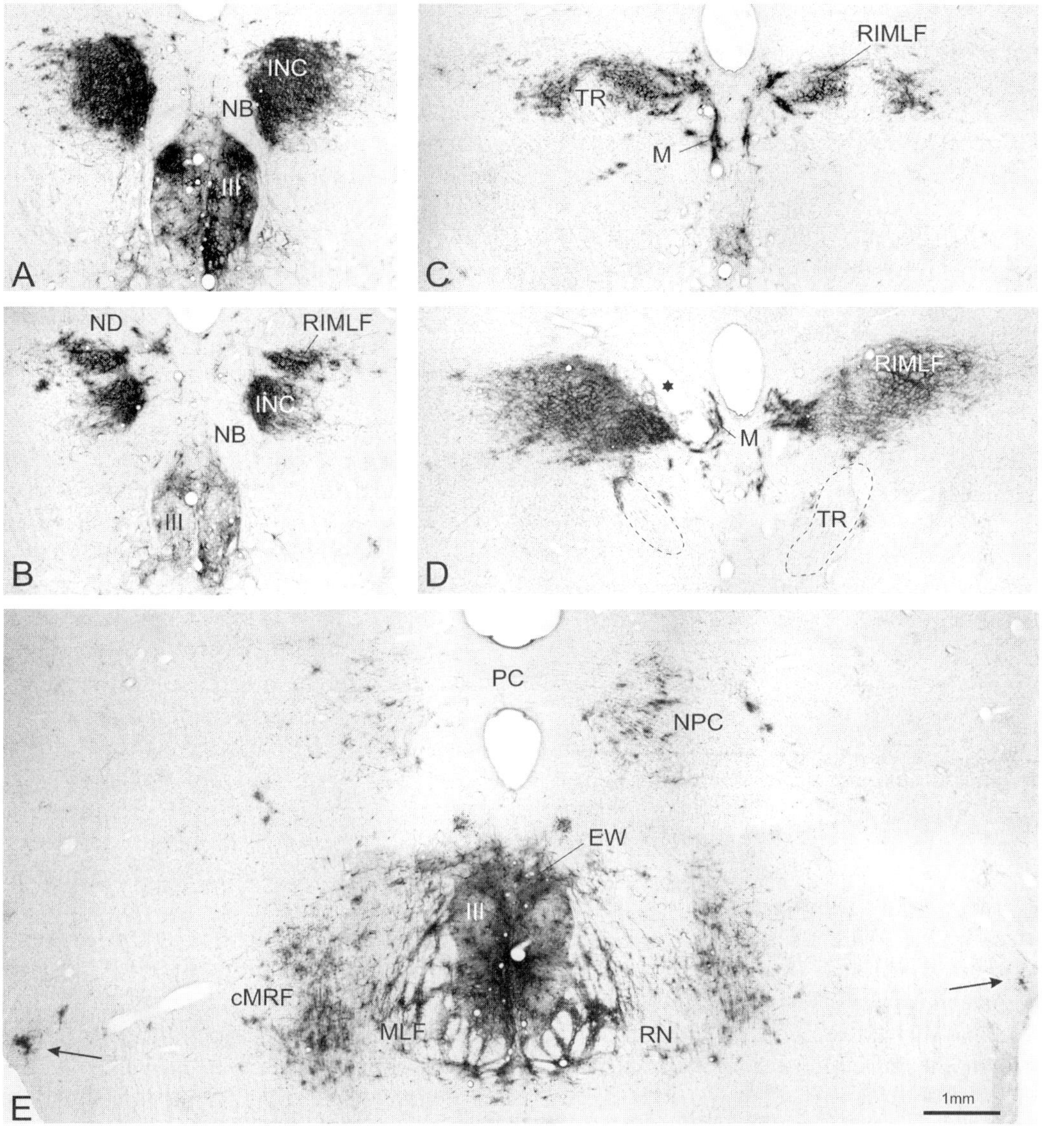

Fig. 3. Labelling in the rostral mesencephalon with rabies virus after injection into lateral rectus and a survival time which allows retrograde transsynaptic transport over 2–3 synapses (LR4). The virus labels vertical burst regions probably as a result of horizontal and vertical cross connections in the vestibular nuclei (Ugolini et al., 2006). Notice the strong bilateral labelling of INC and III in (A). Further rostrally in (B) rostral III, INC, and the caudal most tip of RIMLF is labelled. (C) RIMLF area is crossed by TR, and medially the M group is strongly labelled. In (D) the posterior thalamic subthalamic paramedian artery (asterisk) marks the dorsal border of mid RIMLF. (E) At the level of the PC (caudal to A–D) the III, EW, NPC, and the pretectum are labelled. The cMRF lies lateral to III but only the medial part is labelled here. At the lateral edge of the mesencephalon the accessory optic nuclei (arrows) are also visible.

experiments on monkey and cat, they had all found physiological evidence for a slender column of pause neurons near the midline in the pons, which paused for all saccades, and was thought to provide a high frequency (presumably inhibitory) control of burst neurons. The results were so clear and dramatic that it seemed to me that one must be able to see which cells were the OPNs on brain sections (morphologically), even though up to then no one had been able to find them. I decided then that I would try. We know now that the 'slender column of OPNs' do have a distinctive morphology, with their horizontally oriented dendrites (Fig. 2A, B and D). We first highlighted them by staining PPRF with cytochrome oxidase (Büttner-Ennever et al., 1988), then identified them through their inhibitory transmitter, glycine (Horn et al., 1994). In all mammals so far inspected the OPNs always lie either side of the midline at the same level as the abducens rootlets, as they pass through the dorsal tegmentum (Fig. 2A, B and D). Now after 20 years of studying the OPNs we are beginning to think that the group may not be a homogeneous (Horn, this volume), and that cells lying outside the EBN cluster, lateral to the OPNs may play a role in smooth pursuit (Fig. 2A, B and D) (Keller and Missal, 2003). Whether or not these cells are involved in the 'saccadic latch' is not clear.

The central mesencephalic reticular formation (cMRF) has been defined as an oculomotor structure by lesions, stimulation experiments, single-unit recordings, and tract tracing its neural connections (Waitzman et al., 2002). Its function is complex and currently three different hypotheses of cMRF function are being investigated (Cromer and Waitzman, 2006): (1) it interacts with OPNs to control saccadic triggering. Certainly we have seen monosynaptic projections from cMRF to the OPNs in nucleus raphe interpositus (RIP) using tritiated leucine tract tracer, which is only taken up by neurons in the injection site and not passing axons (personal observation). (2) It participates in the transformation of spatiotemporal coding of SC signals onto EBNs in the pons (feed-forward). (3) It provides an efference copy of saccade velocity, a feedback loop, to higher structures. The cMRF forms the rostral part of nucleus subcuneiformis in monkeys: Chen and May (2000) define it with respect to the tightly coupled reciprocal projections to SC; whereas the results of single unit recording suggest that it may extend rostrally to the Fields of Forel (Waitzman et al., 2000a, b). The rabies experiments labelled cMRF mainly bilaterally after 3 days, but only the medial part of the cMRF region reciprocally connected to SC was labelled in this experiment, implying that a subpopulation of cMRF cells is highlighted here (Fig. 3E).

One of the more recent cell groups that we have described are the non-twitch motoneurons, or, more strictly termed, the motoneurons of multiply innervated muscle fibres (MIFs) of the extraocular muscles. These have become the centre of our attention partly because we do not know to which part of the oculomotor system they belong. Many investigations have shown that extraocular motoneurons in all vertebrates lie in subgroups within the oculomotor, trochlear, and abducens nucleus (III, IV, and VI) (Büttner-Ennever, 2006). They have a phasic and a tonic component of activity, which varies widely in extent, but are considered as a continuum. Only one type of extraocular motoneuron has been recognized in physiological recordings from III, IV, and VI in awake monkeys, and they responded during all types of eye movement. This led to the concept of a 'final common pathway.' However, morphologically at least six different morphological types of striated muscle fibres have been described in extraocular muscles, and these can be divided into two main categories: those that are singly innervated by a central motor endplate (SIFs), and those that are MIFs with endplates along the whole length of the muscle fibre. The MIFs are a very unusual type of muscle fibre, unique to eye muscles, that is they do not occur in the skeletal muscles of mammals: they respond to activation with a relatively slow, graded contraction, not an all-or-nothing twitch like the other fibre types innervated by a single motor endplate (SIFs).

We have found the location of the motoneurons innervating the MIF muscle fibres of the global layer using tract tracers (Büttner-Ennever et al., 2001). They lie around the periphery of the classical III, IV, and VI boundaries and do not intermingle with the SIF motoneurons (Fig. 4). In VI the LR MIFs surrounded the medial aspect of the nucleus; the SO MIFs lay in a dorsal cap over IV; and in III the MIFs of MR and IR are gathered into a small group on the dorsomedial boarder of III (C-group), while those of SR and inferior olive (IO) lay around the midline between the two halves of the III (S-group). In terms of neuroanatomy, when neuronal cell groups lie

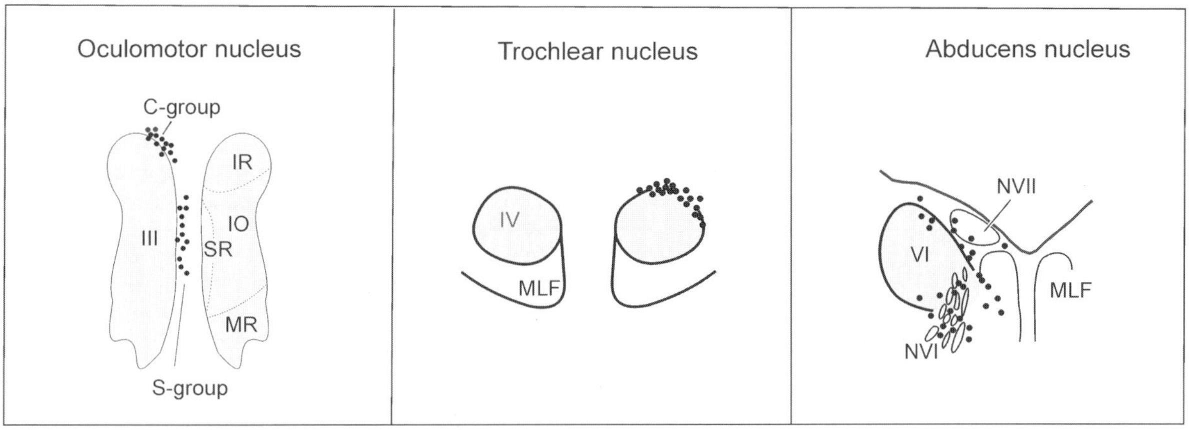

Fig. 4. Drawings of the oculomotor, trochlear, and abducens nuclei to show the location of non-twitch (MIF) motoneurons (black dots) around the periphery of the classical motor nuclei. The position of the motoneuron subgroups innervating medial, superior, and inferior rectus (MR, SR, IR) and inferior oblique muscles (IO) are indicated in III.

separately it is often a sign that they receive different afferent inputs and have different functions. This is indeed the case with SIFs and MIFs, they receive different afferents (Büttner-Ennever et al., 2002; Wasicky et al., 2004; Ugolini et al., 2006). The classical SIF motoneurons have premotor inputs from PPRF and the magnocellular regions of the vetibular nuclei, but these do not directly innervate the MIF motoneurons. The MIFs and their dendrites in III have a close relationship to the Edinger–Westphal complex (EW, near response neurons). The function of MIFs is not clear. Whereas the SIFs are mainly responsible for the rapid eye movements our current hypothesis is that MIFs are involved in more tonic functions such as eye alignment, and may participate in feedback networks which regulate vision and proprioception.

Throughout the search for premotor cell groups of the oculomotor system a major aim has always been to identify the homologous regions in humans. Usually this has been achieved by using additional immunohistochemical techniques. That is, in monkey the motor or premotor groups would be first labelled by tract tracers, and then the histochemical properties of these same labelled neurons would be determined by double-labelling techniques. Next, sections from a likely homologous region in the human brain would be taken, and the material investigated for neurons with the same histochemical properties as the labelled cells in the monkey. These human projects have been led by Anja Horn and have provided a solid scientific basis for the analysis of clinical cases with eye movement disorders (Horn et al., 1994, 1995, 2000; Horn and Büttner-Ennever, 1998). A clear example of this is the recent re-definition of the EW nucleus in man (Horn et al., 2008; May et al., this volume).

Abbreviations

III	oculomotor nucleus
IV	trochlear nucleus
VI	abducens nucleus
ABI	abducens internuclear neurons
cMRF	central mesencephalic reticular formation
EBNs	excitatory burst neurons
EW	Edinger–Westphal complex
IBNs	inhibitory burst neurons
INC	interstitial nucleus of Cajal
IO	inferior olive
M	M-group
MB	mammillary body
MIF	multiply innervated fibre
MT	mammillothalamic tract
MLF	medial longitudinal fasciculus
NVI	abducens nerve

NVII	facialis nerve
NB	nucleus of Bechterew
ND	nucleus Darkschewitsch
NPC	nucleus of the posterior commissure
NRPC	nucleus reticularis pontis caudalis
NRPO	nucleus reticularis pontis oralis
NRTP	nucleus reticularis tegmenti pontis
OPN	omnipause neuron
PC	posterior commissure
PGD	nucleus paragigantocellularis dorsalis
PPRF	paramedian pontine reticular formation
RIMLF	rostral interstitial nucleus of the medial longitudinal fasciculus
RIP	nucleus raphe interpositus
RN	red nucleus
SC	superior colliculus
SIF	singly innervated fibre
TR	tractus retroflexus

Acknowledgements

This work was supported by the Deutsche Forschungsgemeinschaft DFG HO 1639/ 4-2 and the European Union Grant number BIO4-CT98-0546.

References

Baker, R. and Berthoz, A. (1977) Control of gaze by brain stem neurons. Elsevier/North-Holland Biomedical Press, Amsterdam/New York.

Bender, M.B. and Shanzer, S. (1964) Oculomotor pathways defined by electrical stimulation and lesions in the brain stem of the monkey. In: Bender M.B. (Ed.), The Oculomotor System. Harper and Row, New York, pp. 81–140.

Büttner, U., Büttner-Ennever, J.A. and Henn, V. (1977) Vertical eye movement related activity in the rostral mesencephalic reticular formation of the alert monkey. Brain Res., 130: 239–252.

Büttner-Ennever, J.A. (2006) The extraocular motor nuclei: organization and functional neuroanatomy. Prog. Brain Res., 151: 95–125.

Büttner-Ennever, J.A. and Büttner, U. (1978) A cell group associated with vertical eye movements in the rostral mesencephalic reticular formation of the monkey. Brain Res., 151: 31–47.

Büttner-Ennever, J.A., Büttner, U., Cohen, B. and Baumgartner, G. (1982) Vertical gaze paralysis and the rostral interstitial nucleus of the medial longitudinal fasciculus. Brain, 105: 125–149.

Büttner-Ennever, J.A., Cohen, B., Pause, M. and Fries, W. (1988) Raphe nucleus of the pons containing omnipause neurons of the oculomotor system in the monkey, and its homologue in man. J. Comp. Neurol., 267: 307–321.

Büttner-Ennever, J.A., Horn, A.K.E., Graf, W. and Ugolini, G. (2002) Modern concepts of brainstem anatomy. Ann. N.Y. Acad. Sci., 956: 75–84.

Büttner-Ennever, J.A., Horn, A.K.E., Scherberger, H. and D'Ascanio, P. (2001) Motoneurons of twitch and nontwitch extraocular muscle fibers in the abducens, trochlear, and oculomotor nuclei of monkeys. J. Comp. Neurol., 438: 318–335.

Chen, B. and May, P.J. (2000) The feedback circuit connecting the superior colliculus and central mesencephalic reticular formation: a direct morphological demonstration. Exp. Brain Res., 131: 10–21.

Cohen, B. and Henn, V. (1972) Unit activity in the pontine reticular formation associated with eye movements. Brain Res., 46: 403–410.

Cohen, B. and Komatsuzaki, A. (1972) Eye movements induced by stimulation of the pontine reticular formation. Evidence for integration in oculomotor pathways. Exp. Neurol., 36: 101–117.

Cromer, J.A. and Waitzman, D.M. (2006) Neurones associated with saccade metrics in the monkey central mesencephalic reticular formation. J. Physiol., 570: 507–523.

Goebel, H.H., Komatsuzaki, A., Bender, M.B. and Cohen, B. (1971) Lesions of the pontine tegmentum and conjugate gaze paralysis. Arch. Neurol. (Chicago), 24: 431–440.

Horn, A., Eberhorn, A., Härtig, W., Ardelenanu, P., Messoudi, A. and Büttner-Ennever, J. (2008) Perioculomotor cell groups in monkey and man defined by their histochemical and functional properties: a reappraisal of the Edinger–Westphal Nucleus. J. Comp. Neurol., 507: 1317–1335.

Horn, A.K.E. and Büttner-Ennever, J.A. (1998) Premotor neurons for vertical eye-movements in the rostral mesencephalon of monkey and man: the histological identification by parvalbumin immunostaining. J. Comp. Neurol., 392: 413–427.

Horn, A.K.E., Büttner-Ennever, J.A., Gayde, M. and Messoudi, A. (2000) Neuroanatomical identification of mesencephalic premotor neurons coordinating eyelid with upgaze in the monkey and man. J. Comp. Neurol., 420: 19–34.

Horn, A.K.E., Büttner-Ennever, J.A., Suzuki, Y. and Henn, V. (1995) Histological identification of premotor neurons for horizontal saccades in monkey and man by parvalbumin immunostaining. J. Comp. Neurol., 359: 350–363.

Horn, A.K.E., Büttner-Ennever, J.A., Wahle, P. and Reichenberger, I. (1994) Neurotransmitter profile of saccadic omnipause neurons in nucleus raphe interpositus. J. Neurosci., 14: 2032–2046.

Keller, E.L. (1974) Participation of medial pontine reticular formation in eye movement generation in monkey. J. Neurophysiol., 37: 316–332.

Keller, E.L. and Missal, M. (2003) Shared brainstem pathways for saccades and smooth-pursuit eye movements. Ann. N.Y. Acad. Sci., 1004: 29–39.

Luschei, E.S. and Fuchs, A.F. (1972) Activity of brain stem neurons during eye movements of alert monkeys. J. Neurophysiol., 35: 445–461.

Saper, C.B. (2006) Staying awake for dinner: hypothalamic integration of sleep, feeding, and circadian rhythms. Prog. Brain Res., 14: 243–252.

Ugolini, G., Klam, F., Doldan Dans, M., Dubayle, D., Brandi, A.-M., Büttner-Ennever, J.A. and Graf, W. (2006) Horizontal eye movement networks in primates as revealed by retrograde transneuronal transfer of rabies virus: differences in monosynaptic input to "slow" and "fast" abducens motoneurons. J. Comp. Neurol., 498: 762–785.

Waitzman, D.M., Pathmanathan, J., Presnell, R., Ayers, A.S. and DePalma-Bowles, S. (2002) Contribution of the superior colliculus and the mesencephalic reticular formation to gaze control. Ann. N.Y. Acad. Sci., 956: 111–129.

Waitzman, D.M., Silakov, V.L., DePalma-Bowles, S. and Ayers, A.S. (2000a) Effects of reversible inactivation of the primate mesencephalic reticular formation. I. Hypermetric goal-directed saccades. J. Neurophysiol., 83: 2260–2284.

Waitzman, D.M., Silakov, V.L., DePalma-Bowles, S. and Ayers, A.S. (2000b) Effects of reversible inactivation of the primate mesencephalic reticular formation. II. Hypometric vertical saccades. J. Neurophysiol., 83: 2285–2299.

Wasicky, R., Horn, A.K.E. and Büttner-Ennever, J.A. (2004) Twitch and non-twitch motoneuron subgroups of the medial rectus muscle in the oculomotor nucleus of monkeys receive different afferent projections. J. Comp. Neurol., 479: 117–129.

Wickersham, I.R., Lyon, D.C., Barnard, R.J.O., Mori, T., Finke, S., Conzelmann, K.-K., Young, J.A.T. and Callaway, E.M. (2007) Monosynaptic restriction of transsynaptic tracing from single, genetically targeted neurons. Neuron, 53: 639–647.

C. Kennard & R.J. Leigh (Eds.)
Progress in Brain Research, Vol. 171
ISSN 0079-6123

CHAPTER 1.2

Neuronal signalling expression profiles of motoneurons supplying multiply or singly innervated extraocular muscle fibres in monkey

Howard S. Ying[1,*], Karoline Fackelmann[2], Ahmed Messoudi[2], Xiao-fang Tang[2], Jean A. Büttner-Ennever[2] and Anja K.E. Horn[2]

[1]*Ophthalmology, The Johns Hopkins University/Wilmer Eye Institute, Baltimore, MD, USA*
[2]*Institute of Anatomy, Ludwig-Maximilian University, Munich, Germany*

Abstract: Motoneurons of the oculomotor nucleus subserving multiply innervated muscle fibres (MIF) receive different afferent inputs from the motoneurons subserving singly innervated muscle fibres (SIF). We asked whether MIF and SIF motoneurons have different neurotransmitter signalling expression profiles. Adult rhesus monkey extraocular muscles were injected with the retrograde tracer cholera toxin. Sections were then stained for various neurotransmitter-signalling markers. MIF motoneurons showed less glutamate receptor 4 (GluR4) and *N*-methyl-D-aspartate receptor 1 (NMDAR1) immunoreactivity, but showed similar amounts of glutamic acid decarboxylase (GAD) immunoreactive afferent terminals, compared to SIF motoneurons. This difference in excitatory neurotransmitter receptor expression may explain selective oculomotor deficits and allow development of selective pharmacotherapy in the future.

Keywords: non-twitch muscle fibre; twitch muscle fibre; C-group; S-group; GABA; non-NMDA; AMPA

Introduction

Extraocular muscles contain at least six different types of muscle fibres that can be divided into two main categories: multiply innervated fibres (MIF) and singly innervated fibres (SIF). MIF are slowly contracting (non-twitch), fatigue-resistant, innervated by *en grappe* endplates that are distributed along the whole muscle fibre. Their motoneurons likely have a mostly tonic activity pattern, and receive premotor afferents from gaze holding, convergence, or smooth pursuit networks. SIF have a more phasic activity pattern, are fast twitch, innervated by *en plaque* endplates concentrated in a central zone of the muscle fibre, and their motoneurons receive inputs from all ocular motor premotor networks, including: saccade, vestibulo-ocular reflex (VOR), optokinetic reflex (OKN), gaze holding, convergence, or smooth pursuit. SIF motoneurons appear more suited to driving eye movement while MIF motoneurons appear more suited for tonic functions like eye alignment (Ugolini et al., 2006; Büttner-Ennever, 2007). We asked whether SIF and MIF motoneurons could also be distinguished by their

*Corresponding author. Tel.: + (410)502 5383;
Fax: + (410)614 8577; E-mail: hying1@jhmi.edu

DOI: 10.1016/S0079-6123(08)00602-X

neurotransmitter receptor and neuronal signalling expression profiles.

Methods and materials

Adult rhesus monkey superior rectus, medial rectus, or inferior oblique muscles were injected either in the midportion of the muscle belly or in the distal myotendinous junction with cholera toxin subunit B (CTB; 5 µl, 1%, List Biological Laboratories) as previously described (Eberhorn et al., 2005). CTB was detected with goat anti-CTB (List Biological Laboratories) and visualized either with immunoperoxidase methods using diaminobenzidine (brown) or with immunofluorescence using Cy2-tagged anti-goat (Dianova, Hamburg, Germany) (green fluorescence). Glutamic acid decarboxylase (GAD) was visualized with immunoperoxidase methods using a Ni-intensified diaminobenzidine reaction (black). GluR4 and NMDAR1 antibodies (Chemicon, Temecula, CA) were detected with immunofluorescence using Cy3-tagged secondary antibodies (Dianova, Hamburg, Germany) (red fluorescence).

Results

CTB injection into the distal myotendinous junction of the medial rectus or superior rectus or inferior oblique muscle almost exclusively labelled MIF motoneurons in the C-group, dorsomedial to the oculomotor nucleus, or the S-group, in the midline between oculomotor nuclei, respectively. CTB injection into the midportion of the muscle labelled the respective SIF motoneurons in the classic oculomotor nucleus as well as C- or S-group MIF motoneurons. GAD staining was found in synaptic boutons projecting onto both

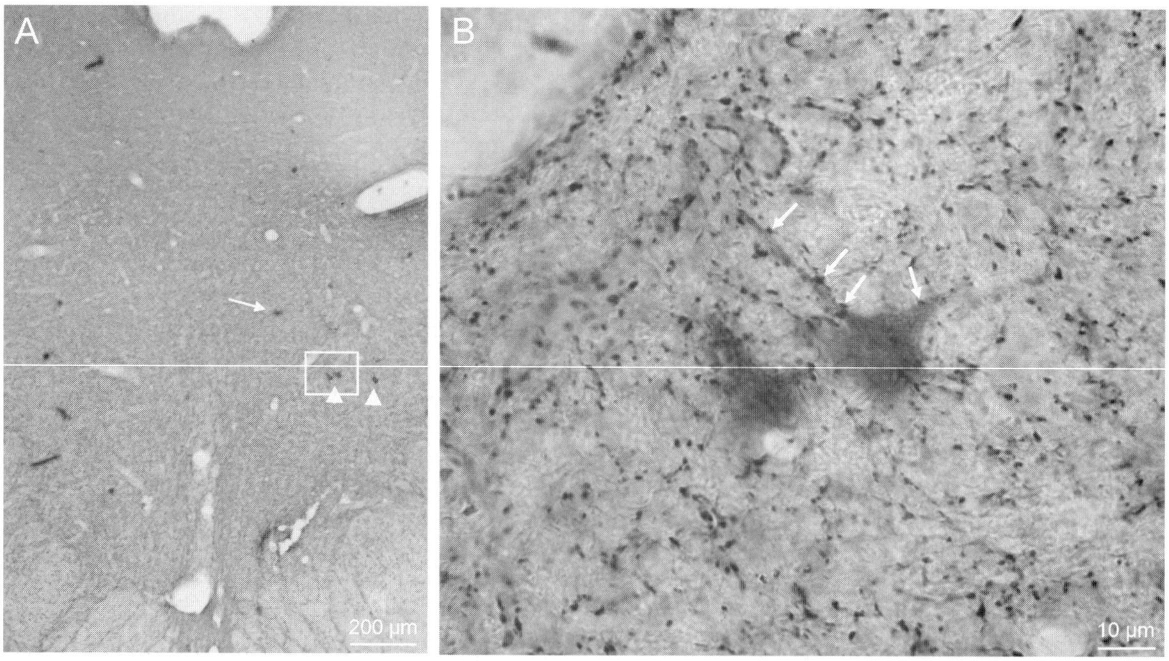

Fig. 1. GAD staining in synaptic boutons (black) was detected adjacent to MIF and SIF motoneurons labelled by cholera toxin (brown). Panel A: Low magnification view of the caudal oculomotor nucleus shows CTB-stained MIF motoneurons (arrow) and SIF motoneurons (arrowheads). Panel B: Higher magnification view of area in white box from panel A shows CTB-stained SIF neurons which are contacted by numerous GAD-positive terminals (arrows). (See Color Plate 1.2.1 in color plate section.)

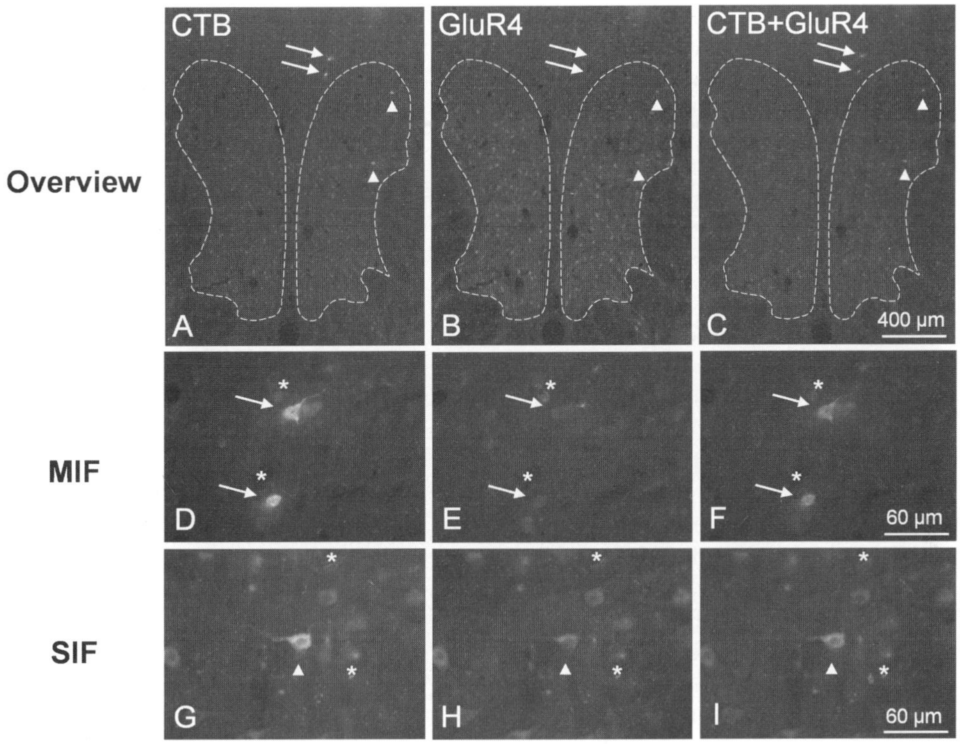

Fig. 2. GluR4 expression (red) was detected in SIF but not MIF motoneurons labelled by cholera toxin (green). Panels A, B, and C show CTB-positive, GluR4-positive, or co-labelled neurons, respectively. The classic oculomotor nucleus is outlined in white. MIF motoneurons are indicated by arrows and SIF motoneurons are indicated by arrow heads. Panels D, E, and F show higher magnification of CTB-stained but GluR4-negative MIF motoneurons. Panels G, H, and I show higher magnification of a CTB-stained and GluR4-positive SIF motoneuron. Blood vessels are indicated by asterisks for reference. (See Color Plate 1.2.2 in color plate section.)

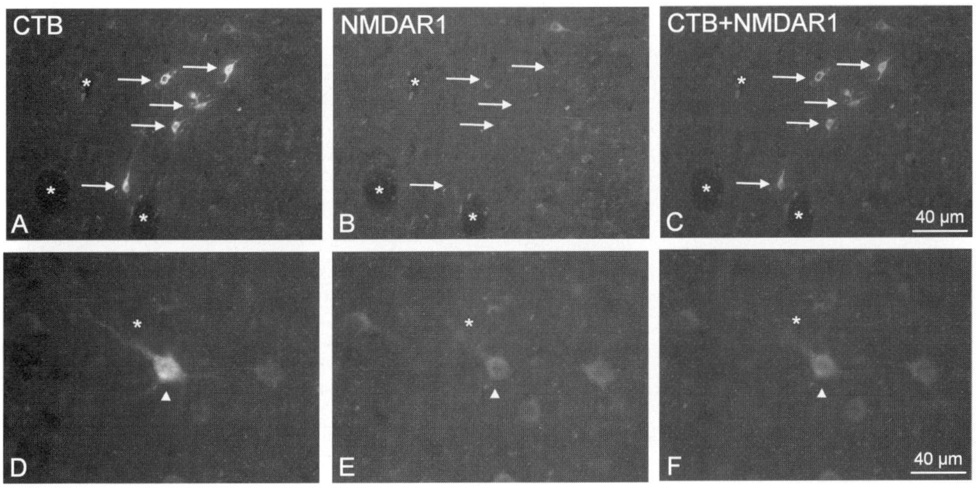

Fig. 3. NMDAR1-immunoreactivity (red) was abundantly detected in SIF but not in MIF motoneurons labelled by CTB (green). Panels A, B, and C show CTB-stained but NMDAR1-negative MIF motoneurons (arrows). Panels D, E, and F show a CTB-stained and NMDAR1-positive SIF motoneuron (arrowhead). Blood vessels are indicated by asterisks for reference. (See Color Plate 1.2.3 in color plate section.)

MIF and SIF motoneurons in similar numbers (Fig. 1). Strong GluR4 immunoreactivity was found in nearly all SIF motoneurons but only in few MIF motoneurons (Fig. 2). NMDAR1 immunoreactivity was also detected in nearly all SIF but few MIF motoneurons (Fig. 3).

Conclusions

MIF motoneurons of the oculomotor nucleus in monkey expressed less NMDA- and non-NMDA receptor immunoreactivity than SIF motoneurons, suggesting that MIF motoneurons were less sensitive to excitatory amino acid input. Similar GAD levels for both populations showed that this decreased glutamate receptor expression was specific. Comparison of MIF and SIF immunofluorescence was confounded by differences in non-specific fluorescence, e.g., from lipofuscin, between populations. If these differences in MIF and SIF neurotransmitter expression are functionally significant, they may explain selective eye alignment or eye movement deficits and allow the development of selective pharmacotherapy in the future.

Acknowledgements

Supported by Research to Prevent Blindness, Wilmer Microscopy and Imaging Core Facility, Knights' Templar Eye Foundation, and National Eye Institute grant NIH EY15025.

References

Büttner-Ennever, J.A. (2007) Anatomy of the oculomotor system. Dev. Ophthalmol., 40: 1–14.

Eberhorn, A.C., Ardeleanu, P., Büttner-Ennever, J.A. and Horn, A.K. (2005) Histochemical differences between motoneurons supplying multiply and singly innervated extraocular muscle fibers. J. Comp. Neurol., 491(4): 352–366.

Ugolini, G., Klam, F., Doldan Dans, M., Dubayle, D., Brandi, A.M., Buttner-Ennever, J. and Graf, W. (2006) Horizontal eye movement networks in primates as revealed by retrograde transneuronal transfer of rabies virus: differences in monosynaptic input to "slow" and "fast" abducens motoneurons. J. Comp. Neurol., 498(6): 762–785.

C. Kennard & R.J. Leigh (Eds.)
Progress in Brain Research, Vol. 171
ISSN 0079-6123

CHAPTER 1.3

Histochemical characterisation of trigeminal neurons that innervate monkey extraocular muscles

K. Fackelmann*, A. Nouriani, A.K.E. Horn and J.A. Büttner-Ennever

Institute of Anatomy, Ludwig-Maximilian University of Munich, Munich, Germany

Abstract: Sensory trigeminal innervation is a consistent feature of extraocular muscles across species, in spite of a variable occurrence of muscle spindles. We studied the histochemical properties of trigeminal ganglion (TG) cells projecting to the extraocular eye muscles to obtain more information about their function. In monkey TG neurons were retrogradely filled by tracer injections (cholera toxin subunit B; wheat-germ agglutinin) into the belly or myotendinous junction of eye muscles; one conjunctival injection served as a control. Retrogradely labelled TG neurons were processed for the presence of parvalbumin (PV), substance P (SP), or nitric oxide synthase (NOS) by double-immunoflourescence. The results indicate that approximately 10% of trigeminal afferents to all parts of the eye muscle are PV-positive, whereas around 20% are SP-positive. Twice as many SP-positive TG projection neurons were counted after a conjunctival tracer injection, presumably relaying nociceptive signals. A surprisingly large population of NOS-positive TG cells (30%) was found only after distal tracer injections. Up to now none of these TG cell groups could be related to the palisade endings located at the myotendinous junction.

Keywords: monkey; trigeminal gang lion; eye muscle; sensory innervation; substance P; parvalbumin; nitric oxide synthase

Introduction

Although anatomical tract-tracing studies have demonstrated that sensory neurons of eye muscles are located in the ophthalmic subdivision of the ipsilateral trigeminal ganglion (TG) (Porter and Donaldson, 1991), there is no information about their function. In general pseudounipolar TG cells innervate mechanoreceptors, thermoreceptors, and nociceptors of the face including the eye and eye muscles. Aside from pain and thermosensation there is evidence for proprioceptive inputs from the eye muscle to the brain (Ruskell, 1999). However, the existence of proprioceptive organs in eye muscles varies widely across species. Besides the well-known muscle spindles present only in few species, and Golgi tendon organs present only in sheep, palisade endings were identified at the myotendinous junction of extraocular muscle of all species so far investigated, and are suited to subserve a proprioceptive function (Büttner-Ennever et al., 2006).

We investigated the histochemical properties of TG cells, which innervate the extraocular muscles to gain insight into their presumed functions.

*Corresponding author. Tel.: +49-89-5160-4876; Fax: +49-89-5160-4802; E-mail: karoline.fackelmann@med.uni-muenchen.de

DOI: 10.1016/S0079-6123(08)00603-1

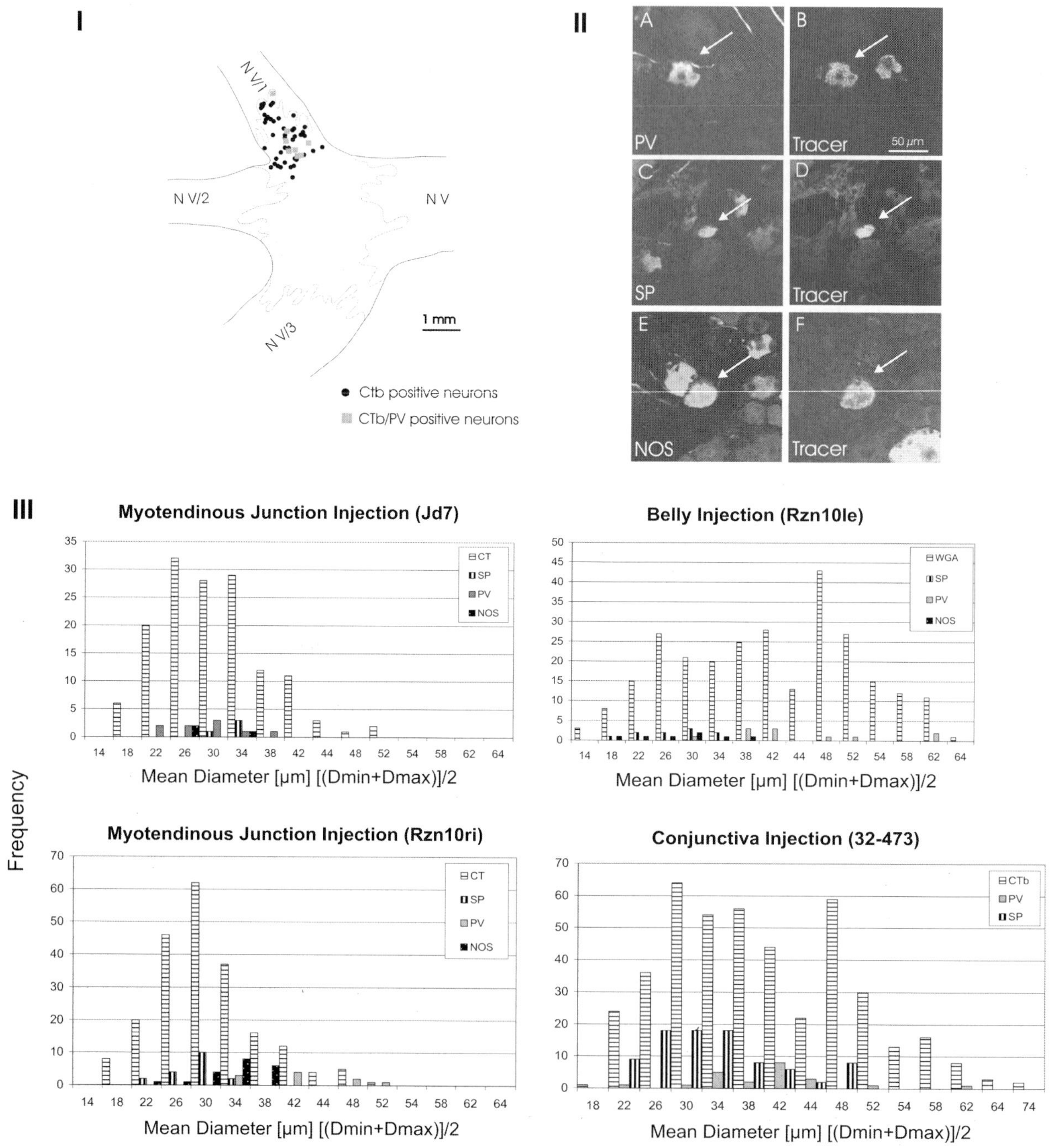

Fig. 1. I: Plot of retrogradely labelled cells (filled circles) some of which are parvalbumin (PV) positive (grey squares). II: Pairs of photographs showing the same section in red fluorescence for the markers PV, substance P (SP), nitric oxide synthase (NOS) (A, C, E), and in green fluorescence for the tracer (B, D, F), respectively. The arrows indicate double-labelled neurons. Scale bar: 50 µm. III: Cell size profiles of the retrogradely labelled neurons and the respective marker associated population in the trigeminal ganglion (TG) after different injection targets in the eye muscle. Note that after tracer injections in the muscle belly or the conjunctiva more large cells were labelled compared to injections into the myotendinous junction. Note that a considerably high number of NOS positive tracer-labelled neurons were found after injection into the myotendinous junction. The highest number of SP positive retrogradely labelled neurons were found after a conjunctival injection.

Methods

A tract-tracer [cholera toxin subunit B (CTB) 5–20 μl, 1%; List Campbell; wheat-germ agglutinin (WGA) 5 μl, 5%; EY Laboratories] was injected either into the belly or the myotendinous junction of an extraocular muscle in anesthetised monkeys. After a 3 day survival time the animals were perfused with 0.9% saline followed by 4% paraformaldehyde. Frozen sections of the brain, the TG, and the eye muscles were processed for the tracer (goat anti-choleragenoid, 1:500; List Biological Laboratories; goat-anti-WGA, 1:25; Axxora) and one of the following markers: Parvalbumin (PV, mouse anti-PV; 1:1000; Swant), substance P (SP, rabbit anti-SP; 1:100; Zymed) or neuronal nitric oxide synthase (nNOS, rabbit anti-NOS; 1:2000; Chemicon). The sections were then reacted with Cy^3-anti-rabbit or Cy^3-anti-mouse (1:200; Dianova) and Alexa 488-anti-goat (1:200; Molecular Probes) for 2 h mounted and coverslipped.

All slides were examined in brightfield or with fluorescence light using appropriate filters for red fluorescent Cy^3 (N2.1) and green fluorescence Alexa 488 (I3). The morphological analysis was performed using Image J software.

Results and discussion

Almost all retrogradely labelled neurons in the TG were found within the ophthalmic division (Fig. 1I). The comparison of different cases where different eye muscles had been injected did not reveal a clear somatotopic organisation in TG.

The morphometric analysis showed that the cell size profile of all retrogradely labelled trigeminal cells resembled that of the complete population seen in Nissl-staining. A small population of medium- and large-sized neurons was PV-positive (2.5–16.9%) (Fig. 1IIA and B; Fig. 1III) and may possibly represent proprioceptive neurons, as described for the dorsal root ganglion (Lazarov, 2002), or the mesencephalic trigeminal nucleus (Shigenaga et al., 1988) where they innervated muscle spindles of the masseter. A considerable population of small to medium-sized neurons expressed SP (10.9–30.1%) (Fig. 1IIC and D; Fig. 1III). SP is a neuropeptide important for pain processing (Levine et al., 1993), and also present in vegetative fibres around blood vessels, presumably involved in vasodilatation (Foreman et al., 1983). Based on this we suggest that the tracer-labelled SP-positive TG neurons represent nociceptive afferents from the eye muscle. At least a subpopulation may play a role in vasodilatation, since SP-positive fibres were found in association with blood vessels in the extraocular muscle (Eberhorn et al., 2005). After tracer injections into the muscle belly a small population of small- to medium-sized neurons NOS-positive neurons (6%) was evident (Fig. 1IIE and F; Fig. 1III). In contrast, a large population of NOS-immunoreactive neurons (42%) was detected after a tracer injection into the myotendinous junction at the distal muscle tip (Fig. 1III). The presence of retrogradely traced NOS-positive TG neurons implicates that nitric oxide is used as a transmitter at sensory nerve endings in the eye muscle (Bredt et al., 1990). The small NOS-positive neurons are presumably involved in pain transmission (Thomsen and Olesen, 1996), whereas the larger NOS-positive cells might play a role in proprioceptive function (Lazarov and Dandov, 1998; Stoyanova and Lazarov, 2005). The large number of tracer-labelled NOS-positive TG neurons filled from the myotendinous junction of the eye muscle, which contains the palisade endings, supports this hypothesis.

References

Bredt, D.S., Hwang, P.M. and Snyder, S.H. (1990) Localization of nitric oxide synthase indicating a neural role for nitric oxide. Nature, 347: 768–770.

Büttner-Ennever, J.A., Konakci, K.Z. and Blumer, R. (2006) Sensory control of extraocular muscles. In: Büttner-Ennever J.A. (Ed.), Progress in Brain Research, Vol. 151: Neuroanatomy of the Oculomotor System. Elsevier B. V., Netherlands, pp. 81–93.

Eberhorn, A.C., Horn, A.K.E., Fischer, P. and Büttner-Ennever, J.A. (2005) Proprioception and palisade endings in extraocular eye muscles. Ann. N.Y. Acad. Sci., 1039: 1–8.

Foreman, J.C., Jordan, C.C., Oehme, P. and Renner, H. (1983) Structure–activity relationships for some substance P-related peptides that cause wheal and flare reactions in human skin. J. Physiol., 355: 449–465.

Lazarov, N.E. (2002) Comparative analysis of the chemical neuroanatomy of the mammalian trigeminal ganglion and mesencephalic trigeminal nucleus. Prog. Neurobiol., 66: 19–59.

Lazarov, N.E. and Dandov, A. (1998) Distribution of NADPH-diaphorase and nitric oxide synthase in the trigeminal ganglion and mesencephalic trigeminal nucleus of the cat: a histochemical and immunohistochemical study. Acta Anat., 163: 191–200.

Levine, J.D., Fields, H.L. and Basbaum, A.I. (1993) Peptides and the primary afferent nociceptor. J. Neurosci., 13: 2273–2286.

Porter, J.D. and Donaldson, I.M. (1991) The anatomical substrate for cat extraocular muscle proprioception. Neuroscience, 43: 473–481.

Ruskell, G.L. (1999) Extraocular muscle proprioceptors and proprioception. Prog. Retin. Eye Res., 18: 269–291.

Shigenaga, Y., Mitsuhiro, Y., Yoshida, A., Qin Cao, C. and Tsuru, H. (1988) Morphology of single mesencephalic trigeminal neurons innervating masseter muscle of the cat. Brain Res, 445: 392–399.

Stoyanova, I.I. and Lazarov, N.E. (2005) Localization of nitric oxide synthase in rat trigeminal primary afferent neurons using NADPH-diaphorase histochemistry. J. Mol. Histol., 36: 187–193.

Thomsen, L.L. and Olesen, J. (1996) The role of nitric oxide in migraine pain. In: Sandler M., Ferrari M. and Harnett S. (Eds.), Migraine Pharmacology and Genetics. Chapman & Hall, pp. 154–166.

C. Kennard & R.J. Leigh (Eds.)
Progress in Brain Research, Vol. 171
ISSN 0079-6123

CHAPTER 1.4

Functional anatomy of the extraocular muscles during vergence

Joseph L. Demer[1,2,3,4,*], Robert A. Clark[1], Benjamin T. Crane[5], Jun-Ru Tian[1], Anita Narasimhan[4] and Shaheen Karim[3]

[1]*Jules Stein Eye Institute, David Geffen Medical School at University of California, Los Angeles, CA, USA*
[2]*Department of Neurology, David Geffen Medical School at University of California, Los Angeles, CA, USA*
[3]*Neuroscience Interdepartmental Program, David Geffen Medical School at University of California, Los Angeles, CA, USA*
[4]*Bioengineering Interdepartmental Program, David Geffen Medical School at University of California, Los Angeles, CA, USA*
[5]*Department of Otolaryngology/Head-Neck Surgery, Johns Hopkins University, Baltimore, MD, USA*

Abstract: Magnetic resonance imaging (MRI) now enables precise visualisation of the mechanical state of the living human orbit, enabling inferences about the effects of mechanical factors on ocular kinematics. We used 3-dimensional (3D) magnetic search coil recordings and MRI to investigate the mechanical state of the orbit during vergence in humans. Horizontal convergence of 23° from a remote to a near target aligned on one eye was geometrically ideal, and was associated with lens thickening and extorsion of the rectus pulley array of the aligned eye with superior oblique muscle relaxation and inferior oblique muscle contraction. There was no rectus muscle co-contraction. Subjective fusion through a 1° vertical prism caused a clockwise (CW) torsion in both eyes, as well as variable vertical and horizontal vergences that seldom corresponded to prism amount or direction. MRI under these conditions did not show consistent torsion of the rectus pulley array, but a complex pattern of changes in rectus extraocular muscle (EOM) crossections, consistent with co-contraction. Binocular fusion during vergence is accomplished by complex, 3D eye rotations seldom achieving binocular retinal correspondence. Vergence eye movements are sometimes associated with changes in rectus EOM pulling directions, and may sometimes be associated with co-contraction. Thus, extraretinal information about eye position would appear necessary to interpret binocular correspondence, and to avoid diplopia.

Keywords: active pulley hypothesis; extraocular muscles; magnetic resonance imaging; pulleys; vergence

Introduction

Some important aspects of ocular kinematics are not specified by explicit neural commands, but are instead determined by mechanical properties of extraocular muscles (EOMs) and orbital connective tissues (Demer, 2004, 2006). With the head stationary, Listing's law (LL) constrains ocular torsion to that specified by rotation from a primary position about a single axis lying in Listing's plane (Tweed and Vilis, 1987). This torsion is not encoded in the discharges of motor

*Corresponding author. Tel.: + (310) 825-5931; Fax: + (310) 206-7826; E-mail: jld@ucla.edu

DOI: 10.1016/S0079-6123(08)00604-3

neurons innervating cyclovertical EOMs (Ghasia and Angelaki, 2005), yet even for abduction evoked by artificial abducens nerve stimulation, torsion conforms to LL (Klier et al., 2006). The ability of EOMs and orbital connective tissues to implement LL is explained by the active pulley hypothesis, which states that pulling directions of rectus EOMs are constrained by connective tissue pulleys that constitute their functional origins (Demer et al., 2000). Insertions of rectus EOM orbital layers on the rectus pulleys actively control their anteroposterior locations to make rectus EOM pulling directions depend on eye position as required to implement LL (Kono et al., 2002a). However, the transverse locations of rectus pulleys are separately under active control, due to insertion upon them of the orbital layers of the inferior oblique (IO) (Demer et al., 2003b) and superior oblique (SO) (Kono et al., 2005) EOMs. Smooth muscle in Tenon's fascia is also positioned for transversely shifting rectus EOM pulleys (Demer et al., 1995, 1997; Kono et al., 2002b; Miller et al., 2003). Magnetic resonance imaging (MRI) demonstrates that the human rectus pulley array counter-rotates during head tilting associated with ocular counter-rolling, a vestibulo-ocular reflex (Demer and Clark, 2005).

Methods

We used MRI to study the functional anatomy of the orbit during vergence, disconjugate ocular rotations bringing images into binocular sensory correspondence. Subjects were young adults confirmed by detailed ophthalmologic examination to have normal binocular alignment and stereo thresholds 40 arcs or better. All subjects gave informed consent according to a protocol approved by an institutional review board. Eye movements were recorded using binocular, dual-winding scleral magnetic search coils applied using topical anaesthesia. Although subjects' heads were firmly restrained, residual head movement was measured and compensated using dual magnetic search coils affixed to a mould of the upper teeth.

Proximal convergence was studied in response to shifting a small lighted target, aligned to one eye, from 500 cm to 15 cm, so that all of the required adduction was in the fellow eye; this was accomplished in an MRI scanner using fibre optics and a half mirror, as elsewhere described (Demer et al., 2003a). Vertical vergence was induced during viewing of a target cross at 20 or 400 cm by placing a 1° base up prism over one eye; ideally, this also requires movement of only the eye viewing through the prism. Orbital MRI was performed in a separate session using a surface coil array in a 1.5 T scanner as previously described (Demer et al., 2003a). Data were analysed only when subjects reported single binocular vision of the target.

Eye positions were represented as quaternions, and converted into Fick angles for display of time sequences and vergence (Crane et al., 2005). Rectus EOM cross sections and pulley locations were determined quantitatively after correction for head positioning and normalisation (Demer et al., 2003a).

Results

Horizontal proximal vergence

As described elsewhere for eight subjects (Demer et al., 2003a), fusion of a target at 15 cm aligned to one eye was associated with 22.4° convergence of the fellow eye, geometrically ideal for bifoveal correspondence. During convergence, there was also thickening of the crystalline lens indicating accommodation. However, there was an extorsional shift in rectus pulley locations, apparently in both the aligned and converging orbits (Fig. 1). This effect was quantified in the aligned orbit for the medial rectus (MR), inferior rectus (IR), and superior rectus (SR) pulleys, all of which shifted significantly in extorsion by 1.5°–2.0° ($P < 0.05$). Due to interference from the lacrimal gland, it was not possible to quantify shift of the lateral rectus (LR) pulley.

Vertical fusional vergence

Preliminary data are illustrated in Fig. 2 for Subject P fusing a 1° base up prism placed over either eye for 10 s intervals alternating with equal

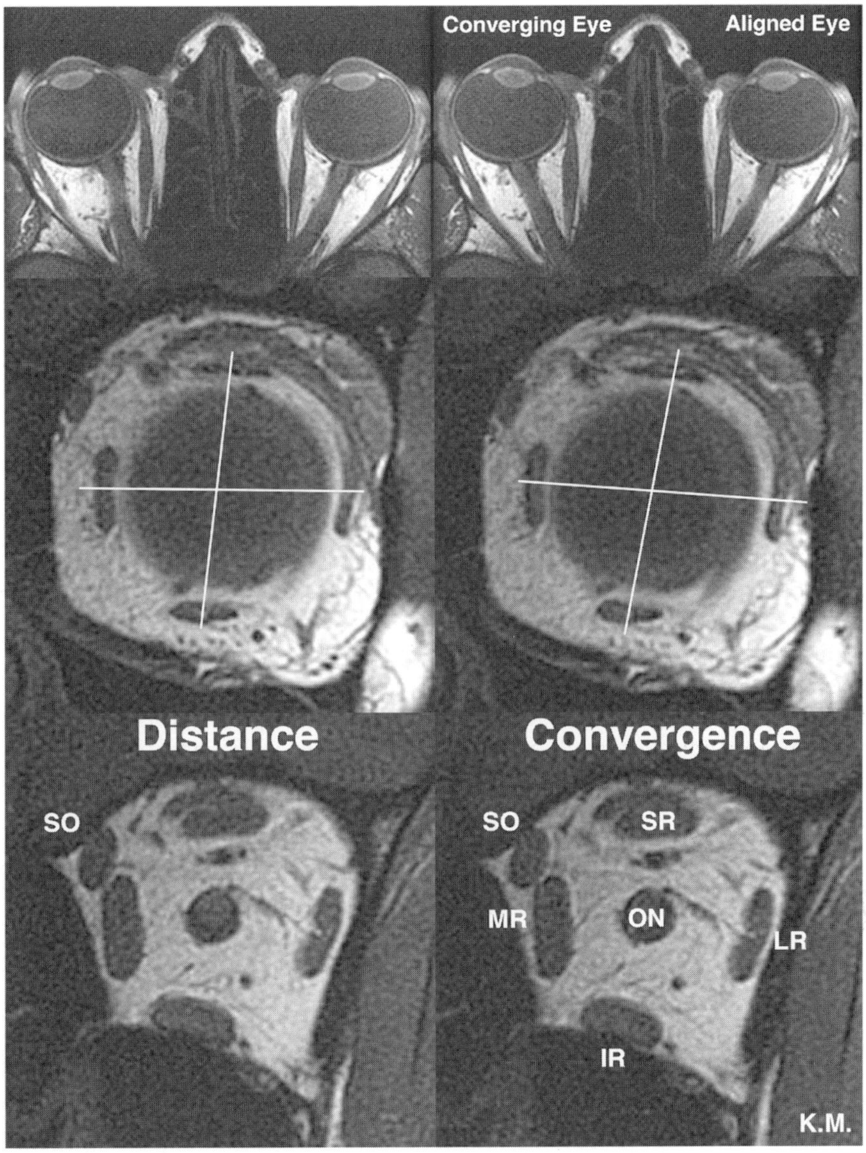

Fig. 1. MRI of orbits showing effects of proximal convergence on transverse pulley location. Target, aligned to left eye, moved from 500 cm (left column) to 15 cm (right column), evoking a 23° convergence movement in right eye only, seen in axial view in upper row. Anterior quasi-coronal images in the middle row at level of pulleys demonstrate transverse, extorsional shift highlighted by white lines through muscle centroids. Posterior quasi-coronal images in lower row suggest reduction in SO cross section in convergence associated with extorsion.

periods of normal vision. Seated upright, this subject reported achieving single binocular vision through the prism within 1–3 s of each transition. Typical of all eight subjects studied, Subject P exhibited a stereotypic pattern of binocular, 3-dimensional (3D) eye movements, regardless of which eye viewed through the prism: 1°–2° right eye infraduction, 2°–3° right eye abduction, and 4°–6° clockwise (CW) cycloversion, as illustrated in Figure 2 for prism placed before the left eye.

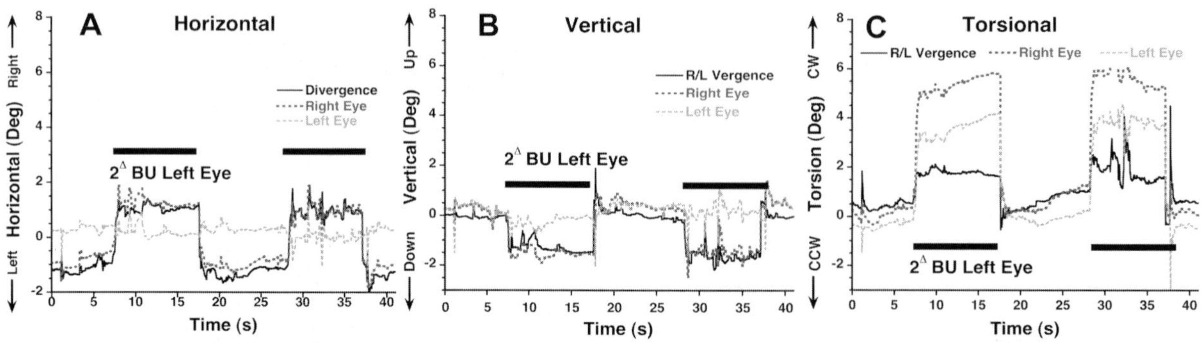

Fig. 2. Eye rotations during vertical fusional vergence in representative subject. A 1° (2Δ) base up prism was placed before left eye for 10 s intervals marked by black bars. Note paradoxical abduction and infraduction of right eye, with CW cycloversion of both eyes, while horizontal and vertical positions of left eye remained paradoxically unchanged despite viewing through prism.

Since a prism deviates gaze towards its base, optically appropriate vertical vergence for left eye base up prism would have required left, but not right, infraduction. Actual vertical vergence paradoxically increased vertical retinal disparity from 1° to ~2.5°, despite subjective binocular fusion. In most trials with base up prism over the left eye in every subject, vertical vergence increased, rather than decreased, objective retinal image disparity, and torsion was similarly directed in both eyes. Findings were similar at 20 and 400 cm viewing distances, and in ±20° horizontal and vertical secondary gaze positions. Subjective tilt was not reported.

Subjects were supine for MRI, and viewed as black cross surrounded by concentric squares 20 cm away inside the scanner. Using T2 fast spin echo, quasi-coronal MRI was performed perpendicular to the long axis of each orbit, and quasi-sagittal MRI transverse to this at 312 μm resolution. Representative images in Subject P showed no consistent torsion of the rectus pulleys associated with subjective fusion through a 1° prism base up over either eye, although the left SO cross section increased with prism over either eye consistent with CW torsion (Fig. 3).

Changes in maximal EOM cross sections represent contractility, and were computed for each EOM under prism viewing and control conditions (Fig. 4). It was sometimes impossible to distinguish the SR from the levator throughout the orbit, but it was treated in the same manner within subjects for all viewing conditions. Consistent with its abduction during prism viewing, right eye MR cross section declined with prism before either eye, with smaller change for the left eye. Correspondingly, LR cross-section generally increased with prism viewing. This suggests that the horizontal vergence response conformed to Sherrington's law of reciprocal innervation. Remaining changes in EOM cross section were frequently paradoxical. Despite the observation that the right eye infraducted during prism viewing, SR cross section generally increased or was unchanged, while IR cross section uniformly increased, with base up prism viewing by either eye. Despite observed CW cycloversion during prism viewing, corresponding to right eye extorsion and left eye intorsion, changes in IO cross section were similar in the two eyes. With left eye base up prism, both the right IO and SO cross sections increased. These paradoxical contractile changes suggest cyclovertical EOM co-contraction during vertical vergence, with all of the cyclovertical EOMs participating in the eye movements.

Discussion

Torsion is considered intrinsic to convergence. Excyclotorsion in convergence has been repeatedly confirmed in humans (Allen and Carter, 1967; Mok et al., 1992; Minken and Van Gisbergen, 1994; Mikhael et al., 1995; Bruno and van den

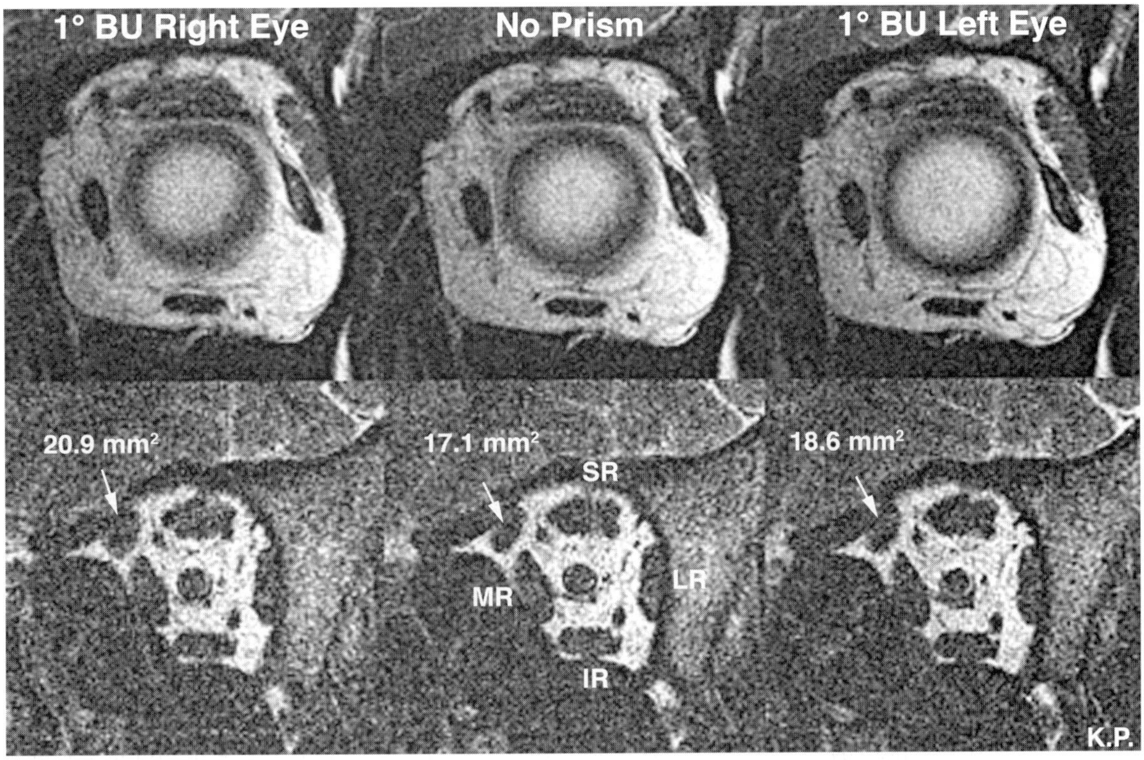

Fig. 3. Quasi-coronal MRI of left orbit of subject during vertical fusional vergence with a 1° prism. Images at the level of the pulleys (top row) suggest no transverse shifts in rectus muscle paths, although the superior oblique cross section (arrows) is increased during prism viewing by either eye (far left and far right columns) as compared with control (middle column).

Berg, 1997; Somani et al., 1998) and monkeys (Misslisch et al., 2001). During asymmetrical convergence, temporal rotation of Listing's plane occurs in both the aligned and converging eyes, independent of eye position (Steffen et al., 2000), corresponding to excyclotorsion in depression and incyclotorsion in elevation (Van Rijn and van den Berg, 1993; Somani et al., 1998; Kapoula et al., 1999). A form of Herring's law of equal innervation has been proposed for the vergence system, such that both eyes receive symmetric version commands for remote targets, and mirror symmetric vergence commands for near targets (Van Rijn and van den Berg, 1993). MRI suggests that convergence is associated with extorsion of the array of rectus EOM pulleys (Demer et al., 2003a), altering pulling directions of the rectus EOMs. This extorsion is apparently mediated by the coordinated actions of the oblique EOMs.

Vertical vergence has received less attention. Enright reported conjugate cyclovergence during vertical vergence, and proposed that vertical vergence may be mediated exclusively by the SO muscles (Enright, 1992). Van Rijn and Collewijn (1994) also observed conjugate cyclovergence whose direction was reversed by reversal of the vertical disparity, but discounted the exclusive role of the SO because the cyclovergence was independent of horizontal eye position. The present data are also incompatible with Enright's proposition, since we also consistently observed right eye infraduction with extorsion (Fig. 2). The SO alone produces infraduction with intorsion. However, in the current study, reversal of the vertical disparity by placing the prism before the opposite eye did not reverse the associated cycloversion, which was consistently CW. Enright also reported mediolateral globe translation (Enright, 1992),

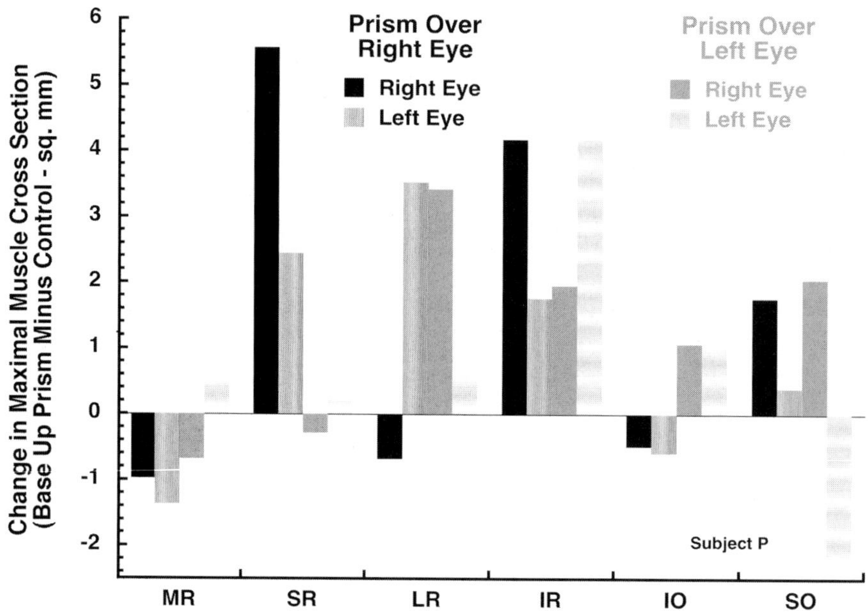

Fig. 4. Effect of fusion of a 1° base up vertical prism on maximal EOM cross sectional areas in Subject P. Note consistent reduction of MR and increase in LR cross section, consistent with reciprocal activity in observed divergence. However, SR and IR, and IO and SO, both frequently increased in cross section, suggesting co-contraction of cyclovertical EOMs.

which may be compatible with the present MRI evidence of co-contraction of multiple EOMs.

Hara et al. (1998) studied adaptation to gradually increasing vertical disparity presented using a virtual reality display. Most of the subjects studied by Hara et al. exhibited vertical movements of both eyes, even though the target was shifted for only one eye, but vertical vergence compensated for about 90% of the visual disparity. Hara et al. (1998) did not observe consistent ocular torsion during vertical vergence. The current study imposed transient vertical disparity using prisms, with strikingly different findings. Although all subjects reported binocular fusion within a few seconds, vertical vergence with base up prism before the left eye was typically misdirected, increasing, rather than decreasing the vertical image disparity. This effect often doubled the prism-induced vertical disparity. While vertical vergence responses exceeding stimulus demand have been reported (Perlmutter and Kertesz, 1982), the present finding of misdirected vertical vergence appears novel. Nevertheless, it is consistent with the recent finding in monkeys with superior oblique palsy that binocular vision increases, rather than decreases, a small vertical phoria (Shan et al., 2007).

Unlike the torsion of the rectus pulley evident by MRI during horizontal proximal convergence to a target aligned on one eye (Demer et al., 2003a), the large cycloversion induced by prism-induced vertical vergence in the present study was not associated with consistent torsion of the rectus pulley array, and was not associated with a subjective sensation of tilt. MRI also demonstrated lack of consistent reciprocal contractile pattern in the rectus and oblique EOMs, and rules out implementation of vertical vergence by any one EOM. Unless the mechanical effect is simply too small to be detected by MRI, absence of a simple mechanical basis for cycloversion during prism-induced vertical vergence suggests novel mechanisms may be playing a role, potentially including co-contraction with globe translation, or differential contraction of subsets of fibres within rectus or oblique EOMs. This possibility deserves further investigation, as does the possibility that the EOM mechanisms of vergence might depend strongly on visual conditions.

Stereopsis presumes internal information of where each eye is pointing in order to interpret retinal image disparity as a veridical sense of depth. In theory, efference copy could serve this purpose if it were accurate. Alternatively, proprioception may permit the interpretation of retinal disparities. Jean Büttner-Ennever has suggested that multiply innervated fibres in the EOMs form part of a proprioceptive system (Büttner-Ennever, 2007). A strong proprioceptive horizontal and vertical eye position signal exists in primary somatosensory cortex (Wang et al., 2007). The observed changes in torsional configuration of rectus pulleys during horizontal vergence, as well as the lack of pulley changes but paradoxical changes in binocular alignment during vertical vergence, argue that the brain probably monitors eye positions and mechanical states of EOMs. Merely knowing the foveal directions of the two eyes apparently does not uniquely determine binocular sensory correspondence, or determine the effect on eye position resulting from change in activity of any single EOM. Conversely, even afferent signals reflecting tension in individual EOMs would require interpretation in context of the tensions and pulling directions of all other EOMs. Given the complex mechanical interactions among EOMs in the orbit, proprioception would seem one essential input both to ocular motor control, and to sensory interpretation of binocular correspondence. In particular, torsional proprioception would seem to be a valuable sensory input to the ocular motor system, yet one unexplored at present.

Abbreviations

3D	three-dimensional
CW	clockwise
EOM	extraocular muscle
IO	inferior oblique
IR	inferior rectus
LL	Listing's law
LR	lateral rectus
MR	medial rectus
MRI	magnetic resonance imaging
SO	superior oblique
SR	superior rectus

Acknowledgements

Supported by National Institutes of Health EY08313. JLD is Leonard Apt Professor of Ophthalmology.

References

Allen, M.J. and Carter, J.H. (1967) The torsional component of the near reflex. Am. J. Optom., 44: 343–349.

Bruno, P. and van den Berg, A.V. (1997) Relative orientation of primary positions of the two eyes. Vision Res., 37: 935–947.

Büttner-Ennever, J.A. (2007) Anatomy of the oculomotor system. Dev. Ophthalmol., 40: 1–14.

Crane, B.T., Tian, J. and Demer, J.L. (2005) Kinematics of vertical saccades during the yaw vestibulo-ocular reflex in humans. Invest. Ophthalmol. Vis. Sci., 46: 2800–2809.

Demer, J.L. (2004) Pivotal role of orbital connective tissues in binocular alignment and strabismus. The Friedenwald lecture. Invest. Ophthalmol. Vis. Sci., 45: 729–738.

Demer, J.L. (2006) Current concepts of mechanical and neural factors in ocular motility. Curr. Opin. Neurol., 19: 4–13.

Demer, J.L. and Clark, R.A. (2005) Magnetic resonance imaging of human extraocular muscles during static ocular counter-rolling. J. Neurophysiol., 94: 3292–3302.

Demer, J.L., Kono, R. and Wright, W. (2003a) Magnetic resonance imaging of human extraocular muscles in convergence. J. Neurophysiol., 89: 2072–2085.

Demer, J.L., Miller, J.M., Poukens, V., Vinters, H.V. and Glasgow, B.J. (1995) Evidence for fibromuscular pulleys of the recti extraocular muscles. Invest. Ophthalmol. Vis. Sci., 36: 1125–1136.

Demer, J.L., Oh, S.Y., Clark, R.A. and Poukens, V. (2003b) Evidence for a pulley of the inferior oblique muscle. Invest. Ophthalmol. Vis. Sci., 44: 3856–3865.

Demer, J.L., Oh, S.Y. and Poukens, V. (2000) Evidence for active control of rectus extraocular muscle pulleys. Invest. Ophthalmol. Vis. Sci., 41: 1280–1290.

Demer, J.L., Poukens, V., Miller, J.M. and Micevych, P. (1997) Innervation of extraocular pulley smooth muscle in monkeys and humans. Invest. Ophthalmol. Vis. Sci., 38: 1774–1785.

Enright, J.T. (1992) Unexpected role of the oblique muscles in the human vertical fusional reflex. J. Physiol., 451: 279–293.

Ghasia, F.F. and Angelaki, D.E. (2005) Do motoneurons encode the noncommutativity of ocular rotations? Neuron, 47: 281–293.

Hara, N., Steffen, H., Roberts, D.C. and Zee, D.S. (1998) Effects of horizontal vergence on the motor and sensory components of vertical fusion. Invest. Ophthalmol. Vis. Sci., 39: 2268–2276.

Kapoula, Z., Bernotas, M. and Haslwanter, T. (1999) Listing's plane rotation with convergence: role of disparity, accommodation, and depth perception. Exp. Brain Res., 126: 175–186.

Klier, E.M., Meng, H. and Angelaki, D.E. (2006) Three-dimensional kinematics at the level of the oculomotor plant. J. Neurosci., 26: 2732–2737.

Kono, R., Clark, R.A. and Demer, J.L. (2002a) Active pulleys: magnetic resonance imaging of rectus muscle paths in tertiary gazes. Invest. Ophthalmol. Vis. Sci., 43: 2179–2188.

Kono, R., Poukens, V. and Demer, J.L. (2002b) Quantitative analysis of the structure of the human extraocular muscle pulley system. Invest. Ophthalmol. Vis. Sci., 43: 2923–2932.

Kono, R., Poukens, V. and Demer, J.L. (2005) Superior oblique muscle layers in monkeys and humans. Invest. Ophthalmol. Vis. Sci., 46: 2790–2799.

Mikhael, S., Nicolle, D. and Vilis, T. (1995) Rotation of Listing's plane by horizontal, vertical and oblique prism-induced vergence. Vision Res., 35: 3243–3254.

Miller, J.M., Demer, J.L., Poukens, V., Pavlowski, D.S., Nguyen, H.N. and Rossi, E.A. (2003) Extraocular connective tissue architecture. J. Vis., 3: 240–251.

Minken, A.W.H. and Van Gisbergen, J.A.M. (1994) A three-dimensional analysis of vergence movements at various levels of elevation. Exp. Brain Res., 101: 331–345.

Misslisch, H., Tweed, D. and Hess, B.J.M. (2001) Stereopsis outweighs gravity in the control of the eyes. J. Neurosci., 21: p. RC126 (online).

Mok, D., Ro, A., Cadera, W., Crawford, J.D. and Vilis, T. (1992) Rotation of Listing's plane during vergence. Vision Res., 32: 2055–2064.

Perlmutter, A. and Kertesz, A.E. (1982) Human vertical fusional response under open and closed loop stimulation to predictable and unpredictable disparity presentations. IEEE Trans. Biomed. Eng., 29: 57–61.

Shan, X., Tian, J., Ying, H.S., Quaia, C., Optican, L.M., Walker, M.F., Tamargo, R.J. and Zee, D.S. (2007) Acute superior oblique palsy in monkeys: I. Changes in static eye alignment. Invest. Ophthalmol. Vis. Sci., 48: 2602–2611.

Somani, R.A.B., Desouze, J.F.X., Tweed, D. and Vilis, T. (1998) Visual test of Listing's law during vergence. Vision Res., 38: 911–923.

Steffen, H., Walker, M.F. and Zee, D.S. (2000) Rotation of Listing's plane with convergence: independence from eye position. Invest. Ophthalmol. Vis. Sci., 41: 715–721.

Tweed, D. and Vilis, T. (1987) Implications of rotational kinematics for the oculomotor system in three dimensions. J. Neurophysiol., 58: 832–849.

Van Rijn, L.J. and Collewijn, H. (1994) Eye torsion associated with disparity-induced vertical vergence in humans. Vision Res., 34: 2307–2316.

Van Rijn, L.J. and van den Berg, A.V. (1993) Binocular eye orientation during fixations: Listing's law extended to include eye vergence. Vision Res., 33: 691–708.

Wang, X., Zhang, M., Cohen, I.S. and Goldberg, M.E. (2007) The proprioceptive representation of eye position in monkey primary somatosensory cortex. Nat. Neurosci., 10: 528–540.

C. Kennard & R.J. Leigh (Eds.)
Progress in Brain Research, Vol. 171
ISSN 0079-6123

CHAPTER 1.5

Induced extraocular muscle afferent signals: from pigeons to people

Paul C. Knox*

Division of Orthoptics, University of Liverpool, Thompson Yates Building, Brownlow Hill, Liverpool, L69 3GB, UK

Abstract: While the muscles that move the eyes, the extraocular muscles (EOM), are well endowed with proprioceptors, afferent signals from these receptors are usually assumed to play little or no role in the control of eye movement. In a series of experiments, a suction contact lens was used to impose movements on one eye, thus inducing afferent signals. Single unit activity was recorded centrally (to examine the interactions between EOM afferent signals and visual or vestibular signals), or the movements of the other eye were measured (to investigate their effects on the output of the oculomotor system). In a model preparation, the decerebrate pigeon, EOM afferent signals modified single unit activity in the medial vestibular nucleus, and the third and sixth motor nuclei, during sinusoidal vestibular stimulation. When one eye was moved to mimic the vestibulo-ocular reflex (VOR), movement faster than required for compensation for a given head velocity reduced the gain of single unit vestibular responses. In awake, alert pigeons the overall output of the VOR, as evidenced by movements of the other eye, was modified. In humans, when one eye was impeded, the saccades and smooth pursuit executed by the other eye were altered. Taken together, these results suggest that EOM afferent signals play a functional role in the shaping of eye movement.

Keywords: extraocular muscles; afferent signals; vestibulo-ocular reflex; saccades; smooth pursuit

Introduction

One of the challenges any organism faces is, given sensory information about the "state of affairs" in the external world, how to shape effective action to improve that "state of affairs" from the organism's point of view. For motor systems, this reduces to the need to monitor and evaluate responsive motor acts. It is of no surprise that most motor systems incorporate various mechanisms for providing feedback about the state of the effectors bringing about movements. One might expect the oculomotor system to be no exception in this regard. Indeed, the muscles which move the eyes in the orbits, the extraocular muscles (EOM), are well endowed with intramuscular receptors of various types (Cooper et al., 1955; Eberhorn et al., 2005). However, the role of proprioception in the oculomotor system has long been controversial (Donaldson, 2000). The sources of EOM proprioceptive signals, the utility and even the necessity for such signals, have all been questioned and debated over a long period.

A number of new developments have prompted a re-evaluation of these issues. Firstly, views of orbital mechanics and control have changed

*Corresponding author. Tel.: +44 (0)151 794 5736; Fax: +44 (0)151 794 5781; E-mail: pcknox@liv.ac.uk

DOI: 10.1016/S0079-6123(08)00605-5

markedly in recent years. The active muscle pulley hypothesis and the notion of separate mechanical and functional roles for, and control of, the global and orbital layers of the EOM (Demer et al., 2000; Ugolini et al., 2006), have led to new suggestions concerning the role of proprioceptive signals from a class of receptor (the palisade ending, Richmond et al., 1984) unique to the EOM and novel structures for modulating the afferent signals arising from them (Buttner-Ennever et al., 2002). Secondly, the demonstration that EOM proprioceptive signals provide somatosensory cortex with an eye position signal (Wang et al., 2007) has raised important functional questions such as the timescale over which EOM afferent signals influence processing. In this review, data from experiments designed to investigate the role of EOM afferent signals in both a model animal preparation (the decerebrate pigeon) and in human subjects will be examined.

Methods

In order to investigate the role of EOM afferent signals, it is necessary to manipulate them. One approach is to remove them either surgically (by cutting the afferent pathway; e.g., Fiorentini et al., 1985, 1986; Buisseret, 1995; Ventre-Dominey et al., 1996) or to block them (Steinbach and Smith, 1981; Wang et al., 2007). However, these approaches rely on knowing where afferent fibres run, and the pathway being accessible; both may be problematic. In order to induce signals, stretch of individual or groups of EOMs (Cooper et al., 1953; Buisseret and Maffei, 1977; Donaldson and Dixon, 1980; Dancause et al., 2007) and single muscle vibration have been used (Allin et al., 1996). Such techniques might provide a highly non-physiological input to central control structures via the afferent pathway. An alternative is to move the whole globe, producing an overall pattern of lengthening and shortening the EOM analogous to that occurring during eye movement (Ashton et al., 1984a; Donaldson and Knox, 1990; Gauthier et al., 1994). A further advantage of this technique is that stimuli may be described parametrically, with reference to the direction, speed, and amplitude of the imposed movement.

An opaque suction contact lens was used to impose movements on one eye in a range of species in single unit recording studies (trout, Ashton et al., 1989; toad, Ashton et al., 1984b; cat, Ashton et al., 1984a). This was done while imparting other stimuli such as vestibular or visual stimuli. A particularly extensive series of experiments was conducted in the pigeon. Pigeons have excellent vision, and a good repertoire of oculomotor behaviour (Nye, 1969; Hodos et al., 1985). The pigeon was also used because both visual and vestibular processing could be explored in a decerebrate, paralysed preparation, free of the complicating effects of general anaesthesia.

Both unparalysed decerebrates and alert animals were also used in behavioural experiments, the results of which revealed that altering the movements of one eye could lead to detectable changes of the movements of the other eye. Therefore in a series of human experiments, a similar lens system (using a fenestrated scleral lens) was used to impede movements of one eye while subjects responded to visual targets with the other eye. Once local anaesthetic was instilled into the right eye, the lens was placed on that eye, and gentle suction applied. The lens carried a stalk that fitted into a fixed holder. By observation it was clear that while the eye continued to move slightly against the lens, its movements were reduced relative to the fellow eye. Lenses were kept in place for short periods of up to 5 min. In these circumstances they were well tolerated.

Single unit results

The primary afferent pathway in the pigeon follows the familiar vertebrate pattern, with primary afferent cell bodies in the ophthalmic division of trigeminal ganglion (Hayman et al., 1995) from which Fahy and Donaldson (1998) recorded their responses to passive eye movement. Single units exhibited both amplitude and velocity sensitivity, and responded to only one direction of eye movement. Sinusoidal oscillation in the horizontal plane was combined with passive eye movement of one eye; the eye was moved rapidly to an eccentric position, held eccentrically for a short period, and then returned to the central position (Donaldson and Knox, 1990). It was clear from unit activity recorded in the medial vestibular nucleus, and the III and VI nuclei, as well as the surrounding reticular formation, that EOM afferent signals conveyed information related to the size, direction, and speed of eye movement (Donaldson and Knox, 1990, 1991). Further, the effects did not build up over a number of trials, but were evident within at least a few tens of milliseconds of the eye being moved. They were thus consistent with the online control of vestibularly generated eye movement in the short term, as well as with longer term, parametric, changes in eye movement control.

As useful as these experiments were, they failed to take account of the functional context. During sinusoidal vestibular stimulation the objective of the oculomotor system is to counter-rotate the eye, with approximately the same peak speed as the head. This suggested a different experiment, which we called the "artificial VOR" (AVOR) experiment (Fig. 1a). We sought to mimic the slow phase of the VOR by imposing a sinusoidal eye movement on one eye 180° out of phase with the head. At an appropriate velocity (in this case 22°/s) the imposed eye movement was compensatory. But we could also introduce functionally specific errors by moving the eye either faster or slower than required for compensation. Note that in our earlier experiments the velocity of the imposed eye movements had been much higher (of the order of 120°/s), although still within the oculomotor range of the pigeon. However, the vestibular responses of single units were sensitive to low velocity errors when induced in this functionally relevant manner (Fig. 1b–i). This sensitivity had an important functional implication. Assuming that the response when the eye velocity was compensatory was the "normal" response, then when the eye was moved more slowly than required for compensation the single unit gain increased; when it was moved more quickly than required, the gain decreased. When the range of eye velocities was held constant, and head velocity was varied over a range, unit firing was also altered (Donaldson and Knox, 1993). This suggests that a comparison is being made between the desired eye velocity or position, and the actual eye velocity and/or position with the resultant error signal altering unit firing.

Behavioural results on the VOR

It is possible that these single unit results do not reflect what is going on in the VOR pathway as a whole. And they were recorded from a reduced (i.e., a decerebrate), paralysed, preparation. However, we found that in an unparalysed, though still decerebrate preparation, during the VOR, EOM afferent signals altered the electromyogram recorded from the muscles (Knox and Donaldson, 1991). We were able to improve on this by repeating the AVOR experiment in awake, alert pigeons (Donaldson and Knox, 2000). The eye movements of head-restrained adult pigeons were recorded from one eye using a chronically implanted scleral search coil, during sinusoidal oscillation in the horizontal plane, while the movements of the other (locally anaesthetised) eye were controlled with a suction contact lens. We observed essentially the same pattern of results as before: when one eye was moved more quickly than required for compensation the gain of the VOR, as measured from the movements of the other eye, was reduced compared to that observed when the eye velocity was compensatory. When the eye was moved more slowly than required, the gain increased. In pilot VOR experiments in human subjects, we found that although the effects did not reach statistical significance, a similar monotonic decline in VOR gain was observed, as the peak speed of imposed eye movement was increased (Knox and Donaldson, 1993).

Experiments on human oculomotor control

The pigeon and human VOR results suggested that by imposing movements on, or interfering with, the movements of one eye, it might be possible to observe effects by measuring the movements of the other eye. We therefore decided to use suction contact lenses on human subjects in order to investigate saccades and smooth pursuit. We were encouraged by the results of Gauthier et al. (1994) who demonstrated that passive deviation of one eye

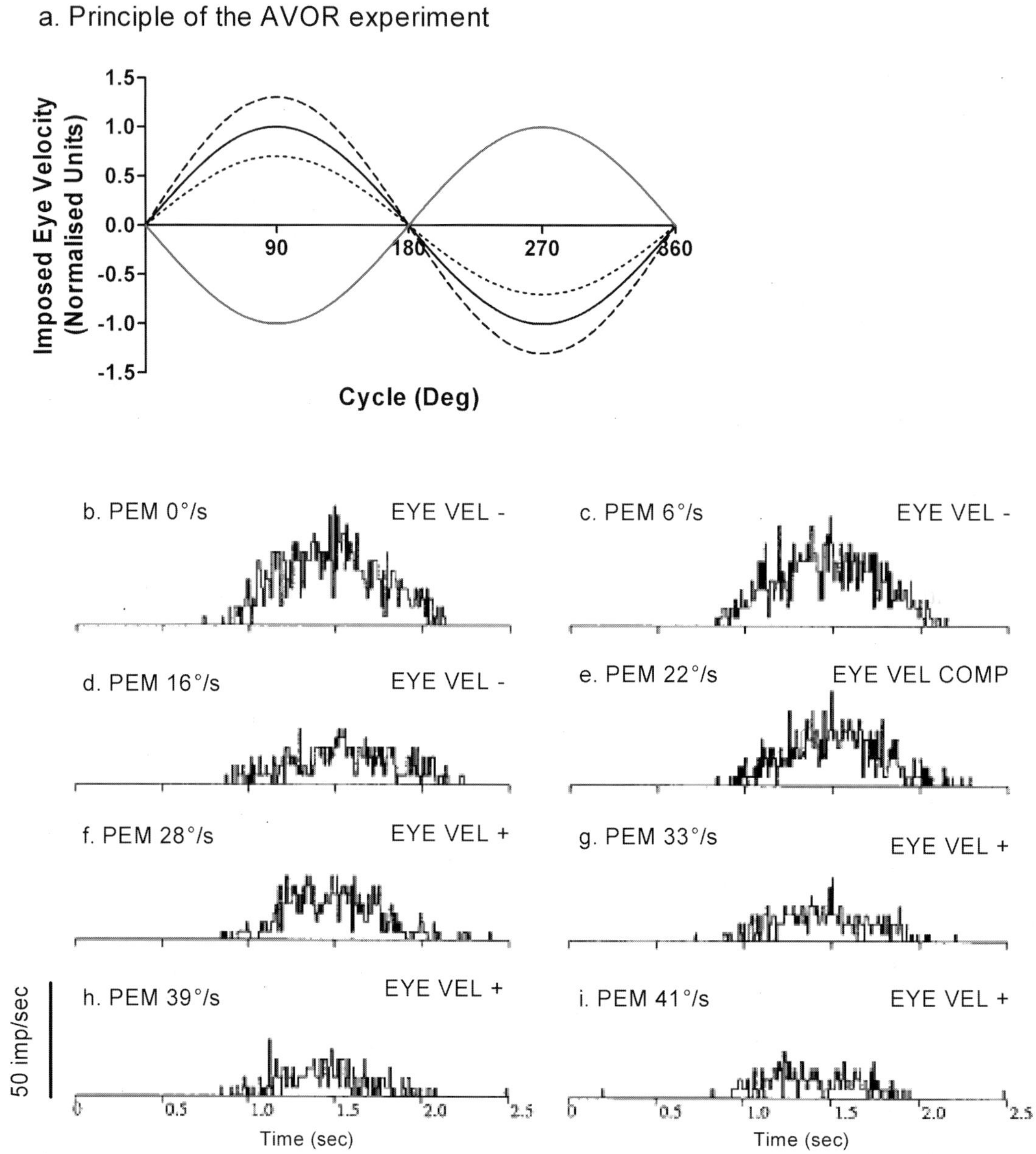

Fig. 1. a: Principle of the "artificial VOR" experiment. Grey solid line: head velocity; black solid line: compensatory eye velocity; dashed line: eye velocity greater than that required for compensation for head velocity; broken line: eye velocity less head velocity. b–i: Peristimulus time histograms illustrating the response of a single unit recorded in the medial vestibular nucleus of the decerebrate pigeon during sinusoidal oscillation in the horizontal plane ($\pm 8^{\circ}$ at 0.4 Hz). In c–i the vestibular stimulus is combined with the passive eye movement of one eye at various velocities as shown. Each panel also shows whether the imposed eye velocity was less (c, d; EYE VEL−) or greater than (f–i; EYE VEL+) the compensatory velocity (e). Adapted with permission from Donaldson and Knox (1993).

in humans for relatively short periods could alter eye alignment in the absence of binocular vision.

Subjects sat facing a stimulus monitor, with their heads stabilised using a chin rest and cheek pads; the movements of the left eye were recorded using infrared oculography. A lens was used to impede the movements of the right eye. In the first experiment, each trial began with a randomised fixation time; a stationary saccade target then appeared (duration 200 ms) at either 5° or 10° to the left or right (Knox et al., 2000). In a second experiment the stimulus was a step-ramp smooth pursuit stimulus (step amplitude 5°, target speed 14°/s; Weir and Knox, 2001).

When the left eye was impeded, the amplitude of saccades executed by the right eye was reduced by 19% averaged across subjects and target positions (Fig. 2). However, the peak velocity and duration of saccades were appropriate for the reduced amplitude. Although the lens was in place for only a few minutes at most (up to approximately 5 min), when it was removed the saccade amplitude returned towards the control values, but remained slightly reduced by approximately 5%. We also examined average saccade amplitude for each target position, for each trial in order, over the three subjects who participated (Fig. 3). This analysis revealed that from the start of the run saccade amplitude was reduced. This is not the same as claiming it was reduced as soon as the lens was in place. It took several tens of seconds to place the lens on the eye, place the stalk into the holder, and initiate the run. During this period, while we sought to keep the subject's vision to a minimum, it is possible that an adaptive effect could have built up. All we can say is that, as Fig. 3 shows, the saccade amplitude was reduced from the first trial in the run. The amplitude effect was relatively consistent throughout the run of trials, neither building up, nor adapting out. As the target presentation time was 200 ms, there was no target present when the reduced amplitude saccades landed. This meant there was no retinal error signal, which might have caused adaptive increases in saccade amplitude. It would be interesting to repeat the experiment with a longer target presentation time in order to establish whether such adaptive effects occur.

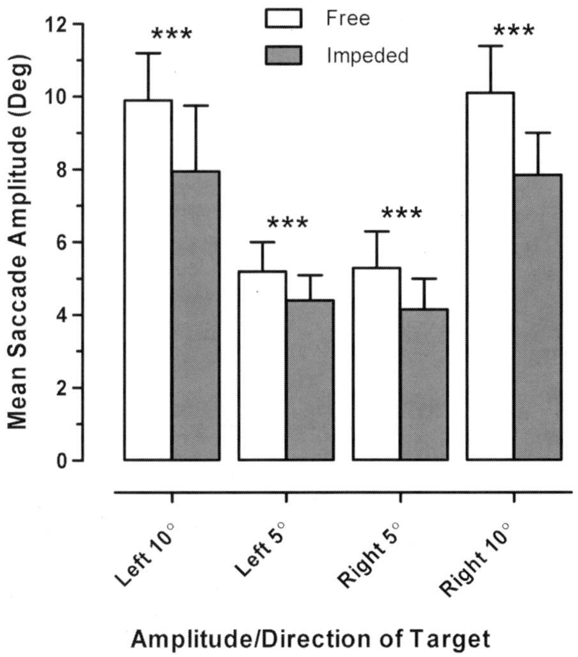

Fig. 2. Pooled mean (±SD) saccade amplitude of the left eye before (open bars "Free") and while (grey bars "Impeded") the right eye was impeded. Data averaged for three human subjects. All column pairs are statistically significantly different (t-test, $p<0.001$).

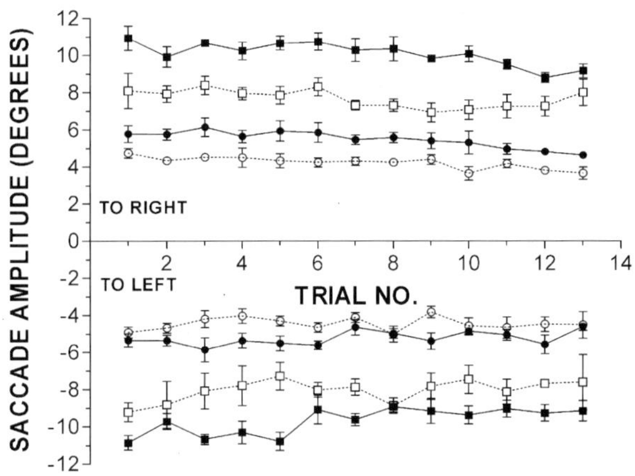

Fig. 3. Effect of impeding one eye on the saccades amplitude of the other eye. Trial-by-trial pooled (±SEM) mean of left eye saccade amplitude before (open symbols) and while (filled symbols) the right eye was impeded. Squares: data from right and left 10° trials; circles: from right and left 5° trials. Data averaged across three subjects. Adapted with permission from Knox et al. (2000).

The smooth pursuit results were generally similar. When one eye was impeded, the initial eye acceleration of the other eye was reduced by approximately 20%, as was the eye velocity 100 ms after pursuit was initiated, i.e., at the end of the open-loop period. When eye velocity was averaged over four 20 ms epochs, with the first epoch beginning at the time of pursuit initiation, the velocity reductions compared to the control (non-impeded) values were evident from early in the pursuit response, and were statistically significant from the third epoch (i.e., 40–60 ms into the response). Again, the velocity reduction effects were present from the first trial, and were unchanged for as long as the other eye was impeded.

These results remain to be confirmed and need to be extended. We have no means of demonstrating independently that the effects which we observed were the result of stimulating intramuscular proprioceptive signals. However, given the existence of various intramuscular receptors in humans, the most parsimonious explanation would be that it is afferent signals from these receptors that mediate the effects on saccades and pursuit. These signals have rapid access to the oculomotor control circuitry. Our results are consistent with both short-term feedback control and rapid and long-term parametric adaptation. Where might such feedback signals interact with the central oculomotor control circuitry? Given that the saccades, although smaller when one eye was impeded, still clearly exhibited the normal main sequence relationships, a site beyond the brainstem is implied. Given also that both saccades and smooth pursuit were modified in similar ways (i.e., we observed gain reductions) a structure involved in both pathways is probably involved. The cerebellum is important for both saccades and pursuit, is involved in gain control, and plays a role in adaptive responses. It would therefore seem to be a likely candidate.

Summary

It has long been generally accepted that the EOMs are endowed with intramuscular receptors of various types. Further, we and others have shown that the signals arising from these receptors, or their effects, can be recorded from a wide range of locations in the central nervous system involved in visuomotor processing. Recently intriguing results have reopened old questions and raised new ones. One recent report has reopened the issue of whether the EOM exhibit stretch reflexes (Dancause et al., 2007) while another has indicated that eye position signals in the primate primary somatosensory cortex appear to be proprioceptive in origin (Wang et al., 2007). Along with the experiments reviewed here, these suggest that the role of EOM afferent signals in visuomotor control remains a conundrum worthy of investigation.

Acknowledgements

All of the experiments I have described were collaborative efforts, and I am grateful to all of those colleagues who have been involved in them. However, particular thanks is due to Iain Donaldson who introduced me to these issues almost 20 years ago, and who for 10 of those 20 years was patient and skilful mentor.

References

Allin, F., Velay, J.-L. and Bouquerel, A. (1996) Shift in saccadic direction induced in humans by proprioceptive manipulation: a comparison between memory-guided and visually guided saccades. Exp. Brain Res., 110: 473–481.

Ashton, J.A., Boddy, A. and Donaldson, I.M.L. (1984a) Directional sensitivity in the responses of units in cat primary visual cortex to passive eye movement. Neuroscience, 13: 653–662.

Ashton, J.A., Boddy, A. and Donaldson, I.M.L. (1984b) Input from proprioceptors in the extrinsic ocular muscles to the vestibular nuclei in the Giant Toad, *Bufo marinus*. Exp. Brain Res., 53: 409–419.

Ashton, J.A., Milleret, C. and Donaldson, I.M.L. (1989) Effects of afferent signals from the extraocular muscles upon units in the cerebellum, vestibular nuclear complex and oculomotor nucleus of the trout. Neuroscience, 31: 529–541.

Buisseret, P. (1995) Influence of extraocular muscle proprioception on vision. Physiol. Rev., 75: 323–338.

Buisseret, P. and Maffei, L. (1977) Extraocular proprioceptive projections to the visual cortex. Exp. Brain Res., 28: 421–425.

Buttner-Ennever, J.A., Horn, A.K.E., Graf, W. and Ugolini, G. (2002) Modern concepts of brainstem anatomy: from extraocular motoneurons to proprioceptive pathways. Ann. N.Y. Acad. Sci., 956: 75–84.

Cooper, S., Daniel, P. and Whitteridge, D. (1953) Nerve impulses in the brain stem of the goat. Responses with long latencies obtained by stretching the extrinsic eye muscles. J. Physiol., 120: 491–513.

Cooper, S., Daniel, P.M. and Whitteridge, D. (1955) Muscle spindles and other sensory endings in the extrinsic eye muscles: the physiology and anatomy of these receptors and of their connexions with the brain-stem. Brain, 78: 564–583.

Dancause, N., Taylor, M., Plautz, E., Radel, J., Whittaker, T., Nudo, R. and Feldman, A. (2007) A stretch reflex in extraocular muscles of species purportedly lacking muscle spindles. Exp. Brain Res., 180: 15–21.

Demer, J.L., Oh, S.Y. and Poukens, V. (2000) Evidence for active control of rectus extraocular muscle pulleys. Invest. Ophthalmol. Vis. Sci., 41: 1280–1290.

Donaldson, I.M.L. (2000) The functions of the proprioceptors of the eye muscles. Philos. Trans. R. Soc. Lond. B, 355: 1685–1754.

Donaldson, I.M.L. and Dixon, R.A. (1980) Excitation of units in the lateral geniculate and contiguous nuclei of the cat by stretch of extrinsic ocular muscles. Exp. Brain Res., 38: 245–255.

Donaldson, I.M.L. and Knox, P.C. (1990) Directionally-specific effects of afferent signals from the extraocular muscles upon responses in the pigeon brainstem to horizontal vestibular stimulation. Neuroscience, 38: 145–161.

Donaldson, I.M.L. and Knox, P.C. (1991) Afferent signals from pigeon extraocular muscles modify the vestibular responses of units in the abducens nucleus. Proc. R. Soc. Lond. B, 244: 233–239.

Donaldson, I.M.L. and Knox, P.C. (1993) Evidence for corrective effects of afferent signals from the extraocular muscles on single units in the pigeon vestibulo-oculomotor system. Exp. Brain Res., 95: 240–250.

Donaldson, I.M.L. and Knox, P.C. (2000) Afferent signals from the extraocular muscles affect the gain of the horizontal vestibulo-ocular reflex in the alert pigeon. Vision Res., 40: 1001–1011.

Eberhorn, A.C., Horn, A.K.E., Eberhorn, N., Fischer, P., Boergen, K.-P. and Buttner-Ennever, J.A. (2005) Palisade endings in extraocular eye muscles revealed by SNAP-25 immunoreactivity. J. Anat., 206: 307–315.

Fahy, F.L. and Donaldson, I.M.L. (1998) Signals of eye position and velocity in the first-order afferents from pigeon extraocular muscles. Vision Res., 38: 1795–1804.

Fiorentini, A., Cenni, M.C. and Maffei, L. (1986) Impairment of stereoacuity in cats with oculomotor proprioceptive deafferentation. Exp. Brain Res., 63: 364–368.

Fiorentini, A., Maffei, L., Cenni, M.C. and Tacchi, A. (1985) Deafferentation of oculomotor proprioception affects depth discrimination in adult cats. Exp. Brain Res., 59: 296–300.

Gauthier, G.M., Vercher, J.L. and Zee, D.S. (1994) Changes in ocular alignment and pointing accuracy after sustained passive rotation of one eye. Vision Res., 34: 2613–2627.

Hayman, M.R., Donaldson, J.P. and Doanldson, I.M.L. (1995) The primary afferent pathway of extraocular muscle proprioception in the pigeon. Neuroscience, 69: 671–683.

Hodos, W., Bessette, B.B., Macko, K.A. and Weiss, S.R.B. (1985) Normative data for pigeon vision. Vision Res., 25: 1525–1527.

Knox, P.C. and Donaldson, I.M.L. (1991) Afferent signals from the extraocular muscles of the pigeon modify the electromyogram of these muscles during the vestibulo-ocular reflex. Proc. R. Soc. Lond. B, 246: 243–250.

Knox, P.C. and Donaldson, I.M.L. (1993) Do extraocular muscle afferent signals play a role in the human vestibulo-ocular reflex? Soc. Neurosci. Abstr., 19: p. 858.

Knox, P.C., Weir, C.R. and Murphy, P.J. (2000) Modification of visually guided saccades by a nonvisual afferent feedback signal. Invest. Ophthalmol. Vis. Sci., 41: 2561–2565.

Nye, P. (1969) The monocular eye movements of the pigeon. Vision Res., 9: 133–144.

Richmond, F.J., Johnston, W.S., Baker, R.S. and Steinbach, M.J. (1984) Palisade endings in human extraocular muscles. Invest. Ophthalmol. Vis. Sci., 25: 471–476.

Steinbach, M.J. and Smith, D.R. (1981) Spatial localization after strabismus surgery: evidence for inflow. Science, 213: 1407–1409.

Ugolini, G., Doldan Dans, M., Dubayle, D., Brandi, A., Büttner-Ennever, J. and Graf, W. (2006) Horizontal eye movement networks in primates as revealed by retrograde transneuronal transfer of rabies virus: differences in monosynaptic input to "slow" and "fast" abducens motoneurons. J. Comp. Neurol., 498: 762–785.

Ventre-Dominey, J., Dominey, P.F. and Sindou, M. (1996) Extraocular proprioception is required for spatial localization in man. NeuroReport, 7: 1531–1535.

Wang, X., Zhang, M., Cohen, I.S. and Goldberg, M.E. (2007) The proprioceptive representation of eye position in monkey primary somatosensory cortex. Nat. Neurosci., 10: 640–646.

Weir, C.R. and Knox, P.C. (2001) Modification of smooth pursuit initiation by a nonvisual, afferent feedback signal. Invest. Ophthalmol. Vis. Sci., 42: 2297–2302.

C. Kennard & R.J. Leigh (Eds.)
Progress in Brain Research, Vol. 171
ISSN 0079-6123

CHAPTER 1.6

Monkey primary somatosensory cortex has a proprioceptive representation of eye position

Mingsha Zhang[1,*], Xiaolan Wang[1] and Michael E. Goldberg[1,2]

[1] *Mahoney Center for Brain and Behavior, Department of Neuroscience, Columbia University College of Physicians and Surgeons, and the New York State Psychiatric Institute, New York, NY, USA*
[2] *Departments of Neuroscience, Psychiatry and Ophthalmology, Columbia University College of Physicians and Surgeons, New York, NY, USA*

Abstract: The visual system is tied to the retina. Because the eyes move in the orbit, and the head moves on the body, accurate location of an object in extrapersonal space cannot simply result from a visual signal. Instead, the retinal signal must be combined with an estimate of where the eyes are in the orbit, and where the head is in space, to calculate where that object is relative to the observer. There is abundant evidence for eye position signals in various areas of the visual cortex. However, the source of that eye position signal is unknown. Estimates of eye position can arise from two different sources. One is *outflow*, an 'efference copy' or 'corollary discharge' which might arise from some eye position signal used to specify eye position for the eye muscles. The second source is *inflow*, a direct proprioceptive signal from the muscles themselves. Nevertheless, neither a proprioceptive representation of eye position nor corollary discharge of a motor command for eye position has ever been demonstrated unambiguously in the cerebral cortex. We recently discovered the neuronal representation of proprioceptive eye position signal in monkey primary somatosensory cortex.

Keywords: extraocular muscle; proprioception; eye position; primary somatosensory cortex; primate; electrophysiology

Introduction

The cerebral cortex must have access to an eye position signal, because humans can report passive changes in eye position in total darkness (Skavenski, 1972). Visual responses in posterior parietal (Sakata et al., 1980; Andersen and Mountcastle, 1983; Mullette-Gillman et al., 2005) and prestriate (Nakamura and Colby, 2002) cortex as well as the superior colliculus (van Opstal et al., 1995) and even the lateral geniculate nucleus (Lal et al., 1990) are modulated by eye position. The source of the cortical and subcortical eye position information is unknown. The general assumption in the literature for decades has been that the cortical eye position signal arises from an outflow signal (Sakata et al., 1980; Andersen and Mountcastle, 1983), an efference copy or corollary discharge signal arising from the eye position component of the motor command, although the possibility that this signal could arise from proprioception has also been raised (Campos et al., 2006). There is a somatotopic representation of the entire body in primary somatosensory cortex, called the

*Corresponding author. Tel.: +1 (212) 543-6931, Ext. 100; Fax: +1 (212) 543-5816; E-mail: mz2132@columbia.edu

DOI: 10.1016/S0079-6123(08)00606-7

'homunculus' in the human. Up until now there has been a hole in the homunculus where the eye muscles should be. The monkey extraocular muscles have a potential sensory structure, the palisade ending or myotendinous cylinder, which resembles a fusimotor receptor in some aspects (Donaldson, 2000; Buttner-Ennever and Horn, 2002; Eberhorn et al., 2005). Neurons in monkey eye muscles project to the spinal trigeminal nucleus and the nucleus cuneatus (Porter et al., 1983), but their thalamic and cortical projections have never been determined. Here we report a sensory representation of eye position in monkey primary somatosensory cortex, dependent on signals from the contralateral orbit (Wang et al., 2007).

Methods

General

Both the New York State Psychiatric Institute and Columbia University Medical Center Institutional Animal Care and Use Committees approved all animal procedures, and certified them to be in compliance with the NIH Guidelines for the Care and Use of Experimental Animals. We prepared monkeys for physiological recording using standard sterile surgical techniques, with ketamine induction and isofluorane endotracheal anaesthesia, with the implantation of a head-holding socket, 2 cm recording chambers positioned at 20 mm A, 27 mm L, and a subconjunctival magnetic search coil in each eye (Judge et al., 1980).

Retrobulbar blocks

Using the technique of Richmond and Wurtz (1980), we held down the animal's eyelid, and passed an Atkinson needle through the upper lateral lid into upper retrobulbar space; we then inject 2 ml of 2% lidocaine without epinephrine and withdrew the needle. If necessary we injected another 0.5 ml of 2% lidocaine into the lower retrobulbar space. The animals tolerated the procedure perfectly well. Every second during the procedure they received a given free liquid reward which they consumed avidly even though they had a needle in their subconjunctival space, and once we withdrew the needle they resumed working immediately. The monkeys worked as efficiently when only one eye moved as they did when both eyes moved. Presumably the lidocaine in the orbit prevented their feeling any possible orbital sensation, and the ptosis prevented diplopia which might have been an uncomfortable result of the retrobulbar block.

Data analysis

We wrote analysis programmes in Matlab, using curve-fitting and statistics toolboxes (MathWorks). Unless otherwise stated, all neural values were calculated from the first 1000 ms after the end of the saccade. Neural latencies were determined using a modified Poisson method (Bisley et al., 2004).

Results

Neurons in somatosensory cortex represent eye position

We trained two Rhesus monkeys (*Macaca mulatta*) to fixate a spot of light and make saccades for liquid reward (Wurtz, 1969). We implanted a 2 cm recording cylinder over somatosensory cortex, and searched in the cylinder until we found multi unit activity with tactile receptive fields on the monkey's brow. We then advanced the electrode deep into the sulcus until we found multiunit activity which had eye position sensitivity, at a depth of 9.5 mm or more. The eye position neurons were in a thin layer, with only one or two cells isolable in each electrode penetration. We found eye position cells in a 4 mm by 4 mm area of the cylinder. Once we found the area we could reliably find eye position neurons in almost every penetration. Not every penetration had obvious tactile receptive fields on the face.

We tested the eye position sensitivity of each neuron by requiring the monkey to fixate nine different points, one at the centre of the screen and the others in a radial array of points 15° from the centre of the screen. An example neuron with a tonic eye position signal for an eye position upward of the centre of gaze is shown in Fig. 1a. We found 88 (33 from monkey C, 55 from monkey W) neurons which informally appeared to have eye position sensitivity. Of these, 81 neurons had a significant ($P<0.05$) main effect of eye position by ANOVA (66 significant at a $P<0.01$ level). Seventy six neurons had a significant difference ($P<0.05$) between maximal and minimal responses by *t*-test (65 significant at $P<0.01$ level).

We did a T1-weighted MRI of a tungsten electrode left in place at the site of an eye position neuron in each monkey (Fig. 1b). The eye position area was in the depth of the central sulcus, in a position consistent with area 3a and the representation of the ophthalmic branch of the trigeminal nerve (Krubitzer et al., 2004). All the cells were tuned for a particular radial direction of movement. Sixty five cells fit a Gaussian distribution for direction of eye position from the centre of gaze with $R^2>0.7$, which is significant at the $P<0.05$ level (Fig. 1c). All directions of eccentric position were represented in a single hemisphere without any apparent directional preponderance. Figure 2a shows the tuning maxima of the Gaussian for all cells which fit a Gaussian with $R^2>0.7$). We were unable to discern any topographic organization or map of orbital position preference.

All position signals increased monotonically with increasing orbital eccentricity. We studied the amplitude tuning of 12 cells using 7 points along their preferred direction. All cells had a measurable baseline activity in the off-direction, and began to increase their discharge as the eye passed near the midline of the orbit (Fig. 2b). We determined this inflection point by eye, and fit different straight lines to each segment. The fit to a straight line in the responsive region was significant for 10/12 cells, with the fits of 8 cells significant at $P<0.0001$. There was no correlation between eye position and signal in the off region.

The signal was tonic, in that it was maintained for as long as we required the monkey to fixate, with the exception of one neuron in which the activity began to decay after 600 ms of fixation. However, roughly 70% of the neurons also exhibited a phasic excitation in addition to the tonic signal when the monkey made a saccade towards the preferred orbital position of the neuron (Fig. 2c), and a phasic suppression of response when the monkey made a saccade away from the preferred position (Fig. 2f). We measured the phasic (0–300 ms from the end of the saccade) and tonic (300–1000 ms from the end of the saccade) responses of 22 neurons as the monkey made saccades from a consistent fixation point. In this sample, 6/22 had no phasic responses. Suppression had a much shorter latency than excitation: the mean latency of the off-suppression was 30 ms from the beginning of the saccade, and the mean latency of the on-excitation was 80 ms from the beginning of the saccade (Fig. 2d). The latencies were correlated (Fig. 2g, $N=22$, $R^2=0.43$, $P=0.0009$). When calculated from end of the saccade the mean latencies were 30 ms before the end of the saccade for the off-suppression, and 5 ms before the end of the saccade for the on-excitation (Fig. 2e). Latencies were correlated with this alignment, too (Fig. 2h, $N=20$, $R^2=0.37$, $P=0.0043$).

We propose that the later generated but longer survived tonic activity reflects the eye position signal more accurately and reliably than the phasic component. We studied the tonic signal during smooth pursuit and during the vestibuloocular reflex (VOR) when we rotated the chair of the monkey whose head was fixed to the chair. In both smooth pursuit and the VOR cases the major Fourier component of the neuronal signal had the same frequency as the chair rotation and the eye position, but lagged the eye position by an average of 94 ms in eight cells. In contrast, there was no correlation between the peak activity of the phasic burst and the maximum velocity of the saccades. Since the signal on the oculomotor neurons (Robinson, 1970) has both a velocity and position signal manifest during saccades, the eye position signal cannot be a simple transduction of the signal on the extraocular muscles.

The eye position signal is proprioceptive and not a motor corollary

The location of the eye position signal, in a region of somatosensory cortex dedicated to muscle proprioception, suggests a proprioceptive origin.

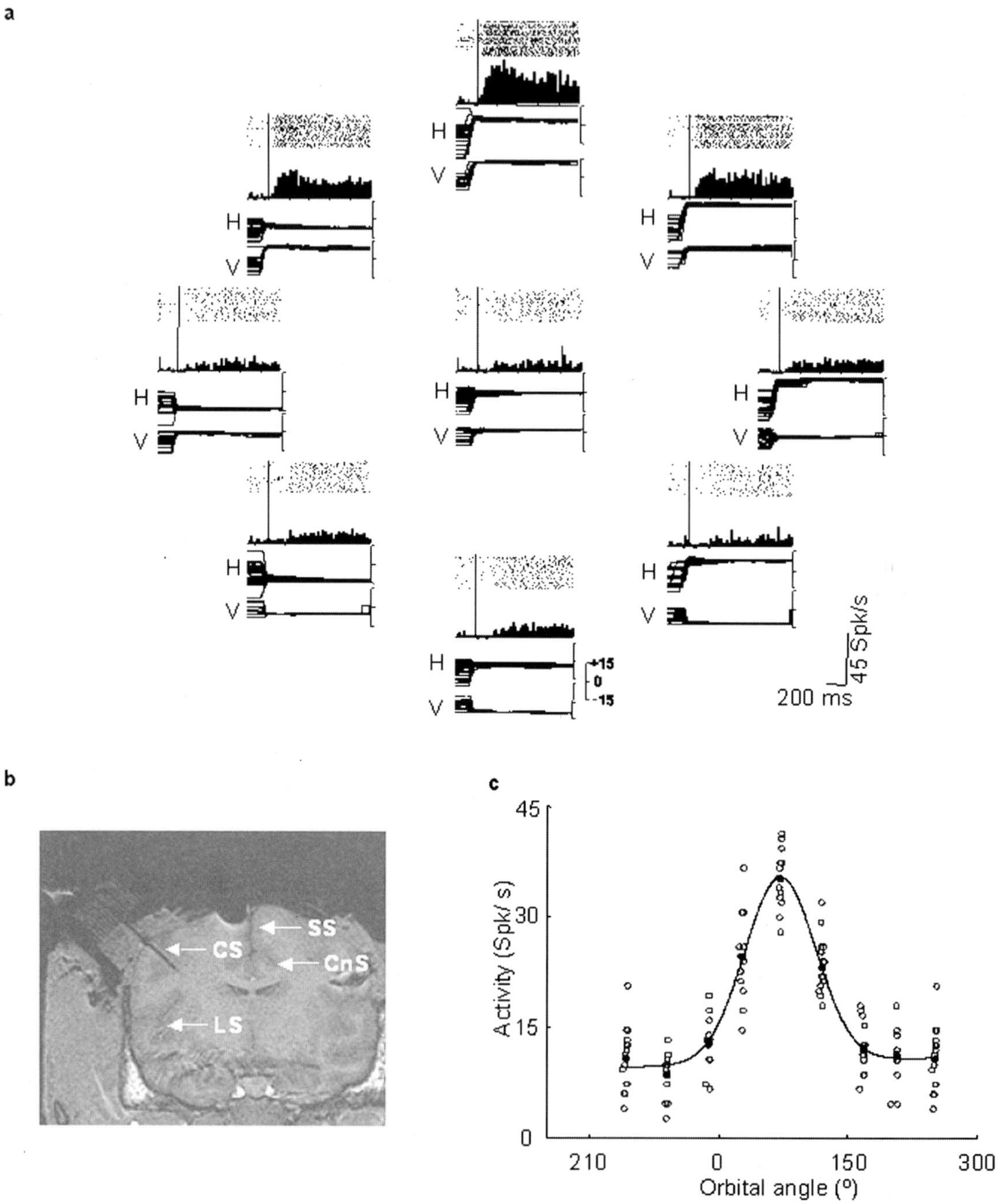

Fig. 1. Activity of a tonic eye position neuron in monkey SI. a: The figure shows nine raster diagrams, one at the centre of the orbit and eight others positioned radially 15 degrees from the centre. The position of the raster in the figure is related to the position of the eye in the orbit. Each tick is an action potential. Each line is a trial. Lines are synchronized on the end of the foveating saccade. Since the trial began with the appearance of the fixation point, the eye position before the saccade was uncontrolled. The histograms beneath each raster average, without smoothing, the activity of the raster above, with a bin width of 25 ms. Horizontal (H) and vertical (V) eye position traces for each trial are shown superimposed beneath each raster. b: T1 MRI showing an electrode (arrow) emerging from the recording grid and travelling to the floor of the central sulcus (CS). The sagittal sulcus (SS), cingulate sulcus (CnS), and lateral sulcus (LS) are also labelled. c: Example of Gaussian tuning of directional activity. Each open symbol is the mean activity of the neuron in one trial in the first 1000 ms after the foveating saccade, and the filled circles are the averages. $R^2 = 0.82$, $P < 0.001$. Adapted with permission from Wang et al. (2007).

However, the eye position signal could also have arisen from an efference copy or corollary discharge of a motor fixation signal. To distinguish between these two alternatives, we temporarily anaesthetized the contralateral orbit by a retrobulbar block while we recorded the activity of an eye position neuron (Richmond and Wurtz, 1980). This procedure caused ptosis, conjunctival and corneal anaesthesia, and a complete or partial paralysis of the eye, but had no effect on the movement of the contralateral eye. The monkey continued to make accurate eye movements in one eye despite the paralysis of the other eye. When the eye movement was completely or nearly abolished, the eye position signal was completely abolished (Fig. 3b). When the block wore off, the eye began to move again, and the eye position signal returned; (Fig. 3c). We performed this experiment on four cells in each of two monkeys. When the eye movement was reduced but not totally abolished, the eye position signal was diminished. For each experiment there was a significant difference in the responses before and during the block ($P = 5.8 \times 10^{-6}$ for the worst case, $P = 6 \times 10^{-9}$ for the median case).

Discussion

Eye position as a somatic sensation

We found the neurons representing eye position in the depth of the central sulcus, in the lateral area dedicated to the representation of the ophthalmic branch of the trigeminal nerve (Krubitzer et al., 2004). This area is consistent with area 3a, the portion of somatosensory cortex in which the skeletal muscle spindles are represented (Phillips et al., 1971). The eye position area is relatively broad, accessible from a surface area 4 mm by 4 mm. Some, but not all of the electrode tracks had easily obtainable tactile receptive fields on the brow. This relatively large area is in keeping with the observation that compared to overlying area 3b, area 3a has a much larger and more diffuse topographic map (Krubitzer et al., 2004).

The signal arises from proprioceptors in the contralateral orbit. When the ocular paralysis was complete, the signal was totally abolished. When the ocular paralysis was partial, the signal was partially abolished. We did not inject the ipsilateral eye, but prior studies of area 3a, even those in anaesthetized monkeys, showed that neurons could be driven exclusively from the contralateral body (Krubitzer et al., 2004). One possibility was that while we were actually performing the block, the monkey became agitated and we lost the neuron. This is unlikely for several reasons: (1) The response never disappeared entirely: instead the neurons maintained a weak background signal. (2) When the block took a few minutes to work, the neurons gradually lost their eye position signal, and gradually redeveloped it as the block wore off. (3) The monkeys tolerated the block well. They avidly consumed free rewards at a high rate during the injection, and resumed working immediately after we left recording booth. Retrobulbar block does not abolish the signal in the abducens nucleus (Richmond and Wurtz, 1980) which projects to the anaesthetized eye, and, because the monkey continued to make eye movements with its normal eye, the block could not have affected any of the more central motor processes associated with eye movements. The only possible interpretation of these data is that the signal arises from some proprioceptive mechanism in the eye, for which the best candidate is the fusimotor myotendinous cylinder.

The function of the proprioceptive eye position signal

Much evidence suggests that oculomotor proprioception is not necessary for on-line processing of visual space for action: monkeys with total ablation of the ophthalmic branch of the trigeminal nerve can perform double-step saccades (Guthrie et al., 1983) and open-loop pointing (Lewis et al., 1998) perfectly well. Corollary discharge, however, must be calibrated, in order for the corollary signal to represent the actual movement reliably. An obvious physiological use for the oculomotor proprioceptive signal which we have discovered is to provide the eye position signal for the eye position modulation of visual responses in the parietal cortex, which could then

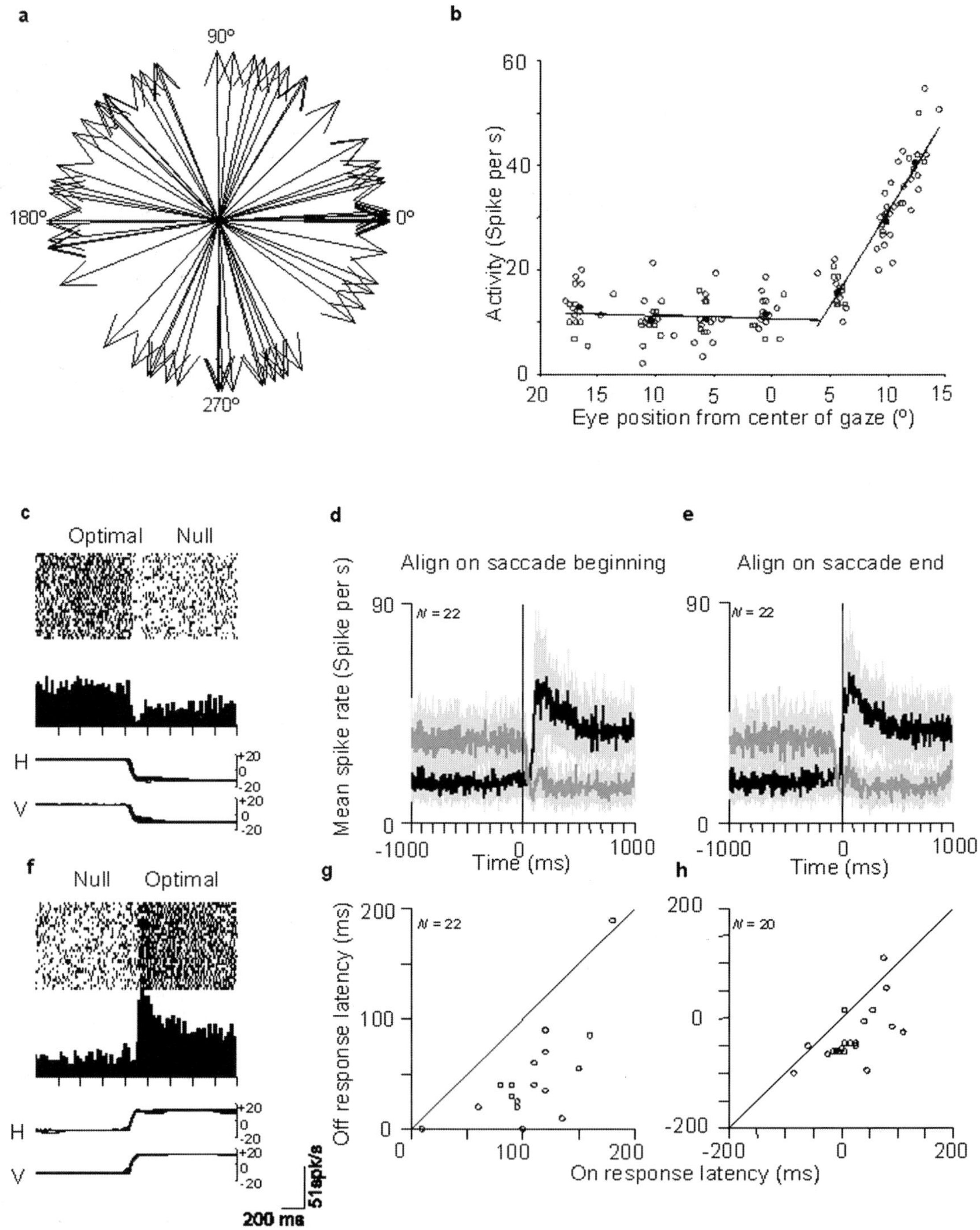
a
90°
180°
0°
270°
b
Activity (Spike per s)
Eye position from center of gaze (°)
c
Optimal
Null
H
V
d
Align on saccade beginning
Mean spike rate (Spike per s)
N = 22
Time (ms)
e
Align on saccade end
N = 22
Time (ms)
f
Null
Optimal
51spk/s
200 ms
g
Off response latency (ms)
N = 22
h
N = 20
On response latency (ms)

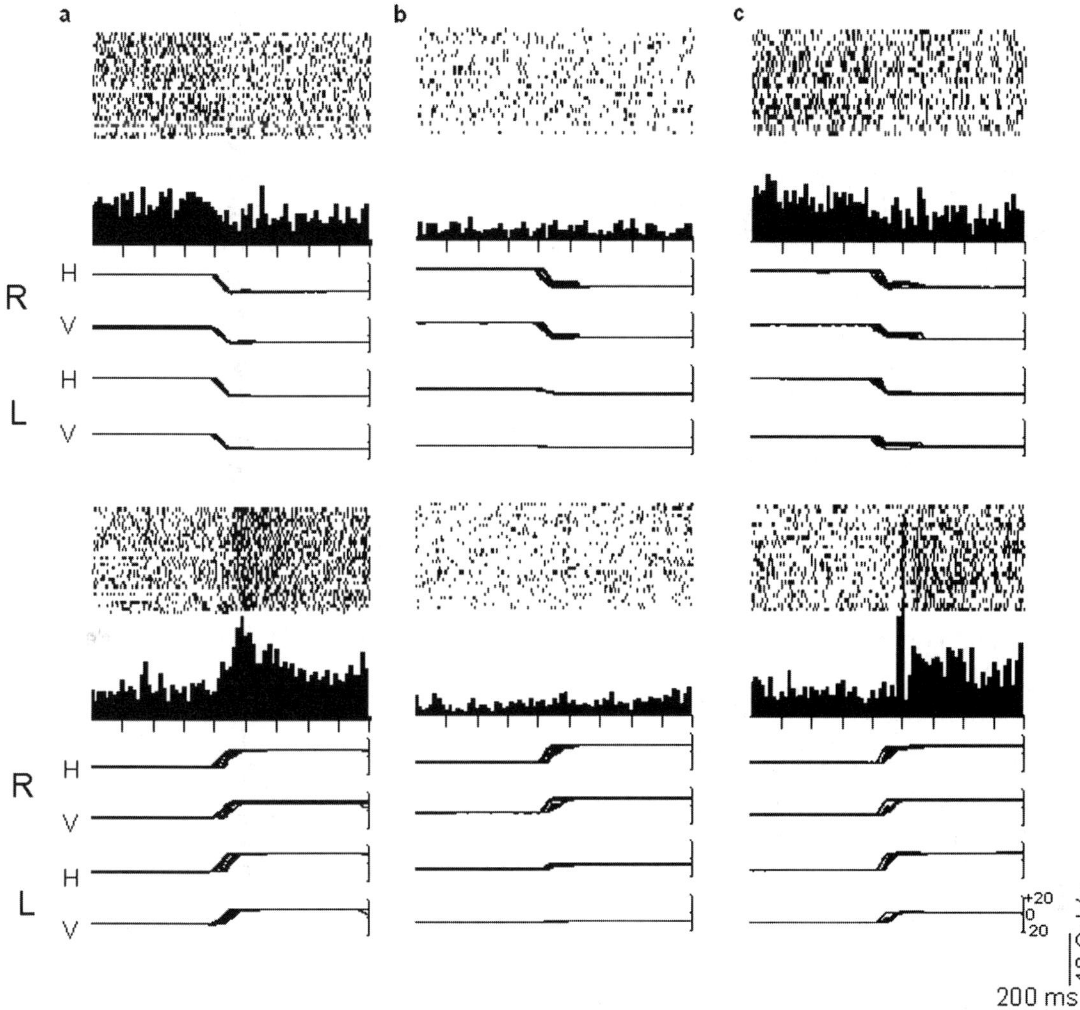

Fig. 3. Effect of retrobulbar block on neural activity. a: Neural activity when the monkey made a saccade from on-region to off-region (top) and off-region to on-region (bottom). Right (R) and left (L) eye traces are shown, with horizontal (H) and vertical (V) eye position. b: Loss of activity after retrobulbar block of the left eye. The left eye moved little, if at all, although the right eye moved normally. c: Recovery of neural activity when the eye regains movement, even though the ptosis was still present. The activity was less than before the block, but the eye had not entirely recovered. Adapted with permission from Wang et al. (2007).

Fig. 2. Tuning, tonic, and phasic response of eye position neurons. a: Optimal position tuning direction, calculated as the peak of the Gaussian to which the neuronal responses (mean of activity in first 1000 ms after the saccade) were fit, for each neuron ($n = 65$) which fit a Gaussian with $R^2 > 0.7$. b: Example of linear tuning with eccentric position of a single cell. $R^2 = 0.8261$, slope = 3.59. c: Single neuron example: The monkey makes a saccade from an optimal to a null orbital position. Note the change in tonic activity and the transient perisaccadic inhibition of response. Horizontal (H) and vertical (V) eye position traces are shown beneath each raster. The rasters were synchronized on the end of the saccades. Note the transient cessation of activity. f: The monkey made a saccade from a null to an optimal position. Here the transient response was excitatory. d: Average responses (dark) and 95% confidence limits (light) of neurons when the monkey made a saccade from null to optimal (black) and optimal to null (grey) orbital positions. Averages calculated in 5 ms bins, and not smoothed, synchronized on the beginning of the saccade. g: Latency of off-response from saccade beginning (ordinate) plotted against on-latency of on-response from saccade beginning. $N = 22$, $R^2 = 0.43$, $P = 0.0009$. e: Average response, displayed as in b, calculated from end of saccade. h: Latency of off-response from saccade end (ordinate) plotted against on-latency of on-response from saccade end. $N = 20$, $R^2 = 0.37$, $P = 0.0043$. For two cells in the sample averaged in c we were unable to determine a satisfactory latency. Adapted with permission from Wang et al. (2007).

be used for calibration (Steinbach, 1986) or for the sensory perception of eye position (Skavenski, 1972). However, if the eye position modulation of the visual responses (the 'gain fields') does arise from proprioception, then gain-field mechanisms could not provide the computational basis for spatially accurate visually guided action which they have been postulated to do (Zipser and Andersen, 1988; Salinas and Abbott, 1995; Pouget et al., 1999).

Another possibility, however, is that the oculomotor proprioceptive signals do not travel in the trigeminal nerve. Tozer and Sherrington (as described by Donaldson, 2000) showed that the palisade endings degenerated when the oculomotor nerves were cut, but not when the trigeminal nerve was cut. They interpreted this to mean that the proprioceptive signals travelled in the oculomotor nerves and not the trigeminal. If this were the case, then section of the trigeminal nerve would not interrupt oculomotor proprioception, and all the work predicated on trigeminal section as an obliterator of oculomotor proprioception would be called into question. Testing if transient inactivation of the trigeminal nerve obliterates the cortical proprioceptive signal would be a way to answer that question.

Acknowledgements

We are grateful to Dr. H. Eggers for teaching us how to perform retrobulbar blocks, to G. Duncan for electronic and computer support, to Y. Pavlova for dedicated animal maintenance, to Drs. M. Osman and G. Asfaw for veterinary care, to S. Dashnaw and Dr. J. Hirsch for MR imaging, and to L. Palmer for facilitating everything. This research was supported by grants from the James S. McDonnell Foundation, the U.S. National Eye Institute (1 R01 EY014978-01 and 1 R24 EY015634-01), and the Whitehall, Keck, Dana, and Kavli Foundations.

References

Andersen, R.A. and Mountcastle, V.B. (1983) The influence of the angle of gaze upon the excitability of the light-sensitive neurons of the posterior parietal cortex. J. Neurosci., 3: 532–548.

Bisley, J.W., Krishna, B.S. and Goldberg, M.E. (2004) A rapid and precise on-response in posterior parietal cortex. J. Neurosci., 24: 1833–1838.

Buttner-Ennever, J.A. and Horn, A.K. (2002) The neuroanatomical basis of oculomotor disorders: the dual motor control of extraocular muscles and its possible role in proprioception. Curr. Opin. Neurol., 15: 35–43.

Campos, M., Cherian, A. and Segraves, M.A. (2006) Effects of eye position upon activity of neurons in macaque superior colliculus. J. Neurophysiol., 95: 505–526.

Donaldson, I.M. (2000) The functions of the proprioceptors of the eye muscles. Philos. Trans. R. Soc. Lond. B Biol. Sci., 355: 1685–1754.

Eberhorn, A.C., Horn, A.K., Fischer, P. and Buttner-Ennever, J.A. (2005) Proprioception and palisade endings in extraocular eye muscles. Ann. N.Y. Acad. Sci., 1039: 1–8.

Guthrie, B.L., Porter, J.D. and Sparks, D.L. (1983) Corollary discharge provides accurate eye position information to the oculomotor system. Science, 221: 1193–1195.

Judge, S.J., Richmond, B.J. and Chu, F.C. (1980) Implantation of magnetic search coils for measurement of eye position: an improved method. Vision Res., 20: 535–538.

Krubitzer, L., Huffman, K.J., Disbrow, E. and Recanzone, G. (2004) Organization of area 3a in macaque monkeys: contributions to the cortical phenotype. J. Comp. Neurol., 471: 97–111.

Lewis, R.F., Gaymard, B.M. and Tamargo, R.J. (1998) Efference copy provides the eye position information required for visually guided reaching. J. Neurophysiol., 80: 1605–1608.

Lal, R., Friedlander, M.J. and Brunet, P. (1990) Effect of passive eye position changes on retinogeniculate transmission in the cat. J. Neurophysiol., 63: 502–522.

Mullette-Gillman, O.A., Cohen, Y.E. and Groh, J.M. (2005) Eye-centered, head-centered, and complex coding of visual and auditory targets in the intraparietal sulcus. J. Neurophysiol., 94: 2331–2352.

Nakamura, K. and Colby, C.L. (2002) Updating of the visual representation in monkey striate and extrastriate cortex during saccades. Proc. Natl. Acad. Sci. U.S.A., 99: 4026–4031.

Phillips, C.G., Powell, T.P. and Wiesendanger, M. (1971) Projection from low-threshold muscle afferents of hand and forearm to area 3a of baboon's cortex. J. Physiol., 217: 419–446.

Porter, J.D., Guthrie, B.L. and Sparks, D.L. (1983) Innervation of monkey extraocular muscles: localization of sensory and motor neurons by retrograde transport of horseradish peroxidase. J. Comp. Neurol., 218: 208–219.

Pouget, A., Deneve, S. and Sejnowski, T.J. (1999) Frames of reference in hemineglect: a computational approach. Prog. Brain Res., 121: 81–97.

Richmond, B.J. and Wurtz, R.H. (1980) Vision during saccadic eye movements. II. A corollary discharge to monkey superior colliculus. J. Neurophysiol., 43: 1156–1167.

Robinson, D.A. (1970) Oculomotor unit behavior in the monkey. J. Neurophysiol., 33: 393–404.

Sakata, H., Shibutani, H. and Kawano, K. (1980) Spatial properties of visual fixation neurons in posterior parietal association cortex of the monkey. J. Neurophysiol., 43: 1654–1672.

Salinas, E. and Abbott, L.F. (1995) Transfer of coded information from sensory to motor networks. J. Neurosci., 15: 6461–6474.

Skavenski, A.A. (1972) Inflow as a source of extraretinal eye position information. Vision Res., 12: 221–229.

Steinbach, M.J. (1986) Inflow as a long-term calibrator of eye position in humans. Acta Psychol. (Amst.), 63: 297–306.

van Opstal, A.J., Hepp, K., Suzuki, Y. and Henn, V. (1995) Influence of eye position on activity in monkey superior colliculus. J. Neurophysiol., 74: 1593–1610.

Wang, X., Zhang, M., Cohen, I.S. and Goldberg, M.E. (2007) The proprioceptive representation of eye position in monkey primary somatosensory cortex. Nature Neuroscience, 10(5): 640–646.

Wurtz, R.H. (1969) Visual receptive fields of striate cortex neurons in awake monkeys. J. Neurophysiol., 32: 727–742.

Zipser, D. and Andersen, R.A. (1988) A back-propagation programmed network that simulates response properties of a subset of posterior parietal neurons. Nature, 331: 679–684.

C. Kennard & R.J. Leigh (Eds.)
Progress in Brain Research, Vol. 171
ISSN 0079-6123

CHAPTER 1.7

Acute superior oblique palsy in the monkey: effects of viewing conditions on ocular alignment and modelling of the ocular motor plant

Christian Quaia[1], Xiaoyan Shan[2], Jing Tian[2], Howard Ying[2], Lance M. Optican[1], Mark Walker[3], Rafael Tamargo[2] and David S. Zee[2,*]

[1]*Laboratory of Sensorimotor Research, National Eye, Institute, Bethesda, MD, USA*
[2]*Department of Neurology, The Johns Hopkins Hospital, 600 N. Wolfe St., Baltimore, MD, USA*
[3]*Case Western Reserve University, Cleveland, OH, USA*

Abstract: We investigated the immediate and long-term changes in static eye alignment with acute superior oblique palsy (SOP) in the monkey. When the paretic eye was patched immediately after the lesion for 6–9 days, vertical alignment slowly *improved*. When the patch was removed and binocular viewing was allowed, alignment slowly *worsened*. In contrast when a monkey was not patched immediately after the lesion vertical alignment did not improve. We also show that a model of the eye plant can reproduce the observed acute deficit induced by SOP, but only by abandoning Robinson's symmetric simplification of the reciprocal innervation relationship within pairs of agonist–antagonist muscles. The model also demonstrated that physiologic variability in orbital geometry can have a large impact on SOP deficits.

Keywords: superior oblique palsy; strabismus; eye plant; ocular motor; adaptation; eye movements

Introduction

In human patients with strabismus, vertical misalignment of the eyes that is more pronounced when the higher eye is in adduction is usually attributed to a palsy of the superior oblique muscle (SOP). In many patients, however, it is difficult to reliably differentiate a 'congenital' from an acquired SOP. Furthermore, orbital imaging suggests that many patients with a presumed congenital SOP do not have a trochlear (IV) nerve palsy but rather an anatomical abnormality in the orbit that mimics SOP (Chan and Demer, 1999). And even when the evidence points to a complete trochlear nerve palsy, the degree and pattern of the static deviations are variable (Graf et al., 2005; Khawam et al., 1967). Inherent variation in the anatomical configuration of the SO muscle and tendon (Fink, 1962; Helveston et al., 1992), secondary changes in the mechanical properties of the palsied muscle and its antagonists (Robinson, 1985; Jampolsky, 1994; Scott, 1994), and central adaptive processes (Maxwell and Schor, 2006) all conspire against a correct diagnosis.

To provide a frame of reference for analysing the clinical presentation of both acute and chronic SOP in humans, and to gather experimental data for developing better models of the eye plant, we

*Corresponding author. Tel.: +1 410 955 3319;
Fax: +1 216 231 3461; E-mail: dzee@dizzy.med.jhu.edu

DOI: 10.1016/S0079-6123(08)00607-9

developed an experimental monkey model of acute SOP using intracranial trochlear nerve section (Shan et al., 2007). We sought to isolate the deficit caused by muscle palsy, and to determine what changes in the alignment of the eyes might occur afterwards. Two surprising findings emerged from this study (Fig. 1). First, when the animals (M1 and M2) wore a patch in front of the palsied eye during the days immediately after the IV nerve section, the vertical misalignment gradually *decreased* over days (left arrow). Second, once the eye patch was removed and binocular vision was allowed, the vertical misalignment *worsened* again, and even exceeded the initial deficit (right arrow).

These findings point to a critical influence of the pattern of habitual viewing (out of one eye or both) on ocular alignment just after the onset of an acute paralytic strabismus. We speculated that the early improvement in alignment during habitual monocular viewing was mediated by signals from the *proprioceptors* in the paretic eye, and that the subsequent deterioration was driven, at least in part, by overriding central mechanisms. Here we present some new findings that confirm the importance of the habitual state of viewing (monocular versus binocular) in the early changes in alignment after acute trochlear nerve palsy. We also present preliminary modelling results that help explain the patterns of misalignment observed with SOP.

Effects of viewing conditions on ocular misalignment

To test the hypothesis that monocular viewing mediates the adaptive mechanism that reduces the vertical misalignment, in a new experiment (monkey M3), instead of patching the paretic eye immediately after the surgery, we allowed uninterrupted binocular vision after the trochlear nerve section (Fig. 2). The result was clear cut: No improvement occurred during the early period after the lesion (left arrow) and, as was the case with the previous animals when binocular viewing was allowed after a period of patching, the misalignment worsened over time (right arrow). These data support the idea that (regardless of mechanism) monocular viewing after a paralytic strabismus is the factor leading to the initial decrease in ocular misalignment, and late increases in misalignment can occur even when the animal has not been patched immediately post-lesion.

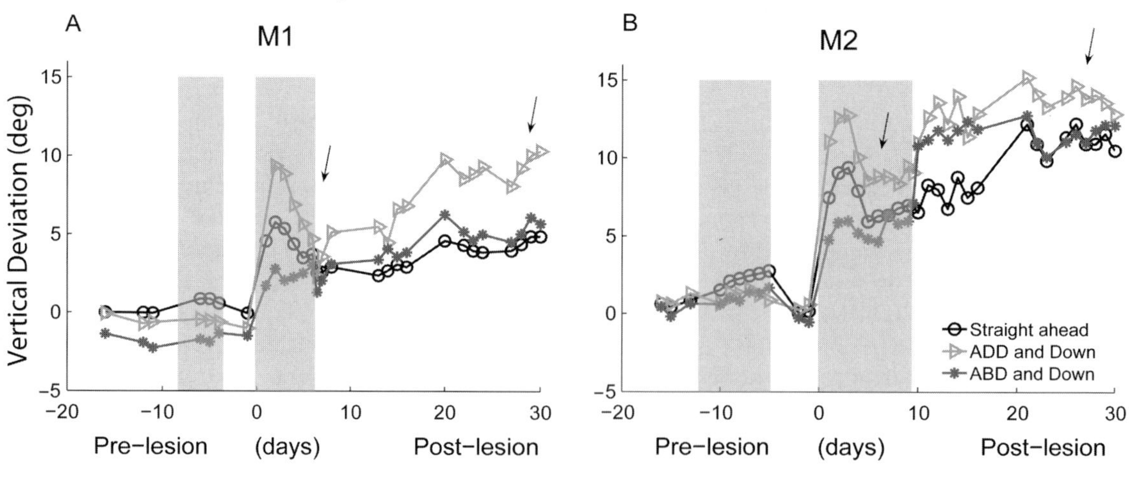

Fig. 1. Time course of vertical deviation with normal eye viewing pre- and up to 30 days post-SOP for M1 (A) and M2 (B). Shaded area: The paretic eye (or to-be-paretic eye) habitually patched. Positive deviation indicates that the paretic eye is relatively higher. Lesion was induced on day zero. ADD: Paretic eye in adduction (normal eye in abduction), ABD: paretic eye in abduction (normal eye in adduction). Left arrow indicates lessening misalignment while the animals were patched and right arrows indicate increasing misalignment after binocular viewing was allowed. Adapted with permission from Shan et al. (2007).

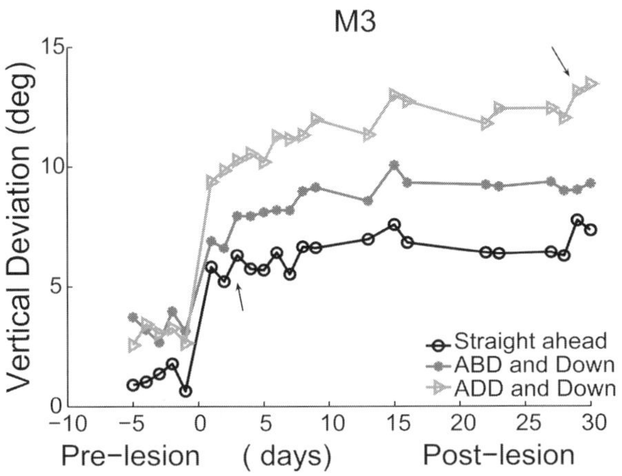

Fig. 2. Vertical deviation of M3 (NEV). Lesion is at day 0. After the lesion about 5 degrees of vertical deviation develops in the straight ahead position, and slightly more in the adducted, down position. Note that prior to the lesion there was a vertical phoria of a few degrees in the down positions probably related to mild restriction in the to-be-paretic eye in down gaze. The upward directed arrow (left) indicates the lack of improvement in alignment during the early days after the lesion and the downward directed arrow (right) indicates that there is some increase in misalignment over time.

Mathematical simulations of ocular motor behaviour

The characterization of the immediate effect of SOP allowed us to effectively validate a mathematical model of the eye plant. To this end we adapted our previously published model of the human eye plant (Quaia and Optican, 2003) to the geometry of the monkey orbit. The measures for the origin and the insertions of the muscles on the eyeball, and the location of the trochlea, were based on Suzuki et al. (1999). The relative strength of each muscle was estimated using two measures: their cross-section (Miller and Robins, 1987) and the number of fibres in the global layer (Oh et al., 2001; Kono et al., 2005). These measures provided us with a fairly tight range of values, but also indicated the importance of future direct experimental measurements.

The functional locations of the pulleys are not known in monkeys, and so we arranged them as follows: the pulleys for the LR and MR were placed at the same coordinates in the coronal plane as their insertions, but they were posterior to the globe centre by 80% of the anterior location of the (respective) muscle insertion. The rationale for this criterion is that in elevation the axis of action of the LR tips back (i.e., gains a torsional component) by a little more than half the elevation angle (Klier et al., 2006), indicating that its inflection point is slightly less posterior than its insertion is anterior (relative to the eye centre). The pulleys for the SR and IR were placed along the path that each muscle would travel with the eye in primary position if there were no pulleys, and at a distance from the insertion that is twice the anterior location of the insertion. This criterion was selected because this appears to be the arrangement in the human orbit (i.e., whereas in primary position there is a fairly clear inflection point for the horizontal recti, the vertical recti largely travel along the shortest path from origin to insertion). We placed the IO pulley: (1) at the same elevation as the IR pulley; (2) laterally between the LR and IR pulleys, but closer to the IR pulley; and (3) half-way between the posterior locations of the LR and IR pulleys. This placed the pulley for the IO between the pulleys for the LR and the IR, as is for humans (Demer et al., 2003). In addition, we have also allowed the pulleys to move as the eye rotates, in accordance with the active pulleys hypothesis (Demer, 2006). The amount of movement of the pulleys for the recti muscles has been inferred from Kono et al. (2002) and Clark et al. (1997), and the amount of motion of the IO pulley is based on Demer et al. (2003). This pattern is compatible with the coordinated pulley hypothesis (Miller, 2007). Because that data refers to a human plant, the amount of motion per degree of eye rotation was scaled down to account for the smaller radius of the eye in monkeys. Using these parameters the length and axis of action of each muscle were computed as a function of the orientation of the eye, and length–tension–innervation equations (Robinson, 1975) were used to compute the muscle force as a function of muscle length and innervation.

Simulating a SOP is then quite simple: first, one needs to solve the *inverse control problem*, i.e., determine the muscle innervations required to hold the orientation of the healthy eye. Then, the innervation to the model superior oblique is forced to zero, thus simulating the effect of a SOP. Since Robinson's original SQUINT model (Robinson, 1975), all mathematical models of the plant have solved the inverse control problem by imposing a specific relationship between the innervation supplied to two muscles in an agonist–antagonist pair. This relationship was always described using a *symmetric* hyperbola. However, when we used this approach the model failed to reproduce the observed results, predicting instead much smaller deficits. No adjustment of the parameters described above (within a physiologic range) improved matters. This might explain why when plant models have been used to replicate the deficit observed in humans with SOP, they had to invoke secondary muscular or adaptive changes (Miller and Demer, 1992; Porrill et al., 2000; Straumann et al., 2003; Haslwanter et al., 2005).

It turns out that the reciprocal innervation formulation is the culprit, as there is no reason to believe that the reciprocal relationship is symmetric. The only reasonable constraint is that it is reciprocal (i.e., if the innervation to one muscle goes up, the innervation to the other goes down, as Sherrington originally proposed). We could actually prove this idea because the length–tension–innervation curve measured by Robinson provides all the information needed to estimate the actual relationship in the horizontal recti (in humans). Using Robinson's original data, the relationship between the innervation in the lateral and medial recti was actually *asymmetric*. Accordingly, we removed the symmetric constraint from our model, and used the data from the SO palsy to directly estimate this relationship for the oblique muscles, as shown in the flow chart. This procedure was run by an optimization algorithm until the best fit was obtained. The final result (Fig. 3) closely fit the behaviour of the two monkeys, and their derived SO/IO innervation relationships (step 6, flow chart) were similar.

In an effort to uncover potential sources of variability in the effects induced by SOP, we have begun examining the sensitivity of our model to changes in orbital parameters. We first examined the angle of insertion of the SO tendon on the globe, a parameter that varies considerably among human subjects (Fink, 1962). We changed the angle by $\pm 30^\circ$ relative to its average insertion (which covers the range reported by Fink in humans), and then applied our simulation algorithm to the modified plant model (Fig. 4). For

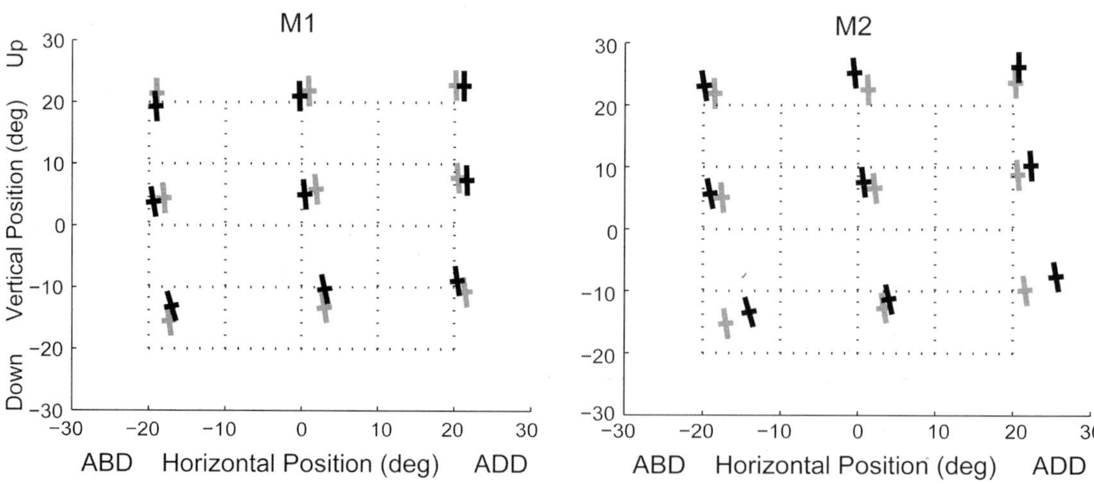

Fig. 3. Simulations for M1 and M2 (left SOP). Grey symbols are simulations and black symbols are monkey data. Shown is paretic eye (PE) position. Fixation points for normal eye are $\pm 20^\circ$ and straight ahead. ADD/ABD refer to PE position. Data are corrected for pre-lesion phorias with same period of patching as post-lesion (2–3 days). Tilt of vertical lines reflects torsional phoria. The simulations capture well the gradients and amplitudes of the vertical and torsional deviations.

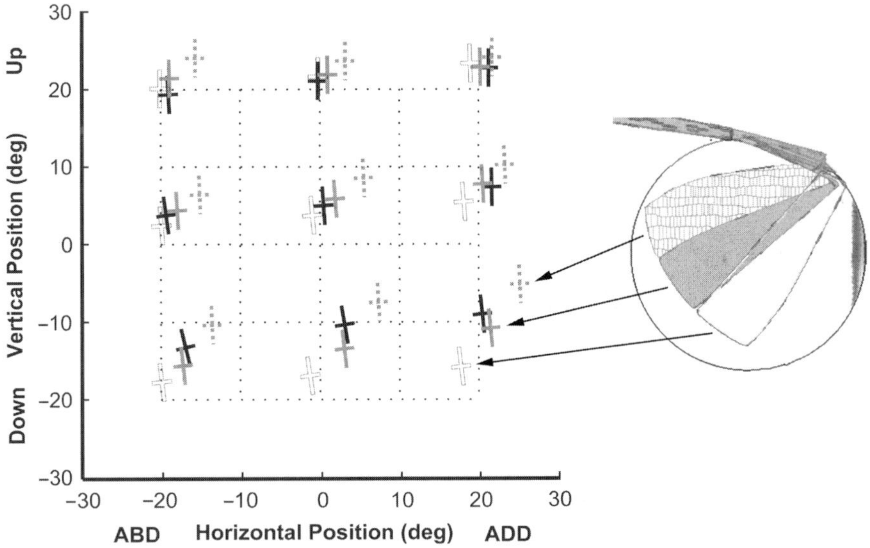

Fig. 4. Simulations of SOP. Black solid crosses: Monkey data (M1). Grey solid crosses: Standard insertion location. Open crosses: $+30^{\circ}$ insertion; dashed crosses: -30° insertion. See right panel for SO tendon path for each insertion location (right eye, seen from above). Note the considerable effect of moving the insertion on the predicted misalignment (vertical, torsional, and horizontal) especially in down gaze. As expected there is a reciprocal relationship between torsion and vertical misalignment, but note also the considerable variation in the horizontal misalignment (even to the point of a small exodeviation).

Table 1. Simulation flow chart

1. Guess the common-mode innervations (i.e., average of the innervations in an agonist–antagonist pair) (asymmetric parabola).
2. Find a set of differential innervations (i.e., difference between innervations in a pair) to keep the intact eye in each of the nine fixation positions.
3. Set the innervation of SO to zero.
4. Change the innervation of the IO to match the positions of the paretic eye.
5. Change the innervation of the SO to match the positions of the normal eye.
6. Fit new innervations for the SO and IO for the nine fixation positions to an asymmetric parabola to estimate reciprocal innervation of SO/IO. This will be unique to each animal.
7. Find a new set of innervations to keep the intact eye in each of the nine fixation positions.
8. Set the innervation of the SO to zero (create a SOP).
9. Compare the alignment of the monkey with the model.

each insertion angle, we repeated the procedure in Table 1. From these simulations we note that in spite of the degrees of freedom of the model, only the original model provides a good fit across the nine positions tested and that orbital geometry can have a large impact on the outcome of the lesion (e.g., compare the difference between the two anatomical variations especially in down gaze). This finding implies that, not surprisingly, orbital geometry can be an important determinant of the variability observed in patients, and suggests that attempts to measure the location of the insertion of the SOP in our animals (e.g., with post mortem anatomical evaluations) will be worthwhile. Also, this suggests that our model and strategy of simulation are reasonable, as a somewhat different geometry would have resulted in a considerably worse fit. In other words, the number of degrees of freedom is not so high as to render the simulations without significance, as would have been the case if changes in the reciprocal innervation patterns had enabled the model to always reproduce the deficit observed, regardless of geometry.

References

Chan, T.K. and Demer, J.L. (1999) Clinical features of congenital absence of the superior oblique muscle as demonstrated by orbital imaging. J. AAPOS, 3(3): 143–150.

Clark, R.A., Miller, J.M. and Demer, J.L. (1997) Location and stability of rectus muscle pulleys: muscle paths as a function of gaze. Invest. Ophthalmol. Vis. Sci., 38(1): 227–240.

Demer, J.L. (2006) Current concepts of mechanical and neural factors in ocular motility. Curr. Opin. Neurol., 19(1): 4–13.

Demer, J.L., Oh, S.Y., Clark, R.A. and Poukens, V. (2003) Evidence for a pulley of the inferior oblique muscle. Invest. Ophthalmol. Vis. Sci., 44(9): 3856–3865.

Fink, W.H. (1962) Surgery of the vertical muscles of the eye. Thomas, Springfield, IL.

Graf, M., Krzizok, T. and Kaufmann, H. (2005) Head-tilt test in unilateral and symmetric bilateral acquired trochlear nerve palsy (in German). Klin. Monatsbl. Augenheilkd., 222: 142–149.

Haslwanter, T., Buchberger, M., Kaltofen, T., Hoerantner, R. and Priglinger, S. (2005) SEE++: a biomechanical model of the oculomotor plant. Ann. N.Y. Acad. Sci., 1039: 9–14.

Helveston, E.M., Krach, D., Plager, D.A. and Ellis, F.D. (1992) A new classification of superior oblique palsy based on congenital variations in the tendon. Ophthalmology, 99(10): 1609–1615.

Jampolsky, A. (1994) A new look at the head tilt test. In: Fuchs A.F., Brandt T.h., Büttner U. and Zee D.S. (Eds.), Contemporary Oculomotor and Vestibular Research: A Tribute to David A. Robinson. Thieme Verlag, Stuttgart, pp. 432–439.

Khawam, E., Scott, A.B. and Jampolsky, A. (1967) Acquired superior oblique palsy. Diagnosis and management. Arch. Ophthalmol., 77(6): 761–768.

Klier, E.M., Meng, H. and Angelaki, D.E. (2006) Three-dimensional kinematics at the level of the oculomotor plant. J. Neurosci., 26(10): 2732–2737.

Kono, R., Clark, R.A. and Demer, J.L. (2002) Active pulleys: magnetic resonance imaging of rectus muscle paths in tertiary gazes. Invest. Ophthalmol. Vis. Sci., 43(7): 2179–2188.

Kono, R., Poukens, V. and Demer, J.L. (2005) Superior oblique muscle layers in monkeys and humans. Invest. Ophthalmol. Vis. Sci., 46(8): 2790–2799.

Maxwell, J.S. and Schor, C.M. (2006) The coordination of binocular eye movements: vertical and torsional alignment. Vision Res., 46(21): 3537–3548.

Miller, J.M. (2007) Understanding and misunderstanding extraocular muscle pulleys. J. Vis., 7(11): 1–15.

Miller, J.M. and Demer, J.L. (1992) Biomechanical analysis of strabismus. Binoc. Vis. Eye Muscle Surg. Quart., 7: 233–248.

Miller, J.M. and Robins, D. (1987) Ocular muscle sideslip and orbital geometry in monkeys. Vision Res., 27(3): 381–392.

Oh, S.Y., Poukens, V., Cohen, M.S. and Demer, J.L. (2001) Structure-function correlation of laminar vascularity in human rectus extraocular muscles. Invest. Ophthalmol. Vis. Sci., 42(1): 17–22.

Porrill, J., Warren, P.A. and Dean, P. (2000) A simple control law generates Listing's positions in a detailed model of the extraocular muscle system. Vision Res., 40: 3743–3758.

Quaia, C. and Optican, L.M. (2003) Dynamic eye plant models and the control of eye movements. Strabismus, 11(1): 17–31.

Robinson, D.A. (1975) A quantitative analysis of extraocular muscle cooperation and squint. Invest. Ophthalmol., 14(11): 801–825.

Robinson, D.A. (1985) Bielschowsky head-tilt test. II. Quantitative mechanics of the Bielschowsky head-tilt test. Vision Res., 25(12): 1983–1988.

Scott, A.B. (1994) Change of eye muscle sarcomeres according to eye position. J. Pediatr. Ophthalmol. Strab., 31(2): 85–88.

Shan, X., Tian, J., Ying, H., Quaia, C., Optican, L., Walker, M., Tamargo, R. and Zee, D.S. (2007) Acute superior oblique palsy in monkeys: I. Changes in static eye alignment. Invest. Ophthalmol. Vis. Sci., 48(6): 2602–2611.

Straumann, D., Steffen, H., Landau, K., Bergamin, O., Mudgil, A.V., Walker, M.F., Guyton, D.L. and Zee, D.S. (2003) Primary position and Listing's law in acquired and congenital trochlear nerve palsy. Invest. Ophthalmol. Vis. Sci., 44(10): 4282–4292.

Suzuki, Y., Straumann, D., Simpson, J.I., Hepp, K., Hess, B.J. and Henn, V. (1999) Three-dimensional extraocular motoneuron innervation in the rhesus monkey. I: muscle rotation axes and on-directions during fixation. Exp. Brain Res., 126(2): 187–199.

C. Kennard & R.J. Leigh (Eds.)
Progress in Brain Research, Vol. 171
ISSN 0079-6123

CHAPTER 1.8

Dynamic aspects of trochlear nerve palsy☆

D. Straumann[1,*], C.J. Bockisch[1,2,3] and K.P. Weber[1,4]

[1]*Department of Neurology, Zurich University Hospital, Zurich, Switzerland*
[2]*Department of ENT, Zurich University Hospital, Zurich, Switzerland*
[3]*Department of Ophthalmology, Zurich University Hospital, Zurich, Switzerland*
[4]*Department of Neurology, Royal Prince Alfred Hospital, Sydney, Australia*

Abstract: Trochlear nerve palsy leads to kinematic aberrations of both the paretic and the unaffected eye. During dynamic head roll, the rotation axis of the covered paretic or unaffected eye deviates inward, while the rotation axis of the viewing paretic or unaffected eye aligns with the line of sight; this convergence of rotation axes increases with gaze moving in the direction of the unaffected eye. During downward saccades, the trajectories of both eyes curve towards the unaffected side; these curvatures increase when the head is rolled to the affected side and gaze directed to the unaffected side. Hence, during both vestibular evoked and saccadic ocular movements, the unaffected eye shows similar kinematic aberrations as the paretic eye. While aberrations of the paretic eye can be explained by decreased force of the superior oblique (SO) muscle, aberrations of the unaffected eye may be due to increased force parallel to the paretic SO in the unaffected eye in accordance with Hering's law. This law, which forms the basis of conjugate eye movements, also seems to govern eye displacements in unilateral eye muscle palsy.

Keywords: superior oblique muscle; vestibulo-ocular reflex; torsion; Hering's law; eye movements; eye muscle palsy; trochlear nerve palsy; saccade

Introduction

For monocular viewing, Hering's law of equal innervation implies that for each position of one eye there is only one corresponding position of the fellow eye, independent of which eye is covered (Hering, 1868). This law is also valid in the presence of extra-ocular muscle palsy, but here the deviation between the three-dimensional (3D) positions of the two eyes increases as gaze is shifted towards the pulling direction of the paretic muscle (Zee et al., 1984; Straumann et al., 2003). Thus, in the static condition with one eye fixing upon a visual target and the other eye covered, the position of the covered eye is always unique relative to the viewing eye, no matter whether the paretic or the non-paretic eye is covered.

While the validity of Hering's law in eye muscle palsy is generally accepted for static eye positions, the law could be violated during eye displacements. For instance, in abducens nerve palsy, clinicians frequently observe an apparent overshooting of the unaffected adducting eye during saccades in the direction of the affected lateral rectus muscle of the

☆Version: 2007-12-11

*Corresponding author. Tel.: +41 44 255 5564;
Fax: +41 44 255 4507; E-mail: dominik@neurol.unizh.ch

DOI: 10.1016/S0079-6123(08)00608-0

other eye. This clinical sign, however, is assessed during binocular viewing, which may lead to an activation of the vergence system and, in turn, to a decrease of the squint angle after the saccade. In other words, clinical observation or measurement of ocular movements with both eyes viewing is not sufficient to dismiss the possibility that Hering's law can be extended to binocular positions during gaze displacements in the absence of vergence movements.

In this paper, we ask whether Hering's law is valid during the vestibulo-ocular reflex (VOR) and during saccades in patients with one-sided eye muscle palsy. We chose to study this question in trochlear nerve palsy, because a weakness of the superior oblique (SO) muscle barely restricts the ocular motor range of the affected eye, which facilitates the kinematic analysis of binocular movements. Here we report on the data from two different paradigms, in which we controlled the movement of the viewing eye, while the fellow covered eye was free to move. First, we rotated patients about their naso-occipital axis during monocular fixations along the horizontal meridian and determined the rotations axes of both eyes. Second, we elicited downward saccades in different static whole-body roll positions and compared the saccadic deviations of the two eyes from a straight trajectory between saccadic onset and offset. Our hypothesis was that the relation between the rotation axes (VOR paradigm) or positional vectors (saccade paradigm) of the two eyes is always in accordance with Hering's law.

Since the pulling direction of the SO muscle includes, besides a vertical component, a considerable torsional component, it is indispensable to record eye movements of patients with trochlear nerve palsy about all principle axes of rotation (horizontal, vertical, and torsional). Here we applied two different technologies. For the VOR paradigm, we used the dual search-coil technique, while for the saccade paradigm we chose 3D video-oculography to bypass the problem of saccadic torsional coil artifacts (Bergamin et al., 2004). Part of the data presented in this paper originates from experiments that were published elsewhere (Weber et al., 2004).

Methods

Subjects

We tested patients with untreated congenital or acquired unilateral trochlear nerve palsy. Informed consent was obtained from the subjects after the experimental procedure was explained. The protocol was approved by a local ethics committee and was in accordance with the ethical standards laid down in the Declaration of Helsinki for research involving human subjects.

Vestibular paradigm

Subjects were seated upright on a turntable with three servo-controlled motor driven axes (prototype built by Acutronic, Jona, Switzerland). The head was restrained with an individually moulded thermoplastic mask (Sinmed BV, Reeuwijk, The Netherlands) and was positioned such that the centre of the interaural line was at the intersection of the three axes of the turntable. Pillows and safety belts minimized movements of the body. Movements of both eyes were recorded in three dimensions with dual search coils manufactured by Skalar (Delft, The Netherlands). A chair-fixed coil frame (side length, 0.5 m) that produced three orthogonal magnetic fields with frequencies of 80, 96, and 120 kHz surrounded the subject's head. The signals were amplified and multiplexed before passing through the turntable slip rings. A high performance 12-bit digital signal processor computed a Fast Fourier transform in real time on the digitized search coil signal to determine the voltage induced in the coil by each magnetic field (system manufactured by Primelec, Regensdorf, Switzerland). Rotation vectors (for positions) and angular velocity vectors (for rotation axes) of both eyes were computed off-line from the coil signals. In patients with left-sided trochlear nerve palsy, the directions of 3D eye position were horizontally mirrored, as if the right eye had been affected by the palsy. Subjects monocularly fixed laser dots projected on a spherical screen at a distance of 1.4 m, whereas the other eye was covered. The dots were located straight ahead and at $\pm 20^{\circ}$ eccentric head-fixed positions. During visual fixation, subjects were oscillated about the naso-occipital (roll) axis with a frequency of 0.3 Hz and an amplitude of ±35°.

Saccade paradigm

In upright, 45° left ear down, and 45° right ear down whole-body roll positions, subjects monocularly tracked a jumping laser target projected on the spherical screen, while the fellow eye was covered. 3D eye movements were binocularly recorded at 200 Hz with a 3D videooculography system (Chronos Eye Tracker Version 1C/2003, Chronos Vision GmbH, Berlin, Germany) mounted on the thermoplastic mask. To optimize pupil tracking, pupils were constricted with pilocarpine 0.5% eye drops 30 min before recording. Vertical saccades with amplitudes of 10° were recorded at −10°, 0°, and 10° horizontal eccentricity. The laser dot jumped to and fro every 3 s without randomization. Raw video data were processed with the iris tracker software (Version 2.1.6.1., Chronos Vision GmbH).

Results

Figure 1 depicts average horizontal angles of ocular rotation axes relative to straight ahead, while subjects were oscillated about the naso-occipital axis and fixed upon horizontal visual targets (circles: straight head; down triangles: 20° left; up triangles: 20° right). The horizontal angles of ocular rotation axes (*x*-axis: paretic right eye; *y*-axis: unaffected left eye) are plotted against each other, while either eye is viewing (filled symbols: right eye covered, left eye viewing; empty symbols: right eye viewing, left eye covered). The example (data from a patient with congenital trochlear

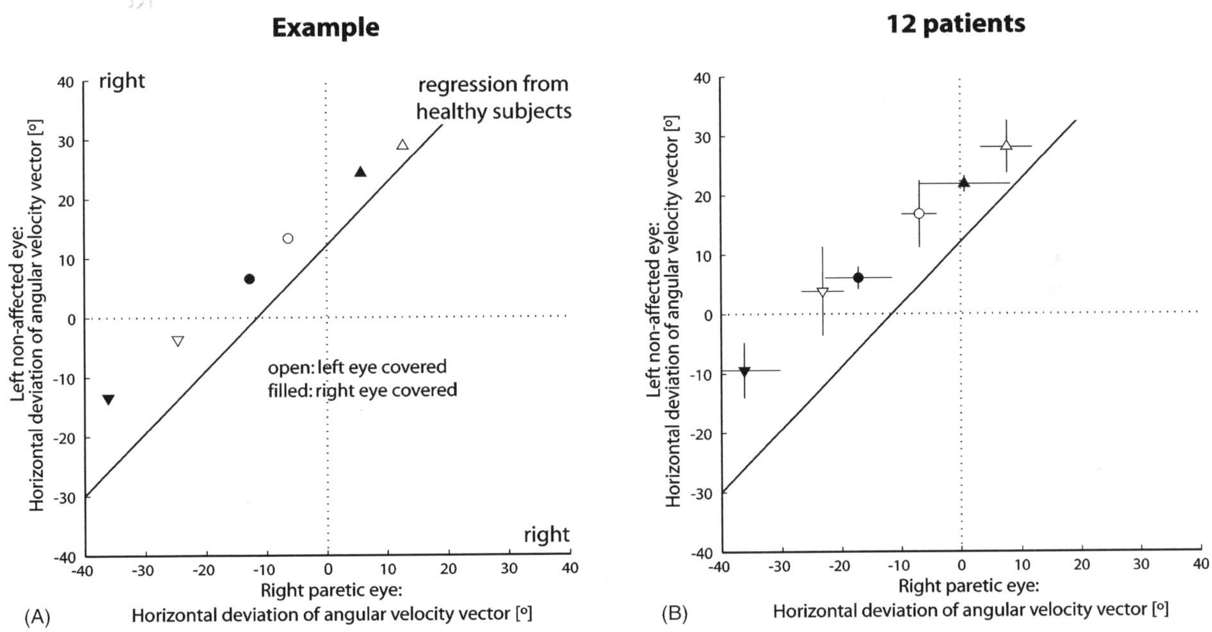

Fig. 1. Average orientation (angle in degrees) of ocular rotation axes in the horizontal plane relative to straight ahead during oscillations about the naso-occipital axis (amplitude: ±35°; frequency: 0.3 Hz). One eye was always viewing, while the other eye was covered. Abscissa: Right paretic eye(s). Ordinate: Left unaffected eye(s). Empty symbols: Right paretic eye(s) viewing. Filled symbols: Right paretic eye(s) covered. Circles: Viewing of straight ahead target. Down triangles: Viewing of 20° left target. Up triangles: Viewing of 20° right target. Dotted lines: Zero angles for right (vertical line) or left (horizontal line) eye. Solid lines: Regression from a population of healthy subjects. Note that data points scatter linearly independent of whether the paretic eye is viewing or covered, but they are shifted to the left relative to the normal regression line. (A) Typical patient with congenital trochlear nerve palsy on the right side. (B) Averages ±1 SD of all 12 patients with trochlear nerve palsy. In patients with left-sided trochlear nerve palsy, the directions of three-dimensional eye position were horizontally mirrored.

nerve palsy) on the left panel of Fig. 1 demonstrates that the correlated horizontal angles between the rotation axes of the two eyes fall almost on a single line which is shifted relative to the line computed from healthy subjects (solid line). Average values (± 1 SD) of all 12 patients with trochlear nerve palsy (paretic eye always on the right side; see Methods) are shown on the right panel. Again, independent of whether the right paretic eyes or the left unaffected eyes were viewing, there was an almost identical linear relationship between the horizontal angles of binocular rotation axes.

Figure 2 shows visually elicited downward saccades in a typical patient with right-sided

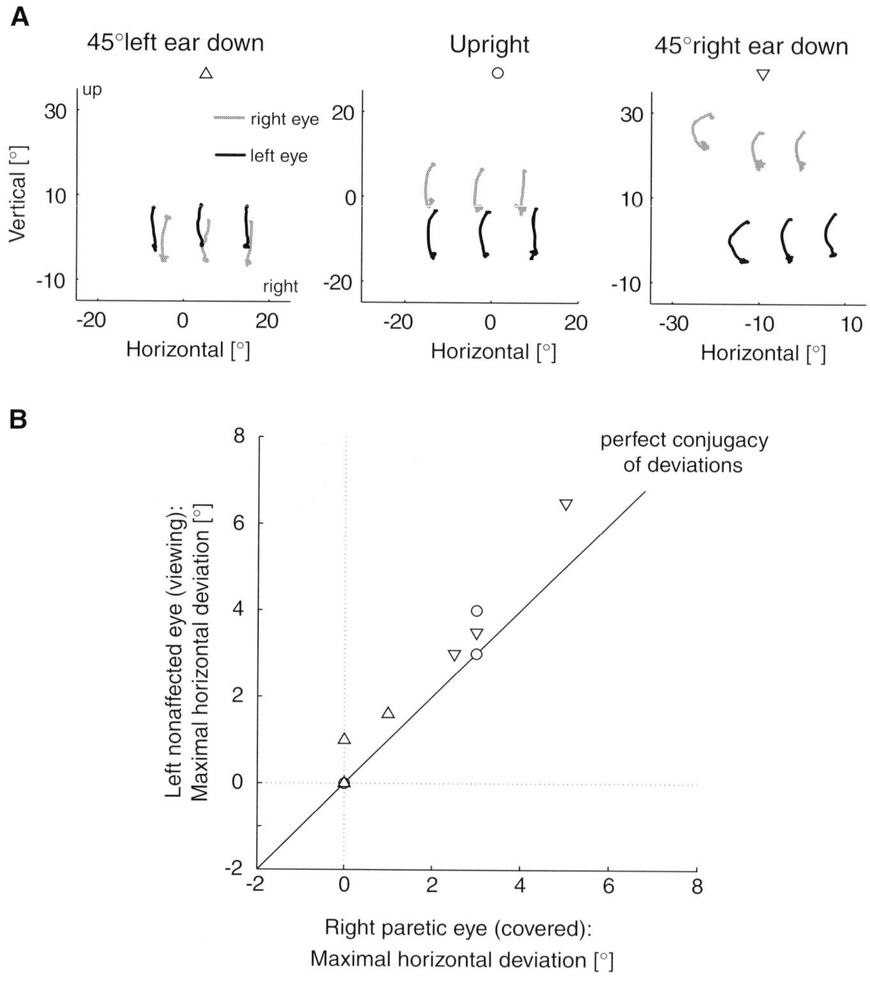

Fig. 2. Mean traces of downward saccades in a patient with acquired right-sided trochlear nerve palsy. Visual targets were presented at $-10°$, $0°$, and $10°$ horizontal eccentricities. Onset targets were $5°$ above and offset targets $5°$ below the horizontal meridian. (A) Horizontal (abscissa; right is positive) and vertical (ordinate; up is positive) eye positions during saccades. Grey traces: Right eye. Black traces: Left eye. Note that trajectories of both eyes deviate to the left before curving back. Curvatures increase in the 45° right ear down whole-body position (right panel; Bielschowsky position) and decrease in the 45° left ear down whole-body position (left panel; anti-Bielschowsky position). (B) Maximal horizontal deviations from straight lines connecting saccadic onsets and offsets. Abscissa: Covered right paretic eye. Ordinate: Viewing left eye. Circles: Data in upright position. Up triangles: Data in 45° left ear down position. Down triangles: Data in 45° right ear down position. Note that data points scatter close to a 45° line through zero, which indicates perfect conjugacy of maximal saccadic horizontal deviations.

acquired trochlear nerve palsy. Visual stimuli were given at different horizontal eccentricities (straight ahead, 10° right, 10° left). The amplitude between stimulus onset and offset was always 10°. Already in the upright position (Fig. 2A, middle panel), the saccades of both the covered paretic right eye and the viewing unaffected left eye curve away from the paretic eye. Curvatures increase in the 45° Bielschowsky whole-body roll position (Fig. 2A, right panel; tilt towards the side of the paretic eye) and decrease in the opposite whole-body roll position (Fig. 2A, left panel). When the maximal saccadic horizontal deviations of the two eyes are plotted against each other (Fig. 2B), data points scatter along the same line for saccades in all three whole-body roll positions (circles: horizontal deviations during vertical saccades at different horizontal eccentricities in upright position; up triangles: in 45° left ear down position; down triangles: in 45° right ear down position). The same pattern, but with individual magnitudes of horizontal saccadic curvatures, was found in the majority of patients with acquired or congenital trochlear nerve palsy (data not shown).

Discussion

In the absence of vergence, the neural signals that drive the extra-ocular muscles of one eye also drive the yoked muscles of the fellow eye to maintain conjugate binocular positions. This "equal innervation" of the two eyes, known as Hering's law (Hering, 1868), is still preserved in the case of monocular eye muscle palsy, although binocular positions become disconjugate. The preservation of Hering's law in one-sided eye muscle palsy can be verified by plotting the positions of the two eyes against each other, while one eye is covered and the other eye is fixing on visual targets at various eccentricities (Zee et al., 1984). Independent of whether the paretic or unaffected eye is covered, the data points fall on the same curve, which is exactly what one expects if the innervation is strictly binocular. In this paper, we demonstrated that Hering's law can be extended to binocular movement trajectories.

During the torsional VOR, the rotation axes of the two eyes are relatively fixed and parallel, when the viewing eye is fixing a target at infinity (Bergamin and Straumann, 2001). While the stability of rotation axes is also preserved in patients with a one-sided trochlear nerve palsy, the rotation axis of the covered eye is rotated inward (Weber et al., 2004). As shown in Fig. 1, the correlated horizontal angles of ocular rotation axes that were determined during monocular fixations of horizontally displaced visual targets scattered close around the same line, independent on whether the paretic or unaffected eye was viewing. This finding, in turn, is compatible with the view that Hering's law also applies to eye movements during the VOR. We anticipate that similar results can be found in other extra-ocular muscle palsies and for different VOR directions.

For downward saccades, both the inferior rectus and the SO muscles contract. If the SO muscle is paretic, we expect that, during the saccade, the affected eye deviates in the pulling direction of the inferior rectus muscle. Indeed, there is dynamic extorsion of the paretic eye during downward saccades as described by Steffen et al. (2007). These authors already noted that the unaffected eyes showed similar torsion in the conjugate direction (intorsion), which casts doubts on whether the saccadic transient deviation of the affected eye is really a direct mechanical consequence of the muscle paresis. In this paper, we concentrated on the saccadic horizontal transients, i.e., curvatures of trajectories during downward saccades in patients with trochlear nerve palsy. These curvatures generally increase with subjects rolling their head towards the side of the affected eye. We found that the magnitude of curvatures was almost the same in the paretic and the unaffected eyes. This finding suggests that aberrant saccadic trajectories cannot be explained by the geometry of a paretic eye muscle alone, but is the consequence of neural adaptation to the paresis and, as a result, is visible on both eyes in accordance with Hering's law.

Acknowledgements

We thank E. Schafflützel, T. Schmückle, and A. Züger for technical assistance. This work is supported by the Swiss National Science Foundation, the Koetser Foundation for Brain Research, and the Center for Integrative Human Physiology, University of Zurich.

References

Bergamin, O., Ramat, S., Straumann, D. and Zee, D.S. (2004) Influence of orientation of exiting wire of search coil annulus on torsion after saccades. Invest. Ophthalmol. Vis. Sci., 45: 131–137.

Bergamin, O. and Straumann, D. (2001) Three-dimensional binocular kinematics of torsional vestibular nystagmus during convergence on head-fixed targets in humans. J. Neurophysiol., 86: 113–122.

Hering, E. (1868) Die Lehre vom binocularen Sehen. Wilhelm Engelmann, Leipzig, Germany.

Steffen, H., Straumann, D., Walker, M.F., Miller, N.R., Guyton, D.L., Repka, M.X. and Zee, D.S. (2007) Torsion in patients with superior oblique palsies: dynamic torsion during saccades and changes in Listing's plane. Graefes Arch. Clin. Exp. Ophthalmol., [epub ahead of print].

Straumann, D., Steffen, H., Landau, K., Bergamin, O., Mudgil, A.V., Walker, M.F., Guyton, D.L. and Zee, D.S. (2003) Primary position and Listing's law in acquired and congenital trochlear nerve palsy. Invest. Ophthalmol. Vis. Sci., 44: 4282–4292.

Weber, K.P., Landau, K., Palla, A., Haslwanter, T. and Straumann, D. (2004) Ocular rotation axes during dynamic Bielschowsky head-tilt testing in unilateral trochlear nerve palsy. Invest. Ophthalmol. Vis. Sci., 45: 455–465.

Zee, D.S., Chu, F.C., Optican, L.M., Carl, J.R. and Reingold, D. (1984) Graphic analysis of paralytic strabismus with the Lancaster red-green test. Am. J. Ophthalmol., 97: 587–592.

C. Kennard & R.J. Leigh (Eds.)
Progress in Brain Research, Vol. 171
ISSN 0079-6123

CHAPTER 1.9

Ocular motor nerve palsies: implications for diagnosis and mechanisms of repair

James A. Sharpe*, Agnes M.F. Wong and Mohammad Fouladvand

Division of Neurology and Department of Ophthalmology and Vision Sciences, University Health Network, University of Toronto, Toronto, Ontario, Canada

Abstract: Measurements of the dynamics of the eyes in ocular motor nerve palsies may aid diagnosis, characterize peripheral and central palsies, and reveal adaptive properties of the brain. Saccadic and vestibulo-ocular reflex (VOR) functions of patients with peripheral and central sixth, and peripheral third and fourth nerve palsies were studied by three dimensional magnetic field search coil oculography. Combined third and fourth cranial nerve microvascular ischaemic palsy in diabetes mellitus produced low ratios of intorsion to adduction amplitudes. Presumed isolated third nerve palsy caused higher ratios of adduction to intorsion and violations of Listing's law. The VOR in third, fourth, and sixth nerve palsies reveals adaptive equilibration of the action of paretic agonist and their non-paretic antagonist muscles in violation of Hering's law during head motion. Saccadic speeds in the field of paretic agonists are repaired in chronic peripheral palsies despite limited ductions, but remain reduced in central palsies. Limited intorsion with third nerve palsy is attributed to concurrent fourth nerve ischaemia in the distribution of the inferolateral trunk of the intracavernous carotid artery. Adaptive repair of the VOR after ocular motor nerve palsies reduces asymmetric retinal image slip and binocular disparity, and repair of saccadic velocity drives both eyes rapidly and simultaneously into the paretic field of motion.

Keywords: ocular motor nerves; abducens nerve; trochlear nerve; oculomotor nerve; saccades; vestibulo-ocular reflex; adaptation; peripheral neuropathy; diabetes mellitus

Introduction

Assessment of paralytic strabismus typically emphasizes static deviations rather than its effects on eye motion. Investigation of the effects of peripheral and central ocular motor nerve palsies on movement can reveal diagnostic features and identify adaptive and repair properties of the central and peripheral nervous systems.

Peripheral third (IIIrd; oculomotor) nerve palsy in diabetes mellitus has an ischaemic microvascular mechanism (Asbury et al., 1970; Weber et al., 1970). The IIIrd nerve palsy may be isolated or associated with involvement of the other cranial nerves (Lapresle and Lasjaunias, 1986). We report simultaneous palsy of the IIIrd and fourth (IVth; trochlear) nerves, propose an explanation of this distinctive combination, and employ three dimensional (3-D) oculography to characterize IVth nerve palsy when it is combined with IIIrd nerve palsy.

Effects of unilateral peripheral sixth (VIth; abducens), IIIrd, and IVth nerve palsies on the

*Corresponding author. Tel.: + 416-603-5950; Fax: + 416-603-5596; E-mail: james.sharpe@uhn.on.ca

DOI: 10.1016/S0079-6123(08)00609-2

angular vestibulo-ocular reflex (VOR) reveal changes in the actions of antagonists to paretic muscles that implicate monocular adaptations to peripheral neuromuscular deficits.

The abducens is the only motor nerve with an ample course both within and outside the brain and it innervates only one muscle, the lateral rectus. VIth nerve palsy affords opportunity to compare recovery after central versus peripheral nerve damage by assessing the dynamics of abduction.

Methods

Patients with unilateral IIIrd, IVth, or VIth nerve palsies were investigated by 3-D search coil oculography using 6 ft diameter field coils (CNC Engineering, Seattle, WA, USA). In each eye, the patient wore a dual-lead scleral coil annulus to detect horizontal, vertical, and torsional gaze positions (Skalar Instrumentation, Delft, Netherlands). Eye position data were filtered with a bandwidth of 0–90 Hz and digitized at 500 Hz.

Three diabetic patients with, pupil sparing, unilateral IIIrd nerve palsies were compared during attempted adduction or depression while fixation with the paretic eye with the normal eye occluded to determine the amplitude of saccades and any associated torsion.

The VOR in darkness and the visually enhanced VOR (VVOR) during fixation were tested with active head on body motion in yaw, pitch, and roll at frequencies of 0.5, 1.0, and 2.0 Hz at one point in the course of 21 patients with VIth nerve palsy (symptom duration ranged from 2 weeks to 96 months, mean duration: 16 months), 10 patients with IIIrd nerve palsies (symptom duration was 1 week to 50 months, mean duration: 18 months), and in 13 patients with IVth nerve palsies (duration 1 week to 132 months, mean duration: 35 months) and expressed as changes from normal subject values, rather than serial intra-subject changes. Any recovery towards normal values was not assessed and abnormalities are interpreted as deficits, or adaptation to those deficits (Wong and Sharpe, 2002; Wong et al., 2002a, b). Results were compared with recordings from 15 normal subjects.

Saccades were measured in 19 patients with unilateral VIth nerve palsy; 14 patients had idiopathic, presumed ischaemic, peripheral palsy, and normal MR or CT imaging. Symptom durations ranged from 3 weeks to 96 months, (mean duration: 21 months). Patients with diplopia of less than 1 month duration were classified as having *acute* palsy; all others were designated here as *chronic*. Five patients were tested acutely (within 1 month of symptom onset). Serial eye movement recordings were performed on these five patients with acute peripheral palsy, first at presentation and then at 2 months after symptom onset. Five of the 19 patients had central (fascicular) VIth palsies identified on MR imaging; two had cavernomas and three multiple sclerosis, involving the abducens nerve fascicle in the pons, but sparing its nucleus. Two of the five patients with central palsies were tested acutely (within 1 month of symptom onset). All measurements were performed for saccades to targets10 degrees left and right of orbital midposition, within the range of limited duction caused by the palsy. Serial recordings were performed on those two patients with acute, central fascicular palsy, at presentation and at 2 months after symptom onset (Sharpe et al., 2005; Wong et al., 2006). Recordings from 10 normal adult subjects provided control data.

Results

Combined IIIrd and IVth versus isolated IIIrd nerve palsies

In a patient with IIIrd palsy attempted adduction, upward, and downward gaze, recordings showed little torsional movements of the right eye, during adduction or depression indicating simultaneous IIIrd and IVth nerve palsies. The ratio of maximum intorsion amplitude to maximums or adduction amplitude was 0.25 (SD 0.03, $n = 10$). Other patients with isolated IIIrd palsies showed intorsion of the palsied eye during attempted adduction and

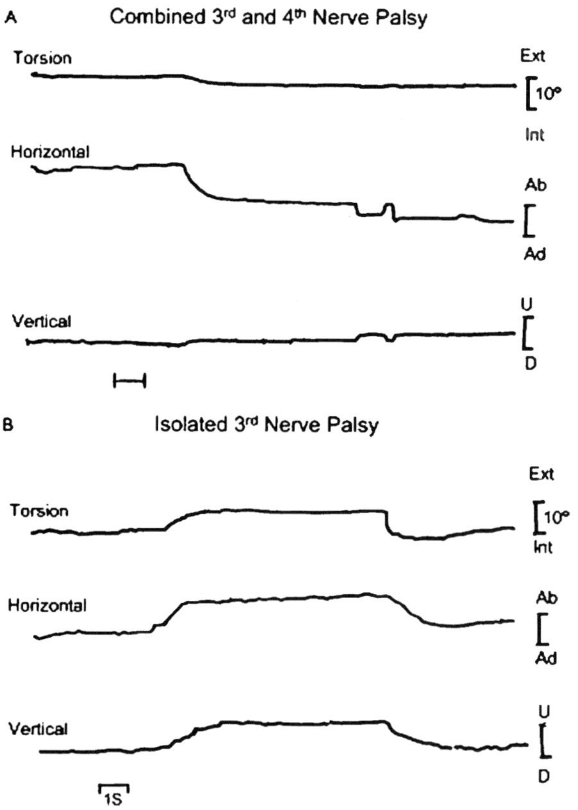

Fig. 1. A. Eye positions of a patient with combined diabetic right IIIrd and IVth nerve palsy has little torsional movement (top trace), during maximal adduction (Ad) saccade (middle trace). B. Recording of a patient with isolated diabetic right IIIrd nerve palsy shows intorsion (Int) (top trace) during adduction saccade (middle trace), indicating intact trochlear nerve and superior oblique muscle function. Slow extorsion (Ext) during abduction (Ab) saccade is explained by paresis of the inferior oblique muscle in comparison to faster intorsion generated by the intact superior oblique muscle. Vertical saccades are slowed upward (U) and downward (D) by paresis of the superior and inferior rectus muscles (bottom trace).

depression and had ratios of intorsion to adduction of 1.16 (SD 0.52) to 0.94 (SD 0.49) during adduction (Fig. 1) in violation of Listing's law.

The VOR in IIIrd, IVth, and VIth nerve palsies

In VIth nerve palsy all 21 patients had decreased horizontal VOR gains in darkness in the paretic eye in both abduction and adduction, but gains remained normal in the non-paretic eye in both directions. In light, horizontal VVOR gains were normal in both eyes in moderate and mild palsy. In severe palsy (with centifugal abduction $<50\%$ of normal range) horizontal VVOR gains were low in the paretic eye in both directions, during viewing with either eye (Fig. 2), while those in the non-paretic eye were higher than normal when the paretic eye viewed.

Similarly, in IIIrd nerve palsy horizontal, vertical, and torsional VOR and VVOR gains of the paretic eye were decreased during abduction and adduction, elevation and depression, and extortion and intorsion. In IVth nerve palsy horizontal, torsional, and vertical VOR gains of the paretic eye of the 13 patients were reduced symmetrically for abduction and adduction, for depression and elevation, and for intorsion and extorsion in darkness. In light, during paretic eye viewing, horizontal torsional VVOR gains remained low in directions of the paretic agonist and intact antagonists, but vertical and horizontal VVOR gains were normal. In both IIIrd and IVth nerve palsies VOR and VVOR gains in the non-paretic eye were normal for torsion and vertical and horizontal head motion. Neither eye showed any significant phase shift from zero in light or in darkness in any direction in IIIrd, IVth, or VIth nerve palsies.

Saccades in central versus peripheral VIth nerve palsies

In acute state of peripheral ischaemic VIth nerve palsies, centrifugal abducting saccades in the paretic eye had reduced peak velocities (Fig. 3A) and longer durations for any given amplitude ($p<0.05$), as anticipated from weakness of the lateral rectus muscle. However, in the chronic state, centrifugal abducting saccadic peak velocities (Fig. 3A) and durations in the paretic eye were within the normal range in the nine chronic peripheral VIth palsies, despite persistent limited abduction. In the five patients with central (fascicular) VIth nerve palsy centrifugal abducting

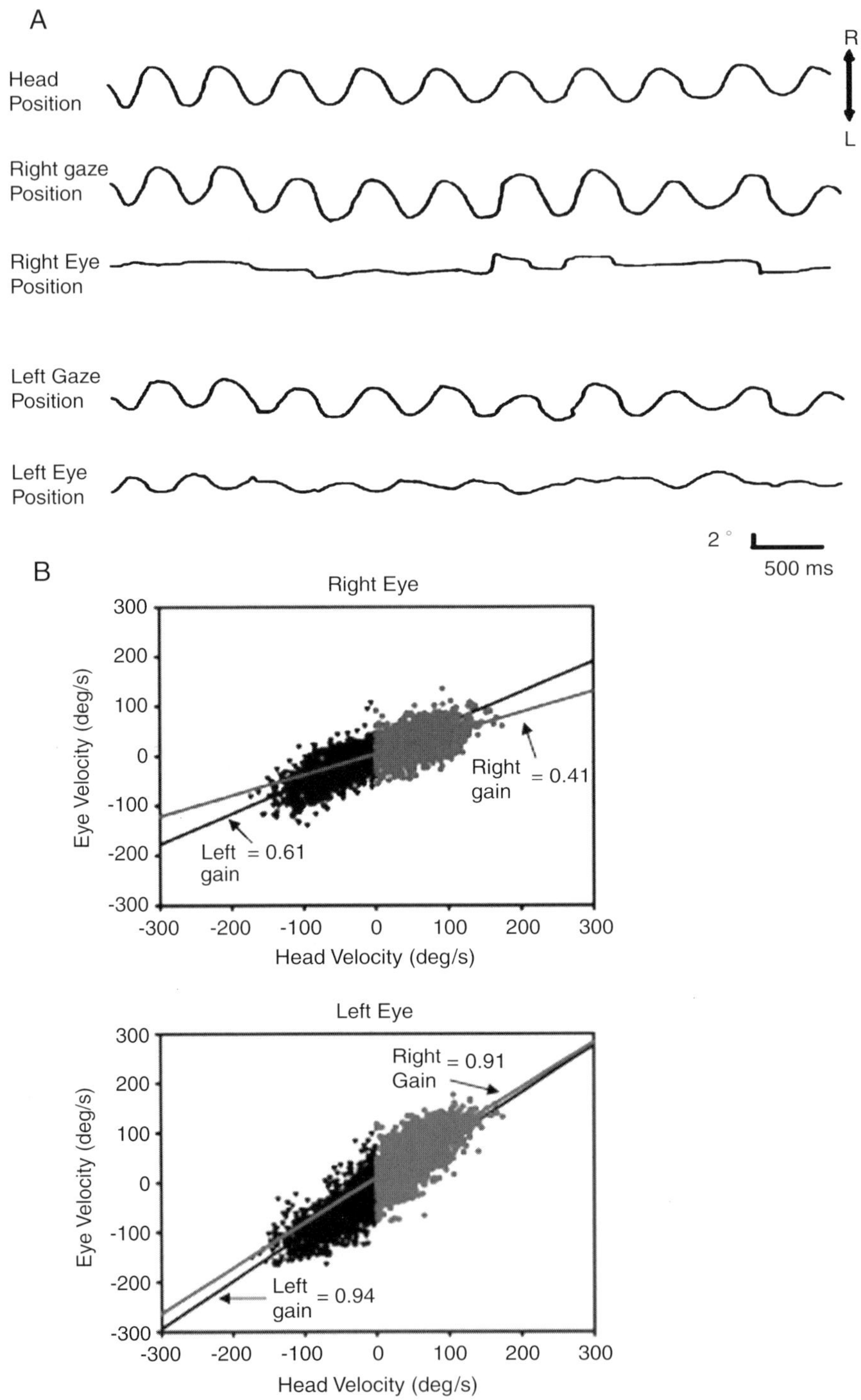

Fig. 2. A. Recording of head, eye, and gaze potions in darkness of a patient with a severe rigth sixth nerve palsy during horizontal head rotation at 2 Hz shows low amplitudes of eye motion to right and left (third trace). The non-paretic eye shows symmetrical higher horizontal amplitude VOR motion (bottom trace). Gaze is measured from coil on right eye (second trace) and left eye (fourth trace). R, rightward; L, leftward. B. VOR gains of right eye (top) and left eye (bottom) in a patient with a right sixth nerve palsy are shown by slopes of best fit lines of data points (dots), plotting head velocity against vestibular smooth eye movement velocity during head rotation at 2 Hz. In the paretic right eye both rightward (abducting) slope (gain 0.41) of and leftward (adducting) slope (gain 0.61) are reduced. For the non-paretic left eye adducting (rightward) VOR gain (0.91) and abducting (leftward) gain (0.94) are symmetrical.

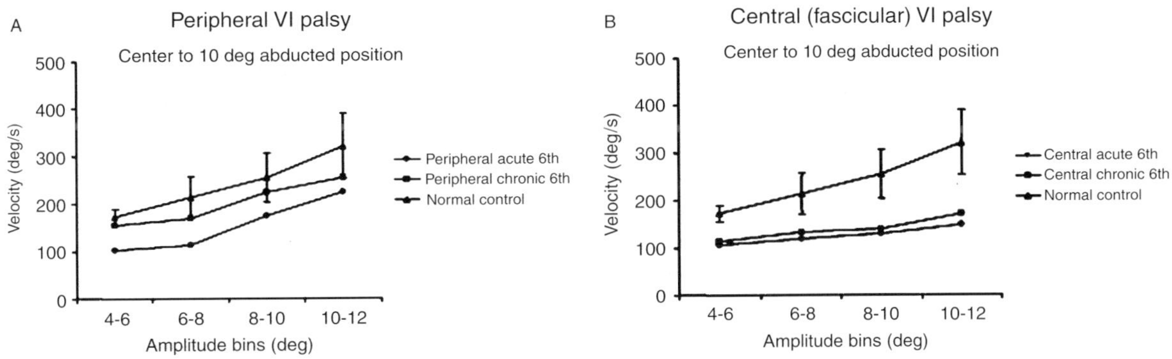

Fig. 3. Mean saccade peak velocities plotted against amplitude show reduced speed of centrifugal abducting saccades in both acute peripheral (A) and central (B) sixth nerve palsies. The paretic eye is used for viewing with the normal eye occluded. In chronic peripheral palsies speed became normal (A) but in chronic central palsies (B) it remained subnormal. Error bars indicate 1 standard deviation from normal group mean.

saccades in the paretic eye had subnormal peak velocities and long durations for any given amplitude, both in the acute and chronic states (Fig. 3B) ($p < 0.05$).

Discussion

Ischaemic ocular motor mononeuropathy versus polyneuropathy

Palsies of the IIIrd cranial nerve from diabetes mellitus or idiopathic presumed ischaemic origin are usually isolated but may be associated with other cranial neuropathies (Annabi et al., 1979; Lapresle and Lasjaunias, 1986). We detected simultaneous IIIrd and IVth nerve palsy in diabetes by 3-D oculography. Their vascular supply to the IIIrd and IVth nerves in their intracavernous region is the inferolateral trunk (ILT), arising from the internal carotid artery. The superior or tentorial branch of the ILT supplies both the IIIrd and IVth nerves, but not other nerves (Lasjaunias et al., 1977; Krisht et al., 1994). Clinical examination of patients with pupil-sparing IIIrd nerve palsy may reveal intorsion during attempted adduction or depression, but the recording (Fig. 1) confirmed the presence of torsion and hence spared IVth nerve function in those patients.

During attempted elevation or depression of the eye innervation of the inferior oblique muscle by the IIIrd nerve must be balanced by innervation of the IVth nerve to restrain torsion and thereby preserve Listing's law (Ferman et al., 1987). In the presence of paresis of the inferior oblique muscle caused by the IIIrd nerve palsy, the unopposed action of the superior oblique muscle is evident, causing intorsion in violation of Listing's law (Fig. 1B), but in the presence of a combined IVth nerve palsy a restraint on torsion is evident (Fig. 1A). After isolated IVth nerve transaction in monkeys Listing's law is largely preserved (Tian et al., 2007). Since both elevation and depression are variably paralysed in IIIrd nerve palsy, but adduction from an abducted position to at least the orbital midposition is preserved, the ratio of intorsion to adduction may be a useful quantitative parameter of assessing paresis or sparing of the IVth nerve, and contribute to localizing the IIIrd nerve palsy to its intracavernous segment when the IVth nerve is also paretic.

Adaptations and deficits in the vestibulo-ocular reflex

Angular VOR gains after unilateral ocular motor nerve palsies are reduced during movement in the directions of actions of paretic muscles as anticipated from their palsy. However gains are also reduced in the fields of actions of their intact antagonist muscles after IIIrd, IVth, or VIth nerve palsies (Wong and Sharpe, 2002; Wong et al., 2002a, b). VOR gains in the non-paretic eye remain normal, implicating a selective adjustment of the paretic eye, specifically to the antagonists of paretic muscles. In light, visual input increases gain of the paretic eye in the motion fields of both paretic agonist muscles and their antagonist muscles, when the paretic eye is used for fixation, while VVOR gains in the non-paretic eye remain normal; this provides further evidence of selective adaptation of innervation to the paretic eye. Torsional dynamic VOR and VVOR gains of the paretic eye are reduced for both extortion and intorsion in IIIrd and IVth nerve palsies (Wong and Sharpe, 2002; Wong et al., 2002a; Sharpe et al., 2003). Motion of the eyes after nerve palsies exemplifies monocular adaptation of the VOR in three dimensions.

After severe VIth nerve palsy, when the centifugal range of abduction was limited to less than 50% of the full range of normal, viewing with the paretic eye increases its abducting and adducting horizontal VVOR gains above VOR gains in darkness but they remain below normal levels in the paretic eye, whereas VVOR gains in the covered non-paretic eye increase to values above unity. In terms adapted from strabismology, the gains measured during fixation with the non-paretic (normal) eye reveal the "primary" VVOR, while gains measured during fixation with the paretic eye reveal the "secondary" VVOR in the non-paretic eye. Elevation of VVOR gain in the non-paretic occluded eye to values above the ideal value of 1.0 during distant fixation with the paretic eye is attributed to increased innervation of yoked muscles of the non-paretic eye, in accordance with Hering's law (see below). However, in mild and moderate palsy vision enhances the VOR in the paretic eye but causes no change in the non-paretic eye, indicating a monocular readjustment of innervation selectively to the paretic eye.

If the VOR gain in the direction of action of antagonists of paretic muscles did not decrease the VOR would be asymmetric in the paretic eye, rotating it further in the direction of action of the intact antagonist with each head rotation, and resulting in increasing position disparity between the two eyes and more diplopia. In the case of a left lateral rectus weakness from a left VIth nerve palsy (for example), any adaptive reduction in innervation to the left medial rectus muscle would be accompanied by reduced innervation to the right lateral rectus muscle, in accordance with Hering's (1868) proposal or law that the brain controls binocular movements conjugately by simultaneuous innervation of yoked pairs of muscles in each eye. However the gain in the non-paretic eye remained normal in both adduction and abduction. Monocular adaptation of VOR and saccadic gain to weakening ocular muscles has been detected in monkeys (Snow et al., 1985; Viirre et al., 1988), but selective effects on paretic muscles or their intact antagonists had not been identified. Selective reduction of VOR gains during action of the antagonist of the paretic muscle by reducing its innervation is apparently the strategy that the brain uses to balance the VOR (Wong and Sharpe, 2002; Wong et al., 2002a, b; Sharpe et al., 2003).

Changes from normal orbital plant mechanics might contribute to the decreased VOR gains in the direction of the antagonists to paretic muscles. The relative contribution of agonist contraction and antagonist relaxation varies with orbital position (Collins, 1975), and it may be altered when one muscle of an agonist–antagonist pair is palsied. Contracture is characterized by muscle shortening and stiffening as a result of decreased number of sarcomeres (Scott, 1994). If the reduction of VOR gains in both directions were due to changes in extraocular muscle mechanics, one would expect VOR gains in the paretic eye to be subnormal during rotation in light (the VVOR) as well as in darkness. However, the results indicate that while VOR gains were decreased, VVOR gains could increase to normal values in light (Sharpe et al., 2003). The results provide evidence that decrease VOR gains in the antagonist of paretic muscles, and equilibration of gains in the directions of paretic agonists and their intact antagonists is not the result of changes in mechanical properties of the orbital plant, but due to a functional, central adaptation to the palsy. Altered proprioceptive inflow (Lewis et al., 1994; Gentle and Ruskell, 1997) from a shortened (slack) antagonist or a palsied muscle might participate in the monocular adaptations after peripheral nerve palsies. Binocular disparity of retinal images that increases during head motion and asymmetry of retinal image slip when the VOR is imbalanced by palsy of a muscle appears to be the visual drive for monocular adaptation to reduce image slip, and diplopia.

Recovery of saccadic velocity in VIth nerve palsy

Patients with *central* palsy, both acute and chronic, had slow abducting saccades but in patients with *peripheral* palsy, abducting saccadic speed became normal in the face of persistent esotropia when tested 2 months after the onset of diplopia, (Sharpe et al., 2005; Wong et al., 2006). Recovery of abducting saccadic velocity in the paretic after peripheral neuropathy may be due to partial regeneration of axons or myelin of the peripheral nerve. The partially recovered peripheral nerve allows the transmission of high firing rates so that saccades are of normal peak velocity. In contrast, tonic signals are impaired resulting in abnormal position (i.e., esotropia) and the limited abduction of the paretic eye, evident in patients in both the acute (<1 month) and chronic states. Disparity between the repair of saccade speed and the persistence of defective ranges of abduction may be due to the selective nature of the nerve fibres damaged or the tempo of their regeneration and remyelination. Large motoneurons within the abducens nucleus innervate singly innervated, twitch muscle fibres, whereas smaller motoneurons around its periphery innervate multiply innervated, non-twitch muscle fibres (Buttner-Ennever et al., 2001, 2003). Normal speed but restricted duction might be explained by either predominant damage to nerve fibres from non-twitch nerve fibres motoneurons just outside the abducens nucleus, or by better repair of nerve fibres from twitch motoneurons within the nucleus. In contrast the absence of recovery saccade speed and range in patients with central palsy indicates limited regeneration of damaged axons and myelin in the glial environment of the brain (Fenrich and Gordon, 2004).

Alternatively, recovery may represent a monocular readjustment of innervation selectively to the paretic eye. A pre-nuclear substrate for possible monocular adjustments of saccade speed in response to peripheral nerve palsy is found in pre-saccadic burst neurons of the caudal paramedian pontine reticular formation (PPRF) which encode monocular movements of either abduction or adduction (Zhou and King, 1998). This organization is contrary to Hering's law. Whether the recovery of saccadic speed in chronic peripheral palsy indicates partial axonal regeneration, remyelination, or monocular adaptation, or all three mechanisms, it allows both eyes to reach targets in the paretic hemifield of motion rapidly and simultaneously.

Acknowledgement

Supported by Canadian Institutes of Health Research (CIHR) Grants MT 15362 and ME 5504.

References

Annabi, A., Lasjaunias, P. and Lapresle, J. (1979) Paralysis of the 3d cranial nerves in diabetes and common oculomotor vascularization. J. Neurol. Sci., 41: 359–367.

Asbury, A.K., Aldredge, H., Hershberg, R. and Fisher, C.M. (1970) Oculomotor palsy in diabetes mellitus: a clinico-pathological study. Brain, 93: 555–566.

Buttner-Ennever, J.A., Eberhorn, A. and Horn, A.K. (2003) Motor and sensory innervation of extraocular eye muscles. Ann. N.Y. Acad. Sci., 1004: 40–49.

Buttner-Ennever, J.A., Horn, A.K.E., Scherberger, H. and D'Ascanio, P. (2001) Motoneurons of twitch and nontwitch extraocular muscle fibers in the abducens, trochlear, and oculomotor nuclei of monkeys. J. Comp. Neurol., 438: 318–335.

Collins, C.C. (1975) The human oculomotor control system. In: Lennerstrand G. and Bach-y-Rita P. (Eds.), Basic Mechanisms of Ocular Motility and Their Clinical Implications. Pergamon Press, Inc., New York, pp. 145–180.

Fenrich, K. and Gordon, T. (2004) Axonal regeneration in the peripheral and central nervous systems: current issues and advances. Can. J. Neurol. Sci., 31: 142–156.

Ferman, L., Collewijn, H. and Berg, A.V. (1987) A direct test of Listing's law-II: human ocular torsion measured under dynamic conditions. Vision Res., 27: 939–951.

Gentle, A. and Ruskell, G. (1997) Pathway of the primary afferent nerve fibers serving proprioception in monkey extraocular muscles. Ophthalmic Physiol. Opt., 17: 225–231.

Hering, E. (1868) Die Lehre vom binokularen Sehen. Wilhelm Englemann, Leipzig, Germany.

Krisht, A., Barnett, D.W., Barrow, D.L. and Bonner, G. (1994) The blood supply of the intracavernous cranial nerves: an anatomic study. Neurosurgery, 34: 275–279, discussion 279.

Lapresle, J. and Lasjaunias, P. (1986) Cranial nerve ischaemic arterial syndromes: a review. Brain, 109: 207–216.

Lasjaunias, P., Moret, J. and Mink, J. (1977) The anatomy of the inferolateral trunk (ILT) of the internal carotid artery. Neuroradiology, 13: 215–220.

Lewis, R.F., Zee, D.S., Gaymard, B.M. and Guthrie, B.L. (1994) Extraocular muscle proprioception functions in the control of ocular alignment and eye movement conjugacy. J. Neurophysiol., 72: 1028–1031.

Scott, A.B. (1994) Change of eye muscle sarcomeres according to eye position. J. Pediatr. Ophthalmol. Strab., 31: 85–88.

Sharpe, J.A., McReelis, K. and Wong, A.M. (2005) Recovery of peripheral versus central nerves identified by saccadic velocity after abducens neuropathy. Ann. N.Y. Acad. Sci., 1039: 417–429.

Sharpe, J.A., Tweed, D. and Wong, A.M. (2003) Adaptations and deficits in the vestibulo-ocular reflex after peripheral ocular motor palsies. Ann. N.Y. Acad. Sci., 1004: 111–121.

Snow, R., Hore, J. and Vilis, T. (1985) Adaptation of saccadic and vestibulo-ocular systems after extraocular muscle tenectomy. Invest. Ophthalmol. Vis. Sci., 26: 924–931.

Tian, J., Shan, X., Zee, D.S., Ying, H., Tamargo, R.J., Quaia, C., Optican, L.M. and Walker, M.F. (2007) Acute superior oblique palsy in monkeys: III. Relationship to Listing's law. Invest. Ophthalmol. Vis. Sci., 48: 2621–2625.

Viirre, E., Cadera, W. and Vilis, T. (1988) Monocular adaptation of the saccadic system and vestibulo-ocular reflex. Invest. Ophthalmol. Vis. Sci., 29: 1339–1347.

Weber, R.B., Daroff, R.B. and Mackey, E.A. (1970) Pathology of oculomotor nerve palsy in diabetics. Neurology, 20: 835–838.

Wong, A.M., McReelis, K. and Sharpe, J.A. (2006) Saccade dynamics in peripheral vs. central sixth nerve palsies. Neurology, 66: 1390–1398.

Wong, A.M.F. and Sharpe, J.A. (2002) Adaptations and deficits in the vestibulo-ocular reflex after third nerve palsy. Arch. Ophthalmol., 120: 360–368.

Wong, A.M.F., Sharpe, J.A. and Tweed, D. (2002a) The vestibulo-ocular reflex in fourth nerve palsy: deficits and adaptations. Vision Res., 42: 2205–2218.

Wong, A.M.F., Tweed, D. and Sharpe, J.A. (2002b) Adaptations and deficits in the vestibulo-ocular reflex after sixth nerve palsy. Invest. Ophthalmol. Vis. Sci., 43: 99–111.

Zhou, W. and King, W.M. (1998) Premotor commands encode monocular eye movements. Nature, 393: 692–695.

C. Kennard & R.J. Leigh (Eds.)
Progress in Brain Research, Vol. 171
ISSN 0079-6123

CHAPTER 1.10

Extraocular proprioception and new treatments for infantile nystagmus syndrome

L.F. Dell'Osso[1,2,3,*] and Z.I. Wang[1,3]

[1]*Daroff-Dell'Osso Ocular Motility Laboratory, Louis Stokes Cleveland Department of Veterans Affairs Medical Center and CASE Medical School, Cleveland, OH, USA*
[2]*Department of Neurology, Case Western Reserve University and University Hospitals Case Medical Center; Cleveland, OH, USA*
[3]*Department of Biomedical Engineering, Case Western Reserve University and University Hospitals Case Medical Center; Cleveland, OH, USA*

Abstract: Our goal is to develop the proprioceptive hypothesis for nystagmus damping; and present the resulting therapies for the treatment of infantile nystagmus syndrome (INS) and acquired nystagmus. Contact lenses, cutaneous stimulation, and neck-muscle vibration damped INS. Four-muscle tenotomy and reattachment was hypothesized as a treatment for INS in 1979 and successfully demonstrated to improve foveation in a canine model of INS and seesaw nystagmus in 1998 and in humans with INS (masked-data, NEI Clinical Trial) in 2003. Subsequently, tenotomy successfully damped acquired pendular nystagmus and oscillopsia in two MS patients and downbeat nystagmus in another. Tenotomy, used in isolation or combination with existing nystagmus and strabismus surgeries, damps different types of nystagmus in their plane of action. Recent neuroanatomical and neurophysiological discoveries support the hypothesis that proprioception is the mechanism for INS damping and allow more realistic models of peripheral ocular motor pathways.

Keywords: ocular motor system; proprioception; nystagmus; treatments; models; control system

Introduction

In the early 1960s, Drs. Larry Stark, Laurence Young, and David Robinson pioneered the use of control systems models to study the ocular motor system (OMS) (Young and Stark, 1963; Robinson, 1964, 1965). In 1967, this approach was extended by studying the behaviour of an OMS with an oscillatory disorder, infantile nystagmus syndrome (INS, also called congenital nystagmus, CN) (Dell'Osso, 1968). An early hypothesis to come out of that extension was that the OMS could *not* function properly using only retinal error position and velocity signals as its inputs. Thus, models that depend solely on retinal inputs cannot duplicate the behavioural responses of subjects with INS, and are also limited in their ability to simulate normal ocular motor behaviour (e.g., pursuit of the imaginary hub of a moving wheel, making saccades in the dark to imagined targets, etc.). Either efference copy of motor output commands or the use of proprioceptive eye-position signals was necessary to reconstruct target information when internal oscillations were present. Figure 1

*Corresponding author. Tel.: + (216) 421-3224; Fax: + (216) 231-3461; E-mail: lfd@case.edu

DOI: 10.1016/S0079-6123(08)00610-9

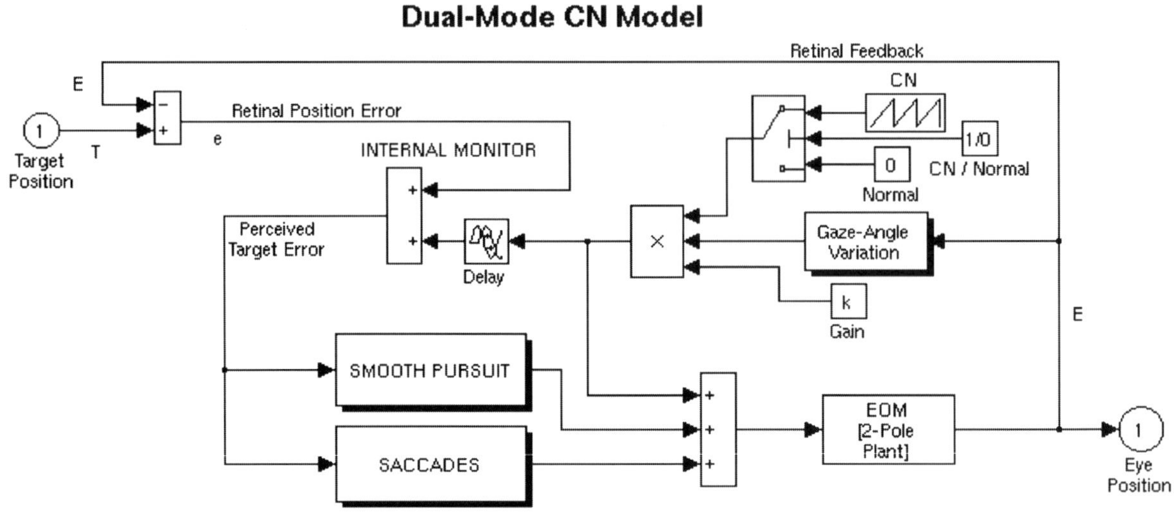

Fig. 1. Block diagram of the first OMS model with INS showing both an internal monitor and proprioceptive feedback signals. The saccadic and pursuit driving signals (Perceived Target Error) are derived from retinal error and efference copy of the nystagmus motor signal, modulated by proprioceptively sensed eye position.

demonstrates how a simple "internal monitor" was used to reconstruct target position in this initial model of INS behaviour. Although a mechanism for generating INS was not hypothesized, a method to simulate the modulation of INS with eye position was; as Fig. 1 shows, it used a proprioceptive measure of the latter. At that time, the use of proprioceptive eye-position information was a reasonable hypothesis since it was known that proprioceptive signals were present in the V cranial nerve as well as in the III, IV, and VI cranial nerves [e.g., Figs. 1–9 of the above thesis (Dell'Osso, 1968)]. Keller and Robinson (1971) published a study that suggested there was no ocular motor stretch reflex and concluded that proprioception did *not* play a role in ocular motor control, at least not for short-term control. In recent years, several lines of evidence have pointed to the importance of extraocular proprioception in ocular motor control (including several articles in this volume).

In this paper, we will present evidence from the studies of the mechanisms of, and therapies for, INS and acquired nystagmus that strongly suggests a significant role for afferent stimuli, both exteroceptive and proprioceptive, in the control of eye movements; we will include that role in a behavioural computer model.

Methods

Recording

Data for our studies were generated using infrared reflection, magnetic search coil, or high-speed digital video systems. Specific details of each may be found in the referenced papers (Wang et al., 2006b).

Protocol

Written consent was obtained from subjects before the testing. All test procedures were carefully explained to the subject before the experiment began, and were reinforced with verbal commands during the trials. Subjects were seated in a chair with headrest and a chin stabilizer, far enough (>5 feet) from either an arc of red LEDs or a reflected laser spot to prevent convergence effects. At this distance the target subtended less than 0.1° of visual angle. The room light could be adjusted from dim down to blackout to minimize extraneous visual stimuli. Experiments consisted of 3–10 trials, each lasting about a minute with time between trials for the subject to rest. Trials were kept short to guard against boredom because INS intensity decreases with inattention.

Analysis

Eye-movement analysis was carried out in the MATLAB environment using software developed and refined in our laboratory. Detailed descriptions of our software and experimental paradigms may be found in the referenced papers and at our web site, http://www.omlab.org. Simulations were carried out in Simulink.

Results

Afferent stimulation and proprioception

One of the first indications that afferent stimulation could affect INS was the observation by J. Lawton Smith in the 1970s that, in some patients, the insertion of contact lenses immediately damped the INS sufficiently for it to be noted clinically. We studied the ocular motor effects of soft contact lenses in INS and found damping across all gaze angles (Dell'Osso et al., 1988). In that paper, we also demonstrated that the damping was immediate by applying light pressure on the eyelid. Topical anaesthetics limited the damping effect by reducing the proprioceptive input from the contact lenses, demonstrating that the effect was not due to the mass of the lenses or the refractive correction. Subsequent studies of other exteroceptive stimuli, including cutaneous stimulation (touch and electrical) of the forehead and neck (including vibration) also demonstrated an immediate damping effect on INS (Dell'Osso et al., 1991; Sheth et al., 1995). We subsequently noted that airflow over the forehead damped INS (unpublished observation).

Four-muscle tenotomy

The four-muscle tenotomy procedure (actually tenotomy, dissection, and resuture at the original muscle insertion) was conceived as a result of studying the ocular motor effects of the Kestenbaum procedure (two-muscle recessions plus two-muscle resections) (Dell'Osso and Flynn, 1979). We noted that in addition to shifting the position of the "null" in INS, several secondary benefits were achieved. They included broadening of the "null" region and an overall damping of the INS at all positions of gaze. We attributed this to "nonlinear changes in the ocular motor plant dynamics." Although conceived in 1979, the tenotomy procedure was not formally hypothesized as a therapy for three classes of INS patients until we had a canine model of INS upon which we could test the procedure (Dell'Osso, 1998).

We successfully demonstrated that the tenotomy procedure damped horizontal canine INS and, 4 months later, vertical canine INS and seesaw nystagmus (Dell'Osso et al., 1999). INS and seesaw nystagmus have different mechanisms, sites, and planes. This led us to hypothesize that tenotomy's therapeutic effects were due to changes in the periphery (the "plant") mediated by a proprioceptive tension-control feedback loop. In an NEI Clinical Trial we demonstrated that the tenotomy procedure successfully damped INS (Hertle et al., 2003, 2004) and in a subsequent study, we demonstrated its "null" broadening effects (Wang et al., 2006a). The beneficial effects of tenotomy were restricted to the small signals responsible for INS slow phases; saccades were not affected (Wang et al., 2006b).

The tenotomy procedure also damped the acquired pendular nystagmus in two patients with MS (Tomsak et al., 2005) and downbeat nystagmus in another patient (Wang et al., 2007b). In the latter paper, tenotomy was combined with nystagmus recessions in the downbeat nystagmus case and strabismus recessions in the INS case.

Convergence

One other means of damping INS (and some forms of acquired nystagmus) is convergence (Dell'Osso et al., 1972; Dell'Osso, 1973). The effect is also immediate (in ms) and results in broadening the "null" in INS (Serra et al., 2006).

Discussion

Top-down, control-systems studies of the normal and dysfunctional OMS have produced the following hypotheses and supporting evidence: (1) the need for a final common neural integrator (Robinson, 1968) and its location (Cannon and Robinson, 1987); (2) the need for an internally resettable saccadic pulse generator (Dell'Osso and Robinson, 1973; Dell'Osso, 1974, both dated entry pages in Dell'Osso's laboratory notebook) (Abel et al., 1978) and its location (Scudder, 1988); (3) the hypothesis that each eye (and, therefore, each eye muscle) is individually controlled (Dell'Osso, 1994) and neurophysiological evidence supporting it (King and Zhou, 2000); and (4) a proprioceptive hypothesis for small-signal gain control (Dell'Osso et al., 1999) and anatomical (Büttner-Ennever et al., 2001, 2002; Eberhorn et al., 2005; Ugolini et al., 2006) and neurophysiological (Wang et al., 2007a) evidence supporting it. Each of these concepts has evolved, often over decades, from an initial hypothesis based on functional analysis of the OMS to a more detailed hypothesis based on anatomical and physiological evidence. Over the past 40 years, the confluence of anatomy, neurophysiology, and control systems analysis has steadily reduced the time required for the transitions between initial hypotheses and the discovery of supporting evidence. The influx of bottom-up, control-systems methods into neuroanatomical and neurophysiological studies has been largely responsible for this progress.

Studies and applications of INS therapies over the past two decades have led to a resurgence of interest in the role of proprioception in ocular motor control. The damping of INS with contact lenses implicated exteroceptive signals carried via the V cranial nerve (Dell'Osso et al., 1988). That is, afferent stimulation via the ophthalmic division of the trigeminal nerve was able to modulate the oscillation; the effect seemed to be immediate. Somehow, signals brought back to the trigeminal (semi lunar) ganglion were able to affect brainstem ocular motor signals. The discovery that other forms of afferent stimulation also damped INS (in ms) reinforced the hypothesis that proprioception played an important role in ocular motor control (Sheth et al., 1995). Finally, the demonstration that the four-muscle tenotomy procedure improved INS waveforms (and eliminated seesaw nystagmus) in a canine, in less than 18 h (Dell'Osso et al., 1999) and in humans (noted immediately post-op by mothers) (Hertle et al., 2003, 2004) provided strong evidence in support of the proprioceptive hypothesis. The success of the tenotomy procedure in many INS patients (Wang et al., 2006a), in the acquired pendular nystagmus of MS (Tomsak et al., 2005), and in downbeat nystagmus (Wang et al., 2007b) further solidified the role of proprioception in nystagmus modulation in each plane, independent of the site or mechanism of the oscillation. Recent research has also identified proprioceptive signals in the cortex (Wang et al., 2007a). We note at this point that convergence also damps INS rapidly and broadens the "null" region — the *same* effects as afferent stimulation and tenotomy (Serra et al., 2006).

Our findings that the tenotomy procedure affects only small signals (e.g., nystagmus slow phases) and not saccades, suggested that current models of the ocular motor plant are inadequate (Wang et al., 2006b). Although the ocular motor neurons (OMN) studied by Keller and Robinson (presumably, the "fast" OMN) did not have a fast stretch reflex, they could still be part of a slower (e.g., a time constant of minutes, hours, or days) tension-control reflex, as is illustrated in Fig. 2. The tension-control loop is independent of the main ocular motor control loop and serves to maintain the resting tension, and therefore (because muscle response is a function of its length) the small-signal gain of the extraocular muscles. Originally, our proprioceptive hypothesis for the mechanism behind the effects of tenotomy was built on the presumption that even in the absence of a short-time-constant stretch reflex, proprioceptive information is used by the OMS for long-term control and adjustment. Because afferent information from the V cranial nerve (e.g., from cutaneous stimulation of the eyelids or forehead) had an *immediate* damping effect on INS, this pathway provides a means of on-line modulation of eye movements. This had presented us with a dilemma; why is there a fast exteroceptive pathway but only a slow proprioceptive pathway? Also, how could cutting a tendon (especially at its distal end), which presumably had no neurological substrate, affect an afferent signal?

Because of the work of Büttner-Ennever and colleagues, we no longer must make a choice

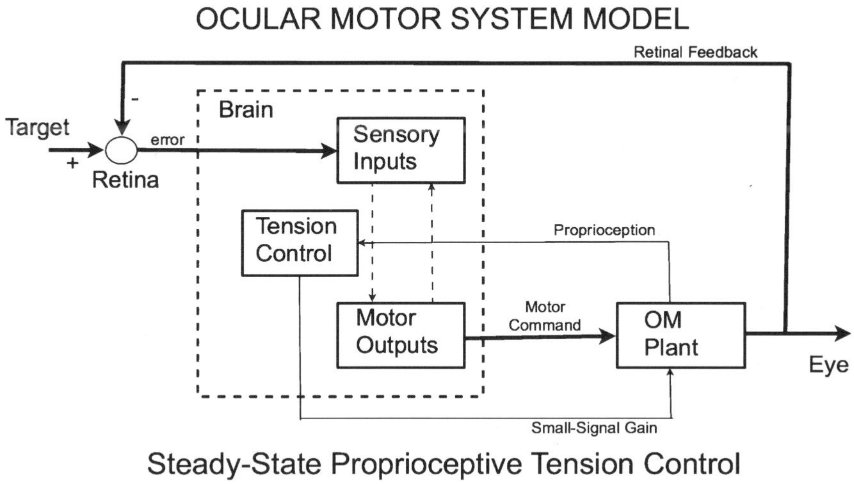

Fig. 2. Block diagram of the OMS showing the addition of a putative proprioceptive tension-control loop. This latter loop was hypothesized to control resting muscle tension and with it, the small-signal gain of the ocular motor (OM) plant. The OMS operates on both retinal error and efference copy of motor commands.

between two separate mechanisms with widely differing time constants. The existence of two types of OMN, two major types of extraocular muscle fibres, palisade endings, and proprioceptive representation in the primary somatosensory cortex not only provided neuroanatomical and neurophysiological substrates for our conclusions but also allowed us to refine and extend our hypothesis (Büttner-Ennever et al., 2001, 2002; Eberhorn et al., 2005; Wang et al., 2007a). In Fig. 3 (top), we show two possible tension-control loops, one for the fast OMN and another for the slow OMN. If the former exists, it would have to be a slow control for maintaining calibration to be consistent with earlier findings (Keller and Robinson, 1971). Although the slow-OMN loop may be similar, it more probably operates on a faster time scale, i.e., more like a skeletal stretch reflex. In Fig. 3 (bottom), we show schematically, based on Büttner-Ennever's work, efferent connections from the various ocular motor subsystems to both fast and slow OMN.

Based on our studies of the effects of tenotomy on INS, we hypothesize that the same on-line modulation of eye movements demonstrated for exteroception will be found for proprioception (i.e., there exists a short-time-constant stretch reflex mediated by the "slow" motor neurons). Given the above-mentioned confluence of bottom-up and top-down research, we expect that evidence supporting this hypothesis will most likely be discovered in the near future by one or more attendees of this meeting. Figure 4 (top) shows a block diagram of the efferent portion of the OMS when two OMN populations are simulated with a simple means to adjust the small-signal gain of the ocular motor plant rather than via a tension-control loop. Although we could have used the latter, at this point too little is known about the time constants of such a loop or its exact function to justify its inclusion. In Fig. 4 (bottom), we show how the two OMN populations and the simplified adjustable-gain plant have been implemented in our behavioural OMS model.

What about convergence? The INS damping from convergence is greater than from gaze angle in most patients and there is some hysteresis present (Serra et al., 2006). We hypothesized that convergence causes a change in muscle-pulley position. However, convergence also produces an imbalance in the muscle lengths of the agonist and antagonist muscles of both eyes (i.e., medial rectus muscles contract and lateral rectus muscles relax). Therefore, we hypothesize that its similar damping and broadening effects as afferent stimulation and tenotomy, employ the same

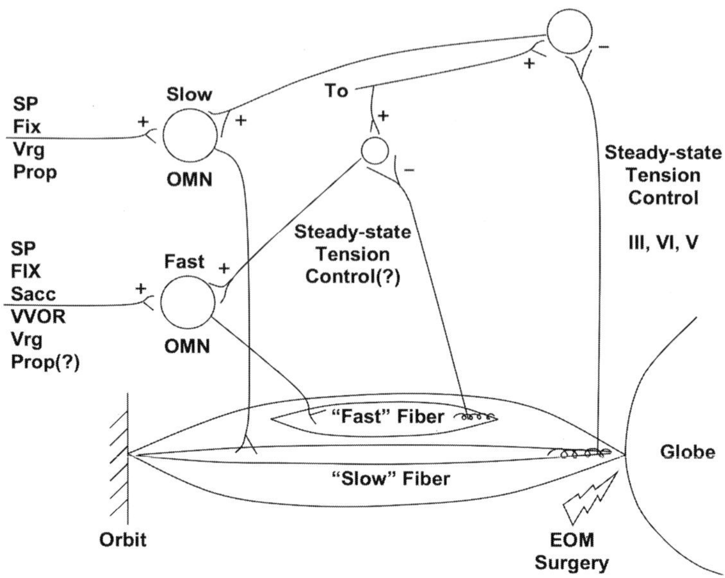

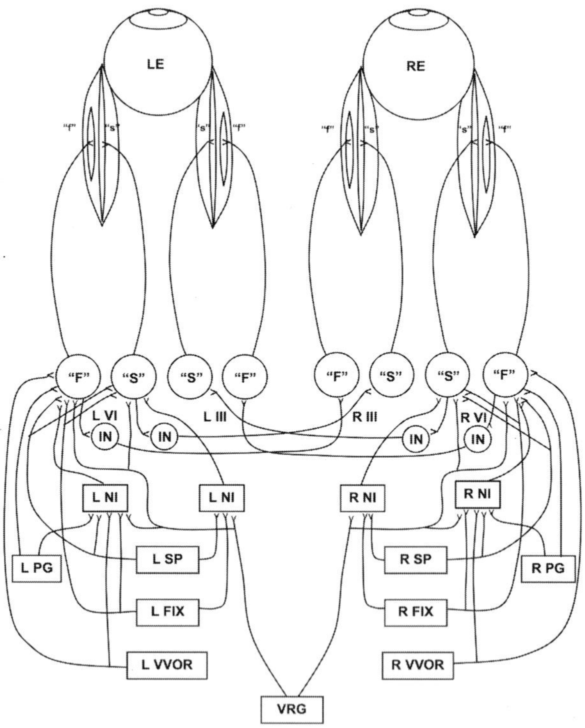

Fig. 3. Top: Putative proprioceptive, tension-control feedback loops for both the "fast" and "slow" OMN and fibres. The "fast" fibres may be involved in slow tension control and the "slow" fibres in a faster, on-line tension control loop. Bottom: Putative wiring diagram for both the "fast" and "slow" OMN and the ocular motor subsystems. Different neural integrator populations on each side of the brain receive inputs from specific OM subsystems and send outputs to their respective OMN populations. The site of extraocular muscle (EOM) surgery is identified, as are the ocular motor subsystems: SP, smooth pursuit; Fix, fixation; Vrg, vergence; Prop, proprioception; Sacc, saccadic; and VVOR, visual vestibuloocular reflex. (R)E, (right) eye; (L)E, (left) eye; "f" or "F", fast; "s" or "S", slow; IN, interneuron; NI, neural integrator; PG, pulse generator.

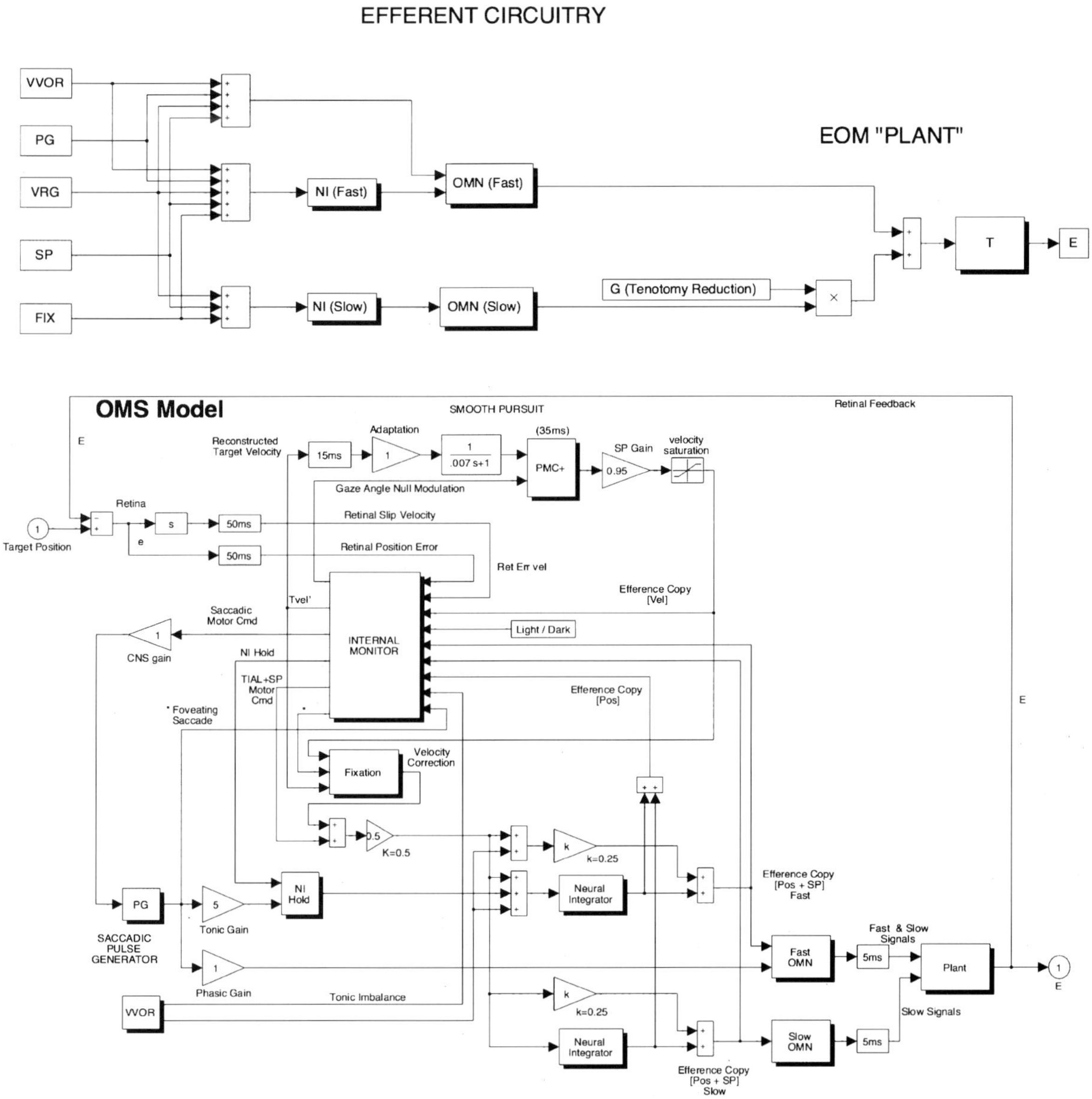

Fig. 4. Top: Efferent block diagram for a unilateral, bidirectional OMS model, including a plant model capable of simulating the effects of tenotomy surgery. Derived from the connections in Fig. 3, this diagram demonstrates how an EOM plant could respond differently to slow signals after tenotomy. Bottom: Incorporation of the two types of OMN and new plant model into an improved behavioural OMS model. This improved OMS model can now allow studies of the tenotomy-induced behavioural changes in INS that were impossible using prior models.

proprioceptive mechanism. That is, contact lenses, convergence, and tenotomy affect an afferent proprioceptive tension-control loop, to produce a damped peripheral ocular-motor response to the nystagmus signal across a broad range of gaze angles.

Although the power and utility of ocular motor proprioception has not been appreciated until recently, proprioceptively based therapies (beginning with the Kestenbaum procedure in 1953) have been used to damp INS and improve visual function for the past five decades. It is often stated that, "proprioception plays no significant role in ocular motor control." That statement is neither credible nor tenable. Top-down studies of ocular motor dysfunction and therapies have produced testable hypotheses that can elucidate the details of how proprioception does affect ocular motor control. Detailed neuroanatomic and neurophysiologic studies of the architecture and signal content of that role are needed to provide more complete understanding of ocular motor control in health and disease and provide the basis for more accurate models of the OMS.

Acknowledgements

The ocular motor and control-system studies of INS cited in this review, and other studies emanating from our Laboratory, were the result of interdisciplinary collaboration of basic and clinical scientists as well as referring physicians from all over the world. They are too numerous to name individually but all deserve recognition for their contributions to the past 45 years of research into the characteristics, mechanisms, and therapies of INS and the insights into normal ocular motor control that this research has provided. Although our approach has been top-down within a control-system framework, we have always been cognisant of, and benefited from, the excellent bottom-up research of Jean Büttner-Ennever and her colleagues. This work was supported in part by the Office of Research and Development, Medical Research Service, Department of Veterans Affairs.

References

Abel, L.A., Dell'Osso, L.F. and Daroff, R.B. (1978) Analog model for gaze-evoked nystagmus. IEEE Trans. Biomed. Eng., BME, 25: 71–75.

Büttner-Ennever, J.A., Horn, A.K.E., Graf, W. and Ugolini, G. (2002) Modern concepts of brainstem anatomy. From extraocular motoneurons to proprioceptive pathways. In: Kaminski H.J. and Leigh R.J. (Eds.), Neurobiology of Eye Movements. From Molecules to Behavior—Ann. N.Y. Acad. Sci., Vol. 956. NYAS, New York, pp. 75–84.

Büttner-Ennever, J.A., Horn, A.K.E., Scherberger, H. and D'Ascanio, P. (2001) Motoneurons of twitch and non-twitch extraocular fibres in the abducens, trochlear and oculomotor nuclei of monkeys. J. Comp. Neurol., 438: 318–335.

Cannon, S.C. and Robinson, D.A. (1987) Loss of the neural integrator of the oculomotor system from brain stem lesions in monkey. J. Neurophysiol., 57: 1383–1409.

Dell'Osso, L.F. (1968) A Dual-Mode Model for the Normal Eye Tracking System and the System with Nystagmus (Ph.D. Dissertation). Electrical Engineering (Biomedical). University of Wyoming, Laramie, WY, pp. 1–131.

Dell'Osso, L.F. (1973) Fixation characteristics in hereditary congenital nystagmus. Am. J. Optom. Arch. Am. Acad. Optom., 50: 85–90.

Dell'Osso, L.F. (1994) Evidence suggesting individual ocular motor control of each eye (muscle). J. Vestib. Res., 4: 335–345.

Dell'Osso, L.F. (1998) Extraocular muscle tenotomy, dissection, and suture: a hypothetical therapy for congenital nystagmus. J. Pediatr. Ophthalmol. Strab., 35: 232–233.

Dell'Osso, L.F. and Flynn, J.T. (1979) Congenital nystagmus surgery: a quantitative evaluation of the effects. Arch. Ophthalmol., 97: 462–469.

Dell'Osso, L.F., Gauthier, G., Liberman, G. and Stark, L. (1972) Eye movement recordings as a diagnostic tool in a case of congenital nystagmus. Am. J. Optom. Arch. Am. Acad. Optom., 49: 3–13.

Dell'Osso, L.F., Hertle, R.W., Williams, R.W. and Jacobs, J.B. (1999) A new surgery for congenital nystagmus: effects of tenotomy on an achiasmatic canine and the role of extraocular proprioception. J. AAPOS, 3: 166–182.

Dell'Osso, L.F., Leigh, R.J. and Daroff, R.B. (1991) Suppression of congenital nystagmus by cutaneous stimulation. Neuro-Ophthalmology, 11: 173–175.

Dell'Osso, L.F., Traccis, S., Abel, L.A. and Erzurum, S.I. (1988) Contact lenses and congenital nystagmus. Clin. Vis. Sci., 3: 229–232.

Eberhorn, A.C., Horn, A.K., Eberhorn, N., Fischer, P., Boergen, K.P. and Buttner-Ennever, J.A. (2005) Palisade endings in extraocular eye muscles revealed by SNAP-25 immunoreactivity. J. Anat., 206: 307–315.

Hertle, R.W., Dell'Osso, L.F., FitzGibbon, E.J., Thompson, D., Yang, D. and Mellow, S.D. (2003) Horizontal rectus tenotomy in patients with congenital nystagmus. Results in 10 adults. Ophthalmology, 110: 2097–2105.

Hertle, R.W., Dell'Osso, L.F., FitzGibbon, E.J., Yang, D. and Mellow, S.D. (2004) Horizontal rectus muscle tenotomy in patients with infantile nystagmus syndrome: a pilot study. J. AAPOS, 8: 539–548.

Keller, E.L. and Robinson, D.A. (1971) Absence of a stretch reflex in extraocular muscles of the monkey. J. Neurophysiol., 34: 908–919.

King, W.M. and Zhou, W. (2000) New ideas about binocular coordination of eye movements: is there a chameleon in the primate family tree? Anat. Rec. (New Anat.), 261: 153–161.

Robinson, D.A. (1964) The mechanics of human saccadic eye movement. J. Physiol., 174: 245–264.

Robinson, D.A. (1965) The mechanics of human smooth pursuit. J. Physiol. (Lond.), 180: p. 569.

Robinson, D.A. (1968) Eye movement control in primates. Science, 161: 1219–1224.

Scudder, C.A. (1988) A new local feedback model of the saccadic burst generator. J. Neurophysiol., 59: 1455–1475.

Serra, A., Dell'Osso, L.F., Jacobs, J.B. and Burnstine, R.A. (2006) Combined gaze-angle and vergence variation in infantile nystagmus: two therapies that improve the high-visual acuity field and methods to measure it. Invest. Ophthalmol. Vis. Sci., 47: 2451–2460.

Sheth, N.V., Dell'Osso, L.F., Leigh, R.J., Van Doren, C.L. and Peckham, H.P. (1995) The effects of afferent stimulation on congenital nystagmus foveation periods. Vision Res., 35: 2371–2382.

Tomsak, R.L., Dell'Osso, L.F., Rucker, J.C., Leigh, R.J., Bienfang, D.C. and Jacobs, J.B. (2005) Treatment of acquired pendular nystagmus from multiple sclerosis with eye muscle surgery followed by oral memantine. DJO, 11: 1–11.

Ugolini, G., Klam, F., Doldan Dans, M., Dubayle, D., Brandi, A.M., Buttner-Ennever, J. and Graf, W. (2006) Horizontal eye movement networks in primates as revealed by retrograde transneuronal transfer of rabies virus: differences in monosynaptic input to "slow" and "fast" abducens motoneurons. J. Comp. Neurol., 498: 762–785.

Wang, X., Zhang, M., Cohen, I.S. and Goldberg, M.E. (2007a) The proprioceptive representation of eye position in monkey primary somatosensory cortex. Nat. Neurosci., 10: 640–646.

Wang, Z., Dell'Osso, L.F., Jacobs, J.B., Burnstine, R.A. and Tomsak, R.L. (2006a) Effects of tenotomy on patients with infantile nystagmus syndrome: foveation improvement over a broadened visual field. J. AAPOS, 10: 552–560.

Wang, Z., Dell'Osso, L.F., Zhang, Z., Leigh, R.J. and Jacobs, J.B. (2006b) Tenotomy does not affect saccadic velocities: support for the "small-signal" gain hypothesis. Vision Res., 46: 2259–2267.

Wang, Z.I., Dell'Osso, L.F., Tomsak, R.L. and Jacobs, J.B. (2007b) Combining recessions (nystagmus and strabismus) with tenotomy improved visual function and decreased oscillopsia and diplopia in acquired downbeat nystagmus and in horizontal infantile nystagmus syndrome. J. AAPOS, 11: 135–141.

Young, L.R. and Stark, L. (1963) A discrete model for eye tracking movements. IEEE Trans. Mil. Elect. MIL, 7: 13–115.

SECTION 2

New Insights into Brainstem Generation of Ocular Motor Commands

C. Kennard & R.J. Leigh (Eds.)
Progress in Brain Research, Vol. 171
ISSN 0079-6123

CHAPTER 2.1

Neural circuits for triggering saccades in the brainstem

Yoshikazu Shinoda*, Yuriko Sugiuchi, Yoshiko Izawa and Mayu Takahashi

Department of Systems Neurophysiology, Graduate School of Medicine, Tokyo Medical and Dental University, Yushima, Bunkyo-ku, Tokyo 113-8519, Japan

Abstract: Here we review the functional anatomy of brainstem circuits important for triggering saccades. Whereas the rostral part of the superior colliculus (SC) is considered to be involved in visual fixation, the caudal part of the SC plays an important role in generation of saccades. We determined the neural connections from the rostral and caudal parts of the SC to inhibitory burst neurons (IBNs) and omnipause neurons (OPNs) in the nucleus raphe interpositus. To reveal the neural mechanisms of triggering saccadic eye movements, we analysed the effects of stimulation of the SC on intracellular potentials recorded from IBNs and OPNs in anaesthetized cats. Our studies show that IBNs receive monosynaptic excitation from the contralateral caudal SC, and disynaptic inhibition from the ipsilateral caudal SC, via contralateral IBNs. Further, IBNs receive disynaptic inhibition from the rostral part of the SC, on either side, via OPNs. Intracellular recording revealed that OPNs receive excitation from the rostral parts of the bilateral SCs, and disynaptic inhibition from the caudal SC mainly via IBNs. The neural connections determined in this study are consistent with the notion that the "fixation zone" is localized in the rostral SC, and suggest that IBNs, which receive monosynaptic excitation from the caudal "saccade zone," may inhibit tonic activity of OPNs and thereby trigger saccades.

Keywords: superior colliculus; inhibitory burst neuron (IBN); omnipause neuron (OPN); abducens motoneuron; fixation neuron; saccade

Introduction

Jean Büttner-Ennever has done much to clarify brainstem anatomy concerned with saccade generation by using innovative tracer and histochemical techniques. Here we summarize our contribution, which consists of studies of the functional anatomy of saccades in the anaesthetized cat. The superior colliculus (SC) contains a motor map that reflects direction and size of saccades (Robinson, 1972; Guitton et al., 1980; McIlwain, 1986). On this map, the rostral SC reflects the foveal region of the retina, and contains neurons that discharge continuously during visual fixation (Munoz and Guitton, 1989, 1991; Munoz and Wurtz, 1993). Conversely, the caudal SC contains movement neurons that start firing before the onset of saccadic eye movement (Munoz and Wurtz, 1995; Stanford et al., 1996). Stimulation of the caudal SC evokes saccadic eye movements to the contralateral side (Robinson, 1972), whereas stimulation of the rostral SC suppresses generation

*Corresponding author. Tel.: +81-3-5803-5152;
Fax: +81-3-5803-5155; E-mail: yshinoda.phy1@tmd.ac.jp

DOI: 10.1016/S0079-6123(08)00611-0

of saccadic eye movements (Paré and Guitton, 1994; Gandhi and Keller, 1999). Therefore, currently, it is often accepted that there are two regions in the SC that have different functions; the "saccade zone" that encodes saccade generation, and the "fixation zone" that encodes visual fixation. More recently, the idea of functional independence of the rostral "fixation zone" has been challenged, since microstimulation of the SC indicated that the saccade size was continuously represented from the caudal to the rostral SC (Gandhi and Keller, 1999). If two different systems of reciprocal functions, i.e., maintaining fixation vs. generating saccades do exist in the rostral and caudal SC, respectively, then the respective parts of the SC should have different neural connections with abducens motoneurons (MNs), inhibitory burst neurons (IBNs), and omnipause neurons (OPNs) in the saccade generator of the brainstem. However, the projections from the rostral and caudal parts of the SC to brainstem saccade generators require clarification.

The OPNs are located in the nucleus raphe interpositus (Büttner-Ennever et al., 1988). There are two important questions concerning the role of OPN in the generation of saccades. First, which excitatory neurons maintain sustained activity of OPNs during steady visual fixation? Second, which inhibitory neurons suppress activity of OPNs so that saccades can be initiated and break fixation? Despite the numerous studies, the neural circuits for triggering saccades by inhibiting OPN activity still remain controversial.

To understand the neural mechanism of generation of saccades and their suppression during visual fixation, we investigated the neural projections from the rostral and caudal SC to IBNs and OPNs in anaesthetized cats by recording intracellular potentials from them. We found different input patterns from the rostral vs. caudal SC to IBNs and OPNs. Thus, IBNs receive monosynaptic excitation and disynaptic inhibition from the caudal SC on the contralateral and ipsilateral side, respectively, and disynaptic inhibition from the bilateral rostral SCs. Most OPNs receive monosynaptic excitation from the rostral SCs and disynaptic inhibition from the caudal SCs. This disynaptic inhibition to OPNs is mediated by IBNs. Possible roles of IBNs in triggering of saccades by actively inhibiting the tonic activity of OPNs will be discussed.

Identification of IBNs and synaptic inputs from the SC to IBNs

To analyse projections from the SCs to IBNs, we searched for neurons in the IBN region about 0.8 mm lateral to the midline. We used the following criteria for identifying penetrated neurons as IBNs; (1) location of a cell in the pontomedullary junction (IBN region), (2) antidromic activation from the contralateral abducens nucleus, (3) monosynaptic excitation from the contralateral caudal SC, and (4) disynaptic inhibition from the ipsilateral caudal SC. To confirm the recorded neurons as IBNs, we injected horseradish peroxidase (HRP) into cell bodies or proximal axons of presumed IBNs that satisfied the criteria listed above, and examined the morphologies of the penetrated cells at the early stages of this series of experiments. The stained cells had morphological features that were similar to electrophysiologically identified IBNs (Yoshida et al., 1982; Strassman et al., 1986).

To investigate the properties of synaptic inputs from the rostral and caudal parts of the SC to IBNs, we recorded intracellular potentials from IBNs and examined effects of stimulation at four rostrocaudal sites in each SC (Fig. 1B). Stimulation of the rostral and caudal parts of the ipsilateral SC evoked inhibitory postsynaptic potentials (IPSPs) in an IBN (Fig. 1B1–4), whereas stimulation of the contralateral SC evoked excitatory postsynaptic potentials (EPSPs) in the same cell (Fig. 1B6–8). The IPSPs and EPSPs usually increased as the stimulation sites moved caudally in the SC. In addition, the most rostral site in the contralateral SC was different from the more caudal sites in that stimulation of the most rostral site evoked IPSPs in the IBN (Fig. 1B5).

Next, we compared the patterns of collicular inputs to IBNs (Sugiuchi et al., 2005) with those to abducens MNs (Izawa et al., 1999). Stimulation of any site of the ipsilateral SC evoked disynaptic IPSPs in abducens MNs (Fig. 1C1–4). This pattern was similar to the pattern described above in IBNs.

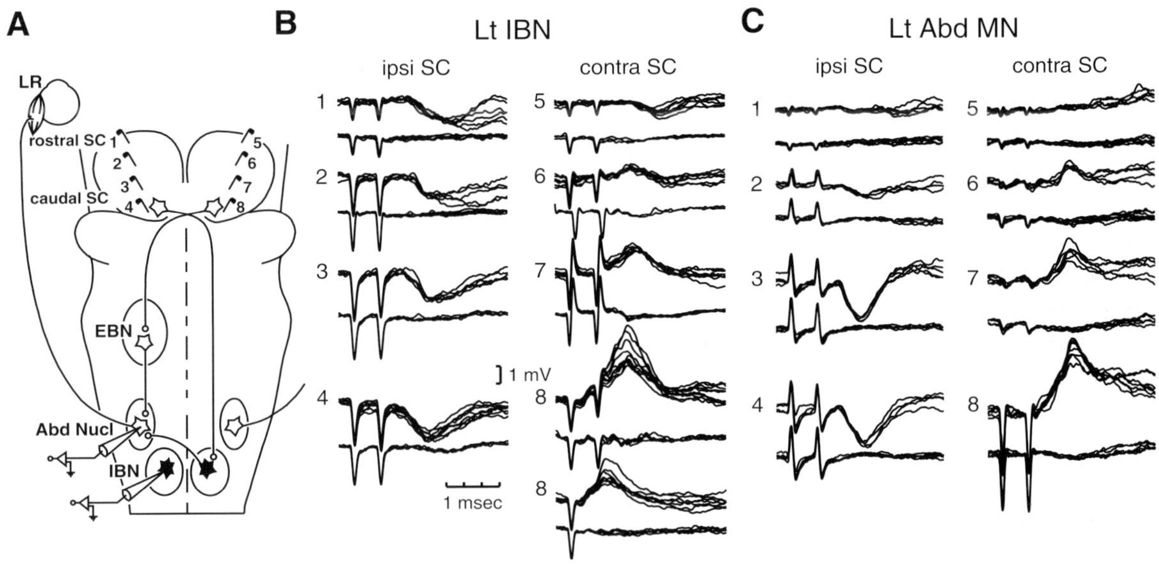

Fig. 1. Comparison of synaptic inputs from the superior colliculi (SCs) on both sides to an inhibitory burst neuron (IBN) (B) and an abducens motoneuron (MN) (C). (A) Experimental setup. LR, lateral rectus muscle; EBN, excitatory burst neuron. (B) Postsynaptic potentials (PSPs) evoked in a left IBN by stimulation of the ipsilateral (1–4) and contralateral SC (5–8) at 500 μA. (C) PSPs evoked in a left abducens motoneuron by stimulation of the ipsilateral (1–4) and the contralateral SC (5–8) at 500 μA. Calibration for (B) applies also to (C). (Adapted with permission from Sugiuchi et al., 2005.)

However, abducens MNs did not usually receive inhibition from the most rostral SC, whereas IBNs always received disynaptic inhibition from the most rostral SC. Stimulation of the contralateral SC evoked excitation in abducens MNs and in IBNs, but the excitation was disynaptic in abducens MNs (Fig. 1C6–8) and monosynaptic in IBNs. Furthermore, stimulation of the most rostral part of the contralateral SC always evoked disynaptic inhibition in IBNs (Fig. 1B5), whereas it never evoked inhibition in abducens MNs (Fig. 1C5). This last finding is the most important difference of the synaptic input pattern between IBNs and abducens MNs.

Pathways from the SC to IBNs

To identify the pathways from SC to IBNs, we made a transverse section in the *right* medial longitudinal fasciculus at a level, about 2 mm rostral to the rostral end of the abducens nucleus, and interrupted the tectoreticular tract connecting between the SC and the OPN region (Fig. 2A). Following sectioning, inhibition from the left rostral and caudal SC disappeared in a left IBN (Fig. 2Bb), but inhibition from the right rostral SC (Fig. 2Bd) remained unaffected in the same IBN. Moreover, stimulation of the caudal part of the left SC did not evoke monosynaptic excitation in a right IBN (Fig. 2Cb), and disynaptic IPSPs evoked by left rostral SC stimulation also disappeared in the same IBN (not shown). However, stimulation of the right rostral SC (Fig. 2Cd) and the right caudal SC (not shown) evoked disynaptic IPSPs in a right IBNs, similar to the control, indicating that the inhibition from the ipsilateral SC was not influenced by sectioning the tectoreticular axons on the same side as the IBNs. It follows that monosynaptic excitation and disynaptic inhibition from the contralateral SC to IBNs were most likely mediated by tectoreticular axons on the same side as the IBNs (Grantyn and Grantyn, 1982), and inhibitory interneurons that mediate this disynaptic inhibitory input were most likely to be located caudal to the level of the section. In contrast, disynaptic inhibition from the ipsilateral SC to IBNs was likely to be mediated by tectoreticular axons on the side opposite to the IBNs.

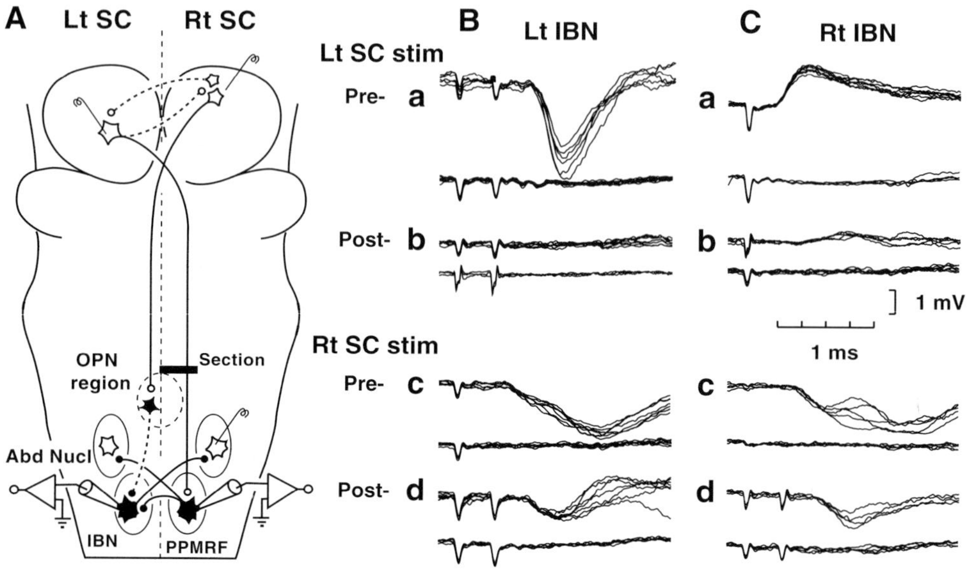

Fig. 2. Effects of the sectioning of tectoreticular axons (thick horizontal bar in A) on the right on SC-evoked PSPs in left (B) and right (C) IBNs. (A) Schematic diagram of the experimental setup. (B–C) Intracellular records from IBNs before (a, c) and after the sectioning (b, d). Note that inhibition from the right rostral SC remained in a left IBN (Bd) and a right IBN (Cd) after the sectioning of the right tectoreticular tract. This finding indicates that the inhibition from the right rostral SC to the left IBN is not due to collicular commissural activation (dotted axons in the SC) of tectoreticular neurons in the left SC. (Adapted with permission from Takahashi et al., 2005.)

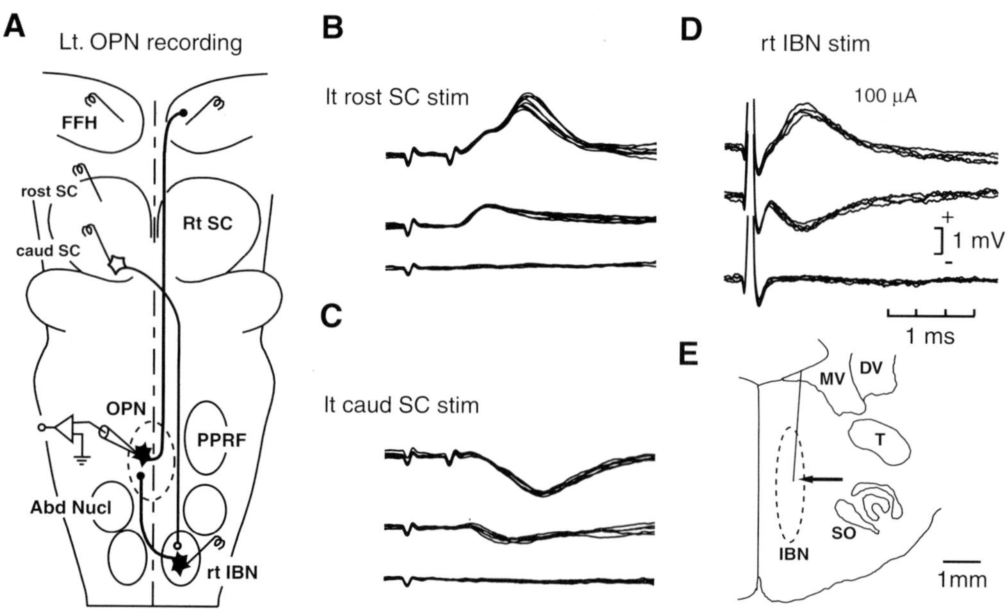

Fig. 3. Synaptic inputs from the SC (B–C) and the IBN region (D–E) to left omnipause neurons (OPNs). (A) Experimental setup. (B–C) EPSPs evoked by stimulation of the left rostral SC (B), and IPSPs evoked by stimulation of the left caudal SC in a left OPN (C). Upper, middle and lower traces; double and single stimulation and field potentials, respectively. (D) Monosynaptic IPSPs (middle traces) and reversed IPSPs after Cl^- injection (upper traces) evoked by stimulation of the right IBN region (arrow in E) in another left OPN. MV, DV, T and SO; medial and descending vestibular nucleus, trigeminal and superior olive nucleus, respectively.

As the stimulation sites in the SC were made more caudally, monosynaptic excitation in the contralateral IBNs increased, and disynaptic inhibition in ipsilateral IBNs and abducens MNs increased. These findings suggest that inhibition in IBNs from the ipsilateral caudal SC might be mediated by contralateral IBNs. To explore further this possibility, we analysed responses in IBNs evoked by stimulation of the contralateral IBN region. Stimulation of the contralateral IBN region evoked IPSPs, with latencies ranging from 0.7 to 1.8 ms. This latter finding showed that contralateral IBNs exerted monosynaptic inhibition on IBNs. We then confirmed that contralateral IBNs mediate disynaptic inhibition from the ipsilateral caudal SC to IBNs, by showing that the monosynaptic inhibition evoked from the contralateral IBN region was facilitated by preconditioning stimulation of the ipsilateral caudal SC.

To confirm further that the inhibition from the ipsilateral caudal SC to IBNs is mediated by contralateral IBNs, we examined the effect of sectioning the midline between the bilateral IBN regions. After a middle section, stimulation of the rostral part of the contralateral SC evoked IPSPs in an IBN. On the other hand, ipsilateral stimulation of the caudal SC did not evoke IPSPs, whereas that of the rostral SC evoked disynaptic IPSPs in the same IBN (Sugiuchi et al., 2005). Taken together, these results confirmed that the disynaptic inhibition from the ipsilateral caudal SC to an IBN is conveyed via contralateral IBNs, and the disynaptic inhibition from the rostral SC on either side is conveyed to IBNs via inhibitory interneurons other than IBNs, most likely OPNs.

Synaptic inputs from the rostral and caudal SC to OPNs

To demonstrate directly that disynaptic inhibition from the rostral SC to IBNs is mediated by OPNs, we recorded intracellular potentials from OPNs (Fig. 3) (Takahashi et al., 2005). OPNs were identified by their antidromic responses to stimulation of the Forel's H area and/or the IBN region. We found that stimulation of the rostral SC evoked EPSPs in an OPN (Fig. 3B). Since OPNs are reported to project to IBNs (Ohgaki et al., 1987; Strassman et al., 1987), our finding indicates that IBNs receive disynaptic inhibition from the rostral SC via OPNs. In contrast, stimulation of the caudal SC evoked disynaptic IPSPs in the same OPN (Fig. 3C). As shown in abducens MNs (Fig. 2), these IPSPs were most probably mediated via contralateral IBNs, because stimulation of the caudal SC could not evoke IPSPs after a transverse lesion in the contralateral medial longitudinal fascicle just rostral to the OPN region. In fact, stimulation of the contralateral IBN region (Fig. 3E) evoked monosynaptic IPSPs with weak stimulus intensity in an OPN (Fig. 3D). Stimulation of the ipsilateral IBN region also induced monosynaptic IPSPs in IBNs. Since IBNs send their stem axons across the midline without giving rise to collaterals on the ipsilateral side (Sugiuchi et al., 2005), this ipsilateral inhibition might be caused by an axon

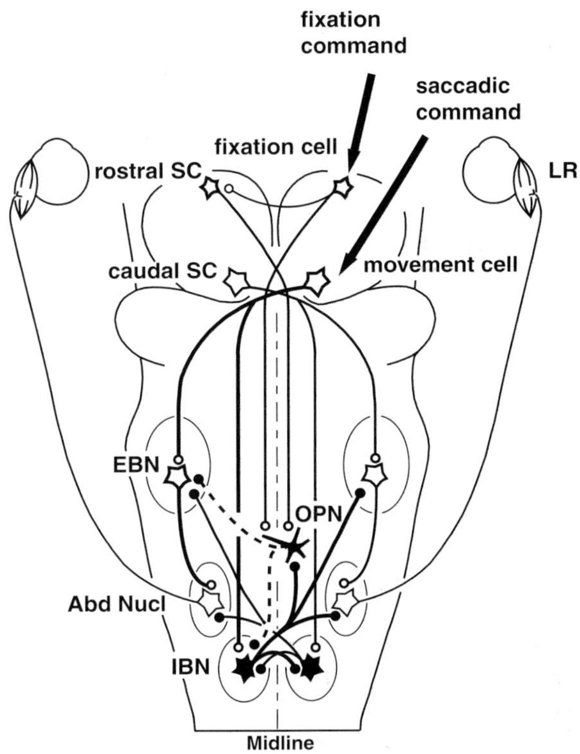

Fig. 4. Summary diagram of neural circuits from the rostral "fixation zone" and the caudal "saccade zone" of the SC to an IBN and an OPN for triggering saccades and visual fixation. Abd Nucl, abducens nucleus; LR, lateral rectus muscle.

reflex of IBN axons. Furthermore, preconditioning stimulation of the caudal SC facilitated the monosynaptic IPSPs evoked by stimulation of the contralateral IBN region (not shown). Therefore, these findings suggest that OPNs receive disynaptic inhibition via contralateral IBNs (Fig. 4).

Conclusions

Our studies have shown that IBNs receive monosynaptic excitation from the contralateral caudal SC, and disynaptic inhibition from the ipsilateral caudal SC via contralateral IBNs. In addition, IBNs receive disynaptic inhibition from the rostral parts of the bilateral SCs via inhibitory interneurons other than IBNs, most probably OPNs.

At saccade onset, SC neurons may activate contralateral excitatory burst neurons (EBNs) and IBNs, and these EBNs excite abducens MNs on the same side. The IBNs inhibit not only abducens MNs, but also IBNs and EBNs on the opposite side, since IBNs give rise to axon terminals in the contralateral abducens nucleus, the paramedian pontine reticular formation (PPRF) and the paramedian pontomedullary reticular formation (PPMRF) (Yoshida et al., 1982; Strassman et al., 1986; Sugiuchi et al., 2005). Therefore, this antagonistic inhibition at the supranuclear as well as at the motoneuronal level assures the suppression of saccade generation towards the opposite side.

Another important finding was that IBNs are disynaptically inhibited by the rostral parts of the bilateral SCs. Since a transverse section of tectoreticular fibres on the right side at the level just rostral to the OPN region eliminated IPSPs evoked by stimulation of the left rostral SC in right IBNs and left IBNs, intervening interneurons for these pathways must be OPNs. Fixation neurons in the rostral SC are antidromically activated by stimulation of the OPN region (Gandhi and Keller, 1997), and rostral SC stimulation activates OPNs (Paré and Guitton, 1994). In fact, intracellular recording from OPNs received monosynaptic excitation from the rostral parts of the bilateral SCs (Takahashi et al., 2005). Paré and Guitton (1994) reported that stimulation of the rostral SC suppresses the generation of saccades in both directions. However, rostral SC inhibition was usually stronger for ipsilateral IBNs than for contralateral IBNs. This result is consistent with the following anatomical data that tectoreticular neurons in the rostral pole of the SC terminate in the OPN region on the opposite site (Büttner-Ennever et al., 1999), and OPNs project mainly to the contralateral PPRF and the IBN region (Ohgaki et al., 1987).

Fixation neurons behaved as if their function was to actively maintain gaze on a target and prevent burst neurons from producing unwanted eye movements (Fuchs et al., 1985). Moreover, by ceasing to fire immediately before saccades away from a fixation target, fixation neurons may partly contribute to triggering of the saccade. Neurons in the "saccade zone" of the caudal SC may receive inputs from the cerebral cortex such as the frontal eye field (FEF) and the lateral intraparietal cortex, whereas fixation neurons in the "fixation zone" of the rostral SC most likely receive inputs from the "suppression area" of the FEF (Izawa et al., 2004). The present results suggest that fixation neurons in the bilateral rostral SCs suppress the initiation of saccades and maintain the direction of gaze by increasing the level of tonic inhibitory input to IBNs and EBNs via OPNs. On the other hand, these fixation neurons stop firing just before a saccade onset. Due to this disfacilitation, and also by active inhibition from the caudal "saccade zone" of the SC, OPNs may stop tonic firing to trigger saccades. Yoshida et al. (2002) showed that stimulation of the SC disynaptically inhibits OPNs. This disynaptic inhibition was considered to be mediated by inhibitory interneurons other than IBNs, most likely inhibitory long-lead burst neurons located in the PPRF (Kamogawa et al., 1996). However, the present result strongly suggests that IBNs are mainly responsible for active inhibition of OPNs at the onset of saccades (Fig. 4). Further study is required to provide anatomical evidence to support a strong projection of IBNs to OPNs.

References

Büttner-Ennever, J.A., Cohen, B., Pause, M. and Fries, W. (1988) Raphe nucleus of the pons containing omnipause neurons of the oculomotor system in the monkey, and its homologue in man. J. Comp. Neurol., 267: 307–321.

Büttner-Ennever, J.A., Horn, A.K.E., Henn, V. and Cohen, B. (1999) Projections from the superior colliculus motor map to omnipause neurons in monkey. J. Comp. Neurol., 413: 55–67.

Fuchs, A.F., Kaneko, C.R. and Scudder, C.A. (1985) Brainstem control of saccadic eye movements. Ann. Rev. Neurosci., 8: 307–337.

Gandhi, N.J. and Keller, E.L. (1997) Spatial distribution and discharge characteristics of superior colliculus neurons antidromically activated from the omnipause region in monkey. J. Neurophysiol., 78: 2221–2225.

Gandhi, N.J. and Keller, E.L. (1999) Comparison of saccades perturbed by stimulation of the rostral superior colliculus, the caudal superior colliculus, and the omnipause neuron region. J. Neurophysiol., 82: 3236–3253.

Grantyn, A.A. and Grantyn, R. (1982) Axonal patterns and site of termination of cat superior colliculus neurons projecting in the tecto-bulbo-spinal tract. Exp. Brain Res., 46: 243–256.

Guitton, D., Crommelinck, M. and Roucoux, A. (1980) Stimulation of the superior colliculus in the alert cat. I. Eye movements and neck EMG activity evoked when the head is restrained. Exp. Brain Res., 39: 63–73.

Izawa, Y., Sugiuchi, Y. and Shinoda, Y. (1999) Neural organization from the superior colliculus to motoneurons in the horizontal oculomotor system of the cat. J. Neurophysiol., 81: 2597–2611.

Izawa, Y., Suzuki, H. and Shinoda, Y. (2004) Suppression of visually and memory-guided saccades induced by electrical stimulation of the monkey frontal eye field. II. Suppression of bilateral saccades. J. Neurophysiol., 92: 2261–2273.

Kamogawa, H., Ohki, Y., Shimazu, H., Suzuki, I. and Yamashita, M. (1996) Inhibitory input to pause neurons from pontine burst neuron area in the cat. Neurosci. Lett., 203: 163–166.

McIlwain, J.T. (1986) Effects of eye position on saccades evoked electrically from superior colliculus of alert cats. J. Neurophysiol., 55: 97–112.

Munoz, D.P. and Guitton, D. (1989) Fixation and orientation control by the tecto-reticulo-spinal system in the cat whose head is unrestrained. Rev. Neurol. (Paris), 145: 567–579.

Munoz, D.P. and Guitton, D. (1991) Control of orienting gaze shifts by the tectoreticulospinal system in the head-free cat. II. Sustained discharges during motor preparation and fixation. J. Neurophysiol., 66: 1624–1641.

Munoz, D.P. and Wurtz, R.H. (1993) Fixation cells in monkey superior colliculus. I. Characteristics of cell discharge. J. Neurophysiol., 70: 559–575.

Munoz, D.P. and Wurtz, R.H. (1995) Saccade-related activity in monkey superior colliculus. I. Characteristics of burst and buildup cells. J. Neurophysiol., 73: 2313–2333.

Ohgaki, T., Curthoys, I.S. and Markham, C.H. (1987) Anatomy of physiologically identified eye-movement-related pause neurons in the cat: pontomedullary region. J. Comp. Neurol., 266: 56–72.

Paré, M. and Guitton, D. (1994) The fixation area of the cat superior colliculus: effects of electrical stimulation and direct connection with brainstem omnipause neurons. Exp. Brain Res., 101: 109–122.

Robinson, D.A. (1972) Eye movements evoked by collicular stimulation in the alert. Vision Res., 12: 1795–1808.

Stanford, T.R., Freedman, E.G. and Sparks, D.L. (1996) Site and parameters of microstimulation: evidence of independent effects on the properties of saccades evoked from the primate superior colliculus. J. Neurophysiol., 76: 3360–3380.

Strassman, A., Evinger, C.R., McCrea, A., Baker, R.G. and Highstein, S.M. (1987) Anatomy and physiology of intracellularly labeled omnipause neurons in the cat and squirrel monkey. Exp. Brain Res., 67: 436–440.

Strassman, A., Highstein, S.M. and McCrea, R.A. (1986) Anatomy and physiology of saccadic burst neurons in the alert squirrel monkey. II. Inhibitory burst neurons. J. Comp. Neurol., 249: 358–380.

Sugiuchi, Y., Izawa, Y., Takahashi, M., Na, J. and Shinoda, Y. (2005) Physiological characterization of synaptic inputs to inhibitory burst neurons from the rostral and caudal superior colliculus. J. Neurophysiol., 93: 697–712.

Takahashi, M., Sugiuchi, Y. and Shinoda, Y. (2005) Synaptic inputs and their pathways from fixation and saccade zones of the superior colliculus to inhibitory burst neurons and pause neurons. Ann. N.Y. Acad. Sci., 1039: 209–219.

Yoshida, K., Iwamoto, Y., Chimoto, S. and Shimazu, H. (2002) Disynaptic inhibition of omnipause neurons following electrical stimulation of the superior colliculus in alert cats. J. Neurophysiol., 85: 2639–2642.

Yoshida, K., McCrea, R.A., Berthoz, A. and Vidal, P.P. (1982) Morphological and physiological characteristics of inhibitory burst neurons controlling horizontal rapid eye movements in the alert cat. J. Neurophysiol., 48: 761–784.

C. Kennard & R.J. Leigh (Eds.)
Progress in Brain Research, Vol. 171
ISSN 0079-6123

CHAPTER 2.2

Brainstem circuits controlling lid–eye coordination in monkey

Anja K.E. Horn* and Jean A. Büttner-Ennever

Institute of Anatomy, Ludwig-Maximilians University of Munich, Pettenkoferstr. 11, Munich, Germany

Abstract: In primate, the M-group is a cell cluster in the rostral mesencephalon which contains premotor neurons for the levator palpebrae (LP) and upward-pulling eye muscles. It is therefore thought to play a role in lid–eye coupling during vertical saccades. To further elucidate its role, the afferents to the M-group and LP motoneurons were studied in monkeys. Anterograde tracer injections were placed in one of the three eye-movement-related areas: 1. superior colliculus (SC), 2. interstitial nucleus of Cajal (INC), and 3. the omnipause neuron (OPN) region. Injections into the medial SC subtending upward saccades led to afferent labelling of the ipsilateral M-group and the adjacent rostral interstitial nucleus of the medial longitudinal fascicle (RIMLF), whereas only RIMLF was labelled after an injection into the lateral SC subtending downward saccades. Both RIMLF and M-group received bilateral projections from INC, but only RIMLF received glycineric inputs from the OPN region. This connectivity pattern supports the hypothesis that the M-group mediates lid–eye coupling during vertical upgaze, but is indirectly driven by collaterals of saccadic burst neurons in the RIMLF during lid saccades. A selective projection from the OPN area to the LP motoneurons, but not to other oculomotor neurons is reported here for the first time. The result is supported by the presence of glycinergic terminals only over LP motoneurons, and implies that a subset of OPNs may directly trigger saccade-related blinks.

Keywords: levatorpalpebrae muscle; vertical saccades; superior colliculus; interstitial nucleus of Cajal; rostral interstitial nucleus of the medial longitudinal fascicle; omnipause neurons

Introduction

In mammals, the eyelids are moved by the interaction of two muscles: the levator palpebrae (LP) muscle elevates the lid and the orbicularis oculi (OO) muscle closes the eye. During vertical eye movements, the upper eyelid closely follows the movement of the globe providing protection, but avoiding obstruction of vision (Becker and Fuchs, 1988; Evinger et al., 1991). The eyelid is elevated by activation of the LP muscle, but lowered exclusively by passive downward forces of connective tissue and a decrease in the LP muscle activity without involvement of the OO muscle (for review, see Sibony and Evinger, 1998). Therefore, the lid–eye coupling is primarily mediated by the synchronized activity of the LP and upward-moving eye muscles, superior rectus (SR) and inferior oblique (IO) muscles (Fuchs et al., 1992). In primates, the LP motoneurons lie

*Corresponding author. Tel.: +49-89-5160-4880; Fax: +49-89-5160-4802; E-mail: Anja.Bochtler@med.uni-muenchen.de

DOI: 10.1016/S0079-6123(08)00612-2

in the central caudal nucleus (CCN) dorsomedial to the trochlear nucleus (nIV) and the caudal pole of the oculomotor nucleus (nIII) (Porter et al., 1989; Van der Werf et al., 1997).

The LP motoneurons receive various afferent inputs, which subserve different functions. In the awake state, the eye is kept open by an excitatory tonic input to LP motoneurons; a decrease of this input would result in lid lowering during drowsiness (Kennard and Glaser, 1964; Helmchen and Rambold, 2007). Superimposed on this tonic input, the LP motoneurons receive excitatory signals from premotor sources, which couple the lid to the eye during vertical eye movements. This synchronicity of eye and lid is only interrupted during blinks, when the LP motoneurons are inhibited and OO muscles are activated (for review, see Helmchen and Rambold, 2007).

In monkey, a premotor cell group, termed the M-group, was identified in the rostral mesencephalon, and shown to project to motoneurons of the LP, those of upward-pulling eye muscles (SR and IO), and motoneurons of the forehead musculature (Fig. 1A–C) (Horn et al., 2000). The M-group lies in close proximity, medial to the premotor burst neurons for vertical saccades in rostral interstitial nucleus of the medial longitudinal fascicle (RIMLF) and exhibits a strong labelling for parvalbumin (PV) or cytochrome oxidase activity, by which it was identified in man as well (Fig. 3A) (Horn et al., 2000; Büttner-Ennever and Horn, 2004). In cat, the homologous group does not form a separate cell group, but is part of the medial RIMLF (Chen and May, 2002). The M-group is thought to play a role in coordinating cranial musculature with vertical eye movements, in particular eye–lid coupling (Horn et al., 2000; Chen and May, 2002). To further elucidate its role, we investigated the afferent connections to the M-group and LP motoneurons from various eye-movement-related areas.

Materials and methods

In anaesthetized monkeys, the anterograde tracer ^{3}H-leucine (0.4–0.8 μl; 100 μCi/μl) was injected into the following saccade-related areas: 1. different locations in the deep layers of the superior colliculus (SC) subserving either up- or downward saccades (Sparks and Hartwich-Young, 1989), 2. the interstitial nucleus of Cajal (INC), and the 3. the nucleus raphe interpositus (RIP) containing saccadic omnipause neurons (OPNs) (Büttner-Ennever et al., 1988). After sacrificing and transcardial perfusion of the animals with 4% paraformaldehyde, the tracer was visualized in transverse brainstem sections by autoradiography. Furthermore, the presence of GABAergic or glycinergic afferent inputs to CCN and the M-group was investigated using antibodies against glutamate decarboxylase (GAD; mouse; 1:5000; Biotrend GC3108) or glycine transporter 2 (GLY-T; sheep; 1:5000; Chemicon AB 1771).

Results

Connections with the RIMLF

The close relationship of the M-group to the RIMLF is clearly seen by the transsynaptic labelling in a monkey case from a previous study, 4 days after a rabies virus injection into an eye muscle (Fig. 1A, B) (Ugolini et al., 2006). At high magnification, tracer labelled fibres can be seen running between the M-group and RIMLF, indicating a direct connection between both areas (Fig. 1C). In addition, these projections are considered as excitatory since no GABAergic or glycinergic neurons have been identified neither in the RIMLF nor in the M-group (Fig. 3D) (Horn et al., 2003).

Connections with the superior colliculus

Dependent on the location in regard to the motor map, ^{3}H-leucine injections in the deep layers of the SC resulted in different afferent labelling patterns in the M-group (Robinson, 1972) (Fig. 1D): whereas only the RIMLF was labelled after a tracer injection in the lateral SC (Fig. 1G) which subserves downward saccades, additional strong terminal staining of the M-group was seen after injections in the medial SC, which subserves upward saccades (Fig. 1E, F). In support of earlier studies, no direct

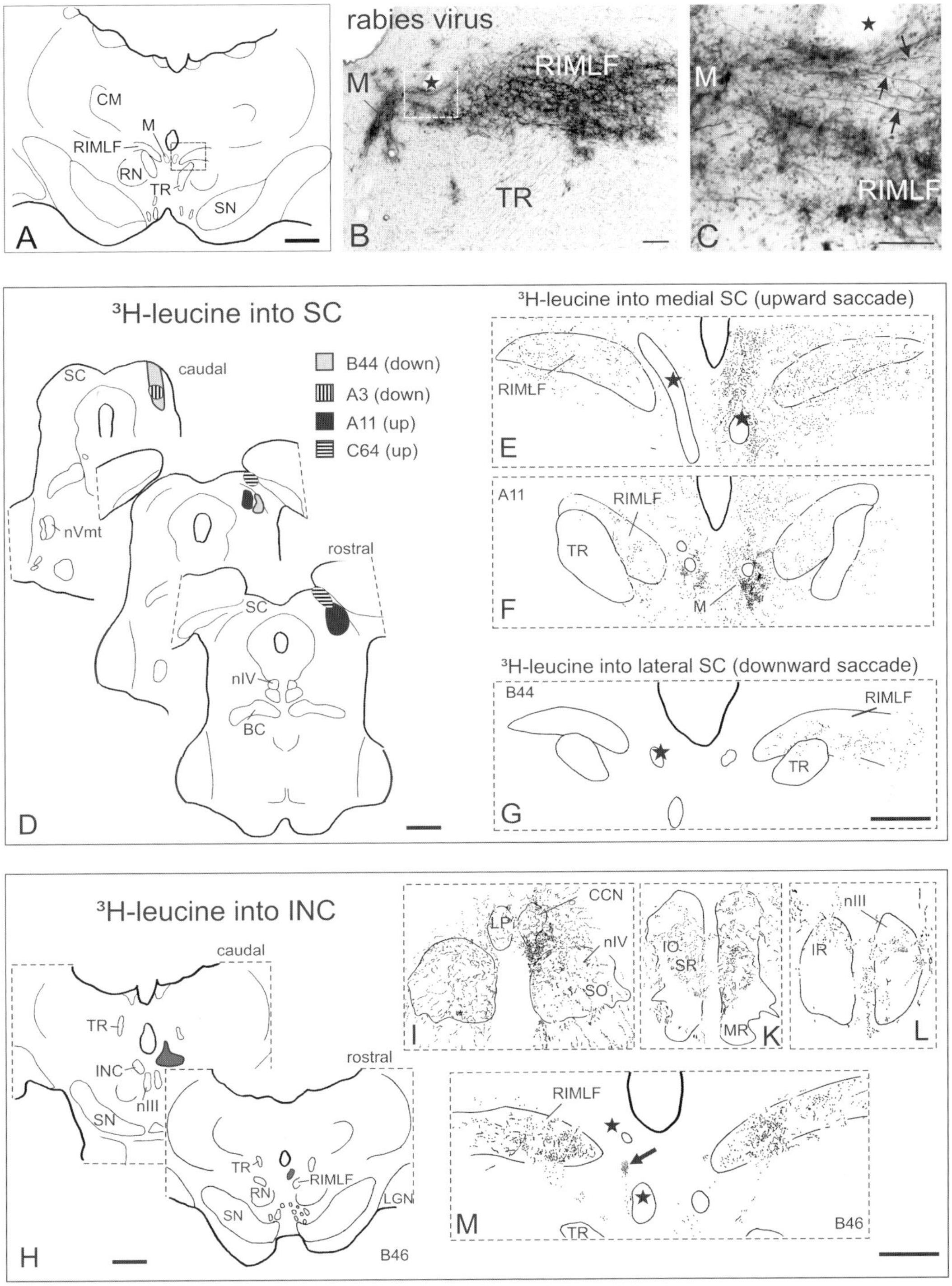

Fig. 1. Location of the M-group in a transverse section of the rostral mesencephalon in monkey (A). Transsynaptically labelled M-group and RIMLF 4 days after Rabies virus injection into an eye muscle (B). Enlargement of A: Tracer-stained fibres may represent collaterals from premotor burst neurons in the RIMLF projecting to the M-group (C, arrows). Reconstructions of ^{3}H-leucine injection sites in the SC (D) and the afferent labelling in RIMLF and M-group after injections into medial SC (E, F) and lateral SC (G). Injection site in INC (H) and terminal labelling in nIV, nIII, CCN, M-group, and RIMLF (I, K, L, M). Scale bar: B: 200 μm; C: 100 μm; E, F, G, I, K, L, M: 1 mm; A, D, H: 3 mm.

projections were found from the SC to extraocular motoneurons (May, 2006).

Connections with the interstitial nucleus of Cajal

In addition to afferent labelling in the motoneurons of vertical-pulling eye muscles in nIII and nIV, the injection of ^{3}H-leucine into the INC led to terminal labelling within the CCN, thereby proving a projection (^{3}H-leucine is not taken up by fibres of passage) from the INC to LP motoneurons described earlier in cat (Fig. 1H, I, K, L) (Chen and May, 2007). The ultra-structural analysis of the INC–CCN projection revealed asymmetric synaptic contacts at LP motoneurons suggestive of excitatory contacts (Chen and May, 2007). The INC is known to act as a velocity-to-position integrator for vertical and torsional eye movements (Fukushima et al., 1992). Thus, the INC neurons are driven by premotor saccadic burst neurons from the RIMLF, and may convey the integrated position signal to motoneurons of vertical-pulling eye muscles as well as to LP motoneurons in the CCN. In addition, the present experiments revealed a considerable projection from INC to the M-group, as well as to the RIMLF of both sides (Fig. 1M). Since the INC contains GABAergic neurons, and the RIMLF as well as the M-group are densely supplied by GAD-immunoreactive terminals, an inhibitory projection from the INC to RIMLF and/or the M-group is probable (Fig. 3B) (Horn et al., 2003).

Connections with the omnipause neuron area

As expected, ^{3}H-leucine injections involving the RIP, which contains saccadic OPNs, resulted in strong afferent labelling of the RIMLF, but no projections were found in the M-group (Fig. 2A, E). This pattern of afferent labelling resembles the terminal GLY-T-immunostaining in the RIMLF and its lack in the M-group (Fig. 3C). A strong projection from the OPN area to the RIMLF and its selective labelling with GLY-T-positive terminals confirms the concept that the OPNs provide a monosynaptic inhibition on saccadic burst neurons, using glycine that is released shortly before and during a saccade (Horn et al., 1994; Kokkoroyannis et al., 1996; Moschovakis et al., 1996). The parallel findings of the absence of GLY-T-positive terminals and the lack of an OPN projection to the M-group supports a concept that, during upward saccades, the M-group receives a copy of the saccadic burst signals via collaterals from up premotor neurons in the RIMLF and does not require its own trigger signal.

In contrast to the lack of any anterograde labelling from OPNs onto motoneurons of extraocular muscles within nIII or nIV, surprisingly a strong anterograde labelling was evident on LP motoneurons within the CCN (Fig. 2B, C, D, F, G). The complementary observation of a strong selective input from GLY-T-positive terminals to LP motoneurons in the CCN, but to no other eye muscle motoneurons, indicates a projection to LP motoneurons from the glycinergic OPNs (Fig. 3E, F) (Horn et al., 1994). The analysis of a case with a wheat germ agglutinin-horseradish peroxidase (WGA-HRP) injection into the CCN revealed several retrogradely labelled neurons in the OPN area, some of which being strongly labelled showing the typical morphological features of OPNs (Fig. 2H, I) (Büttner-Ennever et al., 1988; Horn et al., 1994). However, this population of retrogradely labelled neurons in the OPN area involved only a small fraction compared to those seen after a tracer injection into the RIMLF (Horn et al., 1994), and may represent an unnoticed subgroup of neurons within the OPN area.

Discussion

Based on the present data, the following hypothesis for the lid–eye coordination during vertical saccades can be put forward (Fig. 4): For upward saccades, the up burst neurons in the RIMLF and the M-group receive afferent activating signals from the medial part of the SC, thereby mediating upward eye movements via *t*-projections to SR and IO motoneurons from the RIMLF. The coupling of the eyelid to the upward saccades is mediated by the projection from the M-group to SR, IO, and LP motoneurons. The burst signal for the accompanying lid saccade is relayed through collaterals from excitatory up burst neurons within

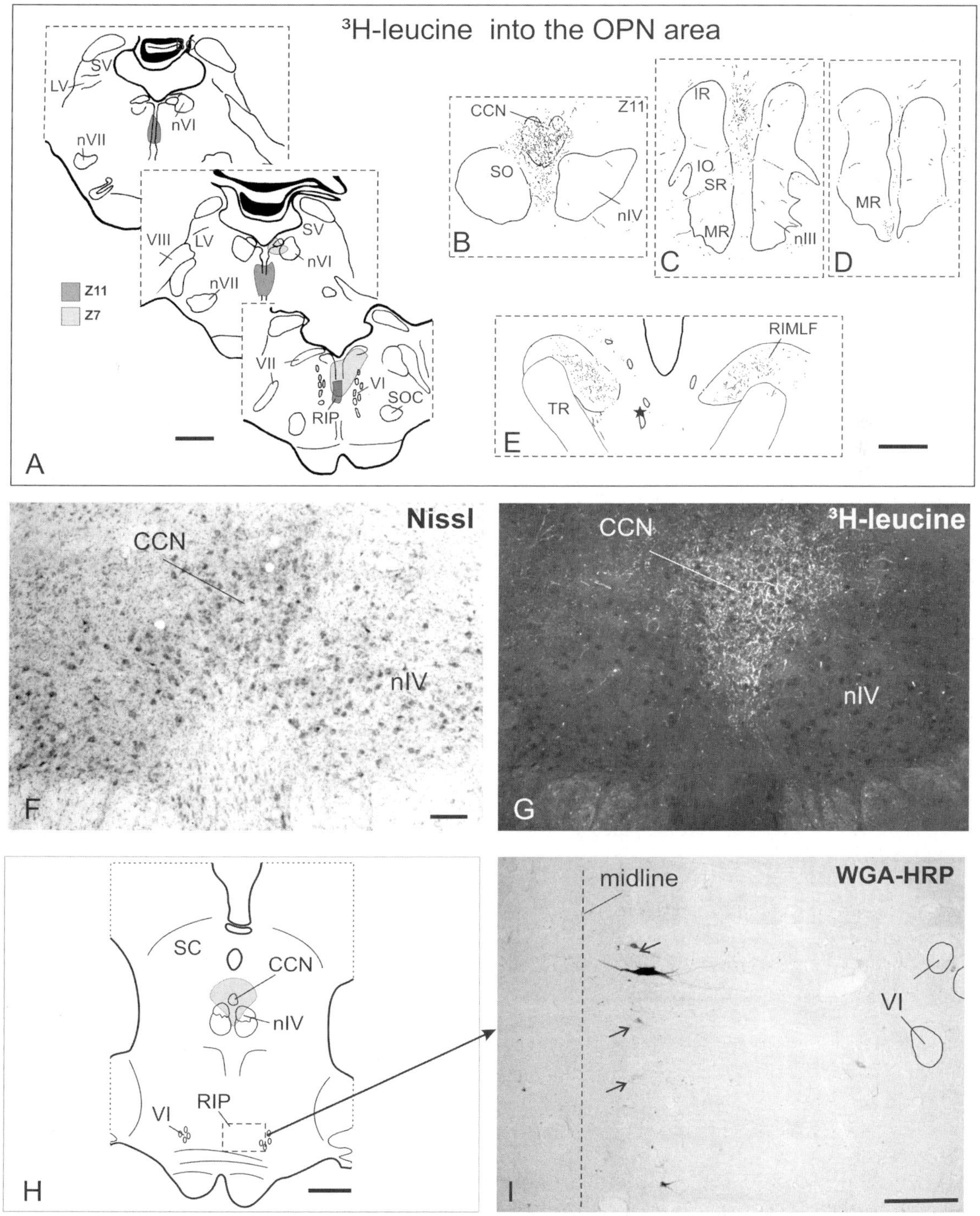

Fig. 2. ³H-leucine injection site in the OPN area (A) and the afferent labelling in nIV, nIII, CCN, M-group, and RIMLF (B, C, D, E). The staining pattern in CCN is shown in darkfield illumination (G), the cytoarchitecture in brightfield of the Nissl-stained section (F). Reconstructions of a WGA-HRP injection into the CCN (H) resulting in retrogradely labelled neurons in the OPN area (I, arrows). Scale bar: A, H: 3 mm; B: 1 mm; F, G, I: 200 μm.

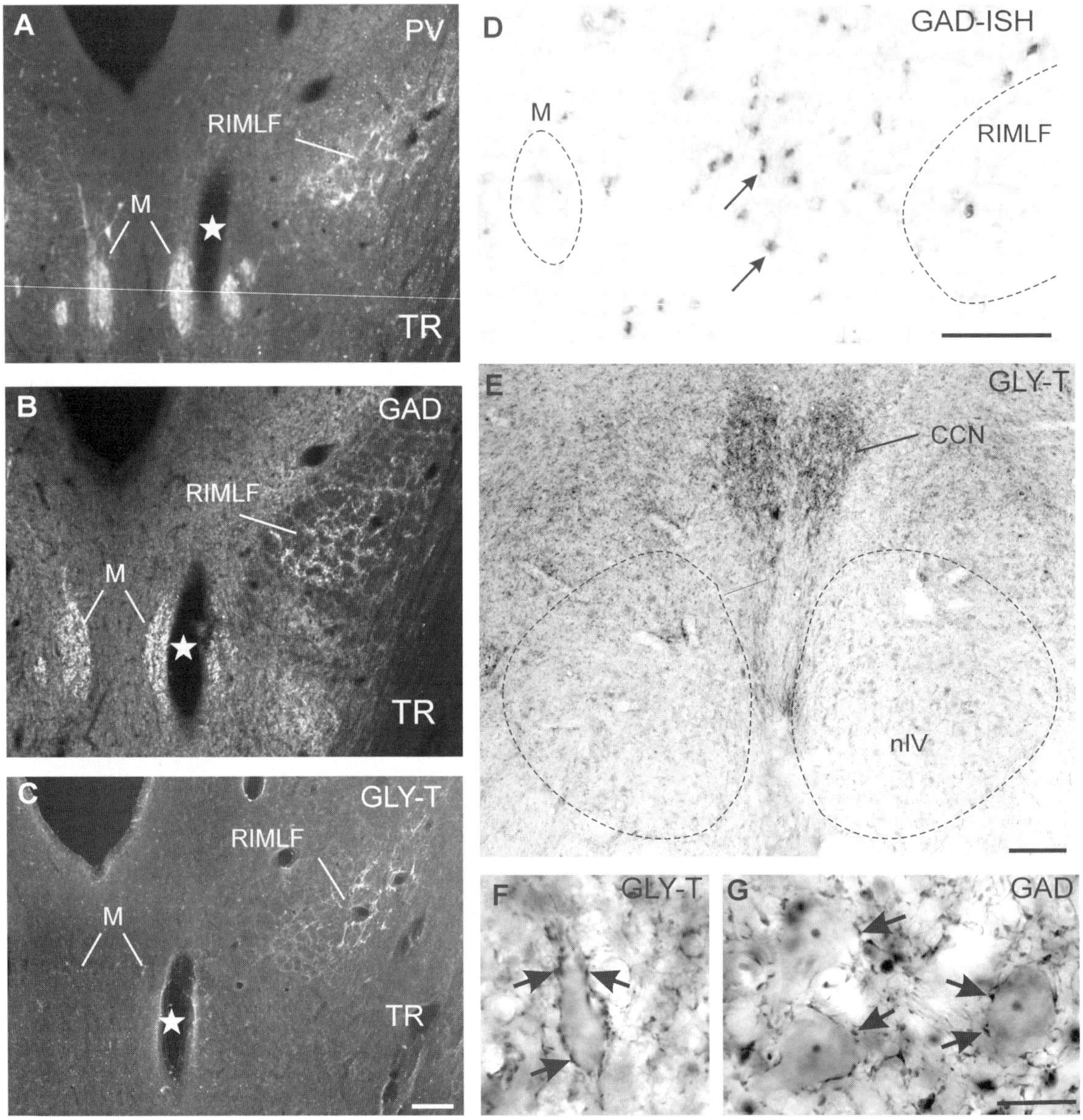

Fig. 3. Immunolabelling for parvalbumin (PV) (A), glutamate decarboxylase (GAD) (B) and glycine transporter 2 (GLY-T) (C) of RIMLF and the M-group. Note that the PV-positive M-group receives numerous GAD-positive terminals, but not GLY-T-positive terminals as opposed to the RIMLF. In-situ hybridization (ISH) for GAD mRNA revealed no GABAergic neurons (arrows) within the M-group (D). Photograph of the CCN demonstrating the selective input of GLY-T-positive terminals (E) on LP motoneurons (F). LP motoneurons are also contacted by GAD-positive terminals (G). Scale bar: A, B, C, D, E: 200 μm; F,G: 30 μm.

the RIMLF to the M-group. Downward lid–eye saccades are generated through the RIMLF, but without involvement of the M-group, which presumably provides only excitatory premotor commands (Chen and May, 2002). Lid lowering is purely mediated by passive downward forces and a decrease of LP muscle activation (Sibony and Evinger, 1998). Therefore, either the M-group or the LP motoneurons must be inhibited during downward eye movements. An inhibitory input

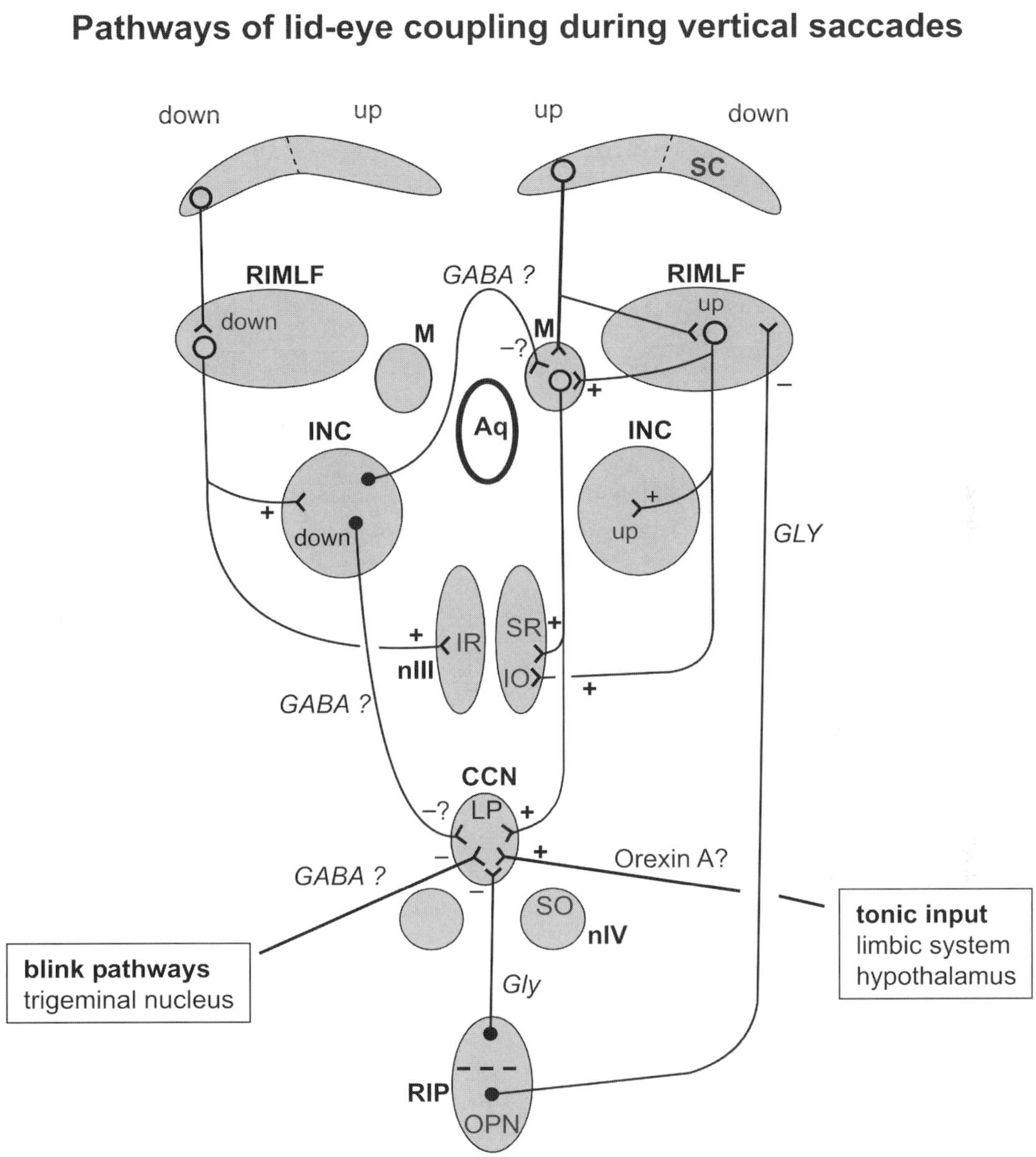

Fig. 4. Summary diagram of proposed connections involved in lid–eye coordination during vertical saccades described in the present paper. Open circles indicate excitatory neurons; filled circles indicate inhibitory neurons.

from neurons in the INC, which are driven by down burst neurons in the RIMLF could provide such a function. The presence of numerous GABAergic neurons in the INC (Horn et al., 2003) and the strong GAD-positive terminal labelling in the M-group (Fig. 3B) could support such an inhibitory projection from the INC to the M-group (Fig. 4). The lid–eye saccade would be triggered by the disinhibition of glycinergic OPNs, which provide a tonic monosynaptic inhibition on saccadic burst neurons in the RIMLF during all non-saccadic eye movements including fixation (Leigh and Zee, 2006). The lack of afferents from the OPNs to the M-group supports the idea that the M-group does not require its own saccadic trigger signal, but is driven from premotor excitatory up burst neurons in the RIMLF. An inhibitory projection from a subset of OPN neurons may contribute to pathways involved in blink-saccade interaction. Close interrelations between blinks and saccades are evident from behavioral and physiological studies (Evinger et al., 1994; Leigh and Zee, 2006; Helmchen and Rambold, 2007): the OPN activity pauses during blinks as well as during saccades, and microstimulation of OPNs suppresses reflex blinks (Mays and Morisse, 1995). Furthermore, voluntary blinks can facilitate abnormally slow saccades in patients (Leigh and Zee, 2006).

The GABAergic terminals on LP motoneuron (Fig. 3G) may represent the inhibitory input for reflexive blinks, either directly from neurons around the principal trigeminal nucleus or via local inhibitory interneurons (van Ham and Yeo, 1996). Orexin A positive terminals found around LP motoneurons might contribute to the tonic activation during the awake state (personal observation; Saper et al., 2001).

In conclusion, afferent pathways have been traced from the medial SC, RIMLF, and INC onto the M-group, which, through direct projections onto LP motoneurons, coordinates the activity of the eyelid with vertical eye movements. The pattern of afferent pathways together with the identification of their neurotransmitters, reflects the activation of the M-group and LP motoneurons during upward eye movements. Furthermore, direct connections were demonstrated for the first time between some OPNs and LP motoneurons, which may be important for the generation of saccade-related blinks.

Abbreviations

VI	abducens nerve
VII	facial nerve
VIII	vestibular nerve
BC	brachium conjunctivum
CCN	central caudal nucleus
CM	centromedian nucleus
GAD	glutamate decarboxylase
GLY-T	glycine transporter 2
INC	interstitial nucleus of Cajal
IO	inferior oblique muscle
IR	inferior rectus muscle
ISH	in-situ hybridization
LGN	lateral geniculate nucleus
LP	levator palpebrae muscle
LV	lateral vestibular nucleus
M	M-group
MR	medial rectus muscle
nIII	oculomotor nucleus
nVI	abducens nucleus
nVII	facial nucleus
nIV	trochlear nucleus
nVmt	motor trigeminal nucleus
OO	orbicularis oculi muscle
OPN	omnipause neuron
PV	parvalbumin
RIMLF	rostral interstitial nucleus of the medial longitudinal fascicle
RIP	nucleus raphe interpositus
RN	red nucleus
SC	superior colliculus
SN	substantia nigra
SO	superior oblique muscle
SOC	superior olive complex
SR	superior rectus
SV	superior vestibular nucleus
TR	tractus retroflexus

Acknowledgement

This work was supported by the Deutsche Forschungsgemeinschaft DFG HO 1639/4-2.

References

Becker, W. and Fuchs, A.F. (1988) Lid–eye coordination during vertical gaze changes in man and monkey. J. Neurophysiol., 60: 1227–1252.

Büttner-Ennever, J.A., Cohen, B., Pause, M. and Fries, W. (1988) Raphe nucleus of the pons containing omnipause neurons of the oculomotor system in the monkey, and its homologue in man. J. Comp. Neurol., 267: 307–321.

Büttner-Ennever, J.A. and Horn, A.K.E. (2004) Reticular formation: eye movements, gaze and blinks. In: Paxinos G.

and Mai J.K. (Eds.), The Human Nervous System. Elsevier Academic Press, Amsterdam, pp. 479–510.

Chen, B. and May, P.J. (2002) Premotor circuits controlling eyelid movements in conjunction with vertical saccades in the cat: I. The rostral interstitial nucleus of the medial longitudinal fasciculus. J. Comp. Neurol., 450: 183–202.

Chen, B. and May, P.J. (2007) Premotor circuits controlling eyelid movements in conjunction with vertical saccades in the cat: II. Interstitial nucleus of Cajal. J. Comp. Neurol., 500: 676–692.

Evinger, C., Manning, K.A., Pellegrini, J.J., Basso, M.A., Powers, A.S. and Sibony, P.A. (1994) Not looking while leaping: the linkage of blinking and saccadic gaze shifts. Exp. Brain Res., 100: 337–344.

Evinger, C., Manning, K.A. and Sibony, P.A. (1991) Eyelid movements. Mechanisms and normal data. Invest. Ophthalmol., 32: 387–400.

Fuchs, A.F., Becker, W., Ling, L., Langer, T.P. and Kaneko, C.R. (1992) Discharge patterns of levator palpebrae superioris motoneurons during vertical lid and eye movements in the monkey. J. Neurophysiol., 68: 233–243.

Fukushima, K., Kaneko, C.R. and Fuchs, A.F. (1992) The neuronal substrate of integration in the oculomotor system. Prog. Neurobiol., 39: 609–639.

Helmchen, C. and Rambold, H. (2007) The eyelid and its contribution to eye movements. Dev. Ophthalmol., 40: 110–131.

Horn, A.K.E., Büttner-Ennever, J.A., Gayde, M. and Messoudi, A. (2000) Neuroanatomical identification of mesencephalic premotor neurons coordinating eyelid with upgaze in the monkey and man. J. Comp. Neurol., 420: 19–34.

Horn, A.K.E., Büttner-Ennever, J.A., Wahle, P. and Reichenberger, I. (1994) Neurotransmitter profile of saccadic omnipause neurons in nucleus raphe interpositus. J. Neurosci., 14: 2032–2046.

Horn, A.K.E., Helmchen, C. and Wahle, P. (2003) GABAergic neurons in the rostral mesencephalon of the Macaque monkey that control vertical eye movements. Ann. N.Y. Acad. Sci., 1004: 19–28.

Kennard, D.W. and Glaser, G.H. (1964) An analysis of eyelid movements. J. Nerv. Ment. Dis., 139: 31–48.

Kokkoroyannis, T., Scudder, C.A., Balaban, C.D., Highstein, S.M. and Moschovakis, A.K. (1996) Anatomy and physiology of the primate interstitial nucleus of Cajal. I. Efferent projections. J. Neurophysiol., 75: 725–739.

Leigh, R.J. and Zee, D.S. (2006) The Neurology of Eye Movements. Oxford University Press, New York.

May, P.J. (2006) The mammalian superior colliculus: laminar structure and connections. Prog. Brain Res., 151: 321–378.

Mays, L.E. and Morisse, D.W. (1995) Electrical stimulation of the pontine omnipause area inhibits eye blink. J. Am. Optom. Assoc., 66: 419–422.

Moschovakis, A.K., Scudder, C.A. and Highstein, S.M. (1996) The microscopic anatomy and physiology of the mammalian saccadic system. Prog. Neurobiol., 50: 133–254.

Porter, J.D., Burns, L.A. and May, P.J. (1989) Morphological substrate for eyelid movements: innervation and structure of primate levator palpebrae superioris and orbicularis oculi muscles. J. Comp. Neurol., 287: 64–81.

Robinson, D.A. (1972) Eye movements evoked by collicular stimulation in the alert monkey. Vision Res., 12: 1795–1808.

Saper, C.B., Chou, T.C. and Scammell, T.E. (2001) The sleep switch: hypothalamic control of sleep and wakefulness. Trends Neurosci., 24: 726–731.

Sibony, P.A. and Evinger, C. (1998) Normal and abnormal eyelid function. In: Miller N.R. and Newman N.H. (Eds.), Clinical Neuro-Ophthalmology. Williams and Wilkins, Baltimore, pp. 1509–1592.

Sparks, D.L. and Hartwich-Young, R. (1989) The deep layers of the superior colliculus. In: Wurtz R.H. and Goldberg M.E. (Eds.), The Neurobiology of Saccadic Eye movements. Elsevier, Amsterdam, pp. 213–255.

Ugolini, G., Klam, F., Doldan Dans, M., Dubayle, D., Brandi, A.-M., Büttner-Ennever, J.A. and Graf, W. (2006) Horizontal eye movement networks in primates as revealed by retrograde transneuronal transfer of rabies virus: differences in monosynaptic input to "slow" and "fast" abducens motoneurons. J. Comp. Neurol., 498: 762–785.

Van der Werf, F., Aramideh, M., Ongerboer de Visser, B.W., Baljet, B., Speelman, J.D. and Otto, J.A. (1997) A retrograde double fluorescent tracing study of the levator palpebrae superioris muscle in the cynomolgus monkey. Exp. Brain Res., 113: 174–179.

Van Ham, J.J. and Yeo, C.H. (1996) Trigeminal inputs to eyeblink motoneurons in the rabbit. Exp. Neurol., 142: 244–257.

C. Kennard & R.J. Leigh (Eds.)
Progress in Brain Research, Vol. 171
ISSN 0079-6123

CHAPTER 2.3

Defining the pupillary component of the perioculomotor preganglionic population within a unitary primate Edinger–Westphal nucleus

Paul J. May[1,*], Wensi Sun[2] and Jonathan T. Erichsen[3]

[1]*Departments of Anatomy, Ophthalmology and Neurology, University of Mississippi Medical Center, Jackson, MS, USA*
[2]*Department of Ophthalmology, Northwest Hospital & Medical Center, Seattle, WA, USA*
[3]*School of Optometry and Vision Sciences, Cardiff University, Cardiff, Wales, UK*

Abstract: The primate Edinger–Westphal nucleus (EW) contains perioculomotor preganglionic ($pIII_{PG}$) motoneurons that control the lens and pupil. Separate subdivisions have been described in EW and termed visceral columns, with the lateral visceral column (lvc) reportedly receiving pretectal inputs for the pupillary light reflex. However, choline acetyl transferase staining reveals a single paired column of cells dorsal to the oculomotor nucleus, suggesting the EW is not subdivided. We investigated this issue by transneuronal retrograde labelling of $pIII_{PG}$ neurons in three monkey species. In all three, $pIII_{PG}$ neurons were contained in a single column. We have also examined which part of the macaque $pIII_{PG}$ population receives pretectal input. Injections of biocytin into the pretectum anterogradely labelled terminals that lay in close association with $pIII_{PG}$ motoneurons retrogradely labelled by ciliary ganglion injections of WGA-HRP. These close associations were concentrated in the ventromedial portion of the middle third of EW, suggesting this $pIII_{PG}$ region mediates pupillary control. In other cases, pretectal WGA-HRP injections, in addition to labelling terminals in the EW, produced a circular field of labelled neurons, and terminals in the periaqueductal grey, dorsolateral to EW. This region may represent the previously designated lvc, but it does not contain $pIII_{PG}$ motoneurons.

Keywords: oculomotor; near response; pupil; para sympathetic

Introduction

We rarely move one joint in isolation. Instead, most actions involve coordination of multiple muscles to produce complex multi-joint movement combinations. Similarly, the extraocular muscles rarely move the eyes in isolation. Since the targets of gaze movements are located within three dimensional space, extraocular muscle action is coordinated with control of the intraocular muscles: the ciliary and sphincter pupillae muscles. Eye movements are thus accompanied by changes in lens accommodation and pupillary diameter to allow the eyes to focus on newly acquired targets (e.g., the near response). The pupils are also regulated with respect to the luminance of each new field of view, to optimize perception of the target in its surroundings. Signals for both extraocular and intraocular muscle contraction flow down the third cranial nerve (IIIn), but in the

*Corresponding author. Tel.: +601-984-1662;
Fax: +601-984-1655; E-mail: pmay@anatomy.umsmed.edu

DOI: 10.1016/S0079-6123(08)00613-4

case of the intraocular muscles, the preganglionic parasympathetic motoneurons that supply the ciliary ganglion lie apart from the somatic motoneurons found in the oculomotor nucleus (III). They are traditionally described as forming a column dorsal to the oculomotor nucleus called the Edinger–Westphal nucleus (EW), which extends into the anteromedian nucleus (AM), rostrally.

While these preganglionic motoneurons were originally described as forming a single paired column in macaque monkeys (Akert et al., 1980), later investigators suggested that they lie in a series of discrete subgroups (Burde and Williams, 1989): AM, the nucleus of Perlia, and the dorsal, medial, and lateral visceral columns (lvcs) of EW. These subdivisions took on added importance when trans-synaptic anterograde tracers were utilized in an attempt to define the pupillary light reflex pathway in the monkey (Kourouyan and Horton, 1997). This study indicated that the lvc represents the portion of EW that receives input from the retina, by way of the olivary pretectal nucleus (OPt). This finding paralleled evidence from conventional anterograde studies that also suggested the pretectum projects to the contralateral lvc (Baleydier et al., 1990; Büttner-Ennever et al., 1996). Our understanding of the structure of EW has been further challenged by studies that indicate that this nucleus also contains neurons projecting to central targets, instead of the ciliary ganglion (cat: Loewy et al., 1978; monkey: Burde, 1988), and that these centrally projecting neurons use neuropeptides, not acetylcholine as their transmitter (Maciewicz et al., 1983). This has recently been reinforced by discovery of a population of neurons that use urocortin, a neuropeptide tied to stress responses and ingestive behaviour, within the rodent EW (Vaughan et al., 1995; Kozicz et al., 2001; Bachtell et al., 2002).

In order to better determine the true identity of EW, we recently investigated the populations of neurons found in the vicinity of the oculomotor nucleus, the perioculomotor (pIII) populations, in the cat and monkey (May et al., 2008). Specifically, we compared the locations of the cholinergic and urocortin peptidergic populations by using antibodies to choline acetyl transferase (ChAT) and urocortin I. The results of this study indicated that there are several pIII populations. In macaque monkeys, $ChAT^{+}$ pIII preganglionic motoneurons ($pIII_{PG}$) lie in the EW and AM, while the perioculomotor population of $urocortin^{+}$ cells ($pIII_{U}$) lies between EW and III, where it partially overlaps with the $ChAT^{+}$ population of the S and C group motoneurons ($pIII_{S\&C}$) that supply multiply innervated fibres (MIFs) in the extraocular muscles. In contrast, the cat EW primarily contains the $pIII_{U}$ population, not the $pIII_{PG}$ population. To clarify the problem posed by EW designating different cell populations in different species, we proposed that these species-specific nuclei be designated EW_{PG} (monkey) and EW_{U} (cat and rodents). However, another striking feature of this work was the observation that the $ChAT^{+}$ $pIII_{PG}$ neurons in the macaque EW_{PG} appeared to form a single-paired column, and were not subdivided into several subgroups, as previously described. These findings concurred with those of Horn et al. (2008), who recently described the pIII populations in monkeys and man.

Consequently, in the present investigation, we utilized conventional and trans-synaptic retrograde tracers to re-examine the distribution of $pIII_{PG}$ motoneurons in several primate species. In addition, we utilized bidirectional tracers, and combinations of anterograde and retrograde tracers to re-examine which $pIII_{PG}$ motoneurons receive input from the OPt, and so are part of the pupillary light reflex pathway in macaques. The procedures used to inject and reveal these tracers have largely been described previously (Chen and May, 2000, 2007; Erichsen and May, 2002).

Results

The distributions of trans-synaptically labelled preganglionic motoneurons ($pIII_{PG}$) in three different species of monkey are shown in Fig. 1. In each case, tracer which crossed the synapses of the ciliary ganglion was revealed by use of tetramethylbenzidene (TMB). Injections of wheat germ agglutinin conjugated horseradish peroxidase (WGA-HRP) into either the aqueous (right) or vitreous (left) of long-tailed macaque monkeys (*Macaca fascicularis*) resulted in the labelling of $pIII_{PG}$ motoneurons in a portion of AM (Fig. 1A),

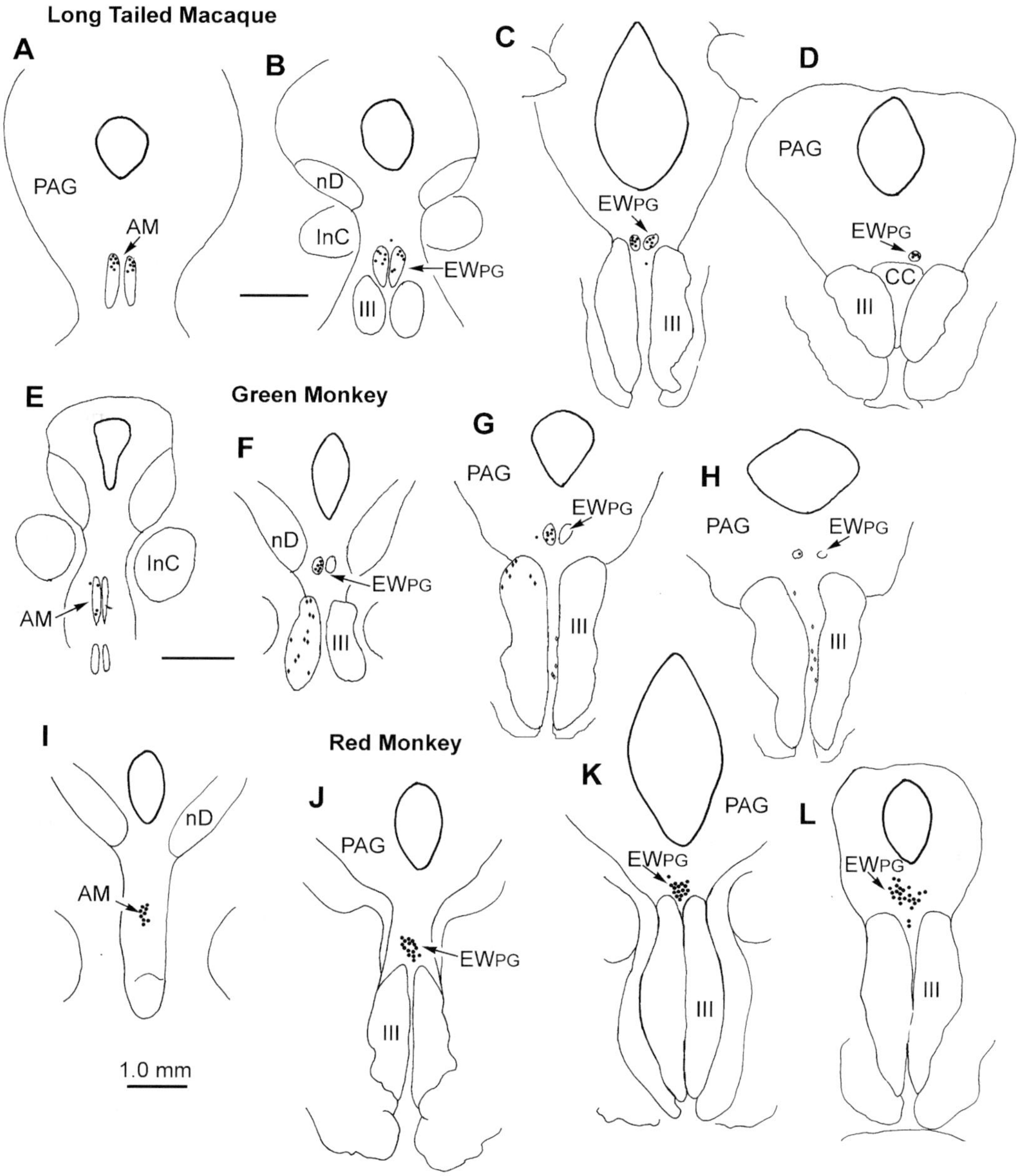

Fig. 1. Pattern of trans-synaptic retrograde label (dots) observed in a rostrocaudal series of sections from three species of monkey following injections of WGA-HRP into the anterior (A–D right and I–L left) and vitreal chambers (A–H left and I–L right) of the eye. Diamonds in E–H represent directly labelled motoneurons. Figure abbreviations: AM, anteromedian nucleus; BC, brachium conjunctivum; CC, caudal central subdivision of III; EW_{PG}, preganglionic Edinger–Westphal nucleus; III, oculomotor nucleus; IIIn, oculomotor nerve; InC, interstitial nucleus of Cajal; lvc, lateral visceral column; MD, mediodorsal nucleus; MRF, midbrain reticular formation; nD, nucleus of Darkschewitsch; nOT, nucleus of the optic tract; nPC, nucleus of the posterior commissure; Opt, olivary pretectal nucleus; PAG, periaqueductal grey; Pt, pretectum; Pul, pulvinar; R, red nucleus; SOA, supraoculomotor area; VP, ventroposterior nucleus.

and in the rest of the EW_{PG} (Fig. 1B–D). The labelled cells lay dorsal to III, and ended above the caudal central subdivision (CC). Although occasional cells were found outside the borders of EW_{PG} (Fig. 1B, C), there was no consistent subdivision of the labelled motoneurons into separate columns. Similar results were observed in rhesus monkeys (*Macaca mulatta*). Evidence from an African green monkey (*Cercopithecus aethiops*), which received an injection of WGA-HRP in combination with HRP into the left vitreal chamber is shown in Fig. 1E–H. Once again, the trans-synaptically labelled $pIII_{PG}$ motoneurons were located in AM (E), and in a column dorsal to III that stretched through EW_{PG} (Fig. 1F–H). The inclusion of HRP in the injection mixture caused some spread to the extraocular muscles, which labelled motoneurons in and between III. However, once again, the segregation of the $pIII_{PG}$ motoneurons into separate subdivisions was not evident. The results from a red monkey (*Erythrocebus patas*) that received WGA-HRP injections into the aqueous (left) or vitreous (right) are illustrated in Fig. 1I–L. As in the other monkeys, trans-synaptically labelled $pIII_{PG}$ motoneurons were located in both AM (Fig. 1I) and EW_{PG} (Fig. 1J–L). However, the cells in this species did not form paired columns. Instead, they formed a single midline column that began rostrally in AM, and extended caudally above III. While occasional outliers were present, there was no consistent subdivision of this population into separate columns.

We then examined the input of the pretectum onto the EW_{PG} population by making an injection of biocytin into the pretectum of a macaque monkey (*M. fascicularis*) in which we had also placed WGA-HRP into the ciliary ganglion. The pretectum was approached on the left side by aspirating the cerebral cortex overlying the midbrain. The ciliary ganglion was approached on the right side by retracting the anterior border of the temporalis muscle to reveal the temporal bone over the lateral wall of the orbit. This bone and the underlying lateral rectus muscle were removed to allow access to the region behind the globe. The ciliary ganglion was then isolated and penetrated by pins with WGA-HRP crystalized on their tips. The tissue was reacted using a dual tracer protocol in which WGA-HRP in the retrogradely labelled neurons was stained brown, and biocytin in the anterogradely labelled axons was stained black (Chen and May, 2007).

The resulting injection site involved portions of all the nuclei in the pretectum (Fig. 2D), but included all the OPt, which is believed to be the source of the luminance signal that drives pupillary preganglionic motoneurons. Anterogradely labelled axons extended ipsilaterally from this injection site through the periaqueductal grey (PAG) towards the pIII region (Fig. 2D). In addition, labelled axons crossed in the posterior commissure and curved into the contralateral pIII region. On both sides, these axons extended rostrally (Fig. 2A–C). As shown in the high magnification illustrations (Fig. 2), retrogradely labelled preganglionic motoneurons were present on the right side, ipsilateral to the injected ciliary ganglion. There was no consistent division of these labelled neurons into different columns, although occasional outliers were present (Fig. 2B). Biocytin labelled pretectal axons arbourized bilaterally within EW_{PG} (Fig. 2B–D), and even in AM (Fig. 2A). They were seen both among the counterstained neurons (left) and retrogradely labelled ones (right). The distribution of terminals was not the same along the course of the $pIII_{PG}$ population; it was densest ventromedially, in middle third of its rostrocaudal extent. It was in this region that close associations between the anterogradely labelled boutons and the somata and proximal dendrites of the retrogradely labelled $pIII_{PG}$ motoneurons were most evident. Examples of these close associations (arrowheads) are shown in Fig. 3A. The biocytin labelled pretectal axons displayed terminal arbour sprays with *en passant* and terminal boutons. These boutons were located adjacent to the proximal dendrites and somata of WGA-HRP labelled $pIII_{PG}$ motoneurons. Due to the limited dendritic filling, contacts with distal dendrites could not be defined. The synaptic contact suggested by these close associations was confirmed ultrastructurally.

Those cells contacted by pretectal inputs represent the pupillary component of the $pIII_{PG}$ population. In contrast to previous reports, they did not lie in the lvc. In fact, no such subdivision

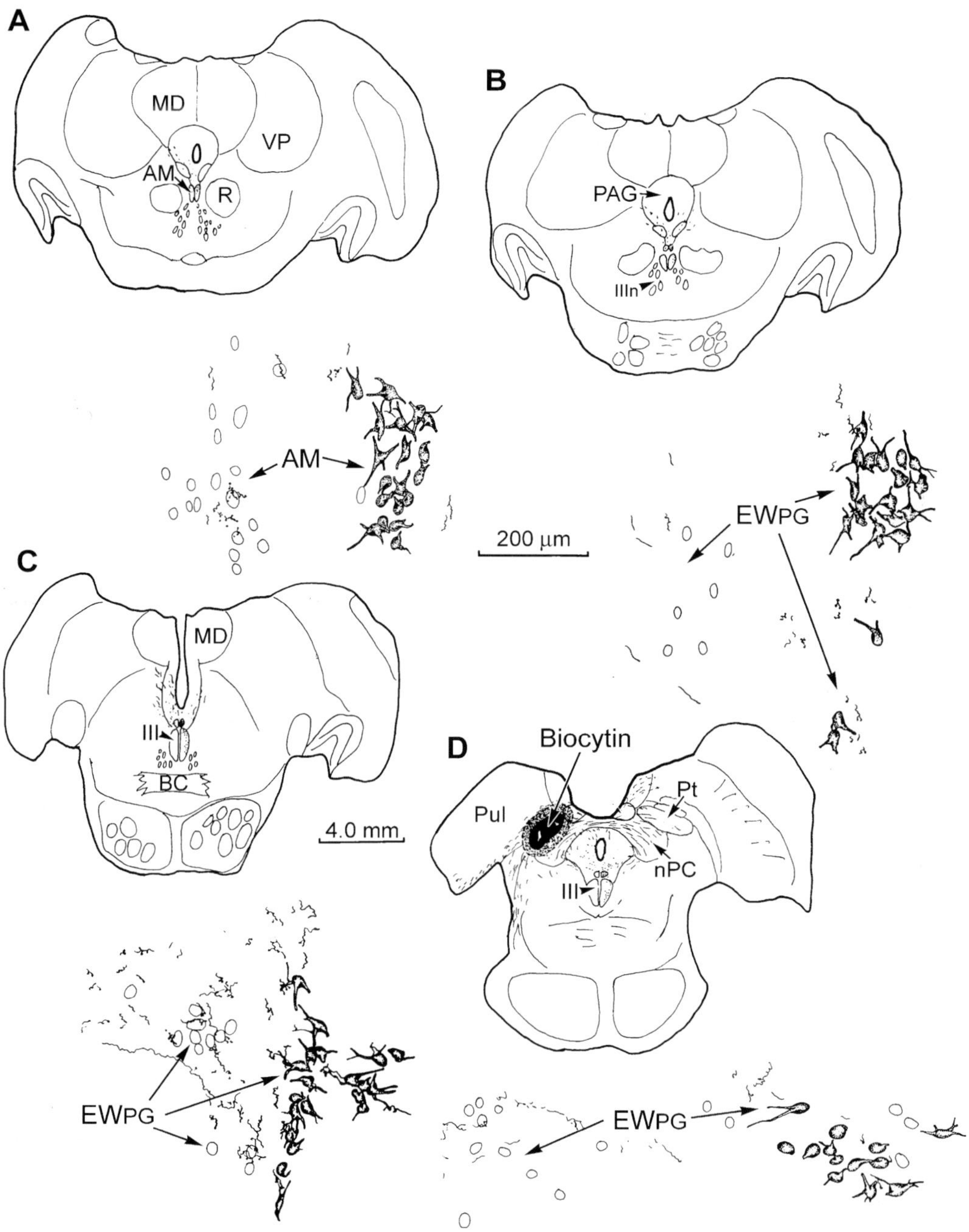

Fig. 2. Pattern of anterograde labelling in axons following a biocytin injection of the pretectum (D), with respect to unlabelled (left) and retrogradely labelled (right) preganglionic motoneurons. The later were labelled following WGA-HRP injection of the right ciliary ganglion. The rostrocaudal level of each illustration is indicated by the accompanying charting.

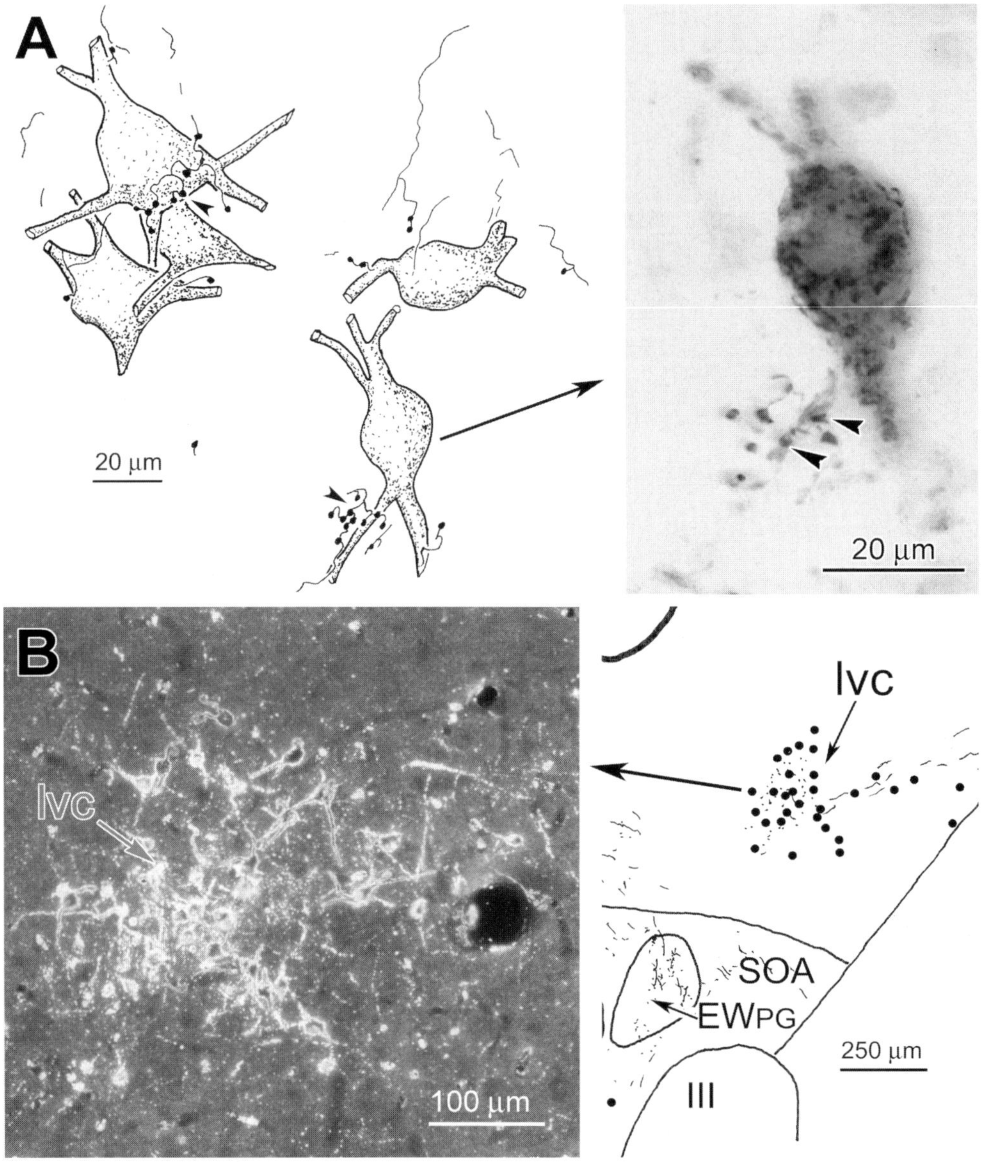

Fig. 3. A: Close associations (arrowheads) between biocytin labelled pretectal axonal boutons and WGA-HRP labelled preganglionic motoneurons. B: WGA-HRP labelled neurons and terminals in the lateral visceral cell column (lvc) following an injection in the pretectum (see Fig. 4). Photomicrograph in B taken with crossed polarizers. (See Color Plate 2.3.3 in color plate section.)

was observed among $pIII_{PG}$ motoneurons. However, evidence for a set of neurons that may represent the previously described lvc came from another experiment in which the macaque (*M. fascicularis*) pretectum was injected with WGA-HRP, and the tissue was reacted using TMB histochemistry. The injection shown in Fig. 4A, B involved portions of all the pretectal nuclei except the medial, and completely filled the OPt. As with the biocytin injection (Fig. 2), labelled axons extended ipsilaterally and, by way of the posterior commissure, contralaterally. Note that the contralateral OPt was largely free of anterograde and retrograde label, but numerous labelled cells (dots), as well as

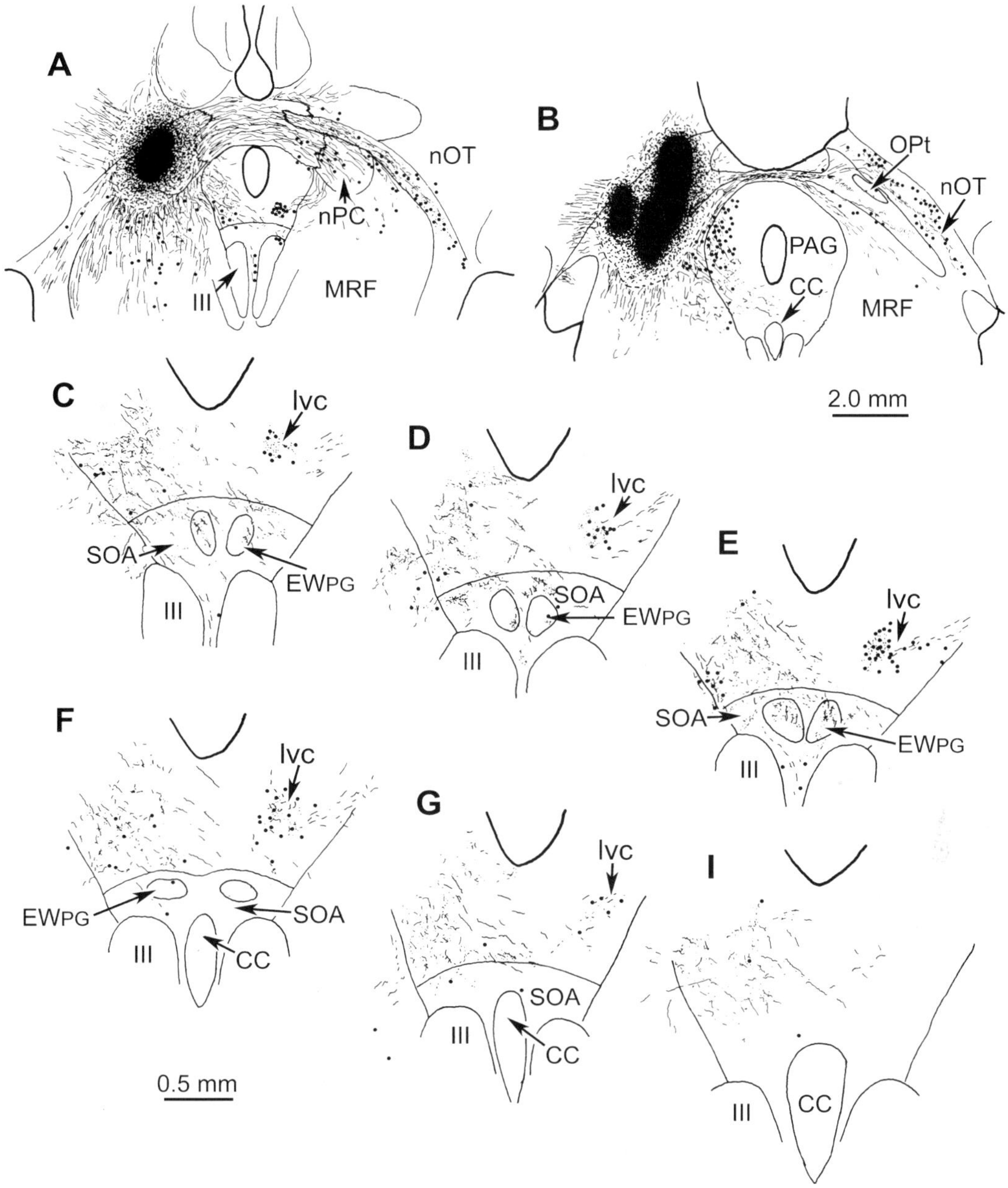

Fig. 4. The pattern of retrograde cell (dots) and anterograde terminal (stipple) label is shown in a rostrocaudal series of illustrations of the pIII region (C–I) following an injection of WGA-HRP into the pretectum (A and B). Note the label in lvc and EW_{PG}.

labelled terminals (stipple), were present in the nucleus of the optic tract (nOT) (Fig. 4B). Labelled axonal arbours were present within the confines of AM (not illustrated) and the EW_{PG} (Fig. 4C–F) on both sides of the midline. It appeared that most of the axons supplying this input arrived via the ipsilateral path. In addition, neurons that were retrogradely labelled with WGA-HRP (dots) were observed contralaterally, in a circular cell group located dorsolateral to EW_{PG} (Fig. 4C–G). We believe this structure is equivalent to the one termed the lvc in previous studies (Baleydier et al., 1990; Büttner-Ennever et al., 1996; Kourouyan and Horton, 1997), and so have continued to use this term to identify it. The lvc appears to be connected to the contralateral pretectum by way of the posterior commissure (Fig. 4B). Punctate labelling, believed to represent labelled terminals (stipple), was also present among these cells, and the same area displayed labelled terminals on the ipsilateral side, as well (Fig. 4C–G). The pattern of contralateral labelling in lvc is further illustrated in Fig. 3B. These lvc cells were slightly smaller than the $pIII_{PG}$ motoneurons in EW_{PG} (Figs. 2 and 3A).

Discussion

The results of these experiments illustrate several important points: (1) The perioculomotor preganglionic ($pIII_{PG}$) population in the primate EW_{PG} is organized into a unitary nucleus, without subdivisions. (2) The projection of the pretectum onto primate $pIII_{PG}$ motoneurons is concentrated ventromedially, in the middle third of the rostrocaudal extent of EW_{PG}. (3) The structure previously considered to be the pupillary subdivision, the lvc, actually contains non-motoneurons that project back on the pretectum.

We did not find evidence that EW_{PG} was separated into subdivisions or "columns" in the primate species investigated here (Fig. 1). These species inhabit a variety of different niches: macaques are semiterrestrial jungle dwellers, green monkeys are arboreal jungle dwellers, and red monkeys roam the plains. Thus, they differ both in the patterns of close and distance work they undertake, and the luminance changes they encounter. Nevertheless, they all exhibited a unitary $pIII_{PG}$ population, which agrees with our previous finding in the prosimian Galago (Sun and May, 1993) and the organization in man (Horn et al., 2008). However, some species differences were observed. The EW_{PG} of the red monkey was organized as a single, midline nucleus, instead of paired columns. This feature was also observed in the Galago (Sun and May, 1993). Evidence for subdivision of the $pIII_{PG}$ population largely comes from a single study (Burde and Williams, 1989), and was not observed by Akert et al. (1980), with the exception of the extension of the $pIII_{PG}$ motoneuron column into AM. We have observed occasional outliers in this population. Perhaps these cells were designated as the visceral columns and nucleus of Perlia in the previous study. It is not surprising that some variation may be present, given that non-primate species do not confine the $pIII_{PG}$ population to EW, and instead scatter these cells dorsal, ventral, and rostral to III (Erichsen and May, 2002; May et al., 2008). However, it appears that among mammals, possession of an EW_{PG} made up of a consolidated preganglionic population is a striking primate feature. The fact that a well defined EW_{PG} is also present in birds (Gamlin et al., 1984) may indicate that a propensity for close work in highly visual species can lead to consolidation of $pIII_{PG}$ motoneurons into a unitary nucleus.

There is considerable evidence that the pathway subserving the light reflex is relayed by way of the OPt (for review, see Loewenfeld, 1993). The OPt primarily contains wide-field luminance units (Pong and Fuchs, 2000; Clarke et al., 2003) that represent the primary source of pretectal input to the pIII region (Steiger and Büttner-Ennever, 1979). Presumably, OPt supplies information on luminance levels to the $pIII_{PG}$ motoneurons that control the sphincter pupillae muscle. Here we have identified a population of motoneurons projecting to the ciliary ganglion that display close associations with boutons labelled anterogradely following pretectal injections (Figs. 2 and 3A). There are reports that vergence units may also be present in the pretectum (Mays et al., 1986), so we cannot be sure that the labelled axons are purely pupil-related. However, in light of the preponderant OPt projection observed by Steiger and Büttner-Ennever (1979), it seems likely that most of the motoneurons with pretectal inputs are part of the light reflex circuit. The ventromedial location of these motoneurons is also in agreement with the distribution of pupil-related units recorded in marmosets (Clarke et al., 1985). The bilateral terminal distribution in EW_{PG} following pretectal injections (Figs. 2 and 4), presumably represents an anatomical substrate for balanced direct and consensual pupillary responses.

Previous anterograde studies of the primate pretectal projections reported terminals in a distinctive nucleus in the ventrolateral PAG, which was termed the lvc (Baleydier et al., 1990; Büttner-Ennever et al., 1996), and was believed to represent a component of EW_{PG} (Burde and Williams, 1989). Trans-synaptic tracer experiments suggested that this nucleus receives retinal input by way of the pretectum (Kourouyan and Horton, 1997). The present study indicates the lvc is not part of EW_{PG}, but is instead a satellite nucleus that receives bilateral pretectal input and projects back onto the contralateral pretectum, by way of the posterior commissure. Based on the findings of Kourouyan and Horton (1997), it must be supplied with visual sensory input from either the OPt or the nOT. The function of this cell group is unknown. However, it is noteworthy that unlike the bird organization where the OPt are interconnected to balance activity levels (Gamlin et al., 1984), there is no evidence of connections between the macaque OPt, even though the nOT are heavily interconnected (Fig. 4). Perhaps this nucleus acts to balance the luminance activity in the two OPt. It will be interesting to determine whether it also projects to EW_{PG}.

Acknowledgement

This work was made possible by NIH Grants EY07166 and EY014263 to PJM.

References

Akert, K., Glicksman, M.A., Lang, W., Grob, P. and Huber, A. (1980) The Edinger–Westphal nucleus in the monkey. a retrograde tracer study. Brain Res., 184: 491–498.

Bachtell, R.K., Tsivkovskaia, N.O. and Ryabinin, A.E. (2002) Alcohol-induced c-Fos expression in the Edinger–Westphal nucleus: pharmacological and signal transduction mechanisms. J. Pharmacol. Exp. Ther., 302: 516–524.

Baleydier, C., Magnin, M. and Cooper, H.M. (1990) Macaque accessory optic system: II. Connections with the pretectum. J. Comp. Neurol., 302: 405–416.

Burde, R.M. (1988) Disparate visceral neuronal pools subserve spinal cord and ciliary ganglion in the monkey: a double labeling approach. Brain Res., 440: 177–180.

Burde, R.M. and Williams, F. (1989) Parasympathetic nuclei. Brain Res., 498: 371–375.

Büttner-Ennever, J.A., Cohen, B., Horn, A.K.E. and Reisine, H. (1996) Pretectal projections to the oculomotor complex of the monkey and their role in eye movements. J. Comp. Neurol., 366: 348–359.

Chen, B. and May, P.J. (2000) The feedback circuit connecting the superior colliculus and central mesencephalic reticular formation: a direct morphological demonstration. Exp. Brain Res., 131: 10–21.

Chen, B. and May, P.J. (2007) Premotor circuits controlling eyelid movements in conjunction with vertical saccades in the cat: II. Interstitial nucleus of Cajal. J. Comp. Neurol., 500: 676–692.

Clarke, R.J., Coimbra, C.J.P. and Alessio, M.L. (1985) Distribution of parasympathetic motoneurons in the oculomotor complex innervating the ciliary ganglion in the marmoset (*Callithrix jacchus*). Acta Anat., 121: 53–58.

Clarke, R.J., Zhang, H.Y. and Gamlin, P.D.R. (2003) Primate pupillary light reflex: receptive field characteristics of pretectal luminance neurons. J. Neurophysiol., 89: 3168–3178.

Erichsen, J.T. and May, P.J. (2002) The pupillary and ciliary components of the cat Edinger–Westphal nucleus: a transsynaptic transport investigation. Vis. Neurosci., 19: 15–29.

Gamlin, P.D.R., Reiner, A., Erichsen, J.T., Karten, H.I. and Cohen, D.P. (1984) The neural substrate for the pupillary light reflex in the pigeon (*Columbia livia*). J. Comp. Neurol., 226: 523–543.

Horn, A.J., Eberhorn, A., Härtig, W., Ardeleanu, P., Messoudi, A. and Büttner-Ennever, J.A. (2008) Perioculomotor cell groups in monkey and man defined by their histochemical and functional properties: a reappraisal of the Edinger–Westphal nucleus. J. Comp. Neurol., 507: 1317–1335.

Kourouyan, H.D. and Horton, J.C. (1997) Transneuronal retinal input to the primate Edinger–Westphal nucleus. J. Comp. Neurol., 381: 68–80.

Kozicz, T., Li, M. and Arimura, A. (2001) The activation of urocortin immunoreactive neurons in the Edinger–Westphal nucleus following acute pain stress in rats. Stress, 4: 85–90.

Loewenfeld, I.E. (1993) The Pupil — Anatomy, Physiology and Clinical Applications. Wayne State University Press, Detroit, MI.

Loewy, A.D., Saper, C.B. and Yamodis, D. (1978) Re-evaluation of the efferent projections of the Edinger–Westphal nucleus in the cat. Brain Res., 141: 153–159.

Maciewicz, R., Phipps, B.S., Foote, W.E., Aronin, N. and DiFiglia, M. (1983) The distribution of substance P-containing neurons in the cat Edinger–Westphal nucleus: relationship to efferent projection systems. Brain Res., 270: 217–230.

May, P.J., Reiner, A.J. and Ryabinin, A.E. (2008) Comparison of the distributions of urocortin containing and cholinergic neurons in the perioculomotor midbrain of the cat and macaque. J. Comp. Neurol., 507: 1300–1316.

Mays, L.E., Porter, J.D., Gamlin, P.D.R. and Tello, C.A. (1986) Neuronal control of vergence eye movements: neurons encoding vergence velocity. J. Neurophysiol., 56: 1007–1021.

Pong, M. and Fuchs, A.F. (2000) Characteristics of the pupillary light reflex in the macaque monkey: discharge patterns of pretectal neurons. J. Neurophysiol., 84: 964–974.

Steiger, H.-J. and Büttner-Ennever, J.A. (1979) Oculomotor nucleus afferents in the monkey demonstrated with horseradish peroxidase. Brain Res., 160: 1–15.

Sun, W. and May, P.J. (1993) Organization of the extraocular and preganglionic motoneurons supplying the orbit in the lesser galago. Anat Rec., 237: 89–103.

Vaughan, J., Donaldson, C., Bittencourt, J.C., Perrin, M.H., Lewis, K., Sutton, S., Chan, R., Turnbull, A.W., Lovejoy, D., Rivier, C., Rivier, J., Sawchenko, P.E. and Vale, W. (1995) Urocortin, a mammalian neuropeptide related to fish urotensin I and to corticotropin-releasing factor. Nature, 378: 287–292.

C. Kennard & R.J. Leigh (Eds.)
Progress in Brain Research, Vol. 171
ISSN 0079-6123

CHAPTER 2.4

Frontal eye field signals that may trigger the brainstem saccade generator

Edward L. Keller[1,*], Byeong-Taek Lee[1,2] and Kyoung-Min Lee[1,3]

[1]*Smith-Kettlewell Eye Research Institute, San Francisco, CA, USA*
[2]*Institute of Psychological Science, Seoul National University, Seoul, Korea*
[3]*Department of Neurology, Seoul National University, Seoul, Korea*

Abstract: Saccades are rapid shifts of gaze that normally place the line of sight on a desired target with a single smooth movement. A number of disease states have been shown to result in saccadic movements that are fragmented, but still end near target position after a multi-step sequence of saccades. Among these disorders are Parkinson's disease and late-onset Tay-Sachs disease (LOTS). We have recently shown that normal human subjects and monkeys also make some two-step saccadic responses in cognitively difficult, choice response tasks. In monkeys we have been able to record neuronal responses as the animals performed a visually guided, choice saccade task. We compared the activity of neurons in the superior colliculus (SC) and the cortical frontal eye field (FEF) during the majority of trials that were accomplished with single-step saccades with those completed with two-step saccades. Several differences in discharge pattern aligned on the first saccade were uncovered. Neurons in the rostral and caudal SC were not modulated at the time of the first saccade, but a class of FEF neurons showed a burst of activity before the first saccade. If these neurons are among those known to project to the saccade generator in the brainstem, they could trigger the onset of a saccade before the remaining machinery in the saccade generator had sufficient activity to sustain the saccade. Overall the results suggest that a delicate balance of triggering and sustaining inputs are required to produce normal single-step saccades. These neural results may also help to clarify the pathology present when fragmented saccades occur in various disease states.

Keywords: saccade; fragmented movements; choice response; superior colliculus; frontal eye fields

Introduction

Saccadic eye movements are rapid shifts of gaze that are normally executed with a single, smooth trajectory that ends near a selected visual target. They are believed to be guided by a brainstem burst generator that delivers a pulsatile burst of activity to the oculo-motoneurons that lasts for the duration of the desired movement (Keller, 1991; Scudder et al., 2002). This step-like, position-altering mechanism is usually thought to be robust in assuring a single movement that finishes near the target. Since vision is degraded during saccades, a single-step arrangement and not multiple steps would optimize visual performance (Stevenson et al., 1986). However, a number of clinical studies have demonstrated that in some pathological states the burst generator loses its robust pulsatile

*Corresponding author. Tel.: +415-345-2102; Fax: +415-345-8455; E-mail: elk@ski.org

drive. In these diseased states saccadic movements become fragmented, but still start in the correct direction and end near the desired target position after a multi-step sequence. Among these disorders are Parkinson's disease (Kimmig et al., 2002) and late-onset Tay-Sachs disease (LOTS) (Rucker et al., 2004). These clinical results suggest that single-pulse saccade signals from the burst generator are subject to a delicate balance of triggering and sustaining activity that may be disrupted by pathological changes in the burst generator network. Possible sites of pathology in the brainstem or cerebellum have been suggested at least for the LOTS patients (Rucker et al., 2004, 2005).

We have recently shown that normal human subjects also make some two-step saccadic responses in cognitively difficult, visually cued, choice response tasks (Lee et al., 2005). Here we report that monkeys also make occasional two-step responses in the same saccade choice task. In monkeys, we have been able to record neuronal responses as the animal performs the choice response task. The behavioural difficulty of this task suggests higher level neural involvement in this type of fragmented saccade. By comparing the activity of neurons recorded in the superior colliculus (SC) and the cortical frontal eye fields (FEF) for the majority of trials that are accomplished with single-step saccades with those having fragmented, two-step saccades, we are able to suggest some of the mechanisms that may be involved in the production of the fragmented movements. These neural results may help to clarify the pathology present when fragmented saccades occur in various disease states.

Methods

Choice response tasks involve the arbitrary pairing of a set of sensory stimuli with a set of motor responses. We designed a saccadic choice response task (see Fig. 1) in which the subjects initially fixated a central grey fixation point. After a delay, an array of coloured disks (the potential targets) appeared. In Fig. 1 the coloured disks are replaced by symbols. The targets in the array were arranged on an imaginary circumference with equal separation of the visual stimuli in physical and colour space. As the subjects remained fixed on the

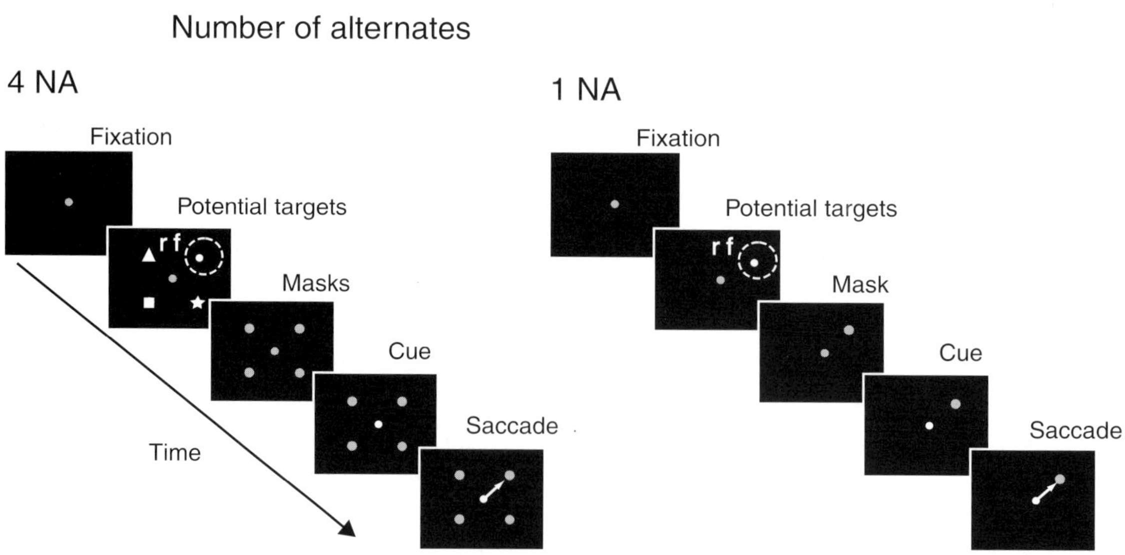

Fig. 1. Schematic illustration of the saccade choice task. The condition with 4 alternative choices (4 NA) is shown on the left and with one choice (1 NA) is shown on the right. Actual arrays of potential targets were coloured disks not symbols. The array of potential targets was arranged so that one of the coloured visual stimuli was positioned in the centre of the response field (rf) of the recorded cell.

central grey disk, grey masks replaced the eccentric coloured disks. After a further random delay the central fixation point changed from its grey colour to one of the colours present in the original eccentric array. The appearance of this colour at the fixation point served as the cue for the subject to make a saccadic eye movement to the remembered location of the former stimulus with the same colour. Response latencies increased dramatically with increases in the number of alternative (NA) potential target/colour parings. Saccade latency in humans was more than 500 ms for the 8 NA condition, suggesting that this task requires extensive cognitive processing for good performance (Lee et al., 2005). In monkey, trials were run in blocks with constant (1, 2, or 4) NA and fixed location/colour associations. In the examples shown in Fig. 1, the red cue signalled that a saccade to the upper right would be counted as correct. The array was positioned so that one of the potential targets was located in the response field (rf) of the neuron being recorded in the monkey.

Results

The extensive cognitive processing required by the saccadic choice response task in the larger NA conditions (4 and 8 NA) resulted in the appearance of fragmented saccades ($\sim 10\%$ of trials) in human and monkey subjects. Figure 2 shows examples of one single-step movement and two, two-step saccades in a monkey in the 4 NA condition. At the completion of the fragmented movements the final positions of the two-step movements were close to the correct target and their initial direction in the first saccade was toward the correct target, but initial eye accelerations are lower than expected for normal movements of the same total size. Saccade reaction times (SRT) in the monkey had a mean latency of 234 ms for the 4 NA condition compared to a mean latency of <180 ms for delayed saccade responses to single targets at the same set of locations. SRT increased logarithmically as a function of NA as expected for this type of associative memory paradigm (Hick, 1952). Although both examples of two-step responses shown in Fig. 2 have longer latencies than the one normal saccade shown, overall there was no significant difference in SRT for the two type of saccades (Wilcoxon rank sum test $p = .45$).

Our first neural observation relates to the activity of fixation neurons (FNs) ($n = 9$) located in the rostral region of the monkey SC. As expected, for one-step saccades, these neurons were completely inhibited during 7°–10° saccades made in either horizontal direction (Munoz and Wurtz, 1993). Furthermore, seven of the nine neurons were also inhibited for small ($\sim 2°$) saccades made in either direction while the other two cells showed a small burst of activity for 2° saccades made in the contralateral direction. When we aligned the mean spike density responses on the second saccade of the fragmented movements made in the same block of trials, these FNs were also inhibited during the second-step movement. However, when we aligned the same responses on the first saccade in the fragmented movements, there was no inhibition of the response, i.e., rostral SC cells continued to fire at the same baseline rate for the first saccades as if they were not aware that the sequence of movements had started.

The second observation relates to the activity of more caudal neurons in the SC. These neurons fired as expected for saccades into their movement fields (Munoz and Wurtz, 1995) (typically 7°–10° in amplitude). We normalized the burst of activity to the peak occurring during single-step movements to centre of the movement field. We averaged the normalized spike density responses across all cells ($n = 19$) to estimate the population activity of saccade-related burst neurons in the SC for single-step saccades (Fig. 3A). We then examined the activity of the same sample of cells for two-step movements made in the same blocks of trials for which the single-step population activity had been calculated. When this activity was aligned on the onset of the second-step saccade it resembled the activity profile present for the single-step movements (Fig. 3B), when the same activity was realigned on the onset of the first step the normal buildup of activity had only just started and no burst of SC activity occurred at the onset of this movement (Fig. 3C). We conclude that like rostral SC neurons, caudal SC cells are not aware that the two-step sequence of movements has begun at the

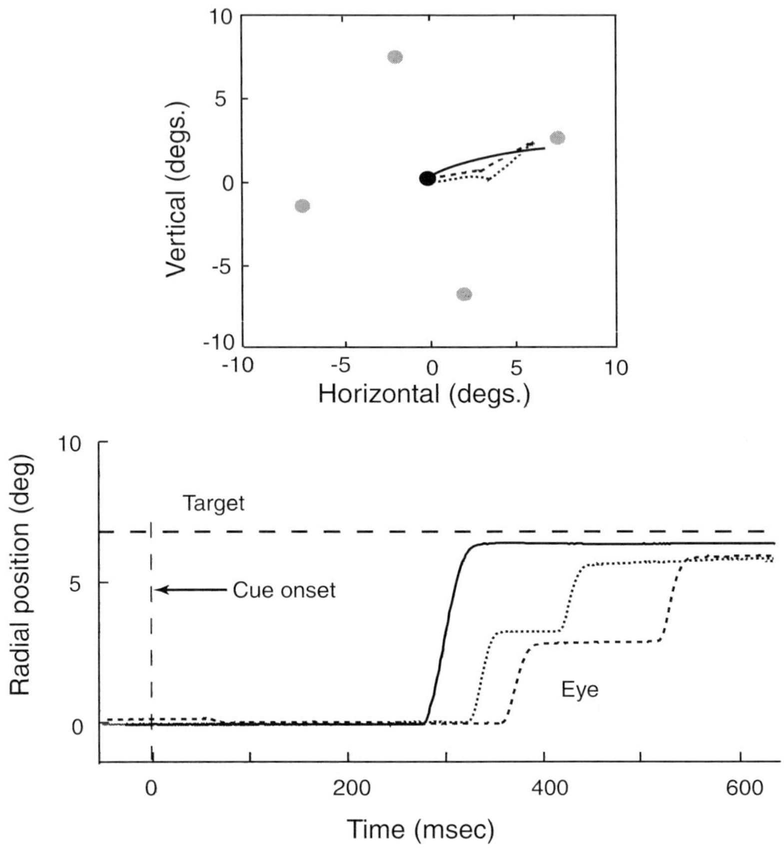

Fig. 2. Spatial and temporal plots showing three example saccades from a block of trials (4 NA condition) in which a red disk had appeared at an amplitude of 7° and 20° to the right and up. In the spatial plot, the array of coloured targets has been replaced by grey masking disks and the fixation point at the centre has turned red (●) as a cue that a saccade should be made to the location of the grey eccentric disk that had formally been red in colour. The eye traces in both the spatial and temporal plots show the radial eye positions for three saccades, one is a normal single-step movement (solid curves) and two are two-step responses (dotted curves). The red colour is shown by a filled circle.

time of the first movement. The normal direct connections of both the rostral (by disfacilitation) and caudal SC (via inhibitory local neurons) to the burst generator that are usually thought to trigger and sustain saccades are not involved in triggering the first movements.

We next examined the activity of saccade-related cells in the FEF of the monkey because this structure, like the SC, has direct inputs to elements of the brainstem burst generator (Segraves, 1992). We found two types of saccade-related cells in the FEF.

The population discharge of one type (called Type 1 here) was similar to that observed in caudal SC neurons ($n = 12$). For single-step saccades to the centre of the recorded cells' movement fields, Type 1 FEF neurons showed a gradual buildup of activity prior to saccade onset (Fig. 4A). This activity peaked just before saccade onset and then declined during the saccadic movement. Figure 5 shows the population activity of another group of FEF cells we called Type 2 ($n = 22$). For single-step saccades made to the centre of the Type 2 cells' movement field there was ramp-like buildup of activity that peaked just before saccade onset and then discharge declined during the movement (Fig. 5A). Thus, the activity pattern for single-step saccades was the same for both Type 1 and Type 2 cells (compare the normalized spike density traces

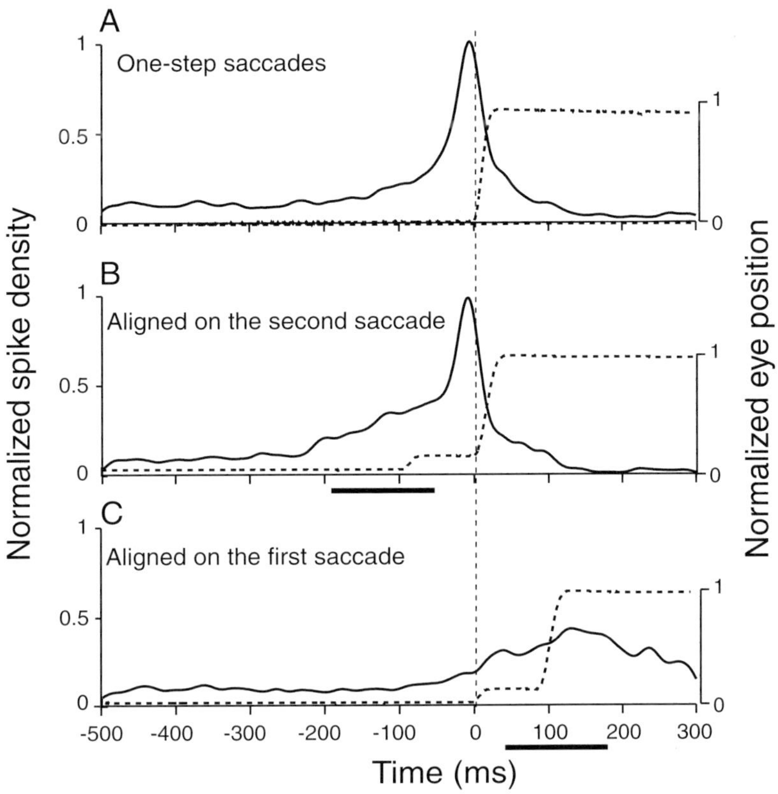

Fig. 3. Population activity of caudal SC saccade-related burst neurons during single- and two-step saccades. In each panel the solid trace is the mean normalized spike density and the dotted trace shows normalized radial eye position for one example saccade. (A) Single-step movements aligned on saccade onset (vertical dashed line). (B) Spike density for two-step movements aligned on the second saccade in the sequence. (C) Spike density for the same set of two-step movements aligned on the first saccade. The solid horizontal bars beneath the abscissas in (B) and (C) show the range of onset times of the first or second movements when alignment is made on the other. The eye movements shown are one representative movement of each type.

in Figs. 4A and 5A) and was similar to that previously reported for visuomotor or motor neurons in FEF (Bruce and Goldberg, 1985). However, when two-step saccades are considered, a different pattern of activity emerges for Type 1 and Type 2 cells. The discharge in Type 1 cells remains aligned with the onset of the second saccade (Fig. 4B). When the activity is aligned on the first saccade, the activity is still low at its onset and peaks well after the first movement (Fig. 4C). This pattern resembles that present for the two-step saccades in cells in the caudal SC. For Type 2 cells, when the activity was aligned on the second saccade it peaked well before second-saccade onset and had declined by the time of this saccade (Fig. 5B). There was another smaller peak that may be associated with the second movement. In contrast, when the activity was aligned on the onset of the first saccade (Fig. 5C) the discharge for this type of FEF cell now peaked before the onset of the first movement and more closely resembled the discharge pattern for normal, single-step saccades (Fig. 5A).

Discussion

The results show that fractionated saccades can be produced in normal subjects performing a difficult choice response task with multiple stimulus/response associations. Furthermore, the neural recordings made in the SC and the FEF of

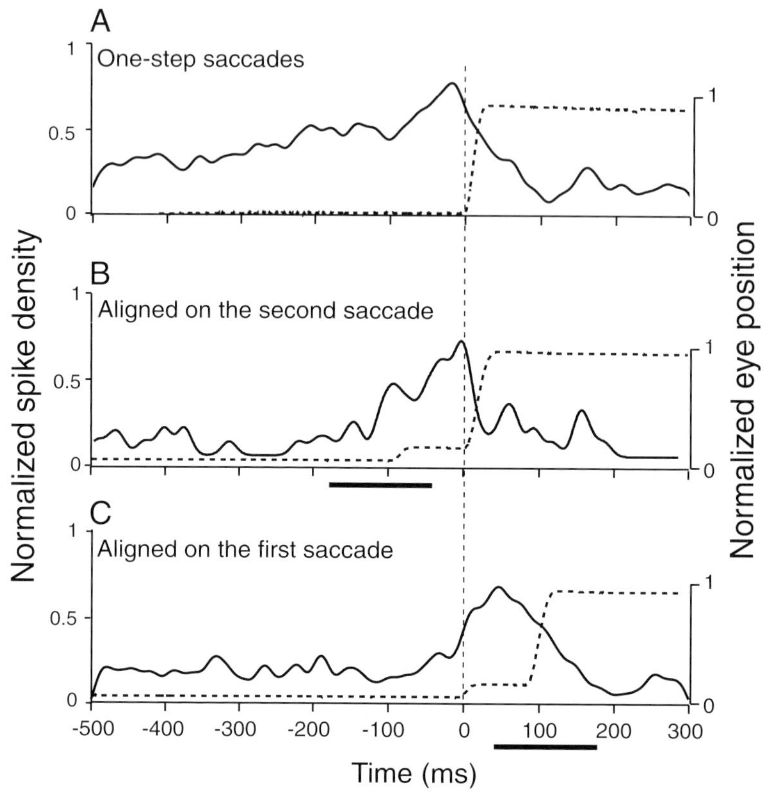

Fig. 4. Population activity of FEF Type 1 cells. Layout is the same as Fig. 3.

monkeys have revealed a previously unknown potential input from the FEF to the brainstem burst generator that may be capable of triggering saccades before activity has built up in the SC. For normal single-step saccades, activity builds up in parallel at matching topographic sites in FEF (both cell types) and caudal SC prior to movement onset. The population activity in the SC has a more clearly demarked inflection point on the spike density trace that separates the buildup phase of activity from the saccade aligned bust of activity and is more sharply peaked just prior to saccade onset. Nevertheless, the temporal course of the population activity in both regions is qualitatively the same, probably reflecting the reciprocal connections between FEF and SC (Sommer and Wurtz, 1998, 2000). This activity pattern is the same for Type 1 and 2 FEF neurons for normal saccades. Our behaviourally difficult saccadic choice task produced very long SRT especially for higher NA conditions. In one monkey trained on the 8 NA condition, mean SRT was 252 ms. In this task for the 4 NA condition two-step saccades begin to appear. If the activity patterns in FEF and SC cells are compared for these movements, a clear difference between SC (and Type 1 FEF) neurons and Type 2 FEF neurons emerges.

The discharge of Type 2 cells was much better aligned with the onset of the first saccade. When the activity was aligned on the second saccade it peaked before second-saccade onset and had declined to a low level by the time of this saccade (Fig. 5B). There was another smaller peak that may be associated with the second movement. In contrast, when the activity was aligned on the onset of the first saccade (Fig. 5C) the discharge of this type of FEF cell now peaked before the onset of the first movement and more closely resembled the discharge pattern for normal, single-step

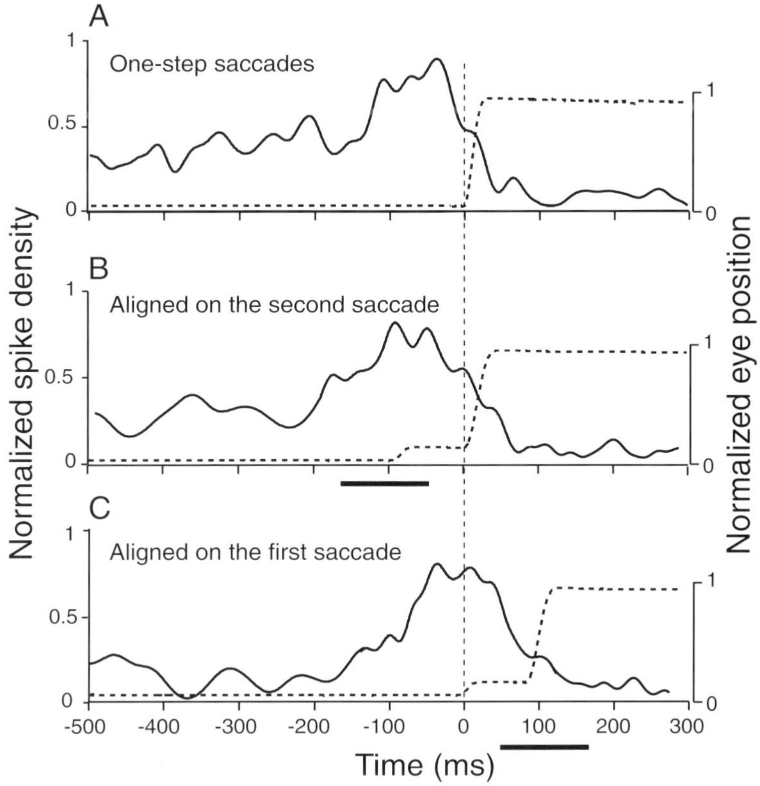

Fig. 5. Population activity of FEF Type 2 cells. Layout is the same as Fig. 3.

saccades (Fig. 4A), and if they selectively project to the burst generator, they may trigger a saccade before caudal SC and Type 1 FEF cells have become fully activated. The latter cells are better aligned with the second saccade.

We hypothesize the following model to explain the mechanisms underlying the production of two-step saccadic movements in our behaviourally difficult task in normal humans and monkeys. The positive drive to excitatory burst neurons (EBN) in the brainstem comes from burst cells in both the SC and the FEF (the "where" signal). Most of the "where" signal from the FEF (Type 1 cells) is routed through the SC. Brainstem omnipause neurons (OPNs) are turned off briefly by an inhibitory trigger signal (the "when" signal) to start saccades. This signal comes jointly from the SC and FEF (Type 2 cells) via local inhibitory interneurons near the OPNs. Normally, the "where" and "when" signals are tightly coupled so that the saccade is triggered only when the full drive signal in the SC (and the FEF) has built up. In cognitively difficult tasks with long SRTs some FEF neurons (Type 2) act as the trigger signal to OPNs before the drive signal has been built up in caudal SC and Type 1 FEF neurons. This cortical signal is sufficient to start, but not sustain the movement. FNs in the rostral SC contribute to the saccade interruption, since they are not turned off for the first saccade in two-step movements as they are in normal saccades. This produces residual excitation in OPNs at the onset of fragmented saccades.

This model for saccade fragmentation is not the same as has been proposed as the cause of interrupted saccades in LOTS patients where the lesion seems to affect oculomotor circuitry in the cerebellum and perhaps the brainstem (see Rucker et al., this volume). Our results suggest that the triggering and maintenance of saccades involves a delicate coordination of the "when" and "where" signals. This coordination may be upset by either brainstem/cerebellar lesions or by cortical mechanisms in difficult tasks with very long SRTs.

Acknowledgements

Supported by NIH Grant EY08060.

References

Bruce, C.J. and Goldberg, M.E. (1985) Primate frontal eye fields. I. Single neurons discharging before saccades. J. Neurophysiol., 53: 603–635.

Hick, W.E. (1952) On the rate of gain of information. Q. J. Exp. Psychol., 4: 11–26.

Keller, E.L. (1991) The brainstem. In: Carpenter R.H.S. (Ed.), Eye Movements. Macmillan Press, London, pp. 200–223.

Kimmig, H., Haussmann, K., Mergner, T. and Lucking, C.H. (2002) What is pathological with gaze shift fragmentation in Parkinson's disease? J. Neurol., 249: 683–692.

Lee, K.M., Keller, E.L. and Heinen, S.J. (2005) Properties of saccades generated as a choice response. Exp. Brain Res., 162: 278–286.

Munoz, D.P. and Wurtz, R.H. (1993) Fixation cells in monkey superior colliculus. I. Characteristics of cell discharge. J. Neurophysiol., 70: 559–575.

Munoz, D.P. and Wurtz, R.H. (1995) Saccade-related activity in monkey superior colliculus. I. Characteristics of burst and buildup cells. J. Neurophysiol., 73: 2313–2333.

Rucker, J.C., Leigh, R.J., Keller, E.L. and Optican, L.M. (2005) Mechanisms of interrupted saccades differ in normal subjects and late-onset Tay-Sachs patients. *Soc. Neurosci. Abstr. CD-ROM:*program 592.3, #592.3.

Rucker, J.C., Shapiro, B.E., Han, Y.H., Kumar, A.N., Garbutt, S., Keller, E.L. and Leigh, R.J. (2004) Neuro-ophthalmology of late-onset Tay-Sachs disease (LOTS). Neurology, 63: 1918–1926.

Scudder, C.A., Kaneko, C.R.S. and Fuchs, A.F. (2002) The brainstem burst generator for saccadic eye movements: a modern synthesis. Exp. Brain Res., 142: 439–462.

Segraves, M.A. (1992) Activity of monkey frontal eye field neurons projecting to oculomotor regions of the pons. J. Neurophysiol., 68: 1967–1985.

Sommer, M.A. and Wurtz, R.H. (1998) Frontal eye field neurons orthodromically activated from the superior colliculus. J. Neurophysiol., 80: 3331–3333.

Sommer, M.A. and Wurtz, R.H. (2000) Composition and topographic organization of signals sent from the frontal eye field to the superior colliculus. J. Neurophysiol., 83: 1979–2001.

Stevenson, S.B., Volkmann, F.C., Kelly, J.P. and Riggs, L.A. (1986) Dependence of visual suppression on the amplitudes of saccades and blinks. Vision Res., 26: 1815–1824.

C. Kennard & R.J. Leigh (Eds.)
Progress in Brain Research, Vol. 171
ISSN 0079-6123

CHAPTER 2.5

The role of omnipause neurons: why glycine?

Lance M. Optican*

Laboratory of Sensorimotor Research, National Eye Institute, NIH, DHHS, Bethesda, MD, USA

Abstract: The anatomy and neurophysiology of the saccadic eye movement system have been well studied, but the roles of certain key neurons in this system are not fully appreciated. Important clues about the functional interactions in the saccadic system can be gleaned from the histochemistry of different saccadic neurons. The most prominent inhibitory neurons in the circuit are the omnidirectional pause neurons (OPN), which inhibit the premotor burst neurons that drive the eye. Most inhibitory neurons in the brain transmit γ-aminobutyric acid (GABA), but OPN transmit glycine (Gly). It is interesting to ask whether the saccadic system would work any differently if OPN were GABA-ergic. Gly and GABA receptors both provide a channel for a hyperpolarizing Cl^- current that inhibits its target neuron. Depolarizing currents that excite the neurons come through several channels, including the NMDA receptor (NMDAR). The NMDAR is unique among receptors in that it has active sites for two different neurotransmitters, glutamate (Glu) and Gly. Gly is a co-agonist that acts to amplify the current produced by Glu. We have proposed a model of the saccadic brain stem circuitry that exploits this dual role of Gly to produce both inhibition of the saccadic circuit during fixation, and to increase its responsiveness, or gain, during movements. This suggests that OPNs act more as a regulator of the saccadic circuit's gain, rather than as a gate for allowing saccades. We propose a new hypothesis: the OPNs play a general role as a modulator of arousal in orienting subsystems, such as saccades, pursuit, head movements, etc.

Keywords: glycine; burst neurons; brainstem; saccades

Introduction

Saccades are rapid, voluntary movements that reorient the retinal region of highest acuity (fovea) towards images of interest. As acuity is greatly impaired during these movements, good vision requires that saccades be fast and accurate. In engineering terms, a fast and accurate controller must have high gain and negative feedback, but having both properties makes a circuit prone to instability. Oscillations around the target at the end of a movement would interfere with vision just as much as a slower, more stable saccade. Thus, the brain must trade-off the speed, accuracy, and stability of the saccade. Indeed, this tradeoff must exist for all orienting subsystems (e.g., eye and head), which suggests the need for a general arousal system that could modulate all orienting behavior to increase sensitivity and maintain stability.

One of the best-studied orienting subsystems, the one for saccadic eye movements, is assumed to have high-gain premotor burst neurons (PBNs, providing the necessary sensitivity to changes) and a class of cells called omnipause neurons (OPN,

*Corresponding author. Tel.: +301-496-9375; Fax: +301-402-0511; E-mail: LanceOptican@nih.gov

DOI: 10.1016/S0079-6123(08)00615-8

cf., Fig. 1A), that fire except during saccades, inhibiting PBNs and ensuring stability of the saccadic system. Here I propose a single function for OPNs that gives them a larger role to play in arousing all systems that orient the animal towards targets of interest.

OPNs were discovered in the brain stem of monkeys (Cohen and Henn, 1972; Luschei and Fuchs, 1972; Keller, 1974) and cats (Evinger et al., 1982). These neurons fire steadily during fixation and pause during saccades. About half the neurons pause for saccades in one direction, and the rest pause for saccades in all directions (i.e., omni-directional pause units, hence the name, omnipause). The OPNs make monosynaptic inhibitory connections (Nakao et al., 1980; Furuya and Markham, 1982; Strassman et al., 1987) to excitatory and inhibitory premotor burst neurons (EBN and IBN) in the brain stem (which connect to horizontal motor neurons) and midbrain (which connect to vertical/torsional motor neurons). The OPNs lie in the midbrain very close to the midline near the rostral pole of the abducens nucleus, in the raphe interpositus nucleus (RIP, Buttner-Ennever et al., 1988; Langer and Kaneko, 1990; Horn et al., 1994).

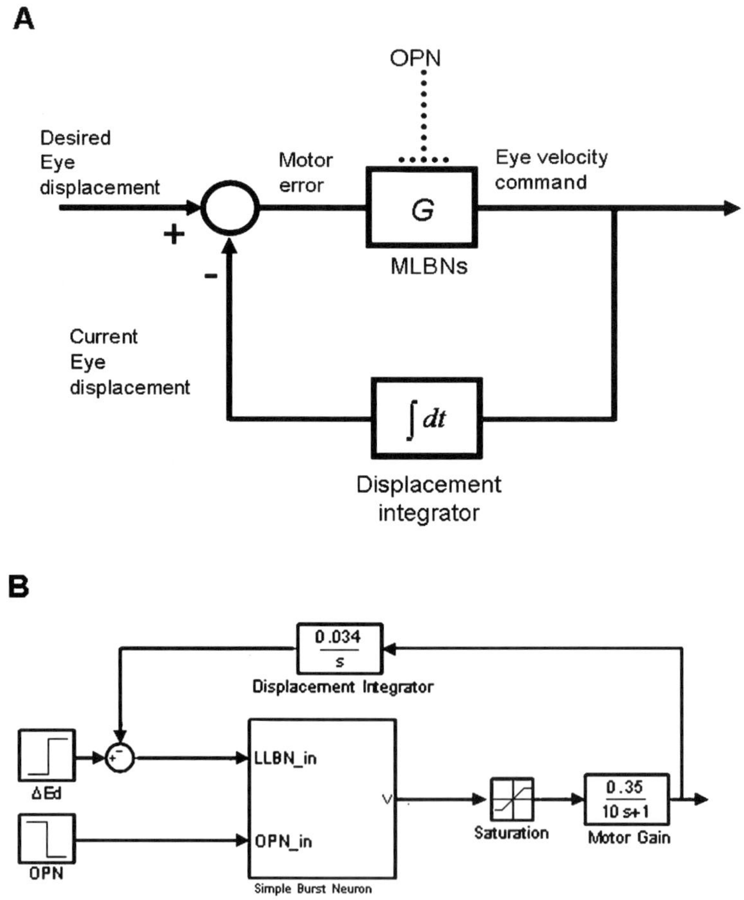

Fig. 1. Central core of local feedback loop pulse generator in a saccadic system. (A) Block diagram showing gating of premotor burst neurons (MLBN) by OPN. (B) Simplified implementation of model in A in SIMULINK. Simple burst neuron is a block that models a single-compartment, spiking neuron. The output of this block is clipped (saturation) and low-pass filtered (motor gain) to produce an eye velocity command. The displacement integrator converts this command into an efference copy of current eye displacement during the saccade, which is subtracted from the desired eye displacement (ΔEd) to compute motor error.

Most inhibitory neurons in the brain transmit either γ-aminobutyric acid (GABA) or glycine (Gly). OPNs happen to be glycinergic (Spencer et al., 1989; Horn et al., 1994). We could ask whether there is any significance to this fact, as the function of OPNs as a gate for saccades would still be possible if they used GABA. Gly differs from GABA in one important aspect. Although both GABA and Gly are inhibitory at their receptors (e.g., the GABAA receptor and the strychnine-sensitive Gly receptor), Gly also has another action at a different receptor. Gly acts as an excitatory modulator of glutamate (Glu) action at the co-agonist sites on the *N*-methyl D-aspartate receptor (NMDAR) (Johnson and Ascher, 1987). Below we explore some of the consequences of this dual action of Gly on the role of the OPNs.

Methods

We simulated a highly simplified model of the saccadic system (Fig. 1B), with essentially one excitatory premotor burst neuron (EBN) and no delays (SIMULINK/MATLAB, The Mathworks, Natick, MA). The EBN was modelled with differential equations for a single-compartment, conductance-based, spiking neuron (Miura and Optican, 2006). The equation for the membrane potential is given by summing the currents through each ionic channel:

$$C\frac{dV}{dt} = -I_L - I_T - I_{\mathrm{Na}} - I_{\mathrm{K}} - I_{\mathrm{Glu}} - \{I_{\mathrm{Gly}} \text{ or } I_{\mathrm{GABAA}}\} \quad (1)$$

In this test, the model EBN had only one or the other of the Gly or GABAA receptors (Figs. 2A, B). Previous studies have demonstrated that the Gly receptors act in a rapid manner when near body temperature. Specifically, a deactivation time constant in the range of 1.3–5.4 ms has been suggested (see discussion in Harty and Manis, 1998). The dynamics of the current, I_{Gly}, are given by:

$$I_{\mathrm{Gly}} = g_{\mathrm{Gly}}x(V - E_{\mathrm{Gly}}) \quad (2)$$

$$\frac{dx}{dt} = \alpha_{\mathrm{Gly}}\mathrm{Gly}_{\mathrm{in}}(1 - x) - \frac{x}{\tau_{\mathrm{Gly}}} \quad (3)$$

where g_{Gly} (1 mS/cm^2) and E_{Gly} (−80 mV) denote the conductance and reversal potential of this channel, respectively, and α_{Gly} and τ_{Gly} are constants. α_{Gly} was set to 0.05, and the decay time constant, τ_{Gly}, was set to 2 ms in the simulations.

The dynamics of the channel for the GABAA receptor are more complicated than the first-order equations for GlyR, and its decay is usually described by two or three exponential functions (Rossi and Hamann, 1998; Dumoulin et al., 2001). For the purposes of this paper, the kinetics were simplified to first order and described by the same equations as the GlyR. The range of time constants for the decay of current in the GABAA channel is quite large (about 40–200 ms in rat cerebellar slice or dorsal root ganglion, Rossi and Hamann, 1998; Dumoulin et al., 2001). Whether or not there are faster subtypes of GABAA receptor proteins in the oculomotor neurons is simply not known. The effects of switching from Gly to GABAA in the OPN model depend almost entirely on the choice of this time constant. For this simulation, we chose a decay time constant of 50 ms; the other parameters were the same as for GlyR.

Results

The first model used a glycinergic OPN and generated a normal saccade (Fig. 3, solid lines). The effects of an OPN lesion were then simulated by turning off the OPN completely, so that no Gly was transmitted (Fig. 3, dotted lines). The effect was to reduce the speed of the saccade, but not its amplitude (because of the feedback loop). In this case, the loss of the Gly co-agonist on the NMDA channel caused the peak speed to drop by 18%. The latency (4.6 ms) was not affected, because it is determined by the threshold of the EBN, not the offset of the OPN (Miura and Optican, 2006).

The effect of an OPN using GABA on a saccade is shown in Fig. 2 (dashed line). First, the peak speed of the movement is greatly reduced (by 54%). This is a much greater drop than can be accounted for by the loss of Gly at the co-agonist

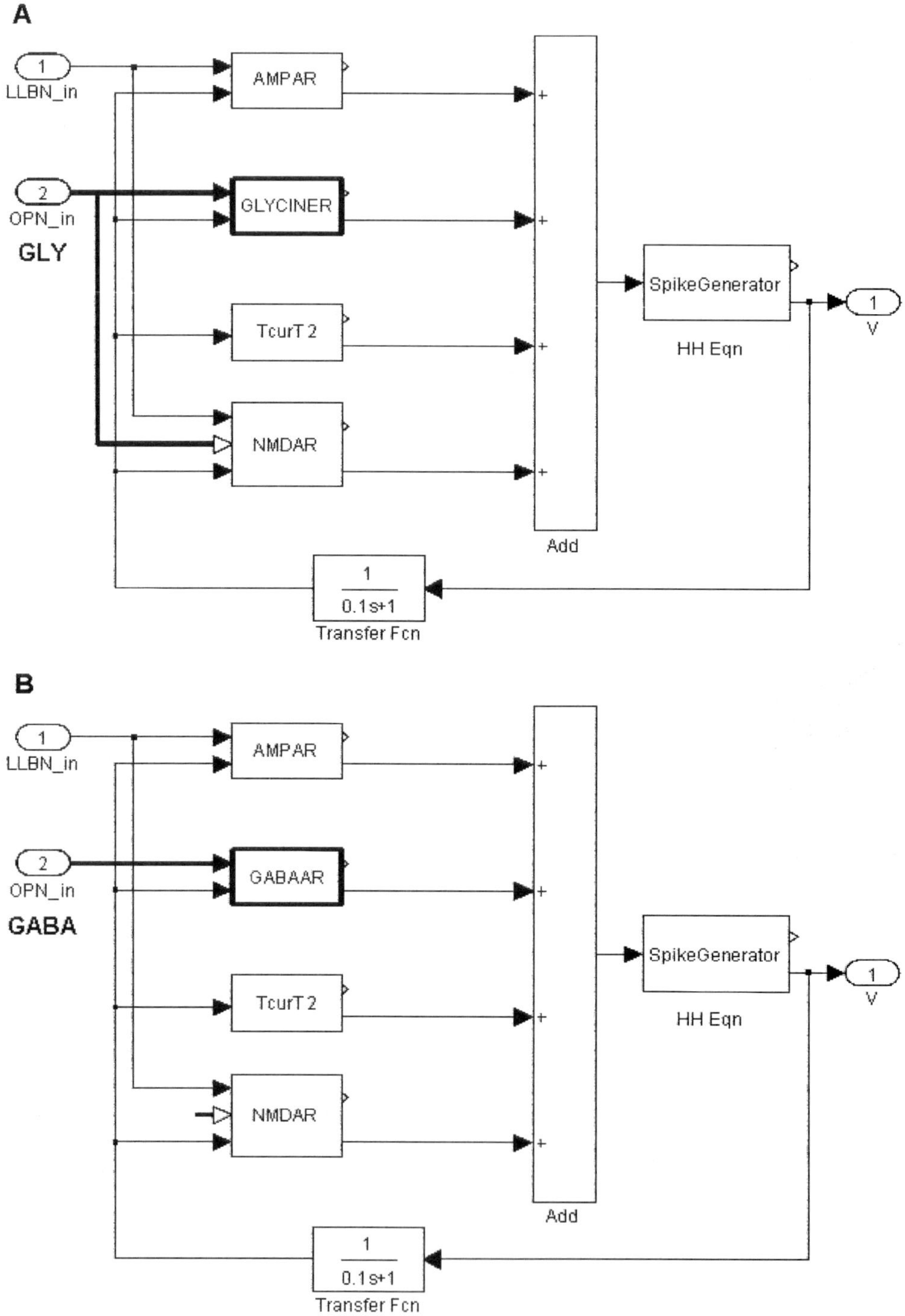

Fig. 2. Inside the simple burst neuron block. (A) A single-compartment neuron represented by adding currents through each ligand-binding receptor (for AMPA, Gly, NMDA, and a T-type Ca^{++} channel) to the Hodgkin–Huxley equations for the voltage-gated channels (for details, see Miura and Optican, 2006). Note that Gly acts at both the strychnine-sensitive glycine receptor (GLYCINER) and the co-agonist cite of the NMDA receptor. (B) Burst neuron simulating GABA-ergic OPNs. Note that OPN output goes only to GABAAR, and not to the co-agonist site on the NMDAR.

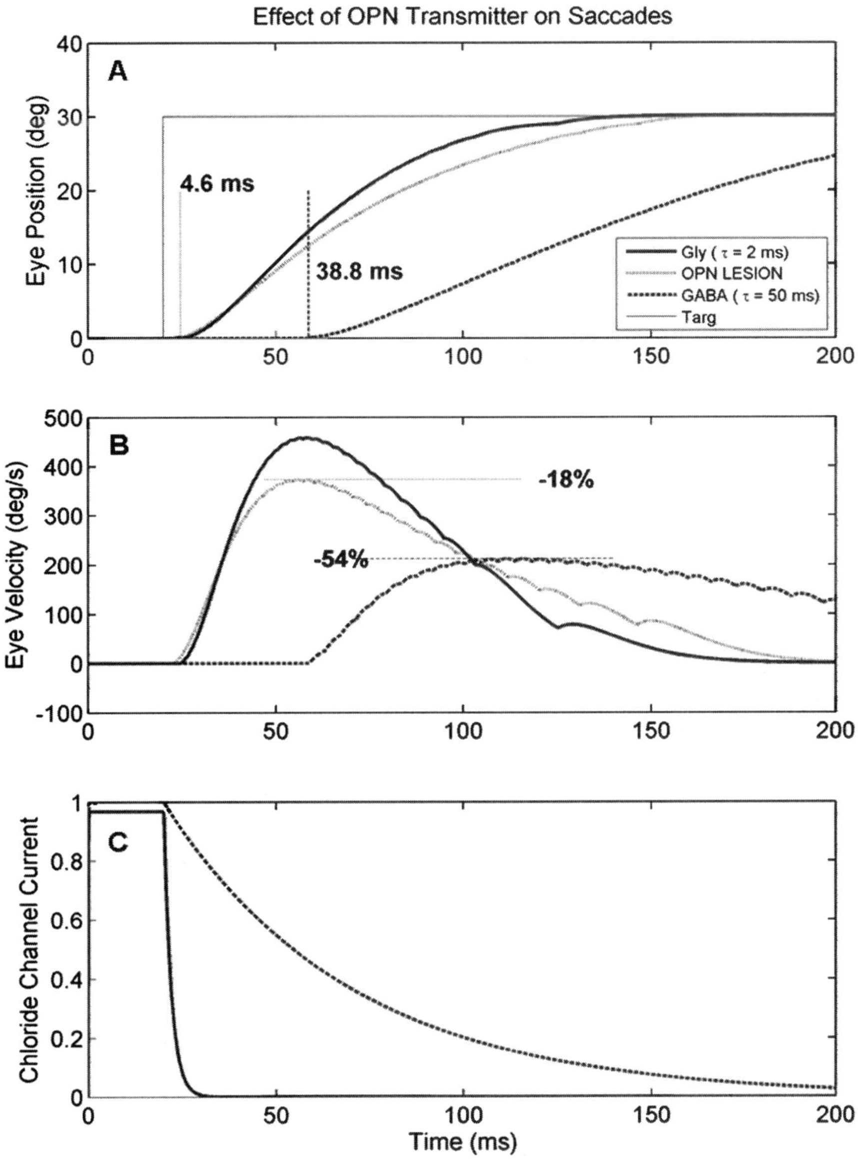

Fig. 3. Effect of OPN transmitter type on saccades. (A) and (B) show simulated position and velocity traces of a single saccade. When OPNs are glycinergic, the saccade (thick line) is fast (peak speed about 450 deg/s), and starts with a short latency (4.6 ms). After OPN lesion (dotted gray line), the saccade latency and accuracy are unaffected, but saccade slows down by about 18% because the loss of the co-agonist at the NMDAR reduces the gain of the Glu input. When OPNs are made GABA-ergic, saccade (dashed line) starts much later (38.8 ms) and is much slower (by 54%). This difference between the GABA traces and the OPN lesion traces are due to the much slower time course of decay of the GABAA receptor (C).

site on the NMDAR (−18%). Thus, the slow recovery of the GABAAR (Fig. 3C) has decreased the acceleration of the EBN firing rate, and thus of the saccade.

Discussion

Here, we propose a new hypothesis about the role and function of OPNs. Evidence shows that OPNs are not simply a gate for saccades, nor are they controlling saccade duration (Busettini and Mays, 2003). Pursuit eye movements slow OPN firing, and OPNs inhibit (but do not stop) pursuit neurons (Missal and Keller, 2002). When OPNs are not active (either during sleep or following lesions of the RIP), saccades are slow, but do not oscillate (Henn et al., 1984; Kaneko, 1996; Soetedjo et al., 2002). We infer from this that OPNs arouse the orienting systems in the brain by increasing their local supply of Gly. Thus, OPN activity signals the brain's expectation that useful information, which may require orientation towards a new target of interest, is about to arrive. When a new orienting movement is ongoing, the OPN signal is reduced (or stopped) because the brain no longer expects any useful information to arrive during the self-motion. This is consistent with the finding that movements are often preceded by a trigger signal and accompanied by a latch signal that prevent OPNs from firing (Yoshida et al., 1999).

How the expectation signal provided by the OPNs is utilized depends upon the downstream neuron. The use of Gly as the OPN transmitter admits two mechanisms. First, the ratio of NMDA to non-NMDA excitatory receptors could be adjusted to allow the OPN signal to modulate the gain (sensitivity) of the downstream neuron. Second, the ratio of strychnine-sensitive to GABA inhibitory receptors could be adjusted to allow the OPN signal to decrease the sensitivity or even gate the activity of the downstream neuron. Presumably, neurons in circuits that are unstable would have more strychnine-sensitive receptors than those that were intrinsically stable. This predicts that saccadic premotor burst neurons would have more strychnine-sensitive receptors than pursuit neurons.

Reduction in OPN activity during pursuit (Missal and Keller, 2002) can then be seen to have two benefits. First, the threshold for a catch-up saccade may be lowered (less Gly at strychnine-sensitive GlyR). Second, pursuit can only follow one object in a complex visual world. Arousal of other subsystems may be lowered (less Gly at NMDA co-agonist sites), preventing distracting stimuli from interfering with the pursuit of the chosen target.

According to this hypothesis, if OPNs were part of a general arousal system for orienting movements, OPNs would be off or slowed during self-motion, because the expectation that a new stimulus would indicate a required reorientation would be low. This suggests that other systems, in addition to pursuit and saccades, should also have an inhibitory effect on OPNs.

Acknowledgement

This work was supported by the Intramural Research Program of the NEI.

References

Busettini, C. and Mays, L.E. (2003) Pontine omnipause activity during conjugate and disconjugate eye movements in macaques. J. Neurophysiol., 90: 3838–3853.

Buttner-Ennever, J.A., Cohen, B., Pause, M. and Fries, W. (1988) Raphe nucleus of the pons containing omnipause neurons of the oculomotor system in the monkey, and its homologue in man. J. Comp. Neurol., 267: 307–321.

Cohen, B. and Henn, V. (1972) Unit activity in the pontine reticular formation associated with eye movements. Brain Res., 46: 403–410.

Dumoulin, A., Triller, A. and Dieudonne, S. (2001) IPSC kinetics at identified GABAergic and mixed GABAergic and glycinergic synapses onto cerebellar Golgi cells. J. Neurosci., 21: 6045–6057.

Evinger, C., Kaneko, C.R. and Fuchs, A.F. (1982) Activity of omnipause neurons in alert cats during saccadic eye movements and visual stimuli. J. Neurophysiol., 47: 827–844.

Furuya, N. and Markham, C.H. (1982) Direct inhibitory synaptic linkage of pause neurons with burst inhibitory neurons. Brain Res., 245: 139–143.

Harty, T.P. and Manis, P.B. (1998) Kinetic analysis of glycine receptor currents in ventral cochlear nucleus. J. Neurophysiol., 79: 1891–1901.

Henn, V., Baloh, R.W. and Hepp, K. (1984) The sleep–wake transition in the oculomotor system. Exp. Brain Res., 54: 166–176.

Horn, A.K., Buttner-Ennever, J.A., Wahle, P. and Reichenberger, I. (1994) Neurotransmitter profile of saccadic omnipause neurons in nucleus raphe interpositus. J. Neurosci., 14: 2032–2046.

Johnson, J.W. and Ascher, P. (1987) Glycine potentiates the NMDA response in cultured mouse brain neurons. Nature, 325: 529–531.

Kaneko, C.R. (1996) Effect of ibotenic acid lesions of the omnipause neurons on saccadic eye movements in rhesus macaques. J. Neurophysiol., 75: 2229–2242.

Keller, E.L. (1974) Participation of medial pontine reticular formation in eye movement generation in monkey. J. Neurophysiol., 37: 316–332.

Langer, T.P. and Kaneko, C.R. (1990) Brainstem afferents to the oculomotor omnipause neurons in monkey. J. Comp. Neurol., 295: 413–427.

Luschei, E.S. and Fuchs, A.F. (1972) Activity of brain stem neurons during eye movements of alert monkeys. J. Neurophysiol., 35: 445–461.

Missal, M. and Keller, E.L. (2002) Common inhibitory mechanism for saccades and smooth-pursuit eye movements. J. Neurophysiol., 88: 1880–1892.

Miura, K. and Optican, L.M. (2006) Membrane channel properties of premotor excitatory burst neurons may underlie saccade slowing after lesions of omnipause neurons. J. Comp. Neurosci., 20: 25–41.

Nakao, S., Curthoys, I.S. and Markham, C.H. (1980) Direct inhibitory projection of pause neurons to nystagmus-related pontomedullary reticular burst neurons in the cat. Exp. Brain Res., 40: 283–293.

Rossi, D.J. and Hamann, M. (1998) Spillover-mediated transmission at inhibitory synapses promoted by high affinity alpha6 subunit GABA(A) receptors and glomerular geometry. Neuron, 20: 783–795.

Soetedjo, R., Kaneko, C.R. and Fuchs, A.F. (2002) Evidence that the superior colliculus participates in the feedback control of saccadic eye movements. J. Neurophysiol., 87: 679–695.

Spencer, R.F., Wenthold, R.J. and Baker, R. (1989) Evidence for glycine as an inhibitory neurotransmitter of vestibular, reticular, and prepositus hypoglossi neurons that project to the cat abducens nucleus. J. Neurosci., 9: 2718–2736.

Strassman, A., Evinger, C., McCrea, R.A., Baker, R.G. and Highstein, S.M. (1987) Anatomy and physiology of intracellularly labelled omnipause neurons in the cat and squirrel monkey. Exp. Brain Res., 67: 436–440.

Yoshida, K., Iwamoto, Y., Chimoto, S. and Shimazu, H. (1999) Saccade-related inhibitory input to pontine omnipause neurons: an intracellular study in alert cats. J. Neurophysiol., 82: 1198–1208.

C. Kennard & R.J. Leigh (Eds.)
Progress in Brain Research, Vol. 171
ISSN 0079-6123

CHAPTER 2.6

Applying saccade models to account for oscillations

S. Ramat[1,*], R.J. Leigh[2], D.S. Zee[3], A.G. Shaikh[3] and L.M. Optican[4]

[1]*Dipartimento di Informatica e Sistemistica, Università degli Studi di Pavia, Pavia, Italy*
[2]*Department of Neurology, Veterans Affairs Medical Center and University Hospitals, Case Western Reserve University, Cleveland, OH, USA*
[3]*Department of Neurology, The Johns Hopkins University, Baltimore MD, USA*
[4]*Laboratory of Sensorimotor Research, IRP, National Eye Institute, Bethesda, MD, USA*

Abstract: Saccadic oscillations are unwanted back-to-back saccades occurring one upon the other that produce a high-frequency oscillation of the eyes (usually 15–30 Hz). These may occur transiently in normal subjects, for example, around the orthogonal axis of a purely horizontal or vertical saccade, during combined saccade-vergence gaze shifts or during blinks. Some subjects may produce saccadic oscillations at will, usually with convergence. Pathological, involuntary saccadic oscillations such as flutter and opsoclonus are prominent in certain diseases. Our recent mathematical model of the premotor circuit for generating saccades includes brainstem burst neurons in the paramedian pontine reticular formation (PPRF), which show the physiological phenomenon of post-inhibitory rebound (PIR). This model makes saccadic oscillations because of the positive feedback among excitatory and inhibitory burst neurons. Here we review our recent findings and hypotheses and show how they may be reproduced using our lumped model of the saccadic premotor circuitry by reducing the inhibitory efficacy of omnipause neurons.

Keywords: saccadic system; saccade models; saccadic Oscillations

Introduction

Saccades are the most frequent voluntary eye movement during everyday life: they rapidly redirect gaze to objects in the environment. Saccades to the sudden appearance of a visual target typically have a latency of ~200 ms, reach peak velocities of ~600 deg/s, last between 30 and 100 ms, and are accurate. The current understanding of the anatomy and physiology of the saccadic system has considerably advanced due to single unit recordings from neurons with activity related to eye movements in the behaving monkey. Experimental studies based on the analysis of the effects of lesions in animals and, vice versa, clinical and postmortem examinations of patients with saccadic disorders, have further clarified the physiology of the saccadic system.

Mathematical models of the generation of saccades have evolved by incorporating these findings, and theoretical studies have generated quantitative hypotheses that sparked new specific questions requiring further experiments. Basic experimental research on the mechanism for generating saccades has interacted with mathematical modelling efforts to advance our understanding of the saccadic system and, more broadly, of motor control in general.

*Corresponding author. Tel.: +39 0382 985745;
Fax: +39 0382 985373; E-mail: stefano.ramat@unipv.it

DOI: 10.1016/S0079-6123(08)00616-X

Recently, the study of saccadic oscillations (ocular flutter) in normal subjects and patients with known lesions has raised new questions that challenge our theories of saccade generation and forced us to update current mathematical models. Here we incorporate these findings into our models to account for saccadic oscillations in normal subjects and in patients with neurological disorders.

Saccadic oscillations

Saccadic oscillations are small, unwanted back-to-back saccades without an intersaccadic interval occurring at a frequency of 15 to 30 Hz. They have been recorded in many neurological conditions (Zee and Robinson, 1979; Ashe et al., 1991; Leigh and Zee, 2006), but some normal subjects can produce similar eye movements at will, as in voluntary nystagmus (Shults et al., 1977). Saccadic oscillations can also be induced in normal subjects by blinks (Hain et al., 1986), by combined saccade-vergence gaze shifts (Ramat et al., 1999), and can occur around the orthogonal axis to a purely vertical or purely horizontal saccade (Zee et al., 1992; Ramat et al., 2005a). Although these oscillations are small (~ 1 deg), they cause blurred vision when they are sustained, as in voluntary nystagmus or in patients with ocular flutter or opsoclonus (Ramat et al., 2005a; Shaikh et al., 2007).

Prior models of saccadic oscillations

In an early model of saccadic oscillations (Zee and Robinson, 1979), Zee and Robinson suggested that the oscillations are due to an increased delay in the local feedback loop controlling saccade amplitude. A delay introduces a phase lag that increases linearly with frequency. This makes the circuit unstable, with oscillations at the frequency at which the loop gain is at least one and the phase lag at least 180 deg. The forward pathway includes a switch representing the action of the omnipause neurons (OPN) that tonically inhibit burst neurons when a saccade is not called for, and a high gain nonlinear block representing the excitatory burst neurons (EBN). The feedback pathway includes a resettable integrator that computes eye position instantaneously during each saccade with a delay of τ s.

The high gain of the burst neurons for small inputs and the delay in the feedback loop make the circuit that generates saccades inherently unstable. Without the OPN gate, which prevents oscillations during fixation, the circuit oscillates at a frequency that depends on the duration of the delay (roughly 0.25/delay) with an amplitude that depends mainly on where the slope of the burst neuron nonlinearity decreases (Zee and Robinson, 1979). This model was initially successful, and later models of saccadic oscillations (Wong et al., 2001) followed the same feedback loop delay idea and combined it with the hypothesis that the fastigial nuclei (FN) is within the feedback loop (Lefevre et al., 1998).

Recent findings in both normal subjects and in patients with saccade oscillations (Ramat et al., 1999, 2005a; Shaikh et al., 2007) have challenged such models and motivated a new hypothesis for saccadic oscillations. First, experimental data showed that individuals can produce oscillations of different amplitudes with little change in frequency, and that oscillations in normal subjects span a large range of frequencies, although they tend to be fixed within a given subject. Both of these observations are not easily explained using the delay model because the range of frequencies requires a physiologically implausible large range of delays in the feedback loop. Second, the finding that patients with surgical ablation of the FN still produce saccadic oscillations argues against both the hypothesis that the delay in the feedback loop may be the mechanism for oscillations, and that such a loop goes through the FN. Third, the report of a familial disorder causing microsaccadic oscillations and limb tremor (μSOLT) is unlikely to be explained by a selective slowing of axonal conduction in just the feedback neurons, because subjects may show oscillations of different amplitudes, but their frequency is mostly constant.

A new model of saccadic oscillations

We recently proposed a new model for saccadic oscillations based on a more detailed representation of the known anatomy of the premotor circuitry for generating saccades, and on the physiological phenomenon of post-inhibitory rebound (PIR) firing (Ramat et al., 2005a).

The premotor saccadic circuitry comprises three types of neurons: excitatory burst neurons (EBN, glutamatergic), inhibitory burst neurons (IBN, glycinergic), and omnipause neurons (OPN, glycinergic). The EBN are responsible for producing the burst of innervation that projects directly to ocular motor neurons and drives the saccade, and to the circuits comprising the ocular motor integrator for generating the tonic signal that holds positions of gaze. EBN are located in the paramedian pontine reticular formation (PPRF) for horizontal saccades (Strassman et al., 1986a) and in the rostral interstitial nucleus of the medial longitudinal fasciculus (riMLF) for vertical and torsional saccades (Horn and Buttner-Ennever, 1998). Horizontal IBN innervate the contralateral abducens nucleus and the contralateral IBN and EBN (Strassman et al., 1986b). Horizontal IBN lie in the medullary reticular formation (medRF), while vertical and torsional IBN lie in the interstitial nucleus of Cajal (INC) (Izawa et al., 2007).

Anatomical studies of the projections of both the inhibitory and the EBN (Strassman et al., 1986b; Buttner-Ennever and Buttner, 1992) show that these neurons are interconnected across the midline and form two positive feedback loops. Since positive feedback can lead to instability, this network of brainstem neurons has the potential to oscillate and has led to a new model of saccadic oscillations (Ramat et al., 2005a). For this model to reproduce the different patterns of saccadic oscillations (Fig. 1, A), we implemented the detailed circuitry of premotor burst neurons (Fig. 1, B) and included the phenomenon of PIR (Enderle and Engelken, 1995). PIR is a property of cell membranes (Perez-Reyes, 2003) that causes the firing of one or more action potentials at the offset of inhibition, due to a particular class of voltage-dependent calcium channels, the T-type calcium channels, that activate at a low membrane potential (Huguenard, 1996; Perez-Reyes, 2003). Recently, Miura and Optican developed a detailed membrane model of burst neurons including T-channels and showed how they may produce rebound depolarization when the OPN are inactivated (Miura and Optican, 2006). A lumped model of PIR behaviour can be obtained to a first approximation using a burst neuron membrane model as illustrated in Fig. 1, B, right, in which a feed forward pathway includes a high-pass filter that causes adaptation (Ramat et al., 2005a). The time constant of adaptation (aTc) determines the duration of PIR; the gain of the adaptation (aGain) determines its amplitude.

The OPN lie in the raphe interpositus nucleus (RIP) and normally inhibit premotor burst neurons bilaterally in both the pons and the midbrain, except when a saccade is called for. Yet, recent studies involving chemical lesions of the RIP (Kaneko, 1996; Soetedjo et al., 2002) have shown that their role is not that of a mere switch enabling or disabling saccades, since their inactivation caused saccades to be slower, although they remained accurate and had a normal latency. It was therefore suggested that the OPN may act as a modulator of the activity of burst neurons and that their incomplete inhibition may be the cause of pathological saccadic oscillations (Ramat et al., 2005b; Shaikh et al., 2007). The OPN use glycine as their neurotransmitter, which acts on burst neurons through two mechanisms: inhibition of strychnine sensitive receptors and activation of NMDA channels (Miura and Optican, 2006).

Experimental results

We recently recorded a second patient, a 57-year-old woman with a surgical resection of a midline cerebellar tumour as a child, producing a midline defect (Fig. 2, CT scan at left) that almost certainly destroyed the posterior portion of the FN. She had marked hypermetria of horizontal saccades (Fig. 2, right top panel), typical of a fastigial nucleus lesion. During large vertical saccades (Fig. 2, right bottom panel), there were small, high-frequency oscillations on the horizontal velocity record. The findings in this patient corroborate the hypothesis that the delay in the feedback loop is not the principal mechanism causing oscillations.

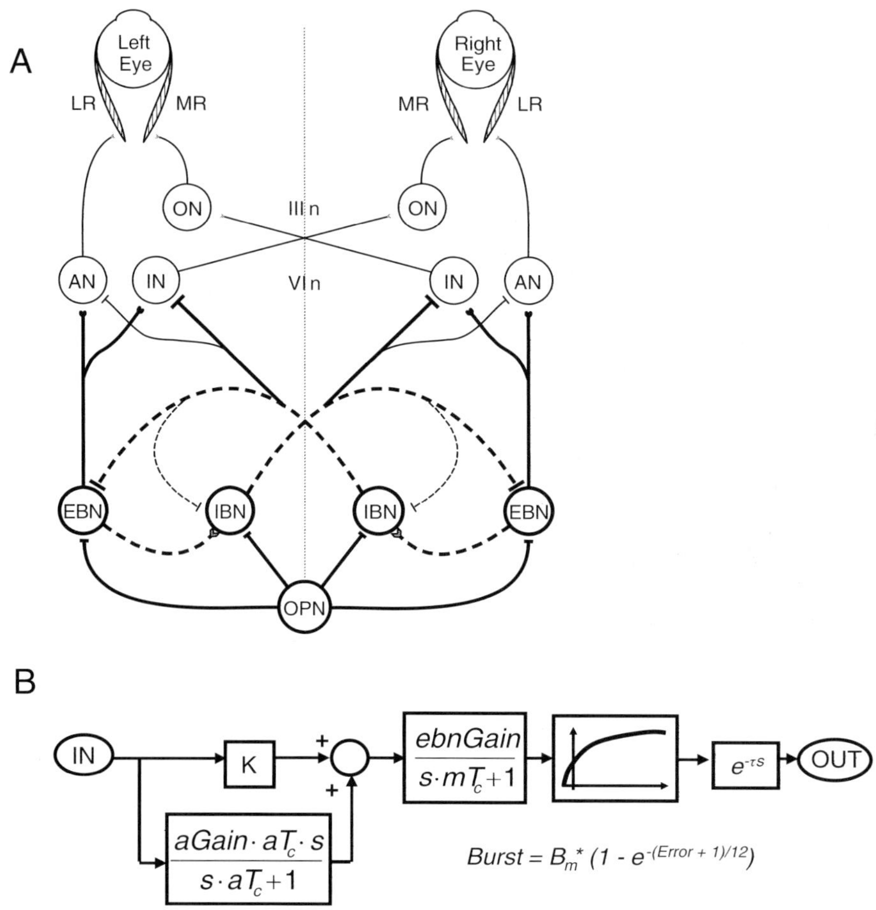

Fig. 1. Panel A: Brainstem circuit for generating saccades. Projections with flat ending are inhibitory, the others excitatory. ML, medial recti; LR, lateral recti; VIn, ipsilateral abducens nucleus; AN, abducens neuron, IN, interneuron projecting to contralateral third nerve (IIIn), ON, oculomotor neuron. EBN, excitatory burst neurons, IBN, inhibitory burst neurons. OPN, omnipause neurons. Panel B: Lumped model of premotor burst neuron used in our model showing adaptation and producing PIR. The high-pass parallel pathway produces PIR with amplitude determined by aGain and duration determined by aTc. The cell membrane then shows a low pass filter (mTc), a nonlinearity (Zee and Robinson, 1979) and a synaptic delay of less than 1 ms. (Adapted with permission from Ramat et al., 2005a.)

Second, we recently recorded a mother and a daughter who each had microsaccadic flutter: an almost continuous saccadic oscillation of the eyes occurring without intersaccadic interval and usually around all three axes (Shaikh et al., 2007). Eye oscillations were also accompanied by limb tremor, and it was suggested that both the eye and the limb oscillations reflected reduced (glycinergic in the case of the eyes) inhibition of an inherently unstable neural circuit (Shaikh et al., 2007).

We recorded the eye movements of the mother with search coils and during fixation she had a virtually continuous saccadic oscillation of both eyes (20.3 ± 0.9 Hz and 7.6 ± 2.2 deg/s horizontal, 21.0 ± 1.0 Hz and 10.3 ± 2.7 deg/s vertical) at about 20 Hz and about 0.1 deg. The first panel in Fig. 3 shows data recorded during fixation followed by a 5 deg saccade after which the oscillations were transiently suppressed for about 200 ms. During large vertical saccades (40 deg), horizontal saccadic oscillations occurred with amplitudes reaching 1.8 degrees (156.2 ± 34.8 deg/s and 21.5 ± 2.3 Hz, as in Fig. 3, second panel). Eye closure and blinks evoked large horizontal oscillations (95 ± 16 deg/s) of significantly lower frequency (18.2 ± 0.9 Hz), as

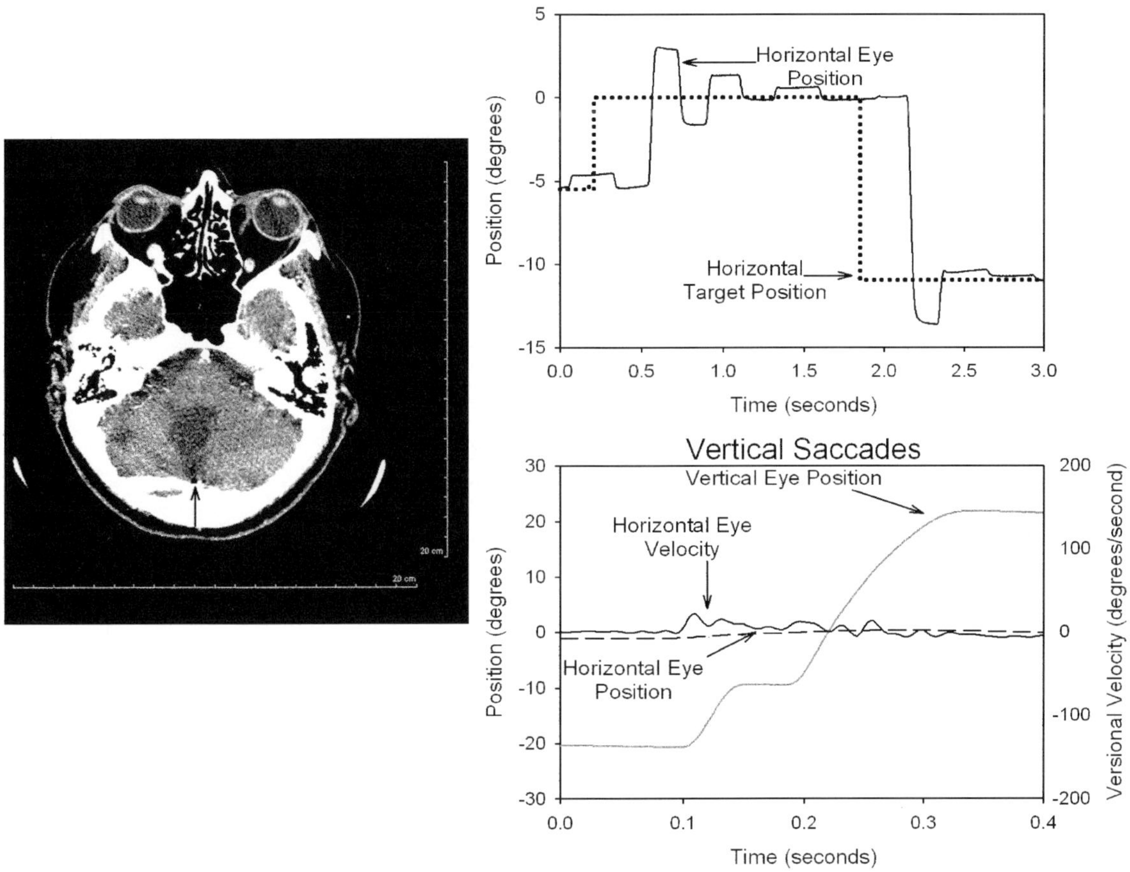

Fig. 2. Finding in a 57-year-old woman, who had undergone resection of a cerebellar tumour as a child leaving a midline defect (indicated by arrow on CT scan at left). She had marked horizontal hypermetria (right, top panel) typical of a fastigial nucleus lesion. When she made large vertical saccades (right, bottom panel), small, high-frequency oscillations were evident on the horizontal velocity record (scale at right).

shown in Fig. 3 third panel. The amplitude of both horizontal and vertical oscillations varied significantly without proportionally affecting their frequency.

The behaviour of this patient was simulated using the lumped model for saccadic oscillations described above, with a population of five burst neurons for each of the horizontal and vertical saccade generators. The values of the parameters representing the membrane characteristics of both EBN and IBN were randomly drawn from a normal distribution with a specified mean and a standard deviation of 10% of such mean. The results of the simulations corresponding to the three conditions shown in Fig. 3 are shown in Fig. 4. The temporary suppression of oscillations at the end of the small amplitude saccade in the left column could be simulated based upon an additional transient inhibition provided by contralateral IBN acting as a "choke" signal. To simulate the patient data we reduced the inhibitory effect of OPN firing on EBN and IBN, reproducing the effect of a pathologically reduced sensitivity to glycinergic inhibition, and increased the strength of PIR (aGain), simulating the increased membrane excitability due to the reduced hyperpolarization of the burst neurons (Shaikh et al., 2007). Reducing the action of glycine in the premotor circuit produces a change in the membrane behaviour of burst neurons that increases the

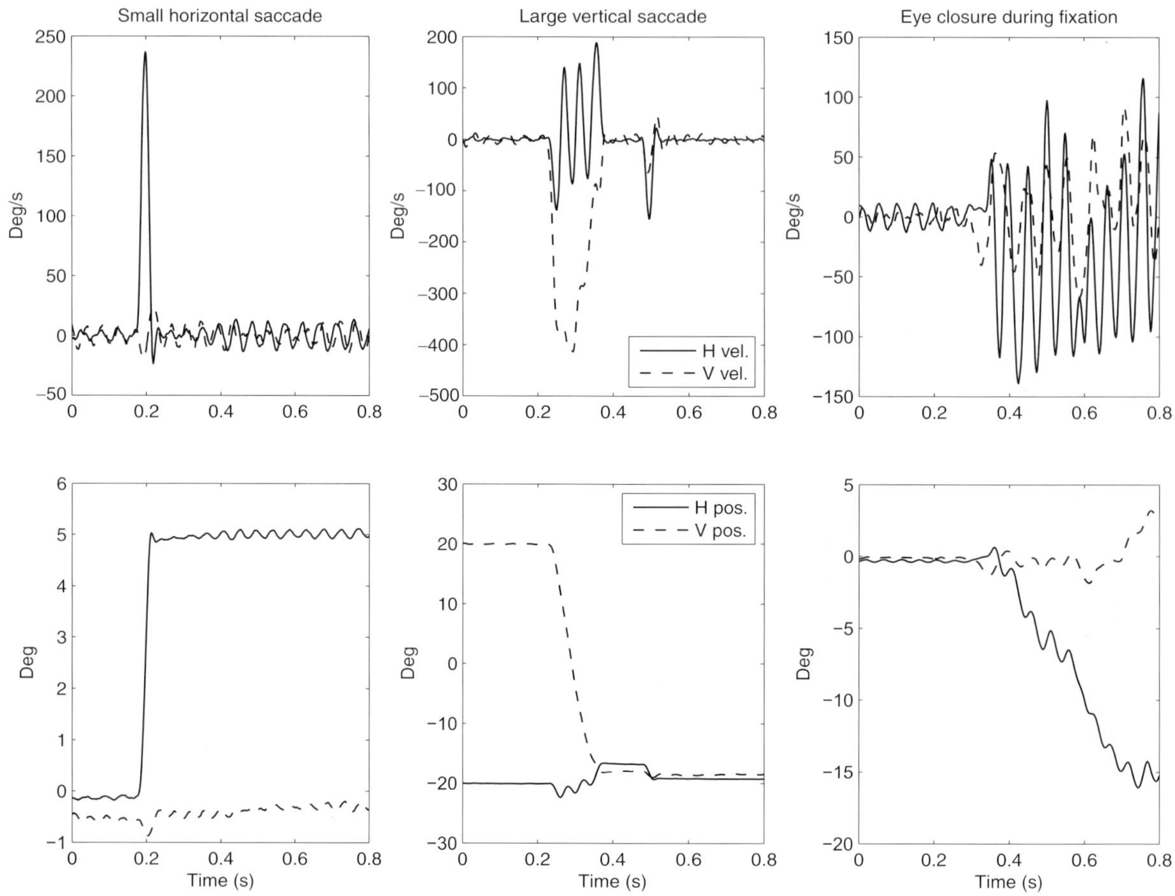

Fig. 3. Representative eye movements recorded in one patient affected by μSOLT. Top row velocity traces, bottom row position traces. Each column represents data recorded in response to different conditions. Left: 5 deg, horizontal saccade; central: 40 deg, vertical saccade; right: eye closure during fixation.

instability of the circuit. Consequently the circuit oscillates during fixation, in spite of some residual OPN firing.

Discussion

Here we have shown further simulations of a recent model for saccadic oscillations (Ramat et al., 2005a; Ramat et al., 2007) which account for the experimental findings in a patient with familial microsaccadic oscillations (Shaikh et al., 2007). The key features of this model are the circuitry representing the detailed interconnections between premotor neurons and the membrane properties of the excitatory and IBN in the brainstem. The model assumes that these high-gain cells show PIR, which causes them to fire a burst of action potentials when their inhibition is removed, even if there is no error signal to trigger a saccade. PIR has been demonstrated in various cells of the nervous system (Perez-Reyes, 2003), and likely produces the high-frequency burst that drives saccades in normal subjects (Enderle and Engelken, 1995; Miura and Optican, 2006). In our model, besides contributing to the drive of normal saccades, PIR is key for explaining saccadic oscillations in normal subjects and in patients (Ramat et al., 2007; Shaikh et al., 2007). Finally, these results further support the idea that

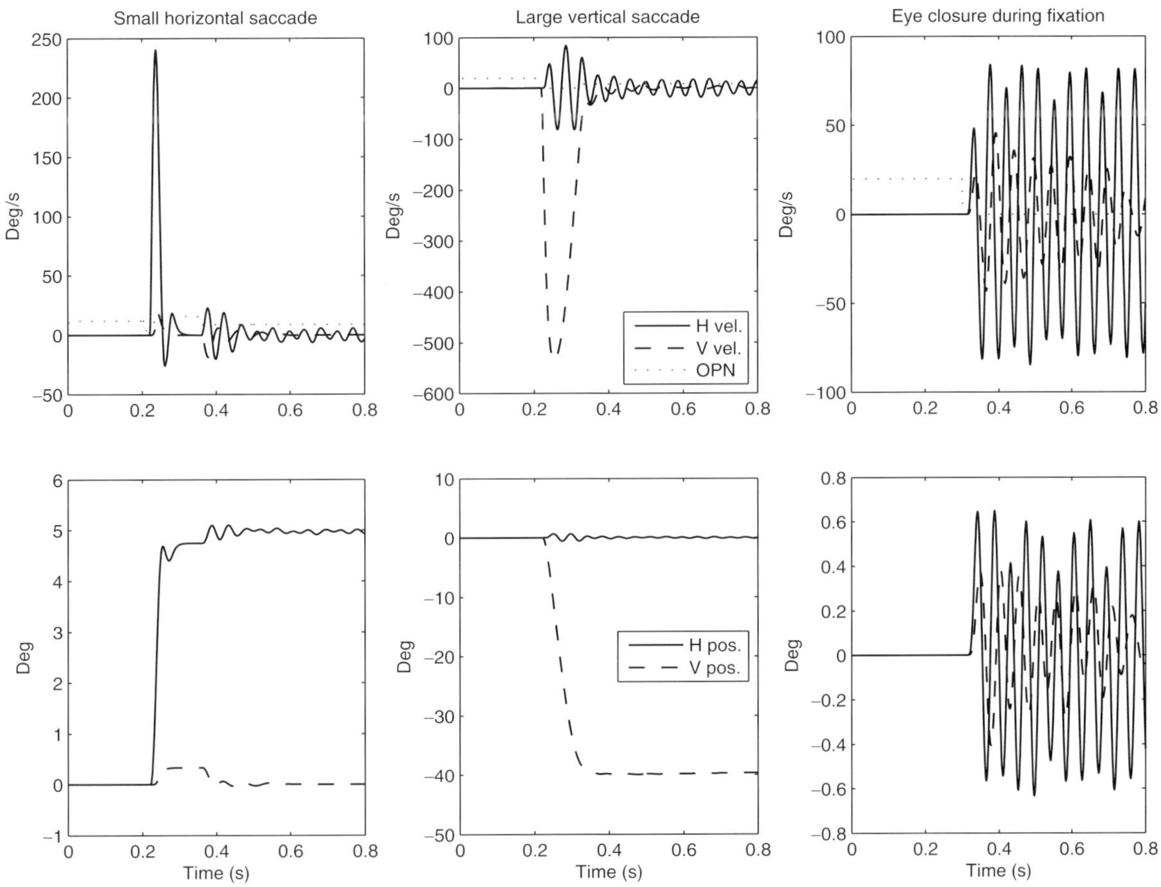

Fig. 4. Model simulations of the patient data in the conditions represented in Fig. 3. The amplitude of the oscillations is mainly determined by the gain of PIR (aGain) and their frequency by the membrane time constants (aTc, mTc). The reduced sensitivity of strychnine-sensitive channels was simulated by lowering the gain of OPN projections onto burst neurons and increasing the gain of PIR in those cells. All patient simulations were performed using the same set of parameters.

pathological saccadic oscillations are caused by a reduction of inhibition in this inherently unstable circuit, which may be due to reduced sensitivity of glycine channels on burst neurons.

References

Ashe, J., Hain, T.C., Zee, D.S. and Schatz, N.J. (1991) Microsaccadic flutter. Brain, 114(Pt 1B): 461–472.

Buttner-Ennever, J.A. and Buttner, U. (1992) Neuroanatomy of the ocular motor pathways. Baillieres Clin. Neurol., 1: 263–287.

Enderle, J.D. and Engelken, E.J. (1995) Simulation of oculomotor post-inhibitory rebound burst firing using a Hodgkin-Huxley model of a neuron. Biomed. Sci. Instrum., 31: 53–58.

Hain, T.C., Zee, D.S. and Mordes, M. (1986) Blink-induced saccadic oscillations. Ann. Neurol., 19: 299–301.

Horn, A.K. and Buttner-Ennever, J.A. (1998) Premotor neurons for vertical eye movements in the rostral mesencephalon of monkey and human: histologic identification by parvalbumin immunostaining. J. Comp. Neurol., 392: 413–427.

Huguenard, J.R. (1996) Low-threshold calcium currents in central nervous system neurons. Annu. Rev. Physiol., 58: 329–348.

Izawa, Y., Sugiuchi, Y. and Shinoda, Y. (2007) Neural organization of the pathways from the superior colliculus to trochlear motoneurons. J. Neurophysiol., 97: 3696–3712.

Kaneko, C.R. (1996) Effect of ibotenic acid lesions of the omnipause neurons on saccadic eye movements in rhesus macaques. J. Neurophysiol., 75: 2229–2242.

Lefevre, P., Quaia, C. and Optican, L.M. (1998) Distributed model of control of saccades by superior colliculus and cerebellum. Neural Netw., 11: 1175–1190.

Leigh, R.J. and Zee, D.S. (2006) The Neurology of Eye Movements. Oxford Press, New York.

Miura, K. and Optican, L.M. (2006) Membrane channel properties of premotor excitatory burst neurons may underlie saccade slowing after lesions of omnipause neurons. J. Comput. Neurosci., 20: 25–41.

Perez-Reyes, E. (2003) Molecular physiology of low-voltage-activated t-type calcium channels. Physiol. Rev., 83: 117–161.

Ramat, S., Leigh, R.J., Zee, D.S. and Optican, L.M. (2005a) Ocular oscillations generated by coupling of brainstem excitatory and inhibitory saccadic burst neurons. Exp. Brain Res., 160: 89–106.

Ramat, S., Leigh, R.J., Zee, D.S. and Optican, L.M. (2007) What clinical disorders tell us about the neural control of saccadic eye movements. Brain, 130: 10–35.

Ramat, S., Somers, J.T., Das, V.E. and Leigh, R.J. (1999) Conjugate ocular oscillations during shifts of the direction and depth of visual fixation. Invest. Ophthalmol. Vis. Sci., 40: 1681–1686.

Ramat, S., Zee, D.S., Leigh, R.J. and Optican, L.M. (2005b) Familial microsaccadic oscillations may be due to alterations in the inhibitory premotor circuit. Abstract Viewer/Itinerary Planner, Program No. 475.15. Society for Neuroscience, Washington, DC.

Shaikh, A.G., Miura, K., Optican, L.M., Ramat, S., Leigh, R.J. and Zee, D.S. (2007) A new familial disease of saccadic oscillations and limb tremor provides clues to mechanisms of common tremor disorders. Brain, 130: 3020–3031.

Shults, W.T., Stark, L., Hoyt, W.F. and Ochs, A.L. (1977) Normal saccadic structure of voluntary nystagmus. Arch. Ophthalmol., 95: 1399–1404.

Soetedjo, R., Kaneko, C.R. and Fuchs, A.F. (2002) Evidence that the superior colliculus participates in the feedback control of saccadic eye movements. J. Neurophysiol., 87: 679–695.

Strassman, A., Highstein, S.M. and McCrea, R.A. (1986a) Anatomy and physiology of saccadic burst neurons in the alert squirrel monkey. I. Excitatory burst neurons. J. Comp. Neurol., 249: 337–357.

Strassman, A., Highstein, S.M. and McCrea, R.A. (1986b) Anatomy and physiology of saccadic burst neurons in the alert squirrel monkey. II. Inhibitory burst neurons. J. Comp. Neurol., 249: 358–380.

Wong, A.M., Musallam, S., Tomlinson, R.D., Shannon, P. and Sharpe, J.A. (2001) Opsoclonus in three dimensions: oculographic, neuropathologic and modelling correlates. J. Neurol. Sci., 189: 71–81.

Zee, D.S., Fitzgibbon, E.J. and Optican, L.M. (1992) Saccade-vergence interactions in humans. J. Neurophysiol., 68: 1624–1641.

Zee, D.S. and Robinson, D.A. (1979) A hypothetical explanation of saccadic oscillations. Ann. Neurol., 5: 405–414.

C. Kennard & R.J. Leigh (Eds.)
Progress in Brain Research, Vol. 171
ISSN 0079-6123

CHAPTER 2.7

Dynamics of saccadic oscillations

Richard A. Clement[1,*], David S. Broomhead[2] and Ozgur E. Akman[3]

[1]*Visual Sciences Unit, Institute of Child Health, University College London, 30 Guilford Street, London WC1N 1EH, UK*
[2]*Department of Mathematics, University of Manchester, P.O. Box 88, Manchester M60 1QD, UK*
[3]*Darwin Building, Centre for Systems Biology, The University of Edinburgh, Kings Buildings, Mayfield Road, Edinburgh EH9 3JR, UK*

Abstract: The brainstem circuitry underlying saccades is symmetrical with respect to the midline. The oculomotor behaviour generated by the circuitry depends on a combination of signals passed along fibre tracts and less easily identifiable connections, such as those across the midline. The midline crossing connections are often affected by developmental disorders which give rise to unstable eye movements (see J. Jen, this volume). The connections at the levels of the colliculus, pause cells, and neural integrator generate different dynamical mechanisms for the development of instabilities, which can be identified in eye movement recordings using phase space analysis techniques.

Keywords: fixed point; phase plane; hopf bifurcation; saccade; nystagmus

Introduction

Disorders of horizontal fixation

The normal saccadic system behaves in a remarkably machine-like fashion, producing 3–4 saccades during every second of our waking lives, all with similar main sequence characteristics. This consistency makes it hard to explain the disorders of normal saccadic behaviour which are found clinically (Ramat et al., 2007). For example, the visual systems of albinos typically show a number of distinctive features which include an underdeveloped retina, abnormal routing of the nerve fibres in the brain, nystagmus, and strabismus. Despite these distinctive features, the eye movement disorders which can be recorded from the patients are relatively varied (Collewijn et al., 1985; Timms et al., 2006), as illustrated in Fig. 1A.

Midline circuitry for horizontal saccades

The saccadic system provides a good starting point for determining how oculomotor disorders arise, because of the extensive knowledge of the neurophysiology of the system (Scudder et al., 2002; Sparks, 2002). The mechanics of the oculomotor plant are heavily damped so that a burst of innervation is required to make a fast eye movement and a tonic change of innervation is required to counter the spring stiffness of the plant. The burst of innervation is delivered by burst cells and a neural integrator converts the burst to a step change in innervation. In addition, the pause cells fire steadily when the eye is fixating the target, and inhibit the burst cells. The behaviour of the normal saccadic system can be characterised by the

*Corresponding author. Tel.: +44 (0)20 7242 9789, Ext. 0284; Fax: +44 207 831 0488; E-mail: r.clement@ich.ucl.ac.uk

DOI: 10.1016/S0079-6123(08)00617-1

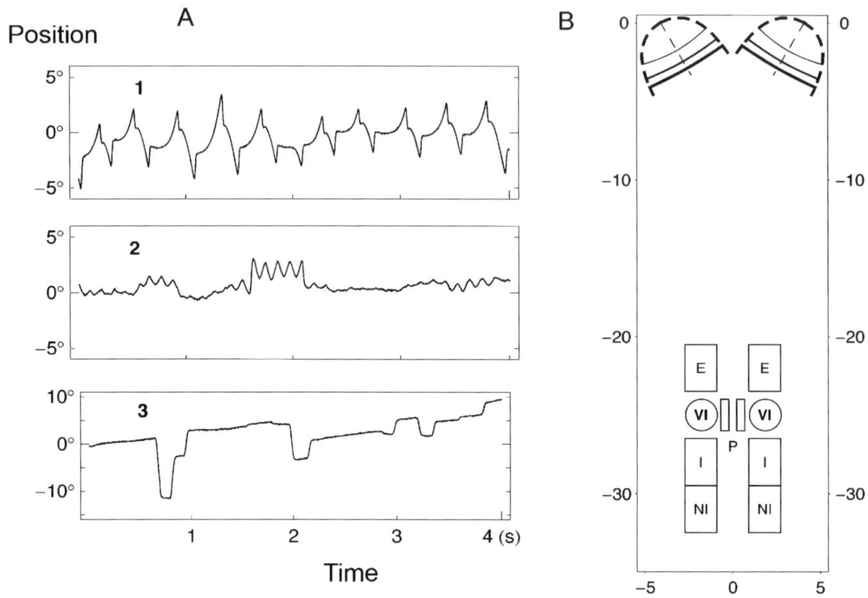

Fig. 1. (A) Three examples of horizontal disorders of fixation recorded from subjects with albinism. (1) Nystagmus, (2) intermittent fine pendular nystagmus, and (3) macrosaccadic oscillations (see Timms et al., 2006 for details of the stimulus and recording technique). (B) Schematic diagram of the relative anatomical proximity of some of the classes of neurons responsible for horizontal saccadic eye movements.

relationships between saccade amplitude and duration and peak velocity, which are referred to as the main sequence.

One noticeable feature of abnormal fixation is that horizontal disorders of fixation are much more common than vertical disorders. The most obvious difference between the horizontal and vertical eye movement systems is that the horizontal system is symmetric with respect to the midline (see Fig. 1B). This requires coordination of signals across the midline via fibre tracts and commisures, and also by fibres crossing the midline. The latter may be more vulnerable to because of the guidance required to ensure successful completion of their connections. Following from this hypothesis, one can go on to ask if different types of oculomotor oscillation are related to failure of different midline crossing connections. This question is addressed with respect to the disorders illustrated in Fig. 1A. Before examining the different disorders the concepts needed to characterise instability are introduced.

Instabilities of fixation

Fixed points

A useful tool for characterising the behaviour of a system is provided by the geometrical technique according to which the state of the system is described by a point in a state space and successive states trace out a curve in state space, known as a trajectory. A trajectory which arrives at a point corresponding to an equilibrium state of the system will stay at the point, which is referred to as a fixed point. In practice, while a trajectory in the neighbourhood of a stable fixed point will be guided towards the point, a trajectory near an unstable fixed point will be guided towards the point in some directions, and repelled from it in other directions in state space. In the context of the oculomotor system, steady fixation corresponds to a stable fixed point, and drifts and oscillations occur when the fixed point becomes unstable. Although it is known that in the case of the oculomotor disorder of nystagmus the state space of fixation is at least three dimensional

(Abadi et al., 1997), the different types of oculomotor instabilities can be revealed by a projection of the state space trajectories into a two dimensional phase plane plot of velocity against position (Dell'Osso et al., 1992). An example of the phase plane trajectory of a single saccade is shown in Fig. 2A. This tool will be use to illustrate how the different oculomotor disorders shown in Fig. 1A can emerge from the normal behaviour of the system.

Neural integrator drift

Stable fixation requires that the eye comes to rest at a particular position and does not move away from that position. On a plot of eye velocity against eye position all stable fixation points will lie on a horizontal line through the origin (Goldman et al., 2002). This relationship has been used to recover the parameters of the neural integrator. The behaviour of the neural integrator can be modelled by a first order, linear differential equation:

$$\frac{\mathrm{d}p}{\mathrm{d}t} = \frac{p}{\lambda} + v_{\mathrm{bias}}$$

where p is the eye position signal, λ the time constant of the integrator, and v_{bias} the constant velocity bias on the input to the integrator. It follows from this equation that the values of both the time constant and the velocity bias can be recovered from a least squares fit to a plot of inter-saccadic velocities against inter-saccadic eye positions (Goldman et al., 2002).

Phase plane analysis can be adapted to identify that the drift shown in Fig. 1A3 is due to an abnormal bias in the neural integrator. If only the position and velocity signals from the intervals between saccades are plotted, then the data is expected to lie on a straight line. The slope of the line is equal to the time constant of the integrator and the offset of the line is equal to the bias on the integrator. The integrator is unstable if the slope is negative and the offset is non-zero.

Burst cell oscillations

The fine pendular nystagmus shown in Fig. 1A2 is an example of a microsaccadic oscillation. Such oscillations have been identified as saccadic

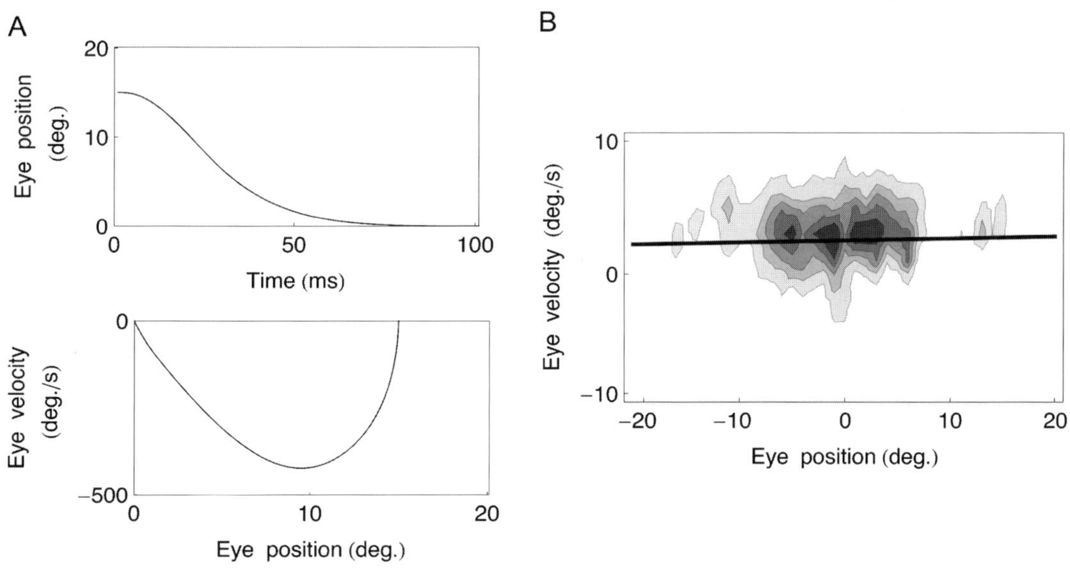

Fig. 2. (A) An example of the phase plane representation of a single saccade. The time course of the saccade is plotted above and the trajectory of the saccade in the phase plane is plotted below. (B) Histogram of the frequency of position and velocity combinations from the phase plane which occur during inter-saccadic intervals. These points were identified by having a velocity of less than 10 degrees/s. The best linear fit to the data is shown by the thick line.

because the particular duration and peak velocity of each movement lie on an extension of the main sequence for small saccades. Both modelling (Ashe et al., 1991; Ramat et al., 2005; Laptev et al., 2006) and experimental results (Ramat et al., 2005) have shown that incomplete inhibition of the burst cells by the pause cells results in high-frequency oscillations.

As illustrated in Fig. 3, the oscillations correspond to closed loops in the phase plane and the dynamical mechanism by which the instabilities develop can be shown to involve a Hopf bifurcation (Laptev et al., 2006). One of the definitive features of the Hopf bifurcation, that distinguishes it from other oscillation generating mechanisms, is that the amplitude of the oscillation grows with the square of the distance from the fixed point. This behaviour can be clearly seen in the simulations plotted in Fig. 3.

Splitting of the pause cells

The final example of an oculomotor instability involves the changes with gaze angle in a nystagmus waveform similar to that shown in Fig. 1A1. The trajectories for nystagmus cycles at different gaze angles can all be plotted on the same phase plane diagram as shown in Fig. 4, which also illustrates the changes in the waveforms. Also plotted in Fig. 4 are the locations of the fixed

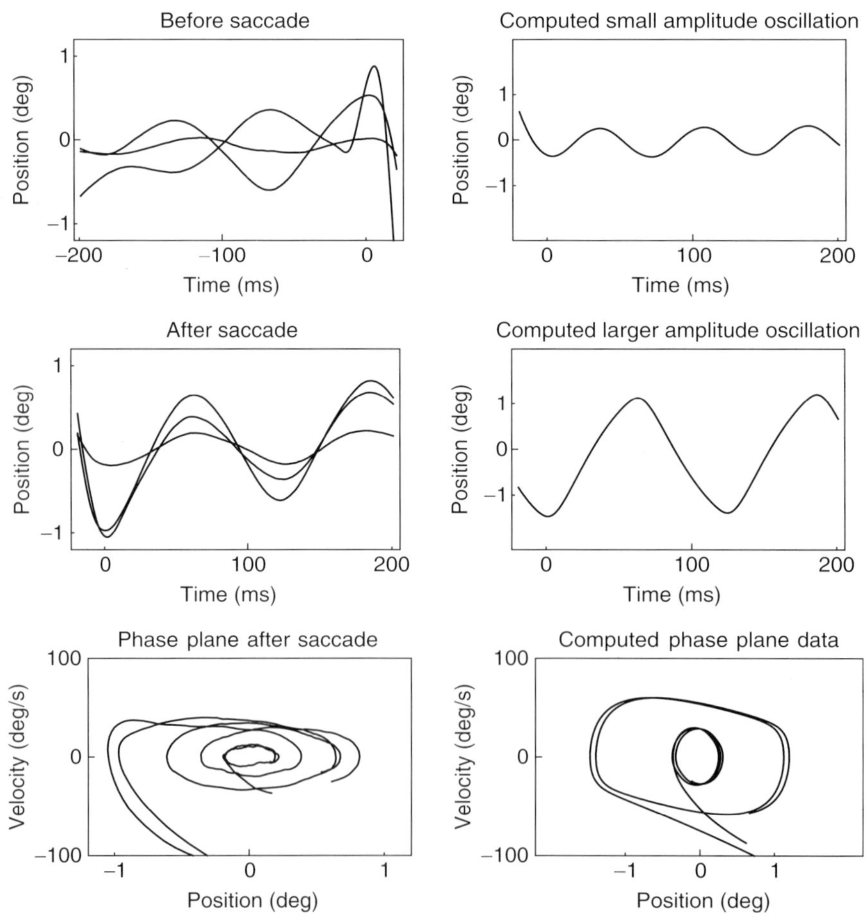

Fig. 3. Measured and modelled microsaccadic oscillations. Both sets of oscillations have their phase reset by voluntary saccades. The simulated waveforms become oscillatory through a Hopf bifurcation mechanism, and in accordance with this mechanism become increasingly sinusoidal as their amplitude decreases.

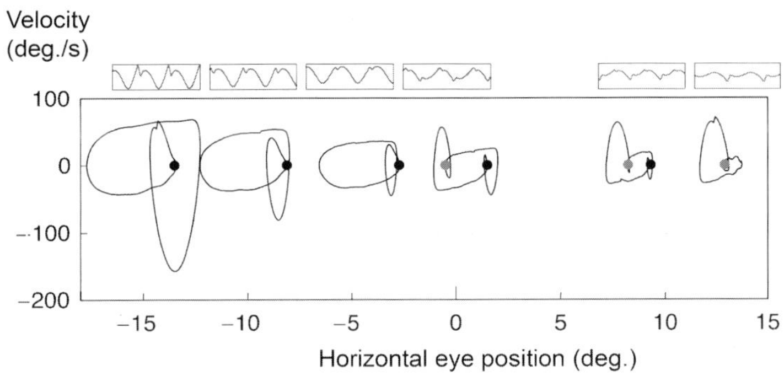

Fig. 4. Examples of nystagmus waveforms recorded from a subject with his head supported by a chin rest. Thirty seconds of fixation was recorded at each target position, which lay in the range from 15 degrees left to 15 degrees right and were located 5 degrees apart. Consistent oscillations were obtained from all target directions except 5 degrees right. Examples of the oscillations are plotted in boxes along the top of the figure. Each box encloses 1 s of eye movement recording and the vertical extent of the box corresponds to 6 degrees of gaze angle. In the phase plane representations the system follows a clockwise path round the loops, drifting away from the fixed point and being brought back by a small saccade. The fixed points of the leftward directed saccades are plotted in black and those of the rightward directed saccades are shown in grey. It is apparent that in the straight ahead direction the two fixed points do not coincide.

points for each waveform. When the target is between the straight ahead direction and 10 degrees to the right of the subject there are two fixed points. If it is assumed that the pause cells fire at the end of a saccade the presence of two fixed points implies that with a single target direction the pause cells can resume firing at two different fixation directions.

Conclusion

A number of horizontal oculomotor disorders appear to be related to failure of the normal development of midline crossing connections in the brainstem. Targets close to the naso-temporal boundary of the retina are represented on both the left and right halves of the colliculus and mutual inhibition between the halves is required to ensure that accurate saccades are generated (van Opstal and van Gisbergen, 1989; Tabareau et al., 2007). Damage to these lateral connections has been shown to result in incessant saccades in the monkey (Carasig et al., 2006). Microsaccadic oscillations occur when activity cycles between the left and right excitatory burst cells. This cycling will only occur with nonlinear mutual inhibition between the burst cells. Although more speculative, it may be that early-onset nystagmus oscillations involve interactions between two groups of pause cells, one of which fires when the target is imaged at a point on the left of the fovea and one of which fires when a target is imaged on the right of the fovea.

The effects of failure of the normal development of midline crossing connections can be investigated experimentally using two recently developed techniques of nonlinear systems analysis. Initially, the experimenter will not know the dimension of the state space of the abnormal fixation system and will have to assume that state space has at least three dimensions since it is known that early-onset nystagmus has at least this dimensionality (Abadi et al., 1997; Akman et al., 2006). A straightforward extension of the phase plane to three dimensions can be effected by plotting not just the position and velocity but also the acceleration of the eye. The fixed points of the system can be found directly from the eye movement recordings by using a technique of nonlinear dynamics which has been adapted for eye movement data by Theodorou and Clement (2007). Once the fixed points have been found, the dimensionality and stability of the system at the fixed points can be estimated by a local principal component method applied to eye movement data by Abadi et al. (1997) and Akman et al. (2006). These techniques enable direct comparison of experimentally measured and simulated eye movement disorders. They can be used to test if the two sets of data have the same number of fixed points and the same dimensionality at each fixed point (Akman et al., 2006).

Once an oscillation is established it is not always possible to establish how it developed. The nonlinear systems approach is based on investigating how the behaviour of the eye movement system changes as a parameter is continuously varied. In many cases, once a patient develops an oscillation they cannot eliminate it by changing a parameter of the system such as gaze angle. So for investigating clinical conditions, the nonlinear dynamics approach will have to be adapted to consist of investigating the dynamics of a range of subjects with a given condition. For example, the hypothesis of a Hopf bifurcation underlying microsaccadic oscillations could be tested by pooling data from many subjects with different sized oscillations to see if the oscillations are more sinusoidal at smaller amplitudes.

Acknowledgement

The approach described in this paper was developed with the help of many contributions from Richard Abadi.

References

Abadi, R.V., Broomhead, D.S., Clement, R.A., Whittle, J.P. and Worfolk, R. (1997) Dynamical systems analysis: a new method of analysing congenital nystagmus waveforms. Exp. Brain Res., 117: 355–361.

Akman, O.E., Broomhead, D.S., Clement, R.A. and Abadi, R.V. (2006) Nonlinear time series analysis of jerk congenital nystagmus. J. Comput. Neurosci., 21: 153–170.

Ashe, J., Hain, T.C., Zee, D.S. and Schatz, N.J. (1991) Microsaccadic flutter. Brain, 114: 461–472.

Carasig, D., Kush, P., Fucito, M., Ramcharan, E. and Gnadt, J.W. (2006) Irrepressible saccades frm a tectal lesion in a Rhesus monkey. Vision Res., 46: 1161–1169.

Collewijn, H., Apkarian, P. and Spekreijse, H. (1985) The oculomotor behaviour of human albinos. Brain, 108: 1–28.

Dell'Osso, L.F., Van der Steen, J., Collewijn., H. and Steinman, R.M. (1992) Foveation dynamics in congenital nystagmus. I. Fixation. Doc. Ophthalmol., 79: 1–23.

Goldman, M.S., Kaneko, C.R.S., Major, G., Aksay, E., Tank, D.W. and Seung, H.S. (2002) Linear regression of eye velocity on eye position and head velocity suggests a common oculomotor neural integrator. J. Neurophysiol., 88: 659–665.

Laptev, D., Akman, O.E. and Clement, R.A. (2006) Stability of the saccadic system. Biol. Cybern., 95: 281–287.

Ramat, S., Leigh, R.J., Zee, D.S. and Optican, L.M. (2005) Ocular oscillations generated by coupling of brainstem excitatory and inhibitory saccadic burst neurons. Exp. Brain Res., 160: 89–106.

Ramat, S., Leigh, R.J., Zee, D.S. and Optican, L.M. (2007) What clinical disorders tell us about the neural control of saccadic eye movements. Brain, 130: 10–35.

Scudder, C.A., Kaneko, C.R.S. and Fuchs, A.F. (2002) The brainstem burst generator for saccadic eye movements. A modern synthesis. Exp. Brain Res., 142: 439–462.

Sparks, D.L. (2002) The brainstem control of saccadic eye movements. Nat. Neurosci. Rev., 3: 952–964.

Tabareau, N., Bennequin, D., Berthoz, A., Slotine, J.-J. and Giraud, B. (2007) Geometry of the superior colliculus mapping and efficient oculomotor computation. Biol. Cybern., 97: 279–292.

Theodorou, M. and Clement, R.A. (2007) Fixed point analysis of nystagmus. J. Neurosci. Methods, 161: 134–141.

Timms, C., Thompson, D., Russell-Eggitt, I. and Clement, R.A. (2006) Saccadic instabilities in albinism without nystagmus. Exp. Brain Res., 175: 45–49.

van Opstal, A.J. and van Gisbergen, J.A.M. (1989) A nonlinear model for collicular spatial interactions underlying the metrical properties of electrically elicited saccades. Biol. Cybern., 60: 171–183.

C. Kennard & R.J. Leigh (Eds.)
Progress in Brain Research, Vol. 171
ISSN 0079-6123

CHAPTER 2.8

Effects of failure of development of crossing brainstem pathways on ocular motor control

Joanna C. Jen*

Department of Neurology, UCLA School of Medicine, 710 Westwood Plaza, Los Angeles, CA, USA

Abstract: Horizontal gaze palsy with progressive scoliosis (HGPPS) is a rare, inherited disorder characterized by a congenital absence of conjugate horizontal eye movement with progressive scoliosis developing in childhood in patients who are otherwise neurologically intact. Detailed structural neuroimaging studies demonstrated abducens nerves and the absence of fibrosis in the extraocular muscles, and a remarkably dysmorphic hindbrain, with hypoplasia and flattened, butterfly-like medulla with deep midline cleft. Diffusion tensor imaging further demonstrated a widespread lack of crossing fibres in the brainstem, supported by evoked potential studies showing uncrossed descending motor and ascending sensory pathways in HGPPS patients. In these patients, we identified homozygous or compound heterozygous mutations in a gene we named *ROBO3*, which shares homology with evolutionarily conserved roundabout genes that are important in neural and vascular wiring. Removal of *Robo3* in mice led to the absence of commissural crossing throughout the spinal cord and hindbrain (and death soon after birth). Therefore, ROBO3 is required for hindbrain axon midline crossing and morphogenesis in both human and mouse. We continue to investigate how ROBO3 mutations lead to massive miswiring in the hindbrain and disruption of conjugate horizontal gaze. Elucidation of the full extent of the anatomical abnormalities in HGPPS awaits improved neuroimaging techniques and detailed pathological studies.

Keywords: horizontal gaze palsy; scoliosis; ROBO3 mutations; roundabout; brainstem maldevelopment; disrupted midline crossing

Introduction

We came to honour Dr. Jean Büttner-Ennever, who has been a pioneer and a giant in elucidating the functional anatomical organization of ocular motor control (Büttner-Ennever, 2007). My laboratory has focused on inherited disorders of imbalance and eye movement control to understand the genetic programmes governing normal neurodevelopment. One of the conditions that we have been studying is horizontal gaze palsy with progressive scoliosis (HGPPS), which is a rare familial disorder characterized by the absence of conjugate horizontal eye movement and progressive scoliosis developing in childhood that often requires surgical intervention early in life. Affected individuals demonstrate preserved vertical gaze and convergence with no other focal neurological deficits. Understanding the molecular and anatomical basis underlying HGPPS may provide insight into the neurodevelopment of

*Corresponding author. Tel.: +310-825-5910; Fax: +310-206-1513; E-mail: jjen@ucla.edu

DOI: 10.1016/S0079-6123(08)00618-3

horizontal eye movement control. Our efforts led to the surprising discovery of a gene that is important in brainstem axon pathway crossing and morphogenesis.

Clinical features of HGPPS

HGPPS was first described in a Chinese family in Jamaica (Crisfield, 1974) and two consanguineous families on the Greek island of Crete (Dretakis and Kondoyannis, 1974). Sharpe et al. (1975) provided detailed descriptions of the eye movement abnormalities in four affected siblings of the previously described Chinese family who moved to Toronto. The patients had absent horizontal gaze, pendular nystagmus, full excursion in vertical gaze, substitution of convergence for attempted lateral gaze without globe retraction, and full abduction by forced duction, suggesting intact midbrain ocular motor function but possible involvement of the paramedian pontine reticular formation (Sharpe et al., 1975). Not all the patients in the earlier reports were examined during infancy to ascertain whether horizontal eye movement was intact. Although the clinical overlap between HGPPS and progressive external ophthalmoplegia raised the possibility of a mitochondrial cytopathy in HGPPS, the presence of consanguinity, the absence of scoliosis or horizontal gaze palsy in antecedent generations, and the equal involvement of both genders are strongly suggestive of an autosomal recessive inheritance pattern.

Genetic basis of HGPPS

We came across two consanguineous but unrelated families with HGPPS. The affected individuals in these families were noted to have no horizontal eye movement since birth, rather than a progressive limitation in horizontal gaze. Genome scanning identified a shared region on chromosome 11 that showed excess homozygosity in affected individuals but not unaffected relatives, suggesting that the disease gene lies in this region (Jen et al., 2002).

Through collaboration with Mac Bosley, Elizabeth Engle, and other clinicians all over the world, we recruited several additional consanguineous families with HGPPS that not only mapped to the same region on chromosome 11 but also helped narrow the candidate region that made candidate gene screening feasible. Annotated genes in the candidate region showed no pathogenic variants. We were finally able to identify in these patients homozygous mutations in a gene that we cloned and named *ROBO3* (Jen et al., 2004). Previously, only a short fragment towards the 3′ end of this gene had been annotated in the genome database as *RBIG*, for homology with a mouse gene *Rig1*, which had been noted to be over-expressed in the brainstem in mice in which the tumour suppressor *Rb* (*Retinoblastoma*) had been eliminated (Yuan et al., 1999). The gene product of *Rig1* is a member of the roundabout (robo) family of proteins that are transmembrane receptors first identified in *Drosophila* mutants in which the axons cross and re-cross the midline (Seeger et al., 1993). In *Drosophila*, Robo proteins bind to chemorepellent Slit to prevent axons from inappropriately re-crossing the midline and direct neuronal migration (Kidd et al., 1998; Brose et al., 1999; Wu et al., 1999).

We found multiple splice isoforms of *ROBO3*, with the longest open reading frame consisting of 28 exons and encoding a 1384 amino acid protein. The ROBO3 protein is predicted to contain an extracellular segment with five immunoglobulin-like domains and three fibronectin III-like domains, a transmembrane segment, and an intracellular segment with three cytoplasmic signalling motifs. We found distinct homozygous mutations in ROBO3 in affected individuals from different HGPPS families. The asymptomatic parents were all heterozygous carriers. We further confirmed that the mutations were not found in ethnically mixed control individuals. The mutations were missense, nonsense, splice site, and frame shift. Different mutations throughout *ROBO3* leading to the same clinical manifestations provide evidence that *ROBO3* is the gene responsible for HGPPS. Furthermore, the requirement that the mutations be present on both alleles for the manifestation of HGPPS to occur suggests that the mutations cause loss of gene function. Indeed, compound heterozygous mutations in non-consanguineous HGPPS families have since been identified (Chan et al., 2006; Jen et al., 2006), including the original Chinese family first described more than 30 years ago (Crisfield, 1974; Sharpe et al., 1975; Jen et al., 2006).

Functional anatomical features of HGPPS

How do mutations in ROBO3 lead to HGPPS? To address this question, we first had to determine where and when ROBO3 is expressed in the developing human brain. We performed in situ hybridization using a ROBO3-specific antisense probe that demonstrated robust signal in the pons in the embryonic human brain. We then performed high-resolution structural magnetic resonance imaging studies in HGPPS patients. Although the neurological deficits appeared restricted to conjugate horizontal gaze, there was marked malformation of the brainstem, with abnormal flattening of the basis pontis and an unusual butterfly-like appearance of the medulla with an unusual midline cleft. Although bulging of the abducens nuclei into the IVth ventricle was absent, the abducens nerves were visualized bilaterally, and the extraocular muscles appeared normal in configuration and size, without any evidence of fibrosis or denervation (Jen et al., 2004), consistent with the previous observation of full abduction on forced duction (Sharpe et al., 1975).

To investigate brainstem function to identify any functional correlates with the abnormal anatomy in HGPPS, we performed evoked potential studies to test the visual, auditory, and sensorimotor pathways. We were astonished to observe ipsilateral sensorimotor projections, suggesting that the long descending corticospinal pathway and the ascending dorsal column-medial lemniscal pathway destined to cross in the medulla failed to do so in HGPPS (Jen et al., 2004; MacDonald et al., 2004). Diffusion tensor imaging demonstrated disruption of other crossing fibres in the brainstem, notably the absence of crossing of the superior cerebellar peduncles and the absence of the trapezoid body (Sicotte et al., 2006). The inferior olives appeared displaced, and there were also more subtle findings of thinning of fibre tracts, including all the cerebellar peduncles and midline fascicles in the brainstem. Detailed functional studies on one of the original Greek families with HGPPS (Dretakis and Kondoyannis, 1974) with genetic confirmation (Jen et al., 2004; Amoiridis et al., 2006) revealed additional findings. Neuropsychological testing showed normal perceptual and cognitive function in two siblings with HGPPS, both left handed. Neurophysiological testing showed uncrossed sensorimotor projections in one but partial crossing in the other subject (Amoiridis et al., 2006), which is different from completely ipsilateral projections in other reports (MacDonald et al., 2004; Jen et al., 2004). Brainstem auditory evoked potentials showed in one subject absent contralateral Wave II on left ear stimulation, suggesting disrupted auditory axon projections consistent with the absence of the trapezoid body; however, the interpretation was complicated by deafness in the right ear. Of note, bilateral sympathetic skin responses to left median nerve stimulation reflected normal crossing of the spinothalamic tract (Amoiridis et al., 2006).

Independent of our investigation in human patients, a mouse model deficient in Robo3 was generated, in which there was a complete lack of crossing axons in the spinal cord and brainstem, with death within hours after birth (Sabatier et al., 2004). It was further demonstrated that, in the Robo3-deficient mice, the precerebellar neurons and their projections fail to cross the midline (Marillat et al., 2004).

Conclusion

The study of HGPPS, which is characterized by a relatively isolated neurological deficit of conjugate horizontal gaze, revealed surprisingly profound structural abnormalities in the brainstem and led to the unexpected discovery of an important molecule in hindbrain axon pathway crossing and morphogenesis. ROBO3 is a divergent member of the roundabout family in that it promotes midline crossing in human and mice. At the genetic level, HGPPS patients harbour biallelic mutations that cause 'knockout' of ROBO3 function, analogous to the situation in the knockout mice. Yet, there are important differences between human and mice. Indeed, despite maldevelopment, there is no evidence of splitting of the brainstem or spinal cord in HGPPS patients, who have normal crossing of the spinothalamic tract as well as normal life span. Hypoplasia and disruption of decussating projections in the medial longitudinal fasciculus and abnormal supranuclear abducens innervation could contribute to the absence of conjugate horizontal gaze. Maldevelopment of the reticular formation involving the central tegmental tract controlling axial tone may lead to progressive scoliosis. Efforts are ongoing to ascertain the molecular and the functional anatomical basis of HGPPS. The ROBO proteins are involved not only in axonal guidance and midline crossing, but also in neuronal migration and fibre tract fasciculation. We anticipate that improved resolution of neuroimaging techniques will reveal new anatomical detail involving the reticular formation, the medial longitudinal fasciculus, and the cerebellar peduncles in HGPPS. Final confirmation awaits the availability of pathological specimen. The study of HGPPS provides us an opportunity to study the development of horizontal eye movement control.

Abbreviation

HGPPS: horizontal gaze palsy with progressive scoliosis

Acknowledgement

This work was supported by the National Institutes of Health R01 EY15311.

References

Amoiridis, G., Tzagournissakis, M., Christodoulou, P., Karampekios, S., Latsoudis, H., Panou, T., Simos, P. and Plaitakis, A. (2006) Patients with horizontal gaze palsy and progressive scoliosis due to ROBO3 E319K mutation have both uncrossed and crossed central nervous system pathways and perform normally on neuropsychological testing. J. Neurol. Neurosurg. Psychiatr., 77(9): 1047–1053.

Brose, K., Bland, K., Wang, K., Arnott, D., W, H., Goodman, C., Tessier-Lavigne, M. and Kidd, T. (1999) Slit proteins bind Robo receptors and have an evolutionarily conserved role in repulsive axon guidance. Cell, 96: 795–806.

Büttner-Ennever, J.A. (2007) Anatomy of the oculomotor system. Dev. Ophthalmol., 40: 1–14.

Chan, W.M., Traboulsi, E.I., Arthur, B., Friedman, N., Andrews, C. and Engle, E.C. (2006) Horizontal gaze palsy with progressive scoliosis can result from compound heterozygous mutations in ROBO3. J. Med. Genet., 43(3): p. e11.

Crisfield, R.J. (1974) Scoliosis with progressive external ophthalmoplegia in four siblings. J. Bone Joint Surg. Br., 56B(3): 484–489.

Dretakis, E.K. and Kondoyannis, P.N. (1974) Congenital scoliosis associated with encephalopathy in five children of two families. J. Bone Joint Surg. Am., 56(8): 1747–1750.

Jen, J., Carr, J., Sharpe, J., Jollands, A., Goudie, D., Mamsa, H., Wan, J. and Baloh, R. (2006) Compound heterozygous ROBO3 mutations cause horizontal gaze palsy with progressive scoliosis in non-consanguineous individuals. Neurology, 66(5): p. A11.

Jen, J., Coulin, C.J., Bosley, T.M., Salih, M.A., Sabatti, C., Nelson, S.F. and Baloh, R.W. (2002) Familial horizontal gaze palsy with progressive scoliosis maps to chromosome 11q23-25. Neurology, 59(3): 432–435.

Jen, J.C., Chan, W.M., Bosley, T.M., Wan, J., Carr, J.R., Rub, U., Shattuck, D., Salamon, G., Kudo, L.C., Ou, J., Lin, D.D., Salih, M.A., Kansu, T., Al Dhalaan, H., Al Zayed, Z., MacDonald, D.B., Stigsby, B., Plaitakis, A., Dretakis, E.K., Gottlob, I., Pieh, C., Traboulsi, E.I., Wang, Q., Wang, L., Andrews, C., Yamada, K., Demer, J.L., Karim, S., Alger, J.R., Geschwind, D.H., Deller, T., Sicotte, N.L., Nelson, S.F., Baloh, R.W. and Engle, E.C. (2004) Mutations in a human ROBO gene disrupt hindbrain axon pathway crossing and morphogenesis. Science, 304(5676): 1509–1513.

Kidd, T., Brose, K., Mitchell, K.J., Fetter, R.D., Tessier-Lavigne, M., Goodman, C.S. and Tear, G. (1998) Roundabout controls axon crossing of the CNS midline and defines a novel subfamily of evolutionarily conserved guidance receptors. Cell, 92(2): 205–215.

MacDonald, D.B., Streletz, L.J., Al-Zayed, Z., Abdool, S. and Stigsby, B. (2004) Intraoperative neurophysiologic discovery of uncrossed sensory and motor pathways in a patient with horizontal gaze palsy and scoliosis. Clin. Neurophysiol., 115(3): 576–582.

Marillat, V., Sabatier, C., Failli, V., Matsunaga, E., Sotelo, C., Tessier-Lavigne, M. and Chédotal, A. (2004) The slit receptor Rig-1/Robo3 controls midline crossing by hindbrain precerebellar neurons and axons. Neuron, 43(1): 69–79.

Sabatier, C., Plump, A.S., Ma, L., Brose, K., Tamada, A., Murakami, F., Lee, E.Y.-H.P. and Tessier-Lavigne, M. (2004) The divergent Robo family protein Rig-1/Robo3 is a negative regulator of Slit responsiveness required for midline crossing by commissural axons. Cell, 117: 157–169.

Seeger, M., Tear, G., Ferres-Marco, D. and Goodman, C.S. (1993) Mutations affecting growth cone guidance in *Drosophila*: genes necessary for guidance toward or away from the midline. Neuron, 10(3): 409–426.

Sharpe, J.A., Silversides, J.L. and Blair, R.D. (1975) Familial paralysis of horizontal gaze. Associated with pendular nystagmus, progressive scoliosis, and facial contraction with myokymia. Neurology, 25(11): 1035–1040.

Sicotte, N., Salamon, G., Shattuck, D., Hageman, N., Rüb, U., Salamon, N., Drain, A., Demer, J., Engle, E., Alger, J., Baloh, J., Deller, T. and Jen, J. (2006) Diffusion tensor MRI shows abnormal brainstem crossing fibers associated with ROBO3 mutations. Neurology, 67(3): 519–521.

Wu, W., Wong, K., Chen, J., Jiang, Z., Dupuis, S., Wu, J.Y. and Rao, Y. (1999) Directional guidance of neuronal migration in the olfactory system by the protein Slit. Nature, 400(6742): 331–336.

Yuan, S., Cox, L., Dasika, G. and Lee, E. (1999) Cloning and functional studies of a novel gene aberrantly expressed in RB-deficient embryos. Dev. Biol., 207: 62–75.

C. Kennard & R.J. Leigh (Eds.)
Progress in Brain Research, Vol. 171
ISSN 0079-6123

CHAPTER 2.9

Neuronal evidence for individual eye control in the primate cMRF

David M. Waitzman[1,*], Marion R. Van Horn[2] and Kathleen E. Cullen[2]

[1]*Department of Neurology, University of Connecticut Health Center, 263 Farmington Avenue, Farmington, CT, USA*
[2]*Department of Physiology, McGill University, McIntyre Medical Sciences Building, 3655 Promenade Sir William Osler, Montréal, Québec H3G 1Y6 Canada*

Abstract: Previous single unit recordings and electrical stimulation have suggested that separate regions of the MRF participate in the control of vergence and conjugate eye movements. Neurons in the supraoculomotor area (SOA) have been found to encode symmetric vergence [Zhang, Y. et al. (1992). J. Neurophysiol., 67: 944–960] while neurons in the central MRF, the cMRF, located ventral to the SOA and lateral to the oculomotor nucleus are associated with conjugate eye movements [Waitzman, D.M. et al. (1996). J. Neurophysiol., 75(4): 1546–1572]. However, it remains unknown if cMRF neurons are strictly associated with conjugate movements since eye movements were recorded with a single eye coil in monkeys viewing visual stimuli at a distance of at least 50 cm. In the current study we addressed whether neurons in the cMRF might also encode vergence-related information. Interestingly, electrical stimulation elicited disconjugate saccades (contralateral eye moved more than the ipsilateral eye) from locations previously thought to elicit only conjugate saccades. Single unit recordings in this same area made in two rhesus monkeys trained to follow visual stimuli moved rapidly in depth along the axis of sight of an individual eye demonstrate that cMRF neurons *do not* simply encode conjugate information during disconjugate saccades; in fact our findings provide evidence that cMRF neurons are most closely associated with the movement of an individual eye. These results support the hypothesis that the midbrain shapes the activity of the pre-motor saccadic neurons by encoding integrated conjugate and vergence commands.

Keywords: saccade; disconjugate eye movement; mesencephalic reticular formation (MRF); superior colliculus (SC); vergence; saccade–vergence interaction; supraoculomotor area (SOA); oculomotor system; disparity

Introduction

The classic assumption that neuronal circuitries generating saccadic and vergence eye movements are largely separate has been challenged by a number of studies that have shown that during disconjugate saccades (i.e., when a vergence movement is combined with a saccade) vergence velocity is sped up and saccade velocities are reduced (Maxwell and King, 1992; Zee et al., 1992; Busettini and Mays, 2005; Kumar et al., 2005). Furthermore, recent work has revealed that a number of commonly assumed "conjugate" saccadic structures in the oculomotor brainstem in fact

*Corresponding author. Tel.: +(860) 679-8011;
Fax: +(860) 679-4446; E-mail: waitzman@nso2.uchc.edu

DOI: 10.1016/S0079-6123(08)00619-5

have monocular tuning (i.e., a combination of conjugate and vergence signals) (McConville et al., 1994; Zhou and King, 1998; Sylvestre and Cullen, 2002; Sylvestre et al., 2003; Van Horn et al., 2008).

The source of vergence-related signals to the pre-motor saccadic neurons remains unknown. Since the bulk of the input to the paramedian portion of the pontine reticular formation (PPRF) originates in the midbrain, specifically the superior colliculus (SC) and the mesencephalic reticular formation (MRF) (Horn, 2005), and the midbrain receives inputs from cortical structures with disparity information [e.g., lateral intraparietal area (LIP) and frontal eye fields (FEF)], we have now begun to re-examine neurons in MRF. The goal in the current study was to establish whether cMRF neurons contribute to the development of neural signals that are suitable for controlling an individual eye or if their discharge is strictly for conjugate control.

Methods

Two rhesus monkeys (*Macaca mulatta*) were prepared for chronic extracellular recording using aseptic surgical procedures described elsewhere (Sylvestre and Cullen, 1999b). The primary difference in the current experiments was the placement of the stainless steel recording chambers which were oriented stereotaxically on the skull towards the oculomotor nucleus. An eye coil was implanted in each eye to allow recordings of binocular eye movements. All procedures were approved by the University of Connecticut Health Center and McGill University Animal Care Committees and were in compliance with the guidelines of the NIH and the Canadian Council on Animal Care.

Behavioural paradigms, data acquisition, and analysis of cMRF neuronal discharges

Data acquisition and behavioural paradigms are identical to those utilized previously and are not repeated for sake of brevity (see Sylvestre and Cullen, 2002). Activity of cMRF neurons was identified as previously described (Waitzman et al., 1996). Horizontal and vertical conjugate saccades were elicited by stepping the target between horizontal and vertical positions, respectively (± 5 to 30 deg). To elicit different types of vergence eye movements an array of 16 computer-controlled red light-emitting diodes (LEDs) was displayed on a board tilted slightly from the horizontal in front of the monkey. To elicit disconjugate saccades LEDs were positioned in a configuration similar to the Müller paradigm (see Ramat et al., 1999) to minimize the movement of one eye. Notably, the eyes are referred to as either ipsilateral or contralateral based on their location relative to the recording site. We also describe eyes movements in terms of conjugate [conjugate = (left eye + right eye)/2] and vergence (vergence = left eye − right eye) coordinates (see Sylvestre and Cullen, 1999b for more details). Analysis of cMRF neuron discharges is precisely the same as those used previously for abducens motoneurons and burst-tonic neurons in the PPRF (Cullen et al., 1996; Cullen and Guitton, 1997; Sylvestre and Cullen, 1999a, b). The dynamic eye position and velocity sensitivities of a neuron during saccades were estimated using linear optimization techniques that have been described in detail elsewhere (Sylvestre and Cullen, 1999b). The dynamic lead time of individual neurons (t_d) was determined during conjugate saccades as described in Sylvestre and Cullen (1999b). For each model parameter in the analysis of disconjugate saccades, we computed 95% confidence intervals using a nonparametric bootstrap approach (Carpenter and Bithell, 2000), and used these confidence intervals to identify non-significant or identical model parameters.

Results

Electrical stimulation

The primary result of electrical stimulation in the cMRF is that both conjugate and disconjugate saccades can be elicited. Examples of the results of stimulating two sites in the cMRF separated by 1 mm in the same monkey are shown in Fig. 1. At the initial cMRF site located approximately 2 mm lateral to the oculomotor nucleus, electrical

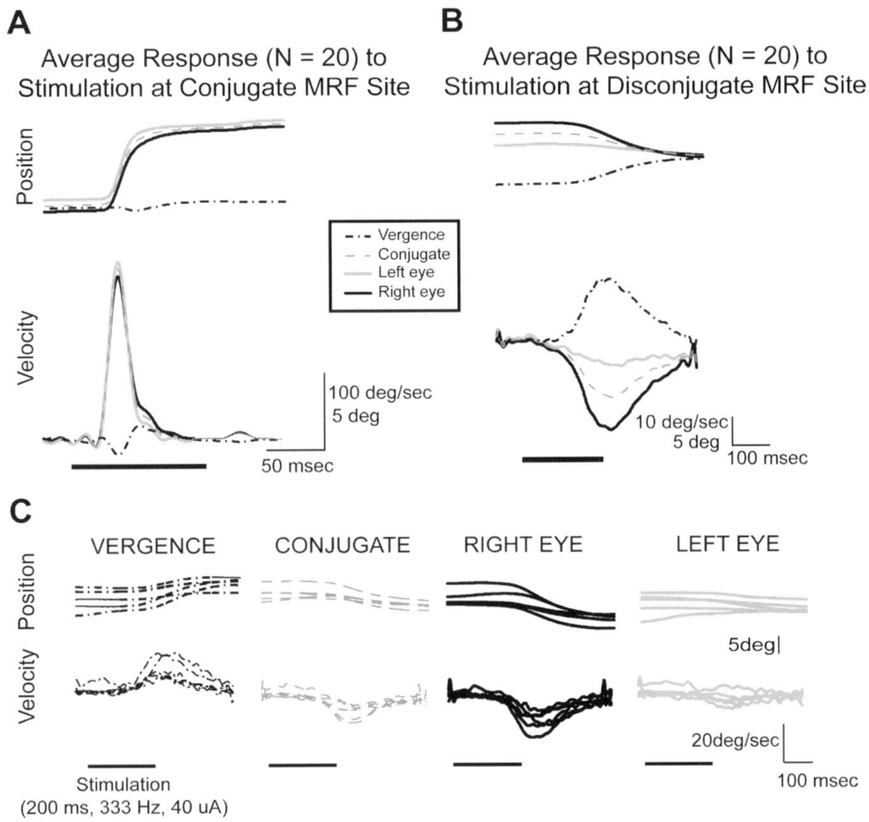

Fig. 1. Generation of conjugate/convergence eye movements following stimulation in the right cMRF of a monkey implanted with two eye coils. (A) Stimulation site in the MRF, 2 mm lateral to the oculomotor nucleus that produced conjugate saccades to the right. (B and C) A MRF site that was 1 mm more *lateral* to the site in A that elicited classic conjugate contralateral saccades. The conjugate signal (thin, dashed grey traces) is primarily the result of leftward movement of the *right* eye (thick, solid black traces) since the left eye moved only slightly (thick, solid grey traces). Average eye position and velocity traces are shown in A and B and individual eye position and velocity traces are shown in C. Note that positive vergence velocities indicate convergence and negative vergence velocities indicate divergence.

stimulation with 30–50 μA produced conjugate saccades to the contralateral side at short latency as has been demonstrated previously (Fig. 1A). Note that both eyes moved exactly the same amount. However in contradiction to previous results (Waitzman et al., 2002) stimulation at a site located 1 mm further *lateral* to this conjugate saccade site (and therefore not in the supraoculomotor area, SOA) elicited disconjugate movements, where the right eye (contralateral to the side of stimulation) moved more than the left (Fig 1B). We have now stimulated more than 40 sites throughout the cMRF and have demonstrated that more than 50% elicit disconjugate saccades.

Single unit recording

One potential criticism of electrical stimulation in the reticular formation is that we could have activated both conjugate and vergence mechanisms via electrical stimulation of axons in passage or antidromically activating regions that might be associated with the generation of vergence movements. Therefore we proceeded to record from single neurons located in the cMRF. An example of one such neuron is shown in Fig. 2. As has been demonstrated previously (e.g., Cromer and Waitzman, 2006), this cMRF neuron fired for contralateral conjugate movements of the eyes.

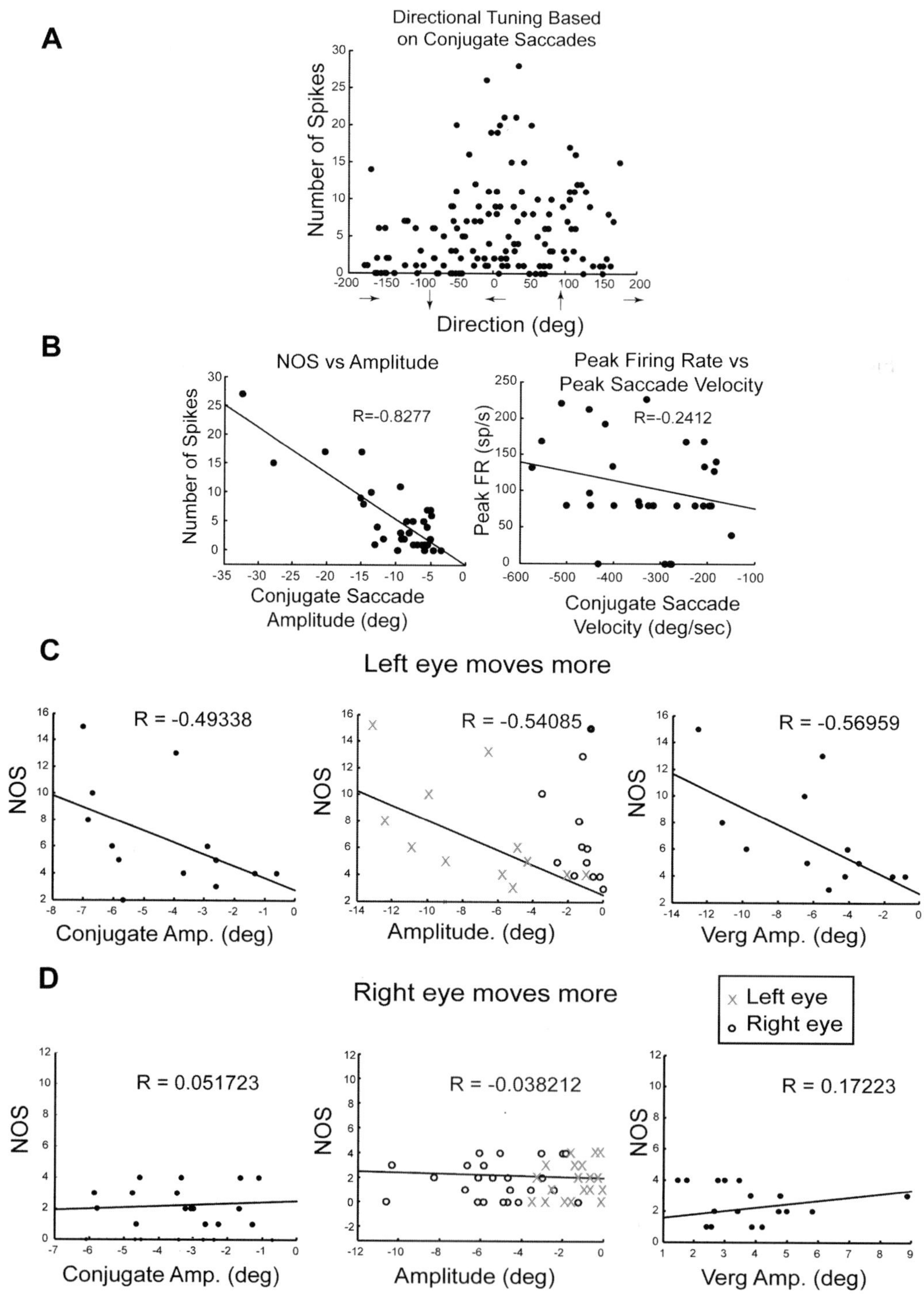

Fig. 2. Comparison of the relationships between number of spikes (NOS) and conjugate or individual eye amplitude and velocity. (A and B) Analysis of conjugate saccades. Note that tuning was best for saccades to the contralateral side. (C and D) Analysis of disconjugate saccades during which the left (C) or right (D) eye moved more. Note the marked improvement in correlation when left eye amplitude (C) was used as opposed to the right eye amplitude (D). Filled circles represent conjugate amplitude, and x's and open circles represent left and right eye amplitude, respectively.

Therefore its movement field consisted of all saccade vectors to the contralateral side (Fig. 2A). In addition there was a monotonic relationship between contralateral saccade amplitude (Fig. 2B, left) and saccade associated spike number, but a weak relationship between peak discharge and peak velocity (Fig. 2B, right).

However, we then analysed the disconjugate saccade trials and segregated them into trials when the left eye moved more (Fig. 2C) or when the right eye moved more (Fig. 2D). It was clear that the neuron fired primarily when the left eye moved to the left (Fig. 2C). The number of spikes in the burst during vergence eye movements was closely associated with vergence amplitude, while movements of the right eye were poorly correlated with the number of spikes regardless of the amplitude measured: conjugate, vergence, or right eye (Fig. 2D). Our sample of cMRF neurons yielded similar results such that during disconjugate saccades the number of spikes was better correlated with individual eye amplitude rather than the amplitude of the conjugate component of the movement.

The number of spikes approach, however, encompasses an inherent assumption that spike number is proportional to amplitude and thus firing rate is proportional to eye velocity. This may not be the case especially for neurons that project directly to the motoneurons and have a direct relationship between eye position and discharge rate (see the Appendix of Sylvestre and Cullen, 2002). Therefore, we used a dynamic method that made no such assumption. The dynamic approach has the added advantage that we can test the prediction of a specific model under a variety of conditions. For instance, we could compare the prediction of the conjugate model with a model that uses the position and velocity of an individual eye. We tested this prediction directly for a neuron located in the cMRF for which we obtained the response during conjugate saccades and hypothesized that the neuronal discharge would be similar during disconjugate saccades. We used the conjugate gaze model:

$$FR(t) = b_{\mathrm{CS}} + k_{\mathrm{CS}} CJ(t-t_d) + r_{\mathrm{CS}} \dot{C}J(t-t_d) \quad (1)$$

where $FR(t)$ is the neuron's instantaneous firing rate, b_{CS}, k_{CS}, and r_{CS} constants and represent the neuron's firing rate at eye position zero, the neuron's conjugate eye position, and eye velocity sensitivities, respectively. However, it was clear that during disconjugate saccades the conjugate gaze model could account for no more 3% of the variance (Fig. 3, top row). We then tested a binocular model that included terms for each individual eye position and velocity:

$$FR(t) = b_{\mathrm{DS}} + k_{\mathrm{i-DS}} IE(t-t_d) + k_{\mathrm{c-DS}} CE(t-t_d) + r_{\mathrm{i-DS}} I\dot{E}(t-t_d) + r_{\mathrm{c-DS}} C\dot{E}(t-t_d) \quad (2)$$

where $FR(t)$ is the neuron's instantaneous firing rate, b_{DS}, $k_{\mathrm{i-DS}}$, $k_{\mathrm{c-DS}}$, $r_{\mathrm{i-DS}}$, and $r_{\mathrm{c-DS}}$ are constants and represent the neuron's bias, ipsilateral and contralateral eye position, and ipsilateral and contralateral eye velocity sensitivity, respectively. The terms $IE(t)$, $CE(t)$, $I\dot{E}(t)$, and $C\dot{E}(t)$ represent the instantaneous ipsilateral and contralateral eye position and eye velocity, respectively. Note that this model follows a similar form to that of the conjugate equation, but now each term is assigned to each eye (i.e., binocular model). The results of the fit to the data with this binocular model were much better and could account for 48% of the variance (Fig. 3A, second row).

Although this observation strongly supported the idea that this neuron did not encode conjugate eye movements, it did not provide enough information to determine if it solely encoded the movements of one eye or a weighted mixture of both eyes' movements. To address this limitation we estimated the 95% confidence intervals in the *binocular* model Eq. (2) using the bootstrap technique (see Sylvestre and Cullen, 2002). Figure 3B shows the bootstrap distributions and the 95% confidence intervals (thick horizontal bars) for the estimated eye position sensitivities. Two important observations can be made from the 95% confidence intervals. First, the parameter values estimated for the ipsilateral ($k_{\mathrm{i-DS}}$) and contralateral ($k_{\mathrm{c-DS}}$) eye position were statistically different (i.e., the confidence intervals did not overlap). Second, neither position parameters overlapped with zero indicating both position terms were significant. We called such neurons "binocular", since they had unequal sensitivity to the position (or the velocity in some cases) of the

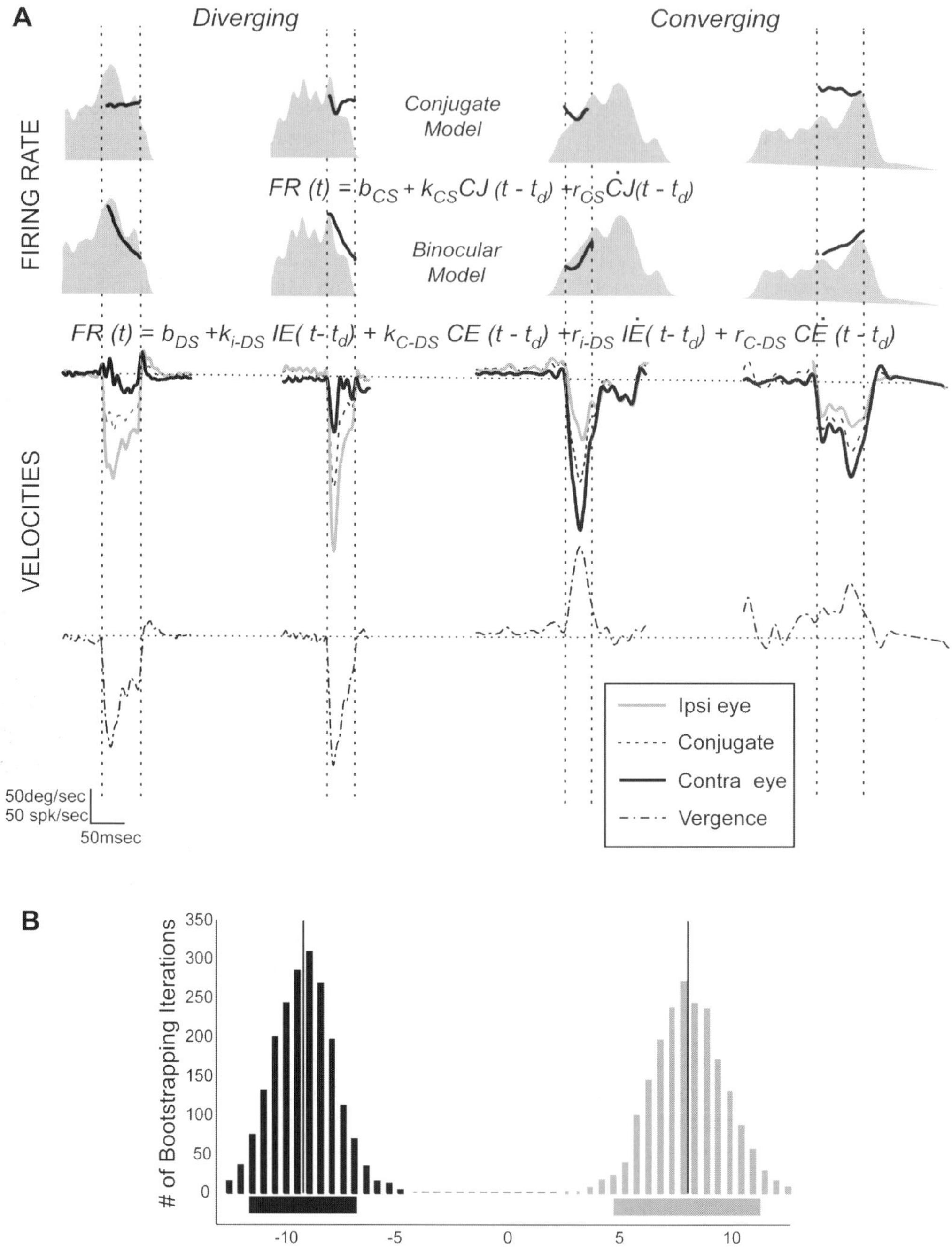

Fig. 3. (A) Model fits for cell ML41_3 during disconjugate saccades (two left columns are diverging saccades and the two right columns are converging saccades). The thick black lines on the firing rates are the model fits using conjugate model parameters (top) and the binocular model (second row). In velocity traces (third row) solid black lines are the contralateral eye and solid grey lines are the ipsilateral eye. Vergence velocity is shown in the fourth row. (B) The histograms resulting from application of the bootstrapping method to the cell shown in A for the position parameters. The 95% confidence intervals for the estimates of position relationship are shown by the heavy horizontal bars below. The black distribution is the contralateral eye and the grey distribution is the ipsilateral eye. Thin vertical black lines indicate the mean for each distribution.

two eyes (ratio of the means of the two position distributions was not 1). Both velocity terms (r_{c-DS} and r_{i-DS}) overlapped with zero indicating the velocity parameters were not important (data not shown). The majority of the neurons in our sample were comparable in that the discharges of the cMRF neurons were best described using individual eye position and/or velocity. Overall, this suggests that cMRF neurons encode the movement of an individual eye rather than solely encoding the conjugate component of a given saccade. Thus, when characterizing the discharge of cMRF neurons it is critical to consider the movement of *both* the ipsilateral and contralateral eye.

Discussion

The findings described in the present study lead to two main conclusions. First, electrical stimulation of the MRF elicits both disconjugate and conjugate saccades. Second, the results of single unit recording from neurons in the cMRF of monkeys have demonstrated that the discharge of MRF neurons can dynamically encode the movement of an individual eye, rather than conjugate eye motion. Taken together, these results support the hypothesis that the midbrain shapes the activity of the premotor saccadic neurons by encoding integrated conjugate and vergence commands.

Role of neurons of the MRF in the control of saccades and vergence

We and others have hypothesized that cMRF neurons could provide a feedback signal from the PPRF or omnipause neurons about the current progress of the saccade to the SC (Waitzman et al., 1996; Soetedjo et al., 2002; Cromer and Waitzman, 2006). It is also likely that neurons in the MRF provide a parallel pathway for descending information to the omnipause region. The projections from the MRF to the omnipause region (Horn, 2005) could inhibit the omnipause neurons and thereby indirectly activate PPRF burst neurons. Notably, our results further suggest that these ascending and descending projections convey not only conjugate but also vergence related information.

Coordinated inputs from the cMRF and intermediate and deep layers of the SC (dSC) to the saccadic pre-motor neurons could serve as an alternative to the previously described cortico-pontine-cerebellar-midbrain loop for the control of vergence (Gamlin et al., 1996). A number of pieces of evidence support this idea. Both the dSC and cMRF receive inputs from disparity sensitive cortical (FEF: Ferraina et al., 2000; LIP: Gnadt and Mays, 1995; Gnadt and Beyer, 1998; Genovesio and Ferraina, 2004) and subcortical regions [e.g., superficial layers of the SC (Mimeault et al., 2004)]. Furthermore, consistent with this idea, are the findings that stimulation of the cMRF (goldfish: Luque et al., 2006; monkey: present study) and SC (Chaturvedi and Van Gisbergen, 1999) have clear affects on vergence. Moreover, neurons in the SC (cat SC: Jiang et al., 1996; monkey SC: Walton and Mays, 2003) *and* the cMRF are modulated during vergence eye movements. Because the modulation of primate SC neurons was observed to be more robust for purely conjugate than disconjugate saccades, it has been suggested that the SC is not tuned in three dimensions (Walton and Mays, 2003). However, the present results combined with other recent findings (Van Horn et al., 2008) that showed neurons in the PPRF were associated with individual eye movement (Zhou and King, 1998; Sylvestre et al., 2003), suggest neurons in the dSC should be re-examined for evidence of an individual eye signal.

Acknowledgements

The authors received support from NIH (Waitzman) (EY009481, EY015356, and NS041224) and NSERC (Cullen).

References

Busettini, C. and Mays, L.E. (2005) Saccade–vergence interactions in macaques. I. Test of the omnipause Multiply Model. J. Neurophysiol., 94(4): 2295–2311.

Carpenter, J. and Bithell, J. (2000) Bootstrap confidence intervals: when, which, what? A practical guide for medical statisticians. Stat. Med., 19: 1141–1164.

Chaturvedi, V. and Van Gisbergen, J. (1999) Perturbation of combined saccade–vergence movements by microstimulation in monkey superior colliculus. J. Neurophysiol., 81: 2279–2296.

Cromer, J.A. and Waitzman, D.M. (2006) Neurones associated with saccade metrics in the monkey central mesencephalic reticular formation. J. Physiol., 570(Pt 3): 507–523.

Cullen, K.E. and Guitton, D. (1997) Analysis of primate IBN spike trains using system identification techniques. II. Relationship to gaze, eye, and head movement dynamics during head-free gaze shifts. J. Neurophysiol., 78(6): 3283–3306.

Cullen, K.E., Rey, C.G., Guitton, D. and Galiana, H.L. (1996) The use of system identification techniques in the analysis of oculomotor burst neuron spike train dynamics. J. Comput. Neurosci., 3(4): 347–368.

Ferraina, S., Pare, M. and Wurtz, R.H. (2000) Disparity sensitivity of frontal eye field neurons. J. Neurophysiol., 83(1): 625–629.

Gamlin, P.D., Yoon, K. and Zhang, H. (1996) The role of cerebro-ponto-cerebellar pathways in the control of vergence eye movements. Eye, 10(Pt 2): 167–171.

Genovesio, A. and Ferraina, S. (2004) Integration of retinal disparity and fixation-distance related signals toward an egocentric coding of distance in the posterior parietal cortex of primates. J. Neurophysiol., 91(6): 2670–2684.

Gnadt, J.W. and Beyer, J. (1998) Eye movements in depth: what does the monkey's parietal cortex tell the superior colliculus? Neuroreport, 9(2): 233–238.

Gnadt, J.W. and Mays, L.E. (1995) Neurons in monkey parietal area LIP are tuned for eye-movement parameters in three-dimensional space. J. Neurophysiol., 73(1): 280–297.

Horn, A.K. (2005) The reticular formation. Prog. Brain Res., 151: 127–155.

Jiang, H., Guitton, D. and Cullen, K.E. (1996) Near-response-related neural activity in the rostral superior colliculus of the cat. Soc. Neurosci. Abstr., 22: 662.

Kumar, A.N., Han, Y., Dell'osso, L.F., Durand, D.M. and Leigh, R.J. (2005) Directional asymmetry during combined saccade–vergence movements. J. Neurophysiol., 93(5): 2797–2808.

Luque, M.A., Perez-Perez, M.P., Herrero, L., Waitzman, D.M. and Torres, B. (2006) Eye movements evoked by electrical microstimulation of the mesencephalic reticular formation in goldfish. Neuroscience, 137(3): 1051–1073.

Maxwell, J.S. and King, W.M. (1992) Dynamics and efficacy of saccade-facilitated vergence eye movements in monkeys. J. Neurophysiol., 68(4): 1248–1260.

McConville, K., Tomlinson, R.D., King, W.M., Paige, G. and Na, E.Q. (1994) Eye position signals in the vestibular nuclei: consequences for models of integrator function. J. Vestib. Res., 4(5): 391–400.

Mimeault, D., Paquet, V., Molotchnikoff, S., Lepore, F. and Guillemot, J.P. (2004) Disparity sensitivity in the superior colliculus of the cat. Brain Res., 1010(1–2): 87–94.

Ramat, S., Das, V.E., Somers, J.T. and Leigh, R.J. (1999) Tests of two hypotheses to account for different-sized saccades during disjunctive gaze shifts. Exp. Brain Res., 129(4): 500–510.

Soetedjo, R., Kaneko, C.R. and Fuchs, A.F. (2002) Evidence that the superior colliculus participates in the feedback control of saccadic eye movements. J. Neurophysiol., 87(2): 679–695.

Sylvestre, P.A., Choi, J.T. and Cullen, K.E. (2003) Discharge dynamics of oculomotor neural integrator neurons during conjugate and disjunctive saccades and fixation. J. Neurophysiol., 90(2): 739–754.

Sylvestre, P.A. and Cullen, K.E. (1999a) Monocularly tuned discharge dynamics of abducens and brainstem inhibitor burst neurons during disconjugate saccades. Soc. Neurosci. Abstr., 25: 1652.

Sylvestre, P.A. and Cullen, K.E. (1999b) Quantitative analysis of abducens neuron discharge dynamics during saccadic and slow eye movements. J. Neurophysiol., 82(5): 2612–2632.

Sylvestre, P.A. and Cullen, K.E. (2002) Dynamics of abducens nucleus neuron discharges during disjunctive saccades. J. Neurophysiol., 88(6): 3452–3468.

Van Horn, M.R., Sylvestre P.A. and Cullen, K.E. (2008) The brainstem saccadic burst generator encodes gaze in three dimensional space. J. Neurophysiol., Mar 12 2008 [Epub ahead of print].

Waitzman, D.M., Pathmanathan, J., Presnell, R., Ayers, A. and DePalma, S. (2002) Contribution of the superior colliculus and the mesencephalic reticular formation to gaze control. Ann. N.Y. Acad. Sci., 956: 111–129.

Waitzman, D.M., Silakov, V.L. and Cohen, B. (1996) Central mesencephalic reticular formation (cMRF) neurons discharging before and during eye movements. J. Neurophysiol., 75(4): 1546–1572.

Walton, M.M. and Mays, L.E. (2003) Discharge of saccade-related superior colliculus neurons during saccades accompanied by vergence. J. Neurophysiol., 90(2): 1124–1139.

Zee, D.S., Fitzgibbon, E.J. and Optican, L.M. (1992) Saccade–vergence interactions in humans. J. Neurophysiol., 68(5): 1624–1641.

Zhang, Y., Mays, L.E. and Gamlin, P.D. (1992) Characteristics of near response cells projecting to the oculomotor nucleus. J. Neurophysiol., 67: 944–960.

Zhou, W. and King, W.M. (1998) Premotor commands encode monocular eye movements. Nature, 393(6686): 692–695.

SECTION 3

Using Eye Movements as an Index of Transformation of Signals by the Cerebellum and Brainstem

C. Kennard & R.J. Leigh (Eds.)
Progress in Brain Research, Vol. 171
ISSN 0079-6123

CHAPTER 3.1

Complex spike activity signals the direction and size of dysmetric saccade errors

Robijanto Soetedjo, Yoshiko Kojima and Albert Fuchs*

Department of Physiology and Biophysics and Washington National Primate Research Center, University of Washington, Seattle, WA, USA

Abstract: The cerebellar oculomotor vermis (OMV) receives inputs from both the superior colliculus (SC) via the nucleus reticularis tegmenti pontis as mossy fibres and the inferior olive as climbing fibres. Lesion studies show that the OMV is necessary for the saccade amplitude adaptation that corrects persistent motor errors. In this study, we examined whether the complex spike (CS) activity due to climbing fibre inputs could serve as an error signal to drive saccade adaptation. When there was an error during behaviourally induced saccade dysmetrias, the probability of CS occurrence depended on the direction and size of the error. If this CS activity actually drives saccade adaptation, we speculate that adaptation should be equally efficient in all directions and that the course of adaptation could have two operating modes.

Keywords: saccades; oculomotor vermis; complex spikes; adaptation; cerebellum; motor learning; monkey

Introduction

The oculomotor system maintains the accuracy of saccadic eye movements until we are well into our 70s (Warabi et al., 1984). It also is able to adapt the hypometric saccades caused by muscle weakness that results either from natural causes (Abel et al., 1978) or experimental manipulations (Optican and Robinson, 1980; Scudder and McGee, 2003). After damage to the midline cerebellum, the oculomotor system no longer can repair such saccade dysmetrias (Barash et al., 1999). We have begun to study the neural mechanisms that could underlie this saccade amplitude adaptation.

After collecting saccade commands from several cortical areas and the basal ganglia, the superior colliculus (SC) distributes them to the eye muscles through two pathways. Our honouree, Jean Büttner-Ennever, has helped elucidate the function of the *direct pathway* to the burst generator (Fig. 1) and then to the motoneurons in a series of seminal papers (Büttner-Ennever and Henn, 1976; Büttner-Ennever et al., 1999, 2001). A second *indirect pathway* (Fig. 1) traverses the nucleus reticularis tegmenti pontis (NRTP), which provides mossy fibres (mf) to vermis lobules VIc and VII (OMV). The Purkinje or P-cells in the OMV inhibit neurons in the caudal fastigial nuclei (CFN), which, in turn, project to the burst generator (for review: Scudder et al., 2002). The oculomotor vermis (OMV) also receives climbing fibre (cf) inputs from the inferior olive (Kralj-Hans et al., 2007).

Marr (1969) and Albus (1971) posited that persistent motor errors cause an increase in cf activity and in the resulting complex spikes (CSs) in their target P-cells. The increased CS activity

*Corresponding author. Tel.: +(206) 543-8014;
Fax: +(206) 685-0305; E-mail: fuchs@u.washington.edu

DOI: 10.1016/S0079-6123(08)00620-1

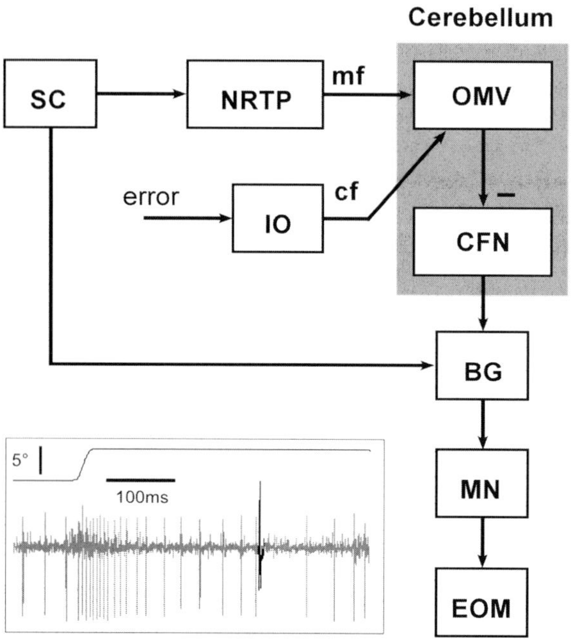

Fig. 1. Projections of the superior colliculus (SC) to the brainstem and cerebellum. The SC projects directly to the saccade burst generator (BG) and indirectly via the nucleus reticularis tegmenti pontis (NRTP) to the oculomotor cerebellum (grey box), including the vermis (OMV) and the caudal fastigial nucleus (CFN). OMV Purkinje cells (P-cells) inhibit CFN neurons, which in turn project to the BG. P-cells receive mossy fibres (mf) from NRTP and climbing fibres (cf) from the inferior olive (IO). Inset: P-cells discharge simple spikes (SSs, grey) at a high rate but only occasional CSs (black). In this P-cell, a SS burst accompanied a 5° saccade. MN: motoneuron; EOM: extraocular muscle.

would alter the simple spike (SS) activity, which, in turn, would adjust a motor pathway in the brainstem until the error was eliminated.

We have begun testing whether CS activity in the OMV could provide an error signal that would be suitable to drive saccade motor learning. To determine whether such a scenario had any promise, we first examined a number of P-cells (like that in Fig. 2) during behavioural amplitude adaptation in only the horizontal direction. In Fig. 2, the 1400 experimental trials are presented from top to bottom in order, each symbol, squares or dots, identifies when a CS occurred, and all trials are aligned on the end of the first saccade. In the first 100 control saccades, the only change in CS activity was a pause, which will remain unchanged throughout adaptation. Between trials 100 and 400, the target jumped forward during the saccade, creating an apparent hypometria (↓, leftward error). When this error was present between the primary saccade and the subsequent corrective saccade, CS activity ceased. Between trials 400 and 1100, we reversed the error direction by stepping the target backwards to produce an apparent hypermetria (↑, rightward error). Now, during the interval between the primary and corrective saccade when there was an error (the error interval), there was a substantial increase in the probability of CS occurrence. However, as adaptation caused the error to become smaller over time (smaller corrective saccades, shorter ↑), the probability of CS occurrence did not appear to decrease. When we again created a left error (↓) by jumping the target forward, CS occurrence again dropped essentially to zero. Such data led us to conclude that CS occurrence reported error direction but not size (Soetedjo and Fuchs, 2006). However, that study examined only horizontal errors with sizes $<5^{\circ}$. Here we test whether CS activity reports errors in all directions and of all sizes.

Methods

Data collection

We recorded extracellular unit activity from P-cells located in vermis lobules VIc and VII while two rhesus monkeys made saccadic eye movements to step changes in target position (see Soetedjo and Fuchs, 2006 for details about the placement of the recording chamber, the measurement of eye movement and the initial processing of unit activity). We identified P-cells by the occurrence of CSs, which, in our recording conditions, had an initial positive component followed by one or more wavelets at ~1 ms intervals and was followed by a brief pause in SS activity (inset Fig. 1). We tested only those P-cells that exhibited phasic SS discharge patterns that were related to saccades, i.e., a burst, a pause, or a combination of the two.

To identify the preferred direction of a CS response, we produced a constant 5° error in eight directions aimed every 45° between 0° (rightward)

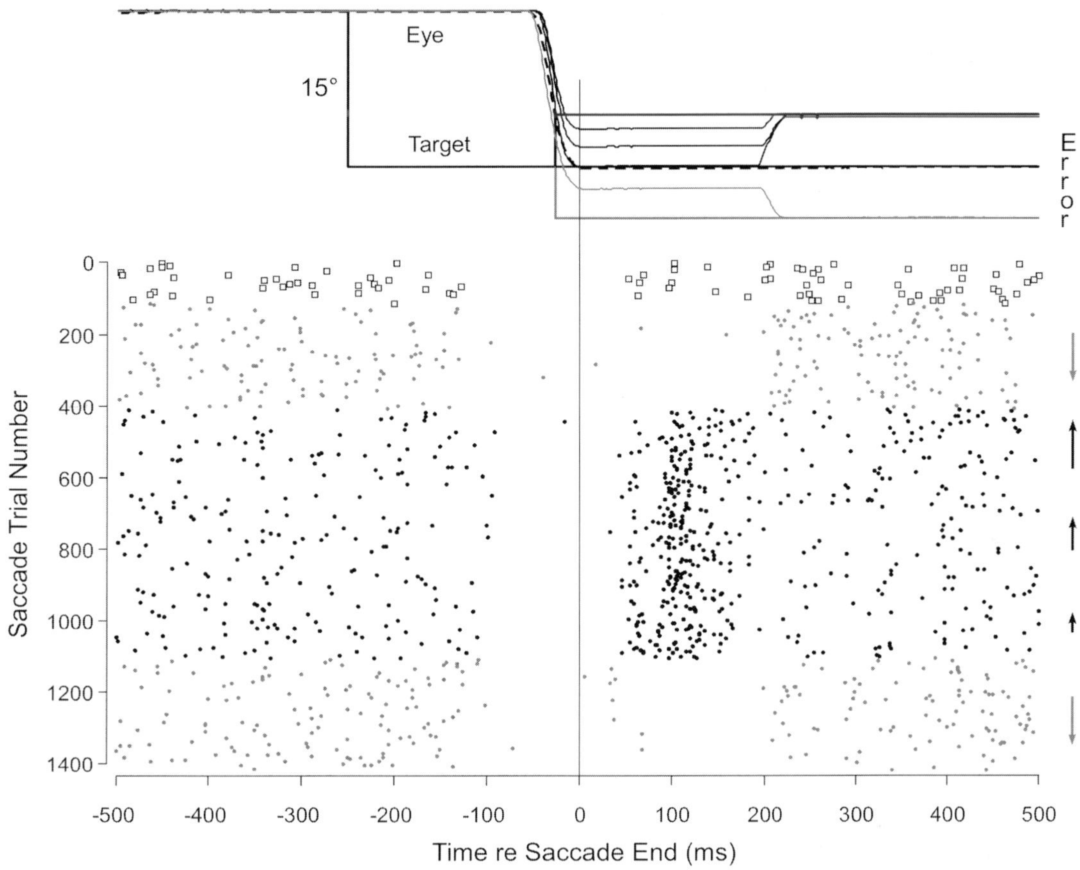

Fig. 2. Occurrence of CSs during adaptation of leftward saccades. Top panel: time courses of representative control saccades (dashed) and saccades in response to forward (leftward, grey) and backward (rightward, black) saccade-triggered adapting target steps. Each row of dots below show the CSs associated with each of 1400 trials plotted from top to bottom in order of occurrence and aligned on the end of the primary saccade. Each dot represents the occurrence of a single CS; CSs do not occur on all trials. Trials 1–100 are controls with no adapt steps (open squares). From trial 100–400, the adapting step was forward (grey dots), so error direction was leftward (↓). From trials ~400 to 1100, the adapting step was backward (black dots), so error direction was rightward (↑). As backward adaptation progressed, the errors became smaller (shorter ↑). From trials 1100 to 1400, the adapt step again was forward (grey dots, leftward error, ↓).

and 315°. Targets jumped by 15° in eight directions randomly from straight-ahead and the targeting saccades triggered 5° backward target (error) displacements (Fig. 3A, grey centripetal arrows). We collected 50 saccades in each of the eight directions and displayed the unit activity associated with each direction, in turn, on two on-line monitors. To estimate the preferred error direction, we counted by eye the CSs within the 200 ms after the primary saccade, which was the average time to the corrective saccade and therefore the approximate end of the error interval. This stimulus produced an error signal that would have caused adaptation after several hundred repetitions in each direction, but little was produced with the 50 repetitions per direction used in this test.

In the remainder of the experiment, we examined CS sensitivity to different error sizes in the direction with the greatest number of CSs by the on-line estimate (the on-direction). We presented different error sizes in either the on (0°, +1.5°, +3°, +5°, +7°, and +9°) or off (−1.5°, −3.0°) directions randomly (Fig. 4A). We collected ~50 saccades for each error size.

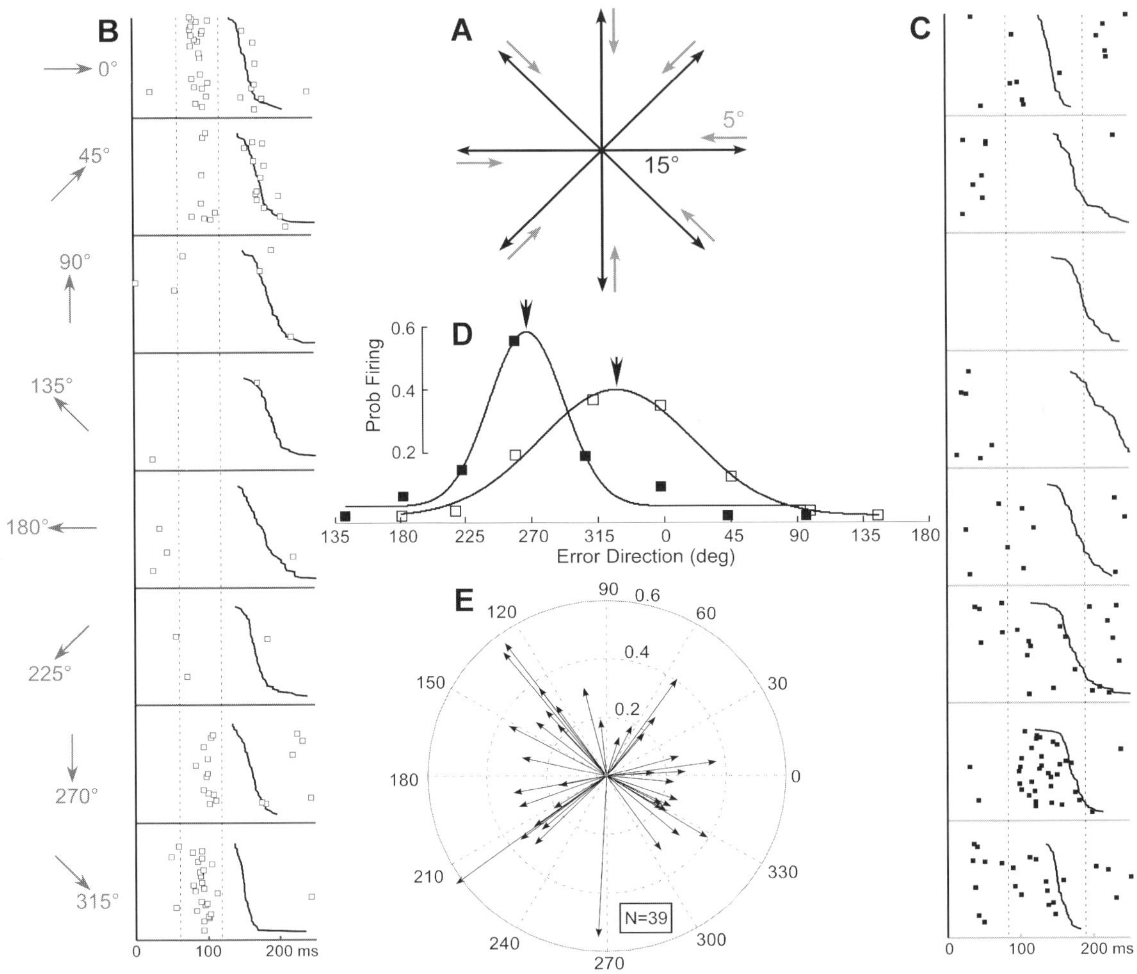

Fig. 3. Directional tuning of CS occurrence. (A) 5° centripetal error steps (grey arrows) triggered by 15° centrifugal saccades in eight random directions. (B) and (C) CS occurrences (open and filled squares, respectively) in each error direction for two units. For each direction (the same directions as grey arrows in A), about 50 trials were ordered from top to bottom according to the latency of the corrective saccade (jagged vertical curves here and in Fig. 4 indicate time of corrective saccade onset). All trials were aligned on primary saccade onset (0 ms). Here and in Fig. 4, vertical dashed lines demarcate the MI windows (see Methods). (D) Gaussian fits of the probabilities of CS occurrence within the MI window against error direction for units in (B) (open squares) and (C) (filled squares). Arrows mark preferred error directions. (E) Distribution of the preferred directions of all units. Arrow length is the peak CS probability.

Data analysis

Because of the stochastic nature of the CS response, we used information theory measures to quantify how CS activity depended on error direction and size. We calculated the mutual information (MI; for review: DeWeese and Meister, 1999) to determine the time window between the primary and corrective saccades in which CSs showed a significant relation with error direction. The mutual information measures the reduction of uncertainty (entropy) of error direction when a CS occurs:

$$\mathrm{MI(Error;CS)} = \mathrm{H(Error)} - \mathrm{H(Error|CS)} \quad (1)$$

H(Error) is the uncertainty of the eight possible error directions measured in bits. The maximum

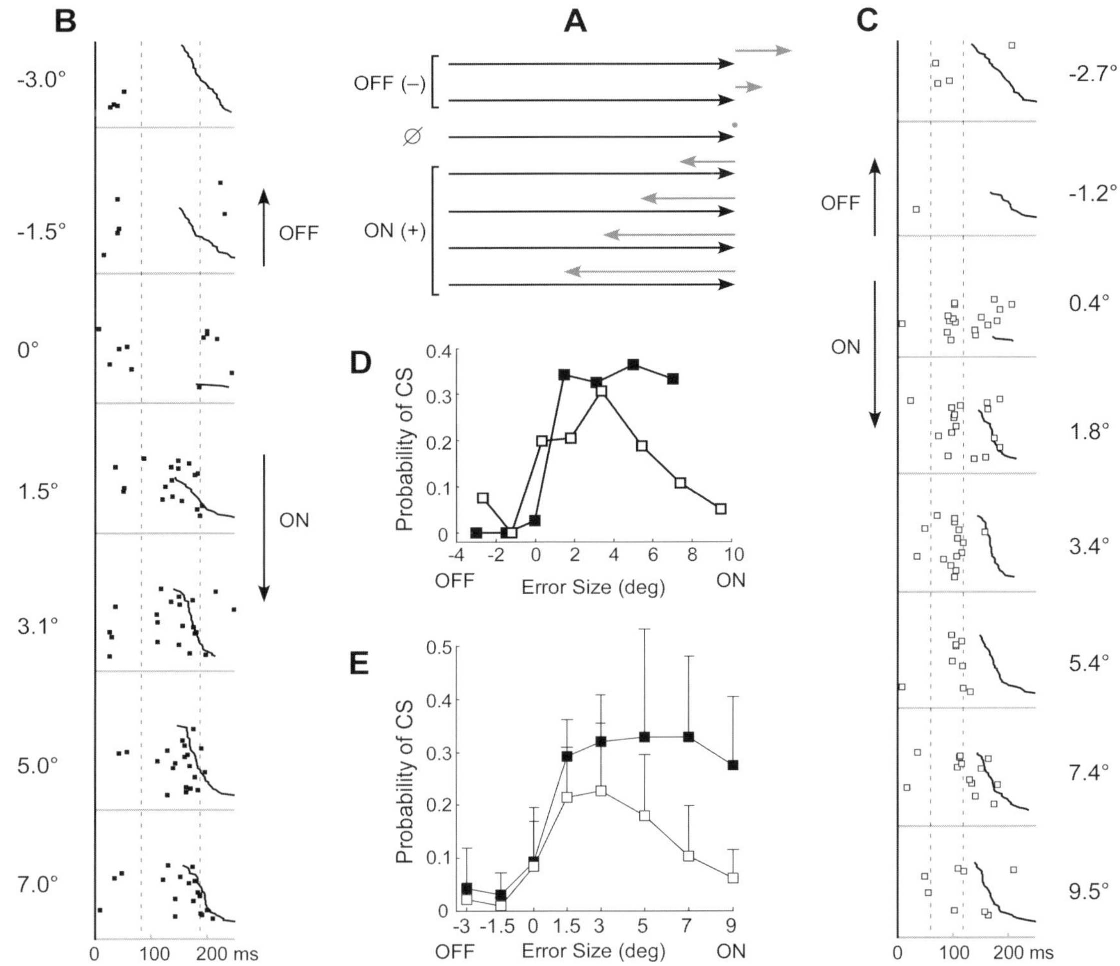

Fig. 4. Sensitivity of CS occurrence to different error sizes. (A) Primary saccades in the preferred direction (here horizontal) triggered either centrifugal (−) or centripetal (+) error steps (grey arrows) of different sizes, including zero (∅, grey dot). (B) (unit in Fig. 3C) and (C) (unit in Fig. 3B), CS occurrences (filled and open squares, respectively) for decreasing average error sizes in the OFF (−) and increasing average error sizes in the ON (+) error directions. For errors near 0, few corrective saccades were made (very short jagged curves). (D) CS probabilities as a function of error size for units in (B) (filled squares) and (C) (open squares). (E) Average CS probability for 13 compact (open squares) and five broad (filled squares) units plotted against error size. Error bars are 1SD.

possible value is 3 bits. If CSs are related to error direction, the occurrence of a CS increases the chance of predicting the error direction correctly. The second term in Eq. (1), H(Error|CS) computes the *remaining* uncertainty of error direction when a CS occurs. The theoretical maximum for H(Error|CS) is 2 bits. Therefore, the maximum possible value of MI is 1 bit.

The MI was computed within a time window of 30 ms that slid every millisecond from the end of the primary saccade until 250 ms later. Within each time window, we counted the number of trials with and without a CS for each direction to produce a 2×8 table. Each of the eight columns indicated an error direction and each row indicated the number of trials with and without a CS for that error direction. This sliding time window created a curve of MI values from the end of the primary saccade until 250 ms later. With G-statistics (Sokal and Rohlf, 1994), we determined a threshold above which the MI curve showed a statistically significant correlation ($p<0.05$) between CS occurrence and error direction. This threshold created a time window (the "MI" interval, dashed vertical line in Fig. 3B, C) within which we computed the probability of CSs for each error direction.

We fitted the probability of CS occurrence in each error direction as a function of error direction using a Gaussian function:

$$prob(\theta) = bias + A \times \exp\left[-\left(\frac{(\theta - B)2}{2 \times SD^2}\right)\right] \quad (2)$$

where θ is the mean error direction and B the unit's preferred direction. The tuning width is 2SD.

Using the same time window established from error direction analysis, we counted the number of CSs (within the MI interval, dashed vertical lines Fig. 4B, C) to compute the probabilities of CS occurrence for each error amplitude.

Results

We recorded CSs in 48 P-cells that exhibited saccade-related phasic discharges of SS activity. The CSs of the majority (39/48) showed directional error tuning. We tested the CS selectivity for error amplitude in 18 of the 39 units.

Direction tuning

Figure 3 shows how CSs were tuned for error direction. The unit in Fig. 3B showed CS responses that were broadly tuned across four error directions (0°, 45°, 270°, and 315°) whereas the unit in Fig. 3C was tightly tuned for 270°. To compare the CS tuning of all cells, we determined the probability of CS occurrence between the dashed lines [the MI interval when CS firing was significant (see Methods)] for each error direction (Fig. 3D). Gaussian fits of those data provided the preferred error direction and the average peak probability of CS occurrence. For the units in Fig. 3B, C, the peak probability of firing occurred at 327.7° (open squares) and 267.1° (filled squares), respectively (arrows) and the widths of the tuning curves (2 × SD) were 103.6° and 50.4°, respectively. Figure 3E plots the preferred error direction and probability for all our units. Most preferred error directions were within ±60° of horizontal; only five units were tuned more to vertical errors. On average, the peak probability across all 39 units was 0.31±0.12, indicating that a CS occurred about once every three trials, on average.

Figure 3 also illustrates that units had two different distributions of CSs within the error interval. For most (24/39), CSs occurred in a *compact* cluster at a short latency after the primary saccade (Fig. 3B; population range from 92 ms to 112 ms). For the others, CSs occurred in a *broad* cluster with a longer latency (Fig. 3C; population range from 83 ms to 183 ms). There was no difference in the distribution of preferred directions between compact and broad units. On average, the widths of the tuning curves of compact units (76.6°±27.2°) were significantly narrower than those of the broad units (100.2°±37.8°).

Amplitude tuning

After we had estimated preferred error directions on-line, 13 compact units and five broad units were still well enough isolated to test their sensitivities to error size in the preferred direction. On average, the on-line estimate of preferred direction differed by only 15.6°±11.1° from that determined by a quantitative MI analysis.

CS responses to error size can be divided into two types as illustrated by the units in Fig. 4B, C. Within the MI interval (dashed vertical lines), the unit in Fig. 4B shows no CSs for off-direction or zero errors but a relatively constant number of CSs for on-direction errors of all sizes. A plot of CS probabilities as a function of error amplitude shows a flat relation for error sizes ≥1.5° (Fig. 4D, filled squares). The relation CS probability with error size for this unit, therefore, resembles that for the unit in Fig. 2. For the unit in Fig. 4C, in contrast, CS occurrence is most probable for errors near 3.4° and less probable for both smaller and larger errors, i.e., it is tuned for error size. Consideration of all the probability vs. error size relations reveals that the 13 compact units were tuned for an average error between 1.5° and 3.0°

(Fig. 4E, open squares) whereas the five broad neurons responded equally well, on average, for all error sizes (Fig. 4E, filled squares).

Discussion

80% of our P-cells showed changes in the probability of CS occurrence after a dysmetric first saccade caused an error and until a corrective saccade eliminated it. Across our entire population, all error directions were represented with direction tuning widths ranging from 77° to 100°. Also, the CSs of some P-cells were tuned for small error sizes in the on-direction whereas the CSs of others responded equally well for all on-direction error sizes.

If this CS activity actually drives saccade amplitude adaptation, we speculate that adaptation should have certain properties. First, adaptation should be equally possible in all directions. There are few studies concerning adaptation in directions other than horizontal. However, preliminary studies in our lab show that adaptation in the monkey indeed is equally robust in all directions. Second, the existence of neurons that either are tuned to small errors or responsive to all sizes suggests that adaptation might have two operating modes. Both neuron types would respond for small errors, but only the un-tuned neurons would report large errors. During growth or ageing, the changes in oculomotor performance are gradual so all units would report the resultant small errors that would drive adaptation. On the other hand, units that reported errors of all sizes, especially large ones, might be responsible for driving the adaptation that compensates for the larger dysmetria caused by injuries. Finally, the combination of directional and amplitude tuning in the majority of our cells might account, in part for adaptation fields where adaptation of saccades of one size and direction transfers poorly to saccades of other sizes and directions (Noto et al., 1999). In particular, the relatively narrow CS direction tuning curves would not drive adaptation of saccades in the orthogonal direction.

In summary, despite their low firing rates, the occurrence of CSs signalled both the direction and amplitude of error. This activity may be used as a teaching signal to drive saccade adaptation (Marr, 1969; Albus, 1971).

References

Abel, L., Schmidt, D., Dell'osso, L. and Daroff, R. (1978) Saccadic system plasticity in humans. Ann. Neurol., 4: 313–318.

Albus, J. (1971) A theory of cerebellar function. Math. Biosci., 10: 25–61.

Barash, S., Melikyan, A., Sivakov, A., Zhang, M., Glickstein, M. and Thier, P. (1999) Saccadic dysmetria and adaptation after lesions of the cerebellar cortex. J. Neurosci., 19: 10931–10939.

Büttner-Ennever, J.A. and Henn, V. (1976) An autoradiographic study of the pathways from the pontine reticular formation involved in horizontal eye movements. Brain Res., 108: 155–164.

Büttner-Ennever, J.A., Horn, A.K., Henn, V. and Cohen, B. (1999) Projections from the superior colliculus motor map to omnipause neurons in monkey. J. Comp. Neurol., 413: 55–67.

Büttner-Ennever, J.A., Horn, A.K., Scherberger, H. and D'Ascanio, P. (2001) Motoneurons of twitch and nontwitch extraocular muscle fibers in the abducens, trochlear, and oculomotor nuclei of monkeys. J. Comp. Neurol., 438: 318–335.

DeWeese, M.R. and Meister, M. (1999) How to measure the information gained from one symbol. Network, 10: 325–340.

Kralj-Hans, I., Baizer, J.S., Swales, C. and Glickstein, M. (2007) Independent roles for the dorsal paraflocculus and vermal lobule VII of the cerebellum in visuomotor coordination. Exp. Brain Res., 177: 209–222.

Marr, D. (1969) A theory of cerebellar cortex. J. Physiol. (Lond.), 202: 437–470.

Noto, C., Watanabe, S. and Fuchs, A.F. (1999) Characteristics of adaptation fields produced by behavioral changes in saccadic gain and direction. J. Neurophysiol., 81: 2798–2813.

Optican, L.M. and Robinson, D.A. (1980) Cerebellar-dependent adaptive control of primate saccadic system. J. Neurophysiol., 44: 1058–1076.

Scudder, C.A., Kaneko, C.S. and Fuchs, A.F. (2002) The brainstem burst generator for saccadic eye movements: a modern synthesis. Exp. Brain Res., 142: 439–462.

Scudder, C.A. and McGee, D.M. (2003) Adaptive modification of saccade size produces correlated changes in the discharges of fastigial nucleus neurons. J. Neurophysiol., 90: 1011–1026.

Soetedjo, R. and Fuchs, A.F. (2006) Complex spike activity of Purkinje cells in the oculomotor vermis during behavioral adaptation of monkey saccades. J. Neurosci., 26: 7741–7755.

Sokal, R.R. and Rohlf, F.J. (1994) Biometry: The Principles and Practice of Statistics in Biological Research (3rd ed.). Freeman, New York, pp. 685–793

Warabi, T., Kase, M. and Kato, T. (1984) Effect of aging on the accuracy of visually guided saccadic eye movement. Ann. Neurol., 16: 449–454.

C. Kennard & R.J. Leigh (Eds.)
Progress in Brain Research, Vol. 171
ISSN 0079-6123

CHAPTER 3.2

Role of the MST-DLPN pathway in smooth pursuit adaptation

Seiji Ono* and Michael J. Mustari

Division of Sensory-Motor Systems, Yerkes National Primate Research Center and Department of Neurology, Emory University, Atlanta, GA, USA

Abstract: The smooth pursuit system is able to adapt to challenges associated with development or injury to maintain pursuit accuracy. The medial superior temporal (MST) cortex and dorsolateral pontine nucleus (DLPN) are major components of the cortico-ponto-cerebellar pathway that carries signals including eye and visual motion that could play a role in smooth pursuit adaptation. To test this possibility, we first studied the effects of unilateral DLPN inactivation on pursuit adaptation in two juvenile rhesus monkeys. In a second series of experiments, we examined the activity of the MSTd neurons during smooth pursuit adaptation. Our results indicate that smooth pursuit adaptation could depend, in part, on eye/visual motion information provided by the MST-DLPN pathway to cerebellum.

Keywords: Pontine; Cortex; Eye movements; Adaptation; rhesus monkeys

Introduction

Smooth pursuit accuracy is maintained by adaptive mechanisms to keep the image of a moving object on the fovea. Previous lesion and single unit recording studies demonstrated that the cerebellum plays a role in ongoing and adapting smooth pursuit (Kahlon and Lisberger, 2000; Takagi et al., 2000). The MST cortex and the dorsolateral pontine nucleus (DLPN) are major components of the cortico-ponto-cerebellar pathway. It is unknown whether MST-DLPN pathway, which delivers eye and visual motion signals to the cerebellum, could also play a role in smooth pursuit adaptation. Here, we have studied the effects of unilateral DLPN inactivation on smooth pursuit adaptation. Because smooth pursuit is driven by volitional motor commands, it seems likely that adaptation might involve cortical circuits. Therefore, we have asked whether MSTd neurons have appropriate response dynamics to support adaptive change in smooth pursuit during double-steps of target speed.

Methods

Adaptive changes of horizontal smooth pursuit were produced by double-steps of target speed that step-up (10°/s–30°/s) or step-down (25°/s–5°/s), similar to prior studies (Fukushima et al., 1996; Kahlon and Lisberger, 1996). In the double-step paradigm, the target begins moving at one speed for first 100 ms and then changes to second speed for the remainder of the trial. Eye movements were measured using the magnetic search

*Corresponding author. Tel.: + 404-727-8128; Fax: + 404-727-9294; E-mail: sono2@emory.edu

DOI: 10.1016/S0079-6123(08)00621-3

coil technique. In muscimol injection experiments, we used a small-diameter (<50 μm) micropipette whose tip was positioned at the depth where smooth pursuit-related neurons in DLPN were recorded. Single unit activity was recorded from neurons in MSTd using customized epoxy-coated tungsten microelectrode (Frederick-Haer Corporation, Brunswick, ME). We verified that our neurons were located in DLPN and MSTd by functional, histological, and MRI (T-1 weighted, fast spin-echo; Siemens, 3-T magnet) criteria (Ono et al., 2004; Ono and Mustari, 2006).

Results

Following unilateral injection of muscimol in DLPN, we observed consistent deficits in the monkey's ability to generate and maintain smooth pursuit of a target moving towards the side of injection (Table 1). We then studied the effects of unilateral DLPN inactivation on smooth pursuit adaptation in a double-step paradigm (Fig. 1). Control testing showed significant adaptive changes of initial eye acceleration after 200 trials ($90.8 \pm 19.5°/s^2$, first 10 trials; $228 \pm 21.3°/s^2$, last 10 trials during adaptation; $P<0.001$, unpaired t-test). Following DLPN inactivation, ipsilesional smooth pursuit adaptation was significantly impaired, as measured after 200 trials ($78.2 \pm 21.7°/s^2$, first 10 trials; $90.4 \pm 20.9°/s^2$, last 10 trials during adaptation; $P=0.22$, unpaired t-test).

In a second series of experiments, we recorded 23 smooth pursuit related MSTd neurons that carry extraretinal signal before, during, and after adaptation. Figure 2A-a illustrates a representative MSTd neuron that showed a significant increase in firing rate (first 100 ms) post-adaptation (28.6 ± 7.2 spikes/s, pre-adaptation; 63.3 ± 25.2 spikes/s, post-adaptation; $P<0.001$, unpaired t-test) correlated with adaptive change in initial eye acceleration ($68.8 \pm 21.5°/s^2$, pre-adaptation; $119 \pm 40.7°/s^2$, post-adaptation; $P<0.01$, unpaired t-test). In contrast, the neuron shown in Figure 2A-b had similar responses across adaptation epochs (18.3 ± 6.5 spikes/s, pre-adaptation; 19.2 ± 7.3 spikes/s, post-adaptation; $P=0.77$, unpaired t-test) even though initial eye acceleration showed significant adaptation ($60.6 \pm 17.9°/s^2$, pre-adaptation; $101.8 \pm 31.4°/s^2$, post-adaptation; $P<0.01$, unpaired t-test). We found that some of our MSTd neurons (11/23) showed significant changes ($P<0.05$, unpaired t-test) in response modulation that correlated with smooth pursuit adaptation (Fig. 2B).

Discussion

Our studies were designed to consider whether eye motion and visual signals carried in the MST-DLPN

Table 1. Smooth pursuit performance and adaptive change following each muscimol injection

Exp. no.	Concentration	Volume	Side	Initial eye acceleration (%)		Adaptive change postlesion (%)			
				Ipsi	Contra	Step-up		Step-down	
						Ipsi	Contra	Ipsi	Contra
W1	5 mg/250 μl (2%)	150 nl	Left	70.9**	96.9	−16.7	103.8**	–	–
W2	5 mg/2.5 ml (0.2%)	200 nl	Left	74.1*	82.8	20.3	–	–	–
W3	5 mg/1.25 ml (0.4%)	200 nl	Left	65.0**	98.6	0.7	–	−56.6**	–
W4	5 mg/1.25 ml (0.4%)	200 nl	Left	60.2**	88.7	–	–	−50.6**	–
W5	5 mg/1.25 ml (0.4%)	200 nl	Left	46.6**	95.4	–	–	−43.3**	−14.5
W6	Saline (0.9%)	200 nl	Left	97.4	98.1	127.0**	–	–	–
H1	5 mg/1.25 ml (0.4%)	200 nl	Right	63.1**	74.9*	1.2	112.4**	–	–
H2	5 mg/1.25 ml (0.4%)	200 nl	Right	71.5**	105.7	4.6	–	–	–
H3	5 mg/1.25 ml (0.4%)	200 nl	Right	58.6**	97.5	14.6	–	−40.1**	–
H4	5 mg/1.25 ml (0.4%)	200 nl	Right	62.7**	85.2	–	–	−39.8**	−15.5
H5	Saline (0.9%)	200 nl	Right	107.5	101.2	89.0**	–	–	–

Note: Adaptive change (%) was calculated as the ratio of eye acceleration early and late in adaptation. Significance: $^{*}P<0.05$, $^{**}P<0.01$, unpaired t-test.

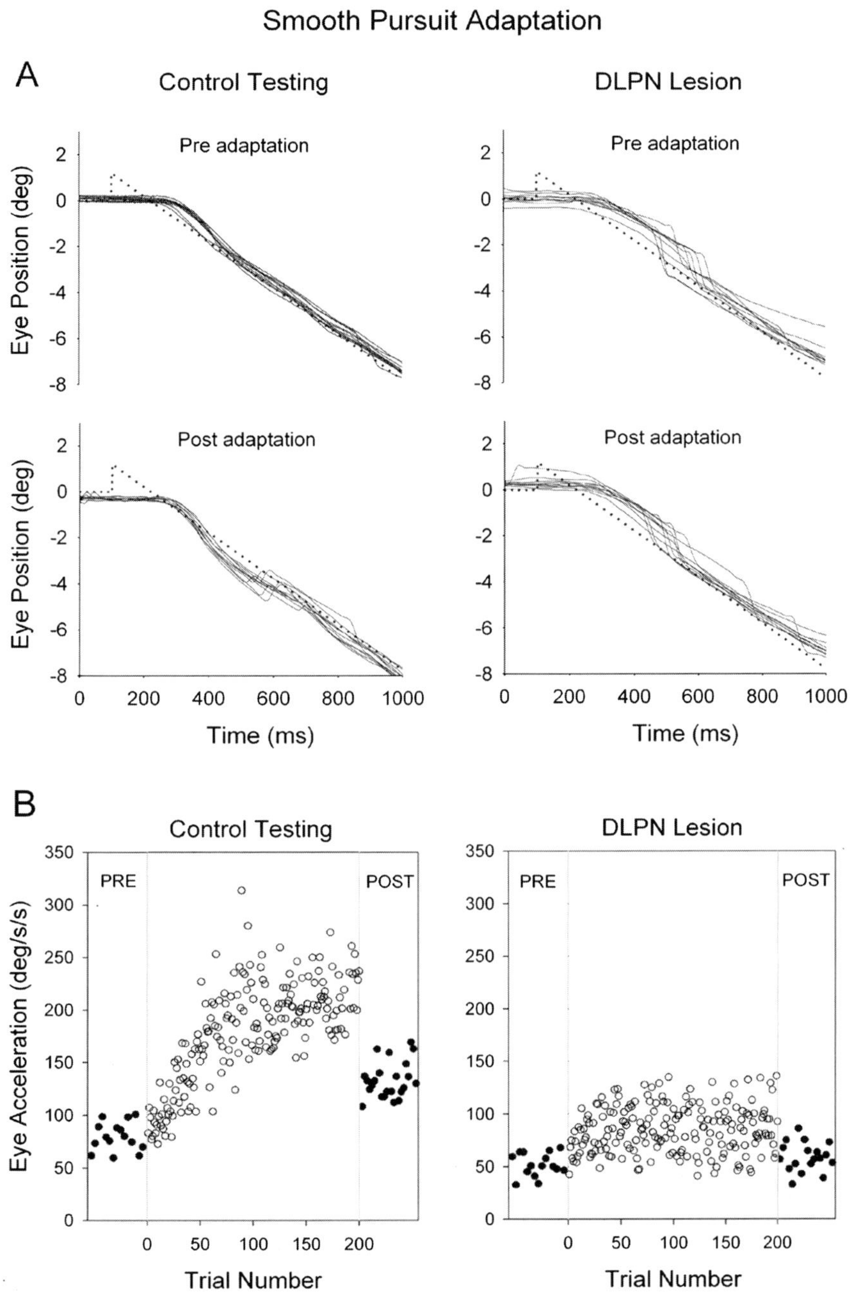

Fig. 1. Representative eye and target position traces in the ipsilesional direction are shown pre- and post-adaptation in control testing and following DLPN lesion during the step-up paradigm (A). Average eye acceleration during the first 100 ms of tracking in the step-up adaptation paradigm shown as a function of trial number in control and post-lesion testing (B).

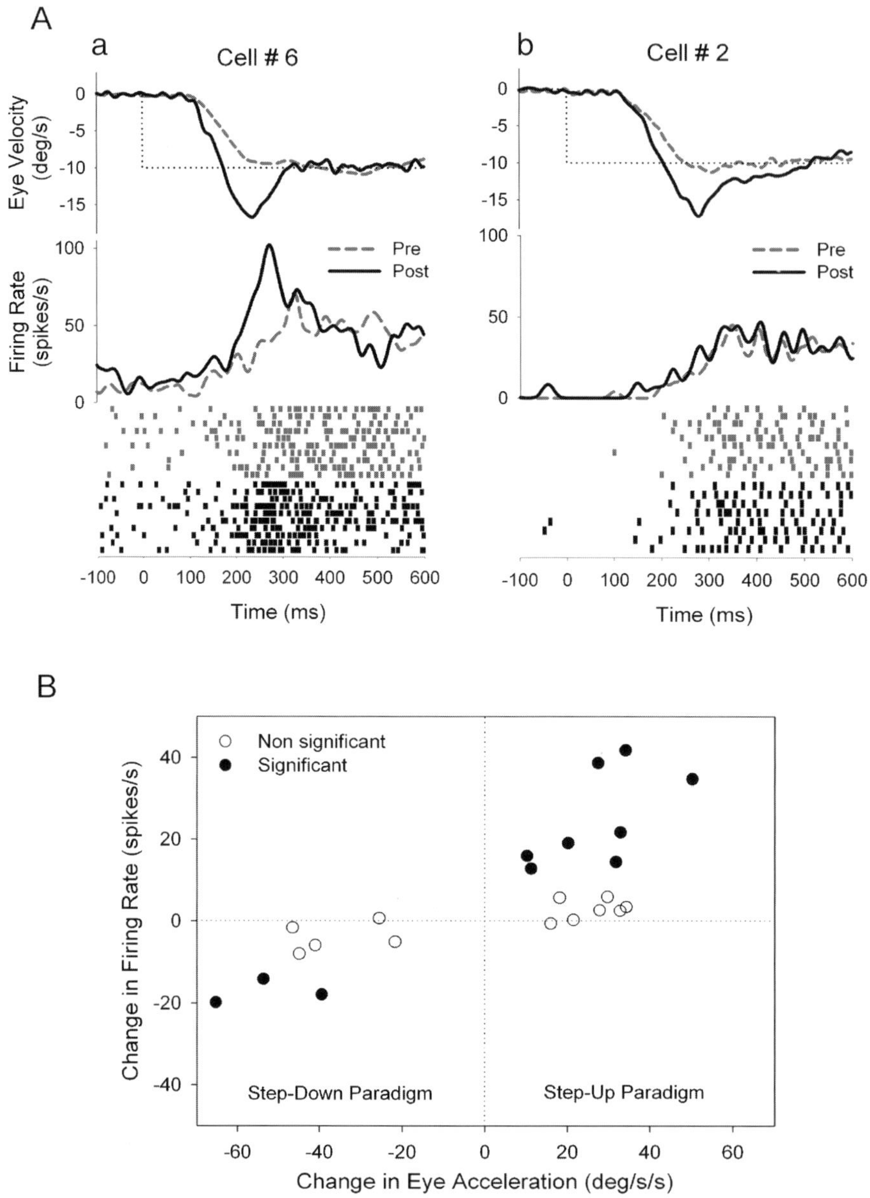

Fig. 2. Eye velocity traces and firing rate are shown for pre- and post-adaptation testing (A). MSTd neuron shows a significant increase in firing rate post-adaptation (A-a). MSTd neuron with no change in firing rate related to adaptive state even though eye velocity shows significant adaptation (A-b). Changes in firing rate of MSTd neurons ($n = 23$) during first 100 ms of pursuit plotted as a function of the concomitant change in initial eye acceleration pre- and post-adaptation (B). Filled circles indicate neurons with significant changes ($P < 0.05$, unpaired t-test) in firing rate post-adaptation.

pathway could play a role in smooth pursuit adaptation. Anatomical studies show that the DLPN is one of the main sources of mossy fibre input to the dorsal/ventral paraflocculus of the cerebellum (see Thier and Möck, 2006 for review). The DLPN also projects to vermal lobules VI and VII. Both the floccular complex (Kahlon and Lisberger, 2000) and oculomotor vermis (Takagi et al., 2000) have been shown to be involved in adaptation of smooth pursuit. During smooth pursuit, DLPN neurons provide eye motion signals (e.g., eye velocity) and foveal/parafoveal visual signals to the cerebellum over mossy fibre pathways. We inactivated the DLPN with muscimol to test its potential role in adaptive changes of horizontal smooth pursuit. Following unilateral DLPN inactivation, smooth pursuit adaptation was significantly impaired in the step-up paradigm for ipsilesional tracking. Therefore, smooth pursuit adaptation could depend, in part, on eye/visual motion information provided by DLPN neurons to cerebellum. The source of pursuit related signals in the DLPN is known to be derived, in part, from the cortical smooth pursuit system, which includes middle temporal (MT), medial superior temporal (MST) and frontal eye field (FEF) and supplementary eye field (SEF) regions of cortex (Distler et al., 2002; see Thier and Möck, 2006 for review). Neurons in area MST are modulated during smooth pursuit and carry visual motion and extraretinal signals (Newsome et al., 1988; Ilg and Thiir, 2003; Ono and Mustari, 2006).We found that some MSTd neurons showed significant changes in response modulation that correlated with smooth pursuit adaptation. Therefore, our findings and those of other laboratories support the suggestion that smooth pursuit adaptation could depend on eye and visual motion information travelling in cortical (MST), pontine (DLPN), and cerebellar pathways. Further studies are needed to determine whether MSTd neurons are essential for smooth pursuit adaptation and whether they are on the afferent or efferent side of the response.

Acknowledgement

This work was supported by National Institutes of Health Grants, NEI, EY13308, and RR00165.

References

Distler, C., Mustari, M.J. and Hoffmann, K.P. (2002) Cortical projections to the nucleus of the optic tract and dorsal terminal nucleus and to the dorsolateral pontine nucleus in macaques: a dual retrograde tracing study. J. Comp. Neurol., 444: 144–158.

Fukushima, K., Tanaka, M., Suzuki, Y., Fukushima, J. and Yoshida, T. (1996) Adaptive changes in human smooth pursuit eye movement. Neurosci. Res., 25: 391–398.

Ilg, U.J. and Their, P. (2003) Visual tracking neurons in primate area MST are activated by smooth-pursuit eye movements of an "imaginary" target. J. Neurophysiol., 90: 1489–1502.

Kahlon, M. and Lisberger, S.G. (1996) Coordinate system for learning in the smooth pursuit eye movements of monkeys. J. Neurosci., 16: 7270–7283.

Kahlon, M. and Lisberger, S.G. (2000) Changes in the responses of Purkinje cells in the floccular complex of monkeys after motor learning in smooth pursuit eye movements. J. Neurophysiol., 84: 2945–2960.

Newsome, W.T., Wurtz, R.H. and Komatsu, H. (1988) Relation of cortical areas MT and MST to pursuit eye movements. II. Differentiation of retinal from extraretinal inputs. J. Neurophysiol., 60: 604–620.

Ono, S., Das, V.E. and Mustari, M.J. (2004) Gaze-related response properties of DLPN and NRTP neurons in the rhesus macaque. J. Neurophysiol., 91: 2484–2500.

Ono, S. and Mustari, M.J. (2006) Extraretinal signals in MSTd neurons related to volitional smooth pursuit. J. Neurophysiol., 96: 2819–2825.

Takagi, M., Zee, D.S. and Tamargo, R.J. (2000) Effects of lesions of the oculomotor cerebellar vermis on eye movements in primate: smooth pursuit. J. Neurophysiol., 83: 2047–2062.

Their, P. and Möck, M. (2006) The oculomotor role of the pontine nuclei and the nucleus reticularis tegmenti pontis. In: Progress in Brain Research, Vol. 151. Elsevier, Amsterdam, pp. 293–320.

C. Kennard & R.J. Leigh (Eds.)
Progress in Brain Research, Vol. 171
ISSN 0079-6123

CHAPTER 3.3

Lesions of the cerebellar nodulus and uvula in monkeys: effect on otolith-ocular reflexes

Mark F. Walker[1,2,4,*], Jing Tian[1], Xiaoyan Shan[1], Rafael J. Tamargo[3], Howard Ying[2] and David S. Zee[1,2]

[1]*Department of Neurology, The Johns Hopkins University, Baltimore, MD, USA*
[2]*Department of Ophthalmology, The Johns Hopkins University, Baltimore, MD, USA*
[3]*Department of Neurosurgery, The Johns Hopkins University, Baltimore, MD, USA*
[4]*Department of Neurology, Case Western Reserve University and Cleveland VAMC, Cleveland, OH, USA*

Abstract: We studied two rhesus monkeys before and after surgical ablation of the nodulus and uvula (Nod/Uv) of the cerebellum. Three-axis eye movements were recorded with the magnetic-field scleral search coil system during a variety of vestibular and ocular motor tasks. Here we describe the effects of the Nod/Uv lesions on dynamic (head translation) and static (head tilt) otolith-mediated vestibulo-ocular reflexes. The main findings were: 1. eye velocity during sinusoidal vertical translation (1.5 Hz) was reduced by 59% in the dark and 36% in the light; 2. eye velocity during steps of horizontal translation was reduced, but only in the dark and more so during the sustained (constant velocity) than the initial (acceleration) part of the response, and 3. there was a torsional nystagmus that depended on the position of roll head tilt, but static ocular counterroll was unchanged. These results suggest new roles for the Nod/Uv in the processing of otolith signals. This is likely important not only for facilitating gaze during linear head motion, but also for maintaining postural stability and one's orientation relative to gravity. The lesions appeared to have a greater effect on responses to vertical motion, particularly in the light (in contrast, responses to interaural translation in the light were nearly normal), suggesting a particular importance of the Nod/Uv in processing signals arising from the sacculi.

Keywords: vestibulo-ocular reflex; translation; torsion; vestibular; cerebellum; utricle; saccule; TVOR; nystagmus

Introduction

The nodulus and the uvula (Nod/UV) of the inferior cerebellar vermis belong to the phylogenetically old vestibulocerebellum. They are densely interconnected with the vestibular nuclei and other brainstem areas involved in eye movements (reviewed in Voogd and Barmack, 2005), and they receive direct afferent input from the vestibular nerve. Prior studies have shown that the Nod/Uv plays an important role in the several key aspects of vestibular reflexes, in particular the ability to relate one's own motion and orientation to the gravitational upright. An example of this is the alignment of the eye-velocity axis with the gravito-inertial vector during sustained off-vertical-axis

*Corresponding author. Tel.: 216-791-3800, Ext. 5225; Fax: 216-231-3461; E-mail: mark.walker@case.edu

DOI: 10.1016/S0079-6123(08)00622-5

rotation; this function depends on both otolith inputs and an intact velocity-storage mechanism and is abolished by lesions of the Nod/Uv (Angelaki and Hess, 1995; Wearne et al., 1996). In fact, it has been suggested that the role of velocity storage in the rotational vestibulo-ocular reflex (rVOR) may be less important than its function in the control of body orientation and posture (Green and Angelaki, 2003).

The sense of orientation depends critically on the proper combination of otolith inputs and semicircular canal signals. Thus, it is expected that the Nod/Uv would play an important role in the processing of otolith information. Although prior studies have shown that diffuse cerebellar diseases can impair the translational vestibulo-ocular reflex (tVOR), the specific role of individual cerebellar areas, including the Nod/Uv, is not well known. In this study, we investigated in two rhesus monkeys the effect of Nod/Uv lesions on the otolith-ocular reflexes, in response to static head tilts (counterroll) and to linear head motion (translation). Here we will focus attention on the role of the Nod/Uv in the control of static and dynamic otolith reflexes. These reflexes both serve vision, by reducing gaze errors induced by head movement, and orientation sense. In the same series of experiments, we also found that the Nod/Uv are critical for normal vertical pursuit but not large-field ocular following (Walker et al., 2008).

Methods

Two juvenile rhesus monkeys were studied under a protocol that was approved by the Institutional Animal Care and Use Committee of the Johns Hopkins University. Eye movements were recorded using the magnetic field scleral search coil system with three magnetic fields and dual coils in each eye. The eye movement calibration method has been previously described (Tian et al., 2007). Lesions of the Nod/Uv were performed by neurosurgical aspiration with direct visualization under inhalation anaesthesia, using full sterile technique and post-operative analgesia. Lesion extent was confirmed with MRI and post-mortem histology.

Motion stimuli

Different systems were used to test horizontal and vertical tVORs. Horizontal translation was performed on a belt-driven linear sled (Acutronic, Zürich, Switzerland). The stimulus consisted of abrupt transient motion (0.4 g to a plateau speed of 40 cm/s), either in the dark or with fixation of a laser target that was back-projected onto a tangent screen in front of the monkey (27 or 70 cm viewing distance). Chair position and velocity were monitored by a linear sensor and fed back to the host computer.

The vertical tVOR was tested on a manually driven, spring-assisted sled that was mounted on tracks to constrain its motion. The chair faced a visual stimulus that was 122 cm away, and recordings were made in both light and complete darkness. Vertical head motion was measured using an accelerometer that was firmly attached to the animal chair, oriented with the vertical axis. To derive head velocity from recorded head acceleration, an optimization algorithm was first used to determine the offset (bias) of the acceleration signal due to gravity. This bias was subtracted from the accelerometer signal and was also used to calibrate the signal (offset = 1 g = 980 cm/s^2). The normalized acceleration signal was numerically integrated (cumtrapz, www.scipy.org) to yield linear head velocity in centimetre per second. Angular eye velocity was determined using standard rotation vector techniques.

Quick phases and saccades were detected and removed from the analysis. The quick-phase detection algorithm was based on acceleration and jerk thresholds (Wyatt, 1998). The right-hand-rule convention is used for all data: positive positions and velocities are leftward, downward, and clockwise.

Results

Vertical tVOR

Smooth eye movements elicited by vertical translation (bob) were substantially reduced by the Nod/Uv lesions (Walker et al., 2008). This decrease was greater for translation in the dark (59% decrease)

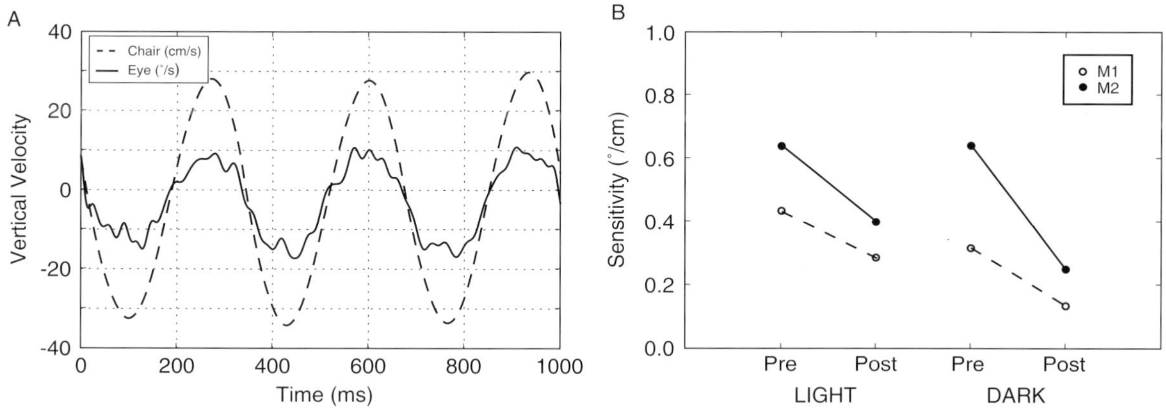

Fig. 1. Responses to vertical translation on a manually driven spring-assisted chair. The stimulus frequency was approximately 1.5 Hz. (A) Example of a response during translation in complete darkness. Head velocity was calculated from the accelerometer signal as described in the Methods section. Head velocity is inverted for easier comparison to eye velocity (vertical component of angular eye velocity). (B) Response sensitivities (eye velocity/head velocity). These were calculated by robust linear regression (rlm in R/rpy, www.cran.r-project.org); the regression slope is the sensitivity.

but was also present during translation in the light (36% decrease) (Fig. 1). Thus, although a visual stimulus was still able to enhance the response after the lesion, this enhancement was not enough to restore eye velocity to its pre-lesion values. Visual enhancement was seen during both upward (85% enhancement) and downward (108% enhancement) motion.

Horizontal interaural tVOR

The interaural tVOR (IA tVOR) was also affected by the Nod/Uv lesions. Figure 2 shows eye velocity during a representative step of leftward motion, before and after the lesion, both in darkness and with lights on. As for the vertical tVOR, the post-lesion response was reduced. The use of abrupt steps of motion allowed us to examine the timing of this change, and we found that the earliest part of the response was less affected than the plateau portion. Elsewhere, we have shown that the response change is consistent with a deficit in the integration of head acceleration to maintain eye speed during constant-velocity head translation (Walker et al., 2007).

Although the Nod/Uv lesions impaired the processing of otolith signals, and hence the tVOR in the dark, they did not prevent visual tracking from compensating for this deficit. In fact, responses in the light had similar magnitudes before and after the lesions (Fig. 2A). Thus, visual mechanisms seemed to play an even greater role for the IA tVOR post-lesion than for the vertical tVOR, although the difference in the stimuli (steps vs. sinusoidal translation) might account for at least some of this difference.

We also examined the binocularity of the IA tVOR. How well is vergence position maintained during prolonged translation, and is this dependent on an intact Nod/Uv? This is important, because a loss of sustained convergence, particularly in the dark, might indicate an impaired ability to maintain the percept of target distance, and hence the response amplitude. We did not find a general loss of horizontal vergence control after the Nod/Uv lesions. For the example of Fig. 2B (M1, 70 cm), there was only a minimal change (<1°) in the vergence angle. Most importantly, there was no loss of vergence during the course of the trials in the dark, either before or after the lesion. Thus, the decrease of the sustained interaural tVOR after the Nod/Uv lesions cannot be attributed to an inability to hold vergence in the dark.

Responses to static roll tilts

The otolith organs not only detect linear head motion, but are also affected by the linear

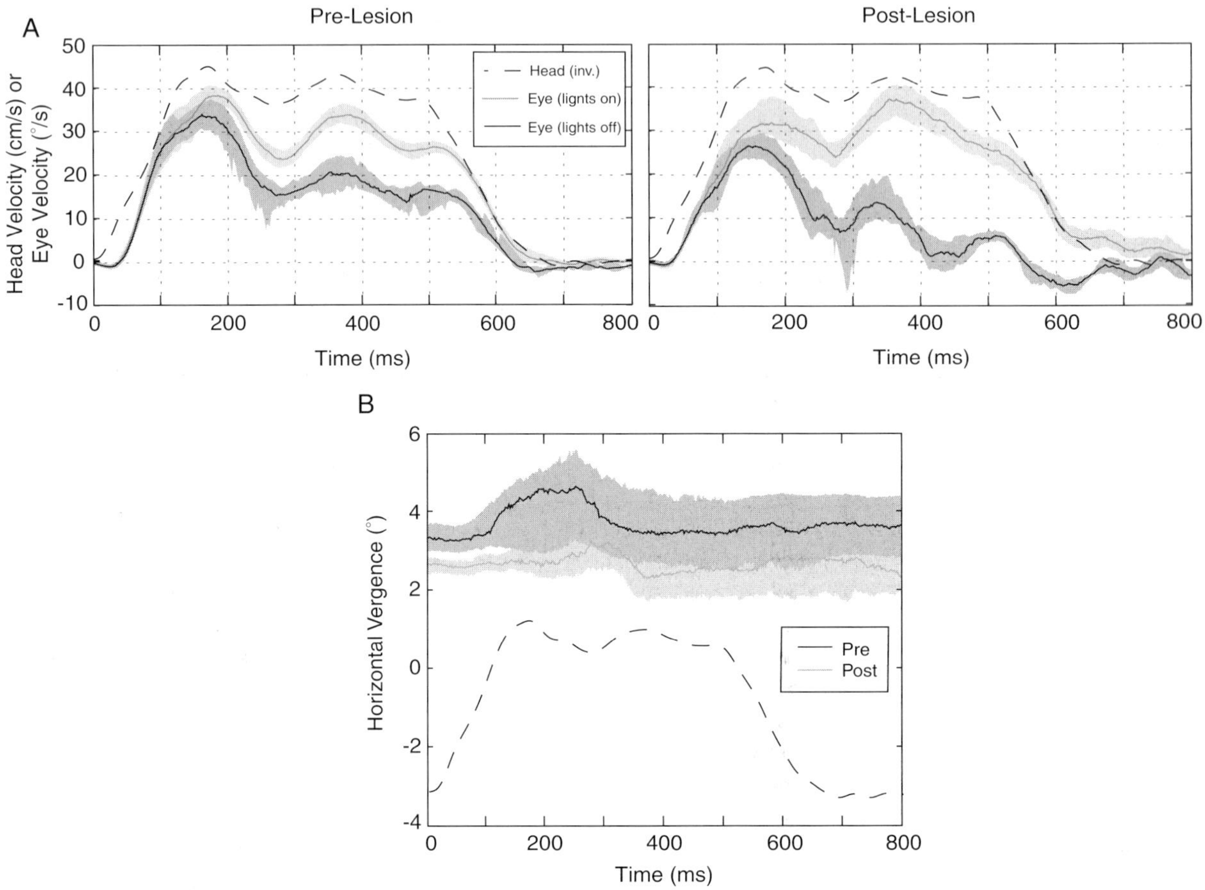

Fig. 2. (A) Responses to rightward interaural translation in M1 (70 cm viewing distance). The left panel shows responses before the Nod/Uv lesion and the right panel responses after the lesion. Solid lines represent the medians of the full set of similar trials, and the corresponding shaded areas include the 25th–75th percentiles of eye velocity. Quick phases and saccades have been excluded. The dashed line shows head velocity (in cm/s) inverted for comparison. Note that this is not the same as the ideal eye velocity, which depends on both head velocity and viewing distance. Each panel compares responses for trials in which the target remained on to those in which the target went out (i.e., translation in the dark). (B) Horizontal vergence during translation in the dark after viewing a target at 70 cm (same data as in A). Convergence is positive. Again, the solid lines show the median values and the shaded areas the two middle quartiles. The head velocity (dashed line) is included for reference only and is not scaled.

acceleration due to gravity, according to the orientation of each otolith organ relative to the gravity vector. Normally, the sensation of head tilt is less important for gaze stability (there is no ongoing head motion) than for maintaining balance and upright posture; nonetheless, the static otolith-ocular reflexes provide a measure of the integrity of the reflexes and of the brain's ability to distinguish a head tilt from a translation that would produce a similar change in linear acceleration. Figure 3 shows the ocular torsion (ocular counterroll, OCR) elicited by roll head tilts in M2 before and after the Nod/Uv lesion (we focused attention on M2, because torsion in M1 may have been affected by cyclovertical muscle weakness related to a separate study of superior oblique paresis (Shan et al., 2007). The lesion had little effect on OCR (15% pre vs. 14% post). Thus, even though Nod/Uv lesions have a large effect on the dynamic roll tVOR mediated by the semicircular canals (Angelaki and Hess, 1994), they do not affect the otolith-mediated response to

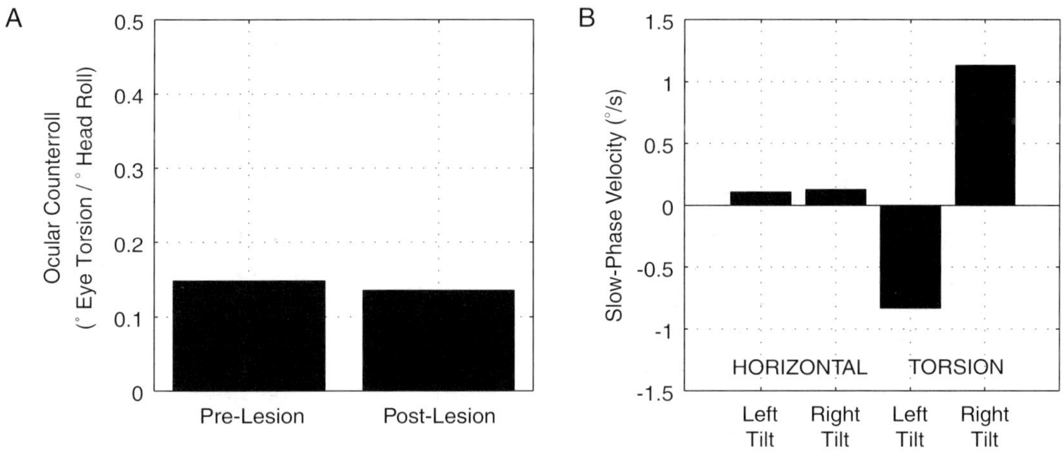

Fig. 3. (A) Ocular counterroll (OCR) before and after the Nod/Uv lesion (M2). OCR was calculated as the difference in the torsional offset of Listing's plane when the head was tilted 15° to the left compared to 15° to the right, divided by the difference in roll head position (30°). OCR was calculated separately for each eye, and the two values which were similar were averaged. (B) Horizontal and torsional slow-phase velocity (SPV) in the dark with the head tilted 15° to the left or right, recorded after the Nod/Uv lesion. Each value is the median eye velocity, after exclusion of saccades, over the full range of spontaneous eye movements. There was no torsional nystagmus before the lesion (SPV 0.02°/s with right tilt and −0.15°/s with left tilt).

static roll. There was no head-tilt-dependent horizontal nystagmus (to suggest that tilt was misinterpreted as translation), although there was a small torsional nystagmus that consisted of clockwise slow phases with the right ear down and counterclockwise slow phases with the left ear down, i.e., the direction of the slow phases was opposite to that of the change in the torsional position of the eyes that characterizes the OTR. The mechanism of the torsional nystagmus is uncertain. Unlike horizontal nystagmus, torsion is never a necessary response to translation, and thus it is unlikely that this is a tilt-translation misinterpretation.

Conclusion

Our study provides important new evidence for a critical role of the Nod/Uv in otolith-ocular reflexes. Not only is the Nod/Uv required for the modulation of the rVOR by otolith information, as others have shown, but here we show it is also important for the processing of otolith signals to drive the tVOR. Ablation of the Nod/Uv impairs the ocular response to vertical translation and reduces eye velocity during sustained horizontal translation along the interaural direction. Thus, processing of both utricular and saccular inputs is under the control of the Nod/Uv. Moreover, our data suggest that the function of the Nod/Uv in the horizontal and vertical tVOR may differ: Nod/Uv lesions reduced the interaural tVOR only in the dark, whereas the vertical tVOR was impaired substantially, even in the light.

Acknowledgements

This research was supported by NIH grant EY001849 (DSZ), the Albert Pennick Fund, and by the Arnold-Chiari Foundation. Dr. Walker was a Pollin Scholar. Adrian Lasker and Dale Roberts provided technical assistance. Measurement of the interaural tVOR was performed in the laboratory of Dr. Lloyd Minor. Histology was performed by Mitchell Smith, slides were scanned by Camilo Pardo, and Dr. Carlos Pardo reviewed the slides. Dr. Peter Barker and Hugh Wall assisted with MRI scans.

References

Angelaki, D.E. and Hess, B.J. (1994) The cerebellar nodulus and ventral uvula control the torsional vestibulo-ocular reflex. J. Neurophysiol., 72(3): 1443–1447.

Angelaki, D.E. and Hess, B.J. (1995) Inertial representation of angular motion in the vestibular system of rhesus monkeys. II. Otolith-controlled transformation that depends on an intact cerebellar nodulus. J. Neurophysiol., 73(5): 1729–1751.

Green, A.M. and Angelaki, D.E. (2003) Resolution of sensory ambiguities for gaze stabilization requires a second neural integrator. J. Neurosci., 23(28): 9265–9275.

Shan, X., Tian, J., Ying, H., Quaia, C., Optican, L.M., Walker, M.F., Tamargo, R.J. and Zee, D.S. (2007) Acute superior oblique palsy in monkeys: I. Changes in static eye alignment. Invest. Ophthalmol. Vis. Sci., 48(6): 2602–2611.

Tian, J., Zee, D.S. and Walker, M.F. (2007) Rotational and translational optokinetic nystagmus have different kinematics. Vis. Res., 47(7): 1003–1010.

Voogd, J. and Barmack, N.H. (2005) Oculomotor cerebellum. In: Büttner-Ennever J.A. (Ed.), Neuroanatomy of the Oculomotor System, Progress in Brain Research, Vol. 151. Elsevier, Amsterdam, pp. 231–268.

Walker, M.F., Tian, J., Shan, X., Ying, H., Tamargo, R. and Zee, D.S. (2007) The cerebellar nodulus and uvula mediate integration of linear head acceleration. Program No. 861.3, Society for Neuroscience, Washington, DC (abstract).

Walker, M.F., Tian, J., Shan, X., Tamargo, R.J., Ying, H. and Zee, D.S. (2008) O1_MRKO1_MRKEffect of cerebellar lesions in monkey on gaze stability. In: Leigh R.J. and Deveraux M.W. (Eds.), Advances in Understanding of Mechanisms and Treatment of Infantile Forms of Nystagmus. Oxford University Press, New York.

Wearne, S., Raphan, T. and Cohen, B. (1996) Nodulo-uvular control of central vestibular dynamics determines spatial orientation of the angular vestibulo-ocular reflex. Ann. N.Y. Acad. Sci., 781: 364–384.

Wyatt, H.J. (1998) Detecting saccades with jerk. Vis. Res., 38(14): 2147–2153.

C. Kennard & R.J. Leigh (Eds.)
Progress in Brain Research, Vol. 171
ISSN 0079-6123

CHAPTER 3.4

Vergence eye movement signals in the cerebellar dorsal vermis

Takuya Nitta[1,2], Teppei Akao[1], Sergei Kurkin[1] and Kikuro Fukushima[1,*]

[1]*Department of Physiology, Hokkaido University School of Medicine, Sapporo 060-8638, Japan*
[2]*Department of Ophthalmology, Hokkaido University School of Medicine, Sapporo 060-8638, Japan*

Abstract: We examined simple-spike activity of Purkinje cells (P-cells) that responded during a search task which required both vergence- and frontal-pursuit. Of a total of 100 responding P-cells, 16% discharged only for frontal-pursuit, 43% only for vergence-pursuit, and 41% for both. Thus, the majority of vermal pursuit P-cells modulated their activity during vergence-pursuit. These P-cells also discharged for vergence eye movements induced by step target-motion in-depth. The majority of vergence related P-cells carried convergence signals with both eye velocity and position sensitivities, and they discharged before the onset of convergence eye movements. Muscimol infusion into the sites where convergence P-cells were recorded resulted in a reduction of peak convergence eye velocity, of initial convergence eye acceleration, and of frontal-pursuit eye velocity. These results suggest specific involvement of the dorsal vermis in vergence eye movements.

Keywords: cerebellum; vermis; Purkinje cell; vergence; pursuit; muscimol

Introduction

The cerebellar dorsal vermis has been known to be related to conjugate eye movements, such as saccade and smooth pursuit in fronto-parallel planes (i.e., frontal-pursuit). However, there is no detailed study on vermal Purkinje cell (P-cell) activity for disconjugate, vergence eye movements. Previous studies indicate that in the upstream pathways, the majority of pursuit neurons in the frontal eye fields (FEF) carry not only frontal-pursuit signals but also vergence signals (Fukushima et al., 2002; Akao et al., 2005). To understand the role of the cerebellar dorsal vermis in vergence eye movements, we examined discharge characteristics of vermal pursuit P-cells.

Materials and methods

Three Japanese monkeys were trained to pursue a virtual target through polarized shutter glasses with their head restrained. Eye movements were monitored using the scleral search coil method. In our search task, a virtual target was moved sinusoidally at 0.5 Hz with oblique trajectories in 3D space. The monkeys pursued the target with the combination of frontal- and vergence-pursuit. P-cells were identified by the existence of complex spikes in the dorsal vermis, and simple spike activity was recorded from single P-cells. Once responsive P-cells were encountered, their responses

*Corresponding author. Tel.: +81 11 706-5038;
Fax: +81 11 706-5041; E-mail: kikuro@med.hokudai.ac.jp

DOI: 10.1016/S0079-6123(08)00623-7

were tested during sinusoidal frontal- and vergence-pursuit, separately. To examine velocity sensitivity, the target was moved at different frequencies (0.3 Hz–1.0 Hz, at ±5°). We also used vergence ramp or step tasks and double target task (Nitta et al., 2007). In the vergence step task, the target was jumped between far point and near points, repeatedly. The monkeys were required to make rapid convergence or divergence eye movements. In a double-target task, monkeys fixated a stationary target and a second spot was sinusoidally moved in depth. The recording sites were confirmed histologically in the cerebellar dorsal vermis (lobules VI and VII). All procedures were performed in strict compliance with the guidelines for the Care and Use of Animals of NIH. Our specific procedures were approved by the Animal Care and Use Committee of Hokkaido University School of Medicine.

Results

We analysed simple-spike discharge of 100 vermal pursuit P-cells that responded during our search task. Of these, 41% responded during both vergence- and frontal-pursuit, 43% responded only during vergence-pursuit, and 16% responded only during frontal-pursuit. Thus, 84% of our pursuit P-cells responded during vergence-pursuit. Figure 1 shows representative responses. This P-cell

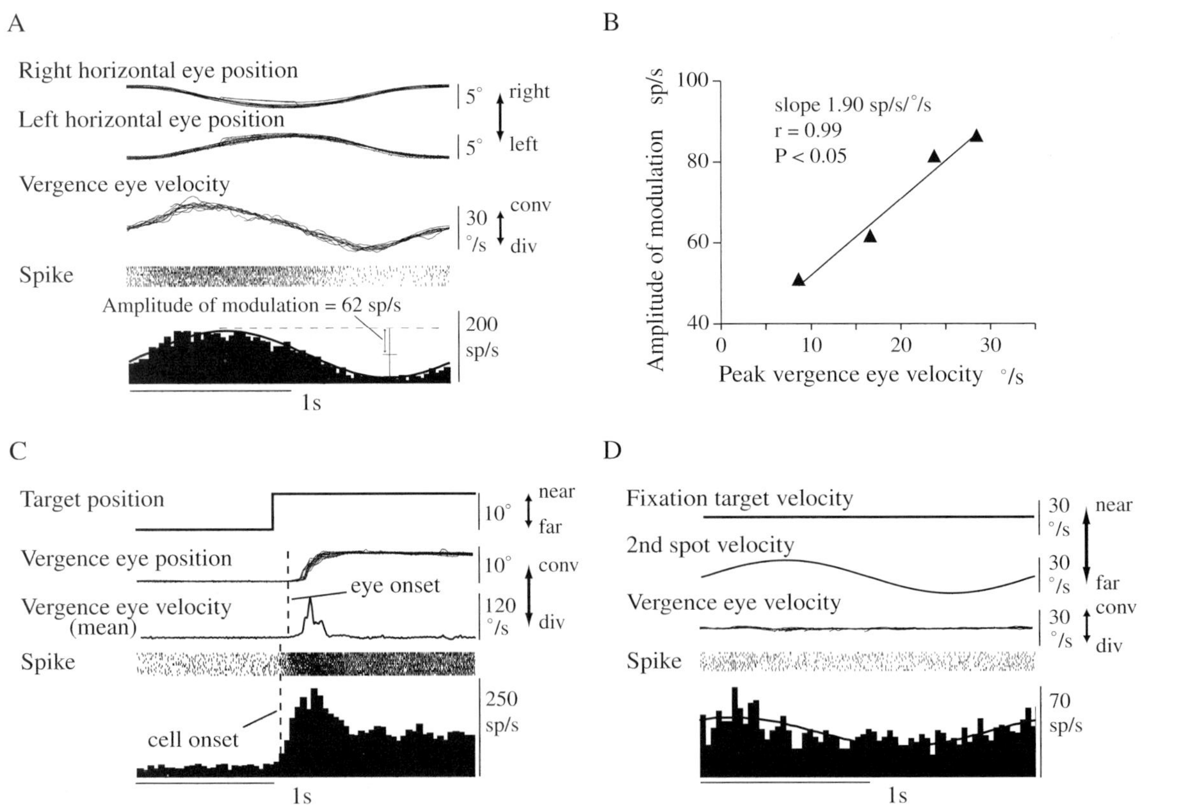

Fig. 1. Responses of a representative vergence-related vermal P-cell. Vergence eye position (velocity) was calculated as the difference between left horizontal eye position (velocity) and right horizontal eye position (velocity). Fitted sine curves are superimposed on the spike histograms (A, D). conv: convergence, div: divergence. Adapted with permission from Nitta et al. (2007).

discharged maximally during peak convergence eye velocity (Fig. 1A, 0.5 Hz, $\pm 5°$). Of the 84 vergence-related P-cells, 45 (54%) discharged during peak convergence eye velocity, 16 (19%) discharged during peak divergence eye velocity, and 23 (27%) discharged during peak vergence eye position. In Fig. 1B, the amplitude of discharge modulation of this P-cell increased as vergence eye velocity increased. Of 42 tested P-cells, the majority ($n = 28$, 67%) had significant positive linear correlation between amplitude of discharge modulation and peak vergence eye velocity.

During target step motion (Fig. 1C), this P-cell showed burst discharge during convergence eye movement followed by tonic discharge during the maintenance of convergence eye position. Of 40 tested P-cells, 8 (20%) showed only burst discharge, 25 (63%) showed both burst and tonic discharge, and 7 (17%) showed only tonic discharge. Thus, the majority ($n = 32$, 80%) had vergence eye position sensitivity. Of the 32 P-cells, 27 (81%) showed tonic discharge for convergence eye position, 5 (19%) showed tonic discharge for divergence eye position. Some vermal P-cells showed visual responses during the double-target task (Fig. 1D). Of 42 tested vergence-related P-cells, 14 (33%) had visual sensitivity in depth. In Fig. 1C, this P-cell discharged before the onset of convergence eye movements. When 40 P-cells were tested during step vergence eye movements toward their preferred directions (33 convergence, 7 divergence), the majority ($n = 29$, 73%) discharged before the onset of vergence eye movements with the median lead time of 16 ms.

After muscimol injection into the right dorsal vermis (Fig. 2A), peak convergence eye velocity and initial convergence eye acceleration decreased. The maintenance of convergence eye movements was not affected. Rightward frontal-plane pursuit eye velocity also decreased. After muscimol injection into the left dorsal vermis (Fig. 2B), the monkey often made rightward saccades near the onset of vergence eye movements. When the monkey made pure vergence eye movements, the onset of

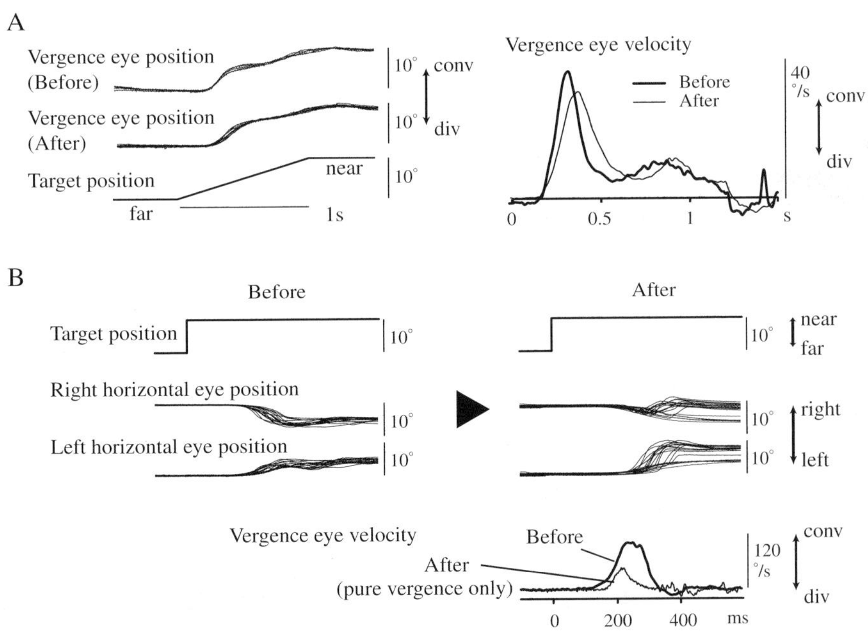

Fig. 2. Muscimol injections (0.5–1.0 μl, 10 μg/1 μl) into the recording sites. Injections were made into the right (A) and left (B) dorsal vermis, respectively. At the bottom (B), mean vergence eye velocities are compared before and after muscimol injection where only the trials without saccade intrusion were averaged. Adapted with permission from Nitta et al. (2007).

convergence eye movement was delayed and peak convergence eye velocity decreased. Divergence eye movements were less affected and ipsiversive saccades became hypometric.

Discussion

Both convergence and divergence signals are found in neurons in the FEF and the nucleus reticularis tegmenti pontis (NRTP) (Gamlin and Clarke, 1995). The NRTP receives projection from the FEF and sends axons to the cerebellar dorsal vermis. Vermal convergence signals that were found in the present study are most likely to be sent from the FEF via NRTP. It is well known that mossy fibres originating in the pontine nuclei send signals not only to the cerebellar cortex but also to the deep cerebellar nuclei with axon collaterals. In the preliminary reports, it was suggested that neurons in the caudal fastigial nucleus, that receives direct inhibitory projections from dorsal vermal P-cells, are also involved in convergence eye movements (Gamlin and Zhang, 1996; Zhang and Gamlin, 1996), suggesting that the dorsal vermis-caudal fastigial pathway may be specifically involved in the initiation of convergence eye movements. To understand the exact neural mechanisms, further studies are needed for discharge characteristics of caudal fastigial neurons.

Acknowledgement

This research was supported by Grant-in-Aid for Scientific Research on Priority Areas (system study on higher-order brain functions) (17022001) and (B) (18300130) from the MEXT of Japan.

References

Akao, T., Kurkin, S.A., Fukushima, J. and Fukushima, K. (2005) Visual and vergence eye movement-related responses of pursuit neurons in the caudal frontal eye fields to motion-in-depth stimuli. Exp. Brain Res., 164(1): 92–108.

Fukushima, K., Yamanobe, T., Shinmei, Y., Fukushima, J., Kurkin, S. and Peterson, B.W. (2002) Coding of smooth eye movements in three-dimensional space by frontal cortex. Nature, 419(6903): 157–162.

Gamlin, P.D. and Clarke, R.J. (1995) Single-unit activity in the primate nucleus reticularis tegmenti pontis related to vergence and ocular accommodation. J. Neurophysiol., 73(5): 2115–2119.

Gamlin, P. and Zhang, H. (1996) Effect of muscimol blockade of posterior fastigial nucleus on vergence and accommodation in the primate. Soc. Neurosci. Abstr., 22: p. 2034.

Nitta, T., Akao, T., Kurkin, S. and Fukushima, K. (2007) Involvement of the cerebellar dorsal vermis in vergence eye movements in monkeys. Cereb. Cortex, doi:10.1093/cercor/bhm143.

Zhang, H. and Gamlin, P. (1996) Single-unit activity within the posterior fastigial nucleus during vergence and accommodation in the alert primate. Soc. Neurosci. Abstr., 22: p. 2034.

C. Kennard & R.J. Leigh (Eds.)
Progress in Brain Research, Vol. 171
ISSN 0079-6123

CHAPTER 3.5

Oculomotor anatomy and the motor-error problem: the role of the paramedian tract nuclei

Paul Dean* and John Porrill

Department of Psychology, University of Sheffield, Sheffield S10 2TP, UK

Abstract: Anatomical evidence indicates that copies of premotor commands to ocular motoneurons are sent to the cell groups of the paramedian tract, whose projections constitute a major mossy-fibre input to the floccular region of the cerebellum. Damage to this relay impairs gaze-holding, so producing retinal slip signals that are also relayed to the flocculus, in this case as climbing-fibre input. These observations suggest that the relation between efference copy and sensory error is important, and might be used by the cerebellum to learn accurate movements. By modelling the flocculus as an adaptive filter using a covariance learning rule, we show that in simulation the cerebellar cortex can in fact learn to decorrelate efference copy from motor command, and thereby compensate for changes to the oculomotor plant. This mechanism appears to be very robust with respect to plant characteristics and noise, and can cope with error-signal delay provided there is a brainstem site of plasticity. Its general significance is that it removes the need for motor-error signals, which are typically unavailable and in any case not conveyed by climbing fibres. This appears to be an example where anatomical findings have helped address a long-standing problem in adaptive control.

Keywords: flocculus; vestibulo-ocular reflex; oculomotor; cerebellum; motor learning; retinal slip; climbing fibre; paramedian tract; decorrelation control

Introduction

Jean Büttner-Ennever has drawn attention to the anatomy of the cell groups of the paramedian tracts (PMT) in a series of reviews (Büttner-Ennever et al., 1989; Büttner-Ennever, 1992; Büttner-Ennever and Horn, 1996). Input to these cell groups appears to be a copy of the inputs sent to the ocular motoneurons, and their outputs are primarily to the floccular region of the cerebellum (flocculus and ventral paraflocculus, hereinafter simply flocculus). In fact, this cerebellar region receives mossy-fibre input from more neurons in the PMT cell groups than from vestibular neurons (Fig. 1A). These anatomical findings raise an intriguing question about the role of the flocculus in oculomotor control: what useful purpose could be served by sending it a massive efference copy of oculomotor commands?

An early clue was provided by clinical observations, suggesting that damage to PMT cell groups might impair gaze-holding (Zee, referred to in Büttner-Ennever et al., 1989, p. 538). Impaired

*Corresponding author. Tel.: +44 (0)114 222 6521;
Fax: +44 (0)114 276 6515; E-mail: p.dean@sheffield.ac.uk

DOI: 10.1016/S0079-6123(08)00624-9

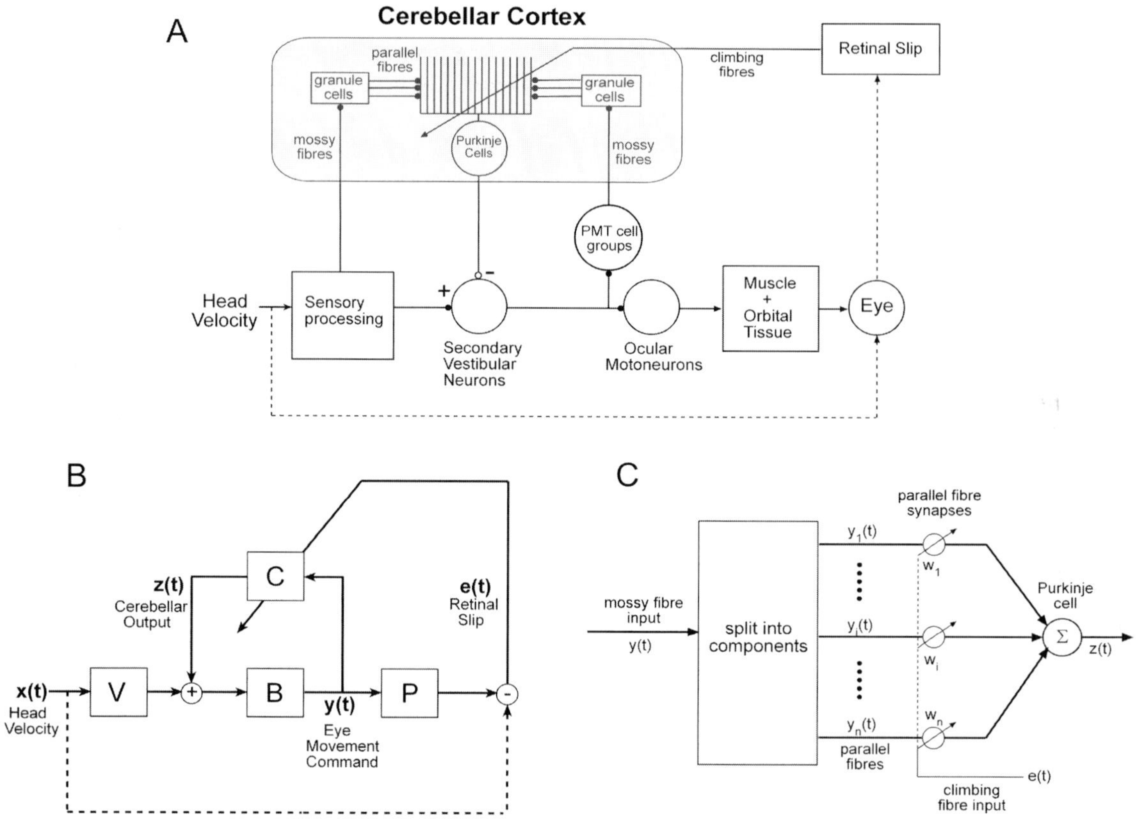

Fig. 1. Circuitry for oculomotor-plant compensation in VOR. (A) Simplified diagram of the circuitry that mediates the horizontal VOR. Head-velocity signals are processed by the semicircular canals and primary vestibular neurons, relayed to secondary vestibular neurons in the brainstem, and then passed to ocular motoneurons. Motor command signals from the motoneurons control the oculomotor plant, i.e., eye muscles plus orbital tissue, in order to produce eye movements that counteract the effects of the head velocity on the retinal image. Inaccurate eye movements produce retinal slip, which is detected by the visual system. A side-loop to the main 3-neuron arc passes through the floccular region of the cerebellum, which receives as mossy-fibre input vestibular information and a copy of the motor command sent to the eye muscles via the PMT cell groups. These mossy-fibre inputs are converted into parallel-fibre signals by granule cells and associated circuitry in the granular layer, and the parallel-fibre signals influence simple spike firing in Purkinje cells. Variation in simple spike firing is transmitted to a subset of secondary vestibular neurons (floccular target neurons) in the brainstem, The flocculus also receives a retinal slip signal as climbing-fibre input, which produces complex spikes. (B) Linearised model of horizontal VOR, derived from the neural circuitry illustrated in panel A. Head velocity $x(t)$ is processed by the filter V, then added to the output $z(t)$ of the adaptive filter C (which corresponds to the floccular region of cerebellum). The summed signal is then passed to the brainstem controller B. The output of B is a motor command $y(t)$, which acts on the plant P. A copy of $y(t)$ is sent back to the adaptive filter C. The command $y(t)$ acts on P to move the eyes, a movement which is added to the head velocity $x(t)$: net image movement is detected as retinal slip $e(t)$ and sent to C. (C) Structure of the adaptive filter shown as C in panel (B). The copy of the eye-movement command $y(t)$ arrives as mossy-fibre input, and is decomposed into components $y_1(t)$ … $y_n(t)$ by the granule cell layer. Each output component $y_i(t)$ is weighted by w_i, corresponding to the efficacy of the corresponding synapse between a parallel fibre and the Purkinje cell. The weighted components are summed by the Purkinje cell and constitute the filter output. The value of each weight w_i is adjusted according to the current value of the correlation between its component $y_i(t)$ and the global retinal slip signal $e(t)$, which arrives as climbing-fibre input. (Adapted from Porrill and Dean, 2007a.)

gaze-holding would result in retinal slip, which so happens to be the major climbing-fibre signal to the flocculus (Simpson et al., 1996). Perhaps then the relation between the two floccular inputs, retinal slip and an efference copy of eye-movement commands, can somehow be used to ensure stable gaze-holding.

The nature of this relationship suggests a possible mechanism. If there is no correlation between movement commands and subsequent retinal slip, then it can be assumed there is no causation: if retinal slip is in fact occurring, then it must be produced by some factor other than inaccurate eye-movement commands (e.g., rotation of an optokinetic drum). If however there is a correlation, then the inference is that inaccurate motor commands *are* causing the retinal slip. Thus, in principle, correlations between mossy-fibre and climbing-fibre inputs to the flocculus could be used to drive motor learning to ensure accurate motor commands and hence stable gaze-holding. Although this idea is consistent in general terms with the influential framework proposed for cerebellar function by early modellers (Marr, 1969; Albus, 1971), the critical question remains of whether it could work in practice. Our first step in addressing this question was to test how the proposed mechanism fared in simulation.

Modelling strategy

The oculomotor plant (that is the extraocular muscles and orbital tissue, Fig. 1A) is primarily viscoelastic (Robinson, 1964). If the plant were a simple viscosity, the force applied to it would be proportional to eye-velocity, so that velocity commands to the eyes would require only a scaling factor for accurate movement. However, the elasticity in the system "diverts" some of the force intended for eye-velocity. For example, at the end of a completed head movement the desired eye-velocity, as specified by the vestibulo-ocular reflex (VOR), is zero. But in the absence of plant compensation the elasticity will pull the eye back to the primary position, which produces eye-velocity when none is required. Impaired gaze-holding is thus just one aspect of inadequate compensation, occurring in the particular circumstances when desired eye-velocity is zero. We therefore modelled gaze-holding as part of the more general process of oculomotor-plant compensation.

The general principles of oculomotor-plant compensation, in the context of velocity commands supplied by the VOR, were first analysed by Skavenski and Robinson (1973). They suggested that it required two pathways to convert the vestibular input to the eye-command output. The first direct pathway was a simple gain, corresponding to the plant's viscosity term. The second indirect pathway was a neural integrator, producing an eye-position command from the desired-velocity input. The two pathways together can exactly compensate for a first-order viscoelastic plant. Although more complex plants (Sklavos et al., 2005) require more complex compensation (Optican and Miles, 1985), the framework proposed by Skavenski and Robinson remains applicable, with the proviso that the neural mechanisms responsible for plant compensation may now be referred to as an "inverse plant model."

To simulate how an accurate inverse plant model could be learnt from the relation between efference copy and retinal slip, we simplified the circuitry illustrated in Fig. 1A in two ways. First, the three main elements (brainstem, cerebellum, and plant) were linearised, to facilitate mathematical analysis of our proposed mechanism. Secondly, we considered only the efference copy mossy-fibre input to the flocculus (Fig. 1B), termed the "recurrent architecture." That was because the problem of plant compensation applies to all eye movements, not just those produced by the VOR, so specific information from the vestibular system is not required. In effect we were modelling the effects on VOR calibration of changing the plant (e.g., muscle weakness), rather than the more familiar effects of changing vestibular processing (magnifying goggles), where indeed the vestibular mossy-fibre inputs to flocculus do become relevant.

In the model, the plant was represented as a first-order system with time constant 0.1 s (further details in Porrill and Dean, 2007a). The properties of the brainstem were deduced from the effects of floccular lesions (Zee et al., 1981), which include a gaze-holding deficit where eccentric gaze returns to the primary position with a time constant of ~1 s. This effect can be simulated by assuming that the brainstem on its own contains a leaky integrator in the indirect pathway; it is the role of the flocculus to assist. Such a role is consistent with Robinson's view of the cerebellum as the "repair shop" of the oculomotor system. In addition, to test the proposed learning mechanism further, the gain of the direct pathway was reduced from its desired value of 1.0 to 0.5.

The flocculus itself was represented as an adaptive filter (Fig. 1C), a development by Fujita (1982) of the Marr–Albus framework alluded to above, and one that is increasingly used in cerebellar modelling. The (mossy-fibre) inputs to the filter are split into components (parallel-fibre signals), which are weighted individually (synapses between parallel fibres and Purkinje cells) then summed to produce the filter's output (Purkinje cell simple spikes). The weights are altered by the error signal (climbing-fibre input) using a learning rule that can be stated qualitatively as follows. If parallel-fibre firing is positively correlated with climbing-fibre firing (which signals retinal slip), reduce the weight (Long Term Depression); if it is negatively correlated, increase the weight (Long Term Potentiation); if uncorrelated, no change. In quantitative form it corresponds to Sejnowski's (1977) covariance learning rule that implements the standard Least Mean Squares algorithm used in adaptive signal processing. Since stability is achieved when the weights cease to change, and this happens when the two input signals are uncorrelated, we have referred to the proposed adaptive mechanism as "decorrelation control."

It is important to emphasise that the brainstem and cerebellar models, and the learning rule, are conventional. The key new feature is the use of the recurrent architecture (Fig. 1B) based on the anatomy of the PMT cell groups (see also Glasauer, 2003) for purposes of adaptive plant compensation.

Modelling results

Before training, all the weights in the simulated cerebellum C were set to zero. The performance of the system thus reflected that of the brainstem alone, with its low gain and leaky integrator. Thus, the "pre" trace in Fig. 2A shows that after a brief head-displacement the compensatory eye movement is too slow and therefore too small, and the globe then returns to the primary position with time constant of 1 s (impaired gaze-holding). This performance is shown in relation to input frequency in Fig. 2B. The "pre" trace here shows the gain of the brainstem alone is roughly constant at 0.5 for frequencies above ~0.3 Hz, but then drops sharply for lower frequencies (reflecting the increasing contribution of plant elasticity). The contribution of the brainstem leaky integrator can also be seen in Fig. 2B: the trace labelled "$B=0.5$" shows the performance of a brainstem with no integrator, very substantially worse that the actual brainstem for frequencies less than ~5 Hz). Finally, Fig. 2C shows the response of the brainstem to a mixed frequency (coloured-noise) head-velocity input in terms of retinal slip.

Training with the decorrelation-control mechanism sharply reduced the retinal slip error (Fig. 2C, D), and gaze-holding became essentially perfect (Fig. 2A). The gain of the VOR became ~1.0 over a wide range of frequencies (Fig. 2C). These results indicate that removing the correlation between retinal slip and an efference copy of the eye-movement command can achieve plant compensation in a simplified system.

Subsequent manipulation of model parameters indicated that this finding was robust. (i) Higher-order plants that require a "slide" as well as a step command (Optican and Miles, 1985) can also be compensated (Dean et al., 2002). (ii) Different methods of splitting mossy-fibre inputs into parallel-fibre components primarily affect speed of learning rather than final convergence (Dean et al., 2002). (iii) The scheme can be readily extended to three dimensions (Fig. 3) (Porrill et al., 2004). (iv) It continues to be effective if the climbing-fibre input signals only the direction of retinal slip, or if its retinal slip signal is delayed (Porrill and Dean, 2007a). This latter result is of particular interest in the context of the VOR, since it showed that a *brainstem* site of plasticity is required for high-frequency VOR performance in the presence of substantial (~100 ms) retinal slip

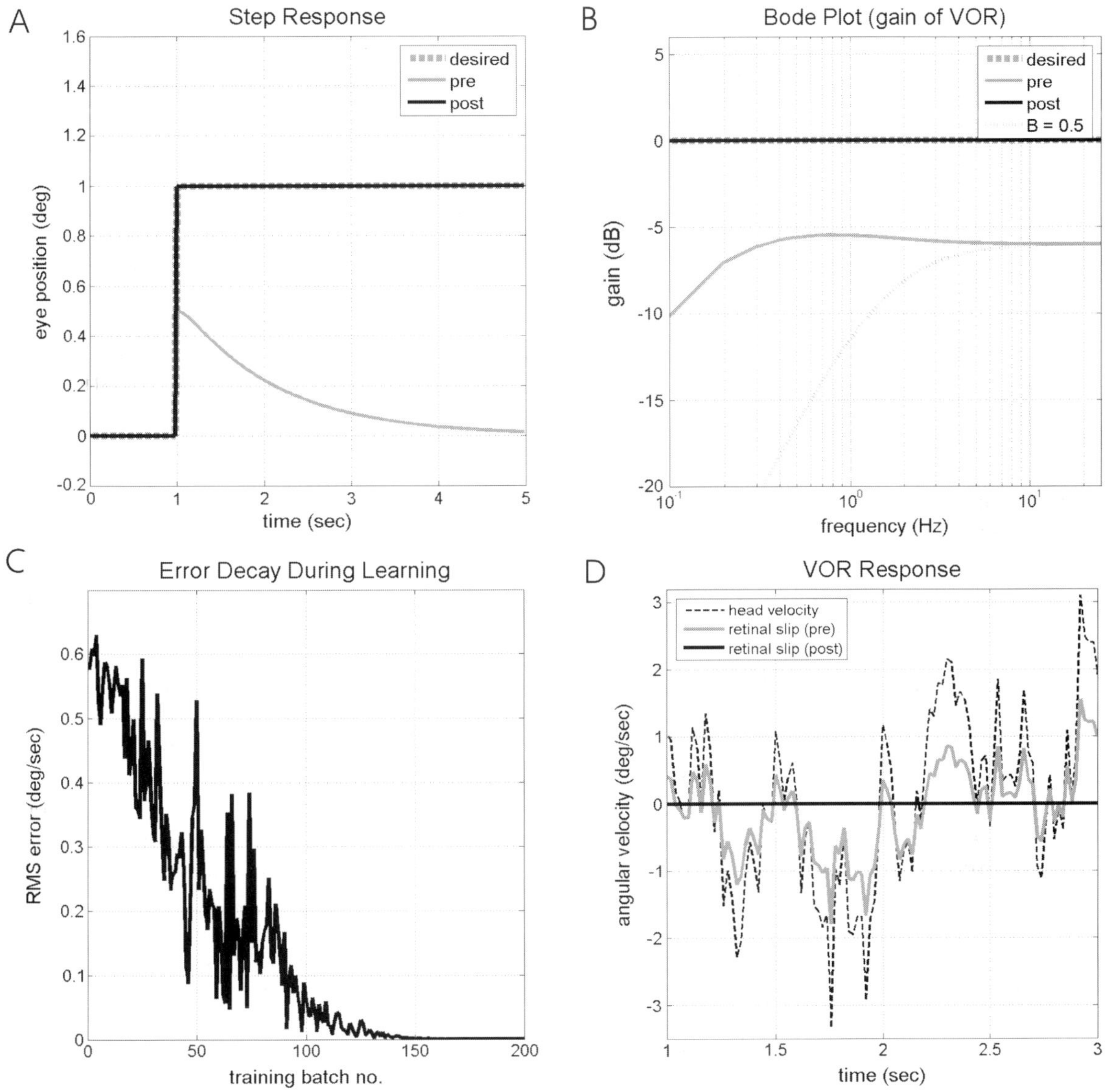

Fig. 2. Model performance before and after training with an undelayed retinal slip signal. (A) Eye-position response to sudden head-displacement. The desired and post-training performances are effectively identical. (B) System gain for sinusoidal input signals as a function of frequency (Bode gain plot). Gain is measured as ratio of eye-velocity amplitude to head-velocity amplitude. Performance before training is shown both for the complete brainstem controller ("pre"), and for the brainstem controller as simple gain ("$B = 0.5$") which corresponds to the direct pathway on its own. After training, the desired and post-training performances overlap. (C) Decline in retinal slip amplitude with training. Root-mean-square (RMS) retinal slip amplitudes, measured over a 5 s training batch, plotted against number of training batches. (D) Example of retinal slip to mixed-frequency head-velocity input before and after training. (Adapted from Porrill and Dean, 2007a.)

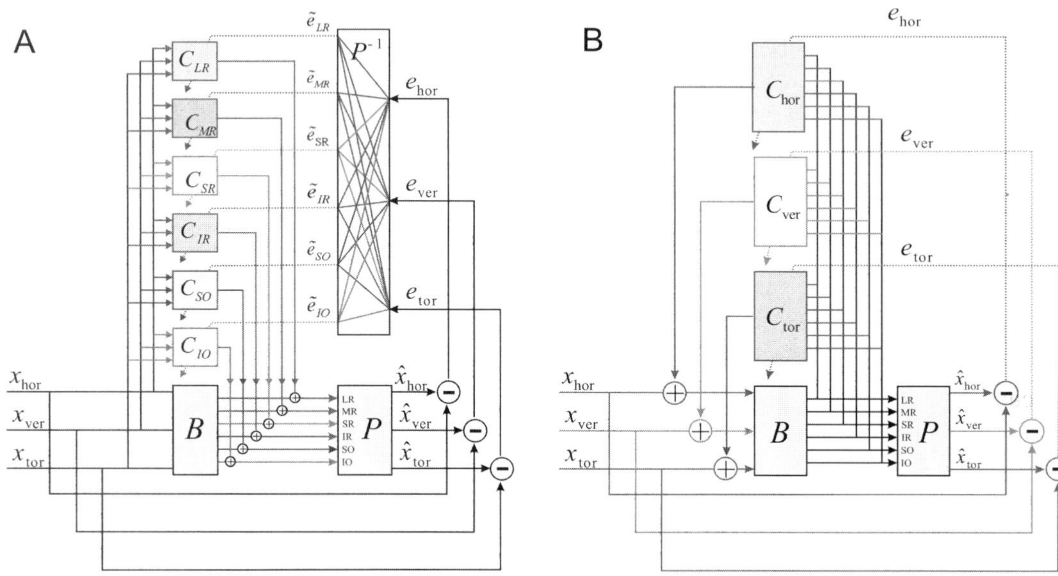

Fig. 3. Alternative architectures applied to simulated VOR in 3D. The vestibular system recovers three components: x_{hor}, x_{ver}, and x_{tor} (horizontal, vertical, and torsional) of head angular velocity. These are processed by the brainstem and cerebellum to produce motor commands to the six extraocular muscles so as to stabilise the eyes rotational position in space. The motor plant is a 3×6 matrix transfer function P and the brainstem contribution to the VOR is a 6×3 matrix transfer function B. (A) Feedback error architecture: the cerebellum takes the three vestibular signals as input and supplies corrections to the six motor commands. Colour is used to highlight the "motor space" modularity. For example, the green component shows those Purkinje cells that contribute to the superior rectus (SR) muscle command, and which require the corresponding motor command error $\tilde{e}_{SR}$ as training signal on their climbing fibres. This signal must be reconstituted from the horizontal, vertical, and torsional components of retinal slip. (B) Recurrent architecture: the cerebellar takes the six motor commands as input and supplies corrections to the three vestibular signals. Colour is used here to highlight the "task space" modularity. For example, the green component shows Purkinje cells contributing to vertical eye motion; these require the vertical component of optic flow as training signal. (Adapted from Porrill et al., 2004.) (See Color Plate 3.5.3 in color plate section.)

delays. Although the existence of brainstem plasticity in VOR adaptation is well-known, its role in the context of Marr–Albus cerebellar learning schemes has hitherto been unclear and contentious (Porrill and Dean, 2007a).

Part of the reason for the scheme's robustness for linearised plant compensation can be demonstrated mathematically (Porrill et al., 2004). The recurrent connectivity implements a partial state feedback controller and using Lyapounov methods it is possible to show that the algorithm can stably learn to adapt to changes in a wide class of motor plants. Thus, the method suggested by the anatomy of PMT cell groups may have applicability beyond eye movements themselves. We consider this first for control of artificial systems including robots, before returning to the role of the recurrent cerebellar architecture in biological sensorimotor processing.

Application to robotics

Adaptive filters and artificial neural nets are often used to implement supervised learning schemes in which weight change is driven by the difference between the device's actual output, and what that output should have been. In the case of motor control, the desired output is the correct command required for the movement in question to be accurate. The difference between this correct command and the actual inaccurate command is sometimes referred to as "motor error."

The correct commands and hence motor error may well be known for a manufactured and unchanging plant. However, if the plant does change in unforeseen ways because of damage or wear, the motor error will not be known. These circumstances apply both to autonomous robots and to biological systems, where the evidence of inaccurate motor commands is *sensory* and concerns the nature of the movement itself. How to use this sensory information to guide learning is a major problem — the distal error problem in adaptive control (Jordan and Wolpert, 2000).

The recurrent architecture suggested by the anatomy of PMT cell groups for eye-movement control may offer a solution to this general problem (Fig. 3). In this architecture the sensory signal denoting inaccurate movement drives learning directly, resulting in a simple modular circuit for multi-dimensional motor control (Fig. 3B). In contrast, the alternative scheme whereby sensory error is converted into an estimate of motor error (Fig. 3A) requires a complex device to extract the relevant combination of retinal slip error for each motoneuron pool. This is equivalent to requiring an estimate of the inverse plant, a potential problem since this is what the system is trying to learn in the first place.

These considerations suggest that the decorrelation-control algorithm may have applications to robotics, and initial tests in a robot head-camera system (Fig. 4A) suggest that it can produce effective gaze-stabilisation in the context of an artificial VOR (Lenz et al., 2007). We have also recently shown how the algorithm can be extended to kinematic control of a simulated two-joint robot arm (Porrill and Dean, 2007b), and this has been applied to the control of a real robot arm by Iyad Obeid (Fig. 4B) who is investigating its potential for control of brain-machine interfaces.

Application to biology

Although the decorrelation-control algorithm arose from biology, and has applications to artificial systems, it still remains only a candidate algorithm for the actual eye-movement control system that inspired it. In particular, specific tests of the algorithm's relevance to floccular function have yet to be carried out. A possible though technically demanding test would be to record from relevant Purkinje cells during simulated plant adaptation, and determine the changes if any in eye-position related firing.

It is however important to note that, notwithstanding this lack, the algorithm is in general consistent with the substantial amount of existing evidence regarding the role of the flocculus in gaze-stabilisation. (i) Inactivating or removing the flocculus abolishes VOR adaptation in general, and adaptation to simulated plant changes in particular (Optican et al., 1986). (ii) Inactivation of PMT cell groups relating to vertical movement in cat "has an effect on the integrator function similar to that found in previous experiments lesioning the flocculus" (Nakamagoe et al., 2000, p. 858). (iii) Electrophysiological evidence suggests that ~75% of primate mossy fibre or other granular-layer input elements modulate their discharge in relation to eye movements (Miles et al., 1980), consistent with a powerful PMT cell group input. (iv) Brainstem cells receiving floccular outputs (floccular target neurons) carry an eye-position signal (Scudder and Fuchs, 1992) that appears to reinforce the function of the nucleus prepositus hypoglossi or brainstem horizontal integrator (Hazel et al., 2002). (v) Some floccular Purkinje cells fire in relation to eye position. The fact that the proportion of such cells is quite low (~20%) in primates (Miles et al., 1980) suggests that plant compensation is not the only function of the flocculus, consistent with the extensive evidence that has established a role for the flocculus in smooth-pursuit.

Finally, the recurrent architecture found for the flocculus appears to be a widespread feature of cerebellar organisation. It has been found for regions of the cerebellum associated with saccades and arm movements (references in Porrill and Dean, 2007b), and it has been suggested that "closed-loop circuits are a fundamental unit of architecture for cerebellar interconnections with the cerebral cortex" (Kelly and Strick, 2003, p. 8441). Interestingly, a "challenge for future studies is to determine the computations that are supported by this architecture" (p. 8443). The results outlined

Fig. 4. Application of decorrelation control to robotics: (A) Eye-Robot, a camera-head rig developed by Bristol Robotics Laboratory (Lenz et al., 2007). Adapted with permission from Bristol Robotics Laboratory, appearing in PLoS Computational Biology, Vol. 3, Issue 10, October 2007. (B) A two-joint arm, developed by Dr. Iyad Obeid, Department of Electrical and Computer Engineering, Temple University, Philadelphia PA, USA.

here suggest a candidate computation, giving any cerebellar area the power to adjust its output appropriately in the face of downstream changes.

Conclusions

The anatomy of the PMT cell groups points to the importance of sending an efference copy of eye-movement commands to the floccular region. We suggest that one of the functions of this signal is to learn accurate compensation of the oculomotor plant, and we describe a procedure where this is achieved by adaptively decorrelating efference copy from retinal slip. The "decorrelation-control" procedure works well in simulation, and can be shown to solve the classic problem of motor error in a principled manner. It may therefore have applications in the field of autonomous robotics, and also to biological motor control of responses besides eye movements, given that recurrent connections are a common feature of cerebellar anatomy.

Abbreviations

PMT	paramedian tract
VOR	vestibulo-ocular reflex

Acknowledgements

This research was supported by the UK Engineering and Physical Sciences Research Council, under the Novel Computation Initiative (GR/T10602/01), and the UK Biology and Biotechnology Research Council under the Integrative Analysis of Brain and Behaviour Initiative (BBS/B/17026).

References

Albus, J.S. (1971) A theory of cerebellar function. Math. Biosci., 10: 25–61.

Büttner-Ennever, J.A. (1992) Paramedian tract cell groups: a review of connectivity and motor function. In: Shimazu H. and Shinoda Y. (Eds.), Vestibular and Brain Stem Control of Eye, Head and Body Movements. S. Karger, Basel, pp. 323–330.

Büttner-Ennever, J.A. and Horn, A.K.E. (1996) Pathways from cell groups of the paramedian tracts to the floccular region. In: Highstein S.M., Cohen B. and Büttner-Ennever J.A. (Eds.), New Directions in Vestibular Research, Vol. 781. New York, New York Academy of Sciences, pp. 532–540.

Büttner-Ennever, J.A., Horn, A.K.E. and Schmidtke, K. (1989) Cell groups of the medial longitudinal fasciculus and paramedian tracts. Revue Neurologique (Paris), 145: 533–539.

Dean, P., Porrill, J. and Stone, J.V. (2002) Decorrelation control by the cerebellum achieves oculomotor plant compensation in simulated vestibulo-ocular reflex. Proc. R. Soc. Lond. B. Biol. Sci., 269: 1895–1904.

Fujita, M. (1982) Adaptive filter model of the cerebellum. Biol. Cybern., 45: 195–206.

Glasauer, S. (2003) Cerebellar contribution to saccades and gaze holding: a modeling approach. In: Brandt T., Cohen B. and Siebold C. (Eds.), Oculomotor and Vestibular Systems: Their Function and Disorders, Vol. 1004. New York, New York Academy of Sciences, pp. 206–219.

Hazel, T.R., Sklavos, S.G. and Dean, P. (2002) Estimation of premotor synaptic drives to simulated abducens motoneurons for control of eye position. Exp. Brain Res., 146: 184–196.

Jordan, M.I. and Wolpert, D.M. (2000) Computational motor control. In: Gazzaniga M.S. (Ed.), The New Cognitive Neurosciences. MIT Press, Cambridge, MA, pp. 601–618.

Kelly, R.M. and Strick, P.L. (2003) Cerebellar loops with motor cortex and prefrontal cortex of a nonhuman primate. J. Neurosci., 23: 8432–8444.

Lenz, A., Balakrishnan, T., Pipe, A.G. and Melhuish, C. (2007) Eye-robot: a biologically inspired adaptive controller. In: Proceedings Towards Autonomous Robotic Systems, Aberystwyth, pp. 231–238.

Marr, D. (1969) A theory of cerebellar cortex. J. Physiol., 202: 437–470.

Miles, F.A., Fuller, J.H., Braitman, D.J. and Dow, B.M. (1980) Long-term adaptive changes in primate vestibuloocular reflex. III. Electrophysiological observations in flocculus of normal monkeys. J. Neurophysiol., 43: 1437–1476.

Nakamagoe, K., Iwamoto, Y. and Yoshida, K. (2000) Evidence for brainstem structures participating in oculomotor integration. Science, 288: 857–859.

Optican, L.M. and Miles, F.A. (1985) Visually induced adaptive changes in primate saccadic oculomotor control signals. J. Neurophysiol., 54: 940–958.

Optican, L.M., Zee, D.S. and Miles, F.A. (1986) Floccular lesions abolish adaptive control of post-saccadic ocular drift in primates. Exp. Brain Res., 64: 596–598.

Porrill, J. and Dean, P. (2007a) Cerebellar motor learning: when is cortical plasticity not enough? PLoS Comput. Biol., 3: 1935–1950.

Porrill, J. and Dean, P. (2007b) Recurrent cerebellar loops simplify adaptive control of redundant and nonlinear motor systems. Neural Comp., 19: 170–193.

Porrill, J., Dean, P. and Stone, J.V. (2004) Recurrent cerebellar architecture solves the motor error problem. Proc. R. Soc. Lond. B. Biol. Sci., 271: 789–796.

Robinson, D.A. (1964) The mechanics of human saccadic eye movement. J. Physiol., 174: 245–264.
Scudder, C.A. and Fuchs, A.F. (1992) Physiological and behavioral identification of vestibular nucleus neurons mediating the horizontal vestibuloocular reflex in trained monkeys. J. Neurophysiol., 68: 244–264.
Sejnowski, T.J. (1977) Storing covariance with nonlinearly interacting neurons. J. Math. Biol., 4: 303–321.
Simpson, J.I., Wylie, D.R. and De Zeeuw, C.I. (1996) On climbing fiber signals and their consequence(s). Behav. Brain Sci., 19: 384–398.
Skavenski, A.A. and Robinson, D.A. (1973) Role of abducens neurons in vestibuloocular reflex. J. Neurophysiol., 36: 724–738.
Sklavos, S., Porrill, J., Kaneko, C.R.S. and Dean, P. (2005) Evidence for a wide range of time scales in oculomotor plant dynamics: implications for models of eye-movement control. Vis. Res., 45: 1525–1542.
Zee, D.S., Yamazaki, A., Butler, P.H. and Gücer, G. (1981) Effects of ablation of flocculus and paraflocculus on eye movements in primate. J. Neurophysiol., 46: 878–899.

C. Kennard & R.J. Leigh (Eds.)
Progress in Brain Research, Vol. 171
ISSN 0079-6123

CHAPTER 3.6

Impulsive testing of semicircular canal function

G. Michael Halmagyi*, Konrad P. Weber, Swee T. Aw, Michael J. Todd and Ian S. Curthoys

Neurology Department, Royal Prince Alfred Hospital, Camperdown, Sydney, NSW, Australia

Abstract: After acute vestibular loss in humans or animals, eye-movement responses to rapid horizontal ipsilesional head rotations ("head impulses") show that there is severe, permanent impairment of the angular vestibulo-ocular reflex. The basis for this appears to be an inhibitory saturation of ipsilesional vestibular nerve, and perhaps vestibular nucleus afferents, in response to high-acceleration, off-direction stimulation of the sole functioning member of any pair of semicircular canals. For the clinician, impulsive testing is an easy and reliable way to identify severe unilateral as well as bilateral impairment of semicircular canal function.

Keywords: head impulse; eye movements; vestibulo-ocular reflex; semicircular canal; catch-up saccade

Introduction

The semicircular canals (SCCs) operate as push–pull pairs due to the bidirectional morphological and physiological polarization of the hair cell bundles on the cupula (Lysakowski and Goldberg, 2004). During any head rotation, displacement of the stereocilia towards the kinocilium depolarizes the hair cell and generates excitation, while displacement of stereocilia away from the kinocilium hyperpolarizes the hair cell and generates inhibition. In response to a head rotation such as a head impulse, one SCC of the push–pull pair is excited while the other is inhibited, generating the total angular vestibulo-ocular reflex (VOR) from direct excitation and indirect disinhibition. In the lateral (horizontal) SCC, ampullopetal endolymph flow causes excitation while ampullofugal flow causes inhibition, whereas in the vertical SCCs ampullofugal flow causes excitation and ampullopetal flow causes inhibition. It is this directional polarization, which produces the on–off directional asymmetry of the VOR generated by a single SCC summarized in what is called Ewald's second law (Halmagyi et al., 1990) and is a means of identifying impairment of individual SCC function.

The smooth eye-movement response to a head impulse is too fast to allow any contribution to the compensatory eye movement from smooth pursuit (Kimmig et al., 2002) or cervico-ocular reflexes (Bronstein and Hood, 1986); it is generated exclusively by the VOR. A head-impulse stimulus should be an unpredictable head rotation of 20°–30° amplitude, 150–300°/s peak velocity, and 2000–5000°/s/s peak acceleration, approximately in the plane of an SCC pair. During the head impulse in normal subjects, with close-to-unity VOR gain, both eyes rotate promptly and smoothly in the compensatory direction (Fig. 2A) and thus keep fixing on target, whereas a patient with reduced VOR gain

*Corresponding author. Tel.: +61-2-9515-7565;
Fax: +61-2-9515-8347; E-mail: michael@icn.usyd.edu.au

DOI: 10.1016/S0079-6123(08)00625-0

cannot keep looking at the fixation target and has to make a, usually observable, catch-up saccade in order to re-fixate the target (Fig. 2C). Graded head impulses can be produced manually with a visual head-velocity feedback signal for the examiner (Black et al., 2005; Weber et al., 2008) (Fig. 3). Head impulses can also be generated with electromechanical devices delivering standardized head-only (Collewijn and Smeets, 2000; Hirvonen et al., 2007) or whole-body stimuli (Tian et al., 2000). The VOR gain is higher, and covert catch-up saccades occur more frequently with active than with passive head impulses (Della Santina et al., 2002; Black et al., 2005).

Individual lateral SCCs can be tested with head impulses about the yaw (Z) axis to excite the ipsilateral and inhibit the contralateral lateral canal (Fig. 1A). Individual vertical SCCs can be tested along their approximate right-anterior and left-posterior (RALP) and left-anterior and right-posterior (LARP) planes (Cremer et al., 1998; Aw et al., 2001).

Recording and analysing the impulsive VOR

The search coil technique (Robinson, 1963; Remmel, 1984; Aw et al., 1996b) is the standard method to measure head and eye position in the impulsive VOR testing. Head and eye positions can be recorded in two dimensions with single-search coils (Halmagyi and Curthoys, 1988; Halmagyi et al., 1990; Aw et al., 1994, 1995) or in three dimensions with dual-search coils (Aw et al., 1996a, b; Cremer et al., 1998; Aw et al., 2001; Lehnen et al., 2004; Weber et al., 2008).

The subject wears search coils on one or both eyes to measure angular eye positions, while another search coil is attached to a dental impression bite-bar to measure the angular head position. Dental impression bite-bar is recommended to prevent head coil slippage. The search coils are pre-calibrated before each recording. The subject is seated with the head in the centre of the magnetic field coil system, which is available in either the two-field or three-field configuration.

The rationale of in vitro calibration is to determine the gains and offsets of the signals from each search coil induced by the magnetic fields. All coils are simultaneously mounted on a Fick gimbal. When a two-field magnetic system is used, the gimbal is moved in yaw, pitch, or roll calibration positions between $\pm 20^\circ$ in 5° steps, and the gains and offsets for each search coil are determined. Maximum errors and cross coupling are less than 2% (Aw et al., 1996b).

When a three-field magnetic system is used, initially offset voltages from undesirable noise pickup and internal amplifier biases are compensated by placing the search coil in a soft iron tube, which isolates it from the magnetic fields, while the amplifier offsets are nulled (Bergamin et al., 2001). Then the annulus is placed on a gimbal system and rotated in six positions, each of which picks up the maximum voltage induced by one magnetic field in one search coil (two coils times three directions for each dual-search coil). These voltage signals are then used to compute the orientation of the dual-search coil in the space-fixed magnetic frame.

In order to obtain good temporal resolution in the head impulse test, it requires a sampling frequency of the search coil signals of at least 500 Hz. The resolution of the analogue to digital converter (ADC) should be at least 16 bits so that digital head or eye velocity derived from its position signal has good signal-to-noise ratio.

Head, gaze, and eye positions are analyzed in three dimensions as rotation vectors or Euler angles (Haslwanter, 1995). Head and gaze positions are the orientations of the head and eye in space-fixed coordinates. Eye position is the orientation of the eye in head-fixed coordinates. Head, gaze, and eye velocities are calculated from their respective positions. A rotation of the coordinate reference frame by 45° about the yaw axis allows re-expression of these vectors as rotations about the approximate preferred axis of individual vertical SCCs (Cremer et al., 1998; Aw et al., 2001) and display of the angular VOR related to the anterior and posterior SCCs. Impulsive VOR gain is analysed in the first 70 ms period after impulse onset, as catch-up saccades can have latencies as short as 70 ms (Tian et al., 2000; Weber et al., 2008).

The latency of the VOR in response to head impulses has been estimated to be about 6–10 ms in

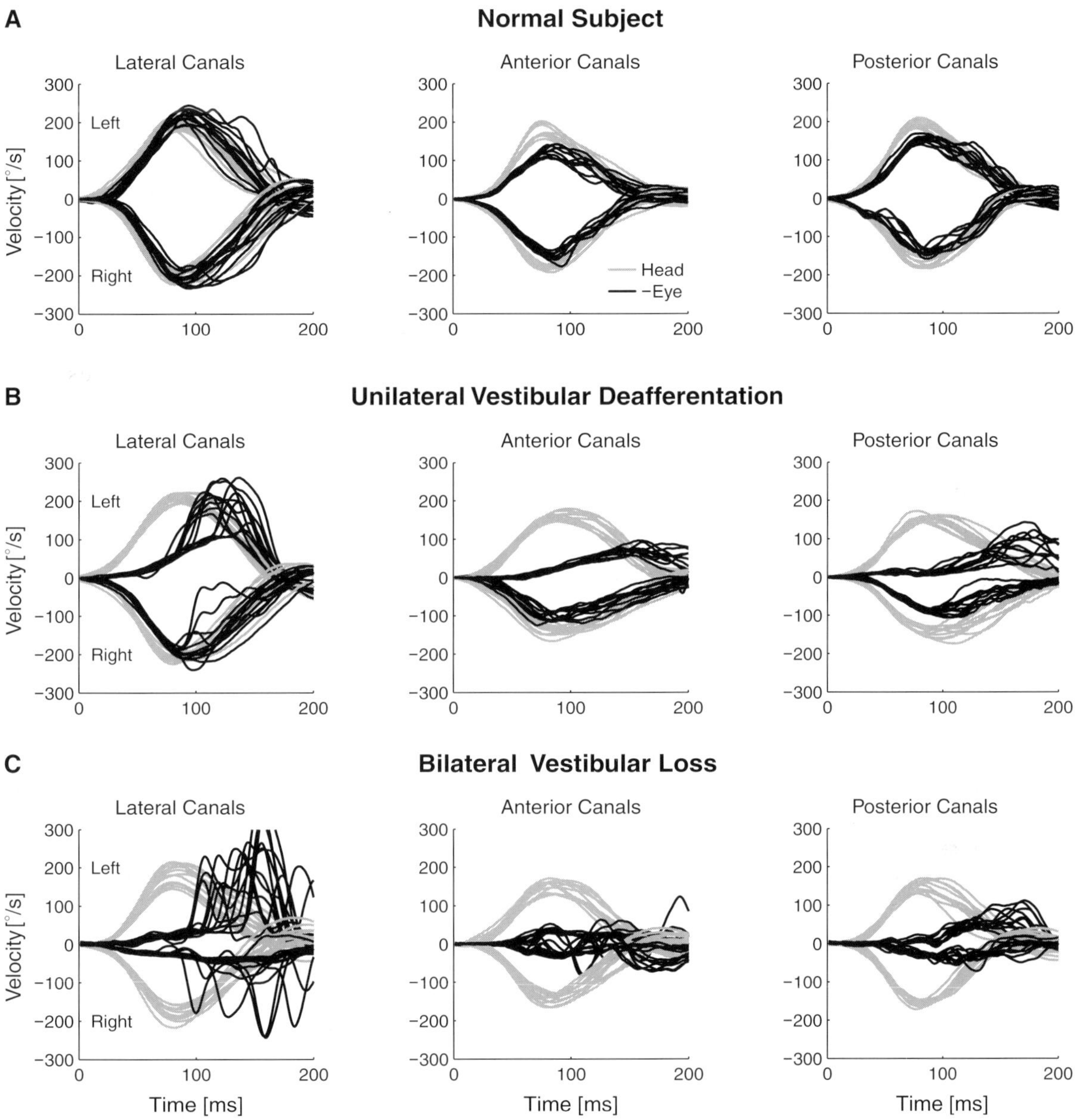

Fig. 1. Superimposed three-dimensional head and eye-velocity data during impulsive testing of individual SCCs in a normal subject (A), a patient with total deafferentation of the left labyrinth (B), and a patient with severe bilateral vestibular loss due to gentamicin toxicity (C). In the normal subject, eye-velocity (inverted for ease of comparison with head velocity) mirrors head velocity for all six SCCs. In contrast, the patient with unilateral deafferentation shows severe impairment of VOR gain for all three SCCs on the left, while the patient with bilateral loss shows severe impairment for gain of all six SCCs. Both patients make covert saccades during horizontal head impulses to their affected sides.

humans (Aw et al., 1996b; Collewijn and Smeets, 2000) and also in animals (Migliaccio et al., 2004). In normal subjects, one method of estimating latency is to shift the eye velocity at 1 ms intervals towards the head velocity, and the least-squares difference between the head and eye velocity is determined after each shift. The latency is the time interval shifted when the least-squares difference

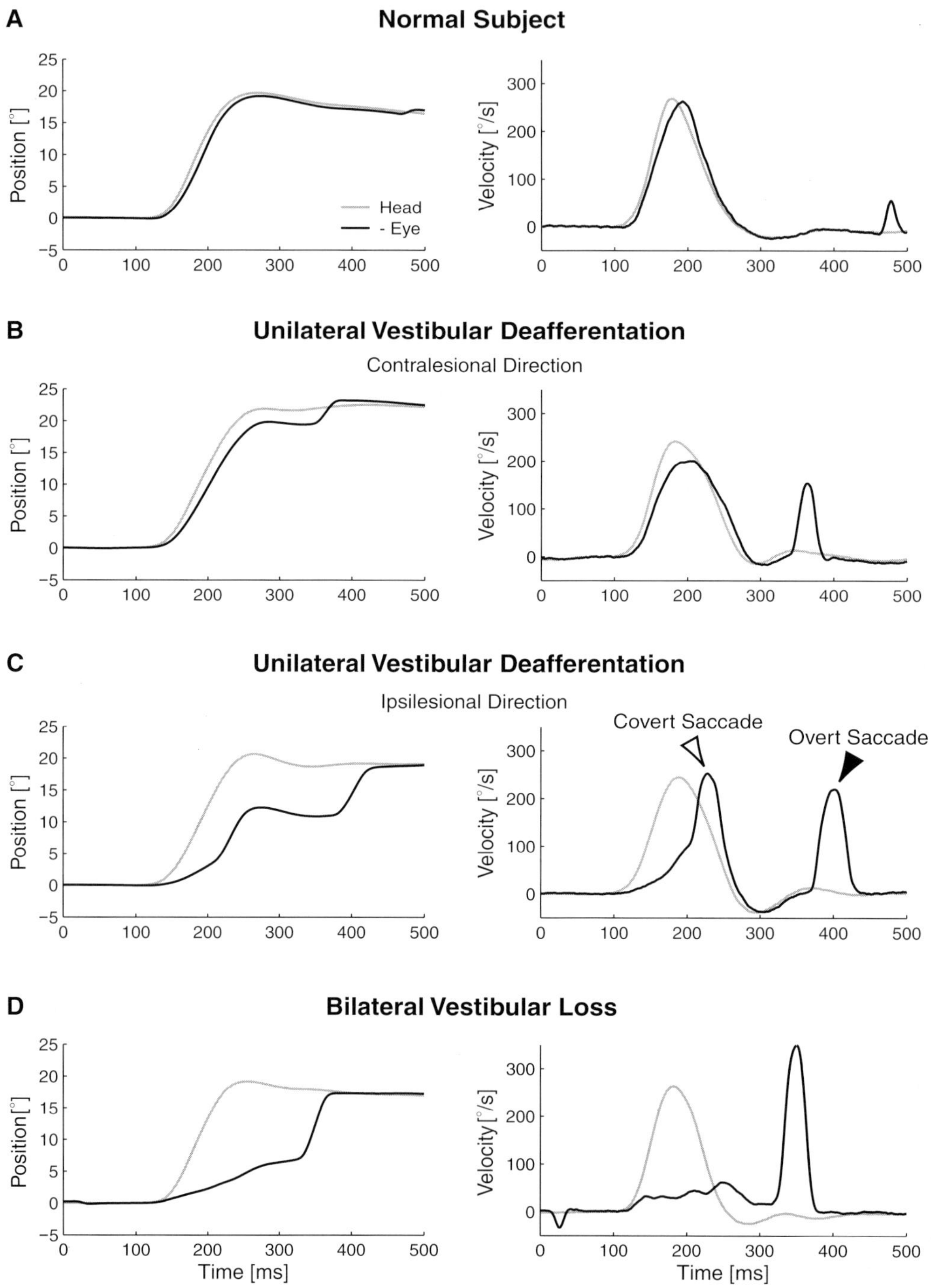

Fig. 2. Horizontal head and eye position (left column) and velocity (right column) during single head impulses in a normal subject (A), a unilateral vestibular deafferentation patient during a contralesional (B) and during an ipsilesional (C) impulse, and in a patient with bilateral vestibular loss (D). Note the low VOR gain in C and D, a covert and then an overt compensatory catch-up saccade in C and the near-normal VOR gain in B with a single small overt saccade.

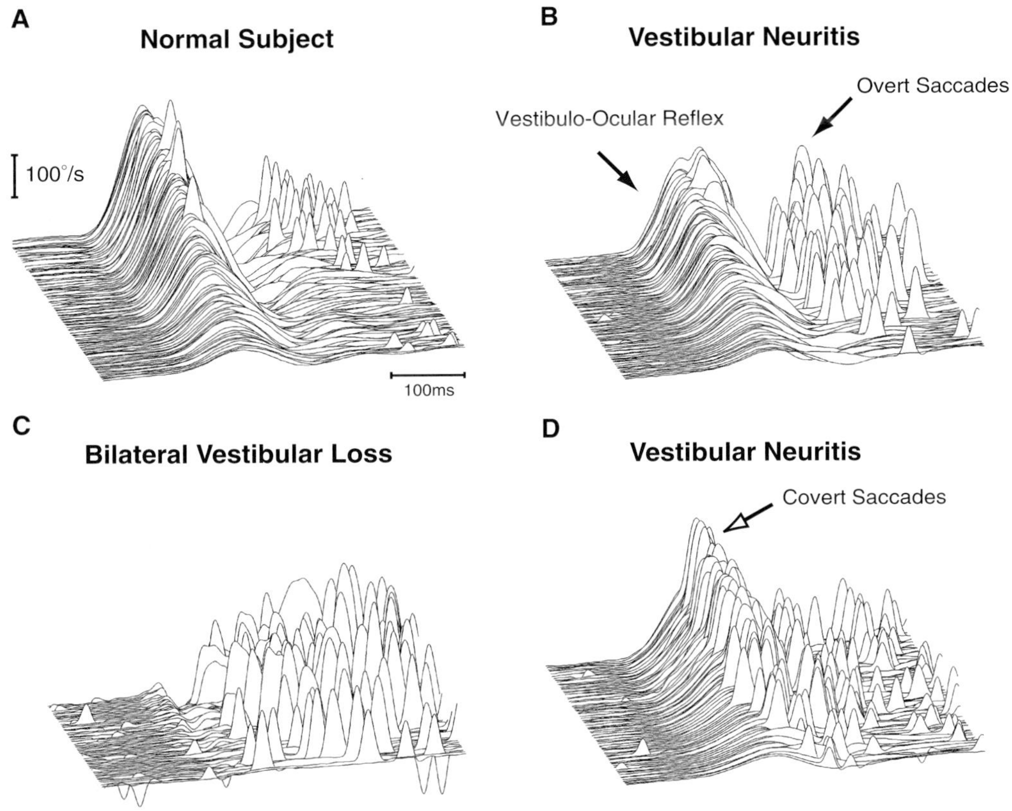

Fig. 3. Horizontal eye-velocity responses to increasing velocity-scaled head impulses (not shown) in a normal subject (A), in a patient with partial loss of lateral SCC function due to vestibular neuritis producing largely overt (i.e., late) saccades which make clinical recognition of the impaired VOR easy (B), in another patient with partial loss of lateral SCC function due to vestibular neuritis producing largely covert (i.e., early) saccades which make clinical recognition of the impaired VOR difficult (D), and in a patient with severe bilateral vestibular loss with a VOR gain less than 0.1 producing large overt catch-up saccades (C).

between the head and eye velocity is minimum (Aw et al., 1996b). Another method is to measure the time interval between the intersections of the linear regressions of (or least square fit) head and eye velocities with the time axis (Collewijn and Smeets, 2000).

VOR gain can be described in at least four different ways:

1. Velocity gain in one dimension referenced to the orthogonal yaw, pitch, and roll axes (Aw et al., 1996a, b; Palla and Straumann, 2004), or to the rotated reference frames of LARP and RALP axes (Cremer et al., 1998; Aw et al., 2001) calculated as instantaneous eye velocity divided by head velocity at or close to peak head velocity (Aw et al., 1996a, b; Collewijn and Smeets, 2000).
2. Acceleration gain calculated as the ratio between the slopes of eye and head velocities for a period prior to peak head velocity (Collewijn and Smeets, 2000; Lehnen et al., 2004; Weber et al., 2008).
3. Three-dimensional speed gain defined as the ratio of eye-velocity magnitude (eye speed) to head-velocity magnitude (head speed) measures the total angular VOR in response to head rotations about a single axis, i.e., yaw, pitch, or roll (Aw et al., 1996b).
4. Impulsive canal paresis is described as three-dimensional gaze instability during a head impulse and defined as the ratio of gaze

velocity to head velocity in SCC coordinates (Aw et al., 2001). Impulsive canal paresis is defined close to peak head velocity, in response to a head rotation towards the on-direction of an SCC, along its SCC plane (Aw et al., 2001). Gaze and head velocities are normalized by dividing each velocity by the magnitude of peak head velocity in each trial and then gaze and head velocities are determined in SCC coordinates.

Based on the VOR gain to each side (g_r, g_l), the percentage of directional VOR gain asymmetry (g_s) can be calculated as follows (Weber et al., 2008):

$$g_s = \frac{g_l - g_r}{g_l + g_r} \times 100$$

The direction of the input–output kinematics of the VOR can be described as a misalignment angle, the instantaneous angle by which the eye rotation axis deviates from perfect alignment with the head rotation axis in three dimensions (Aw et al., 1996b).

Impulsive VOR with unilateral and bilateral vestibular loss

In normal humans (Halmagyi et al., 1990; Aw et al., 1996b; Cremer et al., 1998) and animals (Gilchrist et al., 1998), VOR gain with yaw impulses is close to 1.0 (Fig. 1A). Whereas VOR gain with pitch impulses is also close to 1.0, with roll impulses it is only ~0.6–0.7 (Aw et al., 1996b). With LARP and RALP impulses, head and eye velocity vectors are derived from the pitch and roll components so that LARP and RALP impulsive VOR gain is only ~0.7–0.8 in humans (Cremer et al., 1998; Lehnen et al., 2004) (Fig. 1A) and in animals (Migliaccio et al., 2004).

After total unilateral vestibular deafferentation, the VOR in response to head impulses directed towards each of the three deafferented SCCs is markedly reduced. Figure 1B shows head impulses directed towards each SCC after total unilateral vestibular deafferentation during treatment for a left vestibular schwannoma. The VOR gain during head impulses towards the lesioned left SCCs was only ~0.2–0.3 in all three SCCs, but the responses towards the intact right side are close to normal. Similar deficits occur in guinea pigs (Gilchrist et al., 1998) and monkeys (Lasker et al., 2000; Sadeghi et al., 2006).

In subjects with bilateral vestibular loss, as shown by absent nystagmus responses to caloric and rotational testing, impulsive testing will show VOR gains of 0.2 or less in response to horizontal and vertical head impulses in both directions, indicating defective function of all six SCCs (Fig. 1C).

Mechanism of impulsive VOR asymmetry after unilateral vestibular deafferentation

The simplest explanation for the severe permanent impairment of the VOR with ipsilesional yaw impulses (Fig. 2C) is the necessary off-direction or inhibitory saturation of tonic discharging afferents from the sole functioning lateral SCC. Low-gain irregularly discharging lateral SCC afferents are well-suited to code head acceleration (Hullar et al., 2005) and a mathematical model comprising of a linear VOR pathway coding head velocity and a non-linear pathway coding head acceleration comes close to simulating the results of impulsive testing of the horizontal VOR after unilateral vestibular deafferentation (Lasker et al., 2000).

Clinical head impulse testing and catch-up saccades

If there is reduced impulsive VOR gain, the subject must make a catch-up saccade in order to re-fixate the target. These catch-up saccades can be overt or covert (Weber et al., 2008). Whereas covert saccades occur during head rotation and are imperceptible (Figs. 2C and 3D), overt saccades appear after head rotation and are detectable by the clinician (Fig. 3B). The contribution of both VOR and catch-up saccades on gaze stabilization during head impulses can be estimated graphically with a phase-plane analysis by plotting gaze velocity versus gaze position (Peng et al., 2004).

For the clinician, horizontal (yaw) head impulse testing is a simple and useful way to identify unilateral or bilateral loss of SCC function in patients. With severe unilateral loss of lateral SCC function, head impulses to the affected side will produce a VOR with a gain of 0.4 or less, and the clinician can identify the overt catch-up saccades that the patient must make to re-fixate the target (Fig. 2C). With head impulses to the normal side, VOR gain will be from 0.8 to 0.9 so that the patient will only need to make a small catch-up saccade which might not be obviously abnormal (Fig. 2B). In patients with bilateral vestibular loss, horizontal head impulses to each side will produce catch-up saccades (Fig. 2D). Impulsive testing is useful in distinguishing patients with acute unilateral peripheral vestibular lesions such as neurolabyrinthitis (Fig. 3B, D) from patients with acute central vestibular lesions such as cerebellar strokes (Newman-Toker et al., 2008). This is in contrast to patients with chronic cerebellar degeneration who show impaired modulation of both gain and rotation axis of the impulsive VOR (Crane et al., 2000; Walker and Zee, 2005).

Acknowledgements

The authors acknowledge invaluable technical assistance from Satendra Pratap. This work was supported by the National Health and Medical Research Council, the RPA Neurology Department Trustees, and the Garnett Passe and Rodney Williams Memorial Foundation.

References

Aw, S.T., Fetter, M., Cremer, P.D., Karlberg, M. and Halmagyi, G.M. (2001) Individual semicircular canal function in superior and inferior vestibular neuritis. Neurology, 57: 768–774.

Aw, S.T., Halmagyi, G.M., Curthoys, I.S., Todd, M.J. and Yavor, R.A. (1994) Unilateral vestibular deafferentation causes permanent impairment of the human vertical vestibulo-ocular reflex in the pitch plane. Exp. Brain Res., 102: 121–130.

Aw, S.T., Halmagyi, G.M., Haslwanter, T., Curthoys, I.S., Yavor, R.A. and Todd, M.J. (1996a) Three-dimensional vector analysis of the human vestibulo-ocular reflex in response to high-acceleration head rotations. II. Responses in subjects with unilateral vestibular loss and selective semicircular canal occlusion. J. Neurophysiol., 76: 4021–4030.

Aw, S.T., Halmagyi, G.M., Pohl, D.V., Curthoys, I.S., Yavor, R.A. and Todd, M.J. (1995) Compensation of the human vertical vestibulo-ocular reflex following occlusion of one vertical semicircular canal is incomplete. Exp. Brain Res., 103: 471–475.

Aw, S.T., Haslwanter, T., Halmagyi, G.M., Curthoys, I.S., Yavor, R.A. and Todd, M.J. (1996b) Three-dimensional vector analysis of the human vestibulo-ocular reflex in response to high-acceleration head rotations. I. Responses in normal subjects. J. Neurophysiol., 76: 4009–4020.

Bergamin, O., Zee, D.S., Roberts, D.C., Landau, K., Lasker, A.G. and Straumann, D. (2001) Three-dimensional Hess screen test with binocular dual search coils in a three-field magnetic system. Invest. Ophthalmol. Vis. Sci., 42: 660–667.

Black, R.A., Halmagyi, G.M., Thurtell, M.J., Todd, M.J. and Curthoys, I.S. (2005) The active head-impulse test in unilateral peripheral vestibulopathy. Arch. Neurol., 62: 290–293.

Bronstein, A.M. and Hood, J.D. (1986) The cervico-ocular reflex in normal subjects and patients with absent vestibular function. Brain Res., 373: 399–408.

Collewijn, H. and Smeets, J.B. (2000) Early components of the human vestibulo-ocular response to head rotation: latency and gain. J. Neurophysiol., 84: 376–389.

Crane, B.T., Tian, J.R. and Demer, J.L. (2000) Initial vestibulo-ocular reflex during transient angular and linear acceleration in human cerebellar dysfunction. Exp. Brain Res., 130: 486–496.

Cremer, P.D., Halmagyi, G.M., Aw, S.T., Curthoys, I.S., McGarvie, L.A., Todd, M.J., Black, R.A. and Hannigan, I.P. (1998) Semicircular canal plane head impulses detect absent function of individual semicircular canals. Brain, 121: 699–716.

Della Santina, C.C., Cremer, P.D., Carey, J.P. and Minor, L.B. (2002) Comparison of head thrust test with head autorotation test reveals that the vestibulo-ocular reflex is enhanced during voluntary head movements. Arch. Otolaryngol. Head Neck Surg., 128: 1044–1054.

Gilchrist, D.P., Curthoys, I.S., Cartwright, A.D., Burgess, A.M., Topple, A.N. and Halmagyi, M. (1998) High acceleration impulsive rotations reveal severe long-term deficits of the horizontal vestibulo-ocular reflex in the guinea pig. Exp. Brain Res., 123: 242–254.

Halmagyi, G.M. and Curthoys, I.S. (1988) A clinical sign of canal paresis. Arch. Neurol., 45: 737–739.

Halmagyi, G.M., Curthoys, I.S., Cremer, P.D., Henderson, C.J., Todd, M.J., Staples, M.J. and D'Cruz, D.M. (1990) The human horizontal vestibulo-ocular reflex in response to high-acceleration stimulation before and after unilateral vestibular neurectomy. Exp. Brain Res., 81: 479–490.

Haslwanter, T. (1995) Mathematics of three-dimensional eye rotations. Vision Res., 35: 1727–1739.

Hirvonen, M., Aalto, H., Migliaccio, A.A. and Hirvonen, T.P. (2007) Motorized head impulse rotator for horizontal vestibulo-ocular reflex: normal responses. Arch. Otolaryngol. Head Neck Surg., 133: 157–161.

Hullar, T.E., la Santina, C.C., Hirvonen, T., Lasker, D.M., Carey, J.P. and Minor, L.B. (2005) Responses of irregularly discharging chinchilla semicircular canal vestibular-nerve afferents during high-frequency head rotations. J. Neurophysiol., 93: 2777–2786.

Kimmig, H., Biscaldi, M., Mutter, J., Doerr, J.P. and Fischer, B. (2002) The initiation of smooth pursuit eye movements and saccades in normal subjects and in "express-saccade makers". Exp. Brain Res., 144: 373–384.

Lasker, D.M., Hullar, T.E. and Minor, L.B. (2000) Horizontal vestibuloocular reflex evoked by high-acceleration rotations in the squirrel monkey. III. Responses after labyrinthectomy. J. Neurophysiol., 83: 2482–2496.

Lehnen, N., Aw, S.T., Todd, M.J. and Halmagyi, G.M. (2004) Head impulse test reveals residual semicircular canal function after vestibular neurectomy. Neurology, 62: 2294–2296.

Lysakowski, A. and Goldberg, J.M. (2004) Morphophysiology of the vestibular periphery. In: Highstein S.M., Fay R.R. and Popper A.N. (Eds.), The Vestibular System. Springer Verlag, New York, pp. 57–152.

Migliaccio, A.A., Schubert, M.C., Jiradejvong, P., Lasker, D.M., Clendaniel, R.A. and Minor, L.B. (2004) The three-dimensional vestibulo-ocular reflex evoked by high-acceleration rotations in the squirrel monkey. Exp. Brain Res., 159: 433–446.

Newman-Toker, D.E., Kattah, J.C., Alvernia, J.E. and Wang, D.Z. (2008) Normal head impulse test differentiates acute cerebellar strokes from vestibular neuritis. Neurology, 70: 2378–2385.

Palla, A. and Straumann, D. (2004) Recovery of the high-acceleration vestibulo-ocular reflex after vestibular neuritis. J. Assoc. Res. Otolaryngol., 5: 427–435.

Peng, G.C., Zee, D.S. and Minor, L.B. (2004) Phase-plane analysis of gaze stabilization to high acceleration head thrusts: a continuum across normal subjects and patients with loss of vestibular function. J. Neurophysiol., 91: 1763–1781.

Remmel, R.S. (1984) An inexpensive eye movement monitor using the scleral search coil technique. IEEE Trans. Biomed. Eng., 31: 388–390.

Robinson, D.A. (1963) A method of measuring eye movement using a scleral search coil in a magnetic field. IEEE Trans. Biomed. Eng., 10: 137–145.

Sadeghi, S.G., Minor, L.B. and Cullen, K.E. (2006) Dynamics of the horizontal vestibuloocular reflex after unilateral labyrinthectomy: response to high frequency, high acceleration, and high velocity rotations. Exp. Brain Res., 175: 471–484.

Tian, J., Crane, B.T. and Demer, J.L. (2000) Vestibular catch-up saccades in labyrinthine deficiency. Exp. Brain Res., 131: 448–457.

Walker, M.F. and Zee, D.S. (2005) Cerebellar disease alters the axis of the high-acceleration vestibuloocular reflex. J. Neurophysiol., 94: 3417–3429.

Weber, K.P., Aw, S.T., Todd, M.J., McGarvie, L.A., Curthoys, I.S. and Halmagyi, G.M. (2008) Head impulse test in unilateral vestibular loss: vestibulo-ocular reflex and catch-up saccades. Neurology, 70: 454–463.

C. Kennard & R.J. Leigh (Eds.)
Progress in Brain Research, Vol. 171
ISSN 0079-6123

CHAPTER 3.7

Inter-ocular differences of the horizontal vestibulo-ocular reflex during impulsive testing

Konrad P. Weber[1,*], Swee T. Aw[1], Michael J. Todd[1], Leigh A. McGarvie[1], Satendra Pratap[1], Ian S. Curthoys[2] and G. Michael Halmagyi[1]

[1]*Department of Neurology, Royal Prince Alfred Hospital, Sydney, Australia*
[2]*Vestibular Research Laboratory, School of Psychology, University of Sydney, Sydney, Australia*

Abstract: Impulsive testing, which accurately measures the gain of the vestibulo-ocular reflex (VOR) is mostly recorded from only one eye. We measured the horizontal VOR of both eyes in response to graded velocity head impulses in 14 normal subjects. VOR acceleration gains were similar for each eye in both directions at low accelerations, but as head acceleration increased, gains of the adducting eye exceeded gains of the abducting eye. These inter-ocular VOR gain differences were due to longer latencies but higher accelerations of the adducting eye compared to the abducting eye. Consequently, directional VOR gain asymmetry occurred at high accelerations if the same eye was always analysed — with an average inter-ocular difference of 15.3%. If only the abducting eye was analysed, VOR gains were symmetrical in both directions. For accurate measurements of the VOR at high accelerations, the inter-ocular VOR gain differences require binocular recording.

Keywords: head impulse; vestibulo-ocular reflex; eye movements; binocular

Introduction

Although the head impulse test (HIT) accurately measures the gain of the angular vestibulo-ocular reflex (VOR), little attention has been paid to inter-ocular VOR gain differences because it is mostly recorded monocularly. Since a symmetrical measurement of VOR gain is a prerequisite for diagnosis of unilateral vestibular loss (Weber et al., 2008), we investigated inter-ocular VOR gain differences during HIT and then asked, which eye is best to analyse in order to obtain directional VOR gain symmetry in normal subjects.

*Corresponding author. Tel.: +61 2 9515 7565;
Fax: +61 2 9515 8347; E-mail: konrad.weber@bluewin.ch

Methods

We tested 14 normal subjects between 23 and 65 years (39.9±14 years, mean±SD). Written informed consent was obtained from all subjects; the protocol was approved by a local ethics committee and was in accordance with the Declaration of Helsinki. The subjects fixated a laser dot projected at a distance of 91 cm. About 80 horizontal head impulses to each side were manually applied with graded accelerations (750–6000°/s^2, velocity 50–300°/s, amplitude 5°–25°) using visual feedback to ensure correct head velocity was achieved. Scleral search coils were used to record three-dimensional binocular eye rotations and were mounted on a dental

DOI: 10.1016/S0079-6123(08)00626-2

impression bite bar for measuring head rotations (Aw et al., 2001). Three-dimensional rotation vectors and angular velocity vectors of head, gaze, and eye were derived from coil voltages (Haslwanter, 1995). Acceleration gain was calculated for each eye as the ratio of the slope of a linear regression fitted to eye velocity over head velocity during a 40-millisecond window centred at peak head acceleration (Collewijn and Smeets, 2000). Gain values as a function of head acceleration were smoothed by locally weighted regression using a least squares quadratic polynomial fitting algorithm (robust LOESS, smoothing fraction $f = 0.4$), interpolated for head accelerations between 750°/s^2 and 6000°/s^2 and means ± two-tailed 95% CI of LOESS-functions were determined (Weber et al., 2008). Directional asymmetry (g_s) between left (g_l) and right (g_r) VOR gains was determined as follows (Tabak et al., 1997):

$$g_s = \frac{g_l - g_r}{g_l + g_r} \times 100$$

Results

Figure 1 illustrates the inter-ocular VOR gain difference in HIT of a normal subject. With increasing head acceleration, velocity of the adducting eye showed longer latencies but steeper slopes than velocity of the abducting eye (Fig. 1A). As a result, VOR acceleration gains at low head accelerations were similar in both eyes for both directions, but as head acceleration increased, gains of the adducting eye exceeded gains of the abducting eye (Fig. 1B).

Since VOR gains were different in the two eyes, directional gain asymmetry depended on which eye was used for analysis. If the same eye was always analysed, directional asymmetry biased towards the side of the analysed eye as head acceleration increased (monocular analysis, Fig. 2A). The difference between gain analysis of the left and right eye reached 15.3% at 6000°/s^2 and was significant at head accelerations >3234°/s^2 (paired t-test, $p < 0.05$). If the analysed eye was alternated with HIT direction, symmetry remained stable over all head accelerations (binocular analysis, Fig. 2B). Analysis of the abducting eye (−0.35% ± 1.9; mean ± 95% CI) yielded tighter VOR gain symmetry with smaller confidence intervals than analysis of the adducting eye (3.7% ± 3.5).

Discussion

Our results revealed inter-ocular VOR gain differences in HIT with high accelerations. As a result,

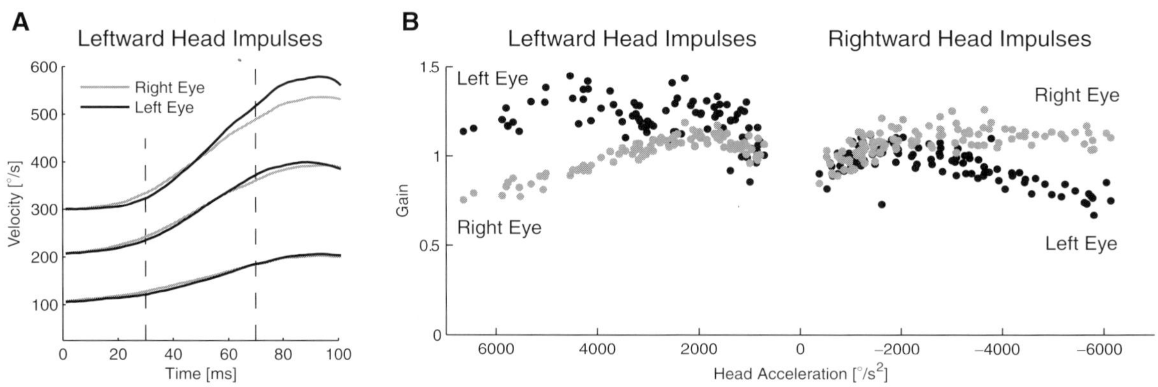

Fig. 1. Inter-ocular VOR difference in HIT of a normal subject. (A) Binocular eye velocity of three leftward head impulses with different head accelerations. Longer latencies but steeper slopes of the adducting left eye compared to the abducting right eye at higher head accelerations. Dashed lines indicate 40 ms time window where acceleration gain was determined centred at peak head acceleration. (B) Binocular acceleration gains of head impulses to both sides. Increasing inter-ocular VOR gain difference at higher head accelerations with higher gains of the adducting eye compared to the abducting eye.

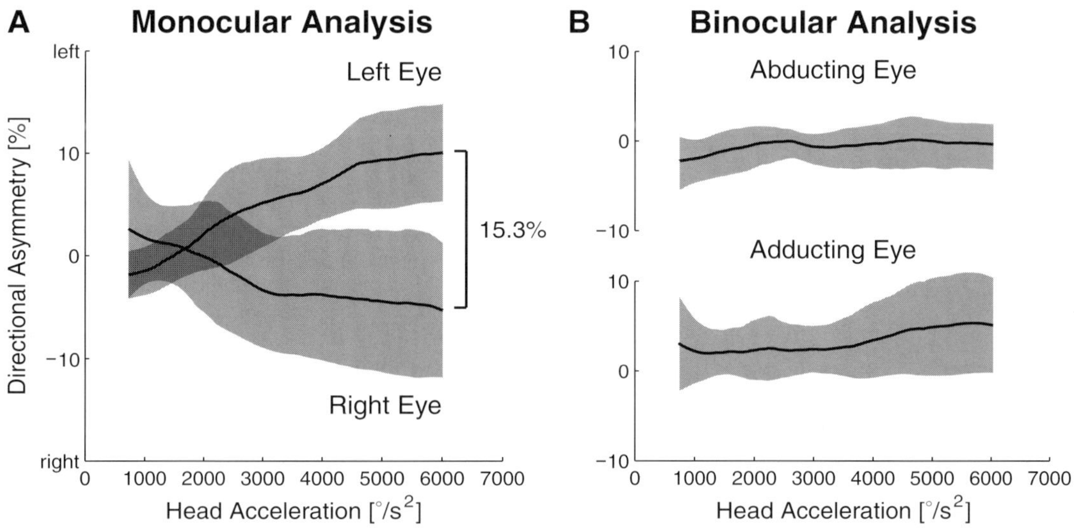

Fig. 2. Directional asymmetry of VOR gains in HIT of 14 normal subjects (mean±95% CI). (A) Monocular analysis: increasing directional asymmetry towards the side of the analysed eye with higher head acceleration. (B) Binocular analysis: tighter VOR gain symmetry with analysis of the abducting eye compared to the adducting eye.

VOR gains in normal subjects showed directional asymmetry if the same eye was always analysed. This directional asymmetry could be cancelled if the analysed eye was alternated with HIT direction and gains were determined always in the abducting eye.

Analysis of the abducting eye provided tighter data than analysis of the adducting eye. This probably reflects the shorter disynaptic pathway from the stimulated horizontal semicircular canal to the lateral rectus muscle of the abducting contralateral eye (Leigh and Zee, 2006). The observed inter-ocular VOR gain difference might be caused by the synaptic delay and the different firing characteristics of the additional abducens internuclear neuron in the pathway from the stimulated horizontal semicircular canal to the medial rectus muscle of the ipsilateral adducting eye (Nakao et al., 1977; Collewijn and Smeets, 2000). Alternatively, different dynamics of the yoked medial and lateral rectus eye muscles may result in inter-ocular VOR gain differences. It has been shown in humans that the maximum active force of the medial rectus muscle was about 26% greater than that of the lateral rectus muscle (Collins et al., 1981).

Mechanical factors such as translation of the adducting eyeball by inadvertent pulling of the skin during manual HIT application may provide another possible explanation. Similarly, artefacts have been reported when the head coil was taped on the forehead instead of mounted on a bite bar (Crane and Demer, 1998). However, inter-ocular VOR gain differences were consistent in our control experiments with different head gripping techniques. On the other hand, previous studies of mechanical HIT attributed artefacts in the initial eye movement response to the inertia of the eyeball itself (Collewijn and Smeets, 2000). Inter-ocular VOR gain differences due to geometrical factors were negligible since the target distance used in our paradigm (91 cm) was large (Viirre et al., 1986).

In conclusion, accurate HIT measurements have to take into account VOR gain differences between the two eyes at high head accelerations and call for binocular HIT recording. Based on the tighter data probably reflecting the shorter disynaptic pathway from the stimulated horizontal semicircular canal to the lateral rectus muscle of the abducting contralateral eye, we recommend analysing the gain of the abducting eye to obtain directional symmetry of VOR gain measurements in normal subjects.

Acknowledgement

Supported by the Garnett Passe and Rodney Williams Memorial Foundation.

References

Aw, S.T., Fetter, M., Cremer, P.D., Karlberg, M. and Halmagyi, G.M. (2001) Individual semicircular canal function in superior and inferior vestibular neuritis. Neurology, 57: 768–774.

Collewijn, H. and Smeets, J.B. (2000) Early components of the human vestibulo-ocular response to head rotation: latency and gain. J. Neurophysiol., 84: 376–389.

Collins, C.C., Carlson, M.R., Scott, A.B. and Jampolsky, A. (1981) Extraocular muscle forces in normal human subjects. Invest. Ophthalmol. Vis. Sci., 20: 652–664.

Crane, B.T. and Demer, J.L. (1998) Human horizontal vestibulo-ocular reflex initiation: effects of acceleration, target distance, and unilateral deafferentation. J. Neurophysiol., 80: 1151–1166.

Haslwanter, T. (1995) Mathematics of three-dimensional eye rotations. Vis. Res., 35: 1727–1739.

Leigh, R.J. and Zee, D.S. (2006) The vestibular-optokinetic system. In: Leigh R.J. and Zee, D.S. (Eds.), The Neurology of Eye Movements, 4th ed., Oxford University Press, New York, pp. 20–107.

Nakao, S., Sasaki, S. and Shimazu, H. (1977) Nuclear delay of impulse transmission in abducens motoneurons during fast eye movements of visual and vestibular origin in alert cats. J. Neurophysiol., 40: 1415–1423.

Tabak, S., Collewijn, H., Boumans, L.J. and van der, S.J. (1997) Gain and delay of human vestibulo-ocular reflexes to oscillation and steps of the head by a reactive torque helmet. I. Normal subjects. Acta Otolaryngol., 117: 785–795.

Viirre, E., Tweed, D., Milner, K. and Vilis, T. (1986) A reexamination of the gain of the vestibuloocular reflex. J. Neurophysiol., 56: 439–450.

Weber, K.P., Aw, S.T., Todd, M.J., McGarvie, L.A., Curthoys, I.S. and Halmagyi, G.M. (2008) Head impulse test in unilateral vestibular loss: vestibulo-ocular reflex and catch-up saccades. Neurology, 70: 454–463.

C. Kennard & R.J. Leigh (Eds.)
Progress in Brain Research, Vol. 171
ISSN 0079-6123

CHAPTER 3.8

Control of ocular torsion in the rotational vestibulo-ocular reflexes

Bernhard J.M. Hess*

Department of Neurology, University Hospital Zurich, CH-8091 Zurich, Switzerland

Abstract: Visual stabilization of the retina during rotational head movements requires that in far vision the eyes rotate about the same axis as the head but in opposite direction with a gain close to unity (optimal strategy). To achieve this goal the vestibulo-oculomotor system must be able to independently control all three rotational degrees of freedom of the eye. Studies of the human rotational vestibulo-ocular reflexes (VOR) have shown that its spatial characteristics are best explained by a strategy that lies halfway between the optimal image stabilization and perfect compliance with Listing's law. Here we argue that these spatial characteristics are fully compatible with an optimal strategy under the condition of a restrained gain of the torsional velocity-to-position integration. One implication of this finding is that the rotational VORs must override the default operation mode of the ocular plant that, according to recent findings, mechanically favours movements obeying Listing's law.

Keywords: eye movements; sensory-motor transformations; vestibular; spatial orientation

Introduction

The rotational vestibulo-ocular reflexes (VOR) stabilize the retinal image by adjusting ocular orientation and gaze in space. In far vision, the entire retinal image can be stabilized against head rotations by rotating the eyes in opposite direction at the same velocity about the same rotation axis as the head. This strategy helps to maintain constancy of the visual field during disturbances of head-in-space orientation and enhances self-motion and orientation (Brandt et al., 1973; Amblard and Carblanc, 1980). In contrast, retinal image stabilization during translational movements requires more selective strategies that stabilize only the foveal and part of the perifoveal image since it cannot be achieved across the entire visual field (Miles, 1999; Angelaki et al., 2003; Angelaki and Hess, 2005). Foveal image stabilization could also be accomplished by anchoring the torsional orientation of the eye to the head and thereby complying with Listing's law (but see Tweed et al., 1998). The peculiar spatial characteristics observed in the human rotational VORs have in fact been interpreted as the result of a compromise strategy halfway between optimal retinal image stabilization and perfect compliance with Listing's law (Misslisch et al., 1994). This law, which underlies all visually guided eye movements, requires that the angular eye velocity depends on eye position in such a way that it always lies in a plane that rotates relative to primary position in the same direction as

*Corresponding author. Tel.: +4144-255-5500;
Fax: +4144-255-4507; E-mail: bhess@neurol.uzh.ch

DOI: 10.1016/S0079-6123(08)00627-4

the gaze line, but only by half the angle (von Helmholtz, 1867; Tweed and Vilis, 1990).

Although the optimal VOR strategy appears to be best suited for visual stabilization, it is not known whether it is in fact the "primum movens" in the evolution and development of the rotational VOR. In the following I will argue that the optimal strategy is indeed underlying the evolutionary old rotational VORs and provide theoretical evidence that its spatiotemporal characteristics follow from first principles of three-dimensional (3D) rotation kinematics and the additional condition of a reduced torsional velocity-to-position integrator.

Independence of gaze position implies torsional rotations in the rotational VORs

While the control strategies underlying the rotational yaw or pitch VOR are not distinguishable in primary gaze direction, clear-cut differences emerge in all other gaze directions. Starting from non-primary positions, rotations of the eye in a horizontal or vertical plane invariably comprise a torsional rotation component in contrast to the torsion-free rotation paths in planes that contain the rotation centre (see the horizontal and vertical great circle in the respective meridional planes in Fig. 1A). Accordingly, the optimal yaw and pitch VOR need to stabilize the eye in planes parallel to the horizontal and vertical meridional planes (see small circles in Fig. 1B, C). As a consequence, the torsional eye position changes the more the VOR drives the eye along these small circular paths. It is interesting to compare the associated angular velocity along the small and the torsion-free great circles in the yaw or pitch VOR (Fig. 1B, C). Interestingly, a difference in angular velocity

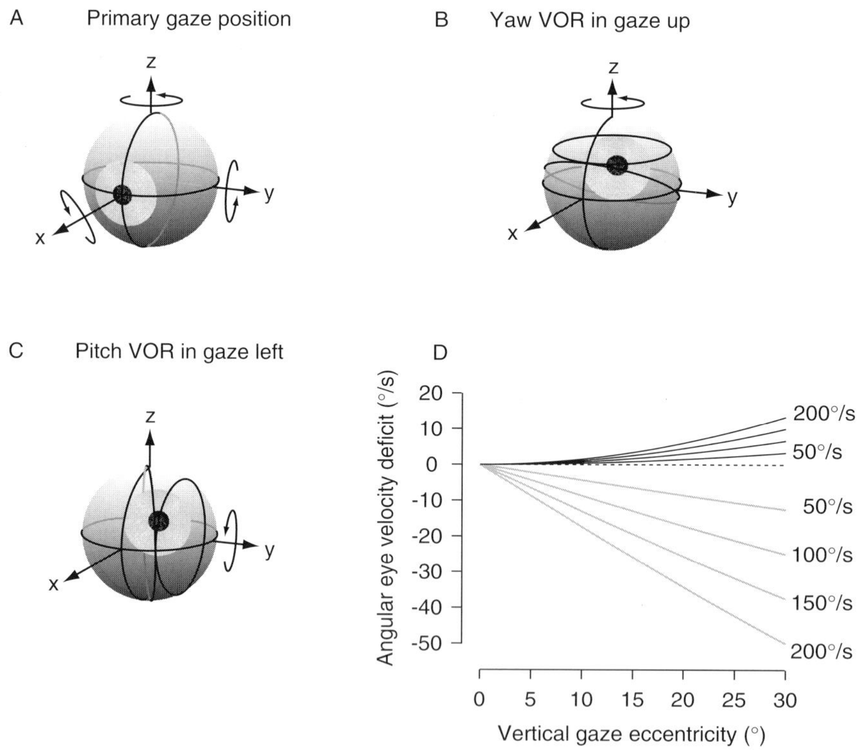

Fig. 1. Optimal vestibulo-ocular reflexes while the eye is (A) in primary or (B, C) in a tertiary position. Small circles indicate the optimal rotation paths; the respective great circles through initial position describe the torsion-free paths. Positional errors (difference between great and small circle paths through initial position shown in B and C) associate with increasing angular velocity deficits (D) as head velocity increases (see labels on the right). *Solid black lines*: horizontal, *dotted black line*: vertical, *solid gray lines*: torsional. *Circular arrows* indicate positive rotation directions around x- (roll), y-(pitch), and z-(yaw) rotation axis.

emerges not only in torsional but also in horizontal or vertical direction. It increases with gaze eccentricity and the magnitude of head angular velocity as illustrated in Fig. 1D for the yaw VOR (optimal strategy with velocity gain = 1). The torsional angular velocity deficit is proportional to the VOR's capacity to change ocular torsion according to the relation: torsional velocity deficit $x(1-k)$ with $k=1$ for optimal performance and $k=0$ for no change in ocular torsion. The reason for the angular velocity difference in horizontal direction is that the velocity associated with the eye movement along the great circle path tilts.

Constraints on the neural implementation of the optimal VOR strategy

A fundamental challenge for the neural implementation of the rotational VORs is related to the fact that the transduced head angular velocity signals from the semicircular canals need to be transformed at least into six different eye velocity commands to the extraocular muscles in such a way that the resulting angular eye velocity at the motor output matches the physical head angular velocity in direction and magnitude. Using the optimal VOR strategy as a model, 3D rotation kinematics provides the guidelines of how the brain might meet this challenge (Tweed and Vilis, 1987). Based on these kinematical principles one can postulate that the neural velocity commands to the eye muscle (formally summarized as eye velocity $\mathrm{d}\hat{E}/\mathrm{d}t$) must follow the proportionality: $\mathrm{d}\hat{E}/\mathrm{d}t \sim \hat{\Omega}_{\mathrm{SCC}} + \hat{\Omega}_{\mathrm{SCC}} \times \hat{E}$, where $\hat{\Omega}_{\mathrm{SCC}}$ represents the transduced head angular velocity and $\hat{E}$ the neural eye position commands for the three pairs of extraocular muscles in the final common pathway. In this proportionality, we neglected quadratic terms in eye position, although they are important in tertiary eye positions. To distinguish neural commands from the respective kinematical variables we use the superscript "^." The crucial element in the mentioned proportionality is the postulated multiplicative interaction of canal angular velocity and eye position signals, which has been implicitly used also in previous models (Tweed, 1997; Smith and Crawford, 1998; Misslisch and Tweed, 2001; Glasauer, 2007). On the motor side, the head rotation imposes a similar fundamental constraint on the neural eye velocity commands by the proportionality: $\mathrm{d}\hat{E}/\mathrm{d}t + \hat{E} \times \mathrm{d}\hat{E}/\mathrm{d}t \sim \Omega_{\mathrm{eye}}$ (where Ω_{eye} is the angular velocity of the eye). These two ancillary conditions, the one on the sensory and the other on the motor output side, set the stage for matching the transduced head velocity to the eye angular velocity at the motor output that results from the neural eye position and velocity commands. A summary of these conditions is presented in Table 1.

Role of the torsional velocity-to-position integrator

Rotational VOR in the yaw and pitch plane

In the optimal VOR strategy, the rotation of the eye has to occur about the same axis as the head independently of the initial ocular orientation.

Table 1. Sensory-motor transformations of head velocity signals in the rotational VOR

Transduced head angular velocity	Eye velocity commands as a function of eye position ($\hat{E}$) and head angular velocity ($\hat{\Omega}$): $\mathrm{d}\hat{E}/\mathrm{d}t = [\hat{\Omega} + \hat{\Omega} \times \hat{E} + (\hat{\Omega} \cdot \hat{E})\hat{E}]/2$	Resulting angular eye velocity at the motor output: $\Omega_{\mathrm{eye}} = 2(\mathrm{d}E/\mathrm{d}t + E \times \mathrm{d}E/\mathrm{d}t)/(1+\lvert E\rvert^2)$
Yaw VOR $\hat{\Omega} = \omega(0,0,1)$	$\mathrm{d}\hat{E}/\mathrm{d}t \approx (\omega/2)(-kE_{\mathrm{ver}}, 0, 1)$	$\Omega_{\mathrm{yawVOR}} \sim \omega[(1-k)E_{\mathrm{ver}}, 0, 1]$
Pitch VOR $\hat{\Omega} = \omega(0,1,0)$	$\mathrm{d}\hat{E}/\mathrm{d}t \approx (\omega/2)(kE_{\mathrm{hor}}, 1, 0)$	$\Omega_{\mathrm{pitchVOR}} \sim \omega[(k-1)E_{\mathrm{hor}}, 1, 0]$
Roll VOR $\hat{\Omega} = \omega(1,0,0)$	$\mathrm{d}\hat{E}/\mathrm{d}t \approx (\omega/2)(k, -E_{\mathrm{hor}}, E_{\mathrm{ver}})$	$\Omega_{\mathrm{rollVOR}} \sim \omega[k, (k-1)E_{\mathrm{hor}}, (1-k)E_{\mathrm{ver}}]$

$E = \mathbf{n}\tan(\rho/2)$: 3D eye position expressed as rotation vector (Haustein, 1989; Hepp, 1990); $\mathbf{n}$ = unit vector along the rotation axis; ρ = angle of rotation about $\mathbf{n}$; $\omega = \omega(t)$: angular velocity profile of head rotation; $k = \Delta E^*_{\mathrm{tor}}/\Delta E_{\mathrm{tor}}$: gain of torsional velocity-to-position integration; $\Delta E^*_{\mathrm{tor}}$: observed change in ocular torsion, ΔE_{tor}: unrestrained kinematic change in ocular torsion predicted by the optimal VOR strategy. Note that $\omega = -g_x\lvert\Omega_{\mathrm{head}}\rvert$ (where the subscript "x" stands for "tor", "ver" or "hor").

This simple requirement is difficult to fulfil without some constraints, because, even in primary position, changes of head position about the roll axis could result in rotations of the eyes that exceed the limits of the torsional oculomotor range. In the optimal yaw and pitch VOR, the eye can acquire large changes in torsional position for eccentric gaze directions as illustrated by the following example. During yaw rotations of the head while fixating an eccentric vertical target, the change in torsional eye position increases linearly with vertical gaze eccentricity (Fig. 2A, left panel), and, correspondingly, also torsional eye velocity (Fig. 2A, middle panel). Both effects are due to the torsional difference of the small circular path that the eye has to follow in the optimal VOR strategy and the torsion-free great circle path (see Fig. 1B). Thus, the eye can only move along the small circle by continuously updating its torsional orientation, which implies perfect integration of the underlying torsional velocity. If the yaw VOR is followed by a pitch VOR, changes in ocular torsion can even accumulate (Tchelidze and Hess, 2008). To further characterize this crucial condition for an optimal performance, we define the gain of the velocity-to-position integration by $k = \Delta E^{*}_{\mathrm{tor}}/\Delta E_{\mathrm{tor}}$, where $\Delta E^{*}_{\mathrm{tor}}$ is the observed change in ocular torsion and ΔE_{tor} is the unrestricted kinematical torsion of the optimal VOR (for more details see Tchelidze and Hess, 2008). When this gain is reduced, say to $k = 0.5$, then the change in torsional eye position and velocity is also reduced by the same factor (see Figs. 1D and 2B, left and middle panel). Note, however, that this does not affect the horizontal eye velocity, implying that the motor commands to the horizontal eye muscles remain unchanged. However, it does affect the spatial orientation of the angular eye velocity at the motor output (Fig. 2B, right panel).

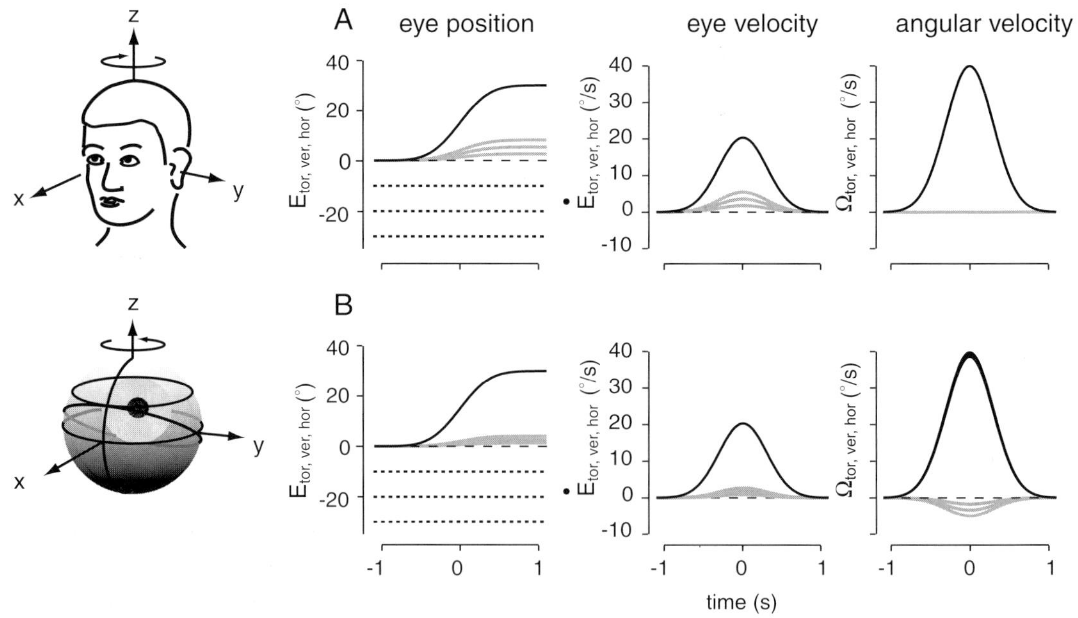

Fig. 2. Simulation of ocular torsion as a function of vertical gaze eccentricity in the yaw VOR. (A) In elevated gaze directions (*dotted lines*), the eye must move also in torsional direction (*grey full lines*) in order to generate an optimal yaw VOR (torsional position gain $k = 1$; velocity gain = 1, head and eye rotation axes parallel, peak head angular velocity = 40°/s). (B) Effect of reducing the torsional position gain to $k = 0.5$. Note reduction of torsional eye position and velocity by half and emergence of torsional angular velocity components (*grey full lines*). *Solid black traces*: horizontal eye position/velocity/angular velocity. *Grey traces*: torsional eye position/velocity/angular velocity. *Dotted black traces*: vertical eye position. Inset: Schematic eye at 30° elevation; small circle = response plane of an optimal yaw VOR; great circle through pupil = plane of constant elevation.

Although its peak horizontal component does change only little (approximately with the cosine of $k\varepsilon/2$, where ε is the angle of vertical gaze eccentricity), there is now a torsional angular velocity component that increases approximately linearly with vertical gaze eccentricity (i.e., with the sine of $k\varepsilon/2$). As a consequence, the resultant angular eye velocity tilts in the pitch plane whenever the gain of torsional velocity-to-position integration is reduced. Similar observations can be made for the pitch VOR in eccentric horizontal gaze directions. In both the non-optimal yaw and pitch VOR, the angular eye velocity tilts always in the same direction as the gaze line with respect to primary gaze direction. The important point is that the tilt amplitude in the yaw as well as in the pitch VOR only depends on the gain of the torsional velocity-to-position integration. It follows approximately the relation $\pm(1-k)\times\varepsilon/2$ where ε is the angle of gaze eccentricity (for details see Table 1 and Tchelidze and Hess, 2008). For $k=1/2$ it results in the so-called quarter angle rule (Misslisch et al., 1994; Glasauer, 2007).

Rotational VOR in the roll plane

In the yaw and pitch VOR, we found that it is the reduced gain of the position integration of torsional eye velocity that limits the torsional compared to the vertical and horizontal oculomotor range. What are the consequences of this limitation on the spatial properties of the roll VOR? Simulations of a hypothetical optimal roll VOR at eccentric gaze directions not only demonstrate exceedingly large torsional eye position changes but also show that, depending on the momentary position along its circular path, eye position changes also in the horizontal and vertical

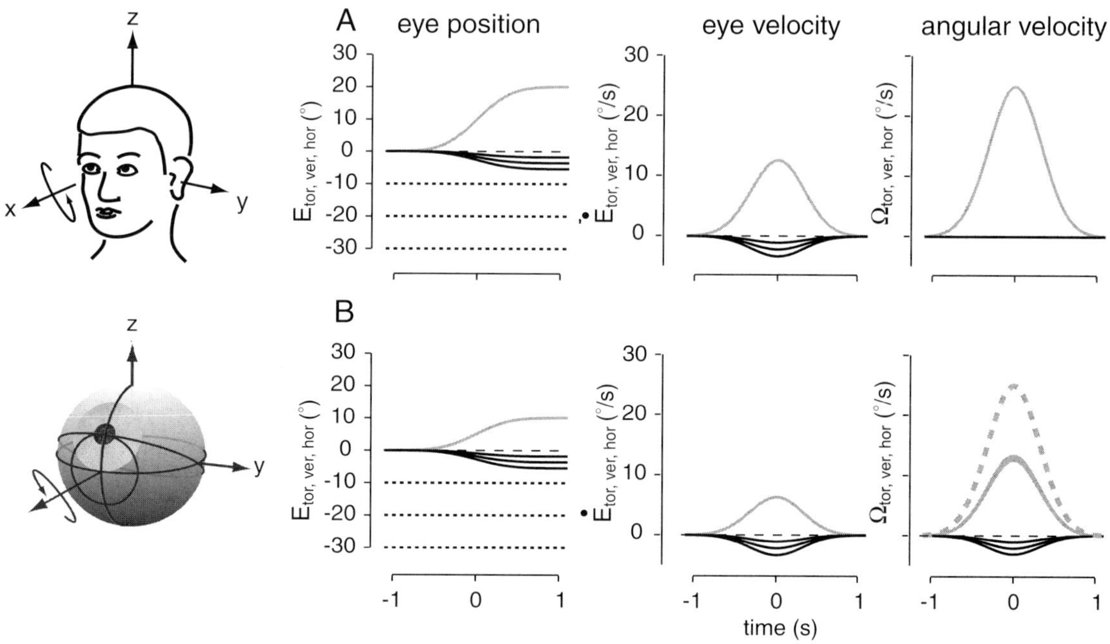

Fig. 3. Simulation of ocular torsion as a function of vertical gaze eccentricity in the roll VOR. (A) In elevated gaze directions (*black dotted lines*), the eye must move also in horizontal direction (*black full lines*) in order to generate an optimal roll VOR (torsional position gain $k=1$; velocity gain = 1, head and eye rotation axes parallel, peak head angular velocity = 25°/s). (B) Effect of reducing the torsional position gain to $k=0.5$. Note reduction of torsional eye position and velocity by half and emergence of horizontal angular velocity components (*black full lines*). *Solid grey traces*: torsional eye position/velocity/angular velocity. *Dashed grey trace*: angular gaze velocity. *Solid black traces*: horizontal eye position/velocity/angular velocity. *Dotted black traces*: vertical eye position. Inset: Schematic eye at 30° elevation; small circle = response plane of an optimal roll VOR; great circle through pupil = plane of constant elevation.

direction. Despite these additional position and velocity components (Fig. 3A, left and middle panel), the angular velocity at the motor output is directed parallel to the head angular velocity at the sensory input side, i.e., along the x-axis (no y- and z-components, Fig. 3A, right panel). Assuming now a torsional velocity-to-position integration gain of $k = 0.5$, which is more realistic (Collewijn et al., 1985; Leigh et al., 1989; Crawford and Vilis, 1991; Tweed et al., 1994; Seidman et al., 1995; Schmid-Priscoveanu et al., 2000), the torsional change in eye position and velocity reduces by the same factor (Fig. 3B, left and middle panel). In contrast to the pitch and yaw VOR, it also reduces the peak torsional angular velocity at the motor output by the same factor. This reduction in amplitude of the angular velocity at the motor output is completely independent of the velocity gain of the roll VOR (assumed to be unity in all simulations). The presented kinematical analysis is based on the following proportionality for the torsional angular velocity of the roll VOR:

$$\Omega_{\text{rollVOR}} \sim g_{\text{roll}}|\Omega_{\text{head}}|[k, (k-1)E_{\text{hor}}, (1-k)E_{\text{ver}}]$$

where $g_{\text{roll}} = \omega/|\Omega_{\text{head}}|$ is the velocity gain (ratio of semicircular canal angular velocity and head angular velocity; Table 1). One conclusion of this analysis is that the overall roll VOR gain ($|(\Omega_{\text{rollVOR}})_{\text{tor}}|/|\Omega_{\text{head}}|$) depends on both the velocity gain (g_{roll}) and the torsional velocity-to-position integration gain k. We assume that the low roll VOR gain during high-acceleration head rotations is entirely due to the restrained gain of the velocity-to-position integration, which under these conditions becomes noticeable only in the roll VOR (Aw et al., 1996). Another consequence of the proposed interaction of neural eye position and velocity signals and a reduced torsional position integrator gain in the roll VOR is the tilt of the angular velocity axis as a function of gaze direction. This feature is experimentally well documented in the human and non-human primate VOR (Crawford and Vilis, 1991; Misslisch et al., 1994; Misslisch and Hess, 2000; Misslisch and Tweed, 2001; Tchelidze and Hess, 2007). In contrast to the yaw and pitch VOR, it follows approximately the relation $\pm(1-k)/k\ \varepsilon/2$ with ε the angle of gaze eccentricity and k the gain of torsional velocity-to-position integration (Tchelidze and Hess, 2008). Note that for $k = 0.5$ this results in the experimentally documented half-angle rule characteristics in the roll VOR (Misslisch et al., 1994; Tchelidze and Hess, 2008).

Dissociation of torsional gaze velocity and ocular torsion

Humans and subhuman primates can perfectly fixate a memorized eccentric target that is presented shortly before onset of a head rotation about the roll axis in complete darkness (Misslisch and Hess, 2000; Tchelidze and Hess, 2008). How can this be if the overall roll VOR gain is as low as 0.5 for moderate head roll velocities? The reason lies in the difference between torsional eye velocity and torsional angular eye velocity. To keep the fovea on an eccentric target during a head movement in the roll plane, it is the torsional eye velocity that matters because it describes the motion of the gaze line (or fovea) along a circular path around the x-axis (see small circle through pupil of schematic eye in Fig. 3B). Kinematical analysis shows that this velocity is a function of the simultaneous modulation of vertical and horizontal eye position and velocity given by the equation $\mathrm{d}E_{\text{tor}}/\mathrm{d}t = (E_{\text{ver}}\mathrm{d}E_{\text{hor}}/\mathrm{d}t - E_{\text{hor}}\mathrm{d}E_{\text{ver}}/\mathrm{d}t)/(E_{\text{ver}}^2 + E_{\text{hor}}^2)$, which only implicitly depends on head angular velocity (see dashed grey curve in Fig. 3B, right panel). This relation adequately describes the neural input pattern to the horizontal and vertical recti muscles that is required for tracking an eccentric target that rotates in a plane parallel to the observer's roll plane. The torsional angular velocity of the roll VOR, i.e., the component $(\Omega_{\text{rollVOR}})_{\text{tor}}$, only matches $\mathrm{d}E_{\text{tor}}/\mathrm{d}t$ if the gain of the torsional velocity-to-position integration is unity. Thus, although the overall roll VOR might be as low as 0.5 due to a reduced k (while the velocity gain g_{roll} is close to unity like in the yaw and pitch VOR) it does not necessarily affect the foveal stability of an eccentric target.

Neural velocity commands and orbital mechanics

The analyses of the sensory-motor transformations in the rotational VOR show that in the optimal strategy of yaw and pitch VOR the axis of the totally commanded eye velocity tilts in the opposite direction as that required for visual guided eye movements that obey Listings law. For example, to minimize the tilt of the angular velocity of the eye during yaw VOR, the eye velocity commands must spin the eye about a forward tilted axis as the gaze line is elevated. Similarly, during the pitch VOR, the neural eye velocity commands must spin the eye about a rightward tilted axis as the gaze line shifts leftward (inverse half-angle rule). The required tilt reduces approximately proportionally to the gain of the torsional velocity-to-position integration (see vertical and horizontal eye position dependence of eye velocity commands in Table 1). If the normal half-angle dependence on gaze eccentricity is the default operation mode of the ocular plant as recent studies suggest (for reviews see Demer, 2007, Miller, 2007) the torsional eye velocity commands in the yaw and pitch VOR need support by an additional torsional signal that overrides the default operation mode.

Acknowledgement

I thank E. Buffone for drawing the figures. Supported by the Swiss National Science Foundation Grant 31-47 287.96, the Hermann Klaus Foundation and the Betty and David Koetser Foundation for Brain Research.

References

Amblard, B. and Carblanc, A. (1980) Role of foveal and peripheral visual information in maintenance of postural equilibrium in man. Percept. Mot. Skills, 51: 903–912.

Angelaki, D.E. and Hess, B.J. (2005) Self-motion-induced eye movements: effects on visual acuity and navigation. Nat. Rev. Neurosci., 6: 966–976.

Angelaki, D.E., Zhou, H.H. and Wei, M. (2003) Foveal versus full-field visual stabilization strategies for translational and rotational head movements. J. Neurosci., 23: 1104–1108.

Aw, S.T., Haslwanter, T., Halmagyi, G.M., Curthoys, I.S., Yavor, R.A. and Todd, M.J. (1996) Three-dimensional vector analysis of the human vestibuloocular reflex in response to high-acceleration head rotations. I. Responses in normal subjects. J. Neurophysiol., 76: 4009–4020.

Brandt, T., Dichgans, J. and Koenig, E. (1973) Differential effects of central versus peripheral vision on egocentric and exocentric motion perception. Exp. Brain Res., 16: 476–491.

Collewijn, H., Van der Steen, J., Ferman, L. and Jansen, T.C. (1985) Human ocular counterroll: assessment of static and dynamic properties from electromagnetic scleral coil recordings. Exp. Brain Res., 59: 185–196.

Crawford, J.D. and Vilis, T. (1991) Axes of eye rotation and Listing's law during rotations of the head. J. Neurophysiol., 65: 407–423.

Demer, J.L. (2007) Mechanics of the orbita. Dev. Ophthalmol., 40: 132–157.

Glasauer, S. (2007) Current models of the ocular motor system. Dev. Ophthalmol., 40: 158–174.

Haustein, W. (1989) Considerations on Listing's law and the primary position by means of a matrix description of eye position control. Biol. Cybern., 60: 411–420.

Hepp, K. (1990) On Listing's law. Commun. Math Phys., 132: 285–292.

Leigh, R.J., Maas, E.F., Grossman, G.E. and Robinson, D.A. (1989) Visual cancellation of the torsional vestibulo-ocular reflex in humans. Exp. Brain Res., 75: 221–226.

Miles, F.A. (1999) Short-latency visual stabilization mechanisms that help to compensate for translational disturbances of gaze. Ann. N.Y. Acad. Sci., 871: 260–271.

Miller, J.M. (2007) Understanding and misunderstanding extraocular muscle pulleys. J. Vis., 7: 10–15.

Misslisch, H. and Hess, B.J. (2000) Three-dimensional vestibulo-ocular reflex of the monkey: optimal retinal image stabilization versus Listing's law. J. Neurophysiol., 83: 3264–3276.

Misslisch, H. and Tweed, D. (2001) Neural and mechanical factors in eye control. J. Neurophysiol., 86: 1877–1883.

Misslisch, H., Tweed, D., Fetter, M., Sievering, D. and Koenig, E. (1994) Rotational kinematics of the human vestibuloocular reflex. III. Listing's law. J Neurophysiol, 72: 2490–2502.

Schmid-Priscoveanu, A., Straumann, D. and Kori, A.A. (2000) Torsional vestibulo-ocular reflex during whole-body oscillation in the upright and the supine position. I. Responses in healthy human subjects. Exp. Brain Res., 134: 212–219.

Seidman, S.H., Leigh, R.J., Tomsak, R.L., Grant, M.P. and Dell'Osso, L.F. (1995) Dynamic properties of the human vestibulo-ocular reflex during head rotations in roll. Vis. Res., 35: 679–689.

Smith, M.A. and Crawford, J.D. (1998) Neural control of rotational kinematics within realistic vestibuloocular coordinate systems. J. Neurophysiol., 80: 2295–2315.

Tchelidze, T. and Hess, B.J. (2008) Non-commutative control in the rotational vestibulo-ocular reflex. J. Neurophysiol, 99: 96–111.

Tweed, D. (1997) Kinematic principles of three-dimensional gaze control. In: Fetter M., Haslwanter T., Misslisch H. and Tweed D. (Eds.), Three-Dimensional Kinematics of Eye,

Head and Limb Movements. Harwood Academic Publishers, Amsterdam, pp. 375–386.
Tweed, D., Haslwanter, T. and Fetter, M. (1998) Optimizing gaze control in three dimensions. Science, 281: 1363–1366.
Tweed, D., Sievering, D., Misslisch, H., Fetter, M., Zee, D. and Koenig, E. (1994) Rotational kinematics of the human vestibuloocular reflex. I. Gain matrices. J. Neurophysiol., 72: 2467–2479.
Tweed, D. and Vilis, T. (1987) Implications of rotational kinematics for the oculomotor system in three dimensions. J. Neurophysiol., 58: 832–849.
Tweed, D. and Vilis, T. (1990) Geometric relations of eye position and velocity vectors during saccades. Vis. Res., 30: 111–127.
von Helmholtz, H. (1867) Handbuch der Physiologischen Optik. Hamburg, Voss.

C. Kennard & R.J. Leigh (Eds.)
Progress in Brain Research, Vol. 171
ISSN 0079-6123

CHAPTER 3.9

Do humans show velocity-storage in the vertical rVOR?

G. Bertolini[1,*], C.J. Bockisch[2], D. Straumann[2], D.S. Zee[3] and S. Ramat[1]

[1]*Dipartimento di Informatica e Sistemistica, Università di Pavia, Italy*
[2]*Department of Neurology, Zurich University Hospital, Zurich, Switzerland*
[3]*Department of Neurology, The Johns Hopkins University, Baltimore, MD, USA*

Abstract: To investigate the contribution of the vestibular velocity-storage mechanism (VSM) to the vertical rotational vestibulo-ocular reflex (rVOR) we recorded eye movements evoked by off-vertical axis rotation (OVAR) using whole-body constant-velocity pitch rotations about an earth-horizontal, interaural axis in four healthy human subjects. Subjects were tumbled forward, and backward, at 60 deg/s for over 1 min using a 3D turntable. Slow-phase velocity (SPV) responses were similar to the horizontal responses elicited by OVAR along the body longitudinal axis, ('barbecue' rotation), with exponentially decaying amplitudes and a residual, otolith-driven sinusoidal response with a bias. The time constants of the vertical SPV ranged from 6 to 9 s. These values are closer to those that reflect the dynamic properties of vestibular afferents than the typical 20 s produced by the VSM in the horizontal plane, confirming the relatively smaller contribution of the VSM to these vertical responses. Our preliminary results also agree with the idea that the VSM velocity response aligns with the direction of gravity. The horizontal and torsional eye velocity traces were also sinusoidally modulated by the change in gravity, but showed no exponential decay.

Keywords: pitch VOR; velocity-storage mechanism; rVOR

Introduction

The horizontal rotational vestibulo-ocular reflex (rVOR) has been investigated extensively, but the vertical rVOR much less so, especially for human subjects. The existence of a velocity-storage mechanism (VSM), which prolongs the time constant of the horizontal rVOR, is unclear for the vertical rVOR. Here we investigated this issue further by recording eye movements in normal subjects to constant-velocity rotation around the pitch axis.

*Corresponding author. Tel.: +39 0382 985352;
Fax: +39 0382 985373; E-mail: giovanni.bertolini@unipv.it

Materials and methods

Vestibular stimulation

Using dual scleral search coils we recorded the three-dimensional eye movements of four healthy subjects monocularly while they were rotated forward and backward about the earth-horizontal interaural axis (pitch rotations) at 60 deg/s in total darkness for over 1 min. The experiments were performed using a motor-driven turntable (prototype built by Acutronic, Switzerland). The turntable was accelerated at 10 deg/s^2, reaching the constant-velocity in 6 s. Details of the coil calibration procedure can be found elsewhere (Marti et al., 2005).

DOI: 10.1016/S0079-6123(08)00628-6

Data analysis

Saccades were interactively removed from the velocity signal, which was linearly interpolated over the intervals corresponding to the removed saccades. The resulting vertical slow-phase velocity (SPV) showed exponentially decaying amplitude, and a residual, otolith-driven, sinusoidal response with a bias. A fit of the sinusoid was estimated from the later part of the response, over 60 s from motion onset, when the rVOR response had presumably vanished, even if the overall rVOR time constant is 20 s. The following sinusoidal function was used to fit the otolith response data

$$y = A * \sin(2\pi f t + \varphi) + c \tag{1}$$

where A is the amplitude, f the frequency, φ the phase, and c the offset. The frequency (f) was kept constant at 0.16 Hz (corresponding to a chair angular velocity of 60 deg/s), while the amplitude (A), the phase (φ), and the offset (c) were iteratively optimized using a nonlinear least-squares algorithm. Once these parameters had been determined, the sinusoidal function was computed for the duration of the whole response, as shown in Fig. 1 panels b–c, and was then subtracted from the SPV data. Finally, the amplitude and the time constant of the rVOR response was estimated by fitting the resulting data with the following exponential function

$$y = A * e^{-t/\tau} \tag{2}$$

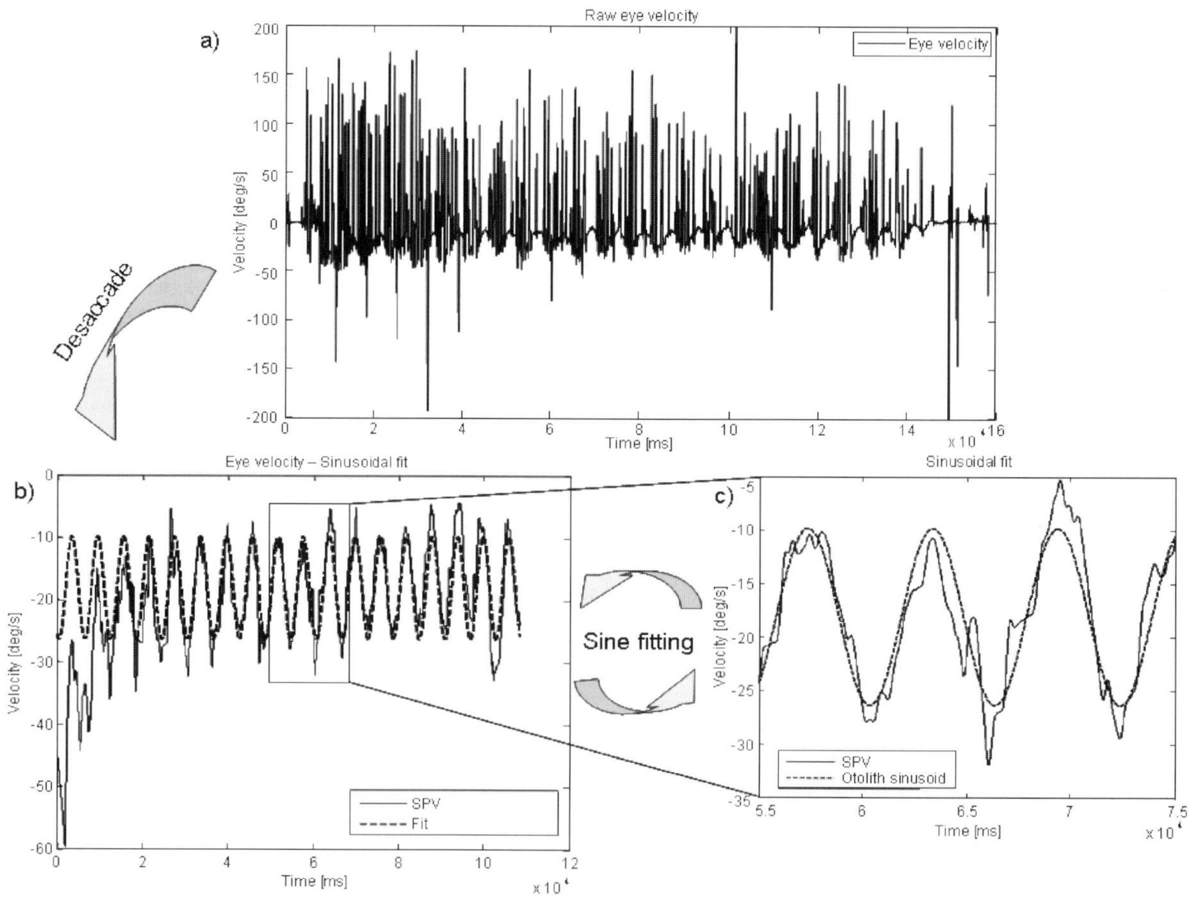

Fig. 1. (a) Calibrated vertical eye velocity trace before removal of saccades. (b) Solid line: Trace after saccades removal. Dashed line: Sinusoidal fit. (c) Detail of the sinusoidal fit.

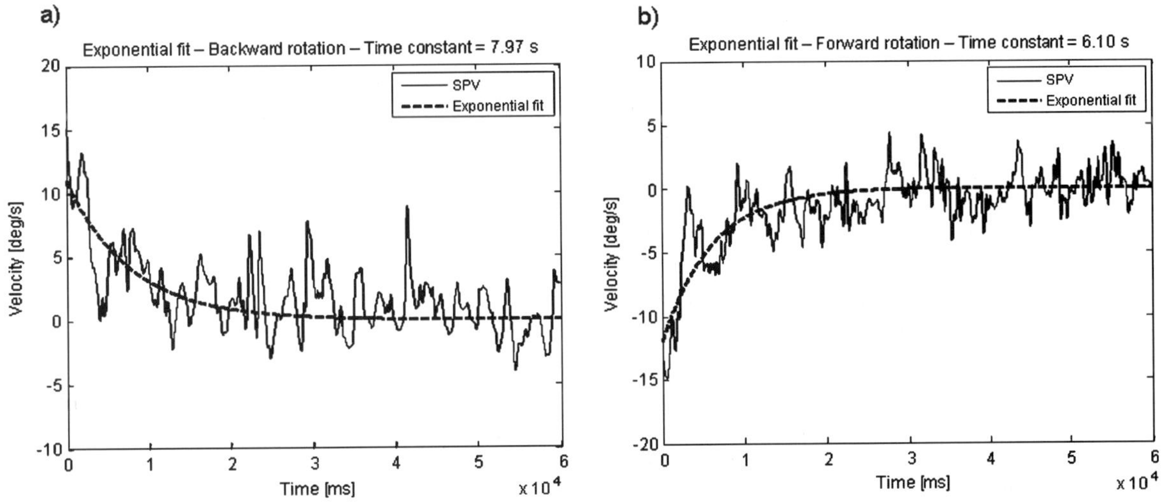

Fig. 2. Representative examples of processed responses to (a) backward rotation, $\tau = 7.97$ s; (b) forward rotation, $\tau = 6.10$ s. Solid line: Processed SPV. Dashed line: Exponential fit.

where A is the amplitude and τ the time constant of the rVOR response.

Results

Our subjects showed an intense vertical nystagmus (Fig. 1a) with SPV amplitude that decayed exponentially to a pure otolith-driven sinusoid with a bias. The inter-subject variability of the responses was considerable, as reflected by the standard deviation of the mean of pooled values. One subject showed markedly larger amplitude of the exponential decay, the sinusoidal modulation and the bias compared to the others, yet the rVOR time constant did not differ from that of the other subjects. Mean values and standard deviations of the data without including that subject are therefore presented within brackets.

The gain of the response prior to subtraction of the otolith-driven sinusoid, estimated as the ratio of eye velocity to chair velocity between seconds two and three of the response (i.e., during the acceleration phase) was 0.34 ± 0.08 (SD) [0.31 ± 0.07]. The sinusoidal response had mean amplitude of 6.19 ± 2.90 [4.72 ± 2.20] deg/s, corresponding to a gain of 0.1 ± 0.05 [0.08 ± 0.04]. The mean of the absolute values of the bias was 5.22 ± 6.67 [3.19 ± 2.95] deg/s and it always introduced a velocity offset that was in the direction compensatory for head rotation. Two examples of the processed data showing the exponentially decaying SPV and the corresponding fit are shown in Fig. 2 (A: backward rotation; B: forward rotation). The mean value of the rVOR time constant over the four subjects was 7.39 ± 1.22 [7.07 ± 1.25] s and its gain was 0.25 ± 0.14 [0.18 ± 0.03]. The time constants were similar for the backward (7.32 ± 1.32 s) and forward (7.46 ± 1.40 s) rotations. Horizontal and torsional eye velocity traces showed a small sinusoidal modulation, but no exponential decay of the response at the onset of constant velocity chair rotation.

Discussion

To assess the contribution of the VSM to the pitch rVOR, we have studied the eye movements of four healthy subjects evoked by constant-velocity pitch rotations at 60 deg/s, in total darkness.

The gain of the pitch rVOR computed from the exponential decay of vertical eye velocity was only about 0.25, which is lower than the typical 0.5 measured for horizontal rVOR evoked by yaw rotations (Barr et al., 1976). Since our data showed no appreciable SPV plateau, we considered a single rVOR time constant corresponding to the original model suggested by Robinson (1977). The mean value of the rVOR time constant was 7.39 ± 1.22 s. These values are lower than the classically considered 20 s, while similar to those reported for vestibular afferents both in monkeys and humans. Considering that recent studies have suggested shorter time constants for both vestibular afferents (between 3.4 s and 4 s) and the rVOR response (between 8 s and 10 s) in humans (Dai et al., 1999; Gizzi and Harper, 2003), our findings confirm the relatively small contribution of the VSM to vertical responses. The horizontal and torsional eye velocity traces were also modulated sinusoidally by the change in gravity, but showed no exponential decay after the onset of chair rotation, thus no VSM contribution. These results agree with the hypothesis that the VSM velocity response is minimal (or absent) when the axis of rotation is not aligned with the gravity acceleration vector (Raphan et al., 1979). The absence of a VSM contribution to torsional and horizontal traces does not conflict with the cross-coupling hypothesis (Raphan and Cohen, 1988) since the direction of gravity with respect to the head continuously changed with our stimulus.

Acknowledgement

Supported by Swiss National Science Foundation, Koetser Foundation for Brain Research, and Center for Integrative Human Physiology, University of Zurich.

References

Barr, C.C., Schultheis, L.W. and Robinson, D.A. (1976) Voluntary, non-visual control of the human vestibulo-ocular reflex. Acta Otolaryngol., 81(5–6): 365–375.

Dai, M., Klein, A., Cohen, B. and Raphan, T. (1999) Model-based study of the human cupular time constant. J. Vestib. Res., 9(4): 293–301.

Gizzi, M.S. and Harper, H.W. (2003) Suppression of the human vestibulo-ocular reflex by visual fixation or forced convergence in the dark, with a model interpretation. Curr. Eye Res., 26: 281–290.

Marti, S., Bockisch, C.J. and Straumann, D. (2005) Prolonged asymmetric smooth-pursuit stimulation leads to downbeat nystagmus in healthy human subjects. Invest. Ophthalmol. Vis. Sci., 46: 143–149.

Raphan, T. and Cohen, B. (1988) Organizational principles of velocity storage in three dimensions. The effect of gravity on cross-coupling of optokinetic after-nystagmus. Ann. N.Y. Acad. Sci., 545: 74–92.

Raphan, T., Matsuo, V. and Cohen, B. (1979) Velocity storage in the vestibulo-ocular reflex arc (VOR). Exp. Brain Res., 35: 229–248.

Robinson, D.A. (1977) Vestibular and optokinetic symbiosis: an example of explaining by modelling. In: Baker R. and Berthoz A. (Eds.), Control of Gaze by Brain Stem Neurons. Elsevier/North-Holland Biomedical Press, Amsterdam, pp. 49–58.

C. Kennard & R.J. Leigh (Eds.)
Progress in Brain Research, Vol. 171
ISSN 0079-6123

CHAPTER 3.10

Preserved otolith function in patients with cerebellar atrophy and bilateral vestibulopathy

S. Marti*, A.A. Tarnutzer, A. Palla and D. Straumann

Department of Neurology, University Hospital Zurich, Zurich, Switzerland

Abstract: Cerebellar degeneration affects vestibular function. For instance, with lesions of the cerebellar flocculus, the ability to adaptively modify the VOR gain is markedly reduced, and cerebellar patients may even demonstrate severe vestibular deficits. We report five patients ($m = 3, f = 2$) with cerebellar disease, in whom search-coil head impulse testing revealed reduced gains of the angular VOR, while sacculus-mediated myogenic potentials were normal. Preserved static ocular counterroll in roll-tilt positions and prominent gravity-dependent modulation of downbeat nystagmus (DBN) along the pitch plane demonstrated the integrity of otolith (OL) function in these patients as well. Probably, at least in some cerebellar patients with marked floccular atrophy, the dissociation between impaired semicircular canal (SCC) function and preserved OL function may be explained by a predilection of the atrophic process for the flocculus and brainstem neurons involved in angular VOR gain control, while structures mediating OL function remain widely spared by the cerebellar degeneration. The exact pathomechanism leading to the vestibular impairment remains unclear: both a primary multi-system-type atrophy involving cerebellar and brainstem vestibular structures as well as a mechanism of secondary retrograde degeneration of floccular brainstem target neurons mediating SCC function seem plausible.

Keywords: cerebellum; bilateral vestibulopathy; downbeatnystagmus

Introduction

Patients with cerebellar disease may demonstrate impaired function of the vestibulo-ocular reflex (VOR) and even bilateral vestibular loss (Migliaccio et al., 2004; Wagner et al., 2007). We recently reported a patient with cerebellar ataxia and prominent downbeat nystagmus (DBN), who demonstrated severe bilateral impairment of semicircular canal (SCC) function, but preserved otolith (OL)-mediated reflexes (patient HH, see Tables 1 and 2; Marti et al., 2008). This patient's brain's magnetic resonance (MR) revealed atrophy predominantly affecting the flocculus, while the nodulus remained relatively preserved. Probably, the dissociated vestibular deficits in this patient might be best explained by the specific pattern of the atrophic process, since the flocculus is particularly involved in angular VOR control (Robinson, 1976), while the nodulus interferes mainly with OL-mediated reflex responses (Hain et al., 1988; Angelaki and Hess, 1995). To address the question whether the dissociated vestibular deficits in this patient represent a rare syndrome or a more common finding in cerebellar disease, we studied SCC- and OL-mediated reflex responses in four other cerebellar patients.

*Corresponding author. Tel.: +41-1-255-3996; Fax: +41-1-255-4380; E-mail: sarah.marti@usz.ch

DOI: 10.1016/S0079-6123(08)00629-8

Table 1. Diagnosis (Dg), most prominent clinical (grading: + mild; + + moderate; + + + severe) and brain MR findings in five cerebellar patients (pat.; initials, age, and sex)

Pat.	Ons.	Dg	DBN	GEN	SP	GA	Dy	LA	Ep	MR
HH, 82, m	60	ILCA	+ + +	+ + +	+ + +	+ + +	–	(+)	–	Severe atrophy of Fl, inf. cerebellar peduncle, inf. V, slight atrophy of N
HN, 81, m	62	ILCA	+ + +	+ + +	+ + +	+ + +	(+)	–	–	Severe atrophy of VestCb
AM, 59, f	62	ILCA	+ +	+ +	+ +	–	–	–	–	Slight atrophy of V
ES, 54, f	35	Probable EA II	+	+	+	(+)	–	–	+ + +	–
EZ, 51, m	30	Probable ADCA III	(+)	+ +	+ +	+ + +	+ +	+	–	Slight atrophy of V and cerebellar hemispheres

Note: Autosomal-dominantly inherited cerebellar ataxia type III (ADCA III); downbeat nystagmus (DBN); dysarthria (Dy); episodic ataxia type II (EA II); episodic symptoms, e.g., vertigo, ataxia of gait and stand, oscillopsia, dysarthria (Ep); flocculus (Fl); gait ataxia (GA); horizontal gaze evoked nystagmus (GEN); limb ataxia (LA); magnetic resonance brain imaging (MR); nodulus (N); age at disease onset (Ons.); idiopathic late-onset cerebellar ataxia (ILCA); impaired horizontal smooth pursuit eye movements (SP); vermis (V); vestibulo-cerebellum (VestCb); male (m); female (f).

Table 2. Quantitative VOR gains assessed by search coil head impulse testing in five cerebellar patients

Pat.	Dg	Hor **R**	Hor **L**	**RA**LP	RA**LP**	**LA**RP	LA**RP**
HH	ILCA	0.11	0.21	0.11	0.14	0.23	0.13
HN	ILCA	0.57	0.42	0.66	0.32	0.5	0.61
AM	ILCA	0.65	0.61	0.55	0.59	0.54	0.53
ES	EA II	0.69	0.76	0.73	0.68	0.66	0.47
EZ	ADCA III	0.73	0.58	0.59	0.34	0.6	0.4

Note: For each of the tested SCC, i.e., horizontal right (Hor R) and left (Hor L), right anterior and left posterior (RALP), left anterior and right posterior (LARP) SCC, the corresponding gain values are indicated. Idiopathic late-onset cerebellar ataxia (ILCA); episodic ataxia type II (EA II); autosomal-dominantly inherited cerebellar ataxia type III (ADCA III).

Case reports

Table 1, summarizes diagnosis and most important clinical and brain MR findings in the five patients with cerebellar disease.

Methods

All patients gave written informed consent to participate in the study. Vestibular-evoked myogenic potentials (VEMPs), i.e., inhibitory potentials in the ipsilateral sternocleidomastoid muscle evoked by brief (4 ms), loud (>90 dB) tone bursts ('pips'), were assessed as part of the standard vestibular testing at the *Center for Vertigo and Balance Disorders* at *Zurich University Hospital.* Quantitative VOR gains were assessed with search coil head impulse testing. Static ocular counterroll in sustained whole-body roll positions and gravity-dependent modulation of DBN along the pitch plain were assessed using the same experimental protocols as described in Marti et al. (2008) and Marti et al. (2002), respectively.

Results

VOR gains measured by head impulse testing were reduced in all five patients (Table 2). Gain reduction was largest in patient HH, who exhibited the most severe floccular atrophy on brain MR, while patient ES, whose brain MR revealed intact vestibulo-cerebellar structures, exhibited the least

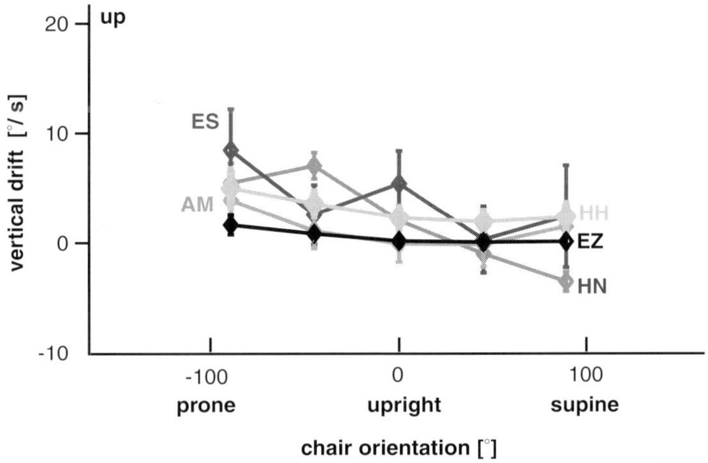

Fig. 1. Gravity-dependent modulation of upward ocular drift (for patients' initials, see legend Table 1). Abscissa: Chair positions, i.e., 90° prone, 45° prone, upright, 45° supine, 90° supine. Ordinate: Vertical ocular velocity (deg/s). Open squares indicate median vertical eye velocities ± 1 SD (error bars) in the respective chair positions. Positive values correspond to upward eye velocities.

gain reduction. Reproducible VEMPs (at 90 dB nHL) were registered in all patients with latencies and amplitudes similar to control data of our laboratory. Static torsional eye position modulated as a function of sustained bilateral whole-body roll tilt in all five patients, similarly as in five healthy subjects (two females, three males, age 66–74 years; data not shown). Upward ocular drift modulated as a function of whole-body pitch increased with nose-down and decreased with nose-up positions in all patients (Fig. 1).

Discussion

We observed a pattern of dissociated vestibular deficits with impaired SCC function (assessed by quantitative head-impulse testing), but largely preserved OL function (demonstrated by intact VEMPs, preserved modulation of static ocular counterroll and preserved gravity-dependent modulation of DBN) in five patients with cerebellar disease. The putative pathomechanisms leading to these dissociated vestibular deficits remain unclear. To postulate a channelopathy (Wagner et al., 2007) or a peripheral vestibulopathy, one would need to explain the selective impairment of SCC function. Probably, at least in some cerebellar patients, the dissociation between impaired SCC function and preserved OL function may be best explained by a predilection of the atrophic process for cerebellar and brainstem structures involved in VOR gain control (Marti et al., 2008). Whether this degenerative process involves primarily both cerebellar and brainstem structures, as in the case of multi-system atrophy (Migliaccio et al., 2004; Wagner et al., 2007), or whether, at least in cases with severe floccular atrophy, they can also be attributed to a mechanism of secondary retrograde degeneration of floccular brainstem target neurons (Smith, 1975), remains unclear. Also, considering the largely preserved OL function in some cerebellar patients, the contribution of the vestibular deficits to the often marked postural imbalance in cerebellar disease needs to be determined (see Chapter 4.22 by Tarnutzer et al. in this volume).

Acknowledgements

This study was financially supported by the Swiss National Science Foundation (3200B0-105434), the Betty and David Koetser Foundation for Brain Research, Zurich, Switzerland, and the Bonizzi-Theler Foundation, Zurich, Switzerland.

References

Angelaki, D.E. and Hess, B.J. (1995) Inertial representation of angular motion in the vestibular system of rhesus monkeys. II. Otolith-controlled transformation that depends on an intact cerebellar nodulus. J. Neurophysiol., 73: 1729–1751.

Hain, T.C., Zee, D.S. and Maria, B.L. (1988) Tilt suppression of vestibulo-ocular reflex in patients with cerebellar lesions. Acta Otolaryngol., 105: 13–20.

Marti, S., Palla, A. and Straumann, D. (2002) Gravity dependence of ocular drift in patients with cerebellar down-beat nystagmus. Ann. Neurol., 52: 712–721.

Marti, S., Tarnutzer, A.A., Schuknecht, B. and Straumann, D. (2008) Dissociation between canal- and otolith-function in cerebellar atrophy. J. Neurol., February 21 [Epub ahead of print].

Migliaccio, A.A., Halmagyi, G.M., McGarvie, L.A. and Cremer, P.D. (2004) Cerebellar ataxia with bilateral vestibulopathy: description of a syndrome and its characteristic clinical sign. Brain, 127: 280–293.

Robinson, D.A. (1976) Adaptive gain control of vestibuloocular reflex by the cerebellum. J. Neurophysiol., 39: 954–969.

Smith, M.C. (1975) Histological findings after hemicerebellectomy in man: anterograde, retrograde and transneuronal degeneration. Brain Res., 95: 423–442.

Tarnutzer, A.A., Marti, S. and Straumann, D. (2008) Gravity perception in cerebellar patients. Abstract for the 'Meeting Using Eye Movements as an Experimental Probe of Brain Function' 2007, Progress in Brain Research, Chapter 4.22.

Wagner, J.N., Glaser, M., Brandt, T. and Strupp, M. (2007) Aetiology of downbeat nystagmus: a retrospective study on 117 patients. J. Neurol. Neurosurg. Psychiatry, September 14 [Epub ahead of print].

C. Kennard & R.J. Leigh (Eds.)
Progress in Brain Research, Vol. 171
ISSN 0079-6123

CHAPTER 3.11

Three-dimensional kinematics of saccadic eye movements in humans with cerebellar degeneration

Matthew J. Thurtell[1,2,3,*], Theodore Raphan[4], Ross A. Black[1], Michael J. Todd[1], R. John Leigh[3] and G. Michael Halmagyi[1,2]

[1]*Department of Neurology, Royal Prince Alfred Hospital, Sydney, NSW, Australia*
[2]*Department of Medicine, University of Sydney, Sydney, NSW, Australia*
[3]*Department of Neurology, University Hospitals Case Medical Center, Cleveland, OH, USA*
[4]*Department of Computer and Information Science, Brooklyn College of City University of New York, Brooklyn, NY, USA*

Abstract: During saccades, the orientation of the eye velocity axis varies depending on the instantaneous eye position. The means by which this eye velocity axis tilting arises remains contentious. Some have argued that muscle pulleys in the orbit implement the tilts, since they cause the pulling directions of the extraocular muscles to change in a manner that depends on instantaneous eye position. Others have suggested that the tilting is centrally programmed. In the current study, three-dimensional eye and head rotation data were acquired, using the magnetic search coil technique, to confirm the presence of eye position-dependent eye velocity axis tilting during saccades. We studied normal humans and humans with inherited or sporadic cerebellar degeneration. While the humans with cerebellar degeneration were noted to have abnormalities in the two-dimensional metrics and consistency of their saccades, the eye velocity axis tilts were similar to those observed in the normal subjects. Our findings suggest that the cerebellum does not encode eye velocity axis tilting during saccades, further supporting the notion that these phenomena arise due to the effects of muscle pulleys.

Keywords: saccades; cerebellar degeneration; kinematics; Listing's law

Introduction

For Listing's law to be obeyed during saccades, the eye velocity axis must tilt by half the angle of eye eccentricity from primary position (Tweed and Vilis, 1990). The mechanism by which this tilting arises remains uncertain. Some have argued that muscle pulleys in the orbit are responsible, since they cause the pulling directions of the extraocular muscles to change in a manner that depends on instantaneous eye position (Demer, 2006). Others have suggested that the tilts are centrally programmed. The cerebellum has been implicated in this regard, since cerebellar disease produces nystagmus slow phases that violate Listing's law (Straumann et al., 2000) and focal cerebellar lesions may lead to violations of Listing's law during saccades (Helmchen et al., 1997). Furthermore, lesions (or stimulation) of the caudal nucleus reticularis tegmenti pontis (which relays

*Corresponding author. Tel.: +1-216-844-4860;
Fax: +1-216-844-3160; E-mail: mj.thurtell@gmail.com

DOI: 10.1016/S0079-6123(08)00630-4

saccadic command signals to the excitatory burst neurons via the cerebellum) in monkey lead to 'instability' of Listing's plane (van Opstal et al., 1996). Lastly, cerebellar disease increases the eye position-dependent eye velocity axis tilts of the angular vestibulo-ocular reflex in response to impulsive stimuli (Walker et al., 2007). While these findings suggest that the cerebellum has some influence on Listing's law for eye movements, it is not known if it is involved in implementing eye velocity axis tilts during saccades. Furthermore, the relative contributions of the cerebellum and muscle pulleys to the implementation of the tilts during saccades have not been differentiated. We therefore studied the kinematics of saccades from humans with cerebellar degeneration and compared them with those from normal humans.

Methods

We studied five normal humans (24–58 years) and four humans (54–68 years) with inherited or sporadic cerebellar degeneration (patients). Each of the patients had downbeat, gaze-evoked, and rebound nystagmus, as well as saccadic dysmetria. Cerebellar degeneration was evident on magnetic resonance imaging of the brain, but a specific genetic cause was not established in any patient, despite an autosomal dominant inheritance pattern in three of them.

Eye and head rotation data were acquired using the magnetic search coil technique. Saccades were recorded while subjects attempted to follow a laser target that moved in 40° horizontal or vertical

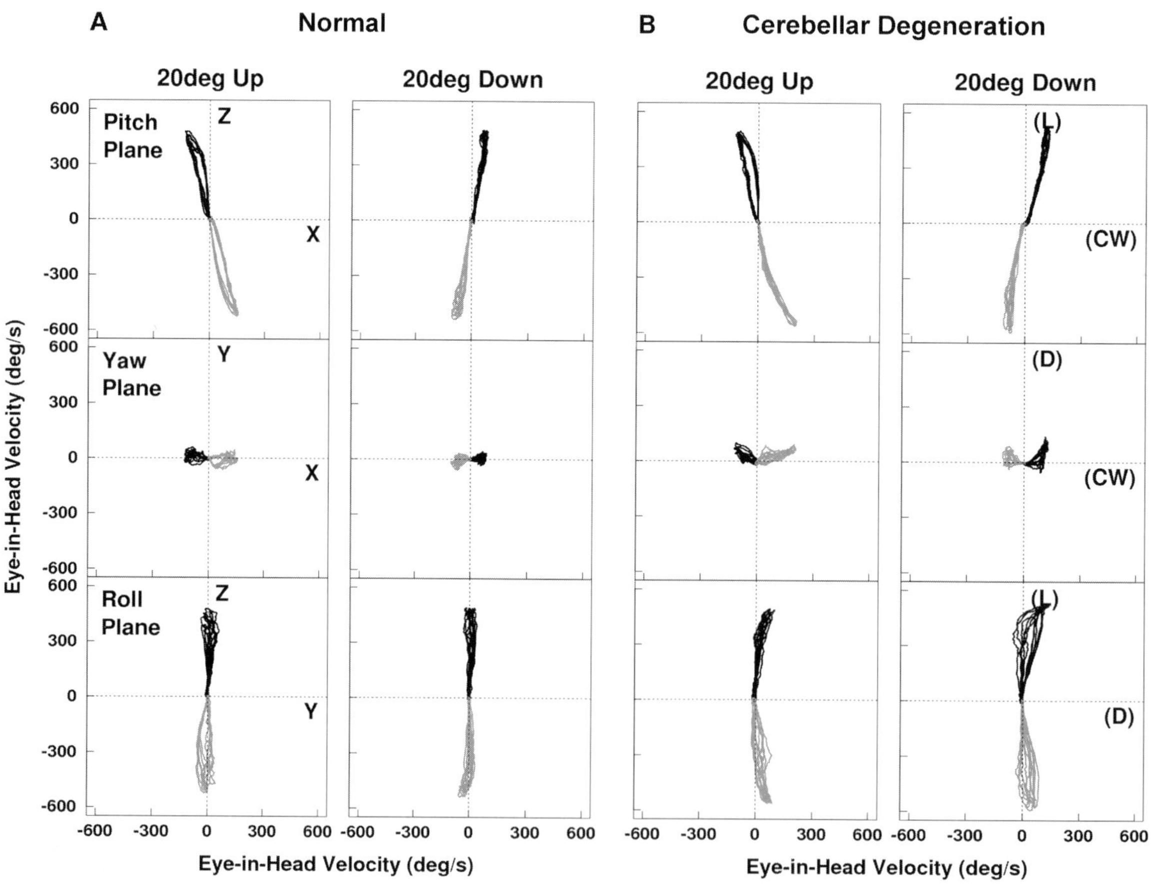

Fig. 1. Eye velocity vectors during horizontal saccades while looking 20° up and 20° down from a normal subject (A) and a patient with cerebellar degeneration (B) are plotted in the pitch, yaw, and roll planes. (X, x-axis; Y, y-axis; Z, z-axis; CW, clockwise; D, down; L, left).

steps between locations 20° up/left, up/right, down/left, and down/right.

The orientation of Listing's plane was determined and all data were rotated into Listing's coordinates. Position data were expressed in axis-angle format (Raphan, 1998), while velocity data were expressed as velocity vectors. Saccades with abnormal amplitudes were not excluded from further analysis, nor were data from subjects unable to hold fixation due to nystagmus. To demonstrate their spatial characteristics, eye velocity vectors were plotted in the pitch, yaw, and roll planes. Eye velocity axis orientation was determined by fitting a line to the spatial data using a linear least-squares algorithm. The angle of axis tilt was calculated as the angle between the fitted axis and the z-axis in the pitch plane for horizontal saccades, and as the angle between the fitted axis and the y-axis in the yaw plane for vertical saccades. To determine how well Listing's law was obeyed during the saccades, the tilt-angle coefficient was calculated as the change in the angle of eye velocity axis tilt divided by the change in eye position eccentricity.

Results

Horizontal saccades made by normal subjects while looking 20° up produced backward eye velocity axis tilt in the pitch plane, while those made while looking 20° down produced forward tilt (Fig. 1A). A similar trend was observed in patients (Fig. 1B). When viewed in the yaw plane, vertical saccades made by normal subjects while

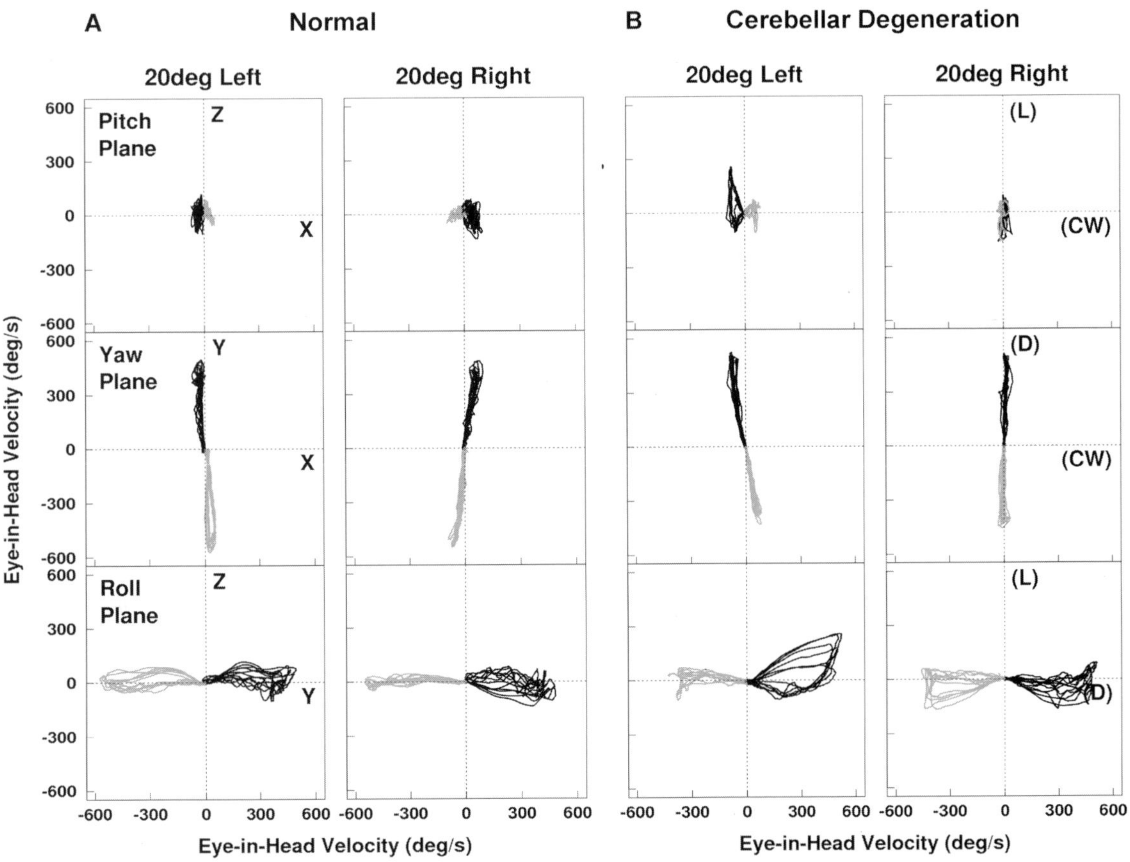

Fig. 2. Eye velocity vectors during vertical saccades while looking 20° left and 20° right from a normal subject (A) and a patient with cerebellar degeneration (B) are plotted in the pitch, yaw, and roll planes. (X, x-axis; Y, y-axis; Z, z-axis; CW, clockwise; D, down; L, left).

looking 20° left produced leftward eye velocity axis tilt, while those made while looking 20° right produced rightward tilt (Fig. 2A). A similar trend was observed in patients (Fig. 2B). Of note, the magnitudes of the axis tilts appeared smaller during vertical saccades than during horizontal saccades in both subject groups.

The tilt-angle coefficients during horizontal saccades were 0.57 ± 0.07 (mean $\pm$ SD) in the normal group and 0.58 ± 0.06 in the patient group. The tilt-angle coefficients of the groups were not significantly different ($P > 0.01$). The tilt-angle coefficients during vertical saccades were 0.30 ± 0.10 in the normal group and 0.19 ± 0.07 in the patient group. Although the tilt-angle coefficients for vertical saccades trended towards being smaller in the patient group, they were not significantly different from those of the normal group ($P > 0.01$). However, when the tilt-angle coefficients for vertical saccades were compared with those for horizontal saccades, they were significantly smaller in both subject groups ($P < 0.01$).

Discussion

Normal humans and patients were found to have similar eye position-dependent eye velocity axis tilts during saccades. Cerebellar degeneration, therefore, does not appear to disrupt the mechanism responsible for producing eye velocity axis tilting during saccades. Rather, our findings indirectly support the hypothesis that muscle pulleys generate the eye velocity axis tilts; a saccadic model incorporating the pulley effect (Raphan, 1998) predicts eye velocity axis tilts similar to those seen in our experimental data.

During horizontal saccades, the tilt-angle coefficient was close to 0.5 in both subject groups, indicating that Listing's law is approximately obeyed. During vertical saccades, however, the tilt-angle coefficient was significantly smaller, consistent with the findings of a prior study (Bruno and van den Berg, 1997). The means by which this disparity arises remains unclear, but could be due to differences in the degree of eye position-dependent change in the tendon pulling directions for the horizontal recti compared with the vertical recti and oblique muscles.

Acknowledgment

Grant DC 05222 (for TR).

References

Bruno, P. and van den Berg, A.V. (1997) Torsion during saccades between tertiary positions. Exp. Brain Res., 117: 251–265.

Demer, J.L. (2006) Current concepts of mechanical and neural factors in ocular motility. Curr. Opin. Neurol., 19: 4–13.

Helmchen, C., Glasauer, S. and Buttner, U. (1997) Pathological torsional eye deviation during voluntary saccades: a violation of Listing's law. J. Neurol. Neurosurg. Psychiatry, 62: 253–260.

Raphan, T. (1998) Modeling control of eye orientation in three dimensions. I. Role of muscle pulleys in determining saccadic trajectory. J. Neurophysiol., 79: 2653–2667.

Straumann, D., Zee, D.S. and Solomon, D. (2000) Three-dimensional kinematics of ocular drift in humans with cerebellar atrophy. J. Neurophysiol., 83: 1125–1140.

Tweed, D. and Vilis, T. (1990) Geometric relations of eye position and velocity vectors during saccades. Vision Res., 30: 111–127.

van Opstal, J., Hepp, K., Suzuki, Y. and Henn, V. (1996) Role of monkey nucleus reticularis tegmenti pontis in the stabilization of Listing's plane. J. Neurosci., 16: 7284–7296.

Walker, M.F., Tian, J. and Zee, D.S. (2007) Kinematics of the rotational vestibuloocular reflex: role of the cerebellum. J. Neurophysiol., 98: 295–302.

C. Kennard & R.J. Leigh (Eds.)
Progress in Brain Research, Vol. 171
ISSN 0079-6123

CHAPTER 3.12

Inferior olive hypertrophy and cerebellar learning are both needed to explain ocular oscillations in oculopalatal tremor

Simon Hong[1], R. John Leigh[2], David S. Zee[3] and Lance M. Optican[1,*]

[1]*Laboratory of Sensorimotor Research, National Eye Institute, Bethesda, MD, USA*
[2]*Veterans Affairs Medical Center and Case Western Reserve University, Cleveland, OH, USA*
[3]*The Johns Hopkins University, School of Medicine, Baltimore, MD, USA*

Abstract: A new model of cerebellar learning explains how the cerebellum can generate arbitrary output waveforms to adjust output timing in the classical delay conditioning. This model can also reproduce the low frequency ocular oscillations seen in oculopalatal tremor (OPT). A novel circuit in the cerebellum uses both interneurons (INs) and Purkinje cells (PC) to control timing. Brain lesions that cause OPT give rise to hypertrophy of the inferior olive (IO) and an increase in conductance through gap junctions among IO neurons. When our model is changed in this way, the heavily coupled IO becomes an oscillator and generates synchronous spike trains at 1–2 Hz. These synchronized spikes do not produce the large amplitude, aperiodic waveforms of OPT. However, the synchronized IO signal goes to the cerebellar cortex (flocculus) directly, on climbing fibres, and indirectly, on mossy fibres from the vestibular nuclei. This creates a pathological association between the IO pulse trains on mossy and climbing fibres in PC. Variable pendular ocular oscillations emerged from the cerebellum model after learning this association. Since electrotonic coupling of IO cells depends on connexin proteins, drugs that block gap junctions, such as anti-malarial agents, might provide a novel therapy for OPT.

Keywords: Purkinje cell; interneuron; interaction; waveform; mGluR; classical conditioning; gap junction

Introduction

The role of the cerebellum in generating timed movements has been demonstrated in animal studies of *classical delay eyeblink conditioning* (e.g., Thompson, 1986). In this paradigm (see Fig. 1A), the animal receives a conditioned stimulus (CS), such as a tone. After a certain delay (inter-stimulus interval, ISI), an unconditioned stimulus (US), such as an air puff directed at the cornea, causes a reflexive blink (unconditioned response, UR). After training, the CS causes the eyelid to blink (conditioned response, CR) at the time when the subsequent US is expected. In this study, we propose a new model of cerebellar timing. The model emphasizes the interaction between the Purkinje cell (PC) and its connected interneurons (INs) in generating modulated waveforms of PC activation (Fig. 1B; Hong and Optican, 2005). It is hypothesized that the delayed

*Corresponding author. Tel.: 301-496-9375;
Fax: 301-402-0511; E-mail: LanceOptican@nih.gov

DOI: 10.1016/S0079-6123(08)00631-6

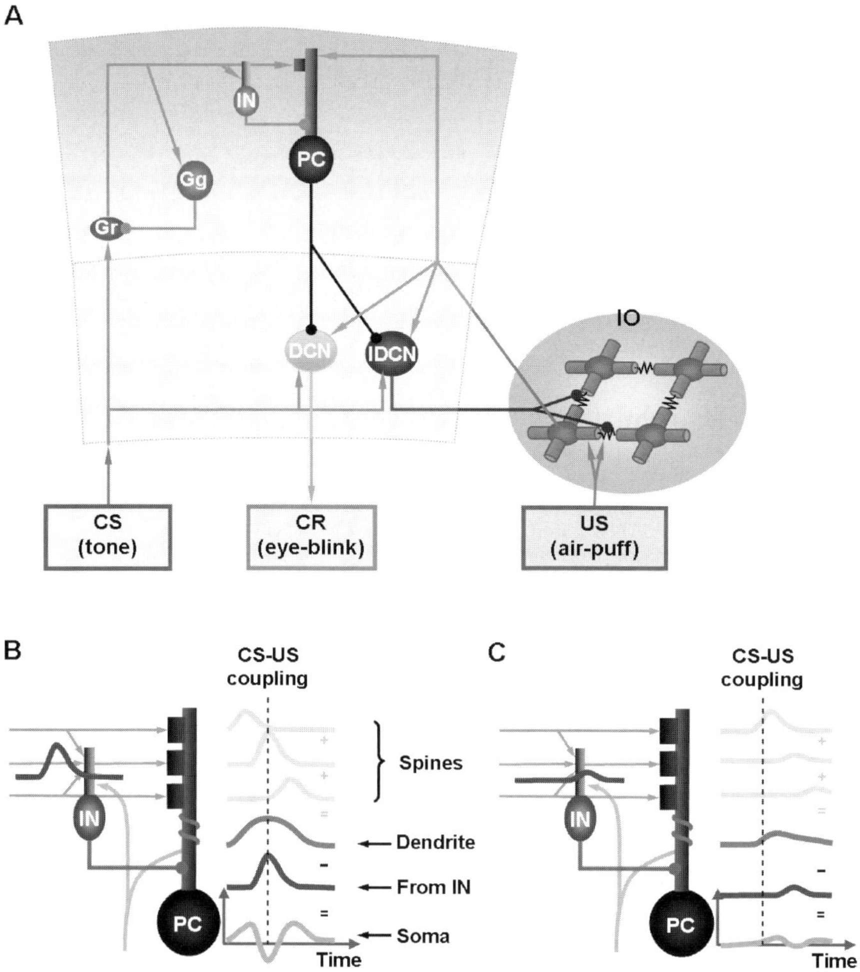

Fig. 1. Schematic of model. (A) Cerebellar modules and IO. See the text for detailed explanations. The zigzags (black) between IO neurons (red) represent gap junctions. Excitatory and inhibitory pathways are represented by triangular heads and round heads, respectively. (B) and (C) PC–IN pairs learn the timing of CS–US coactivation. When the CFs activate after the PF signal with a certain delay, the PC–IN pairs with a matching internal delay timing, such as the one in (B), increase their mGluR-mediated excitability. The "CS–US coupling" with a vertical broken line in (B) and (C) shows the timing. After learning, the increased excitability leads to a big modulation of activity in the PCs of those pairs (the orange wave). This synchronized modulation of a population of PCs leads to the generation of timed CR. For the PC–IN pairs with different internal delay timings, such as the long delay timing in (C), the CS–US coactivation does not change their excitability. Theses pairs remain relatively inactive therefore do not contribute to CR. (See Color Plate 3.12.1 in color plate section.)

coupling of the CS and US delivered via PF and CF, respectively, induces a simultaneous increment of dendritic excitability in INs as well as in the PC, thus leading to an increased modulation of PC activity. This waveform modulation leads to a timed pause of the PC population and makes the deep cerebellar nucleus (DCN) neurons generate the CR signals. We hypothesize that the delayed PC–IN interaction is due to the intracellular activation of the mGluR. This is based on the observations that (1) the decrement of mGluR-mediated excitability in PCs may affect the long-time scale component of motor execution (Coesmans et al., 2003), (2) the group I mGluRs in the INs have similar signalling properties to those in PC (Karakossian and Otis, 2004). More details are provided in Discussion.

Clinical observations also implicate cerebellar circuitry in a variety of motor activities, including eye movements. One clear example is found in patients with oculopalatal tremor (OPT) where the disruption of the deep cerebellar inhibitory projection to the inferior olive (IO) causes a gradual development of involuntary irregular eye movements. The main pathologic finding with OPT is hypertrophy of the inferior olivary nucleus, which may be seen on MRI (Goyal et al., 2000). The olivary nucleus is enlarged, due to hypertrophy of neurons that contain increased acetylcholinesterase reaction product. Such changes begin within a month after the lesion and maximize in about six months, being accompanied by astrocytosis and synaptic and axonal remodelling.

Animal studies have clarified several aspects of how the clinical syndrome develops. An important projection runs from the deep cerebellar nuclei through the superior cerebellar peduncle, decussates, turns caudally at the red nucleus to form the central tegmental tract, and makes inhibitory synapses on inferior olivary neurons (Bengtsson et al., 2004). Disruption of this inhibitory pathway leads to increased activity of inferior olivary neurons. Furthermore, chemical lesions of the deep cerebellar nuclei lead to hypertrophy of inferior olivary neurons with changes in their connectivity (Ruigrok et al., 1990).

In this study, we use progressive development of soma-somatic gap junctions between adjacent inferior olivary neurons and learning in the cerebellar cortex to simulate the involuntary ocular eye movements of OPT.

Methods

This computational study constructs a model of the cerebellar circuit utilizing leaky integrator-type equations for all of the cell types except the IO neurons. IO neurons use more elaborate spiking equations to simulate realistic CF activity including the low frequency (~ 2 Hz) baseline noisy spikes. Figure 1A illustrates our simplified circuit of the cerebellum. There are four major divisions in the circuit: Two inputs (MF and CF), cerebellar cortex, and DCN. MFs provide inputs that represent certain events (CS in Fig. 1), such as a tone signal in the classical eyeblink conditioning paradigm. This input is transmitted to the cerebellar cortex via granule cells and also to the DCN by MF collaterals. CFs, which constitute another major input system, originate in the IO and transmit US signals. The activity of DCN represents the output of the cerebellum and also gates the IO-mediated learning in the cerebellar cortex with its negative feedback via inhibitory DCN→ IO projections.

Normal model

In our model, the timing mechanism is localized to the PC–IN pair. Figure 1B summarizes the key concepts of signal processing in a PC–IN pair. First, it is assumed that repeated CF–PF coupling trains the PC spines and the IN dendrite to increase their excitability. After this training, the increased excitability makes the spines of the PC and the dendrite of the IN increase their potentials upon PF input due to the mGluR-induced long latency calcium activation (the blips in Fig. 1B). The terms *increment* or *decrement of excitability* will be used here for long-time scale plastic changes to differentiate these from the well known LTP and LTD phenomena, which historically refers to the changes in the fast AMPAR pathways. It is assumed that whereas the IN shows a narrow activation profile (blue trace in Fig. 1B), the PC dendrite generates a relatively broad activation profile (red trace) because of the variability in latencies of the slow-acting intracellular calcium components among the dendrite's many spines (three pink traces). At the PC dendrite level, the excitatory potential coming from the PC spines (red trace) and the inhibitory potential coming from the IN (blue trace) interact. This interaction generates a waveform, or temporally modulated pattern, of potential in the PC soma (orange trace in Fig. 1B). We assume that PC–IN pairs have a wide range of mGluR-induced activation latencies that span the possible range of delay timing ($\sim$4 s), similar to the range of PC latencies in the population-based spectral timing model (Fiala et al., 1996). The modulation of the waveforms of PC activations via the PF–CF coactivation happens only in those PC–IN pairs whose latencies match the timing of the coactivation. For example, a PC–IN pair whose internal timing longer than the CS–US coupling timing, as illustrated in Fig. 1C, will not increase the excitabilities, and the PC's activation will not be modulated. This way only the PC–IN pairs having the right internal timing will be recruited by the CS–US coupling leading to a timed decrease of discharge in the PC population. This decrement of inhibition lets the DCN generate the CR.

Model with hypertrophic inferior olive

To further validate this cerebellar model, the circuitry has been adapted to explain OPT which results from an injury that disrupts the DCN→IO inhibitory pathway. Our first hypothesis of the mechanism of these ocular oscillations is as follows: (1) The brain stem or cerebellar injury damages the inhibitory pathway from the DCN to the IO; (2) this causes IO neurons to gradually form soma-somatic gap junctions with their neighbours as part of a progressive hypertrophy (e.g., Ruigrok et al., 1990); (3) the soma-somatic gap junctions increase the communication among the connected IO neurons and as a consequence they start to fire synchronously; (4) the synchronized signals reach the eye movement-related DCN or the vestibular nuclei (VN).

After the implementation of this first hypothesis, which by itself failed to explain the OPT phenomena (see Results), a further modification was made to the model to examine our second hypothesis: The cerebellar cortex acts upon the sequences of pulses coming from the IO and modulates the eye movements. Inputs to PC consist of climbing fibre signals (carrying the synchronized olivary discharge) and parallel fibre signals (via IO→VN→granule cells). Thus, the inferior olivary signal, which projects to both the VN and the cerebellar cortex, could in turn also affect the discharge of PC via a mossy fibre projection from the VN (Zhang et al., 1993). (cf. Fig. 1 in the following article by Liao et al. for the circuit carrying these two signals.) This possibility led us to apply the hypothesis for motor learning in eyeblink conditioning explained above. In the case of OPT, the repeating IO pulses create periodic CF and PF inputs to PC–IN pairs at approximately the same time. The periodic conjunction of the PF–CF signals can train the PC–IN pairs, which after learning can pause after the PF input at the time of the expected CF input. This makes the PCs oscillate with the ongoing IO pulses, which in turn modulate the activity of the VN. The eye movement circuit, which receives its input from the VN, was simulated using a first-order ocular motor plant.

Results

Simulation of classical delay conditioning

Figure 2 shows the simulations of the classical delay eyeblink model with ISIs of 250 ms, 500 ms, and 750 ms. The troughs of the PC population and the peaks of the DCN neuron occur near the arrival time of the US (dashed lines). Individual model PCs often showed a variety of responses (data not shown), often with multiple peaks similar to the patterns in vivo. However, their population response shows a smooth trough around the arrival time of the US (Fig. 2, top).

Simulation of hypertrophic inferior olive in OPT

The lesion of the DCN→IO pathway induces the hyperactivation and ensuing over expression of gap junctions among IO neurons. The model simulates this process by initially selecting a few IO neurons at random as seed points to spread the soma-somatic gap junctions to their nearest neighbours. Figure 3 shows the effects of progression of the hypertrophy in the IO (downward arrows). The large dots indicate the hypertrophic neurons that have soma-somatic gap junctions with their neighbours. Note that as the simulated

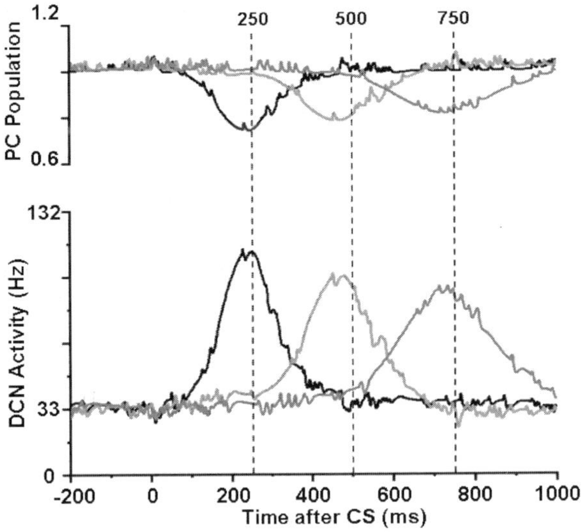

Fig. 2. Simulation of delay eyeblink conditioning. Upper panel: Input given to the simulated DCN neuron from the PC population with different ISIs (black curve: 250 ms, red: 500 ms, blue: 750 ms). Lower panel: DCN activities at different ISIs. PC population reduces the inhibitory output near the arrival of the air-puff US signals (dashed lines). The model DCN peaks near the arrival of the US with appropriate delays because of the reduction of PC inhibition. (See Color Plate 3.12.2 in color plate section.)

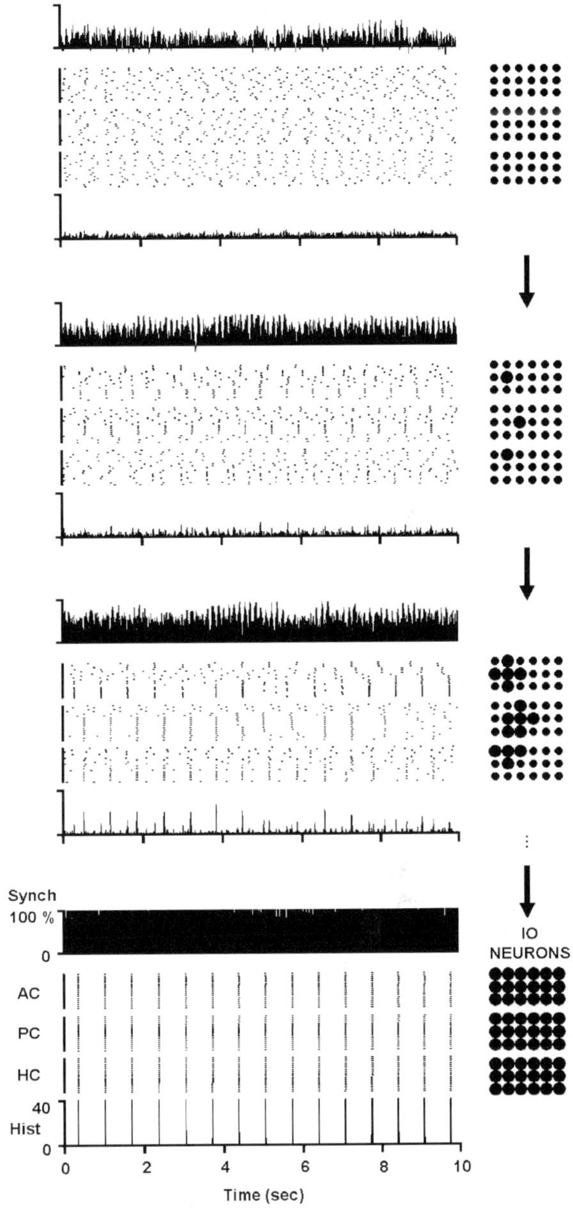

Fig. 3. Results of model IO activity before (top), during (middle two), and after (below) the development of soma-somatic gap junctions. The model IO has three groups of neurons (each group is a 6 × 3 array of dots on the right side of the figure) corresponding to anterior (AC), posterior (PC), and horizontal canals (HC). To simulate the hypertrophy in IO neurons, soma-somatic gap junctions were slowly introduced into the system (from top to bottom). The big dots represent hypertrophic IO neurons. Notice the increasing synchrony (Synch; top trace), the temporal alignment in the spike rasters (middle traces), and the peaks in the spike histogram (Hist; bottom trace).

hypertrophy progresses, the spiking activity among neighbouring IO neurons becomes more synchronized. This finding can also be appreciated in the upper trace of each panel, showing percent synchrony (Synch), which becomes almost 100% in the bottom panel. As a consequence, the spike histogram in the bottom panel (Hist) now shows narrow periodic peaks.

Simulation of OPT without cerebellar learning

Figure 4A shows the result of the simulation when the model does not include learning in the cerebellar cortex. The simulation produced periodic eye movements, but they were small, regular, and jerky, in contrast to ocular oscillations in OPT patients (Fig. 4C). Thus, although our model of the hypertrophied IO accounted for the slow development of OPT and produced oscillations at 1.5 ~ 2 Hz, it could not account for the actual waveform of the ocular oscillations of OPT.

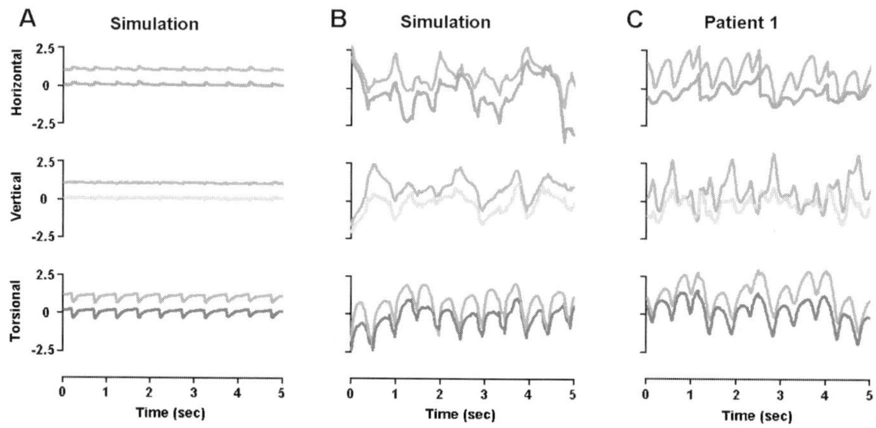

Fig. 4. Three-dimensional eye movement simulations of the OPT without (A) and with (B) the cerebellar cortical learning. (A) The model's simulation with IO hypertrophy but without cerebellar cortical learning shows small, jerky periodic eye movements. (B) When the simulation includes both the IO hypertrophy and the cerebellar cortical learning, the model makes larger, smoother, and more irregular movements, which are qualitatively similar to eye movements recorded in a patient with OPT (C). Gray traces represent right eye, coloured traces represent left eye. (See Color Plate 3.12.4 in color plate section.)

Simulation of OPT with cerebellar learning

When the model includes the cortical circuit that learns the temporal CS–US coactivation as seen in the classical eyeblink condition above, the resulting ocular oscillations (Fig. 4B) are now qualitatively similar to those observed in affected human subjects. Specifically, they are larger, smoother, and more variable than the movements in Fig. 4A. Each of these effects can be attributed to a specific mechanism in the cerebellum. First, the larger amplitude can be attributed to the pause of the PC after the arrival of a mossy fibre pulse, but at the expected arrival time of the next climbing fibre pulse; this, in turn, disinhibits the VN, which fire more vigorously. Second, the waveform is smoother because the distributed PC pauses at around the time of the expected climbing fibre signal. Since the timing of the climbing fibre signal after the parallel fibre signal is variable, the PC population learns the probability distribution of the delay. The variability of activity over the population of PC due to this learning has the effect of smoothing the output of the deep cerebellar nuclei. Third, the irregularity of the waveform can be attributed to mixing of two signals that arrive on PC via mossy fibres: the self-fed mossy fibre signal (via PC to VN to parallel fibres), and the IO signal, also projecting to the VN and then onto the cerebellar cortex. It should be emphasized that the coupling of large CF–PF signals from the IO is a pathological situation created by the abnormal soma-soma electrotonic gap junctions that synchronize many IO neurons. In the normal state the IO is not synchronized and does not generate large, periodic pulses (Fig. 3, top panel).

Discussion

The model described in this article simulates cerebellar timings in classical delay eyeblink conditioning and in the abnormal oscillatory involuntary eye movements in OPT. Unlike most models that assume PCs are the sole player in cerebellar timing, the current model hypothesizes that (1) the interplay between IN and PC generates the PC pause and (2) that the slow-activating mGluR-mediated $[Ca^{2+}]_i$ change in PC (Finch and Augustine, 1998) and in IN (Karakossian and Otis, 2004) determine the timing of the pause. The main reason for us to propose this PC–IN interaction model is the findings by Schreurs et al.

(1997, 1998) pointing out an increased excitability of PC dendrites (in Larsell's lobule HVI) after natural acquisition of classical delay eyeblink conditioning, in contrast to the decreased excitability seen in some in vitro studies. This creates a paradox: if learning increases PC excitability, how can they pause at the right time? Our model suggests a solution to this paradox by proposing a learning mechanism that generates a simultaneous increment of activity in both PC and IN. This concept is a temporal version of the shaping of spatial waveforms in the visual system. For example, in V1 excitatory and inhibitory neurons both increase their activation in response to a relevant input, which shapes the spatial waveforms that define the (centre-surround) receptive fields of visual neurons. We speculate that the modulation of both PCs and INs endow the cerebellar cortex with the versatility to learn almost any temporal output pattern. This point has been demonstrated in the simulation of OPT.

The consistent correlation between the hypertrophic IO and OPT (e.g., Goyal et al., 2000) led us to hypothesize that OPT was caused by the physiological changes in hypertrophic IO neurons. Hypertrophy causes the affected neurons to make extensive electrotonic connections with their neighbours via abnormal soma-somatic gap junctions. This causes a widespread, or even global, synchronization among the IO neurons that induces a behavioural tremor.

We hypothesize that the vestibular circuit is involved in the generation of the OPT. One piece of evidence comes from a recent experiment by Yoshida et al. (2004), who showed that malfunction of the flocculus and connected circuit can generate oscillatory eye movements (albeit of a higher frequency than OPT). Also IO to VN connections are known to exist (e.g., Balaban, 1988). One interesting fact of the vestibular circuit is that the vestibular inputs from head rotation may inhibit the vestibular part of the IO, the dorsal cap of Kooy and the ventrolateral outgrowth, which is known to control eye movements (e.g., Balaban and Beryozkin, 1994). This leads us to suggest that large head rotation could reset the phase of the eye oscillations in the OPT, via the vestibular inhibitory pathway (VN$\rightarrow$IO). The simulation result of this prediction is shown in Fig. 2 of the following article (Liao et al., 2008). The clinical analysis of our study in the following article (Liao et al., 2008) shows that an impulsive head rotation indeed resets the phase of ocular oscillations in OPT patients.

A major insight gained from this study was the pivotal role played by the development of soma-somatic gap junctions on neurons in the inferior olivary nucleus (e.g., Ruigrok et al., 1990). Gap junctions are common in the adult nervous system, and there is debate about their role in dendritic coupling of the normal IO. If development of gap junctions between cell bodies is indeed the key event in turning the IO into a synchronized pulse-generator that drives OPT, then drugs that block gap junctions (e.g., anti-malarial drugs) might have some therapeutic effect. For example, Martin and Handforth (2006) found that the gap junction blocker mefloquine (Cruikshank et al., 2004) suppressed harmaline-induced tremor in rats. However, they did not see any reduction in tremor in their human subjects who took a very low dose of mefloquine. In higher doses, mefloquine has serious side effects, but safer drugs, such as quinine, are also known to block connexin-36 and may be able to reduce tremor. Rash et al. (2004) reported that the connexin-36 labelled gap junctions were localized in proximity with NMDA receptor clusters in IO neurons. This raises a possibility that NMDA receptors may amplify the communication between IO neurons (Du et al., 1997). Thus, a targeted suppression of NMDA receptors may also reduce the synchronization of IO neurons, thereby alleviating the symptoms. This suggests studying the effect of a combination of drugs on tremor, one to block the gap junctions (e.g., anti-malarial drug) and another to block the NMDA receptors (e.g., memantine). This dual action may lessen the synchrony among IO neurons by (1) reducing the initial gap junction currents and (2) disabling the amplification of the currents afterwards. It is possible that once the pulse-generator (hypertrophic IO) is silenced, the learned gratuitous timed pause of the PCs in the cerebellar cortex would undergo extinction, further alleviating the symptoms.

References

Balaban, C.D. (1988) Distribution of inferior olivary projections to the vestibular nuclei of albino rabbits. Neuroscience, 24: 119–134.

Balaban, C.D. and Beryozkin, G. (1994) Organization of vestibular nucleus projections to the caudal dorsal cap of Kooy in rabbits. Neuroscience, 62: 1217–1236.

Bengtsson, F., Svensson, P. and Hesslow, G. (2004) Feedback control of Purkinje cell activity by the cerebello-olivary pathway. Eur. J. Neurosci., 20: 2999–3005.

Coesmans, M., Smitt, P.A., Linden, D.J., Shigemoto, R., Hirano, T., Yamakawa, Y., van Alphen, A.M., Luo, C., van der Geest, J.N., Kros, J.M., Gaillard, C.A., Frens, M.A. and de Zeeuw, C.I. (2003) Mechanisms underlying cerebellar motor deficits due to mGluR1-autoantibodies. Ann. Neurol., 53: 325–336.

Cruikshank, S.J., Hopperstad, M., Younger, M., Connors, B.W., Spray, D.C. and Srinivas, M. (2004) Potent block of C × 36 and C × 50 gap junction channels by mefloquine. PNAS, 101: 12364–12369.

Du, W., Aloyo, V.J. and Harvey, J.A. (1997) Harmaline competitively inhibits [^{3}H]MK-801 binding to the NMDA receptor in rabbit brain. Brain Res., 770: 26–29.

Fiala, J.C., Grossberg, S. and Bullock, D. (1996) Metabotropic glutamate receptor activation in cerebellar Purkinje cells as substrate for adaptive timing of the classically conditioned eye-blink response. J. Neurosci., 16: 3760–3774.

Finch, E.A. and Augustine, G.J. (1998) Local calcium signalling by inositol-1, 4, 5-trisphosphate in Purkinje cell dendrites. Nature, 396: 753–756.

Goyal, M., Versnick, E., Tuite, P., Cyr, J.S., Kucharczyk, W., Montanera, W., Willinsky, R. and Mikulis, D. (2000) Hypertrophic olivary degeneration: metaanalysis of the temporal evolution of MR findings. AJNR Am. J. Neuroradiol., 21: 1073–1077.

Hong, S.Y. and Optican, L.M. (2005) New cellular mechanisms for multiple time-scale adaptation in cerebellum, Program No. 933.7. 2005 Abstract Viewer/Itinerary Planner. Society for Neuroscience, Washington, DC. Online.

Karakossian, M.H. and Otis, T.S. (2004) Excitation of cerebellar interneurons by group I metabotropic glutamate receptors. J. Neurophysiol., 92: 1558–1565.

Liao, K., Hong, Y.S., Zee, D.S., Optican, L.M. and Leigh, R.J. (2008) Impulsive head rotation resets tremor — Examination of a model. Progress in Brain Research (this volume).

Martin, F.C. and Handforth, A. (2006) Carbenoxolone and mefloquine suppress tremor in the harmaline mouse model of essential tremor. Mov. Disord., 21: 1641–1649.

Rash, J.E., Pereda, A., Kamasawa, N., Furman, C.S., Yasumura, T., Davidson, K.G., Dudek, F.E., Olson, C., Li, X. and Nagy, J.I. (2004) High-resolution proteomic mapping in the vertebrate central nervous system: close proximity of connexin35 to NMDA glutamate receptor clusters and co-localization of connexin36 with immunoreactivity for zonula occludens protein-1 (ZO-1). J. Neurocytol., 33: 131–151.

Ruigrok, T.J., de Zeeuw, C.I. and Voogd, J. (1990) Hypertrophy of inferior olivary neurons: a degenerative, regenerative or plasticity phenomenon. Eur. J. Morphol., 28: 224–239.

Schreurs, B.G., Gusev, P.A., Tomsic, D., Alkon, D.L. and Shi, T. (1998) Intracellular correlates of acquisition and long-term memory of classical conditioning in Purkinje cell dendrites in slices of rabbit cerebellar lobule HVI. J. Neurosci., 18: 5498–5507.

Schreurs, B.G., Tomsic, D., Gusev, P.A. and Alkon, D.L. (1997) Dendritic excitability microzones and occluded long-term depression after classical conditioning of the rabbit's nictitating membrane response. J. Neurophysiol., 77: 86–92.

Thompson, R.F. (1986) The neurobiology of learning and memory. Science, 233: 941–947.

Yoshida, T., Katoh, A., Ohtsuki, G., Mishina, M. and Hirano, T. (2004) Oscillating Purkinje neuron activity causing involuntary eye movement in a mutant mouse deficient in the glutamate receptor delta2 subunit. J. Neurosci., 24: 2440–2448.

Zhang, Y., Partsalis, A.M. and Highstein, S.M. (1993) Properties of superior vestibular nucleus neurons projecting to the cerebellar flocculus in the squirrel monkey. J. Neurophysiol., 69: 642–645.

C. Kennard & R.J. Leigh (Eds.)
Progress in Brain Research, Vol. 171
ISSN 0079-6123

CHAPTER 3.13

Impulsive head rotation resets oculopalatal tremor: examination of a model

Ke Liao[1], Simon Hong[2], David S. Zee[3], Lance M. Optican[2] and R.J. Leigh[1,*]

[1]*Daroff-Dell'Osso Laboratory, Veterans Affairs Medical Center and Case Western Reserve University, Department of Neurology, University Hospitals, Cleveland, OH, USA*
[2]*Laboratory of Sensorimotor Research, National Eye Institute, NIH, DHHS, Bethesda, MD, USA*
[3]*Johns Hopkins University, Baltimore, MD, USA*

Abstract: We have described a neuromimetic model of the interaction between the inferior olive (IO) and the cerebellum that accounts for symptomatic oculopalatal tremor (OPT), a disorder characterized by oscillations of the eyes (nystagmus), palate and other branchial muscles. OPT develops months after some brainstem strokes, in association with hypertrophic degeneration of the inferior olivary nucleus (IO). We hypothesized that OPT requires both (1) a pulsatile oscillator created by tighter electrotonic coupling between cells in the IO, and (2) a learned response from the cerebellar cortex that combines with the IO pulses to generate the quasi-pendular oscillations. Since the vestibular nuclei project to both IO and vestibulocerebellum, one prediction of the model is that rapid head rotations could interrupt the oscillator, effectively resetting the timing of the ocular nystagmus. The ocular oscillations in OPT vary in amplitude and phase, making it difficult to determine by Fourier analysis whether head perturbations phase-shift the nystagmus. We applied complex wavelet analysis to data from four patients with OPT and checked whether vestibular stimuli induced a change in phase of the nystagmus. First we calculated a threshold for the spontaneous rate of change of phase of OPT by comparing many segments of nystagmus waveform with their time-shifted versions, bootstrapping these arrays, and computing 95% prediction intervals for each patient. Then we compared the rate of change of phase due to each head perturbation with the threshold for that patient. To minimize the effects of the head perturbation itself on the wavelet analysis, we measured effects in a plane orthogonal to the head rotation, e.g., effects of horizontal head rotations on the torsional component of OPT. In all four patients, the rate of change of phase shift increased sharply at the time of the head perturbation, and in three the change was judged to be statistically significant. Thus, the experimental tests supported the prediction of our model for OPT.

Keywords: inferior olive; wavelets; vestibulo-ocular reflex; gap junctions; clonazepam; memantine

Introduction

The syndrome of oculopalatal tremor (OPT; previously called oculopalatal myoclonus) typically develops over weeks or months following a

*Corresponding author. Tel.: +1 216 707 6428; Fax: +1 216 231 3461; E-mail: rjl4@case.edu

DOI: 10.1016/S0079-6123(08)00632-8

brain stem or cerebellar stroke (Guillain and Mollaret, 1931; Deuschl et al., 1990). The movements of affected muscles are approximately synchronized at a rate of about 2 cycles/s (range 1.8–2.7 Hz) (Deuschl et al., 1994). The ocular oscillations (nystagmus) in OPT are smooth but irregular, usually with prominent, vertical, and torsional components, and variable disconjugacy (Kim et al., 2007). The posterior soft palate, moved by the levator veli palatini muscle, is most often affected, but eyes, facial muscles, pharynx, tongue, larynx, diaphragm, neck, trunk, and extremities may also move in synchrony. The "symptomatic" form of OPT follows a brain stem or cerebellar stroke, and is distinct from either "essential" palatal tremor, which seldom affects the eyes (Deuschl et al., 1994), or OPT arising as a feature of the syndrome of progressive ataxia (Eggenberger et al., 2001; Samuel et al., 2004).

We have developed a computational model of the interaction between the inferior olive (IO) and the cerebellum to account for symptomatic OPT (Leigh et al., 2005). The model we used had been developed to account for classic motor learning of the blink reflex (Hong and Optican, 2005; Hong et al., 2008). Based on experimental studies of inferior olivary hypertrophy (de Zeeuw et al., 1998), we first modified the model so that adjacent IO units progressively developed high-conductance soma-somatic gap junctions; this led to synchronized discharge of the population of IO neurons at ~2 Hz. The model then predicted eye oscillations at 2 Hz, but they were smaller, jerkier, and more regular than those observed in patients. In the next stage in model development (Fig. 1A), pulses from the IO (black spiky curves) are transmitted to the cerebellar cortex via climbing fibres (CFs) and parallel fibres (PFs). The Purkinje cells (PC) react to the incoming signals by discharging the same temporal pattern of spikes as that of the inputs. In the third stage (Fig. 1B) coincidental arrival of PF and CF signals train the cortical PC-IN modules to learn the periodicity of the pulses. After learning, each mossy fibre (MF) pulse prompts the PC population to pause (the red wavy curve) at the expected time of the next CF signal. This periodic pause of the PC population thereby disinhibits the target vestibular neurons. Vestibular neurons now are driven from both the IO (the black spiky curve) and the PC population (the red wavy curve), and generates a signal (the red curve above the eyeball) that is a mixture of the IO pulse and PC modulation. A more complete circuit of the model is illustrated in Fig. 1C, representing a pair of oculomotor circuits; one vestibular axis is shown for simplicity.

This model successfully simulated characteristic features of the ocular oscillations of OPT: their smooth but variable waveform at 1–3 Hz, with vertical, torsional, and horizontal components, and the slow development of oscillations over the course of weeks after the stroke. However, an independent test of the model is necessary to support its validity. As inputs from the labyrinthine semicircular canals project via vestibular nuclei to both IO and the vestibulocerebellum, one prediction of the model is that a high-acceleration (impulse) head rotation would cause a major new input to the system that would shift ("reset") the phase of the ocular oscillations. Figure 2 shows the result of the model's simulation where the relatively periodic torsional component of the OPT shifts its phase after the head rotation. For a better illustration, the five cycles of periodic torsion prior to the head perturbation (in the box) have been displaced (up and to the right) to show what the phase would be if the oscillations were unaffected by the head perturbation. The post-head rotation peaks of the wavelets now do not align with the displaced peaks. We tested this model prediction in OPT.

Because the ocular oscillations of OPT have inherent randomness and non-stationary properties, we could not use classical Fourier analysis to determine phase before and after a head impulse. Accordingly, we used an alternative analysis based on the wavelet transform (WT). Being localized in both time and frequency domains, WT provides detailed information about frequency components at different times, without sacrificing resolution for either frequency or time (Torrence and Compo, 1998).

Subjects and methods

We studied four patients with OPT; their clinical features are summarized in Table 1. We used the

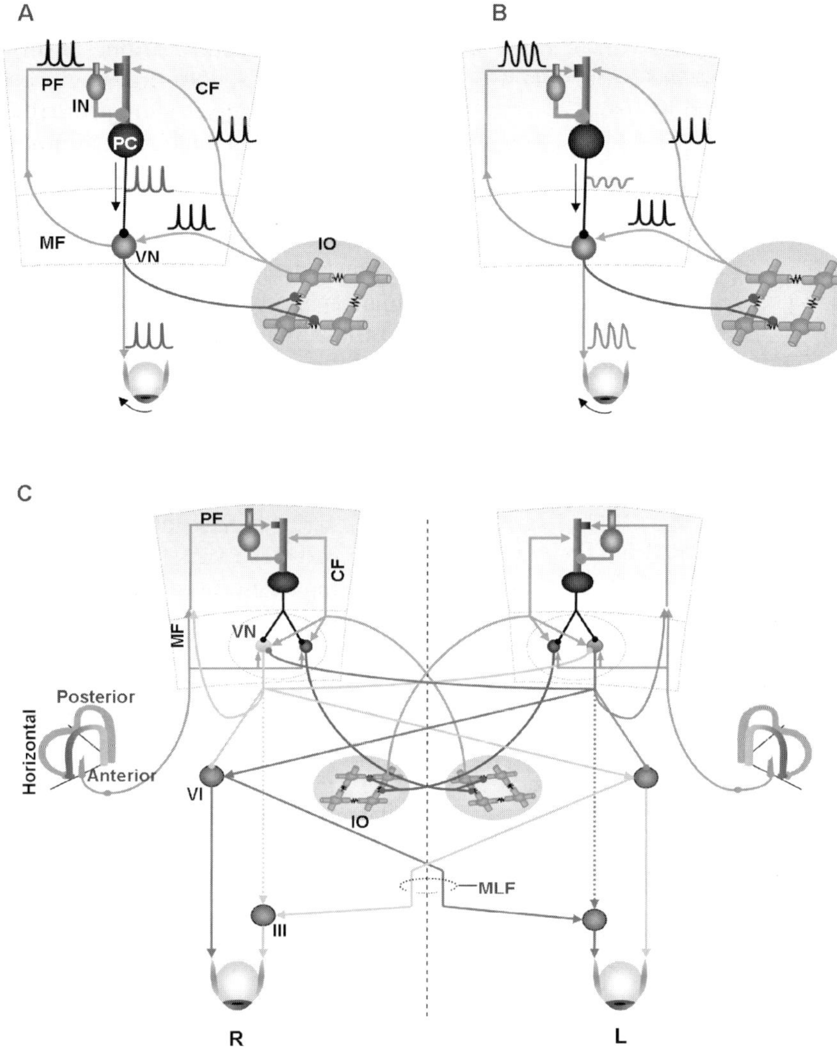

Fig. 1. Summary of the cerebellar circuit that learns the sequences of pulses from the hypertrophied inferior olive. (A) Initial stage of learning. At the beginning of learning, the pulses from the IO (black spiky curves) are simply transmitted to the cerebellar cortex via CFs and PFs. The PC simply reacts to the incoming signals by discharging the same form of spikes as that of the inputs. Note that just one projection from the IO is represented for simplicity. Also for simplicity, just one of three (anterior, posterior, and horizontal) channels is shown. Excitatory and inhibitory projections are indicated by arrow heads and spot heads, respectively. (B) Learned state of the model. The coincidental arrival of PF and CF signals train the cortical PC-IN modules to learn the periodicity of the pulses. After learning, each MF pulse prompts the PC population to pause (the red wavy curve) at the expected time of the next CF signal. This periodic pause of the PC population thereby disinhibits the target vestibular neurons. The vestibular neurons now are driven from both the IO (the black spiky curve) and the PC population (the red wavy curve), and generates a signal (the red curve above the eye ball) that is a mixture of the IO pulse and the PC modulation. (C) A more complete circuit diagram of the model. The diagram has a more detailed representation of the model showing left and right cerebellum and related connections to oculomotor structures. Both excitatory and inhibitory neurons are illustrated in VN. Note that while the inhibitory neuron connects only to the contralateral IO, the excitatory neuron sends its axons to oculomotor neurons and to the cerebellar cortex as a MF. The head rotation triggered phase reset pathway (ear canal→VN→IO) is also illustrated. The indirect projection from VN to the third cranial nucleus (III) via the contralateral sixth cranial nucleus (VI) is assumed to be stronger than the direct projection from the VN to III (dotted line). MLF: medial longitudinal fasciculus. (See Color Plate 3.13.1 in color plate section.)

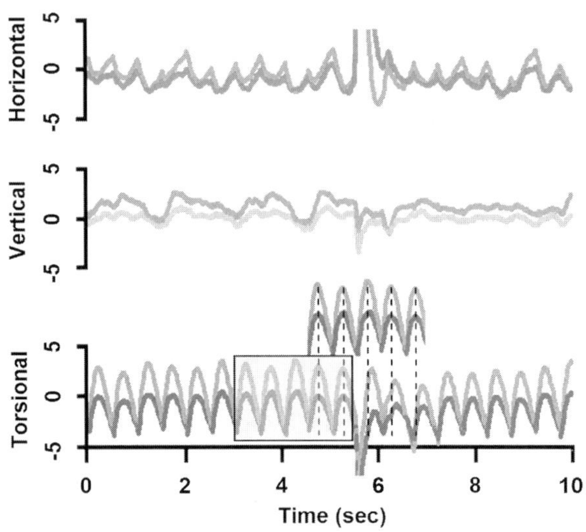

Fig. 2. Simulation of head impulse in OPT. The relatively periodic torsional component of the OPT shifts its phase after the head rotation. For comparison, the five pre-head perturbation wavelets (in box) have been displaced over around the onset of the head perturbation. The simulated head perturbation is a quick horizontal rotation. The overshooting horizontal trace marks the onset of the head rotation. Note that the post-head rotation peaks of the wavelets now do not align with the displaced ones. Grey traces represent right eye, coloured traces represent left eye. Vertical scale: deg/s. (See Color Plate 3.13.2 in color plate section.)

magnetic field/search coil technique to measure 3-D eye and head rotations, as previously described (Steffen et al., 2000). Each subject viewed a central visual target (laser spot projected onto a tangent screen at a distance of 125 cm) with each eye in turn. Coil signals were low-pass filtered (0–150 Hz) prior to digitization at 500 or 1000 Hz. During each experiment, subjects were asked to fixate the central visual target for 15–30 s. Next, the investigator manually applied impulsive head rotations to the subjects approximately every 5 s for 30 s.

We analysed the spectrum information of the signal before and after the head perturbation to evaluate the change in energy. In order to determine the characteristic spectrum of the resting nystagmus, we chose the fixation period of each trial, and picked out a series of saccade-free eye movements (using a criterion of eye velocity less than 40 deg/s). After re-sampling these movements, we obtained a total number of 100 1 s slices, which allowed us to estimate their spectrum and confidence interval. Since the number of the slices is large (1 0 0), we fitted it with a normal distribution. We then calculated the spectrum in a series of 1 s slices after each head perturbation, and compared their spectrum with that of nystagmus during fixation (head stationary).

Table 1. Summary of clinical findings and responses of nystagmus to head impulse stimuli

Age/sex/ duration	Clinical data	Medicines	Peak PSCR at head perturbation[a]	Threshold[a]
P1 50/F/4 year	4 year history of progressive ataxia and visual jumping beginning after a viral infection	Gabapentin atenolol	1032[b]	870
P2 46/M/18 months	Pontine and midbrain haemorrhage; eye symptoms began between 1 and 4 months after haemorrhage	Memantine, clonazepam	917[b]	900
P3 55/M/15 months	Pontine haemorrhage, eye symptoms began shortly after stroke	Norvasc, lopressor, carbidopa and levodopa at bedtime	955[c]	238
P4 57/M/3 year	Bilateral vertebral occlusion causing left anterior medullary and pontine infarction; eye symptoms began 1 year after stroke	Aspirin	95	797

[a]Units of rate of change of phase shift (PSCR) are degrees of phase shift per second. Threshold is 95% prediction interval of phase shift during fixation.
[b]PSCR exceeded threshold.
[c]Although PSCR exceeded threshold, frequent saccades, and blinks also did so.

The easiest way to measure the phase shift of these ocular oscillations is to compare the waveform with its own shifted version. The OPT oscillations usually cycle around a fixed frequency within an epoch of a few seconds, so by shifting the OPT waveform on the time axis by a few cycles, we can predict the phase of the waveform if it were not perturbed. (Testing during the fixation period showed that phase changes between the original and shifted waveforms were relatively small within a unit time period.) Thus, we shifted the ocular oscillation by a few cycles and compared its phase with that of its un-shifted version before and after the head perturbation.

We also calculated the rate of change of phase shift, which is the derivative of the phase difference between the shifted OPT oscillation and its un-shifted version during the head perturbation. Since wavelet coherence analysis results in 2-D data (time and frequency), and the OPT oscillations lie within a certain range of frequencies, we took a circular mean on the frequency axis to convert to 1-D data before we performed a derivative operation. As we observed, the OPT oscillations have their major frequency components at 1–4 Hz, corresponding to a period of 0.25–1.0 s. A circular mean is defined as:

$$a_m = \text{atan}(X, Y) \text{ with } X = \sum_{i=1}^{n} \cos(a_i) \text{ and }$$

$$Y = \sum_{i=1}^{n} \sin(a_i)$$

where a_{m} is the circular mean of a_i ($i = 1$ to n). The circular mean is used here to calculate the mean of trigonometric angles. We calculated a threshold for the spontaneous rate of change of phase of OPT by comparing many segments of OPT waveform with their time-shifted version, bootstrapping these arrays, and computing 95% predictions intervals for each patient. Then we compared the rate of change of phase (PSCR) occurring with each head perturbation with the threshold for that patient.

Results

An example of the ocular oscillations of one patient is shown in Fig. 3A. Note that in this patient the torsional component of the oscillation was the largest. We measured changes in the torsional oscillations in response to orthogonal (horizontal or vertical) head impulses; in this way, artefacts induced by the head rotation were minimized.

After analysing the eye movement with the wavelet decomposition and reconstruction package in Matlab (The Mathworks, Inc.), we found the energy of OPT oscillation only resides from level 6 to level 8 of the wavelet decomposition, corresponding to a frequency range of approximately 1–4 Hz (Liao et al., 2008). The energy that resides in levels 9–12 corresponds to the lower frequency components in the waveform, and the residual energy is of high frequency and lower amplitude, which can be ignored as the noise. Thus, our analysis will be focused on the spectrum of level 6–8 of the wavelet decomposition. A comparison of the energy of resting nystagmus versus that of OPT nystagmus after the head perturbation indicates that the energy of the OPT oscillation was not changed by the head perturbation.

As described above, the original eye movement was shifted by two or more cycles to compare the phase difference between the original and the shifted waveform. Of the four patients studied, two (P1 and P2, Table 1) showed phase difference changes that exceeded the calculated threshold at the moment of the head perturbation. Thus, in Fig. 3B, PSCR is greater than the threshold (870 deg/s) only during the head perturbation, whereas before and after the head perturbation the PSCR is small and well below the threshold. In the two other patients, spikes of PSCR were also evident with head perturbations. However, P3 also exceeded the threshold during frequent saccades and blinks. P4, alone, did not exceed threshold during head perturbations.

Discussion

We set out to test a new model for OPT (Fig. 1) by applying impulsive head rotations. The model predicted that such vestibular stimuli would induce changes in the oscillations that would be evident as phase shifts in the waveform proceeding versus following the head rotation (Fig. 2). Since

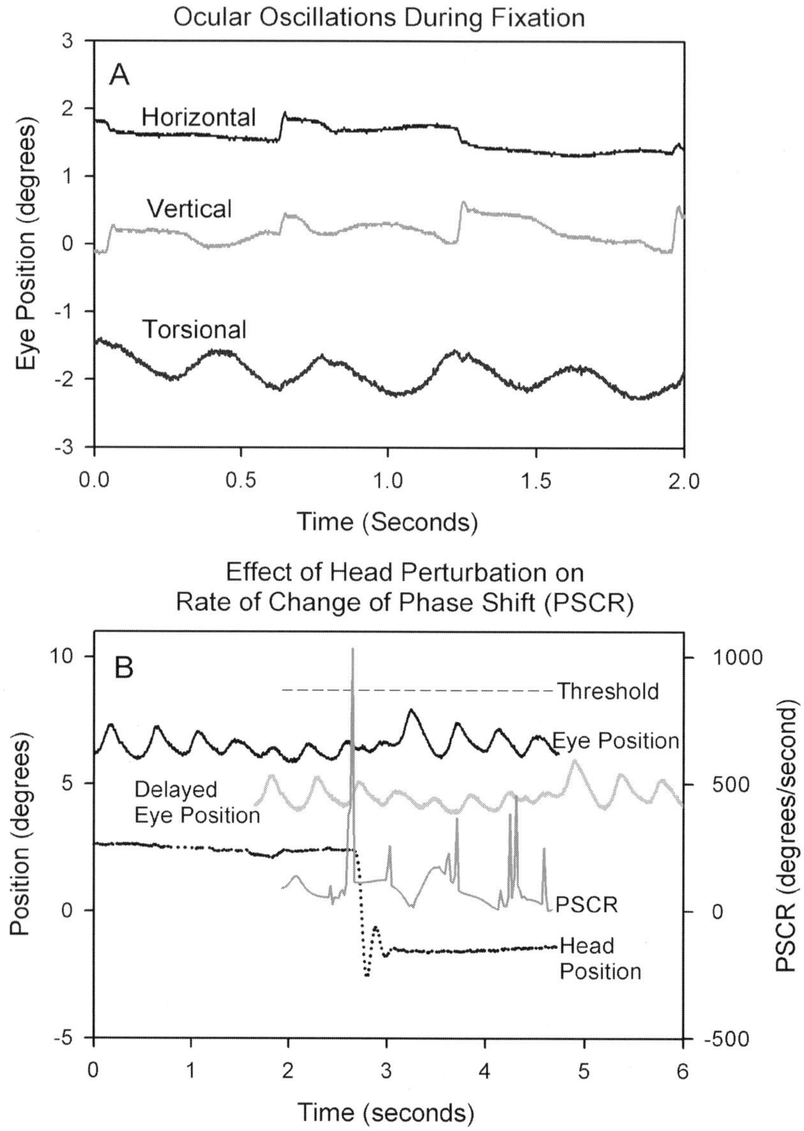

Fig. 3. OPT and impact of head rotation. (A) Example of ocular oscillations of Patient 1; note how the torsional component is the largest; each channel has been offset in position to aid clarity of display. (B) Example of effect of one head impulse on ocular oscillations of Patient 1. The upper black channel represents torsional eye position, whereas the upper gray channel is the same segment of eye position plotted with a delay of 1.65 s and offset of 2 deg in position for ease of comparison. The horizontal head perturbation is shown in the lower black channel. The rate of change of phase shift (PSCR) between the ocular oscillations and their shifted version is shown in the lower gray channel; note separate scale at right. At the time of the head impulse, PSCR exceed a threshold of 95% prediction interval for ocular oscillations during attempted fixation of a stationary target.

OPT has non-periodic and non-stationary properties, it was not possible to simply compare the ocular oscillations with a reference sine wave, as has been done in prior studies of periodic ocular oscillations (e.g., Das et al., 2000). Accordingly, we chose an approach that applied complex wavelet analysis (Torrence and Compo, 1998; Liao et al., 2008).

In all four patients, we detected a substantially greater rate of change of phase during the head impulse stimulus compared with changes of phase that occurred spontaneously during attempted fixation with the head stationary (Fig. 3). In three patients the rate of change of phase was statistically significant (Table 1). In one patient, phase shifts also exceeded threshold during frequent blinks and saccades. It seems possible that the analytic technique that we employed in this study may provide clearer results by asking patients to refrain from blinking and allowing a minimal of five OPT cycles between successive head impulses. Furthermore, this wavelet analysis approach might also be applied to determine whether other eye movements, such as saccades, cause shifts of the OPT waveform, which our results and prior studies (Kim et al., 2007) suggest.

Another way to test our model is to study the effects of drugs (Hong et al., 2008). In occasional patients, OPT is suppressed by a range of medicines, including memantine, which blocks NMDA receptors, and clonazepam, which has GABAergic properties (Leigh and Zee, 2006). In one such patient, with ocular oscillations following pontine haemorrhage, positron emission tomography (PET) demonstrated increased regional cerebral metabolic rate of glucose utilization (rCMRGlu) in the right IO (Yakushiji et al., 2006). Treatment with clonazepam reduced the amplitude of the ocular oscillations, and decreased rCMRGlu in the contralateral inferior cerebellar vermis, but not in the IO. If clonazepam blocked the cerebellar output, the nystagmus would only be driven by the IO output, and so its amplitude would be reduced. A complementary approach would be to study the effects of new drugs that act on the gap junctions of the IO. At present, antimalarial drugs are known to block gap-junctions (Cruikshank et al., 2004), and it seems likely that other agents will become available that could then be subjected to clinical trials for OPT.

To summarize, we have developed a hypothesis for the mechanism underlying the ocular oscillations of the syndrome of OPT, which is associated with inferior olivary hypertrophy. We simulated the ocular oscillations with a mathematical model of IO — cerebellar interactions based on the ideas that (1) hypertrophied olivary neurons become tightly coupled by an increase in gap junctions, leading to periodic synchronized discharge and (2) gratuitous learning of this IO signal by cerebellar cortex modulates the ocular oscillation waveform. An experiment test confirmed model predictions concerning the effects of vestibular stimuli on OPT. The model also makes predictions about potential drug treatments that could be evaluated in clinical trials.

Acknowledgement

This work is supported by NIH grant EY06717; the Office of Research and Department of Veterans Affairs; Evenor Armington Fund; Intramural Division of the National Eye Institute, NIH, DHHS.

References

Cruikshank, S.J., Hopperstad, M., Younger, M., Connors, B.W. and Spray, D.C. (2004) Potent block of Cx36 and Cx50 gap junction channels by mefloquine. Proc. Natl. Acad. Sci. U.S.A., 33: 12365–12369.

Das, V.E., Oruganti, P., Kramer, P.D. and Leigh, R.J. (2000) Experimental tests of a neural-network model for ocular oscillations caused by disease of central myelin. Exp. Brain Res., 133: 189–197.

Deuschl, G., Mischke, G., Schenck, E., Schulte-Monting, J. and Lucking, C.H. (1990) Symptomatic and essential rhythmic palatal myoclonus. Brain, 113(Pt 6): 1645–1672.

Deuschl, G., Toro, C., Valls-Solo, J., Zee, D.S. and Hallett, M. (1994) Symptomatic and essential palatal tremor. 1. Clinical, physiological and MRI analysis. Brain, 117: 775–788.

de Zeeuw, C.I., Simpson, J.I., Hoogenraad, C.C., Galjart, N., Koekkoek, S.K.E. and Ruigrok, T.J. (1998) Microcircuitry and function of the inferior olive. Trends Neurosci., 21: 391–400.

Eggenberger, E., Cornblath, W. and Stewart, D.H. (2001) Oculopalatal tremor with tardive ataxia. J. Neuroophthalmol., 21: 83–86.

Guillain, G. and Mollaret, P. (1931) Deux cas myoclonies synchrones et rhythmées vélo-pharyngo-laryngo-oculodiaphragmatiques: le problèm anatomique et physiolopathologique de ce syndrome. Rev. Neurol. (Paris), 2: 545–566.

Hong, S., Leigh, R.J., Zee, D.S. and Optican, L.M. (2008) Inferior olive hypertrophy and cerebellar learning are both needed to explain ocular oscillations in oculopalatal tremor. Prog. Brain Res. (this volume): 219–226.

Hong, S.Y. and Optican, L.M. (2005) New cellular mechanisms for multiple time-scale adaptation in cerebellum. Soc. Neurosci. Abstr., 933.7.

Kim, J.S., Moon, S.Y., Choi, K.D., Kim, J.H. and Sharpe, J.A. (2007) Patterns of ocular oscillation in oculopalatal tremor imaging correlations. Neurology, 68: 1128–1135.

Leigh, R.J., Hong, S., Zee, D.S. and Optican, L.M. (2005). Oculopalatal tremor: clinical and computational study of a disorder of the inferior olive. Soc. Neurosci. Abstr., 933.8.

Leigh, R.J. and Zee, D.S. (2006) The Neurology of Eye Movements (Book/DVD) (4th edn.). Oxford University Press, New York.

Liao, K., Hong, S., Zee, D.S., Optican, L.M. and Leigh, R.J. (2008). Using wavelet analysis to evaluate effects of eye and head movements on ocular oscillations. In: Leigh, R.J. and Devereaux, M.W. (Eds.), Advances in Understanding Mechanisms and Treatment of Infantile Forms of Nystagmus. Oxford University Press, New York (in press).

Samuel, M., Torun, N., Tuite, P.J., Sharpe, J.A. and Lang, A.E. (2004) Progressive ataxia and palatal tremor (PAPT): clinical and MRI assessment with review of palatal tremors. Brain, 127: 1252–1268.

Steffen, H., Walker, M.F. and Zee, D.S. (2000) Rotation of Listing's plane with convergence: independence from eye position. Invest. Ophthalmol. Vis. Sci., 41: 715–721.

Torrence, C. and Compo, G.P. (1998) A practical guide to wavelet analysis. Bull. Am. Meteorol. Soc., 79: 61–78.

Yakushiji, Y., Otsubo, R., Hayashi, T., Fukuchi, K., Yamanda, N., Hasegawa, Y. and Minematsu, K. (2006) Glucose utilization in the inferior cerebellar vermis and ocular myoclonus. Neurology, 67: 131–133.

SECTION 4

Using Eye Movements as a Probe of Sensory-Motor Processing and Navigation

C. Kennard & R.J. Leigh (Eds.)
Progress in Brain Research, Vol. 171
ISSN 0079-6123

CHAPTER 4.1

Human ocular following: evidence that responses to large-field stimuli are limited by local and global inhibitory influences

B.M. Sheliga*, E.J. FitzGibbon and F.A. Miles

Laboratory of Sensorimotor Research, National Eye Institute, National Institutes of Health, Bethesda, MD, USA

Abstract: Large-field visual motion elicits tracking eye movements at ultra-short latency, often termed ocular following responses (OFRs). We recorded the initial OFRs of three human subjects when vertical sine-wave gratings were subject to horizontal motion in the form of successive 1/4-wavelength steps. The gratings could occupy the full screen (45° wide, 30° high) or a number of horizontal strips, each 1° high and extending the full width of the display. These strips were always equally spaced vertically. In a first experiment, the gratings always had a contrast of 32%. Increasing the number of strips could reduce the response latency by up to 20 ms, so the magnitude of the initial OFRs was estimated from the change in eye position over the initial open-loop period measured with respect to response onset. A single (centred) strip (covering 3.3% of the screen) always elicited robust OFRs, and three strips (10% coverage) were sufficient to elicit the maximum OFR. Increasing the number of strips to 15 (50% coverage) had little impact, i.e., responses had asymptoted, and further increasing the coverage to 100% (full screen image) actually *decreased* the OFR so that it was now less than that elicited with only one strip. In a second experiment, the contrast of the gratings could be fixed at one of the four levels ranging from 8% to 64%, and the OFR showed essentially the same pattern of dependence on screen coverage except that the lower the contrast, the lower the level at which the response asymptoted. This indicated that the asymptote was not due simply to some upper limit on the magnitude of the eye movement or the underlying motion signals. We postulate that this asymptote is the result of normalization due to global divisive inhibition, which has often been described in visual-motion-selective neurons in the cortex. We further suggest that the *decrease* in the OFR when the image filled the screen was due to the increased continuity of the gratings which we postulate would favour the local inhibitory surround mechanisms over the central excitatory ones. This study indicates that robust OFRs can be elicited by much smaller motion stimuli than is commonly supposed and that introducing spatial discontinuities can increase the efficacy of the motion stimuli even while decreasing the area stimulated.

Keywords: ocular following response (OFR); response normalization; surround inhibition

*Corresponding author. Tel.: +301-402-4962;
Fax: +301-402-0511; E-mail: bms@lsr.nei.nih.gov

DOI: 10.1016/S0079-6123(08)00633-X

Introduction

Ocular following responses (OFRs) are the tracking eye movements elicited by large-field visual motion (see Miles, 1998 for review). Two defining characteristics of the initial OFR are its ultra-short latency — less than 80 ms in humans (Gellman et al., 1990) and less than 60 ms in monkeys (Miles et al., 1986) — and its machine-like quality. Recent studies indicate that the OFRs to broadband motion stimuli depend critically on the Fourier composition of the images (Chen et al., 2005; Sheliga et al., 2005), consistent with mediation by local spatio-temporal filters like those in the motion energy model that is commonly used to describe the responses of motion-selective complex cells in the striate cortex (Adelson and Bergen, 1985; Watson and Ahumada, 1985). Further, the fundamental visual properties of the OFR, such as its dependence on contrast and spatial frequency, show remarkably little inter-subject variation, most probably because these characteristics directly reflect the activity of the underlying low-level motion detectors (Sheliga et al., 2005). This has led to the suggestion that the OFR can be used to probe the early cortical processing of visual motion (Kodaka et al., 2007).

Previous studies used large-field stimuli to elicit OFRs and, partly because of this, it has been common to assume that large-field stimuli are necessary. Indeed, the finding that the initial OFRs elicited by patterns moving within stationary circular apertures show strong dependence on the size of the aperture — increasing steadily until the pattern spans 20° or more (Barthelemy et al., 2006) — seem to imply that OFRs are best with large-field stimuli. In the present study, the OFR stimuli were confined to elongated apertures aligned with the motion and indicated that the large-field motion stimuli often used to elicit the OFR are not optimal because they evoke at least two forms of response suppression in the neural networks processing the visual motion. We will argue that one of these suppressive mechanisms involves global inhibitory interactions between the motion-sensitive neurons, which result in divisive normalization, and the other involves local inhibitory surround mechanisms, which render the neurons less responsive to wide-field stimuli. One major consequence of these suppressive mechanisms for the OFR is that some motion stimuli of quite modest proportions are favoured over the usual large-field ones.

Methods

Most of the techniques were very similar to those used previously in our laboratory (Sheliga et al., 2005, 2006). Three subjects participated, and experimental protocols concerned with the use of human subjects were approved by the NEI Institutional Review Board. The horizontal and vertical positions of the right eye were recorded with an electromagnetic induction technique (Robinson, 1963) using a scleral search coil embedded in a silastin ring (Collewijn et al., 1975), as described by Yang et al. (2003). Visual stimuli were presented on a computer monitor that subtended 45° horizontally and 30° vertically. The visual motion stimuli consisted of 1-D vertical gratings with sinusoidal luminance profiles (spatial frequency, 0.25 cycles/°) that underwent successive 1/4-wavelength shifts every frame (i.e., every 10 ms). On any given trial, the grating could occupy the full screen or horizontal strips extending the full width of the display. There could be 1, 3, 7, or 15 strips, each 1° high, with one always at the screen centre and others vertically distributed with equal spacing (see Fig. 1A). The luminance in the spaces between the strips of grating matched the mean luminance of the grating (39 cd/m^2). In Experiment 1, all gratings had the same contrast (32%). In Experiment 2, the gratings could have one of the four contrasts (8, 16, 32, or 64%).

At the beginning of each trial, the grating pattern (randomly selected from a lookup table) appeared together with a central target spot that the subject was instructed to fixate. After the subject's right eye had been positioned within 2° of the fixation target and no saccades had been detected for a randomized period of 800–1100 ms, the fixation target disappeared and the apparent-motion stimulus began. The motion lasted for 200 ms (20 frames), at which point the screen became uniform grey (39 cd/m^2) marking the end of the trial. After an inter-trial interval of 500 ms, a new grating

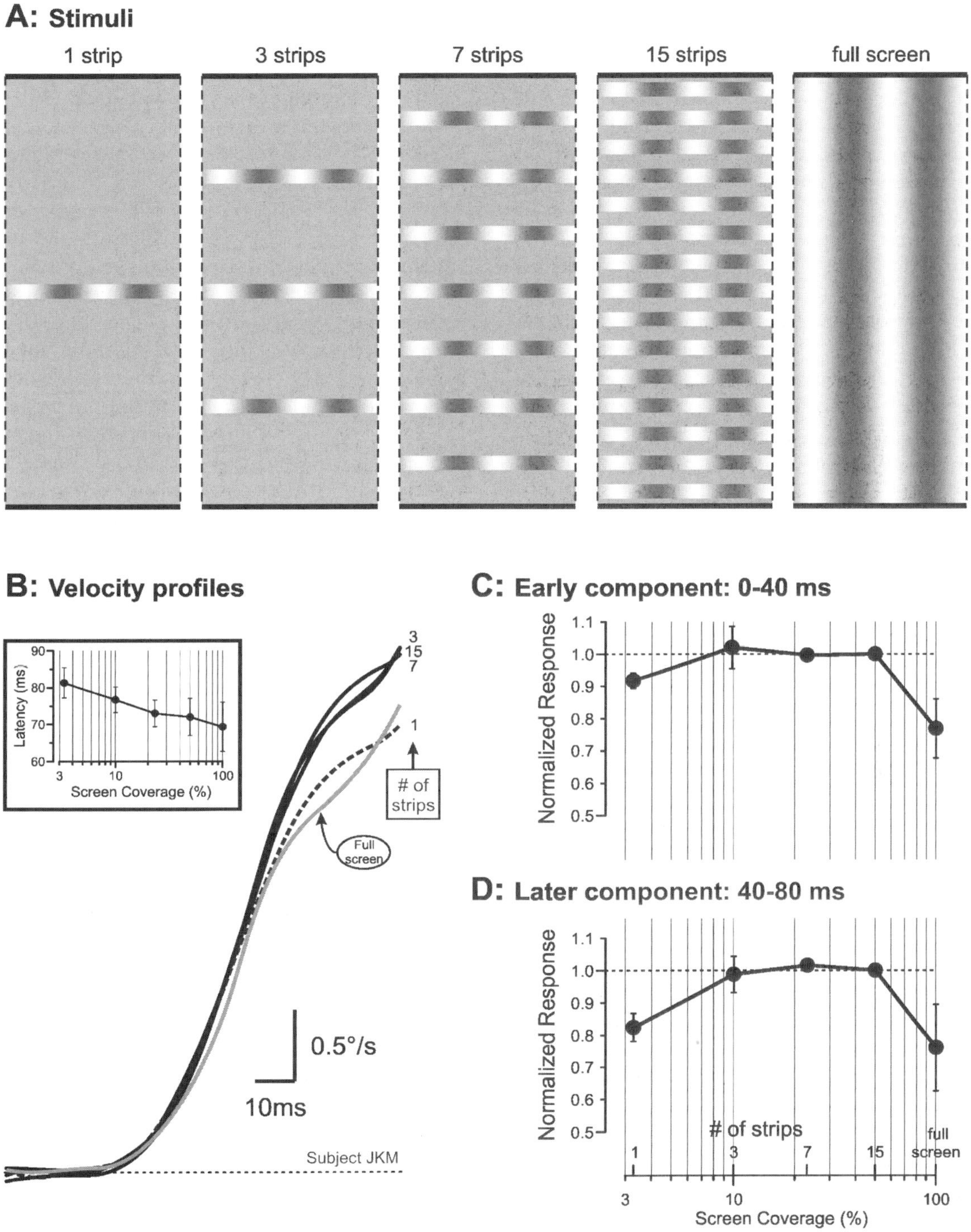

Fig. 1. Dependence of the OFR on screen coverage (determined by the number of strips occupied by the grating). (A) Panels showing the vertical partitioning of the vertical gratings into horizontal strips. (B) Mean horizontal R-L eye-velocity profiles for one subject synchronized to the onset of the responses; inset shows dependence of latency on the screen coverage (means ± SD for three subjects); numbers at end of traces indicate the number of strips making up the grating stimulus; grey trace, grating occupies full screen; dashed trace, grating occupies a single strip. (C) Dependence of the "early component" of the OFR on screen coverage (mean normalized measures ± SD for three subjects). (D) Dependence of the "later component" of the OFR on screen coverage (mean normalized measures ± SD for three subjects).

pattern appeared together with a fixation target, commencing a new trial. The subjects were asked to refrain from blinking or making saccades except during the inter-trial intervals, but were given no instructions relating to the motion stimuli.

The eye-position data were first smoothed with a 6-pole Butterworth filter (3 dB at 45 Hz), and then mean temporal profiles were computed for each subject for all the data obtained for each of the stimulus conditions. To improve the signal-to-noise ratio, the mean horizontal response to each leftward motion stimulus was subtracted from the mean horizontal response to the corresponding rightward motion stimulus: the "mean R-L position responses." By convention, rightward eye movements were positive so that these pooled responses were positive when OFRs were in the forward (i.e., stimulus) direction. Velocity responses were estimated at successive 1-ms intervals by computing the differences between the mean R-L position responses at intervals of 10 ms. The response onset was defined as the time when the mean R-L eye velocity first exceeded 0.2°/s, and the "initial OFRs" were quantified by measuring the changes in eye position over the 60-ms time periods starting with response onset. We also measured the changes in eye position over two 40-ms time periods, one starting with response onset ("early component") and the other 40 ms after response onset ("later component"). For each subject, these various response measures were each normalized with respect to the measures obtained with the 15 strips, and then mean response measures were calculated for each stimulus for the three subjects. The minimum latency of onset was ~70 ms so that the "initial OFR" and "early component" measures were restricted to the open-loop period (i.e., twice the reaction time) but the "later component" measures included some responses after the loop closed.

Results

Experiment 1

Figure 1B shows the mean R-L eye velocity response profiles of a sample subject when the grating stimulus occupied the full screen or was confined to 1, 3, 7, or 15 bands that occupied 3.3, 10, 23, or 50% of the area of the screen, respectively. Note that the latency tended to decrease linearly with the logarithm of the screen coverage, reducing by about 12 ms as the stimulus was increased from a single strip to full screen (see inset graph in Fig. 1B). To facilitate comparison, the traces in Fig. 1B have all been aligned on response onset, and it is immediately apparent that a 30-fold difference in the area of the stimulus — the difference between one strip and the full screen — had only a very modest impact on the initial OFRs. In fact, the profiles during the initial 30 or 40 ms of the responses all look very similar, and then only later do those obtained with the single strip (dashed trace) and the full screen (grey trace) clearly begin to fall progressively below the other profiles. Because these eye-velocity profiles seemed to suggest that the dependence on screen coverage changed ~40 ms into the response, we examined the OFR measures up to this point ("early component") and beyond ("later component") separately (see Methods section). The mean normalized "early component" measures for the three subjects are plotted as a function of screen coverage in Fig. 1C and the mean normalized "later component" measures are similarly plotted in Fig. 1D. (Note the logarithmic abscissas in Figs. 1C, D.) It is now evident that, in fact, the "early" and "later" OFR measures showed qualitatively similar dependencies on the screen coverage and showed only relatively minor quantitative differences. Thus, as the number of strips increased from 1 to 3, the screen coverage increased from 3.3% to 10% and both the "early" and "later" response measures increased to a maximum; both measures then remained at this level until the number of strips reached 15, when the screen coverage was 50%, after which the measures showed a surprising *decrease* of ~24% as the grating filled the screen. Importantly, with just a single strip — only 1° wide and occupying only 3.3% of the screen — the "early" and "later" response measures were, on average, more than 90 and 80%, respectively, of the maxima and were actually *greater* than those elicited by the full screen stimulus by 19 and 8%, respectively.

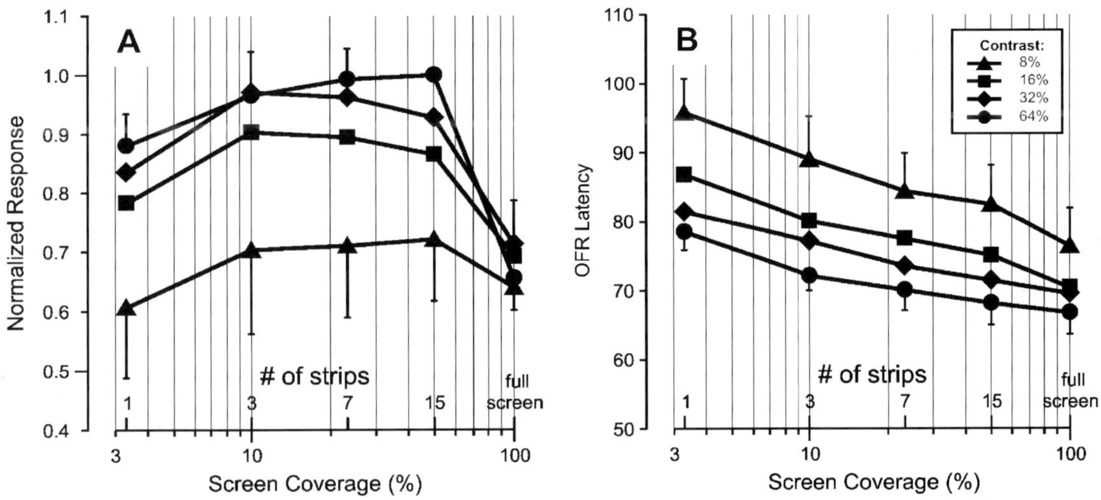

Fig. 2. Dependence of the OFR on screen coverage (determined by the number of strips occupied by the grating): effect of altering the contrast of the gratings. (A) Dependence of the "initial OFR" measures on screen coverage (mean normalized measures ± SD for three subjects) at each of the four contrast levels. (B) Dependence of latency on the screen coverage (means ± SD for three subjects) at each of the four contrast levels. The key indicates the Michelson contrast.

Experiment 2

Figure 2 shows the dependence of the OFR on screen coverage when the gratings had contrasts of 8% (triangles), 16% (squares), 32% (diamonds), and 64% (circles). The mean normalized "initial OFR" measures are plotted in Fig. 2A, and the data obtained at each contrast all show a pattern of dependence on screen coverage very similar to that seen in Fig. 1C, D, except that the lower the contrast, the lower the response level at which responses asymptoted (and the lower the response to the single band). The mean latencies of these responses are plotted in Fig. 2B and show the inverse dependence on screen coverage already apparent in the inset in Fig. 1B. It is now also apparent that the latency was inversely related to the contrast (cf., Miles et al., 1986).

Discussion

Our experiments indicate that robust OFRs can be elicited by much smaller motion stimuli than commonly supposed. Indeed, the initial OFR generally began to level off at its maximum when the grating occupied only 10% of the screen (three strips). In addition, increasing the area of the grating from 50% of the screen to 100%, i.e., from 15 strips to full screen, actually *decreased* the OFR. The net result was that the initial OFRs to a single strip 1° wide that occupied only 3.3% of the screen were actually *greater* than those elicited by the full screen stimulus.

Our finding that the level at which the OFR asymptoted was contrast dependent indicated that this leveling off with screen coverage was not simply due to the passive attainment of some upper limit in the magnitude of the sensory motion signals or the motor response itself. Rather, we postulate an active process like the divisive normalization often described in visual-motion-sensitive neurons in the cortex and generally attributed to some global inhibitory process (Heeger, 1992; Carandini and Heeger, 1994; Carandini et al., 1997; Simoncelli and Heeger, 1998; Britten and Heuer, 1999; Heuer and Britten, 2002). Ideally, the responses of an ocular tracking mechanism to motion of a given speed and direction should be insensitive to the physical characteristics of the moving images, and the current data indicate that, for a given contrast, the initial OFRs are independent of the size of the stimulus over a five-fold range (10–50% coverage).

Over this range, there is clear vector averaging, exactly the sort of behaviour one expects of a system subject to divisive normalization. We suggest that these effects are mediated by the same mechanism that is responsible for contrast gain control whereby the OFR saturates at relatively low contrast, ~30% (Masson and Castet, 2002; Sheliga et al., 2005).

We further postulate that the *decrease* in the OFR when the screen coverage increased from 50 to 100% was due to the increased continuity of the gratings as the image filled the screen, and we invoke local inhibitory surround mechanisms to explain it (cf., Barthelemy et al., 2006). Direction-selective neurons with powerful inhibitory surrounds are commonplace in cortical area MT, which is a major source of the motion signals reaching MST, a region known to be critical for the genesis of the OFR (Takemura et al., 2007). Some MT neurons have antagonistic surrounds whose preferred direction of motion is the same as that at the centre, rendering these neurons sensitive to local-motion contrast and insensitive to wide-field motion (see Born and Bradley, 2005 for recent review). We postulate that it is because of such neurons that introducing spatial discontinuities increases the OFR — even while decreasing the area stimulated by motion — by reducing the activation of the antagonistic surrounds.

A crucial feature of the stimuli in the present study was that they were in effect seen through elongated apertures aligned with the axis of motion and hence were inherently broadband. Moving images confined to stationary circular apertures, as in the study of Barthelemy et al. (2006), become increasingly high-pass when the aperture is reduced in diameter, compromising the low spatial frequencies that are preferred by the OFR. Thus, the effects of the aperture here are less to do with its area than with its spatial-frequency bandwidth, which depends on the length of the aperture along the axis of motion. Many other studies have examined the so-called smooth pursuit tracking responses to single, small moving spots that are obviously not confined to a stationary window. These pursuit responses have latencies that are generally at least twice that of the OFR (Heinen and Watamaniuk, 1998).

Acknowledgement

This research was supported by the intramural programme of the National Eye Institute at the National Institutes of Health.

References

Adelson, E.H. and Bergen, J.R. (1985) Spatiotemporal energy models for the perception of motion. J. Opt. Soc. Am. A, 2: 284–299.

Barthelemy, F.V., Vanzetta, I. and Masson, G.S. (2006) Behavioral receptive field for ocular following in humans: dynamics of spatial summation and center-surround interactions. J. Neurophysiol., 95: 3712–3726.

Born, R.T. and Bradley, D.C. (2005) Structure and function of visual area MT. Ann. Rev. Neurosci., 28: 157–189.

Britten, K.H. and Heuer, H.W. (1999) Spatial summation in the receptive fields of MT neurons. J. Neurosci., 19: 5074–5084.

Carandini, M. and Heeger, D.J. (1994) Summation and division by neurons in primate visual cortex. Science, 264: 1333–1336.

Carandini, M., Heeger, D.J. and Movshon, J.A. (1997) Linearity and normalization in simple cells of the macaque primary visual cortex. J. Neurosci., 17: 8621–8644.

Chen, K.J., Sheliga, B.M., FitzGibbon, E.J. and Miles, F.A. (2005) Initial ocular following in humans depends critically on the Fourier components of the motion stimulus. Ann. N.Y. Acad. Sci., 1039: 260–271.

Collewijn, H., Van Der Mark, F. and Jansen, T.C. (1975) Precise recording of human eye movements. Vision Res., 15: 447–450.

Gellman, R.S., Carl, J.R. and Miles, F.A. (1990) Short latency ocular-following responses in man. Vis. Neurosci., 5: 107–122.

Heeger, D.J. (1992) Normalization of cell responses in cat striate cortex. Vis. Neurosci., 9: 181–197.

Heinen, S.J. and Watamaniuk, S.N. (1998) Spatial integration in human smooth pursuit. Vision Res., 38: 3785–3794.

Heuer, H.W. and Britten, K.H. (2002) Contrast dependence of response normalization in area MT of the rhesus macaque. J. Neurophysiol., 88: 3398–3408.

Kodaka, Y., Sheliga, B.M., Fitzgibbon, E.J. and Miles, F.A. (2007) The vergence eye movements induced by radial optic flow: some fundamental properties of the underlying local-motion detectors. Vision Res., 47: 2637–2660.

Masson, G.S. and Castet, E. (2002) Parallel motion processing for the initiation of short-latency ocular following in humans. J. Neurosci., 22: 5149–5163.

Miles, F.A. (1998) The neural processing of 3-D visual information: evidence from eye movements. Eur. J. Neurosci., 10: 811–822.

Miles, F.A., Kawano, K. and Optican, L.M. (1986) Short-latency ocular following responses of monkey. I. Dependence on temporospatial properties of visual input. J. Neurophysiol., 56: 1321–1354.

Robinson, D.A. (1963) A method of measuring eye movement using a scleral search coil in a magnetic field. IEEE Trans. Biomed. Eng., 10: 137–145.

Sheliga, B.M., Chen, K.J., FitzGibbon, E.J. and Miles, F.A. (2005) Initial ocular following in humans: a response to first-order motion energy. Vision Res., 45: 3307–3321.

Sheliga, B.M., Chen, K.J., FitzGibbon, E.J. and Miles, F.A. (2006) The initial ocular following responses elicited by apparent-motion stimuli: reversal by inter-stimulus intervals. Vision Res., 46: 979–992.

Simoncelli, E.P. and Heeger, D.J. (1998) A model of neuronal responses in visual area MT. Vision Res., 38: 743–761.

Takemura, A., Murata, Y., Kawano, K. and Miles, F.A. (2007) Deficits in short-latency tracking eye movements after chemical lesions in monkey cortical areas MT and MST. J. Neurosci., 27: 529–541.

Watson, A.B. and Ahumada, A.J., Jr. (1985) Model of human visual-motion sensing. J. Opt. Soc. Am. A, 2: 322–341.

Yang, D.S., FitzGibbon, E.J. and Miles, F.A. (2003) Short-latency disparity-vergence eye movements in humans: sensitivity to simulated orthogonal tropias. Vision Res., 43: 431–443.

C. Kennard & R.J. Leigh (Eds.)
Progress in Brain Research, Vol. 171
ISSN 0079-6123

CHAPTER 4.2

Short-latency disparity vergence eye movements: dependence on the preëxisting vergence angle

H.A. Rambold* and F.A. Miles

Laboratory of Sensorimotor Research, National Eye Institute, National Institutes of Health, Bethesda, MD, USA

Abstract: We recorded the vergence eye movements that are elicited at ultra-short latencies when binocular disparities are applied to large-field patterns (Busettini, C., Miles, F.A. and Krauzlis, R.J. (1996). J. Neurophysiol., 75: 1392–1410) and determined their dependence on the *preëxisting vergence angle* (*PVA*). The search coil technique was used to record the movements of both eyes in four healthy subjects (two with presbyopia). Using dichoptic viewing, the two eyes saw identical images each consisting of a fixation cross at the centre of a random-dot pattern in a circular aperture. The subject fixated the crosses and then the images (crosses, random dots, windows) moved horizontally (1.5°/s) in opposite directions so as to bring the eyes to the desired horizontal vergence position without changing the accommodation demand. After a further 800–1200 ms to permit fusion at this new vergence angle (now, the *PVA*), a disparity step was applied and, 200 ms later, the screen changed to uniform grey, marking the end of the trial. The disparity steps could have one of six magnitudes and four directions (crossed, uncrossed, right-hyper, left-hyper) while the *PVA* was varied systematically. The horizontal and vertical disparity vergence responses (DVRs) of one of the presbyopes consistently showed robust linear dependence on the *PVA* ($r^2 > 0.96$). The horizontal DVRs of the other three subjects showed no sensitivity to the *PVA* and their vertical DVRs showed only very weak dependence. The experiment was repeated on one of the non-presbyopes after cycloplegia, but the outcome was the same, indicating that the negative findings were not due to the influence of the vergence-accommodation response. Our data indicate that the DVRs can be scaled by the *PVA*, but most subjects do not show this effect, perhaps because they relied on other distance cues that are uninformative in our experimental situation.

Keywords: disparity vergence eye movements

Introduction

When the images seen by the two eyes are slightly misaligned (binocular disparity) vergence eye movements are elicited at ultra-short latencies in humans and monkeys (Busettini et al., 1996, 2001; Masson et al., 1997; Takemura et al., 2001; Sheliga et al., 2006, 2007). The direction of these disparity vergence responses (DVRs) is as expected of a negative feedback control system that uses binocular parallax to eliminate vergence errors: convergent with crossed disparities, divergent with uncrossed disparities, left sursumvergent with left-hyper disparities, and right sursumvergent

*Corresponding author. Tel: +301-496-9080; Fax: +301-402-0511; E-mail: ramboldh@nei.nih.gov

DOI: 10.1016/S0079-6123(08)00634-1

with right-hyper disparities (Busettini et al., 2001). The DVR is one of three oculomotor reflexes that share much in common (for review, see Miles, 1998). One of these other reflexes, the radial flow vergence response (RFVR), has a gain that is a linear function of the preëxisting vergence angle (*PVA*) in humans (Yang et al., 1999). The other reflex, the ocular following response (OFR), has a gain that is inversely related to viewing distance in monkeys (Busettini et al., 1991) but modulation with viewing distance was very variable in humans unless subjects were placed on a sled to encourage the expectation of ego-motion (Busettini et al., 1994). In the case of the OFR it was felt that the modulation with viewing distance was linked to *perceived distance* with possible contributions from efference copies of the vergence angle and the accommodative state, but other depth cues (e.g., texture, size) and context also seem to play an important role in humans (Busettini et al., 1991, 1994). In the present study we recorded the initial open-loop DVRs that were elicited when binocular disparities were applied to random-dot patterns (RDs) and examined their dependence on the *PVA*. The data indicate that only one of four subjects used the *PVA* to scale the DVR.

Methods

We recorded the initial DVRs elicited by disparities applied to large textured patterns. Some of the techniques were very similar to those used previously (Yang et al., 2003; Sheliga et al., 2006, 2007) and will be described only in brief here. Four subjects participated in this study: One was an author (FAM), two were experienced in eye-movement recordings but uninformed about the purpose of the experiments (BMS, DSZ), and the fourth was both inexperienced and uninformed (ST). All subjects had normal or corrected-to-normal vision. Experimental protocols were approved by the NEI Institutional Review Board concerned with the use of human subjects. Two of the subjects were presbyopic (FAM and DSZ) and two were emmetropic (BMS and ST). Otherwise, the eye examination in all subjects was normal.

The horizontal and vertical positions of both eyes were recorded with the electromagnetic induction technique (CNC Engineering, Seattle, USA) using scleral search coils embedded in silastin rings (Skalar, Delft, The Netherlands), and each eye was sampled at 1 kHz (Robinson, 1963; Collewijn et al., 1975). At the beginning of each recording session a calibration procedure was performed for each of the two eyes independently with defined fixations targets as previously described (Sheliga et al., 2006, 2007).

Dichoptic stimuli were presented using a Wheatstone mirror stereoscope (Sheliga et al., 2006, 2007). The visual stimuli consisted of RDs with centred grey fixation crosses (width 5°, height 5°, thickness 0.1°) that were identical at the two eyes. The RDs were presented inside a circular aperture (diameter, 30°) with a black surround. The individual dots were circular (diameter, 0.5°), covering 25% of the pixels, and were either black or white (contrast 70%) on a grey background (18.7 cd/m^2). At the start of each trial a randomly chosen pair of identical RDs (from a lookup table) was presented, creating a single binocular image at the monitor distance (53 cm; approximate vergence angle, 6.5°). Both half images moved at 1.5°/s horizontally in opposite directions until the appropriate *PVA* was reached (range, 0°–18°). After a further 800–1200 ms to permit fusion of the two half images, a disparity step was applied and then, after 200 ms, the screen changed to uniform grey (18.74 cd/m^2) marking the end of the trial. After an inter-trial interval of 500 ms the next pair of binocular images appeared signalling the start of another trial. The subjects were asked to fixate the centre of the crosses whenever present and to refrain from blinking or making saccades after a beep tone was presented indicating the end of the ramp movement and the beginning of the fixation period before the disparity step was applied. If no saccades were detected during the requisite period the data were stored on a hard disc; otherwise, the trial was aborted and repeated later. Each block of trials consisted of 10–50 randomly interleaved stimulus presentations, and each block was repeated at least 50 times. There were 10 different patterns of dots in the RDs and one was randomly selected for each trial. The RDs, including the circular aperture, were corrected for the tangent error so that the retinal images did not vary with the *PVA*.

The horizontal and vertical eye-position measures obtained during the calibration procedure were each fitted with second-order polynomials whose parameters were then used to linearize the corresponding eye-position data recorded during the experiment proper. The linearized eye-position measures were smoothed with a 6-pole Butterworth filter (3 dB at 45 Hz) and mean temporal profiles were computed for each stimulus condition. Trials with saccadic intrusions (that had failed to reach the eye-velocity threshold of 18°/s during the experiment) were deleted. The horizontal (vertical) vergence angle was computed by subtracting the horizontal (vertical) position of the right eye from the horizontal (vertical) position of the left eye. In our convention, rightward and upward eye movements were positive so that the DVRs were positive when convergent or left sursumvergent and negative when divergent or right sursumvergent. The initial horizontal and vertical DVRs (HDVRs, VDVRs) in each stimulus condition were quantified by measuring the changes in the horizontal and vertical vergence position measures over the 70-ms time periods commencing 90 ms after the onset of the disparity stimuli. The minimum latency of the DVRs was ~80 ms from the first appearance of the disparity stimuli so that these vergence-response measures were restricted to the initial open-loop period (i.e., twice the reaction time).

Results

Figure 1A shows sample HDVR temporal profiles obtained from one subject (FAM) when crossed and uncrossed disparities of 0.4° were applied to RDs at different *PVA*s as indicated by the numbers superimposed on the traces. Crossed disparities (X) elicited convergent (CONV) and uncrossed disparities (UX) elicited divergent (DIV) eye movements that increased in amplitude with increases in the *PVA*. With high *PVA*s (fixations at near) the vergence velocity traces showed oscillations that were especially prominent with crossed disparities (Fig. 1A). The associated response measures, based on the change in vergence angle over the interval 70–160 ms from stimulus onset, are plotted in Fig. 1B and show clear linear dependence on the *PVA*, with regression coefficients for the HDVRs to crossed disparities of 0.090 (intercept) and $0.006^{\circ-1}$ (slope) with an r^2 value of 0.999; similarly, the coefficients for the HDVRs to uncrossed disparities were −0.045 (intercept), $-0.009^{\circ-1}$ (slope) with an r^2 value of 0.998.

The VDVRs showed a very similar dependence on the *PVA*. Figure 1D shows sample VDVR temporal profiles obtained from the same subject when left-hyper and right-hyper disparities of 0.2° were applied to RDs at different *PVA*s. Left-hyper disparities (LH) elicited left sursumvergent (LSSV) and right-hyper disparities (RH) elicited right sursumvergent (RSSV) eye movements that increased in amplitude with increases in the *PVA*. The associated response measures plotted in Fig. 1E show clear linear dependence on the *PVA*, with linear regression coefficients for the VDVRs to left-hyper disparities of 0.047 (intercept) and $0.0026^{\circ-1}$ (slope) with an r^2 value of 0.999; similarly, the coefficients for the VDVRs to right-hyper disparities were −0.035 (intercept), $-0.037^{\circ-1}$ (slope) with an r^2 value of 0.998.

To determine if the dependence of the DVRs on the *PVA* was evident with all disparity stimuli we obtained disparity tuning curves at two different *PVA*s (3° and 15°). The two tuning curves for the HDVRs are shown in Fig. 1C and for the VDVRs are shown in Fig. 1F (subject FAM). All curves were well fitted by Gabor functions (Busettini et al., 2001), which explained over 96% of the variance (Table 1). The scale factors of the best-fit Gabor functions, *G*, were significantly greater when the *PVA* was 18° than when it was 3°: 13.1 vs. 7.2 (HDVR) and 5.5 vs. 3.8 (VDVR). Other parameters of the best-fit Gabor functions — offset (*A*), width (σ), and phase shift (θ) — were not significantly different for the two *PVA*s.

We also explored the effect of horizontal disparity offsets (0–2.5°, crossed and uncrossed disparities) on the VDVR elicited by given vertical disparity steps (0.2° right-hyper, 0.2° left-hyper) at two *PVA*s (3°,15°) for subject FAM. The orthogonal (horizontal) disparity offsets were applied at the

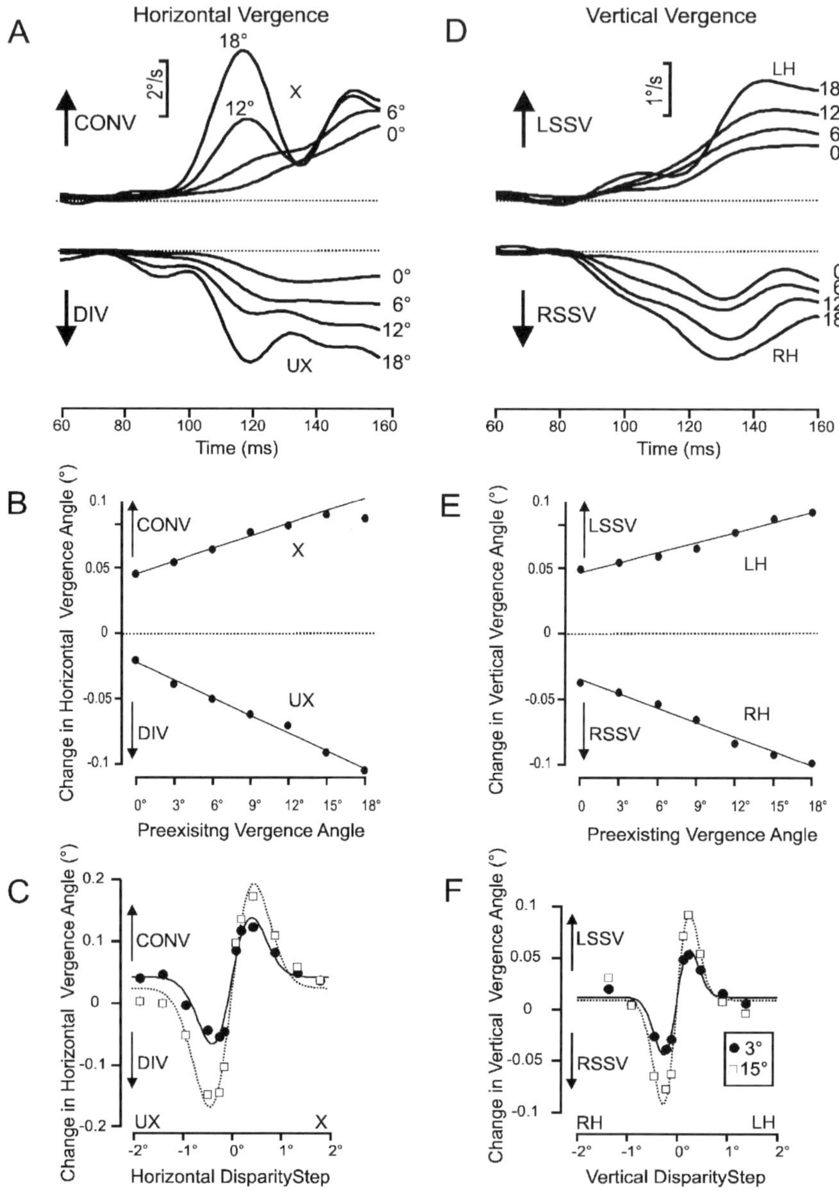

Fig. 1. Dependence of the DVRs on the *PVA* (subject FAM). (A) Mean horizontal vergence velocity profiles; *numbers* on traces indicate the *PVA*. (B) Mean changes in horizontal vergence angle as a function of the *PVA*; straight lines are linear regressions. (C) Horizontal disparity tuning data when the *PVA* was 3° (filled symbols) and 15° (open symbols); curves are best-fit Gabor functions (parameters in Table 1). (D) Mean vertical vergence velocity profiles; *numbers* on traces indicate the *PVA*. (E) Mean changes in vertical vergence angle as a function of the *PVA*; straight lines are linear regressions. (F) Vertical disparity tuning data when the *PVA* was 3° (filled symbols) and 15° (open symbols); curves are best-fit Gabor functions (parameters in Table 1). Means based on data from 49 to 55 trials. Abbreviations: X, crossed disparities; UX, uncrossed disparities; RH, right-hyper disparities; LH, left-hyper disparities; CONV, convergence; DIV, divergence; LSSV, left-sursumvergence; RSSV, right-sursumvergence.

Table 1. Parameters of the best-fit Gabor functions in Fig. 1C, F (subject FAM)

PVA	B	G	d^F	σ	f	ϕ	r^2	
3°	0.042	7.16	0.026	0.40	0.010	−90	0.966	HDVRs
15°	0.025	13.06	0.019	0.45	0.010	−90	0.970	
3°	0.012	3.76	−0.025	0.26	0.014	−90	0.976	VDVRs
15°	0.009	5.53	−0.029	0.26	0.017	−90	0.971	

Note: The Gabor function had the form, $A + G \times e^{[-(D-d^F)^2/(2\sigma^2)]} \times \cos[2\pi f(D-d^F)+\phi]$, where A is the y-offset, G the gain, d^F the x-offset, D the applied disparity, σ the standard deviation, ϕ the phase in degrees, and f the spatial frequency in cycles/°.

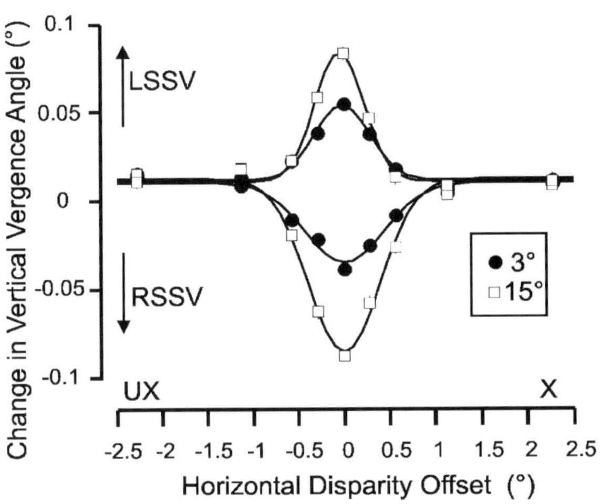

Fig. 2. The effect of horizontal disparity offsets on the mean VDVRs to right-hyper and left-hyper disparities of 0.2° with two different *PVA*s: 3° (filled symbols) and 15° (open symbols). Curves are best-fit Gaussian functions (parameters in Table 2). Subject, FAM. Abbreviations same as for Fig. 1.

same time as the vertical disparity steps, i.e., the disparity vector was oblique. Figure 2 shows the dependence of the mean VDVRs (±SE) on the horizontal disparity offset for the two *PVA*s (filled symbols, 3°; open symbols, 15°). The dependence on the horizontal offset was always well described by a Gaussian function (r^2: 0.989±0.010) as previously reported by Yang et al. (2003). The parameters of the best-fit Gaussian functions for the DVRs to left-hyper (right-hyper) disparities with the two *PVA*s differed significantly only in the scale factor, g (Table 2).

Table 2. Parameters of the best-fit Gaussian functions in Fig. 2 (subject FAM)

PVA	A	g	μ	σ	r^2	
3°	0.012	0.04	−0.01	0.31	0.994	LSSV
15°	0.011	0.07	−0.04	0.27	0.989	
3°	0.010	−0.05	0.01	0.44	0.974	RSSV
15°	0.012	−0.10	0.00	0.39	0.993	

Note: The Gaussian had the form, $A + g \times e^{[-(d-\mu)^2/(2\sigma^2)]}$, where A is the y-offset, g the scale factor, σ the width, μ the x-offset of the peak, and d the applied disparity. All units are in degrees.

The HDVRs of the other three subjects showed no significant dependence on the *PVA* and the VDVRs showed only a minor increase with increases in the *PVA*. In fact, this effect was rather subtle and only visible in the velocity traces. Linear regression of the DVRs to the *PVA* had r^2 values >0.99 (mean±SD, slope of HDVRs to X: −0.0041±0.0044, UX: −0.0038±0.0051; VDVRs to RH: 0.0013±0.0204, LH: −0.013±0.0287).

In our setup, the changes in the *PVA* were not accompanied by changes in the accommodation demand and the non-presbyopes occasionally reported problems maintaining focused images, presumably due to the influence of vergence-accommodation (Fincham and Walton, 1957). The experiments were repeated on one of the non-presbyopes (BMS) after cycloplegia (Cyclopentolate, 1% solution, Cyclogyl®), but there were no significant changes in this subject's data, suggesting that vergence-accommodation was not responsible for the lack of dependence of this subject's DVRs on the *PVA* (slope: HDVRs to X: 0.0003, UX: −0.0004; VDVRs to RH: 0.0000, LH: −0.0004).

Discussion

The data in Figs. 1 and 2 from one subject clearly indicate that horizontal and vertical DVRs can be scaled (linearly) by the *PVA*. This is in line with the unpublished observations of C. Busettini, G.S. Masson, and F.A. Miles [cited by Miles (1998)] who did essentially the same experiments as in the present study but used a somewhat different setup from the present one. In fact, these workers used the same setup that Yang et al. (1999) subsequently used to demonstrate that the initial RFVR was a linear function of the *PVA*. The only difference between those two studies was that one used radial-flow steps and the other used disparity steps. The RDs in those studies were somewhat larger in extent than in the present one (90° × 90°) and no correction was made for the slight changes in the retinal images of the patterns as the *PVA* changed (due to the tangent error). The scaling of the DVRs by the *PVA* in Fig. 1 is very similar to that previously reported for the RFVR by Yang et al. (1999), who postulated that radial-flow vergence operates synergistically with the translational vestibulo-ocular reflex (TVOR) during fore-aft motion of the observer to maintain binocular alignment on objects in the scene ahead and that the two reflexes share a pathway whose gain modulates with vergence angle. The data in Fig. 1 suggest that the DVR might also share this same pathway and operate in synergy with the RFVR and TVOR mechanisms. Interestingly, Busettini et al. (1991) had earlier invoked a similar synergy between the OFR and TVOR to compensate for lateral and vertical accelerations of the head and showed that, in monkeys, both reflexes have gains that modulate linearly with vergence angle, an effect that they also attributed to shared pathways and neural gain elements. These explanations imply that the modulation arises rather late in the neural pathways, i.e., on the motor side, and our finding that the *PVA* merely scales the disparity tuning curves (Fig. 1C, F) and the curves describing the dependence on orthogonal disparity offsets (Fig. 2) is consistent with this.

Unlike the RFVR, the DVR showed clear dependence on the *PVA* in only one of our four subjects, casting doubt on the notion that the modulation arises from a gain element shared by the two reflexes. The subject showing the dependence was one of two presbyopes, but since the other did not show this dependence, presbyopia does not seem to be the determining factor. One problem in trying to rationalize these differences between subjects is that it is not clear what purpose is served by modulating the DVR gain with the *PVA*. In particular, what advantage is gained from increasing the sensitivity of the DVR for near fixations? High gains can cause instability and this might be the reason for the oscillations that we observed with near viewing (Fig. 1A). [Note that the period of these oscillations was less than half the response latency of the DVR and hence must involve internal feedback loops rather than the external visual feedback loop.] Busettini et al. (1991) pointed out that the OFR shows progressive saturation with image speed, the peak eye velocity being inversely related to the viewing distance, indicating that the speed saturation occurs upstream of the dependence on viewing distance. These workers postulated that, under normal viewing conditions, this speed saturation would tend to offset the dependence on viewing distance because the retinal slip speeds experienced by the moving observer tend to vary inversely with viewing distance, resulting in a greater tendency for the OFR to saturate with near viewing. It is known that the DVR saturates with retinal image speed at a point in the visual pathways before the inputs from the two eyes are combined (Masson et al., 2002), opening up the possibility that under normal viewing conditions this saturation will tend to offset the DVR's dependence on the *PVA*.

Why would the DVR of most subjects in our study not show this dependence on the *PVA*? One factor might be cue redundancy. There are many potential cues to viewing distance in addition to the vergence angle — accommodation, size, perspective, overlay, motion parallax (Howard and Rogers, 2002) — and different subjects might assign different weights to these cues at different times depending on the context. In our experiment, the depth indicated by these other cues is generally fixed and in conflict with the vergence angle so that any subject who used them would not show dependence on the vergence angle. Similar arguments were advanced to explain inter-subject variability in a study on the dependence of the human OFR on viewing distance (Busettini et al., 1994). This last study also found contextual effects, whereby dependence of the OFR on viewing distance became much clearer when the subject was placed on a sled, raising the possibility that the visual motion stimulus could result from ego-motion. This all implies that there might be much more flexibility in the deployment of these reflexes than generally supposed and, in our present study, this might be responsible for the failure to see a dependence on the PVA in three of our four subjects.

Acknowledgements

The authors thank B.M. Sheliga and E. Fitzgibbon for technical and experimental support and B.M. Sheliga, S. Tanabe, and D.S. Zee for taking part in the experiments. This study was supported by the Alexander von Humboldt Foundation (Germany) and the Intramural Program of the National Eye Institute of the National Institutes of Health.

References

Busettini, C., Fitzgibbon, E.J. and Miles, F.A. (2001) Short-latency disparity vergence in humans. J. Neurophysiol., 85: 1129–1152.

Busettini, C., Miles, F.A. and Krauzlis, R.J. (1996) Short-latency disparity vergence responses and their dependence on a prior saccadic eye movement. J. Neurophysiol., 75: 1392–1410.

Busettini, C., Miles, F.A. and Schwarz, U. (1991) Ocular responses to translation and their dependence on viewing distance. II. Motion of the scene. J. Neurophysiol., 66: 865–878.

Busettini, C., Miles, F.A., Schwarz, U. and Carl, J.R. (1994) Human ocular responses to translation of the observer and of the scene: dependence on viewing distance. Exp. Brain Res., 100: 484–494.

Collewijn, H., van der Mark, F. and Jansen, T.C. (1975) Precise recording of human eye movements. Vision Res., 15: 447–450.

Fincham, E.F. and Walton, J. (1957) The reciprocal actions of accommodation and convergence. J. Physiol., 137: 488–508.

Howard, I.P. and Rogers, B.J. (2002) Seeing in depth. I. Depth perception. Porteous, Toronto, Canada.

Masson, G.S., Busettini, C. and Miles, F.A. (1997) Vergence eye movements in response to binocular disparity without depth perception. Nature, 389: 283–286.

Masson, G.S., Yang, D.S. and Miles, F.A. (2002) Version and vergence eye movements in humans: open-loop dynamics determined by monocular rather than binocular image speed. Vision Res., 42: 2853–2867.

Miles, F.A. (1998) The neural processing of 3-D visual information: evidence from eye movements. Eur. J. Neurosci., 10: 811–822.

Robinson, D.A. (1963) A method of measuring eye movement using a scleral search coil in a magnetic field. IEEE Trans. Biomed. Eng., 10: 137–145.

Sheliga, B.M., Fitzgibbon, E.J. and Miles, F.A. (2006) Short-latency disparity vergence eye movements: a response to disparity energy. Vision Res., 46: 3723–3740.

Sheliga, B.M., Fitzgibbon, E.J. and Miles, F.A. (2007) Human vergence eye movements initiated by competing disparities: evidence for a winner-take-all mechanism. Vision Res., 47: 479–500.

Takemura, A., Inoue, Y., Kawano, K., Quaia, C. and Miles, F.A. (2001) Single-unit activity in cortical area MST associated with disparity-vergence eye movements: evidence for population coding. J. Neurophysiol., 85: 2245–2266.

Yang, D., Fitzgibbon, E.J. and Miles, F.A. (1999) Short-latency vergence eye movements induced by radial optic flow in humans: dependence on ambient vergence level. J. Neurophysiol., 81: 945–949.

Yang, D.S., Fitzgibbon, E.J. and Miles, F.A. (2003) Short-latency disparity-vergence eye movements in humans: sensitivity to simulated orthogonal tropias. Vision Res., 43: 431–443.

C. Kennard & R.J. Leigh (Eds.)
Progress in Brain Research, Vol. 171
ISSN 0079-6123

CHAPTER 4.3

MSTd neurons during ocular following and smooth pursuit perturbation

U. Büttner[1,*], S. Ono[2], S. Glasauer[1], M.J. Mustari[2] and U. Nuding[1]

[1]*Department of Neurology and Bernstein Center for Computational Neuroscience, Ludwig-Maximilians-University, Munich, Germany*
[2]*Yerkes National Primate Research Center and Department of Neurology, Emory University, Atlanta, GA, USA*

Abstract: MSTd neurons in the behaving monkey were investigated during step-ramp smooth pursuit eye movements (SPEM), short perturbations of the small visual target during ongoing pursuit, and large-field visual stimulation inducing ocular following responses (OFR). Neurons responded with short latencies to visual motion during OFR. In contrast the non-retinal responses during SPEM and perturbations followed the eye movements by 100–150 ms and were in the opposite direction to the OFR response. Often neurons were not modulated by the perturbation. Although, both the OFR and the perturbation response are involuntary eye movements due to visual motion, it seems very unlikely that these MSTd neurons with non-retinal responses are involved in their direct control. Based on these responses, we suggest that our MSTd neurons may code for gaze direction in space based on visual estimates of self-motion and extraretinal estimates of eye-in-head motion.

Keywords: MST; smooth pursuit eye movements; eye movement perturbation

Introduction

Moving visual stimuli can lead to slow tracking eye movements. Depending on the stimulus conditions different types of movements can be distinguished.

1. Smooth pursuit eye movements (SPEM) are used to track voluntary small moving visual objects. They occur only in foveate species and are mostly investigated in humans and monkeys. They can reach velocities above 100 deg/s (monkey: Lisberger et al., 1981; man: Simons and Büttner, 1985). The latency for SPEM initiation is 100–150 ms (Robinson, 1965).
2. Optokinetic response (OKR). These compensatory eye movements occur in response to large-moving visual fields. During continuous stimulation the eyes are reset by saccades (fast phases of optokinetic nystagmus). In afoveate animals without SPEM, OKR is solely determined by the so called "indirect" response, which causes a gradual increase in slow-phase velocity during continuous stimulation. The "indirect" response can best be demonstrated during optokinetic after-nystagmus (OKAN), i.e., the nystagmus, which continues in the dark after the light has been turned off. The "indirect" response is related to the "velocity storage" mechanism and manifests itself with activity changes in the vestibular nuclei (Boyle et al., 1985; Waespe

*Corresponding author. Tel.: +49-89-70954823; Fax: +49-89-70955561; E-mail: Ulrich.buettner@med.uni-muenchen.de

DOI: 10.1016/S0079-6123(08)00635-3

and Henn, 1987). In foveate species there is also a "direct" response, which causes a rapid increase in slow phase eye velocity. Particularly in monkeys the "direct" and "indirect" response add together leading to slow phase velocities above 180 deg/s (Cohen et al., 1977; Büttner et al., 1983).

3. Ocular following response (OFR). These are involuntary eye movements in response to large-moving visual fields with very short latencies (60–70 ms). Thus, the short latencies, the stimulus size, and the involuntary character distinguish OFR from SPEM (Kawano et al., 1994; Kawano, 1999). OFR only considers the initial 100–150 ms (open-loop) phase of the eye movements, which afterwards continue as "direct" OKR.
4. Smooth pursuit perturbation response (SPPR). If during ongoing pursuit the moving visual target is perturbed (i.e., one cycle of a high frequency sinusoidal oscillation), the eyes follow involuntarily. Interestingly the response amplitude depends on the on-going pursuit velocity. It is small during fixation and increases monotonically with higher constant pursuit velocities. This reflects a nonlinear mechanism and has been called dynamic gain control (Schwartz and Lisberger, 1994; Keating and Pierre, 1996).

Several structures at the cortical, brainstem, and cerebellar levels are involved in the generation of these eye movements. Major cortical structures are the middle temporal (MT) visual area and the medial superior temporal (MST) visual area located at the parieto-temporo-occipital junction and the smooth pursuit-related region of the frontal eye field (FEF) (Büttner and Büttner-Ennever, 2006).

These areas project to the brainstem, mainly to the dorsolateral pontine nuclei (DLPN) and to the nucleus reticularis tegmenti pontis (NRTP). NRTP receives its input mainly from FEF and projects predominantly to the oculomotor vermis (OV), whereas for DLPN the input mainly derives from MT/MST with cerebellar projections to OV and the paraflocculus. Thus, there is some evidence for two parallel pathways from the cerebral cortex via

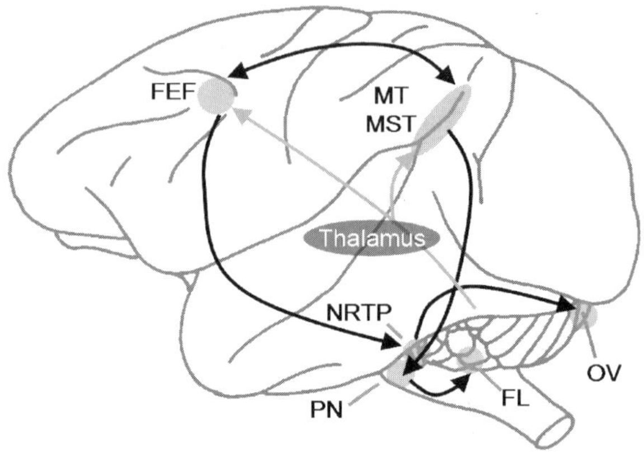

Fig. 1. Schematic drawing of smooth pursuit related pathways shown on a lateral view of the monkey brain. Both cortical areas (FEF, MT/MST) project to different brainstem areas (PN, NRTP), which in turn send afferents to distinct cerebellar structures (OV, FL). Also shown is a pathway for feedback signals via the thalamus.

the brainstem to the cerebellum (Fig. 1), although these pathways are certainly not completely separated with ample reciprocal connections at the cortical level and the brainstem areas not exclusively projecting to one cerebellar structure.

The specific functions of the different pathways are still poorly understood. In the following mainly neuronal activity in MST will be considered, particularly in relation to SPPR and OFR. In MST two areas (MSTl, MSTd) are distinguished. Anatomically MSTl lies next to MT, where neurons have mainly visual properties. There are a number of features, which allow a functional distinction of areas MSTl and MSTd. Neurons in both areas have visual motion sensitive receptive fields (RFs), which are smaller in MSTl (average 15 deg) (Ilg et al., 2004) compared to MSTd, where RF sizes often exceed 20 deg (Komatsu and Wurtz, 1988), RF's in MSTl are in the contralateral hemifield and often include the fovea (68%) (Ilg et al., 2004), but generally do not extend in the ipsilateral hemifield. In the MSTd RFs are also in the contralateral hemifield, but are often eccentric and can include both contralateral and ipsilateral visual field components (Komatsu and Wurtz, 1988; Churchland and Lisberger, 2005).

Many neurons in MSTd and MSTl are also modulated during SPEM. A major difference is the neuronal response latency in relation to eye movement onset. While MSTl neurons lead the eye movement (average −26 ms; see Ilg et al., 2004), SPEM-related activity in MSTd starts after the eye movement, often by 100–150 ms (Newsome et al., 1988). Furthermore, most MSTl neurons have the same preferred direction for visual motion and SPEM (Ilg et al., 2004) whereas MSTd neurons have opposite preferred directions (Komatsu and Wurtz, 1988).

Large-field (LF) visual motion induces OFR (Kawano, 1999). Neuronal responses start on average about 10 ms before eye movement onset and latencies are similar for MSTd and MSTl neurons (Kawano et al., 1994). Only few neurons so far have been investigated during OFR and SPEM. This applies particularly for MSTd neurons with opposite preferred directions for OFR and SPEM (Kawano et al., 1994). From these single-unit studies in MST it has been concluded, that MST (and also DLPN) mainly act as a relay of visual information for OFR and that only in the floccular region more motor-related signals are encountered (Kawano, 1999). On the other hand, partially based on the SPEM-related activity, MST has also been considered to play a more direct role in smooth pursuit performance, particularly for maintenance of SPEM (Krauzlis, 2004; Thier and Ilg, 2005).

Thus, the functional differences and the underlying neuronal mechanisms for SPEM, OFR, and SPPR are not fully understood. It was pointed out that SPEM are voluntary eye movements in contrast to SPPR and OFR, which are involuntary. One major difference between SPPR and OFR would be then the stimulus size with OFR reflecting the response to large-moving visual fields and SPPR the response to small visual stimuli. This raises the question, whether OFR and SPPR share common neuronal mechanisms or reflect different features. To further elaborate these questions single-unit recordings in MSTd were performed during OFR and SPPR.

Methods

Monkeys (*Macaca mulatta*) were chronically prepared for eye movement and single-unit recordings. Under anaesthesia they received a head stabilization post and a stereotaxically implanted recording chamber. For eye position recordings a scleral search coil was inserted underneath the conjunctiva of one eye. All procedures were approved by the Institutional Animal Care and Use Committee of Emory University. For more details of the surgical procedure, see Ono and Mustari (2006). During the experiments monkeys were seated in a primate chair with their head fixed in a completely dark room. Visual stimuli were projected on a tangent screen 57 cm away. For responsive neurons the optimal stimulus direction was determined by moving a target in eight cardinal directions, separated by 45 deg. For SPEM a step-ramp tracking task was applied.

The following stimulus conditions were applied:

1. SPEM. A small target (0.2 deg) first stepped away from the central position and then moved at constant velocities (10–20 deg/s).
2. Smooth pursuit perturbation response (SPPR). During ongoing pursuit the small target was perturbed by one sinusoidal cycle (5 Hz ± 10 deg/s) with the first half cycle increasing the stimulus velocity (peak-first perturbation).
3. OFR. For this a LF (35 × 35 deg) random dot pattern was used as a visual stimulus. The monkey fixated the small target spot located at the centre of gaze and the centre of the LF stimulus. It was turned off, when the LF started to move at constant velocity (10–20 deg/s) for 2 s. This stimulus consistently led to following eye movements with a gain close to one.
4. Blinking. The non-retinal origin of the SPEM response for MSTd neurons was tested by blinking the target during ongoing pursuit for 100–200 ms. Trials with and without blinking were randomly interleaved (Ono and Mustari, 2006).

Single-unit activity was recorded in MSTd using customized epoxy-coated tungsten microelectrodes (Frederick-Haer, Brunswick, ME). Location of recording sites of MSTd neurons was verified by using structural MRI (T1-weighted, fast spin echo, Siemens, 3 T magnet).

Results

General characteristics

Neurons in MSTd responded directionally specific to LF visual motion. In addition many neurons also responded during SPEM (Komatsu and Wurtz, 1988). There was no obvious difference in the visual motion response between those neurons, which responded during SPEM and those, which did not. Neurons responding only during SPEM were very rare. For neurons responding during SPEM and LF motion the preferred direction was in nearly all instances in the opposite direction. The non-retinal origin of the response for neurons modulated with SPEM was tested by target blinking (100–200 ms) during ongoing pursuit, which showed continuous response during the blink (Ono and Mustari, 2006). All neurons responding to visual motion had large RF in the contralateral hemifield, in most instances considerable larger than 20 deg. For some neurons the RF expanded by more than 20 deg in the ipsilateral hemifield.

Smooth pursuit eye movements (SPEM)

Figure 2 shows a typical example of a MSTd neuron in response to a step-ramp stimulus eliciting SPEM. Except for the moving target the monkey was in complete darkness. The target step was adjusted to prevent an initial catch-up saccade. After the step the eyes start to move with a latency of 130 ms. Constant eye velocity is reached after another 170 ms.

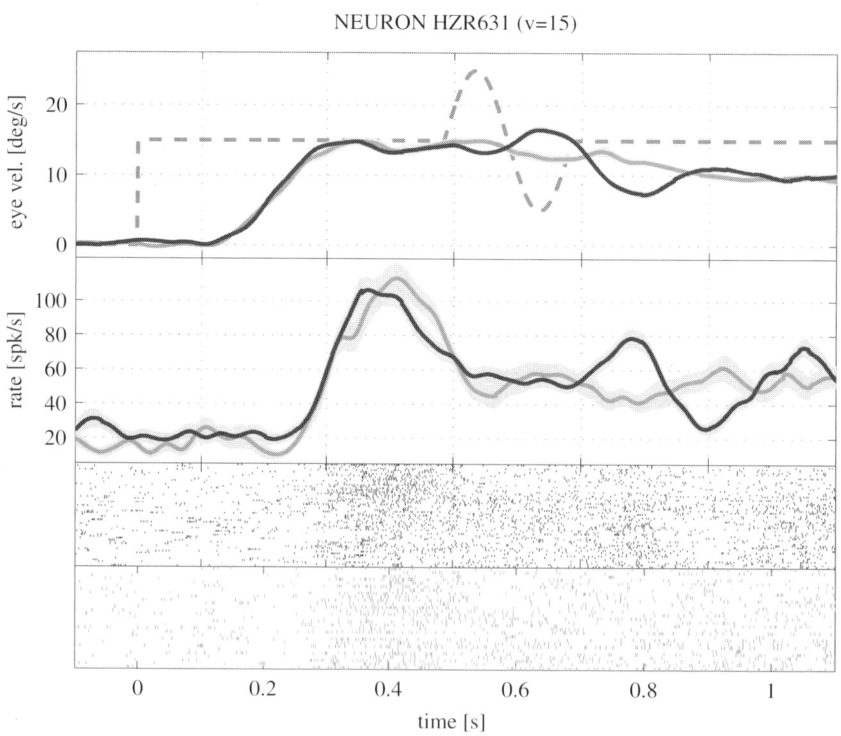

Fig. 2. MSTd neuron during smooth pursuit eye movement (step-ramp) with and without perturbation (one cycle 5 Hz ± 10 deg/s). Upper traces show stimulus velocity (grey broken line) and eye velocity with (black solid line) and without (grey solid line) perturbation. The middle trace shows the related neuronal activity average of the individual trials shown below (upper half with, lower half without perturbation). The perturbation leads to a change in eye velocity. Neuronal activity starts ca. 120 ms after eye movement onset. Activity changes related to SPPR also occur ca. 120 ms later, which is more than 200 ms after stimulus perturbation.

In MSTd activity started to increase for most neurons after the beginning of SPEM (Newsome et al., 1988). For the neuron shown in Fig. 2 the increase of neuronal activity was 120 ms after eye movement onset. For some neurons (Fig. 2) higher activity was observed during the initial acceleration phase, but in general the activity changes during SPEM related to the constant velocity period.

Smooth pursuit perturbation response (SPPR)

The perturbation (one cycle 5 Hz ± 10 deg/s) during ongoing pursuit reliably led to a change in eye velocity (Figs. 2 and 3). The latency of the eye movement response (SPPR) was often shorter than the smooth pursuit onset during ramp stimulation. For the neuron shown in Fig. 2 it was 100 ms. There was always a clear modulation of the eye movement response (SPPR), which increased with higher ongoing smooth pursuit velocities. As described earlier this reflects dynamic gain control, but will not be discussed further here. For the example shown in Fig. 2 with an ongoing velocity of 15 deg/s the gain was 0.47 for the SPPR.

For many neurons the SPPR was not accompanied by neuronal activity changes (Fig. 3). But for more than 35% of the neurons such activity changes occurred (Fig. 2). Noticeable neuronal changes occurred after the SPPR, often with a delay of more than 100 ms. For the neuron shown in Fig. 2 the delay between peak of SPPR and peak of activity changes was 120 ms, which is the same time difference as seen after SPEM onset.

Ocular following response (OFR)

Also the LF visual motion led consistently to eye movements (OFR). The onset was shorter than

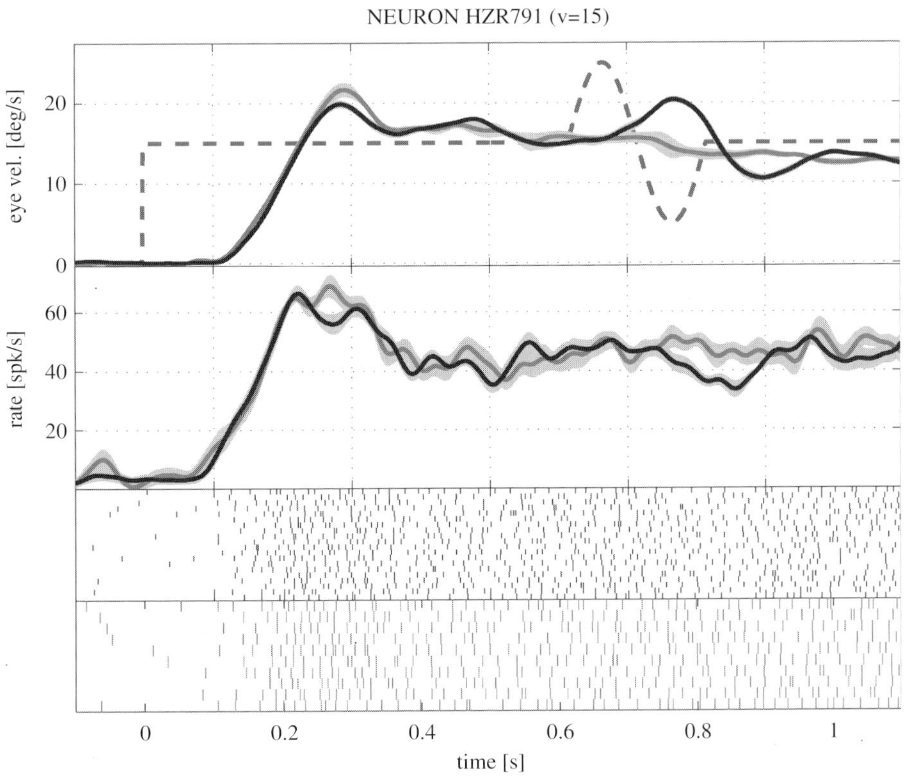

Fig. 3. MSTd neuron during step ramp smooth pursuit, which is not modulated during SPPR despite a clear eye movement modulation after the perturbation. Traces as in Fig. 2. Note that this is one of the rare MSTd neurons, in which the neuronal activity starts to increase slightly before eye movement onset.

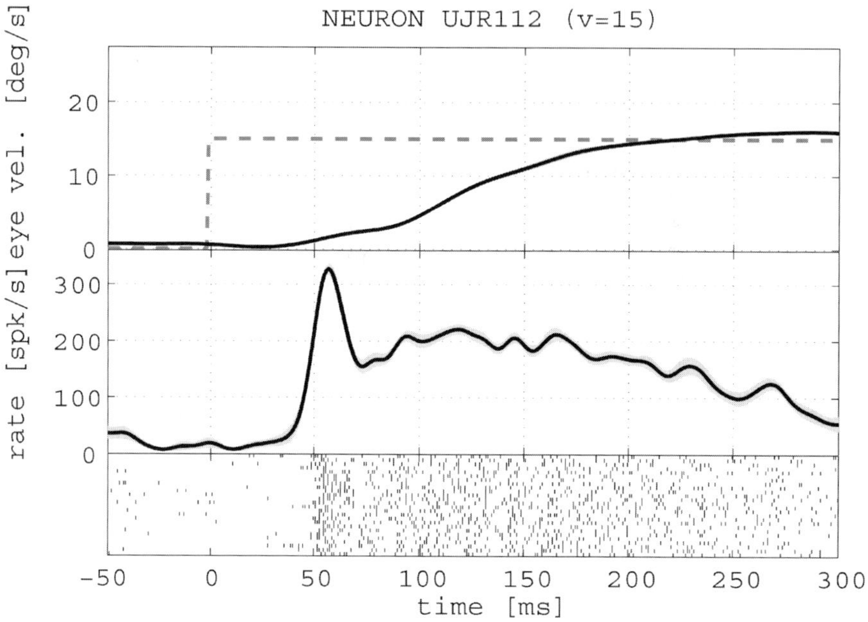

Fig. 4. Response of a MSTd neuron to large field (LF) visual motion, which elicits an ocular following response (OFR). Upper trace shows LF stimulus velocity and eye velocity; middle trace averaged neuronal activity of individual trails shown below. Note the different time scale in comparison to Figs. 2 and 3. The LF starts to move at 0, leading to eye movements (OFR) ca. 50 ms after stimulus onset. The neuronal activity increases at the same time. It starts to decrease when eye velocity is close to stimulus velocity.

during SPEM. For the example shown in Fig. 4 it was 50 ms. After initiation eye velocity increased continuously and reached a gain of 0.5 after about 65 ms. MSTd-neurons generally responded vigorously to LF visual stimulation. As pointed out earlier, the LF responses were nearly always in the opposite direction to the SPEM responses. Neurons started to respond 40–50 ms after stimulus onset and in contrast to the SPEM and SPPR responded always with or before the beginning of the eye movements. For the neuron shown in Fig. 4, activity increased with eye movement onset. In general activity changes under these conditions were related to retinal-slip velocity, which was high at the beginning and decreased with increasing eye velocity (Fig. 4).

Discussion

Neurons were recorded in MSTd during SPEM (step-ramp), smooth pursuit perturbation (SPPR), and ocular following OFR. Whereas SPEM initiation is a voluntary act, SPPR and OFR are involuntary, visually induced eye movements, the latter being distinguished by the stimulus size, which is small for SPPR and large for OFR.

Neurons in MSTd respond to large-moving visual fields. Many of these neurons are also modulated during SPEM eye movements. The smooth pursuit related activity generally starts 100–150 ms after eye movement onset, which has been described previously (Newsome et al., 1988). Perturbation (one cycle 5 Hz $\pm$ 10 deg/s) during ongoing pursuit reliably led to eye velocity changes. For many MSTd neurons these changes are not accompanied by neuronal changes. For those neurons which are affected by the SPPR the activity changes occur after the eye movements, often up to 100–150 ms. Based on these findings it can be ruled out that these neurons in MSTd participate in the generation of the SPPR. These extraretinal signals probably provide an eye velocity feedback signal, which might be transmitted via the thalamus (Tanaka, 2005). Besides being involved in the maintenance of SPEM

(Krauzlis, 2004; Thier and Ilg, 2005), they could play a role in estimating target velocity in space (Shenoy et al., 2002; Page and Duffy, 2003; Thier and Ilg, 2005).

MSTd neurons also respond to LF visual motion, which lead to OFR. OFR has very short latencies (see Fig. 4) (Kawano, 1999). The neuronal responses often precede the eye movements (Kawano et al., 1994) and can be related to the visual (retinal) component of the MSTd neurons. Thus the neuronal relationship between SPPR and ORF is quite different, one (SPPR) reflecting an extraretinal signal and the other (OFR) a retinal (visual) signal. Furthermore the preferred directions are opposite. This makes a simple role of this group of MSTd neurons in the direct control of OFR and SPPR unlikely.

In order to propose a hypothesis about the variable coded by our MSTd neurons, the main findings are recapitulated: (1) responses to LF visual motion occur with short delay and brisk onset, (2) responses to SPEM are in opposite direction and occur clearly after eye movement onset, and (3) responses to SPPR are similar to SPEM responses.

We suggest that such MSTd neurons may code for an estimate of *gaze velocity in world-centred coordinates*, just like MSTl neurons code for object motion in world-centred coordinates (Ilg et al., 2004). We argue as follows: if rightward LF visual motion (relative to the observer) is interpreted as self-motion, then *gaze in space* moves *to the left*, as long as no eye movements occur. In contrast, rightward target motion, not being interpreted as self-motion, causes SPEM to the right, and thus movement of *gaze in space* is also *to the right*. Consequently, assuming that our MSTd neurons code for gaze in space not only explains the difference in OFR and SPEM preferred directions, but also the difference in latencies to LF motion and SPEM response. However, there are certainly several distinct functional groups of MST neurons. It has been described above that MSTl neurons have quite different response characteristics than MSTd neurons. It is also known that only a small percentage of MST neurons project to the DLPN. Further studies are required using a combination of different paradigms to elucidate the role of different groups of MST neurons in the control of visually driven eye movements (SPEM, SPPR, OFR).

Acknowledgement

This work is supported by BMBF (BCCN Munich 016Q0440), NEI EY06069, and RR00165.

References

Boyle, R., Büttner, U. and Markert, G. (1985) Vestibular nuclei activity and eye movements in the alert monkey during sinusoidal optokinetic stimulation. Exp. Brain Res., 57: 362–369.

Büttner, U. and Büttner-Ennever, J.A. (2006) Present concepts of oculomotor organization. Prog. Brain Res., 151: 1–42.

Büttner, U., Meienberg, O. and Schimmelpfennig, B. (1983) The effect of central retinal lesions on optokinetic nystagmus in the monkey. Exp. Brain Res., 52: 248–256.

Churchland, A.K. and Lisberger, S.G. (2005) Relationship between extraretinal component of firing rate and eye speed in area MST of macaque monkeys. J. Neurophysiol., 94: 2416–2426.

Cohen, B., Matsuo, V. and Raphan, T. (1977) Quantitative analysis of the velocity characteristics of optokinetic nystagmus and optokinetic after-nystagmus. J. Physiol., 270: 321–344.

Ilg, U.J., Schumann, S. and Thier, P. (2004) Posterior parietal cortex neurons encode target motion in world-centered coordinates. Neuron, 43: 145–151.

Kawano, K. (1999) Ocular tracking: behavior and neurophysiology. Curr. Opin. Neurobiol., 9: 467–473.

Kawano, K., Shidara, M., Watanabe, Y. and Yamane, S. (1994) Neural activity in cortical area MST of alert monkey during ocular following responses. J. Neurophysiol., 71: 2305–2323.

Keating, F. and Pierre, A. (1996) Architecture of a gain controller in the pursuit system. Behav. Brain Res., 81: 173–181.

Komatsu, H. and Wurtz, R.H. (1988) Relation of cortical areas MT and MST to pursuit eye movements. III. Interaction with full-field visual stimulation. J. Neurophysiol., 60: 621–644.

Krauzlis, R.J. (2004) Recasting the smooth pursuit eye movement system. J. Neurophysiol., 91: 591–603.

Lisberger, S.G., Miles, F.A., Optican, L.M. and Eighmy, B.B. (1981) Optokinetic response in monkey: underlying mechanisms and their sensitivity to long-term adaptive changes in vestibuloocular reflex. J. Neurophysiol., 45: 869–890.

Newsome, W.T., Wurtz, R.H. and Komatsu, H. (1988) Relation of cortical areas MT and MST to pursuit eye movements. II. Differentiation of retinal from extraretinal inputs. J. Neurophysiol., 60: 604–620.

Ono, S. and Mustari, M.J. (2006) Extraretinal signals in MSTd neurons related to volitional smooth pursuit. J. Neurophysiol., 96: 2819–2825.

Page, W.K. and Duffy, C.J. (2003) Heading representation in MST: sensory interactions and population encoding. J. Neurophysiol., 89: 1994–2013.

Robinson, D.A. (1965) The mechanics of human smooth pursuit eye movements. J. Physiol. Lond., 180: 569–591.

Schwartz, J.D. and Lisberger, S.G. (1994) Initial tracking conditions modulate the gain of visuo-motor transmission for smooth pursuit eye movements in monkeys. Vis. Neurosci., 11: 411–424.

Shenoy, K.V., Crowell, J.A. and Andersen, R.A. (2002) Pursuit speed compensation in cortical area MSTd. J. Neurophysiol., 88: 2630–2647.

Simons, B. and Büttner, U. (1985) The influence of age on optokinetic nystagmus. Eur. Arch. Psychiatry Neurol. Sci., 234: 369–373.

Tanaka, M. (2005) Involvement of the central thalamus in the control of smooth pursuit eye movements. J. Neurosci., 25: 5866–5876.

Thier, P. and Ilg, U.J. (2005) The neural basis of smooth-pursuit eye movements. Curr. Opin. Neurobiol., 15: 645–652.

Waespe, W. and Henn, V. (1987) Gaze stabilization in the primate. The interaction of the vestibulo-ocular reflex, optokinetic nystagmus, and smooth pursuit. Rev. Physiol. Biochem. Pharmacol., 106: 37–125.

C. Kennard & R.J. Leigh (Eds.)
Progress in Brain Research, Vol. 171
ISSN 0079-6123

CHAPTER 4.4

Neural activity in cortical areas MST and FEF in relation to smooth pursuit gain control

Ulrich Nuding[1,*], Seiji Ono[2], Michael J. Mustari[2], Ulrich Büttner[1] and Stefan Glasauer[1]

[1]*Department of Neurology and Bernstein Center for Computational Neuroscience, Ludwig-Maximilians-University, Munich, Germany*

[2]*Yerkes National Primate Research Center and Department of Neurology, Emory University, Atlanta, GA, USA*

Abstract: Two cortical areas that crucially contribute to the generation and maintenance of smooth pursuit eye movements (SPEM) are the medial superior temporal area (MST) and the pursuit area of the frontal eye fields (FEF). They both project to the brainstem premotor structures via different parallel pathways. A special property of the pursuit system is the increased sensitivity to retinal image motion for increasing pursuit velocities (dynamic gain control), which might be attributed to the FEF. On the other hand, MST is regarded as the main site for estimating the target velocity in space and therefore for maintaining an ongoing SPEM. However, neural activities in these areas during visually driven SPEM have not been fully understood. We present a simple model of smooth pursuit gain control that comprises population activities, as well as single unit sensitivities with respect to eye and retinal image motion in MST and FEF. We show that the nonlinear computation necessary for dynamic gain control may be accomplished by units that exhibit a curvilinear sensitivity to retinal image motion and eye velocity.

Keywords: sensory reweighting; dynamic gain control; spike probability; single-unit recordings

Introduction

Several cortical areas are involved in the control of smooth pursuit eye movements (SPEM; for review see Lisberger et al., 1987; Krauzlis, 2004). Specifically, the medial superior temporal area (MST) and the frontal eye fields (FEF) are crucially involved in the generation and maintenance of SPEM (Newsome et al., 1988; MacAvoy et al., 1991). A noteworthy property is the parallel pathway architecture from higher visual cortical to lower motor areas in the brainstem. For example, the pathway originating in MST projects to the cerebellar floccular complex (e.g., ventral paraflocculus, vPF) via the dorsolateral pontine nuclear region (DLPN), whereas the other major pathway originates in FEF and projects strongly to the cerebellar dorsal vermis (DV) via the nucleus reticularis tegmenti pontis (NRTP). The MST pathway is considered to be mainly responsible for estimating the target movement in world-centred coordinates and for the compensation of self-induced visual motion (Page and Duffy, 2003; Thier and Ilg, 2005), while the FEF pathway is commonly viewed as being crucial for the initiation of visually guided pursuit (MacAvoy et al., 1991). Another major property of the pursuit system is the so-called dynamic gain control (DGC) which can also be

*Corresponding author. Tel.: +49-89-70954833; Fax: +49-89-70954801; E-mail: UNuding@nefo.med.uni-muenchen.de

DOI: 10.1016/S0079-6123(08)00636-5

attributed to the FEF (Tanaka and Lisberger, 2001). This mechanism dynamically adjusts the feed-forward gain, i.e., the sensitivity to retinal image motion, as a function of ongoing tracking velocity. For example, the eyes are largely insensitive to perturbations of a stationary target during fixation. If, however, the target's trajectory is perturbed during an ongoing SPEM, a clear motor response to that perturbation can be observed. Nevertheless, a direct connection between the functional properties of these pathways and their neuronal activities is still lacking. Here, we review a simple model of smooth pursuit that features the dual pathways and DGC. Basic response properties of single units during step-ramp pursuit, as well as more complex response behaviours during tracking of a randomly moving target can be explained qualitatively.

A dual pathway model for smooth pursuit gain control

The basic model for one-dimensional eye movements is shown in Fig. 1(a). The eye plant (modeled as a first-order lowpass with time constant $\tau_E = 200$ ms) is controlled by a feedback loop with two complementary branches. Both utilize eye velocity information as supplied by the efference copy of the eye motor command (note that in Fig. 1 we assume a perfectly accurate internal eye plant model to reconstruct eye velocity from the motor command, therefore the efference copy is directly derived from the actual eye velocity). The MST–DLPN–vPF pathway essentially adds estimated eye velocity to retinal slip, which yields an estimate of current target velocity in extrinsic space. The FEF–NRTP–DV pathway implements the DGC mechanism. As shown in several studies (e.g., Keating and Pierre, 1996; Churchland and Lisberger, 2002), the feedforward gain depends linearly on tracking velocity, therefore retinal slip is multiplied by a rectified linear function of the internal estimate of eye velocity (rectification is justified by the fact that the gain is independent of the direction of target motion). Figure 1(b) shows typical responses associated with DGC which are reproduced correctly by the proposed model. Based on a detailed analysis of the model, which is omitted for brevity, MST carries mainly eye velocity related signals, whereas FEF is mostly associated with eye acceleration.

Neural activities in MST and FEF

To test for the above predicted signal attributes we recorded single units in awake monkeys during

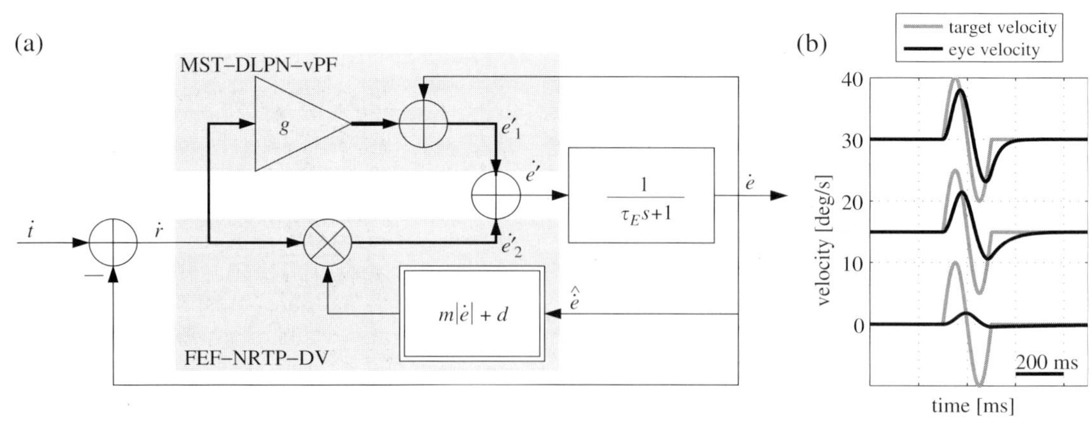

Fig. 1. (a) Scheme of the dual-pathway model for smooth pursuit gain control. The MST–DLPN–vPF pathway receives eye velocity feedback and serves for estimating the target velocity in space by adding ongoing eye velocity to retinal slip input. The FEF–NRTP–DV pathway regulates the gain by weighting the retinal input with a rectified linear function of eye velocity. Parameters were: $m = 0.2$; $d = 0.25$; $g = 0.25$; $\tau_E = 0.2$. (b) Typical responses associated with gain control were reproduced correctly with the model: increasing tracking velocities lead to a greater sensitivity to high-frequency perturbations.

step-ramp smooth pursuit. We analysed the firing rates of eye movement related neurons in the dorsal aspect of MST (MSTd) and in the smooth eye movement region of FEF (FEFsem). (For further details of the procedure, see Büttner et al., this volume.) As can be already conceived from the responses depicted in Fig. 2(a), MST shows a sustained activity during smooth pursuit, whereas FEF activity is characterized by a more transient behaviour. To further quantify this finding, we analysed the firing rates by means of a standard linear regression to explain the neuronal responses as a linear combination of eye velocity $\dot{e}$ and acceleration $\ddot{e}$:

$$r(t) = \beta_0 + \beta_1 \dot{e}(t - T) + \beta_2 \ddot{e}(t - T)$$

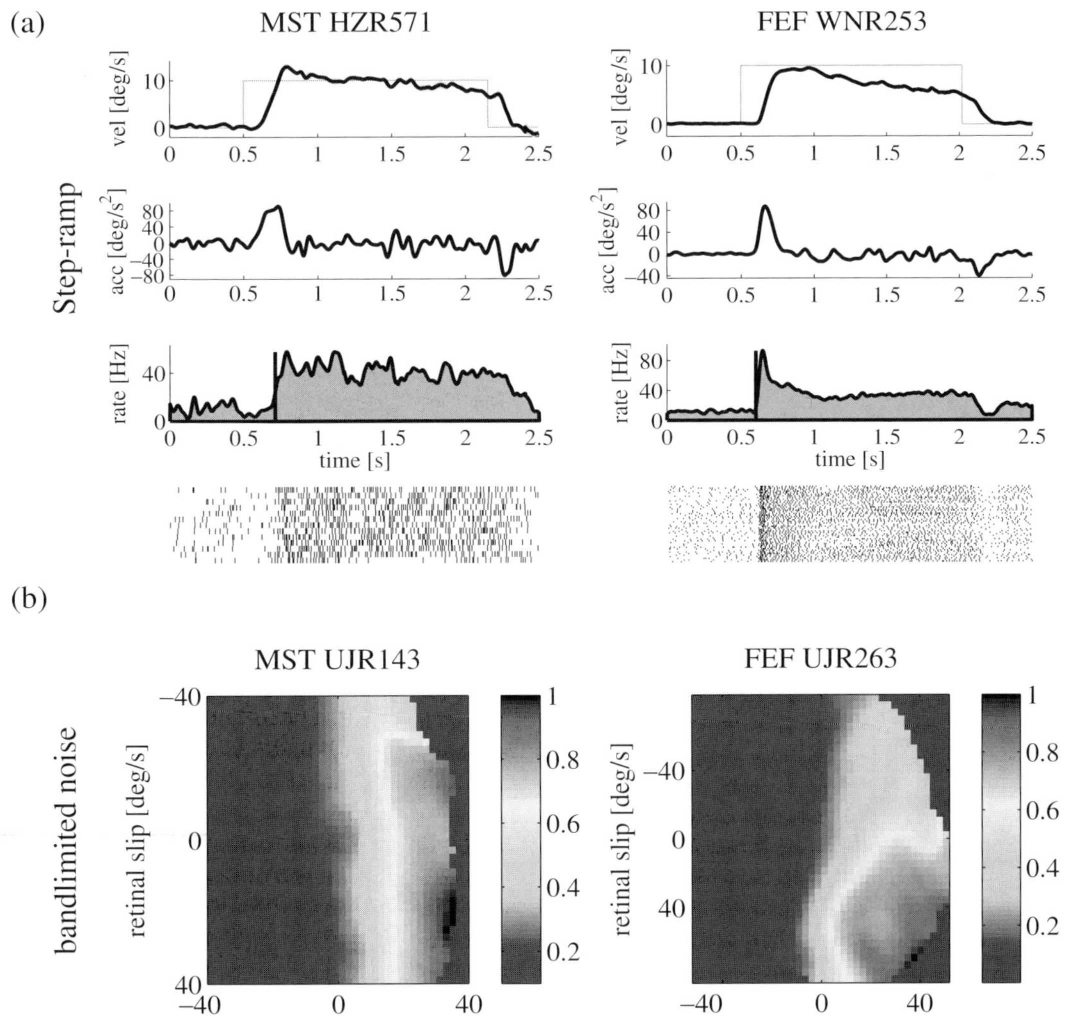

Fig. 2. (a) Typical response behaviour of MST and FEF neurons during step-ramp smooth pursuit. From top to bottom: eye/target velocity, eye acceleration, mean firing rate, spike raster plot. (b) Spike probability (normalized) against eye velocity and retinal slip during pursuit of a bandlimited Gaussian noise stimulus (details see text). Only the FEF neuron shows a curvilinear sensitivity in stimulus space. (See Color Plate 4.4.2 in color plate section.)

where the assumed time delay T between neuronal response and eye motion was optimized to yield the maximal coefficient of determination. We calculated the partial R^2 values for each regressor and found that for 12/15 MSTd neurons a higher partial R^2 value was assigned to eye velocity, whereas for 9/13 FEFsem neurons eye acceleration dependence was more pronounced (with individual R^2 values up to 0.9).

A distinctive feature of the model is the multiplicative interaction between eye velocity and retinal slip in FEF. These sort of nonlinear interactions are likely to be realized in neural substrate by so-called gain fields, as has been suggested by Nuding et al. (2006). As a consequence, FEF units should show a curved sensitivity with respect to retinal slip and eye velocity. This has been tested experimentally in awake monkeys during pursuit of a horizontally moving target whose trajectory was given by bandlimited Gaussian noise with limited maximal position and velocity. From records lasting several minutes, we estimated the spike probability in stimulus space, i.e., the probability of eliciting a spike for each combination of retinal slip and eye velocity. Preliminary results shown in Fig. 2(b) suggest indeed a curved sensitivity of FEF neurons in stimulus space. Furthermore, MST neurons do not exhibit this kind of gain field behaviour, as computation therein is essentially linear with respect to retinal slip and eye velocity (cf. Fig. 1).

Conclusion

A simple model of smooth pursuit which consolidates the parallel pathway architecture and DGC is able to explain the very basic extraretinal response properties of MST and FEF neurons, namely, the relation to eye velocity and acceleration, respectively. Furthermore, the assumed non-trivial multiplicative interaction between eye velocity and retinal slip in FEF can be realized by gain field units whose existence still remain to be proven, but first evidence is provided by our single-unit recordings during pursuit of a Gaussian noise stimulus. Although it seems that FEF is crucial for performing the nonlinear computation associated with DGC, MST may be also substantially involved due to the massive recurrent interactions with FEF.

Acknowledgements

Study supported by BMBF (BCCN Munich 01GQ0440), NEI EY06069, RR00165.

References

Churchland, A.K. and Lisberger, S.G. (2002) Gain control in human smooth pursuit eye movements. J. Neurophysiol., 87: 2936–2945.

Keating, E. and Pierre, A. (1996) Architecture of a gain controller in the pursuit system. Behav. Brain Res., 81: 173–181.

Krauzlis, R.J. (2004) Recasting the smooth pursuit eye movement system. J. Neurophysiol., 91: 591–603.

Lisberger, S.G., Morris, E.J. and Tychsen, L. (1987) Visual motion processing and sensory-motor integration for smooth pursuit eye movements. Annu. Rev. Neurosci., 10: 97–129.

MacAvoy, M., Gottlieb, J. and Bruce, C. (1991) Smooth-pursuit eye movement representation in the primate frontal eye field. Cereb. Cortex, 1: 95–102.

Newsome, W., Wurtz, R. and Komatsu, H. (1988) Relation of cortical areas MT and MST to pursuit eye movements. II. Differentiation of retinal from extraretinal inputs. J. Neurophysiol., 60: 604–620.

Nuding, U., Ono, S., Mustari, M.J., Büttner, U. and Glasauer, S. (2006) Smooth pursuit gain control: network simulations and experimental analysis. Program no. 735.15. In: Abstract Viewer and Itinerary Planner. Society for Neuroscience, Atlanta, GA.

Page, W.K. and Duffy, C.J. (2003) Heading representation in MST: sensory interactions and population encoding. J. Neurophysiol., 89: 1994–2013.

Tanaka, M. and Lisberger, S.G. (2001) Regulation of the gain of visually guided smooth pursuit eye movements by frontal cortex. Nature, 409: 191–194.

Thier, P. and Ilg, U.J. (2005) The neural basis of smooth-pursuit eye movements. Curr. Opin. Neurobiol., 15: 645–652.

C. Kennard & R.J. Leigh (Eds.)
Progress in Brain Research, Vol. 171
ISSN 0079-6123

CHAPTER 4.5

Eye position and cross-sensory learning both contribute to prism adaptation of auditory space

Qi N. Cui[1], Laura Bachus[1], Eva Knoth[1], William E. O'Neill[1,2] and Gary D. Paige[1,2,3,*]

[1]*Department of Neurobiology and Anatomy, University of Rochester School of Medicine and Dentistry, Rochester, NY, USA*
[2]*Center for Navigation and Communication Sciences, University of Rochester School of Medicine and Dentistry, Rochester, NY, USA*
[3]*Center for Visual Science, University of Rochester School of Medicine and Dentistry, Rochester, NY, USA*

Abstract: Optical prisms shift visual space, and through adaptation over time, generate a compensatory realignment of sensory-motor reference frames. In humans, prism-induced lateral shifts of visual space produce a corresponding shift in sound localization. We recently reported that sound localization shifts towards eccentric eye position, approaching ~40% of gaze over several minutes. Given that eye position affects sound localization directly, prism adaptation may well reflect contributions of both eye position and sensory adaptation; while the visual world is shifted by the prisms, the eyes must also shift simply to gaze ahead. To test this new concept of prism adaptation, 10 young (18–27 year) adults localized sound targets before and after 4 h of adaptation to base-right or base-left prisms that induced an 11.4° shift left or right, respectively. In separate sessions subjects were exposed to: (1) natural binaural hearing; (2) diotically presented inputs devoid of meaningful spatial cues; or (3) attenuated hearing to simulate hearing loss. These preliminary results suggest that the prism adaptation of auditory space is dependent on two independent influences: (1) the effect of displaced mean eye position induced by the prisms, which occurs without cross-sensory experience; and (2) true cross-sensory learning in response to an imposed offset between auditory and visual space.

Keywords: sound localization; prism; adaptation; eye movement; spatial perception; gaze; multi-sensory

Introduction

Optical prisms shift visual space, and this in turn induces an adaptive realignment of sensory-motor reference frames (Redding et al., 2005). For auditory–visual coordination in particular, prism-induced lateral shifts of visual space produce a corresponding shift in sound localization in the barn owl (Knudsen and Knudsen, 1985; Brainard and Knudsen, 1995). Although prism adaptation of auditory space has also been studied in humans, it is unclear whether the resultant shift in spatial localization can be attributed to an adaptive response in the auditory system (Lackner, 1976). Recently, we reported that sound localization shifts by simply maintaining eccentric eye position. This shift is robust, time-dependent, and spatially broad; it develops exponentially over

*Corresponding author. Tel.: 585-275-6395/275-2591; Fax: 585-442-8766; E-mail: gary_paige@urmc.rochester.edu

DOI: 10.1016/S0079-6123(08)00637-7

minutes in the direction of ocular deflection and approaches ~40% of eye eccentricity (Razavi et al., 2007).

Given that eye position itself affects sound localization, prism adaptation of auditory space may well reflect contributions of both eye position and cross-sensory adaptation. Previous examinations of auditory–visual coherence did not take into account the effect of optical prisms on ocular deviation. Specifically, while the visual world is shifted in the direction dictated by the prisms, the eyes must also shift in the same direction simply to fixate the same field of targets. Even though the eyes are free to move, *average* eye position will shift over time, and this alone, apart from any shift due to auditory–visual interactions, will cause a corresponding shift in sound localization.

More formally, we hypothesize that the prism adaptation of auditory space is comprised of two components: (1) a *physiologic adaptation* of auditory space in direct response to a new average eye position; and (2) cross-sensory *adaptive plasticity*, an experience-dependent learning phenomenon that recalibrates auditory to visual space over time. In this preliminary experiment, we revisited the paradigm of prism adaptation to test this hypothesis directly and to quantify the efficacy of these two components. The effect of eye position was examined in isolation by eliminating meaningful binaural localization cues normally present in auditory signals while wearing prisms, thereby disrupting cross-sensory re-calibration. By comparing this adaptation paradigm to that with normal binaural hearing, the contribution of eye position to the overall phenomenon was quantified, while the remaining difference presumably constitutes a cross-sensory learning effect.

Methods

Subjects

Ten normal human subjects (4 male, 6 female; 18–27 years old) participated in this study.

Apparatus and stimulus

Subjects sat in a dark, echo-attenuated room facing the centre of a cylindrical screen of black speaker cloth at 2 m distance. The head was aligned with the horizontal plane using Reid's baseline, and fixed in place using a personalized bite-bar. Acoustic targets were presented using an 8-cm-diameter two-way coaxial speaker mounted on a two-axis servo-controlled robotic arm hidden behind the screen. The setup enabled rapid positioning of the speaker in cylindrical coordinates and provided an unlimited array of possible targets over the range of $\pm 65^\circ$ azimuth (Az) and $\pm 25^\circ$ elevation (El). Spatially diffuse Gaussian white noise (65 dB sound pressure level, SPL), delivered through two stationary loudspeakers, masked potential predictive positional cues during speaker movements between localization trials. Auditory targets consisted of 150 ms bursts (10 ms rise–fall time) of broadband (0.1–20 kHz) Gaussian noise (equalized to compensate speaker frequency response), repeating at 5 Hz, and randomly varied between 70 dB and 75 dB SPL from trial-to-trial (Razavi et al., 2007).

Experimental paradigm and response measures

Subjects manually localized stationary auditory targets using a red laser-LED pointer mounted on a 2-axis cylindrical joystick, aligning its beam with the perceived sound locations. For each target presentation, subjects registered response endpoint with a key press, and the target and pointer positions were recorded. Subjects were instructed to localize quickly but accurately. Auditory localization was studied before, during, and after 4 h of adaptation to either base-right (R) or base-left (L) prisms (20 prism-diopters) that induced an 11° shift L or R, respectively. For each session, a normal baseline of sound localization was first established without prisms or other devices. Subjects then donned the prisms for 4 h, during which time they engaged in normal active behaviour in and around the University of Rochester Medical Center (noise level $\leq$ 90 dB SPL), and returned to the laboratory for repeat testing after 1 and 4 h. Testing occurred without prisms or other devices (always removed or restored while subjects were on the bite-bar with eyes closed). In separate sessions, subjects were exposed to: (1) natural *binaural* hearing; (2) *diotic* hearing (portable Lavalier microphone-amplifier presented the same monaural signal to both ears through earphones); and (3) sound-attenuation (44 ± 5 dB SPL) using earmuffs and earplugs (*near-deafness*). Twenty-six randomly distributed target locations were tested (repeated measures design; Fig. 1a). To complete all three acoustic conditions under both prism conditions (L and R), six sessions on different days at least 2 days apart were required per subject.

To effectively separate eye position effects from those of cross-sensory interaction, it is important to control for the availability of auditory localization cues. Control sessions examined the efficacy of the lavalier microphone–amplifier (*diotic* condition) and the earplugs–earmuffs (*near-deafness* condition) in eliminating effective auditory spatial cues. Linear regression was performed to quantify *spatial gain* (slope) of response relative to target positions across horizontal space. As would be expected following the elimination of useable cues, spatial gains in Az fell to near zero for both *diotic* (0.05 ± 0.07) and *near-deafness* (0.03 ± 0.21) conditions, in contrast to a normal binaural baseline of 1.19 (± 0.12, or a 19% overshoot of target positions).

Data analysis

Data were sorted by prism direction (L or R shift), auditory input (*binaural*, *diotic*, or *near-deafness*), and adaptation duration (baseline, 1, and 4 h). *Accuracy*, the error between response and target locations, was normalized to baseline localization for each target. The effect of prism adaptation was quantified as *shift magnitude* (Δ accuracy in Az) between R and L prism directions across all targets.

Results

Visual prisms consistently shifted sound localization in the direction of visual deviation. The localization accuracy at 1 h differed significantly from baseline for both *diotic* and *binaural* conditions ($p < 0.01$). The shift magnitude (Δ accuracy in Az; R minus L) in the normal *binaural* condition increased over time between 1 h ($5.92 \pm 2.32°$, mean ± SD) and 4 h ($10.4 \pm 2.60°$, or ~ half the visual shift) of prism adaptation (ANOVA, $p < 0.01$). In contrast, sound localization also shifted under *diotic* conditions, but did not significantly increase between 1 h ($4.62 \pm 2.65°$) and 4 h ($6.52 \pm 2.94°$; $p = 0.1$) of prism adaptation. Interestingly, the small augmentation in shift magnitude between the two time points of the *diotic* condition was restricted to the central region of head-centred auditory space, demonstrating a spatial selectively not present in the *binaural* condition (Fig. 1b). Note that the shift magnitudes were similar between the two conditions after 1 h of prism adaptation ($p = 1.00$), but differed significantly after 4 h ($p = 0.02$; Fig. 1b, c). Additionally, the shift magnitude in the *near-deafness* condition ($3.89 \pm 3.39°$) was comparable to that of both *diotic* ($p = 1.00$) and *binaural* ($p = 0.63$) conditions following 1 h of prism adaptation.

Six subjects (2 male, 4 female; 21–27 years old) also participated in an *alternating fixation* paradigm (Razavi et al., 2007). In this paradigm, head-fixed subjects maintained ocular fixation on one of three red laser-LED spots projected on the screen (centre, L20°, or R20° Az; 0° El), and used peripheral vision to guide the pointer to localize auditory targets. Two sessions (on different days) of 161 trials were parsed into 5 separate but contiguous epochs (Fig. 2). Sessions began and ended with an epoch of central fixation, interjected by a sequence of three epochs of ocular eccentricities that reversed order between sessions. The paradigm measured the shift's dynamics in response to a ±20° (e.g., epochs 1→2, 4→5) as well as a ±40° (e.g., epochs 2→3, 3→4) change in eye position. The time course and amplitude of the shift in sound localization in response to a change in eye position was parameterized using the first-order exponential equation: $y(t) = y_0 + a(1 - e^{-1/\tau})$. Interestingly, results demonstrated that the amplitude of the shift (a, from starting point to asymptote) for a given change in eye position correlated closely with the 1 h shift magnitude in the *diotic* condition under prisms (0.88; $p = 0.02$).

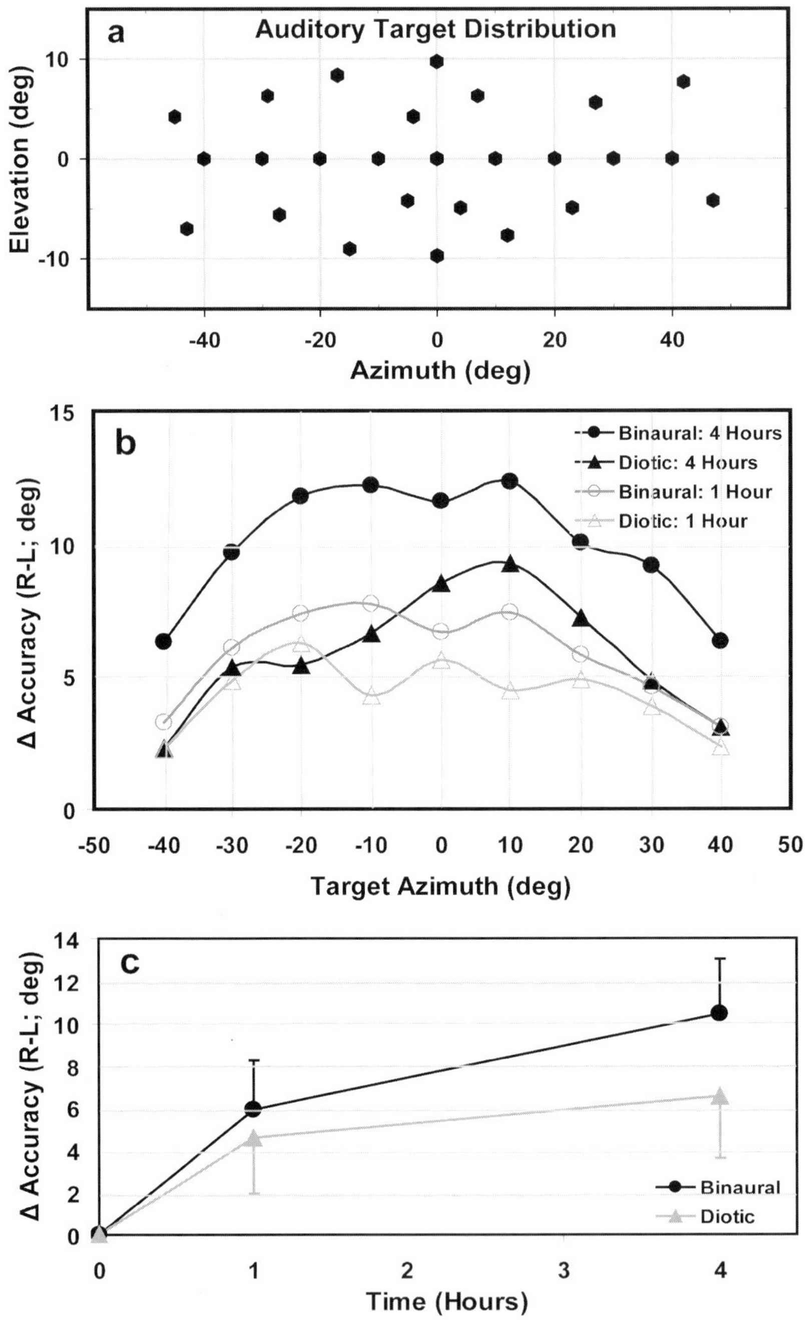

Fig. 1. (a) Auditory target distribution. (b) Mean change in Az localization accuracy (shift magnitude) for both *diotic* and *binaural* hearing conditions following 1 and 4 h of prism adaptation (R and L combined), binned in 10° intervals of target Az. (c) Average shift magnitude across subjects, pooled for all target locations (error bars are SDs).

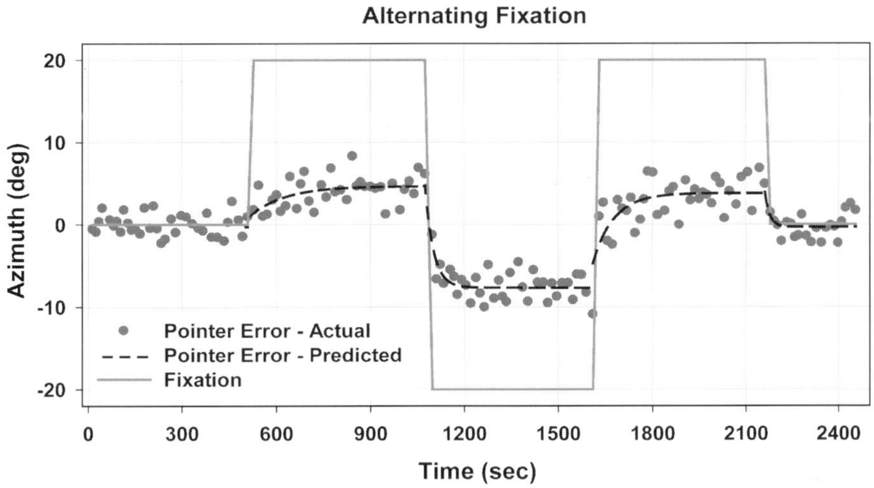

Fig. 2. Sound localization accuracy (error) across target locations in one subject. Fixation (solid trace) started at centre and then alternated between R and L 20° in Az. The dashed trace reflects the exponential model. Mean shift across epochs (including session 2) measured 7.96° in this subject.

Discussion

The phenomenon of visual prism adaptation was re-evaluated in a context that includes a newly described phenomenon of a gaze-dependent auditory localization shift (Razavi et al., 2007). Preliminary findings support the notion that the prism adaptation of auditory space is dependent on two independent influences: (1) the effect of displaced mean eye position induced by the prisms, which occurs without cross-sensory experience; and (2) true cross-sensory learning in response to an imposed offset between auditory and visual space.

We have shown previously that the shift in sound localization in response to eye position increases exponentially over time at a highly variable rate among subjects, eventually approaching ~40% of gaze in a matter of minutes. Because the temporal dynamics of the gaze-dependent responses are unknown beyond 30–40 min, we are uncertain whether the small increase in the *diotic* shift magnitude for longer periods is attributable to additional drift in sound localization towards gaze or the change in shape across space (central progression only). Nevertheless, at 4 h, adaptation of auditory space under *binaural* conditions showed continued wide-field growth far exceeding that for *diotic* conditions, suggesting that cross-sensory learning accounts for the additional magnitude and progression of adaptation, beyond that for diotic hearing. Additional support for our hypothesis stems from the close correlation between the shift in sound localization after 1 h of prism adaptation and the shift amplitude during the alternating fixation paradigm. The latter constitutes a pure assessment of the auditory spatial shift induced by ocular eccentricity.

Acknowledgements

We thank Babak Razavi, Marina Dobreva, Martin Gira, and John Housel for technical assistance and valuable insights. This work was supported by National Institutes of Health (NIH) — National Institute on Aging Grant R01-AG16319 and NIH — National Institute on Deafness and Other Communication Disorders (NIDCD) Grant P30-DC05409. Q. N. Cui was supported by training grants from the NIH — NIDCD (F30-DC009372) and NIH — National Institute of General Medical Sciences (T32-GM07356).

References

Brainard, M.S. and Knudsen, E.I. (1995) Dynamics of visually guided auditory plasticity in the optic tectum of the barn owl. J. Neurophysiol., 73(2): 595–614.

Knudsen, E.I. and Knudsen, P.F. (1985) Vision guides the adjustment of auditory localization in young barn owls. Science, 230(4725): 545–548.

Lackner, J.R. (1976) Influence of abnormal postural and sensory conditions on human sensorimotor localization. Environ. Biol. Med., 2(3): 137–177.

Razavi, B., O'Neill, W.E. and Paige, G.D. (2007) Auditory spatial perception dynamically realigns with changing eye position. J. Neurosci., 27(38): 10249–10258.

Redding, G.M., Rossetti, Y. and Wallace, B. (2005) Application of prism adaptation: a tutorial in theory and method. Neurosci. Biobehav. Rev., 29(3): 431–444.

C. Kennard & R.J. Leigh (Eds.)
Progress in Brain Research, Vol. 171
ISSN 0079-6123

CHAPTER 4.6

Hysteresis effects of the subjective visual vertical during continuous quasi-static whole-body roll rotation

A. Palla*, M. Tatalias and D. Straumann

Neurology Department, Zurich University Hospital, Zurich, Switzerland

Abstract: Healthy human subjects, when roll tilted in darkness, make systematic errors in estimating subjective visual vertical (SVV). Typically, roll tilt underestimation occurs at angles beyond 60° (A-effect). At smaller tilt angles, overestimation may occur (E-effect). At ~135° whole-body roll tilt, Kaptein and Van Gisbergen (2004, 2005) found an abrupt A/E transition, the exact location of which depended on the preceding rotation direction indicating hysteresis. Since this was observed using relatively fast roll velocity, it remains unclear whether the described hysteresis is dynamic or static. To clarify this uncertainty, we continuously rotated nine healthy subjects about the earth-horizontal naso-occipital axis, while they performed SVV adjustments every 2 s. Starting from the upright position, three full quasi-static constant velocity rotations (2°/s) were completed in both directions (CW: clockwise; CCW: counterclockwise). SVV deviation from earth-verticality was plotted as a function of whole-body roll position. A bimodal Gaussian distribution function was fitted to SVV differences between CW and CCW rotations. A-effects (peaks at 88° and 257° chair position) at identical whole-body positions were larger after rotations from upside-down than after rotations from upright (average peak difference: 26°). These results demonstrate static hysteresis for SVV estimation.

Keywords: spatial orientation; visual vertical; earth vertical; vestibular system; vestibular psychophysics

Introduction

Healthy human subjects, when tilted along the roll plane in darkness, make systematic errors in estimating visual verticality. Typically, roll tilt underestimation occurs at angles beyond 60° (Aubert, 1861), i.e., the subjective visual vertical (SVV) deviates from the 'true' gravitational vertical in the direction of the subject's head or whole-body tilt (A-effect). At smaller roll angles, overestimation, i.e., SVV deviation from 'true' vertical in the opposite direction of head or whole-body roll (E-effect), may occur (Müller, 1916). Contrary to previous descriptions of a gradual increase and decrease of the A-effect peaking at ~130° whole-body roll (Mittelstaedt, 1983; Van Beuzekom and Van Gisbergen, 2000), Kaptein and Van Gisbergen (2004) recently reported of an abrupt transition from the A- to the E-effect approximately at 135°. Remarkably, the transition zone occurred at different whole-body tilting positions depending on the direction of the preceding rotation, which

*Corresponding author. Tel.: +41-1-255-5500;
Fax: +41-1-255-4507; E-mail: antpalla@access.unizh.ch

DOI: 10.1016/S0079-6123(08)00638-9

suggests hysteresis properties of SVV (Kaptein and Van Gisbergen, 2005). Since, however, this phenomenon was observed using relatively fast roll velocities (30°/s), it remains unclear whether the hysteresis is dynamic (i.e., velocity dependent) or static (i.e., position dependent). The present study was undertaken to clarify this ambiguity by an SVV paradigm during very slow, i.e., quasi-static, continuous whole-body roll rotations.

Methods

Nine healthy human subjects (4 females; 23–45 years old) participated in this study. Informed consent of all subjects was obtained after full explanation of the experimental procedure. Subjects were seated upright on a turntable with three servo-controlled motor-driven axes (prototype built by Acutronic, Jona, Switzerland). An individually moulded thermoplastic mask (Sinmed BV, Reeuwijk, The Netherlands) restrained the head. Subjects were positioned so that the intersection of the interaural and naso-occipital axes was at the intersection of the three axes of the turntable. Experiments were performed in a dark environment. A chair-fixed laser projected a visual line with an arrowhead (length: 500 mm; width: 3 mm) onto the centre of a sphere located 1.5 m away from the subject's eyes. Chair position and line orientation signals were digitized at 200 Hz per channel with 12-bit resolution, and stored on a computer hard disc for offline processing. We used the same turntable experimental protocol as in our previous study on hysteresis of static ocular counterroll (Palla et al., 2006). Starting from the upright position, subjects were rotated about their earth-horizontal naso-occipital axis counterclockwise (CCW, positive torsional direction) or clockwise (CW, negative torsional direction) at a constant angular velocity of 2°/s. CCW and CW were defined from the subject's viewpoint. The velocity plateau was reached by turntable acceleration of $0.05°/s^2$, which is below the detection threshold of the semicircular canals (Shimazu and Precht, 1965; Diamond et al., 1982). A total of three consecutive 360° turntable rotations were performed. Subjects were instructed to rapidly (≤ 2 s) adjust the orientation of the visual line earth-vertical, while turntable was rotated at constant angular velocity. The roll orientation of the visual line was pseudo-randomly offset every 2 s either CW or CCW (random offset range: 0°–360°). Chair and visual line signals were processed with interactive programmes written in MATLAB™ (MathWorks, Natick, MA). SVV deviation from the earth-vertical was plotted as a function of CCW and CW whole-body roll position. SVV deviation from 'true' gravitational vertical was defined as positive when it deviated in the CCW direction, and as negative when it deviated in the CW direction from the subject's viewpoint. The following bimodal Gaussian distribution function was fitted to SVV differences between CW and CCW rotations:

$$y = a_1 e^{(x-b_1)^2/(2c_1)^2} + a_2 e^{(x-b_2)^2/(2c_2)^2} + d$$

where a_1, a_2 are the amplitudes, b_1, b_2 the centres, and c_1, c_2 the widths of the two Gaussian peaks, and d the offset.

Results

Figure 1 shows response errors of the visual line setting to the SVV plotted against whole-body roll position in one subject (M.T.). The three traces represent the responses for each of the three turntable rotation cycles performed in CCW and CW directions. Note that the definition of SVV deviation (see Methods section) implies that the A-effect from upright is positive for CCW and negative for CW rotations. Likewise, the A-effect from upside-down is negative for CCW and positive for CW rotations. For both CCW (upper panel) and CW (lower panel) turntable rotations, the following pattern was observed: 1. a gradual increase of the A-effect during the first 90° of whole-body roll; 2. a gradual decrease of the A-effect during the last 90° (beyond 270°) of whole-body roll; 3. a zone of instability with switches between A- and E-effects between 90° and 270°, i.e., in head-down positions. If SVV traces were identical for CCW (upper panel from left to right) and CW (lower panel from right to left) rotations, the trace of differences between these two traces should be a horizontal zero line.

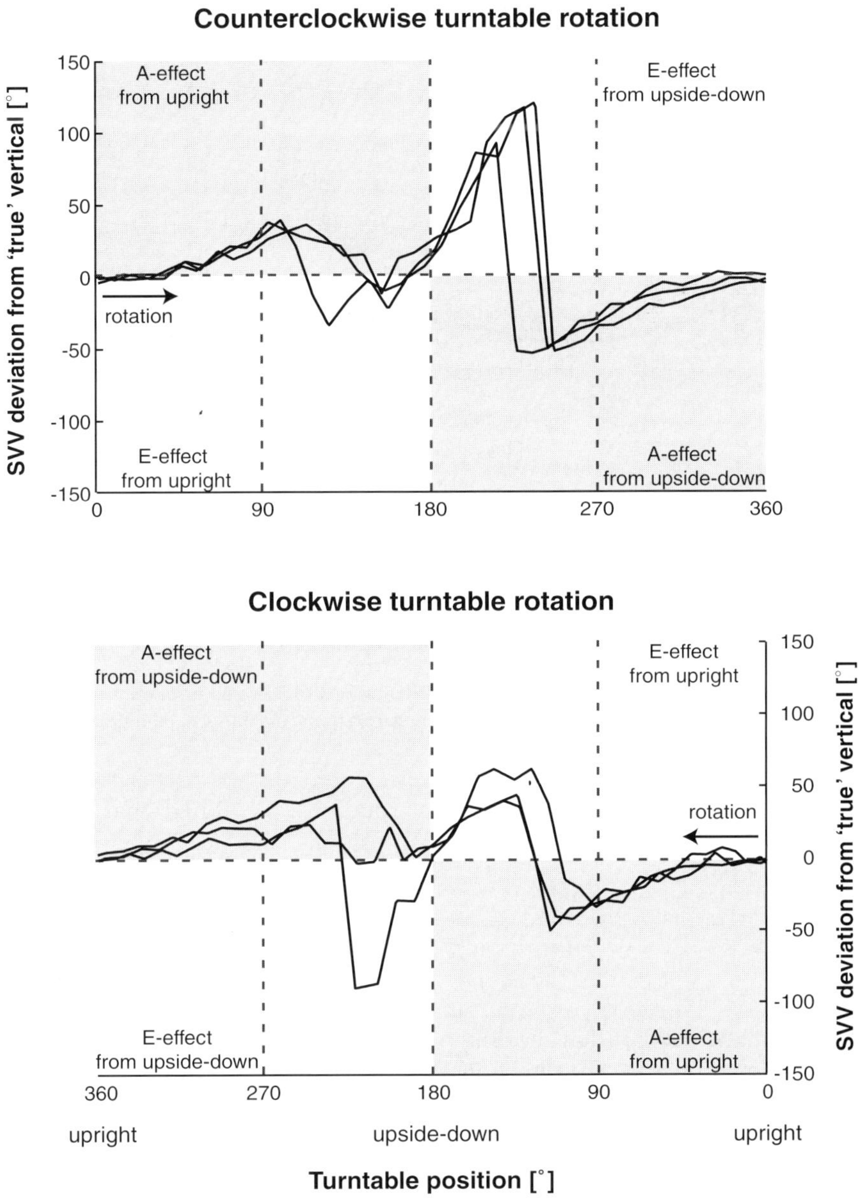

Fig. 1. Example of SVV deviation from 'true' vertical plotted as a function of whole-body roll position in a typical subject (M.T.). Each trace ($N = 3$) represents interpolated SVV adjustments during one full rotation cycle. Upper panel: CCW turntable rotation; lower panel: CW rotation. CW traces are flipped (read from right to left) for comparison of identical CCW and CW whole-body positions.

Figure 2 plots the differences between SVV deviations during CCW and CW rotations (circles: individual averages of 10° wide bins from three rotations) as a function of whole-body roll position in all nine subjects. A bimodal Gaussian distribution function (see Methods section) was fitted to SVV differences (black trace). The two dips at 88° and 257° indicate that A-effects at identical whole-body positions were larger after rotation from upside-down than after rotation

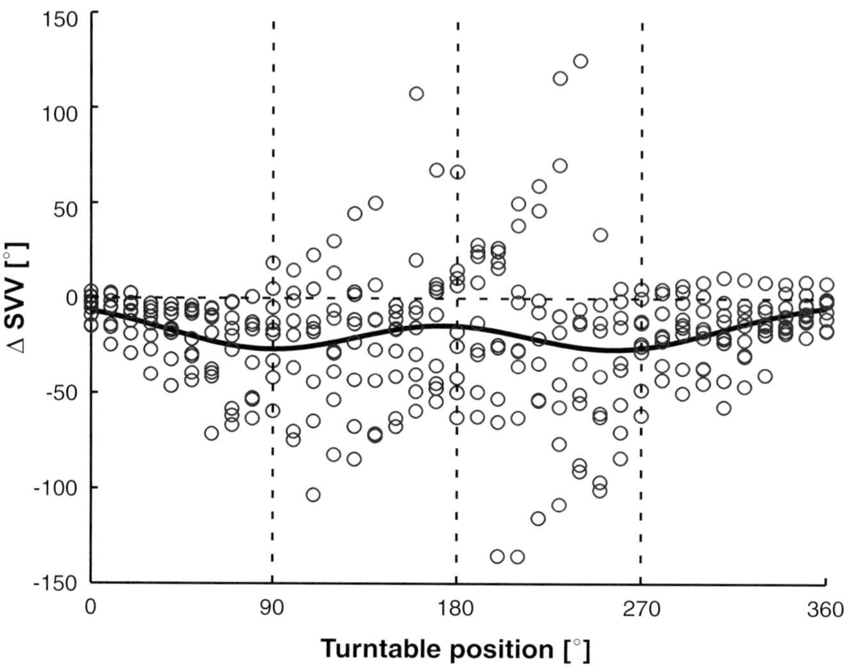

Fig. 2. Differences between SVV deviations during CCW and CW rotations (ΔSVV) as a function of whole-body roll position in subjects ($N=9$). Circles: individual averages of 10° wide bins from three rotations; black trace: bimodal Gaussian distribution function fitted to the data.

from upright. The average peak difference between SVV traces during CCW and CW rotations was 26° [CI (35.14, 18.07), $p<0.05$].

Discussion

We analysed the systematic errors that healthy human subjects make when estimating the direction of gravity by setting a visual line (SVV) during constant 'quasi-static' velocity whole-body rotations about the naso-occipital axis.[1] The majority of subjects showed an instability zone around the upside-down position between 90° left-ear and 90° right-ear down where rapid switches between tilt underestimation (A-effect) and tilt overestimation (E-effect) occurred. When comparing SVV deviations measured during CCW and CW turntable rotations, we found that at identical whole-body positions the A-effect was larger when passing through upside-down than by 'direct' rotation from upright (peaks in the side positions). Because rotations were performed 'quasi-statically,' these results demonstrate that there is hysteresis for SVV estimation that is independent of roll velocity. Moreover, this static SVV hysteresis is already present at roll angles below 90°, where the A-effect dominates. We speculate that the hysteresis found by Kaptein and Van Giesbergen (2004, 2005) for the switching between A- and E-effects around 135° is a direct result of static hysteresis for the A-effect. In the instability zone, the direction-dependent magnitude of the A-effect may influence the roll position for switching between A- and E-effects.

[1]'Quasi-static' is defined as low constant-velocity whole-body rotations (2°/s) about the earth-horizontal naso-occipital axis with the initial acceleration below the detection threshold of the semicircular canals (0.05°/s^2).

Acknowledgements

This study was supported by the Swiss National Science Foundation, the Koetser Foundation for Brain Research, the Baasch Medicus Foundation, and the Center for Integrative Human Physiology, University of Zurich, Switzerland.

References

Aubert, H. (1861) Eine scheinbare bedeutende Drehung von Objekten bei Neigung des Kopfes nach rechts oder links. Arch. Pathol. Anat., 20: 381–393.

Diamond, S.G., Markham, C.H. and Furuya, N. (1982) Binocular counterrolling during sustained body tilt in normal humans and in a patient with unilateral vestibular nerve section. Ann. Otol. Rhinol. Laryngol., 91: 225–229.

Kaptein, R.G. and Van Gisbergen, J.A. (2004) Interpretation of a discontinuity in the sense of verticality at large body tilt. J. Neurophysiol., 91: 2205–2214.

Kaptein, R.G. and Van Gisbergen, J.A. (2005) Nature of the transition between two modes of external space perception in tilted subjects. J. Neurophysiol., 93: 3356–3369.

Mittelstaedt, H. (1983) A new solution to the problem of the subjective vertical. Die Naturwissenschaften, 70: 272–281.

Müller, G.E. (1916) Über das Aubertsche Phänomen. Z. Sinnesphysiol., 49: 109–246.

Palla, A., Bockisch, C.J., Bergamin, O. and Straumann, D. (2006) Dissociated hysteresis of static ocular counterroll in humans. J. Neurophysiol., 95: 2222–2232.

Shimazu, H. and Precht, W. (1965) Tonic and kinetic responses of cat's vestibular neurons to horizontal angular acceleration. J. Neurophysiol., 28: 991–1013.

Van Beuzekom, A.D. and Van Gisbergen, J.A. (2000) Properties of the internal representation of gravity inferred from spatial-direction and body-tilt estimates. J. Neurophysiol., 84: 11–27.

C. Kennard & R.J. Leigh (Eds.)
Progress in Brain Research, Vol. 171
ISSN 0079-6123

CHAPTER 4.7

Perception of self motion during and after passive rotation of the body around an earth-vertical axis

N. Sinha[3], N. Zaher[4], A.G. Shaikh[1], A.G. Lasker[1], D.S. Zee[1,*] and A.A. Tarnutzer[1,2]

[1]*The Johns Hopkins University, School of Medicine, Baltimore, MD, USA*
[2]*Zurich University Hospital, Zurich, Switzerland*
[3]*The Physiological Laboratory, University of Cambridge, Cambridge, England*
[4]*Tishreen University, Lattakia, Syria*

Abstract: We investigated the perception of self-rotation using constant-velocity chair rotations. Subjects signalled self motion during three independent tasks (1) by pushing a button when rotation was first sensed, when velocity reached a peak, when velocity began to decrease, and when velocity reached zero, (2) by rotating a disc to match the perceived motion of the body, or (3) by changing the static position of the dial such that a bigger change in its position correlated with a larger perceived velocity. All three tasks gave a consistent quantitative measure of perceived angular velocity. We found a delay in the time at which peak velocity of self-rotation was perceived (2–5 s) relative to the beginning or to the end of chair rotation. In addition the decay of the perception of self-rotation was preceded by a sensed constant-velocity interval or plateau (9–14 s). This delay in the rise of self-motion perception, and the plateau for the maximum perceived velocity, contrasts with the rapid rise and the immediate decay of the angular vestibuloocular reflex (aVOR). This difference suggests that the sensory signal from the semicircular canals undergoes additional neural processing, beyond the contribution of the velocity-storage mechanism of the aVOR, to compute the percept of self-motion.

Keywords: vestibular; vestibulo-ocular reflex; semicircular canals; velocity storage; perception; self-motion; rotation; sensation

Introduction

Compared to the angular vestibuloocular reflex (aVOR), relatively little is known about the mechanisms underlying vestibular perception. A constant-velocity rotation around an earth-vertical axis in darkness leads to an exponential decay of horizontal slow-phase velocity, typically with a time constant of ~15–20 s (Cohen et al., 1981). This time constant is considerably larger than that estimated from the physical properties of the SCC (~4 s, Gizzi and Harper, 2003) or from recordings of activity in the vestibular nerve in monkeys (~7 s, Büttner and Waespe, 1981). It is thought that this process is carried out by a central 'velocity-storage' integrator that perseverates the ocular motor response to a constant-velocity head rotation and so improves the low-frequency response of the aVOR (Raphan et al., 1979). A key question is whether perception of rotation and the aVOR share the same velocity-storage mechanism. Okada et al. (1999) recorded both the aVOR and perceived angular velocity

*Corresponding author. Tel.: +1 410 955 3319;
Fax: +1 216 231 3461; E-mail: dzee@dizzy.med.jhu.edu

DOI: 10.1016/S0079-6123(08)00639-0

simultaneously during constant-velocity yaw-axis rotation and found that the shape (fast rise followed by an exponential decay) and time constant of both responses were similar, suggesting a common vestibular signal generated in a single brainstem velocity-storage integrator. In contrast, we have found that the dynamic characteristics of the aVOR and perceived rotation are different, and suggest that the perception of angular rotation undergoes additional neural processing beyond the velocity-storage mechanism involved in generating the aVOR (unpublished data). Because of this inconsistency we investigated subjects using three methods for reporting their perception of rotational velocity.

Methods

Fifteen naïve healthy human subjects (8 males, 7 females, age 22–55 years) participated. They were seated upright on a single-axis motorized rotating chair with the head restrained. All experiments were performed in total darkness. A disc equipped with a potentiometer placed on the right armrest was used to record perceived velocity of rotation. Next to the disc a button was used to report the occurrence of certain events. Subjects were accelerated to a velocity of 100 deg/s within 1.3 s and their per-rotatory responses were recorded for 75 s. Then the chair was stopped within 1.2 s and post-rotatory responses were recorded. In Experiment A, subjects reported each of four different time points by pressing a button at their: (i) first sensation of motion, (ii) sensation of peak velocity, (iii) first sensation of a decrease in velocity, and (iv) sensation that the chair had stopped rotating (see Fig. 1A). In Experiment B, subjects reported their sensation of spinning by rotating a disc just as they felt their body was rotating (see Fig. 1B). Subjects increased the displacement of the disc from the zero position as they felt their perceived velocity increasing. Both the time to peak disc velocity and the time constant, *tau*, of the decay in velocity were calculated in Experiment B. In Experiment C subjects reported their perceived rotation using a 'speedometer.' The maximum deflection of the disc represented the subject's sensed peak rotational velocity (see Fig. 1C). We calculated the disc deflection values corresponding to 10 and 90% of this maximum. The time interval needed to increase from the 10th percentile to the 90th percentile of sensed peak rotational velocity was taken as a quantitative measure of the perceptual 'rise time.' The interval between increasing above the 90th percentile and decreasing below the 90th percentile of sensed peak rotational velocity was taken as a measure of perceptual 'plateau time.' Finally, the interval to decrease from the 90th percentile below the 10th percentile (decay time) was determined as the 'decay time'.

Results

In Experiment A, first sensation of motion (i.e., reaction time) was 0.4 ± 0.2 s (both per- and post-rotatory) on average after the chair began moving (see Fig. 2A). Peak velocity (i.e., perceptual rise time) was sensed after 4.7 ± 2.3 s (per-rotatory) and 3.9 ± 2.0 s (post-rotatory), respectively. First decrease in perceived velocity was significantly delayed in per- compared to post-rotatory trials (18.8 ± 6.8 s vs. 13.2 ± 4.6 s; $p < 0.001$). A perceptual 'plateau time' (time from peak to first decay in perceived rotation) was 14.2 ± 5.9 s (per-rotatory) and 9.4 ± 3.7 s (post-rotatory) and sensed rotation stopped after 37.3 ± 10.8 s (per-rotatory) and 33.1 ± 8.3 s (post-rotatory). Whereas the rise-time in Experiment B was similar for both per- and post-rotatory trials (2.6 ± 1.6 s vs. 2.3 ± 1.3 s), time constant *tau* was significantly longer for per- compared to post-rotatory trials (18.2 ± 4.8 s vs. 12.6 ± 4.5 s, $p < 0.001$), as seen in Fig. 2B. Overall, post-rotatory *tau* values were in the range of those reported by Okada et al. (1999) (16 s on average). Compared to Experiment B, average rise-times (per-rotatory vs. post-rotatory) in Experiment C were similar (3.6 vs. 2.6 s, $p > 0.05$). For Experiment C plateau time was calculated as the time between increasing above the 90th percentile and decreasing below the 90th percentile of sensed peak rotational velocity (Fig. 2C). Plateau time was significantly longer in per- compared to post-rotatory trials (13.8 ± 4.4 s vs. 8.5 ± 5.3 s, $p < 0.001$) whereas there was no significant

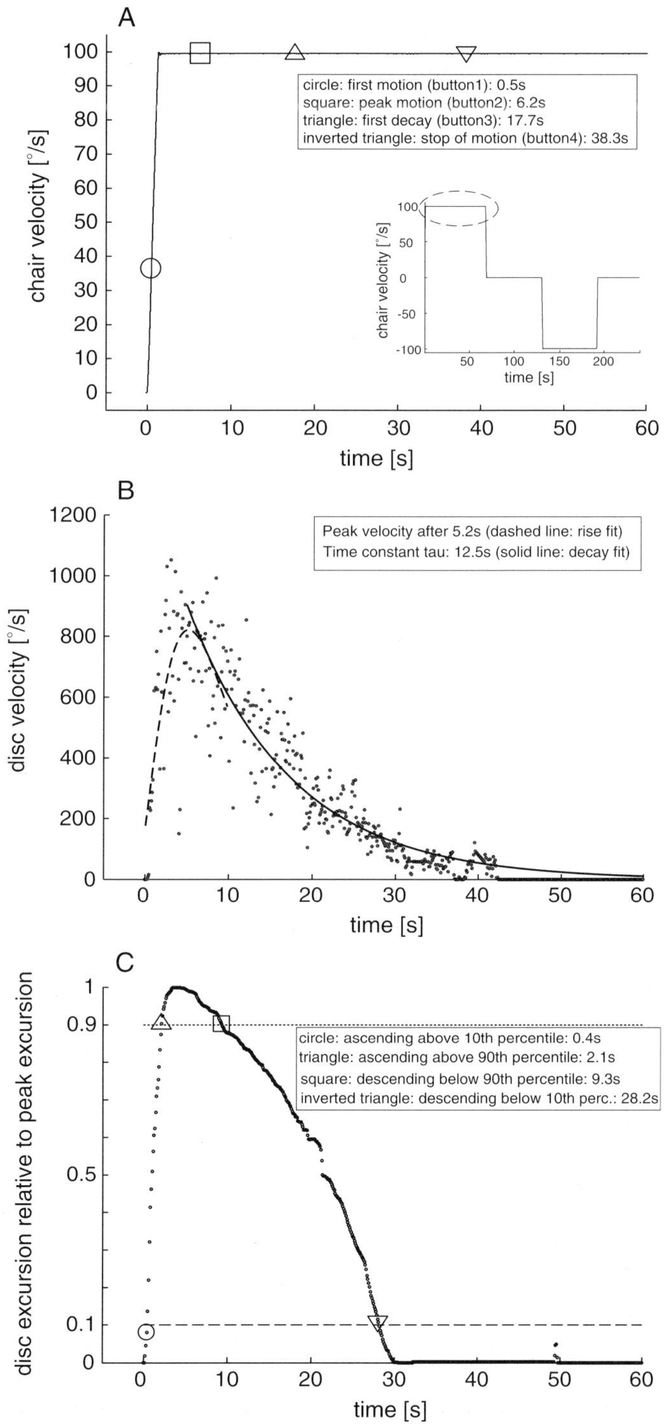

Fig. 1. Sample trials in a single subject. In Experiment A, button presses (as defined in the inlet) are plotted against chair velocity (A). The dashed ellipse in the inlet (representing the entire velocity stimulus) refers to the time window shown in A. Disc velocity (dots) is plotted against time for Experiment B (B), determined time to peak (dashed curve) and the exponential decay function (solid line) are shown. In Experiment C (C), disc excursion relative to peak excursion (dots) is plotted against time. The 10th (dashed line) and the 90th percentile (dotted line) are illustrated and time points when crossing these limits are marked.

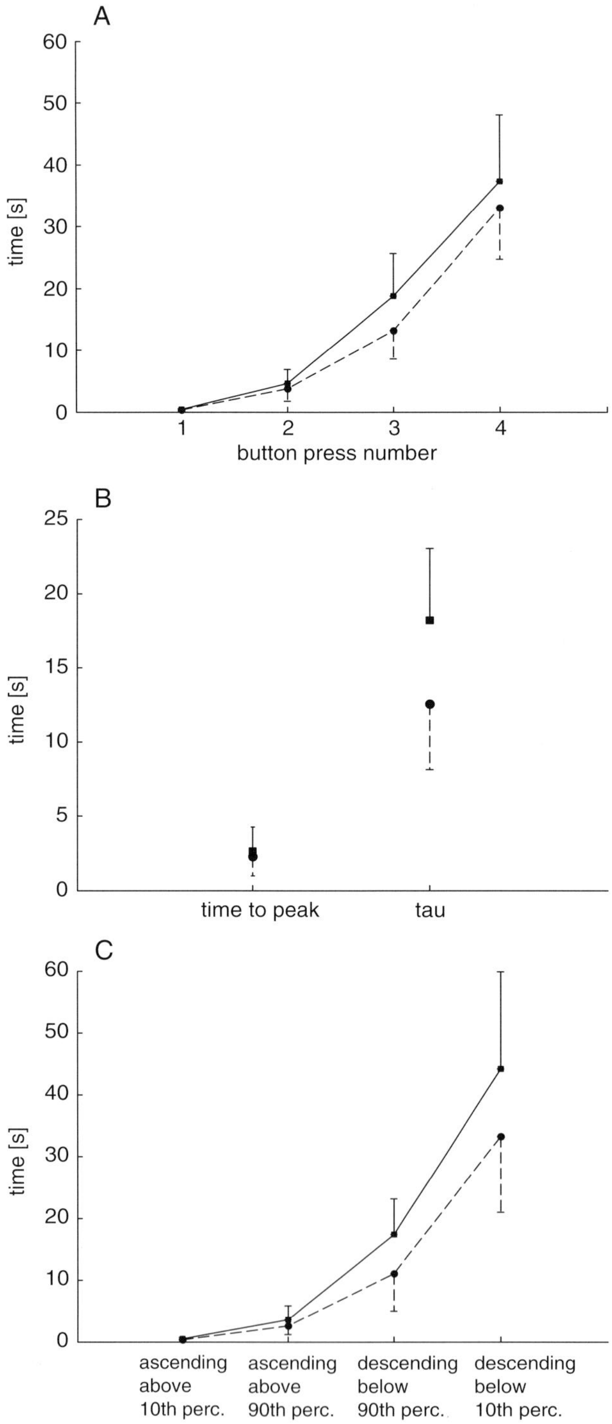

Fig. 2. Mean ± 1 SD of individual mean values for all three experiments. In A, the delays of the four button presses relative to onset are shown both for per-rotatory (squares, connected by solid lines) and post-rotatory (circles, connected by dashed lines) trials in Experiment A. In B, time to rise and time constant *tau* for both per-rotatory (squares, solid error bars) and post-rotatory (circles, dashed error bars) trials are illustrated for Experiment B. Time-points when crossing the 10th and 90th percentile are shown in C for Experiment C, again showing per-rotatory (squares, connected by solid lines) and post-rotatory (circles, connected by dashed lines) trials separately.

difference between per- and post-rotatory decay times (26.7 ± 15.2 s vs. 22.2 ± 9.4 s, $p > 0.05$).

Discussion

Our results show that the time course of vestibular perception of angular velocity is different from the time course of the velocity output of the aVOR. We found a consistent delay in the time to perceived peak velocity of self-rotation in all experiments. A plateau of perceived rotation was found in both Experiments A and C. One interpretation of this difference between the time course of vestibular perception and the time course of the velocity output of the VOR is that the sensory signal from the semicircular canals undergoes additional neural processing, beyond the velocity-storage mechanism, to compute the perception of self rotation. Another consistent observation across the three different experiments was that most measures of the time course of vestibular perception were prolonged during rotation compared to after rotation, suggesting that additional sensory cues (e.g., vibration from the chair, airflow) may also contribute to the perception of self-rotation. Simultaneous recordings of both per- and post-rotatory aVOR and sensed rotation will be necessary for further analysis of ocular motor and perceptual responses to rotation.

Acknowledgement

This study was financially supported by The Wellcome Trust as a Student Elective Prize.

References

Büttner, U. and Waespe, W. (1981) Vestibular nerve activity in the alert monkey during vestibular and optokinetic nystagmus. Exp. Brain Res., 41(3–4): 310–315.

Cohen, B., Henn, V., Raphan, T. and Dennett, D. (1981) Velocity storage, nystagmus, and visual-vestibular interactions in humans. Ann. N.Y. Acad. Sci., 374: 421–433.

Gizzi, M.S. and Harper, H.W. (2003) Suppression of the human vestibulo-ocular reflex by visual fixation or forced convergence in the dark, with a model interpretation. Curr. Eye Res., 26(5): 281–290.

Okada, T., Grunfeld, E., Shallo-Hoffmann, J. and Bronstein, A.M. (1999) Vestibular perception of angular velocity in normal subjects and in patients with congenital nystagmus. Brain, 122(7): 1293–1303.

Raphan, T., Matsuo, V. and Cohen, B. (1979) Velocity storage in the vestibulo-ocular reflex arc (VOR). Exp. Brain Res., 35(2): 229–248.

C. Kennard & R.J. Leigh (Eds.)
Progress in Brain Research, Vol. 171
ISSN 0079-6123

CHAPTER 4.8

The freezing rotation illusion

Max R. Dürsteler*

Department of Neurology, University Hospital Zurich, CH-8091 Zurich, Switzerland

Abstract: The "freezing rotation illusion" refers to decrease in perceived speed of a continuously rotating central region when a swaying surround co-rotates. We observed the following effects for rotations: First, when the centre and its surround are turning in the same direction, and their velocities are distinguishable, the perceived speed of the centre is lower than its physical speed. Second, when the surround and the centre are counter-rotating, the perceived speed of the centre is enhanced. Third, even when the surface size of the centre matches the surface size of the surround, swaying of the centre is unable to induce significant changes in the perceived speed of the surround, whereas swaying of the surround still induces changes in the perceived speed of the centre.

Keywords: motion illusion; motion perception; figure-ground; induced motion

Introduction

Perception of an object's motion depends on the properties of its background. Thus, Duncker (1929) reported how illusionary motion of a central disc, induced by rotation of a larger background disc, could be nulled if subjects counter-rotated the central disc. Pavard and Berthoz (1977) reported a freezing motion phenomenon of presumably vestibular origin: a translating pattern projected on the virtual side windows of a cart appeared stationary when the cart moved abruptly forward. A similar visuo-visual illusion emerged when a stationary subject observed a visual pattern scrolling at a fixed velocity across a mobile monitor: when the monitor started moving, the scrolling pattern appeared to stop (Mesland and Wertheim, 1996; Wertheim and Reymond, 2007). In our laboratory, Stefan Hegemann and I perceived freezing of a rotating scene on either on head-mounted or laptop display while we rolled our heads side-to-side or turned the display back and forth around its roll axis (Dürsteler, 2005). To demonstrate the purely visual origin of this freezing rotation illusion, I borrowed Duncker's experimental design, with a small central disc on a larger disc, each of which could rotate independently. Video clips of the freezing motion illusion can be found on the internet (e.g., http://www.youtube.com/watch?v = bZbVTi3Ibmw).

Methods

Three observers (ages 27, 27, and 59) with normal, or corrected-to-normal vision, looked at a 19 in Samsung SyncMaster 912 T liquid crystal display (33.7° × 25.9°) connected to an HP DC7100 PC. In the first experiment, the test stimulus was composed of a surround with black und white random dots (diameter 16.5°, luminance contrast 93.8%) rotating at 37.7°/s, and a centre with blue and red dots

*Corresponding author. Tel.: +41-44-2555542;
Fax: +41-44-2554429; E-mail: max.duersteler@usz.ch

DOI: 10.1016/S0079-6123(08)00640-7

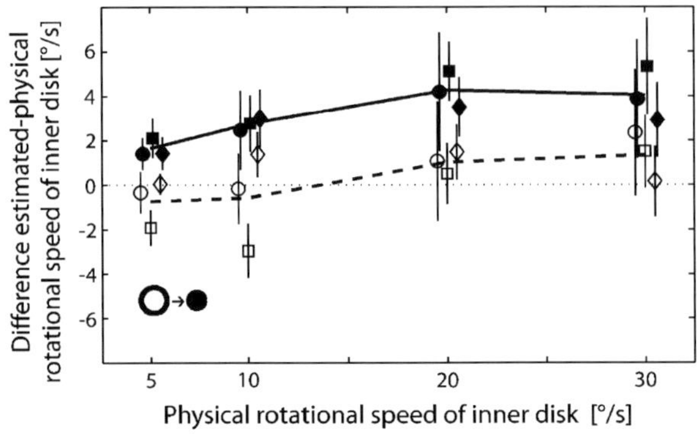

Fig. 1. Influence of surround rotation on perceived speed of inner disc. The mean values and 95% confidence intervals of the difference between estimated and physical rotational speed of the inner disc are plotted as a function of the physical speed of the inner disc for conditions with equal (open symbols for the means of single observers, stippled line for the mean of all three observers) or opposing (filled symbol for the means of single observers, continuous line for the mean of all observers) rotational directions. In all trials the rotational speed of the surrounding annulus was set to 37.7°/s.

(diameter 11.7°, luminance contrast 41.8%) rotating at 5, 10, 20, or 30°/s. The probe stimulus was a centre with the black and white Julesz type random dots and a grey surround. The task of the observers was to indicate if the probe's centre velocity was faster or slower than the velocity of the centre in a preceding test stimulus. The accelerated stochastic approximation method (Kesten, 1958) was used. In the second experiment, the test stimulus was composed of a centre with a white and black random dot pattern turning at a speed of 37.7°/s and a surround with blue and red dots turning at speeds of 5, 10, 20, or 30°/s. The probe stimulus was composed of a surround painted with the black and white random dot pattern and a grey centre. In both experiments, I performed a two-way ANOVA with speed of the induced test stimulus and the relative direction of inducing and induced stimulus element (opposite or same) as factors, and their interaction. The figures show means with their 95% confidence limits adjusted by Tukey's honestly significant difference (HSD) criterion for multiple comparisons.

Results

First experiment

The first experiment examined the effects of the surround rotation direction on the perceived speed of a continuously turning disc. In all three observers the perceived speeds were on average significantly higher in trials with opposite rotation directions than in trials with the same directions (Fig. 1). In a two-way ANOVA (speed × relative rotation direction), the rotation direction (opposite or same) was significant for all observers at a $p<0.0025$. There was no significant interaction ($p>0.49$). The differences in rotational speed estimation did not depend in a systematic way on the speed of the centre. The result of the first experiments shows that that perceived speed is enhanced in trials with opposing rotation directions.

Second experiment

The second experiment tested the effect of the inner disc rotation direction on the perceived speed of the surround (Fig. 2). Two observers did not show significant differences in their speed estimation for conditions with opposite and the same rotation direction ($p>0.06$), one observer (diamonds) showed a significant overall difference together with a significant interaction of speed and relative direction ($p<0.001$), indicating that the differences were significant at some speeds only. The second experiment showed that for the same speeds of the induced stimulus, where we saw significant effects of the surround rotation on the perceived disc rotation, rotations of the inner disc

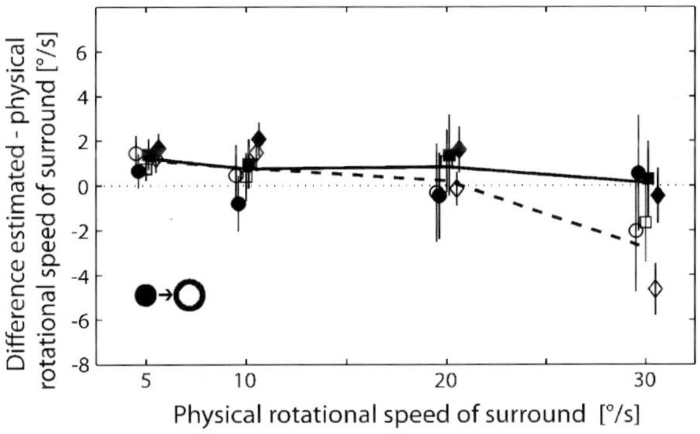

Fig. 2. Influence of figure rotation on perceived speed of surround. The mean values and 95% confidence intervals of the difference between estimated and physical rotational speed of the surrounding ring are plotted as a function of the physical speed of the surround for conditions with equal (open symbols for the means of single observers, stippled line for the mean of all three observers) or opposing (filled symbol for the means of single observers, continuous line for the mean of all observers) rotational directions. In all trials the rotational speed of the inner disc was set to 37.7°/s.

did not have a significant effect on the perceived rotation of the surround.

Discussion

The freezing rotation illusion suggests that the rotation of an included centre is referenced to the rotation of the surround, but that the rotation of the surround is not referenced to the rotation of the included centre. The rotational speed of the centre is increased by an individually different amount, when its direction is opposite to the direction of the surround, and decreases by a fixed amount, when its direction is the same as the surround's direction. By focusing attention on the stimulus element whose speed one had to estimate, the surround rotation may have been less effective as when focusing on the whole scene, when looking at freezing rotation illusion in the video.

Neuronal correlates of the freezing motion may be found in area MST in the monkey (Duffy and Wurtz, 1991) or part of the MT+ complex in humans (Dukelow et al., 2001). After chemical lesions in area MST including the adjoining floor of the STS in macaques, a directional pursuit deficit was found, where the monkey underestimated the speed of a small target moving towards the lesion side and overestimated the speed of a target moving way from the lesion side (Dürsteler et al., 1987), reminiscent to the effects of surround rotation directions on the perceived centre's rotational speed in the freezing rotation illusion.

References

Duffy, C.J. and Wurtz, R.H. (1991) Sensitivity of MST neurons to optic flow stimuli. II. Mechanisms of response selectivity revealed by small-field stimuli. J. Neurophysiol., 65(6): 1346–1359.

Dukelow, S.P., DeSouza, J.F., Culham, J.C., van den Berg, A.V., Menon, R.S. and Vilis, T. (2001) Distinguishing subregions of the human MT+ complex using visual fields and pursuit eye movements. J. Neurophysiol., 86(4): 1991–2000.

Duncker, K. (1929) Über induzierte Bewegung (Ein Beitrag zur Theorie optisch wahrgenommener Bewegung). Psychol. Forsch., 12: 180–259.

Dürsteler, M.R. (2005) Eine neuartige Bewegungstäuschung: Duncker Illusion und "Motion Capture" im Widerstreit. Ophta, 3: 19–26.

Dürsteler, M.R., Wurtz, R.H. and Newsome, W.T. (1987) Directional pursuit deficits following lesions of the foveal representation within the superior temporal sulcus of the macaque monkey. J. Neurophysiol., 57(5): 1262–1287.

Kesten, H. (1958) Accelerated stochastic approximation. Ann. Math. Stat., 29: 41–59.

Mesland, B.S. and Wertheim, A.H. (1996) A puzzling percept of stimulus stabilization. Vis. Res., 36(20): 3325–3328.

Pavard, B. and Berthoz, A. (1977) Linear acceleration modifies the perceived velocity of a moving visual scene. Perception, 6(5): 529–540.

Wertheim, A.H. and Reymond, G. (2007) Neural noise distorts perceived motion: the special case of the freezing illusion and the Pavard and Berthoz effect. Exp. Brain Res., 180(3): 569–576.

C. Kennard & R.J. Leigh (Eds.)
Progress in Brain Research, Vol. 171
ISSN 0079-6123

CHAPTER 4.9

Geometrical considerations on canal–otolith interactions during OVAR and Bayesian modelling

Jean Laurens*, Bernhard J.M. Hess and Dominik Straumann

Vestibulo-Oculomotor Laboratory, Zürich University Hospital, Frauenklinikstr. 26, CH-8091 Zürich, Switzerland

Abstract: During constant-velocity rotation about a tilted axis (OVAR), the VOR and the rotation perception last indefinitely, but show a striking dependency on tilt angle. We show that, during OVAR, a variety of motions can account for the head motion relative to gravity. Some of these are in conflict with canal signals, but correspond to a lower angular velocity; we suggest that the brain performs a trade-off in order to select the best motion. We show that this theory explains the effect of tilt angle on velocity estimation during OVAR.

Keywords: off-vertical axis rotation; modelling; Bayesian gravity

Introduction

During a constant velocity rotation about an axis tilted with respect to gravity (also called off-vertical axis rotation, OVAR), the angular velocity signal originating from the canals decays away. However, the orientation of the head relative to gravity changes constantly, which has been shown to give rise to a continuous perception of rotation, provided that the tilt angle is large enough (Guedry, 1965). Previous modelling work has proposed that the brain constructs an estimate of motion in space from otolith signals, which matches sensory signals (Bos and Bles, 2002). During OVAR, a sustained estimate of rotation can account for head reorientation relative to gravity, and is therefore coherent with sensory signals. Nonetheless, modelling motion perception and vestibulo-ocular reflexes (VORs) during OVAR remains a challenging task, mainly due to the complexity of the tri-dimensional motion and the multiplicity of sensory sources involved.

In the present report, we show how to simplify the problem of angular motion estimation during OVAR, assuming that the head orientation is perceived correctly. Specifically, we point out and formalise two aspects of motion estimation during OVAR: (1) that there is a variety of head motions that lead to the same otolith stimulation as during OVAR and (2) that the information provided by the semicircular canals plays a role in the estimation process.

Otolith signals

During OVAR, the angular velocity of the head can be represented by a vector Ω_0, which is aligned with head-fixed Z-axis (see Fig. 1a). In an egocentric frame of reference, this rotation causes the gravity vector G to rotate around Z, according to $G' = -\Omega_0 \times G$, where G' is the time derivative of G and 'x' represents the vector cross product (see Fig. 1b). A fundamental observation is that

*Corresponding author. Tel.: +41 44 255 5564;
Fax: +41 44 255 4507; E-mail: jean.laurens@gmail.com

DOI: 10.1016/S0079-6123(08)00641-9

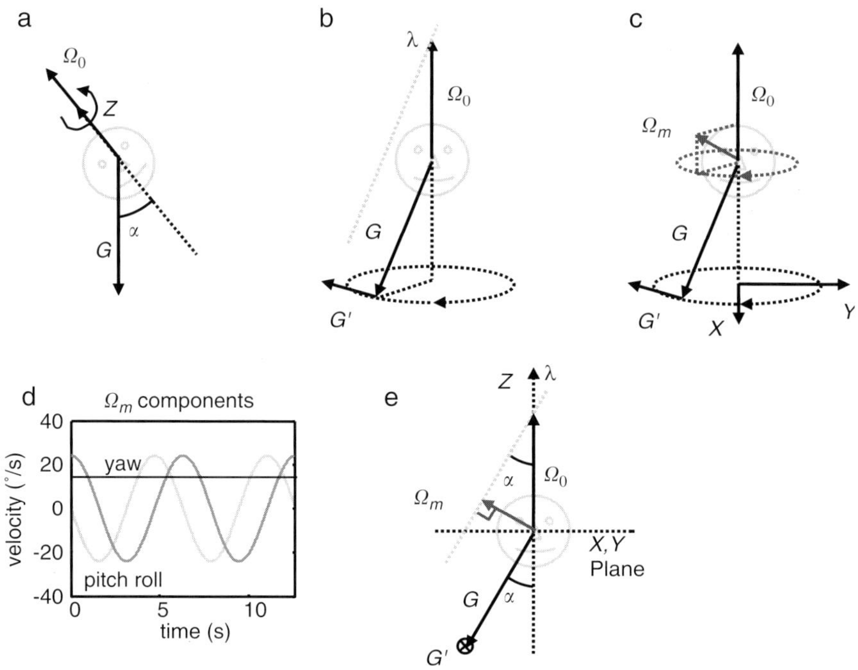

Fig. 1. Geometrical aspects of OVAR. (a) Motion of the head in a geocentric reference frame. (b) Displacement of gravity (G) in an egocentric reference frame, and ensemble of possible rotation vectors. (c) Motion of the vector Ω_m relative to the head. (d) Instantaneous head velocity corresponding to Ω_m. (e) Reformulation of the diagrams b and c in two dimensions.

any angular velocity vector Ω which satisfies the equation $G' = -\Omega \times G$ (i.e., $\Omega \times G = \Omega_0 \times G$) can explain the displacement of G, and is therefore coherent with the otolithic input (also see Hess, 1992). The ensemble of possible Ω vectors is simple to compute. If we decompose Ω as the sum of Ω_0 and an additional vector Ω_1 (i.e., $\Omega = \Omega_0 + \Omega_1$), we obtain $(\Omega_0 + \Omega_1) \times G = \Omega_0 \times G$, i.e., $\Omega_1 \times G = 0$. This means that the additional velocity vector has to be parallel to G. The ensemble of possible velocity vectors is represented in Fig. 1b as a line λ (which passes at Ω_0 and is parallel to G).

Another observation is that some velocity vectors have a smaller magnitude than Ω, which means that they correspond to a smaller angular velocity. The vector with the smaller magnitude is $\Omega_m = -G \times G'/|G|^2$. We previously presented the hypothesis that the brain favours motion estimates with a lower angular velocity (Laurens and Droulez, 2007). Accordingly, we would expect that, on the basis of otolith signal only, the perceived motion correspond to Ω_m.

Semicircular canal signals

In a steady state, Ω_0 does not vary over time in an egocentric frame of reference. It corresponds to a constant-velocity rotation about the head vertical axis, which is only detected at the beginning of rotation by the semicircular canals. This peripheral input typically fades away over a few seconds. In contrast, the other angular velocity vectors continuously rotate around Z. For instance, the trajectory of the Ω_m vector is illustrated in Fig. 1c. The projection of Ω_m on the Z axis is constant, whereas its projection on the (X, Y) plane is rotating around the origin. Therefore, the projection on the X and Y axis varies sinusoidally over time (Fig. 1c). In other words, Ω_m corresponds to the summation of a constant-velocity rotation in yaw and of pitch and roll oscillations (as illustrated on Fig. 1d). In terms of sensory inputs, a motion corresponding to the vector Ω_m would not activate the horizontal canals, but the dynamic pitch and roll components would activate the

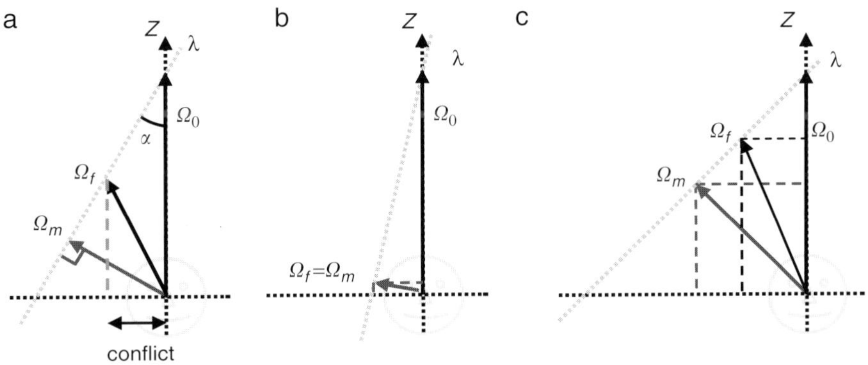

Fig. 2. Optimal estimation of motion in a general case (a), for tilt angles of 15° (b), and 45° (c). See text for details.

vertical canals. As these canals are not activated during the real OVAR, the motion corresponding to Ω_m is in conflict with their signal. For a given vector Ω, this conflict is proportional to the amplitude of the pitch and roll oscillations, which is represented by the projection of Ω on the X, Y plane. As pitch and roll oscillations are equivalent for the purpose of our demonstration, the Fig. 1c can be reduced to a two dimensional diagram (Fig. 1e), in which the abscissa represents the amplitude of the pitch and roll oscillations, and the ordinate the constant yaw velocity component.

This diagram allows capturing the issues discussed above in a simple geometrical representation. The ensemble of possible motions forms a line λ, passing through the Ω vector with an angle α relative to the ordinate. The Ω_m vector is obtained by orthogonal projection of the origin on line λ. Each possible motion is represented by a vector Ω. The length of Ω is equal to the angular velocity of this possible motion, and its projection on the abscissa represents the magnitude of the conflict with the vertical canals.

Bayesian modelling

In a previous work, we implemented a general Bayesian model of self-motion perception (Laurens and Droulez, 2007), as well as a simplified model dedicated to motion estimation during OVAR (Laurens, 2006). The latter uses the constraints described above. We will briefly describe the principles of this estimation by using the diagram in Fig. 2a. The model assumes that motion perception during OVAR can be modelled as a trade-off between the minimisation of the angular velocity and the minimisation of the conflict with canals signal. Therefore the perceived motion (Ω_f) is expected to fall between the vectors Ω_m (which minimises the angular velocity) and the vector Ω_0 (for which there is no conflict). This trade-off can easily be visualised for two angles of tilt on the Fig. 2b, c. For a small angle of tilt ($\alpha = 15°$, Fig. 2b), the vector Ω_m has a very small magnitude compared to Ω_0. Furthermore, its projection on the abscissa is small, which means that it corresponds to a small conflict. Therefore the optimal motion is Ω_m. This explains the absence of yaw rotation perception at small angles of tilt (Denise et al., 1988; Vingerhoets et al., 2006). In contrast, for a tilt angle of 45° (Fig. 2c), Ω_m has a higher yaw component. As it also corresponds to a higher conflict, Ω_f falls between Ω_m and Ω_0. As the tilt angle approaches 90°, Ω_f gets closer to Ω_0. This explains the close to veridical motion perception during OVAR with a large angle of tilt (Guedry, 1965).

Conclusion

It is widely accepted that the brain can derive an angular velocity estimate from the otolith input during OVAR. We have formalised this process and we show that a variety of motion is compatible with the otolith signal. We also emphasise the role of the information provided by the canals during OVAR. In particular, we demonstrate that, although this signal fades away during OVAR, it contributes to deducing that the head is rotating at a constant velocity.

Acknowledgements

This study was supported by the Swiss National Science Foundation, the Betty and David Koetser Foundation for Brain Research, Zurich, Switzerland, and the Center of Integrative Human Physiology, University of Zurich, Switzerland.

Abbreviations

OVAR	off-vertical axis rotation
VOR	vestibulo-ocular reflex

References

Bos, J.E. and Bles, W. (2002) Theoretical considerations on canal–otolith interaction and an observer model. Biol. Cybern., 86(3): 191–207.

Denise, P., Darlot, C., Droulez, J., Cohen, B. and Berthoz, A. (1988) Motion perceptions induced by off-vertical axis rotation (OVAR) at small angles of tilt. Exp. Brain Res., 73(1): 106–114.

Guedry, F.E. (1965) Orientation of the rotation-axis relative to gravity: its influence on nystagmus and the sensation of rotation. Acta Otolaryngol., 60: 30–48.

Hess, B.J. (1992) Three-dimensional head angular velocity detection from otolith afferent signals. Biol. Cybern., 67(4): 323–333.

Laurens, J. (2006) Modélisation Bayésienne des interactions visuo-vestibulaires. Ph.D. Université Paris VI.

Laurens, J. and Droulez, J. (2007) Bayesian processing of vestibular information. Biol. Cybern., 96(4): 389–404.

Vingerhoets, R.A., Medendorp, W.P. and Van Gisbergen, J.A. (2006) Time course and magnitude of illusory translation perception during off-vertical axis rotation. J. Neurophysiol., 95(3): 1571–1587.

C. Kennard & R.J. Leigh (Eds.)
Progress in Brain Research, Vol. 171
ISSN 0079-6123

CHAPTER 4.10

Listing's plane and the otolith-mediated gravity vector

Andrew H. Clarke*

Vestibular Research Lab, ENT Department, Campus Benjamin Franklin, Charité Universitaetsmedizin Berlin, Germany

Abstract: The orientation of Listing's plane (LP) was examined under one-g and zero-g conditions during parabolic flight and during prolonged spaceflight. Ten healthy subjects participated in the parabolic flight study. In zero-g, the orientation of LP was consistently altered. LP elevation was tilted backwards by approximately 10° ($p = 0.003$). The azimuth angles of the left and right eyes also diverged in zero-g, with a statistically significant change ($p = 0.04$) in the vergence angle between 6.1° and 11.8°. A discernible dissociation in torsional eye position was also observed, which proved to be statistically significant ($p = 0.03$). The thickness of LP was found to be of the order of 1°, and was not significantly altered by the transitions between one-g and zero-g. The results of the spaceflight measurements demonstrate that this LP re-orientation persists over 6-month periods in microgravity, and that re-adaptation after returning to one-g earthbound conditions proceeds over several days to weeks. The findings demonstrate that in contrast to repositioning of the head in one-g, the elimination of gravity represents a qualitative change for the vestibular and oculomotor systems. It appears that given the lack of voluntary control of ocular torsion, the tonic otolith inputs are instrumental in the stabilization of torsional eye position and consequently of LP. The observed torsional divergence also provides support for the so-called otolith asymmetry hypothesis.

Keywords: Listing's plane; gravity; microgravity; otoliths; eye movements; video oculography-spaceflight

Introduction

Under one-g conditions, the orientation of Listing's plane (LP) has been found to be dependent on the otolith-mediated gravity vector in the monkey (Haslwanter et al., 1992; Hess and Angelaki, 1997), and to a much lesser extent in the human (Bockisch and Haslwanter, 2001; Furman and Schor, 2003). The present study examines the behaviour of LP in the complete absence of gravity during the zero-g phases of parabolic flight and during prolonged spaceflight on the International Space Station (ISS). In contrast to the previous Earthbound studies, where despite changes in head position the gravity vector persists, the absence of gravity during parabolic flight and spaceflight represents a qualitatively distinct condition for the sensorimotor systems.

Methods

Eye and head movements were sampled at 100 Hz using a binocular, three-dimensional eye-tracking device (ETD, Chronos Vision, Berlin). A detailed description of this device has been published previously (Clarke et al., 2000). Calibration of eye

*Corresponding author. Tel.: +49 30 8445 2434;
Fax: +49 30 8445 4119; E-mail: Andrew.Clarke@charite.de

DOI: 10.1016/S0079-6123(08)00642-0

movements was performed by evaluation of a sequence of fixations to target spots set at fixed angles, which were generated using a head-mounted laser diode. Ten subjects were tested during consecutive one-g and zero-g phases of 4–8 parabolas, during which the subject performed sequences of saccades to randomly placed target points at approximately 1.5-m distance, over a period of 20 s. The spaceflight tests were performed by trained cosmonauts on the ISS using identical equipment and protocol as for the parabolic flight experiments. Training and baseline data collection were performed during the 4 weeks prior to launch. During their 6-month stay on the ISS, the cosmonauts performed the experiment protocol at 3-week intervals, yielding eight data points in zero gravity. Postflight testing was then performed every second day for 2 weeks and finally 60 days after returning to Earth.

The recorded eye-image sequences were analysed offline after completion of the flights. Three-dimensional eye position was calculated in Fick co-ordinates, which then transformed into quaternions. A principal components' analysis was employed to calculate the best-fit plane for each data set. This provides the elevation, azimuth, and thickness of the displacement plane. Details of the methods are described in Clarke and Haslwanter (2007).

The examination and test procedures employed were in accordance with the ethical standards of the Ethics Committee of the Charite' Medical School, Berlin and the Medical Board of the European Space Agency on human experimentation and with the Declaration of Helsinki.

Results

The results of the parabolic flight experiments are summarized in Fig. 1. Change in LP elevation on transition from one-g to zero-g was consistently in one direction (tilting back). The median change in LP elevation for the right and left eye was 10.3° and 8.9°, respectively ($p = 0.02$, Wilcoxon matched pairs test).

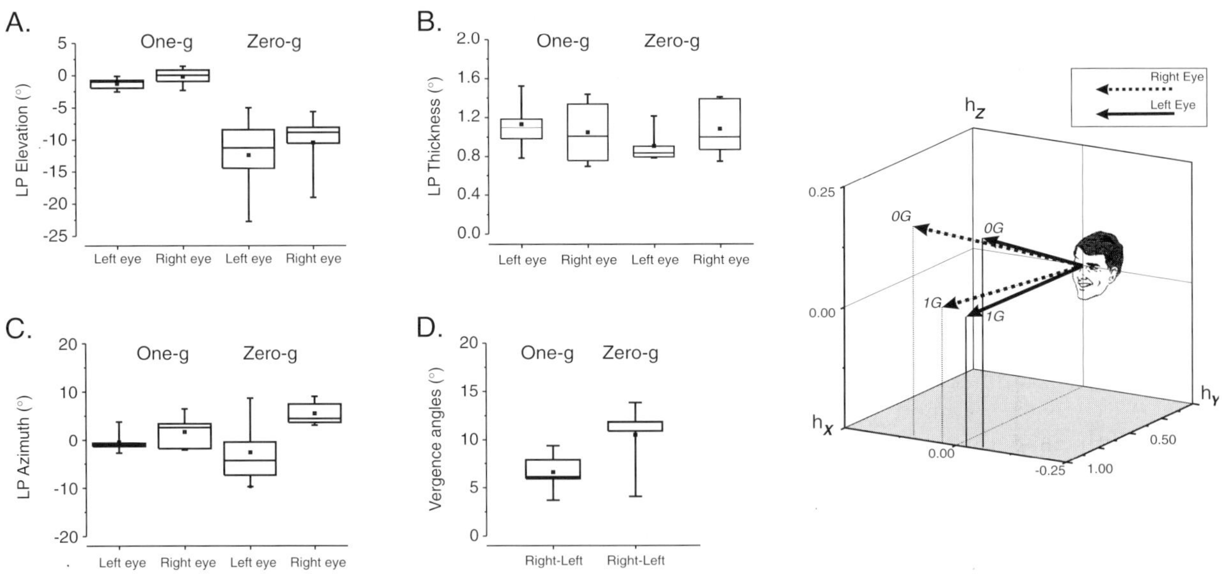

Fig. 1. Left panel: Each graph depicts the median value (horizontal bar) 25–75% range (box), 5–95% range (whiskers), and the mean value (small square). The calculated parameters are arranged for comparison of one-g and zero-g conditions. (A) Elevation angle for right and left eyes; (B) LP thickness for left and right eyes; (C) LP azimuth angle for each eye; and (D) vergence angle determined as the difference in azimuth angles of the right and left eyes. The one-g results are equivalent to those from laboratory control measurements using the same equipment and protocol (i.e., median for LP elevation = 2.37, variability of LP thickness = 1.24°) Right panel: Primary position during parabolic flight. Each vector depicts the median value calculated from the data of five subjects, demonstrating the tilt in LP elevation and the divergence of azimuth in zero-g conditions.

No significant change ($p = 0.12$) in LP thickness was found between one-g and zero-g conditions. Median thickness in one-g amounted to 0.90° (max. 1.62°, min. 0.26°), and in zero-g to 0.78° (max. 2.12°, min. 0.32°).

The differences in left-eye azimuth angle between one-g and zero-g conditions (Fig. 1C) proved to be statistically significant (Wilcoxon for matched pairs, $p = 0.04$). Comparison of the vergence angles ($R_{azimuth} - L_{azimuth}$) between one-g and zero-g (Fig. 1D) also yielded a statistically significant difference (Wilcoxon for matched pairs, $p = 0.02$). The 3D primary position vector for each eye was also calculated for the one-g and zero-g conditions. The median vectors are shown in Fig. 1 (right panel), illustrating the consistent change in elevation and divergence between one-g and zero-g conditions.

The preliminary results from one subject tested over a 6-month stay on the ISS are shown in Fig. 2. These demonstrate the consistent backward tilt of LP elevation over the period in zero gravity. The postflight measurements conducted during the 2 weeks after landing demonstrate a clear re-adaptive modification in LP orientation. Recovery to preflight baseline value appears to take longer than this 2-week interval. Testing after 60 days indicates a comparable value to preflight.

Discussion

The analysis of the parabolic flight and spaceflight recordings reveal a consistent gravity-dependent change in the elevation and azimuth of LP between microgravity and normal gravity. This involves changes in elevation of the order of 10°, and an increase in divergence between the right- and left-eye azimuth angles of LP. The parabolic flight recordings also reveal a small shift of LP along the x-axis, i.e., corresponding to a torsional deviation in the resting eye position. These findings indicate that transitions between one-g and zero-g induce discrete, qualitative changes in the circuitry responsible for the maintenance of LP.

The observation that the 3D orientation of LP is changed consistently by these transitions strongly supports the idea that the otolith-mediated gravity reference is the driving signal for this effect. It could be argued that LP orientation is simply modulated by the direction and magnitude of the gravito-inertial vector. This may explain the relatively small modulation of LP orientation found previously in humans during static pitch re-positioning under one-g conditions. However, it does not accommodate the very much larger changes in 3D orientation observed in the present study. The more pronounced re-orientation of LP

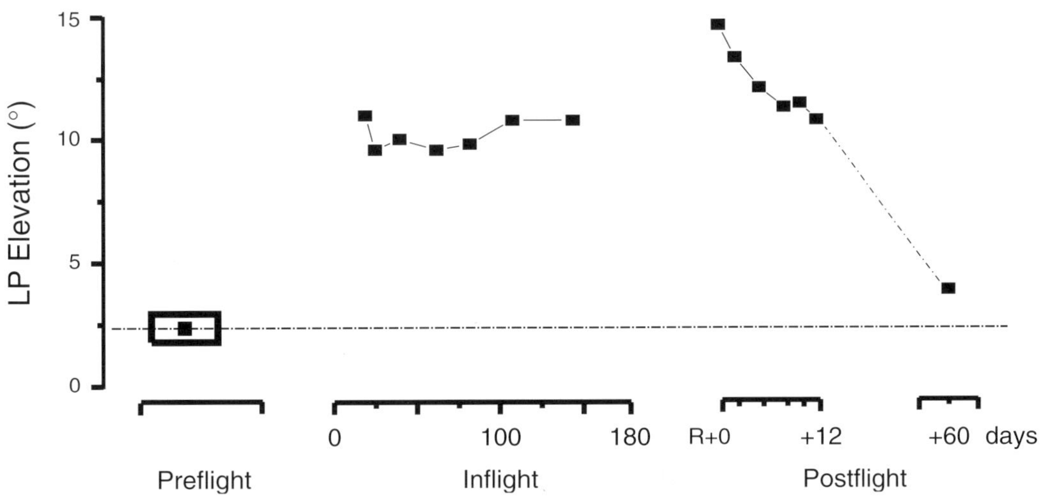

Fig. 2. LP elevation measured in one subject during the preflight, inflight, and postflight intervals of a 6-month period on the International Space Station (ISS). The backward tilt of LP elevation persists throughout to the 6-month interval in microgravity. Postflight re-adaptation proceeds over several days or weeks. The R + 60 measurement result is comparable with preflight baseline.

can be understood as resulting from the radical loss of the otolith-mediated gravity reference. Further evidence of such a coupling between the gravity reference and the oculomotor control circuitry is the gravity-dependent bias component on neural signals described by Frens et al. (1998).

The individual changes in ocular torsional position observed under zero-g conditions support the so-called "otolith asymmetry hypothesis" (Egorov and Samarin, 1970; von Baumgarten and Thumler, 1978), according to which any inherent otoconial mass asymmetry is centrally compensated under normal one-g conditions; however, on transition to zero-g this compensatory re-weighting would persist, and manifest as the observed deviation or bias in torsional eye position.

Taken together, it appears that given the lack of voluntary control of ocular torsion, the tonic otolith afferences and the resultant generation of an internal gravity reference are instrumental in the stabilization of torsional eye position, and consequently of LP.

Acknowledgement

The research was supported by grant WB0429 from the German Space Agency (DLR).

References

Bockisch, C.J. and Haslwanter, T. (2001) Three-dimensional eye position during static roll and pitch in humans. Vision Res., 41: 2127–2137.

Clarke, A.H. and Haslwanter, T. (2007) The orientation of Listing's plane in microgravity. Vision Res., 47: 3132–3140.

Clarke, A.H., Steineke, C. and Emanuel, H. (2000) High image rate eye movement measurement: a novel approach using CMOS sensors and dedicated FPGA devices. In: Lehmann T. (Ed.), Bildverarbeitung in der Medizin. Springer, Berlin, pp. 398–402.

Egorov, A.D. and Samarin, G. (1970) Possible changes in the paired operation of the vestibular apparatus during weightlessness. Kosm. Biol. Aviakosm. Med., 4(2): 85–86.

Frens, M.A., Suzuki, Y., Scherberger, H., Hepp, K. and Henn, V. (1998) The collicular code of saccade direction depends on the roll orientation of the head relative to gravity. Exp. Brain Res., 120: 283–290.

Furman, J.M. and Schor, R.H. (2003) Orientation of Listing's plane during static tilt in young and older human subjects. Vision Res., 43: 67–76.

Haslwanter, T., Straumann, D., Hess, B.J. and Henn, V. (1992) Static roll and pitch in the monkey: shift and rotation of Listing's plane. Vision Res., 32: 1341–1348.

Hess, B.J. and Angelaki, D.E. (1997) Kinematic principles of primate rotational vestibulo-ocular reflex. II. Gravity-dependent modulation of primary eye position. J. Neurophysiol., 78: 2203–2216.

von Baumgarten, R.J. and Thumler, R.A. (1978) A model for the vestibular function in altered gravitational states. Life Sci. Space Res., 17: 161–170.

C. Kennard & R.J. Leigh (Eds.)
Progress in Brain Research, Vol. 171
ISSN 0079-6123

CHAPTER 4.11

A reinterpretation of the purpose of the translational vestibulo-ocular reflex in human subjects

Ke Liao[2], Mark F. Walker[1], Anand Joshi[2], Millard Reschke[3], Zhong Wang[2] and R. John Leigh[1,2,*]

[1]*Department of Neurology, Daroff-Dell'Osso Laboratory, Veterans Affairs Medical Center and University Hospitals, Case Western Reserve University, Cleveland, OH, USA*
[2]*Department of Biomedical Engineering, Daroff-Dell'Osso Laboratory, Veterans Affairs Medical Center and University Hospitals, Case Western Reserve University, Cleveland, OH, USA*
[3]*Neurosciences Laboratories, Johnson Space Center, Houston, TX, USA*

Abstract: In a prior study we reported that the human translational vestibulo-ocular reflex (tVOR) in response to vertical (bob) 2 Hz oscillations generated eye rotations of only 60% of those required to keep the eyes pointed at a stationary visual target, whether located at near (~17 cm) or far (2 m). Best responses occurred in ambient illumination and we concluded that relative image motion between the target and background was an important determinant of tVOR behaviour. To investigate further how visual conditions influenced tVOR, we measured responses as subjects binocularly viewed the bridge of their own nose in a mirror at ~8.5 cm, a visual condition that required similar convergence to viewing the near target, but cancellation of tVOR. Median tVOR cancellation gain [(near-viewing response−mirror viewing response)/near-viewing response] was 0.81 (range 0.55–0.97), which was substantially greater than the gain of smooth visual tracking of a large visual display moving at 2 Hz (median gain 0.27, range 0.09–0.42). Thus, visual inputs other than smooth tracking must contribute to tVOR cancellation. We then compared tVOR response to 2 Hz bob as subjects fixed upon a visual target at 17 cm and viewed a large textured background at 1.5 m that was either stationary or moving at 2.1 Hz. Vertical eye rotations waxed and waned as a function of the difference between platform and background oscillations. These findings support our hypothesis that tVOR evolved not to stabilize the image of the target on the fovea, but rather to minimize retinal image motion between objects lying in different planes, in order to optimize motion parallax information. A geometrically based optimization function is proposed to account for tVOR responses at different target distances.

Keywords: locomotion; vergence angle; moving platform; retinal slip

Introduction

In a recent study, we measured translational vestibulo-ocular reflex (tVOR) during vertical (bob) translations in ambient light, and also in combination with horizontal (yaw) rotations (Liao et al., 2008). Our goal was to determine whether these test conditions, which approximated components of head perturbations during locomotion, would increase the overall tVOR response. We found that tVOR increased only slightly during translation–rotation in ambient light compared with reported

*Corresponding author. Tel.: + 216-844-3190; Fax: + 216-231-3461; E-mail: rjl4@case.edu

DOI: 10.1016/S0079-6123(08)00643-2

values (Gresty et al., 1987; Israël and Berthoz, 1989; Paige, 1989; Gianna et al., 1997; Ramat and Zee, 2003; Ramat et al., 2005). Although the velocity of eye movements increased by almost a factor of 10 between far versus near viewing, compensation gain (eye rotational velocity/required eye rotational velocity to maintain foveal target fixation) remained at 0.55–0.60. Based on responses during binocular or monocular viewing, and ambient or reduced illumination, we concluded that relative image motion between the target and background was a critical determinant of tVOR behaviour.

The goals of the present study were to investigate further the influence of vision on tVOR performance during two test paradigms: (1) to compare tVOR cancellation versus smooth visual tracking. (2) To determine whether motion of the visual background influenced tVOR during viewing of a near target.

Methods

Subjects

We studied 13 healthy subjects (6 females, median age 60 years, range 25–72), who were subjects in a prior study of tVOR (Liao et al., 2008). All gave informed, written consent, in accordance with the Declaration of Helsinki and the Institutional Review Board of the Cleveland Veterans Affairs Medical Center.

Vestibular stimuli

Subjects sat in a chair on a Moog 6DOF2000E electric motion platform (East Aurora, New York) that could move with 6° of rotational and translational freedom. Each experimental run started with three cycles of bob at 0.2 Hz (typical amplitude ±5.6 cm). Responses to this 0.2 Hz stimulus held subjects' eyes continuously on target and served as a calibration check. Then, we applied bob translations at 2 Hz (typical amplitude ±1.5 cm) for 12 s.

Comparison of cancellation of tVOR and smooth tracking

For the first set of experiments, for which all 13 subjects participated, there were three visual conditions, each employed for one experimental run. (1) Subjects binocularly viewed a laser spot projected on a wall at a distance of 2 m. (2) Subjects binocularly viewed an earth-stationary near target (reflective ball, diameter 1 cm) suspended at a distance of ~17 cm in front of their left eye. (3) Subjects viewed a small mark on the bridge of their own nose through an earth-fixed mirror at a distance of ~8.5 cm. The actual positions of the near target and the mirror, for each subject, were measured directly. During these experiments, room lights were turned on so that natural visual cues, such as motion parallax and relative size, were available. We compared the cancellation of tVOR during mirror viewing with smooth visual tracking of a moving visual stimulus (Amsler grid), subtending 25.6° horizontally and 18.6° vertically with a central dot, at a target distance of 110 cm. The stimulus moved sinusoidally, in the vertical plane: (A) through ±5.6° at 2.0 Hz (peak velocity 70°/s) or (B) through ±2.8° at 2.0 Hz (peak velocity 35°/s). The first moving visual stimulus imposed the same requirements on eye movements as those imposed by the translation stimuli, if there were no vestibulo-ocular responses; the second stimulus corresponded to the remaining visual motion if tVOR compensated for half of that required.

Investigation of background motion on tVOR

For the second set of experiments, subjects viewed the near target at ~17 cm with a background consisting of horizontal stripes or a photograph of a park, displayed on a large flat screen at 1.5 m, which subtended 50° horizontally and 30° vertically. Room lights were turned out, but the earth-stationary near target could be easily seen by its reflected light, which emanated from the large screen; however, other background landmarks in the room could not be seen. The two visual conditions were then (1) near target against the background which was stationary; (2) near target against the background that moved sinusoidally at 2.1 Hz. Three subjects took part in these experiments (age range 25–60).

Measurement of eye and head movements and data processing

Three-dimensional eye rotations were measured using the magnetic search coil technique; details of the system and processing of coils signals have been previously described (Liao et al., 2008). Linear and rotational movements of the chair frame and subject's head were monitored by an infrared reflection system (Vicon Motion Systems, Los Angeles, CA), and its signals were used to calculate the movements of the subject's head, the coil frame, and "ideal eye rotations" to hold gaze (corresponding to the line of sight) on the visual target (Liao et al., 2008). We carried out Fourier transforms of eye and head velocity, measuring the response at the frequency of the stimulus. We quantified the responses with two parameters, which we term the responsivity and the compensation gain. We define the responsivity (output/input) of tVOR as eye rotational velocity/head translational acceleration, which has units of degrees/second of eye rotation per meters/second^2 of head translation (hereafter stated as deg $\times$ s/m). We also calculated compensation gain as eye rotational velocity/required eye rotational velocity to maintain foveal fixation of the visual target (Ramat et al., 2005).

In the case of mirror viewing, ideally, tVOR should be negated. Thus, as an index of tVOR cancellation, we compared tVOR responses during mirror viewing with responses during viewing of the near target. The measured viewing distance of the near target (16.9$\pm$0.9 cm) and virtual image in the mirror (17.7$\pm$1.0 cm) were similar but not identical and, accordingly, we made a geometric correction to the near-target responses. We then calculated tVOR cancellation ratio (CanR) from [(near-viewing response−mirror viewing response)/near-viewing response]. The gain (eye velocity/target velocity) and phase lag of smooth tracking with respect to visual target motion were calculated by desaccading eye velocity data, and computing Fourier transforms. To test whether smooth tracking was the only factor that cancelled the tVOR response during mirror viewing, we calculated the expected response during mirror viewing (MR) by subtracting smooth pursuit response (SP) from tVOR response during near viewing

$$\mathrm{MR} = \mathrm{tVOR} - \mathrm{SP}$$

The predicted cancellation gain or ratio (CanR) can then be calculated as:

$$\mathrm{CanR} = \frac{A_{\mathrm{tVOR}} - A_{\mathrm{MR}}}{A_{\mathrm{tVOR}}}$$

where A_{tVOR} is the measured amplitude of tVOR, and A_{MR} the calculated amplitude of the MR. In this way we could compare values of predicted CanR with measured results.

Results

Comparison of cancellation of tVOR and smooth visual tracking

Representative records from one subject during the three visual test conditions are shown in Fig. 1A; note that, apart from vergence, individual traces have been offset to aid clarity of display. The magnitude of tVOR (responsivity) increases from far to near viewing, but is largely negated during mirror viewing, although the vergence angle is similar to during near viewing.

Results from all 13 subjects are summarized in Fig. 1B. For the group of subjects during mirror viewing, mean vergence angle was 20.0° and median tVOR cancellation gain was 0.81 (range 0.55−0.97). The median gain (range) for smooth tracking for the full amplitude stimulus (located at 110 cm) was 0.27 (0.09−0.42) with mean phase lag of 58.6°$\pm$15.6 for measured eye velocity with respect to ideal eye velocity required to follow the target; for the half amplitude stimulus, median gain was 0.18 (0.06−0.38), with a phase lag of 56.4° ($\pm$13.7). As shown in Fig. 1B, the cancellation gain (CanR) calculated from either full or half amplitude smooth

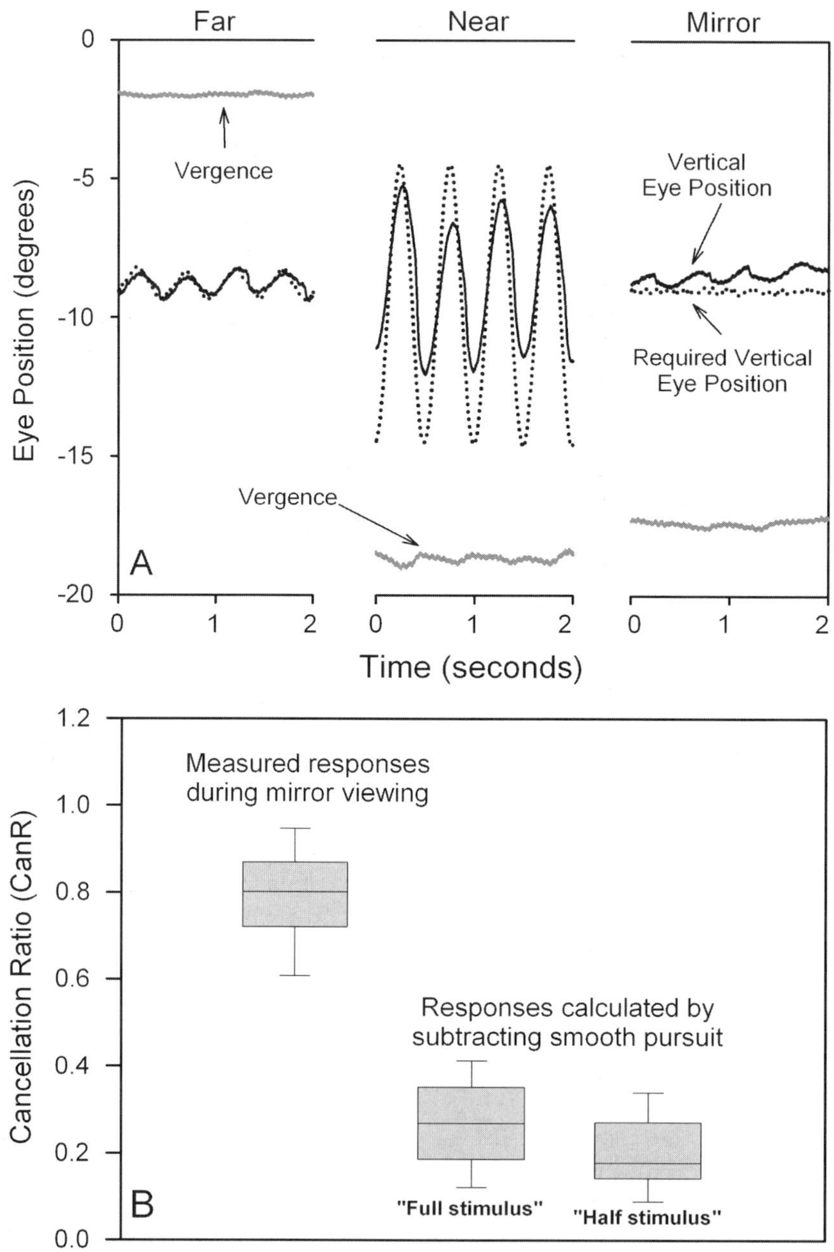

Fig. 1. (A) Representative records from the left eye of one subject illustrating typical responses during viewing of the far target, near target, and mirror viewing. Note that, except for vergence, individual traces have been offset in position to aid clarity of display. Note how tVOR (vertical eye rotation) increases during near viewing compared with far, but is largely cancelled during mirror viewing (visual target moves with subject). Required eye rotations were computed from measured head movements (Liao et al., 2008). (B) Comparison of measured CanR during mirror viewing and estimated CanR based on smooth-tracking performance (see text for details); measured CanR values are significantly greater than values calculated from either "full amplitude" or "half amplitude" smooth tracking responses.

pursuit response is significantly smaller than the measured CanR ($p < 0.001$). Furthermore, a paired comparison for each subject (Wilcoxon rank-sum test) showed a significantly lower value of calculated CanR than measured CanR ($p < 0.001$). Thus, smooth visual tracking could not account for cancellation of tVOR responses during near mirror viewing.

Effect of moving visual background on tVOR

Representative responses from one subject are shown in Fig. 2. As the difference between platform and background motion decreased, so did the tVOR response compared with responses with a stationary background (for whom median peak eye velocity is indicated by horizontal dotted lines). Subsequently, as the difference between platform and background increases, tVOR started to increase. It is evident that the effect on tVOR lags the changing stimulus by 2 s. This behaviour is addressed further in the Discussion.

Discussion

We found that visual cancellation of tVOR, as each subject viewed the bridge of their nose in a near mirror, was achieved much more successfully than could be accounted for by visual tracking mechanisms such as smooth pursuit. These results are consistent with a prior study of visual cancellation of tVOR in bob in two human subjects (Paige, 1989). Thus, similar to visual cancellation of aVOR (Leigh et al., 1989; Huebner et al., 1992; Das et al., 1998), mechanisms other than superposition of visual tracking appear to contribute. During smooth visual tracking at 2 Hz, large phase lags of the eye occur with respect to the target. In contrast, the phase lag of tVOR at 2 Hz

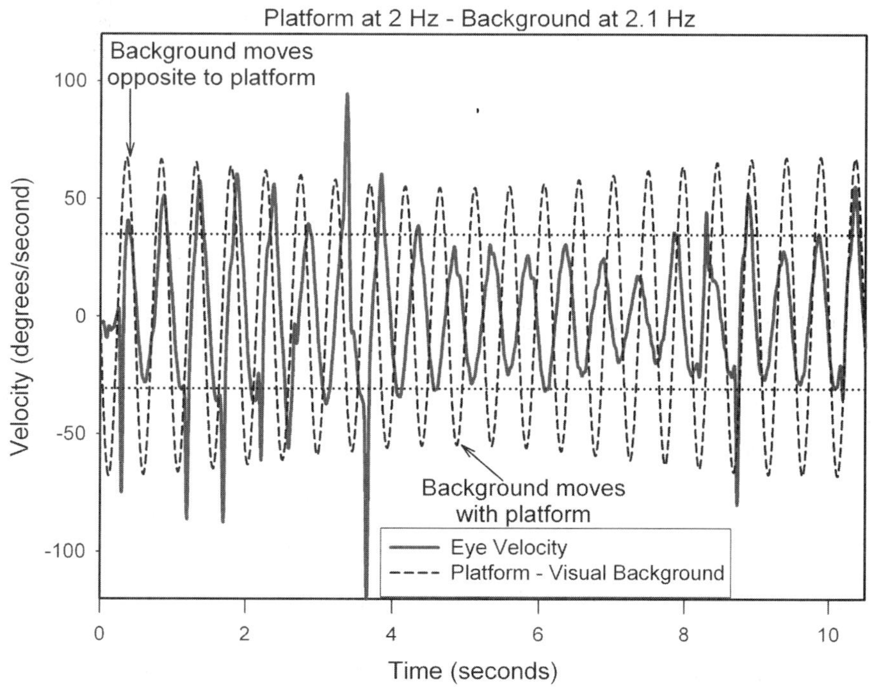

Fig. 2. Effects of tVOR as a subject fixed upon a small earth-stationary target at 17 cm against a moving background (horizontal gratings) at a viewing distance of 1.5 m. The subject bobbed sinusoidally at 2.0 Hz and the background moved at 2.1 Hz. The subject's vertical eye velocity and the difference between platform and background, which was constantly changing, are plotted. Horizontal dotted lines correspond to median values of peak eye velocity while this subject viewed the same near target against a stationary background.

encountered in our prior experiments was about 18°, irrespective of the viewing condition (Liao et al., 2008), which could be largely accounted for by the reported tVOR latency of about 25 ms (Gresty et al., 1987; Ramat and Zee, 2003). Thus, one conclusion is that although visual stimuli are important for setting tVOR responsivity to an appropriate level, smooth visual tracking (such as smooth pursuit) appears to play little role if any.

Our second experiment addresses the issue: Why should the compensation ratio of tVOR be systematically smaller than the amount of eye rotation that is required to keep the fovea pointed at the near target? During translation of the observer, relative motion of objects located at different distance is inevitable, for which eye movements cannot compensate. This relative motion of retinal images provides a cue to the distance of objects in the environment. Motion detection follows Weber's law, such that discrimination of relative motion is better at lower velocities of retinal image motion (Nakayama, 1985). We have postulated that tVOR responses are set to minimize retinal image speed (RIS) for both the target and the visual background (Liao et al., 2008). In this study, we found that motion of the visual background may influence tVOR performance compared with viewing a stationary background. Thus, when the platform and target were moving in opposite directions (180° phase shift), assuming the subject's gaze remained on the near target, then retinal image due to background motion should be minimized and we postulate that tVOR responsivity should be maximized. Conversely, when the platform and target were moving in the same direction (0° phase shift), then assuming the subject's gaze remained on the near target, retinal image should be maximized, and we postulate that tVOR responsivity should be minimized. The responses shown in Fig. 2 generally conform to these predictions, but more studies are needed to confirm them.

In our prior study, based on measured tVOR responses during viewing of targets at three viewing distances (2 m, 40 cm, 17 cm), we were able to calculate geometric predictions of retinal image velocity (Schwarz and Miles, 1991), assuming a compensation gain of 0.6. The predictions of this simple geometric model generally fitted the measured values of RIS well (Fig. 3 — curve and squares). If tVOR has a compensation gain of 0.6, it will hold peak RIS below 5°/s (a threshold for clear vision) for target distances greater than 90 cm. Note that measured peak RIS of the background lying at 200 cm (diamonds in Fig. 3), which might provide motion parallax information, is similar to that of the image of the fixation target, although opposite in direction.

Our present finding that background motion may influence tVOR led us to develop a more general optimization function that minimizes retinal slip due to both the visual target and the background. RIS, when the eye is stationary, depends on the distance (D) between target and eye, the amplitude of the head movement (A), and the frequency of oscillations (f) in hertz

$$\mathrm{RIS} = \frac{A}{D} \times f \times 360 \tag{1}$$

RIS calculated with the eye stationary also corresponds to the speed of the required eye movement to keep the fovea pointed at the target. To calculate RIS when the eye is moving with velocity $\dot{\theta}$, then if $C = A \times f \times 360$

$$\mathrm{RIS} = \dot{\theta} - \frac{C}{D} \tag{2}$$

We postulate that tVOR is set to minimize the sum of squares of RIS of both the background ($\mathrm{RIS}_{\mathrm{bk}}$) and the foreground target ($\mathrm{RIS}_{\mathrm{tar}}$). Then, the optimization function (F_{opt}) is:

$$F_{\mathrm{opt}} = \mathrm{RIS}_{\mathrm{bk}}^2 + \mathrm{RIS}_{\mathrm{tar}}^2 \tag{3}$$

Substituting $F_{\mathrm{opt}} = (\dot{\theta} - C/D_{\mathrm{bk}})^2 + (\dot{\theta} - C/D_{\mathrm{tar}})^2$ and its minimum occurs when

$$\dot{\theta} = \frac{C}{2} \times \left(\frac{1}{D_{\mathrm{bk}}} + \frac{1}{D_{\mathrm{tar}}} \right) \tag{4}$$

Taking into account the different effect of background RIS and target RIS, we assign a weight coefficient to each. Then the optimized eye velocity becomes:

$$\dot{\theta} = \frac{K_1 C/D_{\mathrm{bk}} + K_2 C/D_{\mathrm{tar}}}{K_1 + K_2} \tag{5}$$

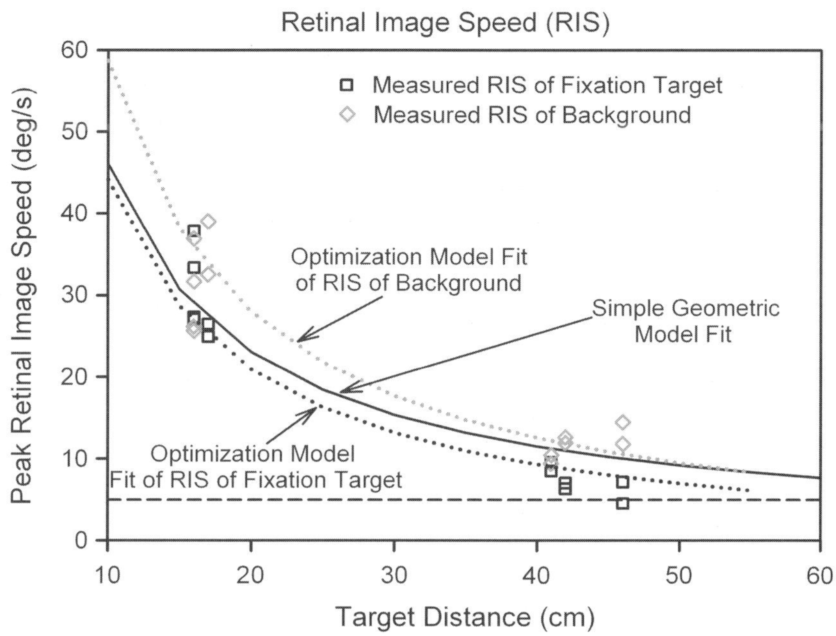

Fig. 3. Geometry of peak retinal image speed (RIS) as a function of target distance for three subjects versus their measured peak retinal image speeds. Squares indicate measured values of RIS of the fixation target at each of the target distances for each subject. Diamonds indicate measured values of RIS of the background at 200 cm, which is opposite to that of the target image motion, but is similar in magnitude The black curve defines responses based on a simple geometric model (Schwarz and Miles, 1991) with a mean compensation gain of 0.6. The dotted curves summarize the fit of the optimization model (see text) for RIS of target and background, which generally fit the data better than the simple geometric model. The dashed horizontal line corresponds to a retinal image speed of 5°, above which visual acuity for high spatial frequencies will decline.

where K_1 is the weight for the background, and K_2 the weight for the target. To determine the value of K_1 and K_2, we fit the model to measured responses from three subjects for target distances of 40 and 17 cm, and a background distance of 2 m by minimizing the error of predicted eye velocity to the measured eye velocity with a least-squares method. Thus we obtain

$$K_1 = 0.43, K_2 = 0.57$$

We then use this optimization model to fit the RIS of targets (black dotted line) and RIS of the background (grey dotted line) in Fig. 3. The optimization model generally gave better fits of the measured RIS of subjects for target or background than the curve based on the simple geometric model with a compensation gain of 0.6. Experiments are required to test the predictions of this model, such as varying the background distance. This model proposes a new interpretation of the purpose of tVOR: to control retinal image motion of objects at different distances so that motion parallax information is optimized.

Acknowledgements

This study was supported by NASA/NSBRI NA00208, Office of Research and Development, Medical Research Service, Department of Veterans Affairs, NIH grant EY06717, and the Evenor Armington Fund. We are grateful to Drs. Harold Bedell, David Zee, and Gary Paige, for their helpful advice, and to Ulrich Büttner for critical review of the manuscript. The work reported in this paper constitutes research performed by Ke Liao as part of requirements for his Doctoral Dissertation.

References

Das, V.E., DiScenna, A.O., Feltz, A., Yaniglos, S.S. and Leigh, R.J. (1998) Tests of a linear model of visual-vestibular interaction using the technique of parameter estimation. Biol. Cybern., 78: 183–195.

Gianna, C.C., Gresty, M.A. and Bronstein, A.M. (1997) Eye movements induced by lateral acceleration steps. Exp. Brain Res., 114: 124–129.

Gresty, M.A., Bronstein, A.M. and Barratt, H. (1987) Eye movement responses to combined linear and angular head movement. Exp. Brain Res., 65: 377–384.

Huebner, W.P., Leigh, R.J., Seidman, S.H., Thomas, C.W., Billian, C., DiScenna, A.O. and Dell'Osso, L.F. (1992) Experimental tests of a superposition hypothesis to explain the relationship between the vestibuloocular reflex and smooth pursuit during horizontal combined eyehead tracking in humans. J. Neurophysiol., 68: 1775–1792.

Israël, I. and Berthoz, A. (1989) Contribution of the otoliths to the calculation of linear displacement. J. Neurophysiol., 62: 247–263.

Leigh, R.J., Maas, E.F., Grossman, G.E. and Robinson, D.A. (1989) Visual cancellation of the torsional vestibulo-ocular reflex in humans. Exp. Brain Res., 75: 221–226.

Liao, K., Walker, M.F., Joshi, A., Reschke, M.F. and Leigh, R.J. (2008) Vestibulo-ocular responses to vertical translation in normal human subjects. Exp. Brain Res., 185: 553–562.

Nakayama, K. (1985) Biological image motion processing: a review. Vision Res., 25: 625–660.

Paige, G.D. (1989) The influence of target distance on eye movement responses during vertical linear motion. Exp. Brain Res., 77: 585–593.

Ramat, S., Straumann, D. and Zee, D.S. (2005) The interaural translational VOR: suppression, enhancement and cognitive control. J. Neurophysiol., 94: 2391–2402.

Ramat, S. and Zee, D.S. (2003) Ocular motor responses to abrupt interaural head translation in normal humans. J. Neurophysiol., 90: 887–902.

Schwarz, U. and Miles, F.A. (1991) Ocular responses to translation and their dependence on viewing distance. I. Motion of the observer. J. Neurophysiol., 66: 851–864.

C. Kennard & R.J. Leigh (Eds.)
Progress in Brain Research, Vol. 171
ISSN 0079-6123

CHAPTER 4.12

Dynamics of binocular fixation of targets during fore-aft motion

Sergei. B. Yakushin[1,*], Dmitri A. Ogorodnokov[1], Michael Kunin[2], Bernard Cohen[1] and Theodore Raphan[1,2]

[1]*Department of Neurology, Mount Sinai School of Medicine, New York, NY, USA*
[2]*Department of Computer and Information Science, Brooklyn College of the City University of New York, Brooklyn, NY, USA*

Abstract: When subjects move in the fore-aft direction, version and vergence are generated through the linear vestibulo-ocular reflex (lVOR) and pursuit systems to maintain binocular fixation. While the dynamics of pursuit and the lVOR have been studied for monocular gaze, the frequencies over which these systems operate to maintain binocular fixation are not known. The frequency characteristics of pursuit and the lVOR and their interaction were determined in two monkeys that were trained to fixate targets at vertical, lateral, and oblique positions on a computer screen. Animals were oscillated at 0.05 Hz to 4 Hz along their naso-occipital axes while fixating the targets (lVOR + pursuit). Animals were also stationary and fixated targets on the screen moving horizontally, vertically, or obliquely over $\approx 3^\circ$ at various frequencies up to 1 Hz (pursuit). Eye movements were recorded with dual scleral search coils, and Fick angles were computed for each eye and compared to target positions. When animals pursued targets moving on the screen at low frequencies (0.05 Hz–0.5 Hz), the gains were close to unity, and the eyes were in-phase with the target. Pursuit gains decreased to ≈ 0.5 and phases lagged by $\approx 35^\circ$ as the frequencies of target oscillation increased above ≈ 0.5 Hz. When fixating the targets during fore-aft oscillation (lVOR + pursuit), the gain was close to 1.0, and phases were within 10°–15° for frequencies up to 3 Hz. Thus, although pursuit alone does not contribute to visual following at higher frequencies, the interaction of binocular pursuit and the lVOR supported accurate binocular fixation for frequencies up to 3 Hz.

Keywords: monkey; pursuit; translation; linear VOR

Introduction

Binocular fixation of targets while in motion requires the coordination of both eyes by the angular and linear vestibulo-ocular reflexes (aVOR, lVOR) as well as ocular pursuit. When moving in the fore-aft direction, convergence movements also need to be generated, dependent on the binocular gaze point in three dimensions (Paige and Tomko, 1991a, b; McHenry and Angelaki, 2000). There has been considerable work to show that the eye velocity sensitivity is a function of gaze direction and viewing distance, and that it generally follows predictions based on geometric dependencies required for the maintenance of foveal visual acuity (Paige and Tomko, 1991a, b; McHenry and Angelaki,

*Corresponding author. Tel.: + 212 241 7068;
Fax: + 212 831 1610; E-mail: sergei.yakushin@mssm.edu

DOI: 10.1016/S0079-6123(08)00644-4

2000). However, the nature of the interaction between binocular pursuit, the lVOR, version, and vergence is not well understood. For example, the aVOR, lVOR, and pursuit, which presumably coordinate the compensatory versional and vergence eye movements, have different frequency characteristics.

The vestibular system can provide compensatory eye movements for angular and linear head movement up to 10 Hz–20 Hz (Minor et al., 1999; Ramachandran and Lisberger, 2005), but the gain of the compensatory lVOR, while high at high frequencies and viewing close targets, is small at low to midband frequencies (Paige and Tomko, 1991a, b; Schwarz and Miles, 1991). Pursuit complements the aVOR and lVOR. When subjects pursue sinusoidal targets moving at small amplitudes in one or two dimensions, the eyes follows the target with a 3-dB cut-off at 1 Hz (Fender and Nye, 1961; Wyatt and Pola, 1983). Above 1 Hz, ocular pursuit is diminished, and the saccadic system corrects for eye position errors. At higher frequencies visual fixation is obscured when pursuit is diminished and the saccadic refractory period prevents the eye from following the target smoothly (Robinson, 1965; Fuchs, 1967). Here, we investigated the effectiveness of 3-D binocular gaze fixation as a function of frequency for lVOR-pursuit interactions and when vergence was dynamically modulated to maintain binocular fixation of targets during fore-aft linear motion.

Methods

The experiments, which were performed on two rhesus monkeys (*Macaca mulatta*, M0652, M3082), conformed to the Guide for the Care and Use of Laboratory Animals and were approved by the Institutional Animal Care and Use Committee. Surgical procedures were done in two stages under anaesthesia in sterile conditions. First, a head mount was implanted on the skull to provide painless head fixation in stereotaxic coordinates during testing (Yakushin et al., 2000). Two weeks later, two coils were implanted in both eyes. One coil measured the horizontal and vertical components of eye position (Robinson, 1963; Judge et al., 1980). Another coil, placed approximately orthogonal to the frontal coil (Cohen et al., 1992), was used to measure the torsional component of eye position.

Eye movements recording

Eye movements were recorded using a rotating magnetic field system (Ogorodnikov et al., 2007), which only required the calibration of the primary position for each eye in order to compute eye orientations. Eye movements were calibrated by presenting a target on a computer screen located 276 mm in front of the animal. Animals first fixated with the right eye with the left eye covered to establish the reference position signals and corresponding rotation matrix. A similar procedure was used to obtain the reference position and corresponding rotation matrix for the left eye. Movement of each eye was obtained by computing the rotation matrix relative to that at the reference position. Vergence of the eyes was computed as a relative rotation matrix between the two eyes, from which roll, pitch, and yaw components were derived as Fick angles.

Saccades were identified and removed using a maximum likelihood ratio criterion (Singh et al., 1981). The residual signals were visually inspected and eye positions were compared to actual visual target positions for the left and right eyes separately. If either eye deviated more that 1° from expected target location, the data for both eyes in this portion of the record were eliminated from further analyses.

The gain of the oculomotor responses was computed based on the amplitudes of the roll, pitch, and yaw components of eye movement and target movement in head coordinates. Figure 1 shows a typical example of data obtained when the animal visually fixated a stationary target while the chair was oscillated fore-aft (left column) or while pursuing the visual target with the head stationary in space (right column). The Fick coordinates of the visual target for the left and right eyes as well as the degree of convergence necessary to fixate the target binocularly were computed (Fig. 1D–I, grey traces). The desaccaded eye position trace was fit with a sinusoid at the stimulus frequency. The gain of the yaw component of the movement of the left eye, for example, was determined as the amplitude

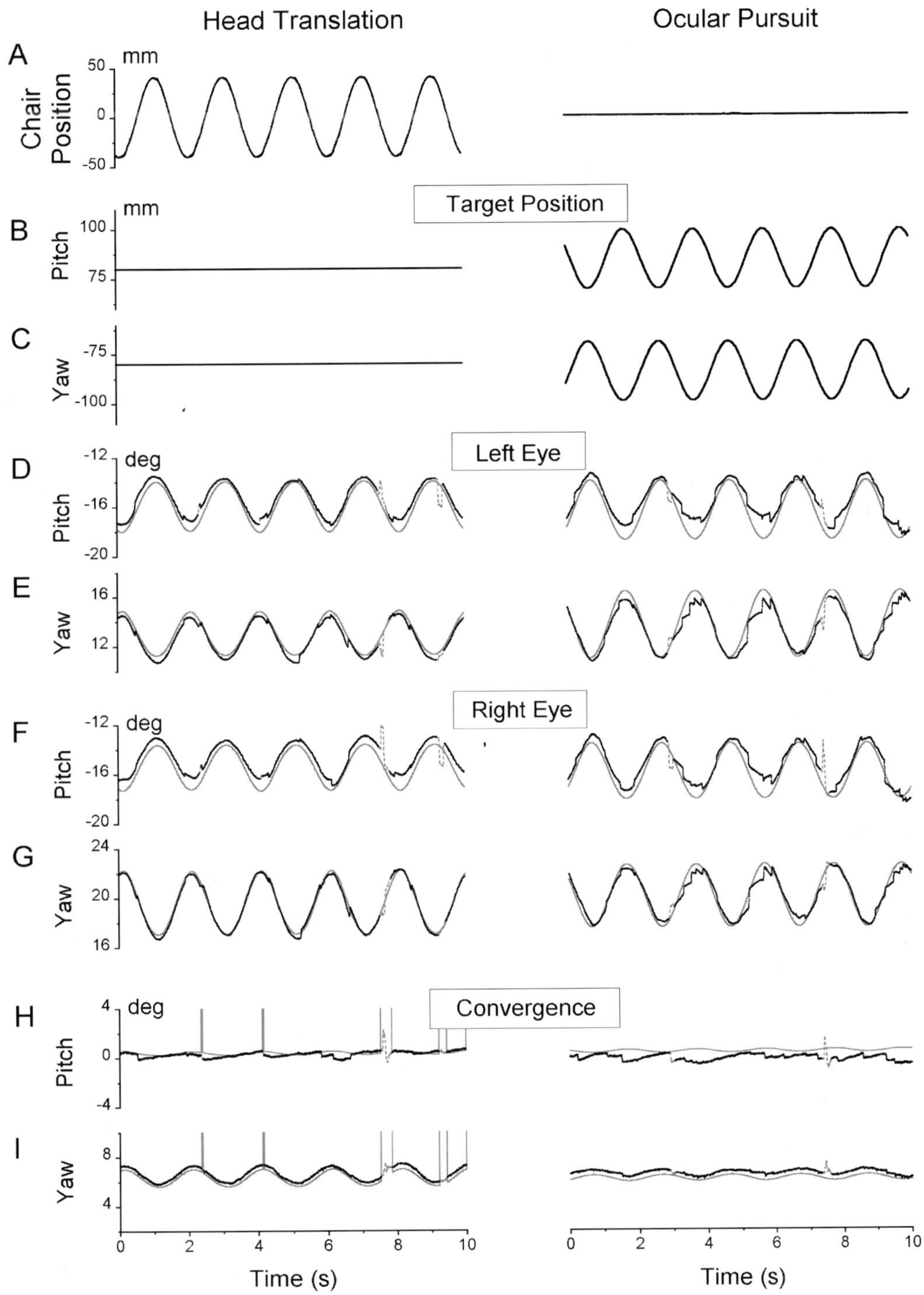

Fig. 1. Left column, Eye movements evoked by fore-aft oscillation of the animal at 0.5 Hz while visually fixating target located in the upper left corner of the screen (Target #3). Right column, Sinusoidal pursuit of the target (Target #3, see Fig. 2A, B) oscillated diagonally with 4° amplitude, about the same location as on the left plot.

of a sinusoidal fit through the yaw eye position, divided by the amplitude of the sinusoidal fit through the computed yaw target position. Similarly, the gain was computed for the pitch component of each eye separately. At least four cycles of oscillation were taken to compute the gain at 0.05 Hz and more than 10 cycles to compute gains at all other frequencies.

Stimuli

Linear oscillations

Animals sat in an upright position visually fixating a target in one of nine locations (Fig. 2A, crosses #1–#9): straight ahead or 17° left, right, up, down from this location or 24° (17°/sin 45°) diagonally (Fig. 2C, D). Animals were oscillated along their naso-occipital axes at frequencies that ranged from 0.05 Hz to 4 Hz. The linear acceleration was negligible at the 0.05 Hz oscillation but was 0.11g at 1 Hz. Accelerations were 0.27g, 0.38g, and 0.35g for oscillations at 2, 3, and 4 Hz, respectively. The amplitude of the oscillations varied from ±50 mm at 0.05 Hz to ±6 mm at 4 Hz.

Pursuit

Visual targets were presented at the same nine locations as above (Fig. 2A, B). The target was then oscillated with a 3° amplitude horizontally when it was at locations #1, #2, and #6; vertically when it was at locations #4 and #8 and 4° obliquely (3°/sin 45°) when it was at locations #3, #5, #7, and #9. The amplitudes and directions of eye movements were similar during pursuit to those evoked when monkeys fixated stationary targets during linear oscillations (Fig. 2E, F). Peak target velocities varied from 0.9°/s to 18.8°/s for horizontal or vertical target movements and from 1.3°/s to 26.7°/s for oblique target motions over a frequency range of 0.05 Hz to 1 Hz.

Results

When the target was placed in front of one eye and the animal was oscillated fore-aft from 0.05 Hz to 4 Hz, that eye remained stationary, and the other eye, which was laterally displaced relative to the target, verged to maintain binocular fixation on the target. The gain of the eccentric eye was close to unity at all tested frequencies ($p<0.05$). There was a phase shift that increased with frequency to ≈10° at 4 Hz, and there were no significant differences for data obtained for the left and right eyes. Targets were then presented in 2° increments from the straight ahead position for each eye up to ±10° horizontally and vertically. The target oscillation in pitch (Fig. 3A, B) and yaw relative to the head (Fig. 3C, D) was linearly related to the vertical (Z) and horizontal (Y) directions of the target relative to each eye in head coordinates. There was no pitch modulation of the target relative to the head when it was positioned horizontally at the $Z=0$ level. Similarly there was no yaw modulation when the target was positioned vertically at the $Y=0$ level. The amplitude of the pitch component of the target movement increased linearly as the visual target deviated from the $Z=0$ position, regardless of the Y position (Fig. 3A, B).

Similarly, the amount of yaw linearly increased as the eye deviated from the $Y=0$ position, regardless of its orientation in pitch (Fig. 3C, D). The movements of each eye at 1 Hz from M3082 (Fig. 3E–H) and M0652 (Fig. 3I–L) were correlated with the motion of the target relative to that eye (Fig. 3A–D). These data show that the eye moves to maintain binocular gaze accurately on target when the animals were oscillated at 1 Hz. The amount of yaw and pitch was determined by the visual target location in head coordinates.

When M0652 was oscillated at 3 Hz (±10 mm, 0.38g), the amplitude of the target displacement was less than that when oscillating 1 Hz (±30 mm, 0.11g). The amplitude of pitch (Fig. 3M, N) and yaw (Fig. 3O, P) were smaller, reflecting the reduction in target movement amplitude. Thus, binocular gaze fixation of targets in space, which could be evoked by the visual pursuit or lVOR, was accurate while translating along the naso-occipital axis over a wide range of frequencies.

We next compared how binocular fixation was maintained on target when the animal was

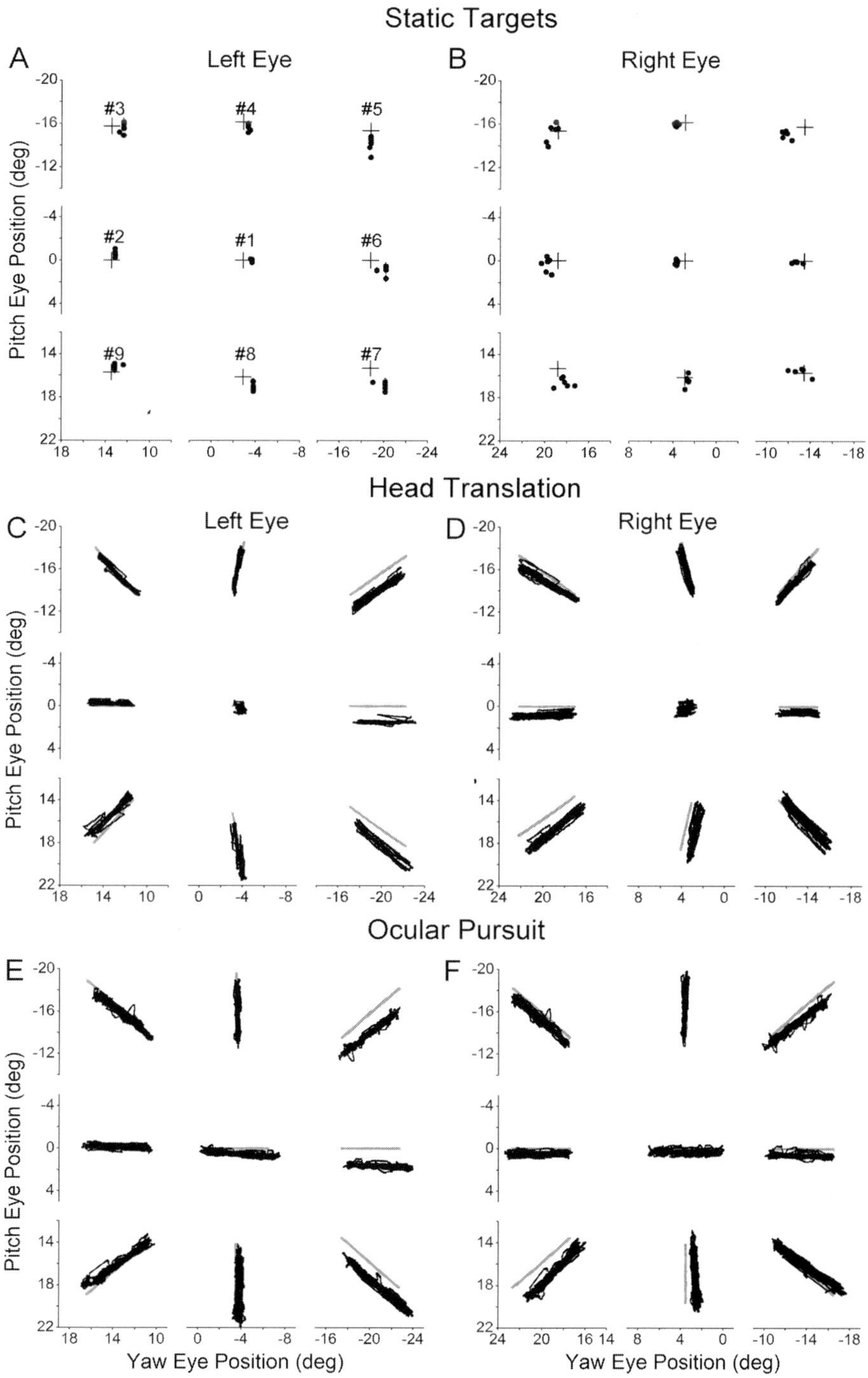

Fig. 2. (A, B) Visual fixation of static targets (#1–#9). The locations of each target (crosses) were computed for left (A) and right (B) eye. Dots about each target represent the average eye position during fixations obtained on six different experimental days. (C, D) Eye movements (black) obtained during visual fixation of each target while the animal was sinusoidally translated fore-aft at 0.5 Hz $\pm$ 50 mm. Relative movements of the target in head fixed coordinate systems (grey) associated with the left (C) and right (D) eyes, respectively. (E, F) Eye movements evoked during sinusoidal oscillation of the target in different locations. It should be noted that eye and target positions were derived as Fick angles, with leftward and downward positions as positive values according to a right hand rule.

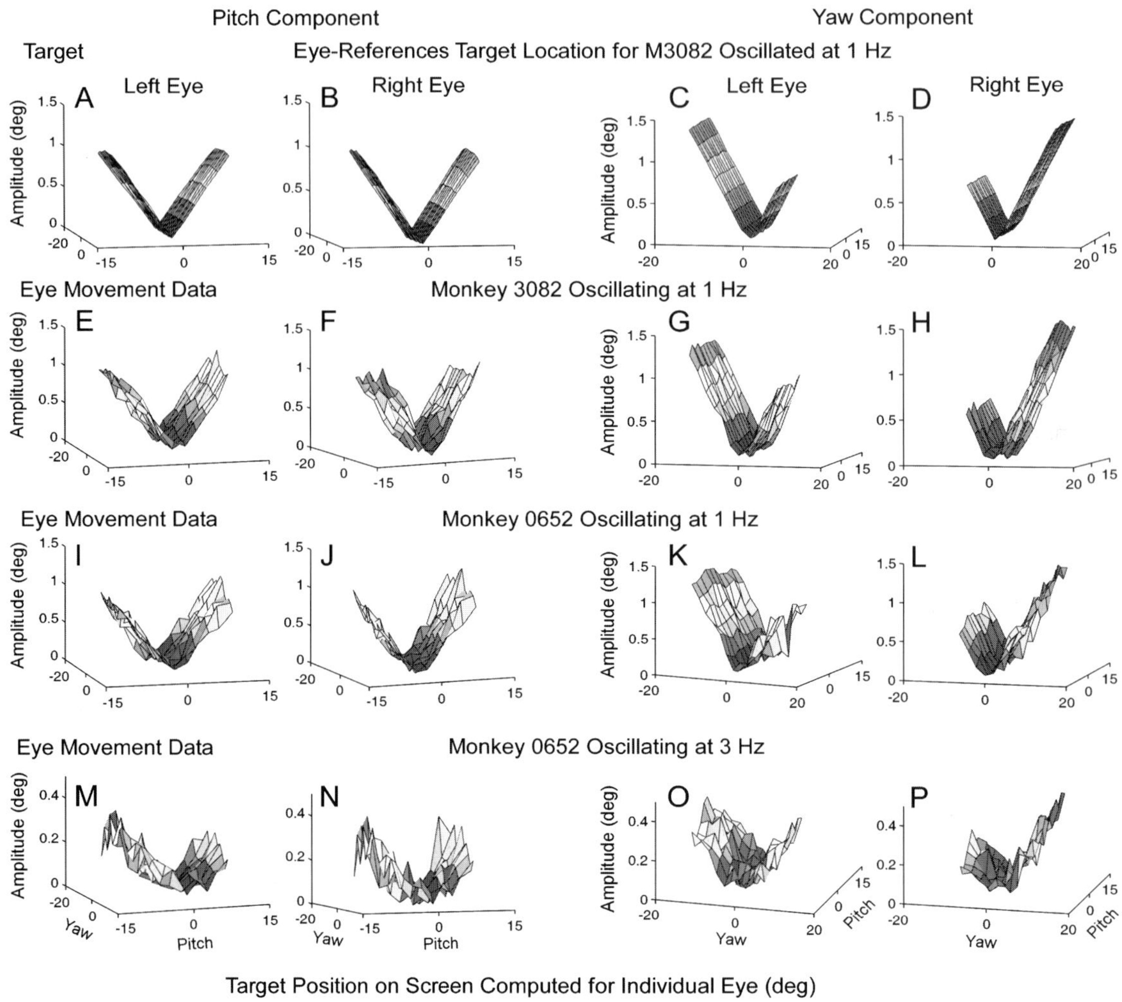

Fig. 3. Amplitude of target (A–D) and eye position (E–P) oscillation computed for each eye, when animals were sinusoidally translated in fore-aft direction while visually fixating stationary targets presented in straight ahead position or deviated left, right, up, and down up to 10° in 2° increments. Three-dimensional plots were obtained using spline interpolation techniques.

translated or was stationary and pursued eccentric visual targets. Stationary visual targets were first presented for 3 s in one of nine locations in random order. Each target was presented five times and the average pitch and yaw eye position was computed. The black dots in Fig. 2A, B represent the average eye position obtained in six experiments performed on different days. There was some disparity between the target location and average eye position when monkeys fixated the target, and this disparity was the same at different times. The disparity between target position and eye position was different in the two animals. We then tested whether the difference between the expected and measured eye positions differed for the different tests. Eye positions were obtained when animals were translated sinusoidally at 0.5 Hz (Fig. 2C, D) and when animals pursued the visual target at 0.5 Hz while the head was stationary (Fig. 2E, F). The disparity between expected and actual eye positions was the same for the two behavioural paradigms (Fig. 2), and for different test frequencies (not shown). Therefore, data from different paradigms could be compared.

Frequency response for sinusoidal pursuit

Both animals were able to pursue the target moving with $\pm 3^{\circ}$ amplitude at low frequencies binocularly, regardless of the target location on the screen. That is, when the target moved at 0.05 Hz the amplitude of eye movement corresponded to the target displacement, so the gain was ≈ 1 for each component, and there was no phase difference between eye and target locations. There were individual differences between two animals, however, as the frequency of pursuit increased. One animal (M3082) had a pursuit gain close to unity even when stimulus frequency reached 1 Hz and the target moved purely horizontal (targets #1, #2, #6) or vertical (target #4 and #8) (Fig. 4A, C, grey symbols). When the target moved obliquely (targets #3, #5, #7, #9) at the frequencies above 0.5 Hz, the gain of the yaw (Fig. 4B, grey symbols) and pitch components (Fig. 4D, grey symbols) decreased to 0.7–0.8 as stimulus frequency reached 1 Hz. The gain of the yaw and pitch pursuit of the second animal (M0652) was only accurate for lower frequencies of target motion (Fig. 4E, G, grey symbols). Gains gradually declined to 0.5 as the stimulus frequency approached 0.6 Hz–0.7 Hz. Similar but more dramatic changes were observed for both components of the oblique pursuit (Fig. 4F, H, grey symbols). Regardless of the gain differences between the two animals, the phase of the pursuit

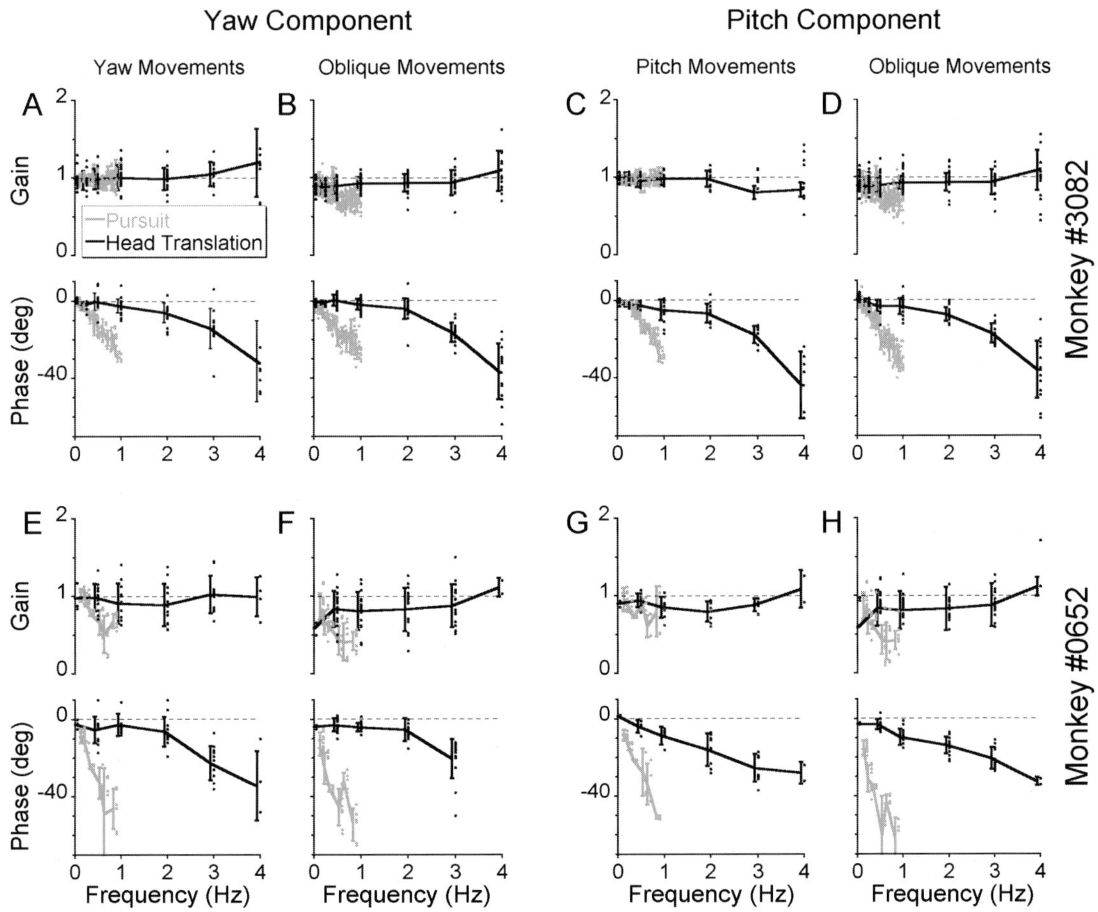

Fig. 4. Gains and phases of the eye movement evoked during sinusoidal pursuit (grey) and fore-aft oscillations (black) as a function of oscillation frequency.

was only compensatory at 0.05 Hz (Fig. 4A–H, grey symbols), and the phase lag gradually increased to $\approx 30^\circ$ as stimulus frequency approach 1 Hz. These data suggest that binocular pursuit performance is accurate only for frequencies below ≈ 0.5 Hz.

Fixation frequency responses to naso-occipital linear translation

When the animals were oscillated along the X-axis, while visually fixating a spatially fixed target, both animals had gains of ocular fixation close to 1.0 over the frequency range from 0.05 Hz to 4 Hz, regardless of whether the evoked eye movement were in yaw, in pitch, or were oblique (Fig. 4, black symbols). The phase lag was less than 20° even when the animals visually fixated the target while they were oscillated at 3 Hz, but the phase lag increased to $\approx 30^\circ$–40° at 4 Hz. Thus, animals were able to accurately fixate the visual target binocularly while translating up to 3 Hz.

Discussion

This study shows that there is accurate binocular fixation of targets located on and off centre while linearly translating in the fore-aft direction over a wide range of frequencies (0 Hz–3 Hz). This is in contrast to binocular pursuit of targets that move over the same amplitudes relative to the head. Eye movements induced by the pursuit stimulus were only accurate at frequency ranges below 1.0 Hz. When both amplitude and phase of the binocular fixation were considered, differences between movement of the animal and the target were even more striking. The phase of ocular pursuit while the animal was stationary was accurate only for frequencies below 0.5, while when moving, the phase accuracy was within 15° for frequency ranges up to 3 Hz.

It is known that the compensatory lVOR has high frequency characteristics that cut-in at approximately 0.5 Hz and that the gain is low even at higher frequencies of head oscillations (Paige and Tomko, 1991a, b; McHenry and Angelaki, 2000). Pursuit itself, however, is known to be compensatory with a gain close to 1.0 at frequencies below 1 Hz and falls off at higher frequencies (Fender and Nye, 1961; Robinson, 1965; Fuchs, 1967; Wyatt and Pola, 1983). Our data indicate that there is interaction between pursuit and the lVOR that improves ocular performance even at higher frequencies. While the mechanism for the enhancement of the gain during movement by pursuit is not known, it could be related to high frequency oscillatory pursuit mechanisms that have been identified at greater than 1 Hz (Robinson, 1965; Krauzlis and Lisberger, 1989). It also may be related to the pursuit in depth that is required by fore-aft translation compared to the tangential movement for pure pursuit. These considerations are beyond the scope of this study, but must be considered when analysing binocular pursuit in three dimensions.

The present study also verified that binocular fixation during fore-aft movement follows closely the geometrical constraints imposed on binocular eye movements by the fixation point relative to the head over a wide range of frequencies. When testing animals with 1 Hz and 3 Hz oscillations along the naso-occipital axis and fixating targets located up to 10° laterally or vertically from straight ahead position in 2° increments, the minimal ocular response of each eye occurred when the target was located in front of that eye. The amount of yaw oscillation did not depend on the pitch component of eye position and conversely the amount of pitch was only a function of pitch and not eye orientation in yaw. This is in accordance with predictions based on geometrical predictions (Paige and Tomko, 1991a, b), although some small discrepancies from exact geometrical predictions have been noted (McHenry and Angelaki, 2000).

Our experimental paradigm also tested whether animals could follow targets that demanded dynamic changes in vergence as the eyes fixated the target both during pure pursuit and during fore-aft linear motion, extending the iso-vergence paradigm used by McHenry and Angelaki (2000). Our results indicated that the requirement that animals dynamically change their vergence during both pursuit and fixation during fore-aft translation, did not compromise their ability to maintain binocular fixation at a wide range of off-centre fixation points and frequencies of oscillation.

Binocular fixation when there was head movement in the fore-aft direction was phase shifted within 10° relative to target motion. This is considerably smaller than the much larger phase shift (30°–40°) when the animals were stationary and the target was moved, generating only binocular pursuit. Thus, head movement enhances the ability to track targets binocularly, which is beyond the capabilities of just superposing pursuit and the lVOR.

Abbreviations

aVOR	angular vestibulo-ocular reflex
lVOR	linear vestibulo-ocular reflex

Acknowledgements

SUPPORT: EY04148; DC04996; EY11812; DC05204; EY01867.

References

Cohen, H., Cohen, B., Raphan, T. and Waespe, W. (1992) Habituation and adaptation of the vestibuloocular reflex: a model of differential control by the vestibulocerebellum. Exp. Brain Res., 90: 526–538.

Fender, D.H. and Nye, P.W. (1961) An investigation of the mechanisms of eye movement control. Kybernetik, 1: 81–88.

Fuchs, A.F. (1967) Saccadic and smooth pursuit eye movements in monkey. J. Physiol., Lond., 191: 609–631.

Judge, S.J., Richmond, B.J. and Chu, F.C. (1980) Implantation of magnetic search coils for measurement of eye position: an improved method. Vision Res., 20: 535–538.

Krauzlis, R.J. and Lisberger, S.G. (1989) A control systems model of smooth pursuit eye movements with realistic emergent properties. Neural Comp., 1: 116–122.

McHenry, M.Q. and Angelaki, D.E. (2000) Primate translational vestibuloocular reflexes. II. Version and vergence responses to fore-aft motion. J. Neurophysiol., 83: 1648–1661.

Minor, L.B., Lasker, D.M., Backous, D.D. and Hullar, T.E. (1999) Horizontal vestibuloocular reflex evoked by high-acceleration rotations in the squirrel monkey. I. Normal responses. J. Neurophysiol., 82: 1254–1270.

Ogorodnikov, D., Tarasenko, S., Cohen, B., Yakushin, S.B. and Raphan, T. (2007) Rotating magnetic field approach for recording binocular 3-D eye movements in monkeys. Program no. 179.176/LL172, Online. In: Society for Neuroscience, San Diego, CA.

Paige, G.D. and Tomko, D.L. (1991a) Eye movement responses to linear head motion in the squirrel monkey. I. Basic characteristics. J. Neurophysiol., 65: 1170–1182.

Paige, G.D. and Tomko, D.L. (1991b) Eye movement responses to linear head motion in the squirrel monkey. II. Visual-vestibular interactions and kinematic considerations. J. Neurophysiol., 65: 1183–1196.

Ramachandran, R. and Lisberger, S.G. (2005) Normal performance and expression of learning in the vestibulo-ocular reflex (VOR) at high frequencies. J. Neurophysiol., 93: 2028–2038.

Robinson, D.A. (1963) A method of measuring eye movement using a scleral search coil in a magnetic field. IEEE Trans. Biomed. Eng., 10: 137–145.

Robinson, D.A. (1965) The mechanics of smooth pursuit eye movement. J. Physiol., Lond., 180: 569–591.

Schwarz, U. and Miles, F.A. (1991) Ocular responses to translation and their dependence on viewing distance. I. Motion of the observer. J. Neurophysiol., 66: 851–864.

Singh, A., Thau, G.E., Raphan, T. and Cohen, B. (1981) Detection of saccades by a maximum likelihood ratio criterion. Proc. 34th Ann Conf Eng Biol, Houston, TX 136.

Wyatt, H.J. and Pola, J. (1983) Smooth pursuit eye movements under open-loop and closed-loop conditions. Vision Res., 23: 1121–1131.

Yakushin, S.B., Reisine, H., Büttner-Ennever, J., Raphan, T. and Cohen, B. (2000) Functions of the nucleus of the optic tract (NOT). I. Adaptation of the gain of the horizontal vestibulo-ocular reflex. Exp. Brain Res., 131: 416–432.

C. Kennard & R.J. Leigh (Eds.)
Progress in Brain Research, Vol. 171
ISSN 0079-6123

CHAPTER 4.13

Differential coding of head rotation by lateral-vertical canal convergent central vestibular neurons

Julia N. Eron[1,*], Bernard Cohen[1], Theodore Raphan[1,2] and Sergei B. Yakushin[1]

[1]*Department of Neurology, Mount Sinai School of Medicine, New York, NY, USA*
[2]*Departments of Computer and Information Science, Brooklyn College, City University of New York, New York, NY, USA*

Abstract: Convergent inputs from the lateral and vertical semicircular canals (LC and VC) to 31 central vestibular-only (VO) and vestibular-plus-saccade (VPS) neurons were determined by oscillating monkeys about a spatial vertical axis while the head was tilted forward and backward up to 90°. Activity of each neuron varied as a function of head tilt. Seven neurons had maximal activation when the head was tilted ≈30° forward (spatial phase), indicating convergent inputs from the LC, while peak activation of 10 units occurred with the head tilted back ≈50°, indicating VC input. Fourteen neurons had spatial phases that deviated more that 15° from the LC and VC planes, indicating convergent inputs from LC and VC. Seven of these neurons had a spatial phase less than 15° forward and 35° back, indicating canal inputs from both sides. Seven other neurons had spatial phases more that 45° forward and 65° back, indicating inputs from canals located on the same side. Thus, there are two groups of central vestibular neurons: one group responds maximally when the head is rotated about a spatial vertical axis in an upright position, declining as the head is tilted away from this position. Another group responds minimally to rotation in an upright head orientation, increasing as the head is tilted away from the upright. A majority of the cells also had convergent otolith input. The otolith and canal inputs superposed when the animals were rotated about roll and pitch axes from an upright position. This insured that these neurons would respond over a broad frequency range from very low to high frequencies.

Keywords: central vestibular neurons; semicircular canals; vestibular nuclei; monkey

Introduction

Head movements in three-dimensions during natural behaviour rarely occur about a single axis. Therefore, coding of head orientation must require interaction of inputs from the lateral and vertical canals (LC and VC) and from the otolith organs. Central vestibular neurons are known to receive convergent projections from various semicircular canals and the otolith organs (Curthoys and Markham, 1971; Baker et al., 1984; Dickman and Angelaki, 2002; Uchino et al., 2005). It is also known that neurons that receive convergent inputs from LC and VC located on different sides of the

*Corresponding author. Tel.: +212 241 7068;
Fax: +212 831 1610; E-mail: julia.eron@mssm.edu

DOI: 10.1016/S0079-6123(08)00645-6

head will respond maximally when the head is rotated about a spatial vertical axis in the upright position (Yakushin et al., 2006). Units with convergent LC and VC inputs from the same side should respond minimally to rotation about the spatial vertical with the animal upright. They should respond maximally when oscillated about a spatial vertical axis when the head is tilted. Such neurons have only been identified with orthodromic stimulations of the canal nerves (Uchino et al., 2005), but their spatial characteristics have not been studied. The purpose of this study was to characterize such units over a wide range head orientations to determine their convergent inputs quantitatively and compare them to units that have maximal response, when upright.

Methods

During experiments animals sat in the primate chair placed in the rotator with the head fixed in the stereotaxic horizontal plane (Fig. 1D) with the LC tilted up by $\approx 30^\circ$ (Yakushin et al., 1998). Single cells were recorded extracellularly in the vestibular nuclei. Details of the head holder, eye coil implantation, and unit recording techniques are described in details elsewhere (Yakushin et al., 2006). All experimental procedures were approved by the Institutional Animal Care and Use Committee of the Mount Sinai School of Medicine.

Thirty-one vestibular-only (VO) and vestibular-plus-saccade (VPS) neurons were recorded extracellulary in superior and rostral medial vestibular nuclei in two cynomolgus monkeys (*Macaca fascicularis*). To identify the canal convergent inputs, the animal's head was first sinusoidally oscillated about a spatial vertical axis at 0.2 Hz, peak velocity 60°/s, while the head was upright or tilted forward and backward up to $\pm 90^\circ$ in 15° increments. The temporal sensitivities of the unit firing rates (FR) were plotted as a function of head tilt and fit with a cosine: $y = S \times \cos(x + b)$, where S is the maximal sensitivity and b is the spatial phase (Fig. 1A–C). Positive phase values corresponds to forward and negative values to backward head tilts. We assumed that the lateral canal-related units were maximally activated when the head was tilted forward about 30°, while vertical canal-related neurons were maximally activated when the animals were tilted 50° backward (Yakushin et al., 1998, 2006). We also assumed that the units received convergent input from a single canal if the spatial phases of the response did not deviate more than $\pm 15^\circ$ from the canal plane (Reisine et al., 1988; Yakushin et al., 2005).

Static otolith-related input was determined as a projection of the otolith polarization vector onto the head horizontal plane (Schor et al., 1984). This was determined by tilting the head 30° from the upright in various head orientation in yaw (Fig. 2A, B). Units that had inputs from body proprioceptors and eye movement-related sensitivity were excluded from the analyses.

Results

Typical modulations in FR during verification of the convergent canal inputs are shown in Fig. 1A. Seven neurons had maximal FR modulation with the head was tilted $\approx 30^\circ$ ($27 \pm 12^\circ$, Fig. 1E, filled circles), indicating input from the LC while 10 neurons were maximally activated $-56 \pm 10^\circ$, indicating input from the VC (Fig. 1E, filled triangles).

The predicted spatial sensitivity for neurons with LC and VC related input located on opposite sides and on the same side of the head are shown in Fig. 1F, G, respectively. Seven neurons (23%) had a spatial phase greater than the upper limits for the LC (45° forward) and VC (65° backward) planes, respectively. The FR of such units had little modulation in the upright position, but the FR was in-phase with head velocity when the animal was tilted forward and out-of-phase when tilted backward (Fig. 1A, B). The temporal sensitivity and phases of the unit plotted as a function of head tilt had a spatial phase of -76° (Fig. 1C, thick line). This could be explained if the neuron received inputs from the LC and VC located on the same side (Fig. 1G). The spatial gain of the unit was $0.82\,\text{imp} \times \text{s}^{-1}/\text{deg} \times \text{s}^{-1}$ (Fig. 1C). The VC's are minimally activated when the head is tilted forward 40° and only LC could contribute to the response (Yakushin et al., 1998, 2005). From these data, the LC spatial sensitivity was estimated as $0.27\,\text{imp} \times \text{s}^{-1}/\text{deg} \times \text{s}^{-1}$ (Fig. 1C). Similarly the

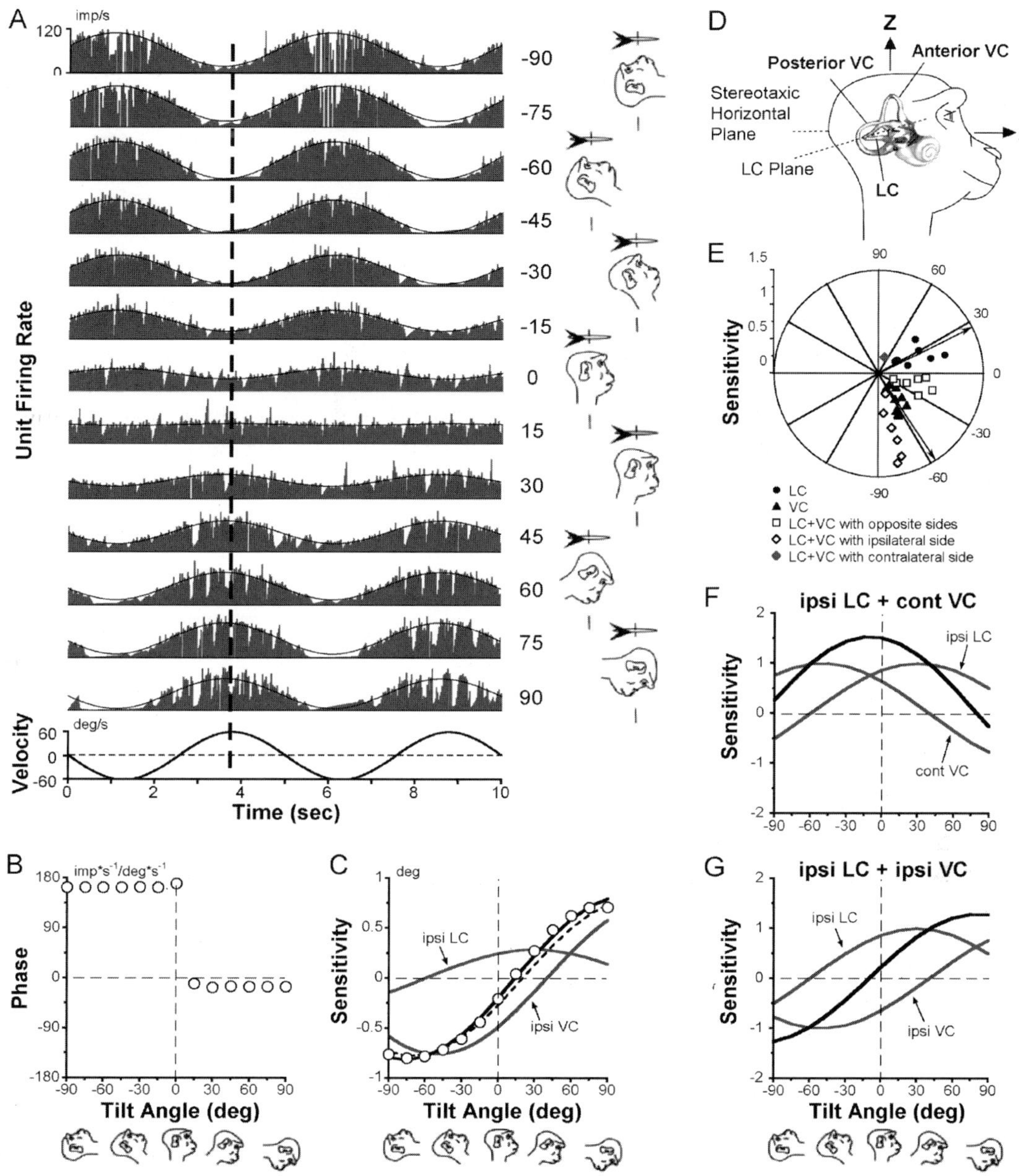

Fig. 1. (A) Modulation of the firing rate (FR) of a vestibular neuron during rotation about a spatial vertical axis with the head tilted forward and backward in 15° increments. (B) Temporal sensitivities and phases (C) of the neuron plotted as function of head orientation and fitted by a cosine (dark line). Grey lines are predictable changes of sensitivity for LC and VC-related inputs (arrows). (D) Coordinate system used in this study and the position of right labyrinth of a monkey. (E) Polar plots of the spatial sensitivities and phases of all tested units. Arrows are the average planes of LC and VC related units. (F, G) Models of convergent inputs from semicircular canals located on opposite side (F) and same side (G). Grey lines are ideal models of the sensitivity for lateral and vertical canals (see arrows), where spatial sensitivity (S) is assumed as 1 for each canal; spatial phase (b) is 30° for LC and −50° for VC. Dark line is a sum of two model sinusoids for LC and VC.

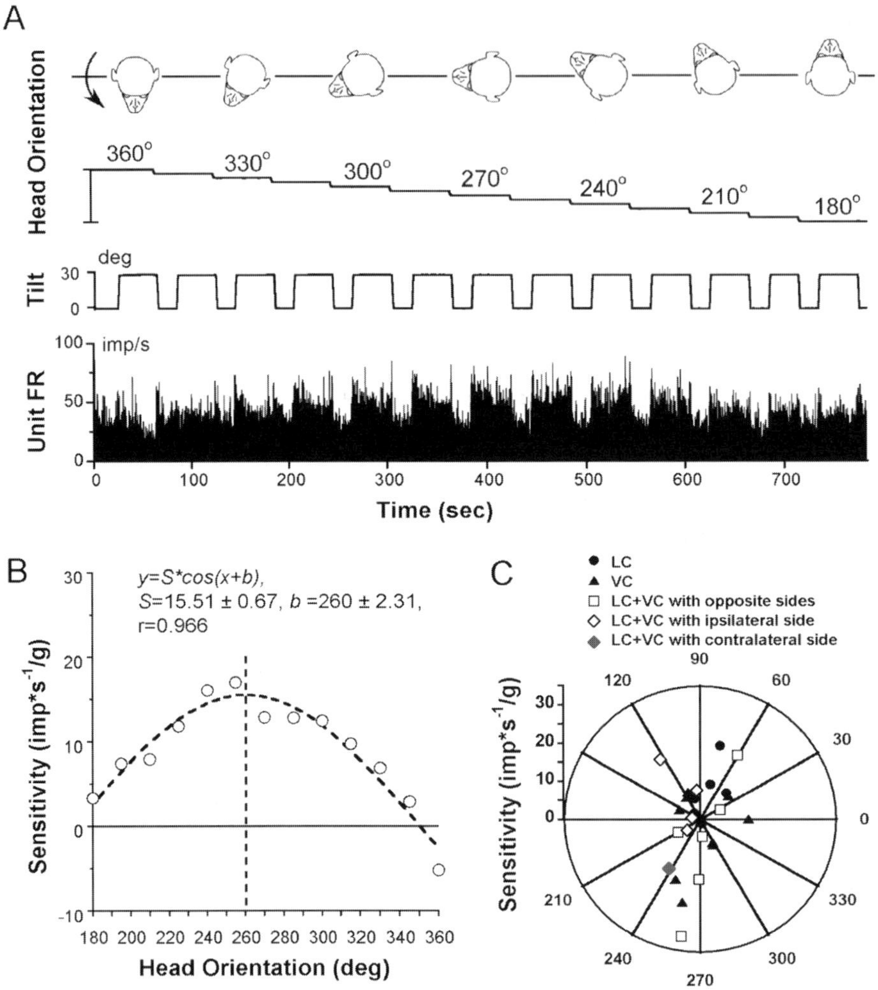

Fig. 2. Determination of otolith inputs. (A) Effect of static tilts of 30° on the firing rate of a VO neuron. The head was oriented about a yaw axis in 15° increments over 180°. (B) Temporal sensitivity plotted as a function of head orientation in yaw. (C) Polar plot of the spatial sensitivities and phases for the otolith input of units.

VC sensitivity was estimated as $0.78\,\mathrm{imp} \times \mathrm{s}^{-1}/\mathrm{deg} \times \mathrm{s}^{-1}$ from the data with the head tilted back 60° (Fig. 1C). When the two sinusoids were summated, the predicted (dotted line) and measured (dark line) sensitivities did not differ significantly ($p > 0.05$). Analysis of the remaining neurons indicates that six neurons received inputs from the ipsi-LC and VC, and one from the contra-LC and VC. The average spatial phase of neurons with ipsilateral canal-convergent inputs was $-76 \pm 4°$. Neuron with contra-LC and VC related inputs had spatial phase of 69°.

The spatial phases of the other seven neurons were less then 15° and greater than −35° ($-19 \pm 9°$, Fig. 1E, open squares), indicating the convergent inputs from LC and VC were from opposite sides of the head (Fig. 1F). The FR of these neurons was maximally modulated when the head was oscillated about the spatial vertical, close to the upright position.

Most of the tested neurons (27/31, 87%) had a significant static otolith input. For each unit, the spatial sensitivity and phase were determined by the amplitude (S) and phase (b) of the cosine fit through the data (Fig. 2B). When individual sensitivities were plotted versus phase in a polar plot (Fig. 2C), there was no predominant orientation of the otolith input for either class of units.

Discussion

This study showed that there are VO and VPS neurons that have convergence of ipsilateral LC and VC input, which had previously been suggested from experiments based on orthodromic stimulation (Uchino et al., 2005). Seven of the 31 (23%) neurons reported in this study did not respond to oscillation about a spatial vertical axis when the animals were close to the upright position, but had substantial modulation when the head was close to the prone and supine positions. Based on model predictions (Yakushin et al., 1998, 2006), we suggest that these neurons receive convergent input from the LC and VC in the labyrinth located on the same side. These units have not been reported previously, possibly due to difficulties in identifying them, since oscillation near the upright position does not modulate their activity to reveal their vestibular nature. In the present study, however, the searching stimulus was random oscillation in the each of three-canal planes. All of these neurons were recorded in the vicinity of the superior vestibular nuclei. The large number of these neurons (23%) indicates, that they were functionally important. That is one class of these units are silent when the head is rotated about spatial vertical axis in upright, but this class responds as the head is deviated in any direction from the upright.

Seven other neurons (23%) received convergent inputs from LC and VC located on the opposite sides, optimally responding to head rotation about the yaw (*Z*) axis. However, when the head was rotated about a spatial vertical axis, their activation was just opposite to that of the neurons described above. That is they were maximally activated in animals, which were close to the upright position, and the FR's decreased as the head was tilted in any direction. Thus, neurons with LC and VC inputs from the same and from opposite sides form two populations of units (46%) that provide information about head rotation about a spatial vertical at various orientations relative to gravity.

These data demonstrate that different classes of VO and VPS neurons may have different functional roles in terms of how they code head perturbations. For example, neurons that have their maximal FR when the head is upright, reduce the modulation of their FR minimally when the head is tilted by a small angle. On the other hand, neurons that have a minimal response to yaw rotation when the head is upright have a greater change in temporal sensitivity to tilt perturbations. Since locomotion generates both roll, pitch and yaw head perturbations, a greater change in sensitivity to yaw rotation is required when the head is pitched or rolled. These latter types of units could be important for gaze stabilization, giving the angular vestibulo ocular reflex the appropriate sensitivity to optimize compensation for yaw rotation during locomotion.

Abbreviations

AVOR	angular vestibulo ocular reflex
FR	firing rate
LC	lateral canal
VC	vertical canal
VO	vestibular-only
VPS	vestibular-plus-saccade

Acknowledgement

This work is supported by DC04996; EY11812; EY04148; DC05204; and EY01867.

References

Baker, J., Goldberg, J., Hermann, G. and Peterson, B. (1984) Spatial and temporal response properties of secondary neurons that receive convergent input in vestibular nuclei of alert cats. Brain Res., 294: 138–143.

Curthoys, I.S. and Markham, C.H. (1971) Convergence of labyrinthine influences on units in the vestibular nuclei of the cat. I. Natural stimulation. Brain Res., 35: 469–490.

Dickman, J.D. and Angelaki, D.E. (2002) Vestibular convergence patterns in vestibular nuclei neurons of alert primates. J. Neurophysiol., 88: 3518–3533.

Reisine, H., Simpson, J.I. and Henn, V. (1988) A geometric analysis of semicircular canals and induced activity in their peripheral afferents in the rhesus monkey. Ann. N.Y. Acad. Sci., 545: 10–20.

Schor, R.H., Miller, A.D. and Tomko, D.L. (1984) Responses to head tilt in cat central vestibular neurons. I. Direction of maximum sensitivity. J. Neurophysiol., 51: 136–146.
Uchino, Y., Sasaki, M., Sato, H., Bai, R. and Kawamoto, E. (2005) Otolith and canal integration on single vestibular neurons in cats. Exp. Brain Res., 164: 271–285.
Yakushin, S.B., Raphan, T., Büttner-Ennever, J.A., Jun-Ichi Suzuki, J.-I. and Cohen, B. (2005) Spatial properties of central vestibular neurons of monkeys after bilateral lateral canal nerve section. J. Neurophysiol., 94: 3860–3871.
Yakushin, S.B., Raphan, T. and Cohen, B. (2006) Spatial properties of central vestibular neurons. J. Neurophysiol., 95: 464–478.
Yakushin, S.B., Raphan, T., Suzuki, J.-I., Arai, Y. and Cohen, B. (1998) Dynamics and kinematics of the angular vestibulo-ocular reflex in monkey: effects of canal plugging. J. Neurophysiol., 80: 3077–3099.

C. Kennard & R.J. Leigh (Eds.)
Progress in Brain Research, Vol. 171
ISSN 0079-6123

CHAPTER 4.14

Cyclovergence evoked by up–down acceleration along longitudinal axis in humans

I. Olasagasti[1,*], C.J. Bockisch[1,2,3], D.S. Zee[4] and D. Straumann[1]

[1]*Department of Neurology, Zurich University Hospital, Zurich, Switzerland*
[2]*Department of ENT, Zurich University Hospital, Zurich, Switzerland*
[3]*Department of Ophthalmology, Zurich University Hospital, Zurich, Switzerland*
[4]*Department of Neurology, Johns Hopkins Hospital, Baltimore, MD, USA*

Abstract: We present results of a study of torsional eye movements evoked by earth-vertical accelerations along the subject's longitudinal axis. The earth-vertical stimulus leads to a gravito-inertial acceleration vector that changes magnitude but not direction. It can therefore be viewed as a dynamic change of the gravity level. Up–down oscillations induced relatively symmetric cyclovergence (0.6–2.2° peak-to-peak). Eyes intorted/extorted for higher/lower effective gravity. The phase of this modulation was small relative to chair acceleration. We contrast this behaviour to the dynamics of cycloversion in response to interaural acceleration, which shows a considerably larger phase lag. This strikingly different dynamics suggest a different processing of otolith signals during interaural and longitudinal stimulation.

Keywords: disconjugate torsion; orienting responses; linear acceleration; gravity; otoliths

Introduction

Earth-horizontal acceleration along the longitudinal axis (z-axis) elicits ocular cycloversion (conjugate torsion of the eyes) in side position (Merfeld et al., 1996) and cyclovergence (oppositely directed torsion) in supine position (de Graaf et al., 1996; Merfeld et al., 1996). While the former could be interpreted as a compensatory response to the changing orientation of the gravito-inertial force vector (GIFV) in the frontal plane, the latter awaits an explanation. So far, studies did not give a detailed characterization of this response, and their stimuli varied not only the projection along the longitudinal axis but also generated a swing of the gravito-inertial vector in the sagittal plane. Reports of static cyclovergence to quasi-static changes of gravity have also been reported during parabolic flights (de Graaf et al., 1996).

Methods

On a motion simulator with six degrees of freedom, healthy human subjects seated upright were oscillated along an earth-vertical (to induce cyclovergence, $N=9$) or earth-horizontal line (to induce cycloversion, $N=3$) with the head (thermoplastic mask) and body (4-point seatbelts, evacuation pillows) restrained. Chair motion was sinusoidal with varying frequencies and peak accelerations (earth-horizontal: 0.1–0.4g, 0.5–1.25 Hz; earth-vertical: 0.4–0.7g, 1–2 Hz; leftward and upward chair

*Corresponding author. Tel: +41 44 255 39 96;
Fax: +41 44 255 45 07; E-mail: miren.olasagasti@usz.ch

DOI: 10.1016/S0079-6123(08)00646-8

motions are positive). Experiments were performed in the dark. Subjects fixated a platform-fixed LED 1.4 m straight ahead, which suppressed horizontal and vertical, but not torsional eye movements. Eye positions were recorded binocularly with dual search coils. During earth-horizontal accelerations, the motion of the head within the coil frame was monitored with a third coil placed on a bite-piece.

Coil signals were transformed into rotation vectors (positive signs: leftward, downward, and clockwise from the subject's viewpoint). Angular velocities were calculated for both eyes and the head. Finally, head angular velocity was subtracted from eye angular velocity to obtain the angular velocity of the eyes in the head. We defined cycloversion as the average torsion of both eyes and cyclovergence as the difference in torsion (right eye minus left eye). With this definition, positive cyclovergence corresponds to extorsion of the eyes (provided both eyes show equal movement).

For a given trial of sinusoidal chair motion, angular velocity signals were overlaid for consecutive chair-triggered cycles. The median torsional velocity of each eye was calculated and fit to a sinusoidal function that included the first two harmonics. We then analysed the gain and phase corresponding to the first harmonic.

Results

We found that during up–down acceleration, eyes intorted/extorted for higher/lower effective gravity (Fig. 1).

Cyclovergence evoked by earth-vertical acceleration (example on Fig. 2, right panel) showed a very small phase lag with respect to the driving acceleration for all stimulus frequencies. On average, the phase lag at 1 Hz stimulation frequency was $0° \pm 11$ SD. This contrasts with cycloversion evoked by interaural acceleration (example on Fig. 2, left panel), which showed a considerable phase lag. At 1 Hz, the average phase lag of cycloversion relative to chair position was $79° \pm 20$ SD).

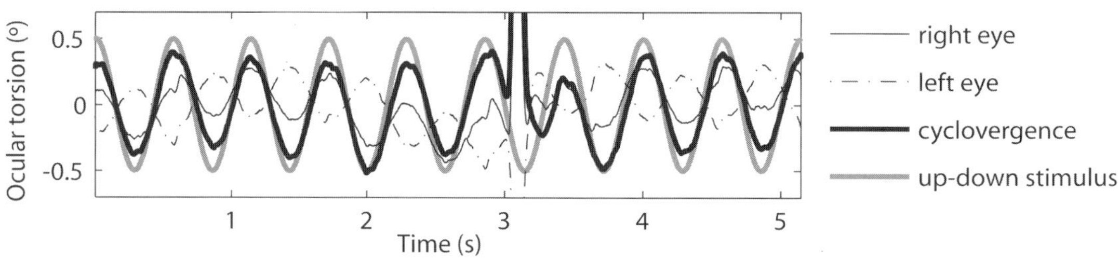

Fig. 1. Torsional component of rotation vectors for an up–down trial at 1.75 Hz and 0.46g peak acceleration. Thin solid line: torsional right eye position; dashed line: torsional left eye position; black solid line: cyclovergence; thick grey line: chair position in arbitrary units (positive upwards).

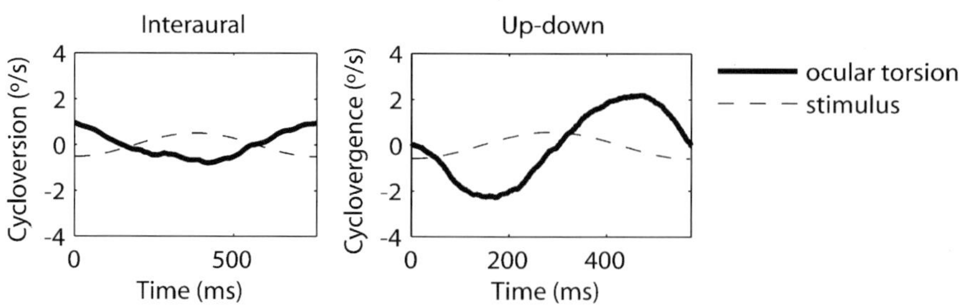

Fig. 2. Examples of binocular torsional movements (thick continuous lines: median angular velocity of nine cycles in degree per second) evoked by linear oscillation of the chair (dashed lines: in units of gravity). Left: Cycloversion during interaural acceleration. Right: Cyclovergence during up–down acceleration.

Discussion

An earlier report of eye torsion during parabolic flights found that the right eye extorted during 1.8g periods and intorted during 0g periods (de Graaf et al., 1996). This is the opposite of what we find here under dynamic conditions. Presently, we do not know the reason for this discrepancy.

Our main finding is the striking phase difference between cycloversion during interaural acceleration and cyclovergence during earth-vertical acceleration, which suggests a different dynamic processing of the interaural and longitudinal components of gravito-inertial stimulation. The phase of cyclovergence relative to chair position, which had not been reported before, remains small for the range of frequencies explored here. However, the phase lag of cycloversion relative to chair position is significantly different from zero, consistent with previous studies (Merfeld et al., 1996).

What could be the origin of the cyclovergence elicited by up–down acceleration? Cycloversional movements observed during earth-horizontal linear acceleration directed along the interaural axis can be understood as an orienting response to the reorientation of the gravito-inertial vector in the head frame, since those responses can be observed during actual reorientations of the head that lead to similar changes in the gravito-inertial vector. We cannot think of an equivalent picture for cyclovergence. If cyclovergence has some functional significance, there could be a dedicated cyclovergence system. It is also possible that cyclovergence just reflects an asymmetry between the ipsilateral and contralateral pathways descending from the otoliths to the motor neurons in the cycloversion system. Due to the overall left–right symmetry, when otolith sensors modulate symmetrically, as is the case during up–down acceleration in upright position, they send opposing torsional commands to the eyes. The only way to prevent the eyes from intorting or extorting is by having an exact balance in the inputs from the left and right otoliths. When the sensors on both sides modulate in opposite directions, as is the case during interaural acceleration, the same circuitry delivers a conjugate torsional command, and signals from both otoliths reinforce each other. Thus, an imbalance (asymmetry) between ipsilateral and contralateral projections could explain a small cyclovergence in response to linear accelerations that stimulate right and left otolith sensors symmetrically.

One has to ask whether this picture is consistent with the dynamic difference between cycloversion (evoked by interaural acceleration) and cyclovergence (evoked by earth-vertical acceleration), i.e., if the circuitry is shared, is it possible to obtain different phase lags? We think that the answer is yes. Similar to eye-velocity commands for horizontal and vertical eye movements, torsional commands from the otoliths go through parallel direct and indirect pathways. Since these structures are bilateral, the signal that arrives at a given ocular motor neuron results from a combination of four signals: the direct and indirect signals from the ipsilateral otoliths and the direct and indirect signals from the contralateral otoliths. Since the relative weight of direct and indirect pathways is what determines the phase lag and gain, conjugate and disconjugate components will, in general, show different dynamics.

Recently Maruta et al. (2008) reported that in rabbits the component of the gravito-inertial vector along the head's vertical axis modulated the strength of orienting responses. Such a modulation has also been observed in humans for ocular counterroll, which is larger during centrifugation than during static head tilt for the same level of interaural stimulation (MacDougall et al., 1999). We cannot exclude that cyclovergence is the result of such a modulation, since it is plausible that asymmetries in the gain modulation of the tonic activities could lead to cyclovergence.

Abbreviation

g magnitude of gravity ($9.81\,m/s^2$)

Acknowledgements

This study was supported by the Swiss National Science Foundation, the Betty and David Koetser Foundation for Brain Research, Zurich,

Switzerland, and the Center of Integrative Human Physiology, University of Zurich, Switzerland.

References

de Graaf, B., Bos, J.E. and Groen, E. (1996) Saccular impact on ocular torsion. Brain Res. Bull., 40(5–6): 321–326.

MacDougall, H.G., Curthoys, I.S., Betts, G.A., Burgess, A.M. and Halmagyi, G.M. (1999) Human ocular counterrolling during roll-tilt and centrifugation. Ann. N.Y. Acad. Sci., 871: 173–180.

Merfeld, D.M., Teiwes, W., Clarke, A.H., Scherer, H. and Young, L.R. (1996) The dynamic contributions of the otolith organs to human ocular torsion. Exp. Brain Res., 110(2): 315–321.

Maruta, J., Raphan, T., Simpson, J.I. and Cohen, B. (2008) Vertical (*Z*-axis) acceleration alters the ocular response to linear acceleration in the rabbit. Exp. Brain Res., 185: 87–99.

C. Kennard & R.J. Leigh (Eds.)
Progress in Brain Research, Vol. 171
ISSN 0079-6123

CHAPTER 4.15

Oblique gaze shifts: head movements reveal new aspects of component coupling

Edward G. Freedman* and Aaron L. Cecala

Department of Neurobiology and Anatomy, University of Rochester, Rochester, NY, USA

Abstract: When the head is prevented from moving, it has been clearly demonstrated that the horizontal and vertical components of oblique saccades are not independently produced. The duration of the smaller of the two components is stretched in time to match the duration of the larger component. Several hypotheses have been proposed and each can account for the observed interaction between horizontal and vertical saccade components. When the head is free to move, gaze shifts can be accomplished by combining eye and head movements. During repeated gaze shifts of the same amplitude, as head contribution increases, saccade amplitude declines but saccade duration increases. Thus, the expected relationship between duration and amplitude of saccadic eye movements can be reversed. We have used this altered relationship to determine whether the duration of the vertical saccade component is affected by the amplitude or the duration of the horizontal component. We find that the relative amplitudes of horizontal and vertical saccades cannot account for the observed temporal stretching: vertical component duration increases despite a decrease in the amplitude of the horizontal component. These results are likely inconsistent with models that rely on calculating the vector or relative component amplitudes to account for component stretching.

Keywords: eye-head coordination; oblique saccades; component coupling

Introduction

When the head is prevented from moving, saccades are characterized by a set of stereotyped relationships between peak velocity, duration, and amplitude. These stereotyped relationships (sometimes referred to as the "main sequence") have been described in detail during horizontal, vertical, and oblique movements in a variety of species (Bahill et al., 1975; Baloh et al., 1975; Guitton and Mandl, 1980; Evinger et al., 1981; vanGisbergen et al., 1985; Yee et al., 1985; King et al., 1986; Collewijn et al., 1988a, b; Becker and Jürgens, 1990; Smit et al., 1990). During oblique saccades, the duration of the smaller of the two components is stretched in time to match closely the duration of the larger component (Guitton and Mandl, 1980; Evinger et al., 1981; vanGisbergen et al., 1985; King et al., 1986; Becker and Jürgens, 1990). Although various mechanisms accounting for component stretching have been proposed, nearly all hypotheses rely on a calculation of the relative amplitudes of the vertical and horizontal components (or equivalently saccade direction) in order to scale the velocity of the two components (vanGisbergen et al., 1985; Grossman and Robinson, 1988; Becker and Jürgens, 1990; Smit et al., 1990; vanOpstal and

*Corresponding author. Tel.: +585 273 1892;
Fax: 585 756 5334; E-mail: ed_freedman@urmc.rochester.edu

DOI: 10.1016/S0079-6123(08)00647-X

Kappen, 1993; Nichols and Sparks, 1996; Quaia and Optican, 1997).

When the head is free to move, and gaze shifts are made to targets displaced both vertically and horizontally, the question of component coupling remains. However, although the eyes and head move together to accomplish large changes in the direction of the line of sight, the trajectories along which they proceed are not necessarily the same. In general the eyes accomplish a larger proportion of the vertical component of oblique gaze shifts compared to their contribution to the accomplishment of the horizontal component (Tweed et al., 1995; Freedman and Sparks, 1997b). This raises questions about component coupling and the type of neural interactions that lead to the temporal stretching of saccade components. For instance, are horizontal and vertical components coupled at the level of gaze control signals? Or are gaze commands separated into eye and head signals before the coupling of horizontal and vertical components? Few of these issues have been addressed and the majority of gaze control models have been restricted to simulation of gaze shifts along the horizontal meridian. Nonetheless, just as understanding the interactions of horizontal and vertical saccades constrained hypotheses related to the neural control of saccadic eye movements (vanGisbergen et al., 1985; Grossman and Robinson, 1988; Becker and Jürgens, 1990), so too will a clearer understanding of oblique gaze shifts constrain hypotheses designed to describe the neural control of coordinated eye–head movements in two dimensions.

Investigating oblique gaze shifts provides insight into the neural control of eye–head coordination. In addition, because of the ways in which accompanying head movements alter ongoing saccades, some of the fundamental assumptions regarding the brainstem control of saccadic eye movements may need to be reassessed. In this report, conditions are identified in which the usual saccadic duration–amplitude relationship is reversed; saccade duration *declines* as amplitude *increases*. This violation of one of the fundamental rules defining saccades is a direct consequence of two previous observations. The first is that during gaze shifts of similar amplitude made to the same target location, the relative amplitudes of the eyes and head vary inversely as a function of the starting positions of the eyes in the orbits (Delreux et al., 1991; Volle and Guitton, 1993; Freedman and Sparks, 1997b; Freedman, 2001; Stahl, 2001; Freedman, 2005). For example, a 35° gaze shift can be accomplished by combining a 10° saccadic eye movement and a head movement that contributes 25° to the change in the direction of the line of sight. A 35° gaze shift can also be accomplished with a 35° saccadic eye movement. Head movements occurring on trials like this begin after the gaze shift is complete and therefore do not contribute to its accomplishment. The second important observation is that as head movement amplitude and velocity increase, the velocity of the concurrent saccadic eye movement declines (Tomlinson and Bahra, 1986; Phillips et al., 1995; Freedman and Sparks, 1997b; Freedman and Sparks, 2000). Thus, during a series of 35° gaze shifts, when the eyes begin deviated away from the direction of the movement, the head movement will be small. The resulting saccadic eye movement will be large and fast. In contrast, when the eyes begin deviated in the direction of the gaze shift, the head contribution will be large. The concurrent saccadic eye movement will be smaller and due to the effects of the large, fast head movement the velocity of this saccade will be reduced. We will show that under these conditions the duration–amplitude relationship for saccades has a negative slope. This uncoupling of saccade duration and amplitude allows us to address the question of component coupling during oblique saccades in a new way. It is now possible to assess the differential effects of horizontal component duration and amplitude on the duration of a constant amplitude component amplitude increased, component duration also increased. The resulting hypotheses about the nature of component coupling rely on calculation of the relative amplitudes (or direction) of saccades and use this to scale the velocities of the two components. As shown in this report, the duration of the constant amplitude vertical saccade component is not altered as a function of the amplitude of the horizontal component. Temporal stretching of the vertical component duration is correlated with changes in horizontal component duration and is independent of amplitude. These data may require a new set of hypotheses to account for the interaction of horizontal and vertical components during oblique saccades.

Methods

Data from one female rhesus monkey (*Macaca mulatta*; 5 kg) are reported here. A scleral eye coil was implanted in one eye (Judge et al., 1980) and a small head-restraint post secured to the skull. Surgical procedures were carried out using general inhalation anaesthesia (isoflourane) and aseptic techniques. After full recovery, the subject was trained to make gaze shifts to visual targets for a juice reward. All surgical and experimental procedures were approved by the University of Rochester Committee on Animal Resources and comply with the NIH guide for the Care and Use of Animal.

The subject sat in a primate chair that prevented movements of the hips and shoulders but permitted unrestricted horizontal and vertical movements of the head. The chair was positioned such that the head of the monkey was located at the centre of a 1.2 m cube that housed three pairs of magnetic field coils (CNC Engineering, Seattle, WA). The four vertical faces of the cube (front–back and left–right) contained two pairs of Helmholtz coils in spatial and phase quadrature (Collewijn, 1977). The current induced in the implanted scleral coil (and similar head-mounted coil) varied linearly (within 2%, over ~360°) as a function of horizontal angle between the eye and head coils and the front vertical face of the cube. The two horizontal faces of the cube (top and bottom) contained a third pair of Helmholtz coils for measurement of the vertical angle of the scleral and head-mounted coils. A 5-pole low pass Bessel filter with a cutoff frequency of 3 kHz was used to remove the coil system carrier frequency. A second low-pass filter (1-pole, $\tau = 0.3$ ms) conditioned coil signals before digitizing. This filter will introduce no more than a 0.2 ms delay in signals with frequency content between 0 and 530 Hz.

A lightweight cam-lock device was secured to the head restraint post. This device carried a coil similar to that implanted in the eye in order to measure head position. In addition, three small laser diodes (red: 650 nm) were daily secured to the head. The central laser was aligned with the mid-sagittal plane of the subject, the second laser was directed ~18° to the left of centre, and the third was directed ~18° to the right.

Each trial began with the illumination of one of the three head-mounted lasers and a central fixation light. The subject was trained to look at the central light and also to align the head-mounted laser spot with the central fixation spot. When the central head-mounted laser was illuminated, this served to align the eyes and head. When one of the other head-mounted lasers was illuminated the eyes began deviated either in the direction of the ensuing gaze shift or in the opposite direction. Initial eye positions ranged from 19.6° in the direction of the movement to 23.2° in the opposite direction. After a variable delay the gaze shift target was illuminated. The subject was required to maintain fixation and head-laser alignment for a variable delay period that ended with the turning off of both the central fixation spot and the head-mounted laser. This was the cue to initiate a gaze shift to the still illuminated target. Movements to many different target locations were randomly interleaved. For this report only movements to two targets are presented. Data along the horizontal meridian directed at a target displaced 35° to the left, and oblique gaze shifts directed at a target displaced 35° left and 35° down.

Coil signals were digitized at 1 kHz and stored for off-line analyses (Matlab, Natick, MA). All behavioural contingencies were executed in real time with 1 ms resolution using custom data acquisition and behavioural control software.

Results

As stated in the Introduction, during gaze shifts of similar amplitude and direction, when the eyes begin in different orbital positions, the linear relationship between saccade duration and amplitude can be reversed. An example of this type of reversal is shown in Fig. 1. This figure plots the duration of horizontal saccades as a function of their amplitude. Gaze shifts were made from the fixation target (0,0) to a target 35° to the left (note that absolute value of horizontal saccade amplitude is plotted along the abscissa in this figure). During 526 gaze shifts (these data were extracted

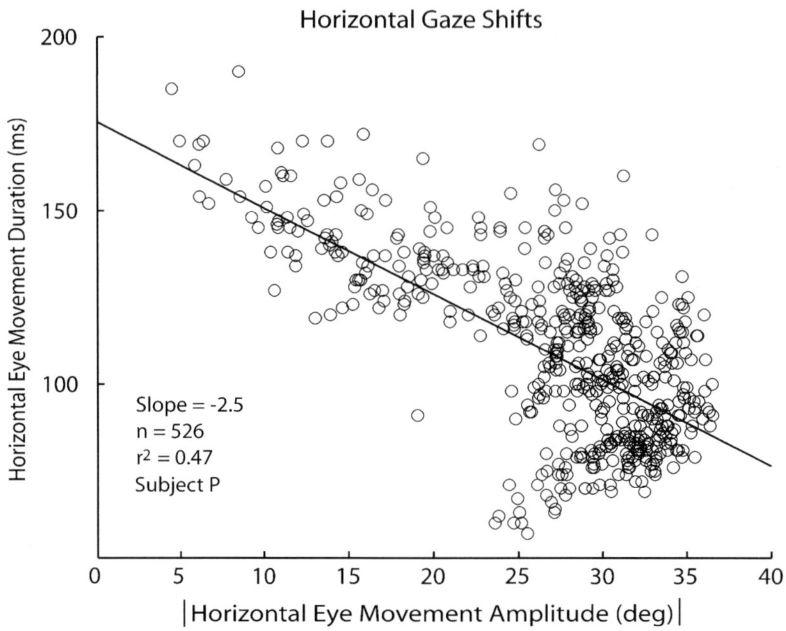

Fig. 1. Horizontal saccade duration plotted as a function of the absolute value of horizontal saccade amplitude. This set of movements occurred during head unrestrained gaze shifts to the same target displaced 35° to the left along the horizontal meridian. Initial orbital positions of the eyes varied as a result of the head-alignment requirements of the task. Consequently the amplitude of saccades varied (from 5° to 37°) during constant amplitude gaze shifts. As shown the line of best fit to these data has a negative slope; the duration–amplitude relationship was reversed.

from multiple data collection sessions; a variety of target locations were randomly interleaved during any particular session) the initial orbital positions of the eyes varied from ~20° in the direction of the gaze shift to ~20° in the opposite direction. As a result the amplitude of the saccadic eye movements that made up these gaze shifts varied from 5° to >35°. Because gaze shift amplitude varied very little during these movements (mean (SD) = 35.2° (1.6)), when saccade amplitudes were small the associated head movements were necessarily large. Because saccade velocities are reduced when associated with large, fast head movements (Freedman and Sparks, 2000), as saccade amplitude declined, head amplitude and velocity increased leading to a decline in saccade velocity. This results in the negative slope of the line of best fit for the duration–amplitude relationship for saccades made under these conditions (Fig. 1; slope = −2.5).

During oblique gaze shifts to a target located 35° to the left and 35° down relative to the central fixation target, the vertical positions of the eyes at the beginning of movements did not vary. However, as a result of the head-alignment requirements of the task, the horizontal positions of the eyes varied over the central 40° relative to the head. The vertical amplitude of eye movements made during gaze shifts to these targets varied very little. However, under these conditions, the amplitude of the horizontal component of saccades varied as did the duration of the horizontal component. And, as in Fig. 1, the slope of the duration–amplitude relationship was negative. The effects of this altered duration–amplitude relationship are illustrated in Fig. 2, during three example movements. Horizontal and vertical saccade positions are plotted as functions of time for three gaze shifts (Fig. 2A). One of these movements began with the eyes and head nearly aligned (dark grey), a second movement began with the horizontal positions of the eyes deviated away from the direction of the movement (black), and the third began with the eyes deviated in the direction of the ensuing saccade (light grey). Note that the initial vertical positions

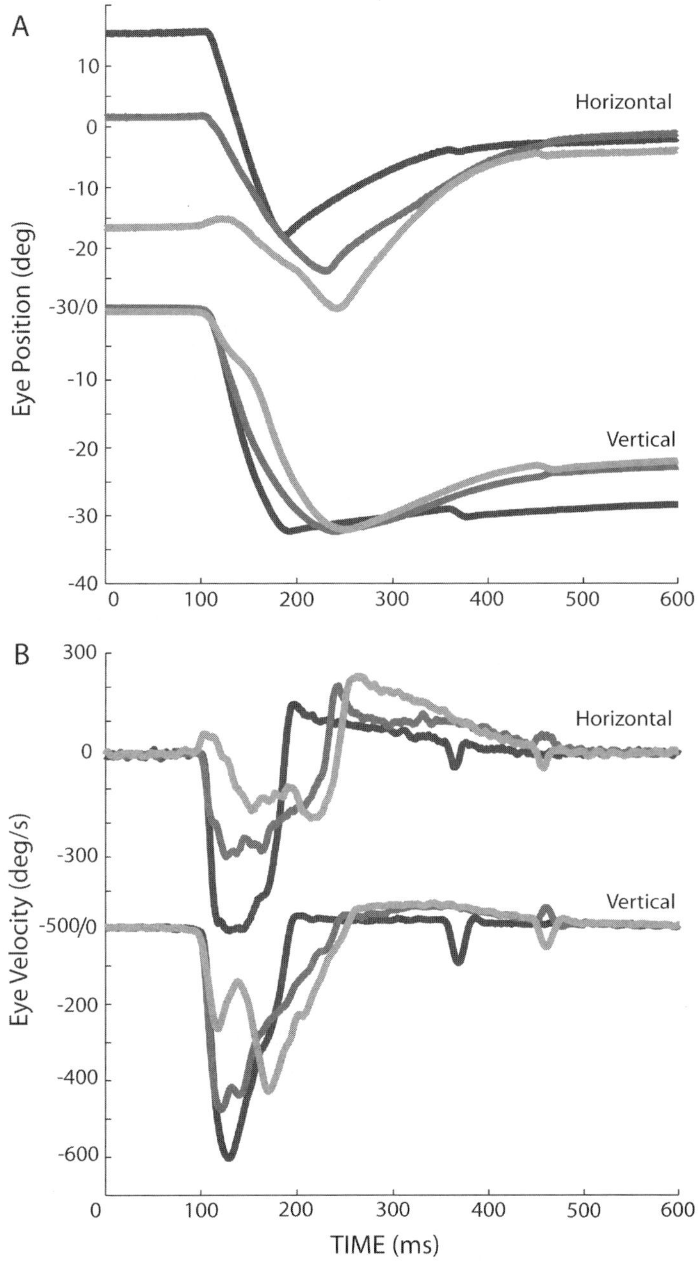

Fig. 2. Three single trial examples of oblique saccades made during constant amplitude gaze shifts to a target displaced to the left 35° and down 35°. (A) Eye position plotted as functions of time for horizontal and vertical components of saccades. The initial vertical positions of the eyes in the orbits were 0 (centred). Initial horizontal positions varied and were 15° to the right (away from the direction of movement (Black)), 17° to the left (light grey), and aligned with the head (centred in the orbits: dark grey). Horizontal and vertical velocities are plotted as functions of time in panel B.

of the eyes were zero in all cases. In Fig. 2B the corresponding component velocities are plotted as functions of time. It is clear from these examples that when the eyes began deviated in the direction of the movement, the amplitude of the horizontal component was small (~12°; light grey) and the velocity was low (peak horizontal velocity <200°/s). In contrast, when the eyes began deviated in the opposite direction (black) the horizontal component amplitude was large (33°) and velocity was high (peak velocity 500°/s). Although the amplitudes of the vertical saccade components were similar during each of these trials, the duration increased when the 35° vertical component was associated with a 12° horizontal component (light grey) compared to the vertical component duration when coupled with a larger horizontal component (black).

During 214 similar movements to the same target location, the duration of the constant vertical component is plotted as a function of the horizontal

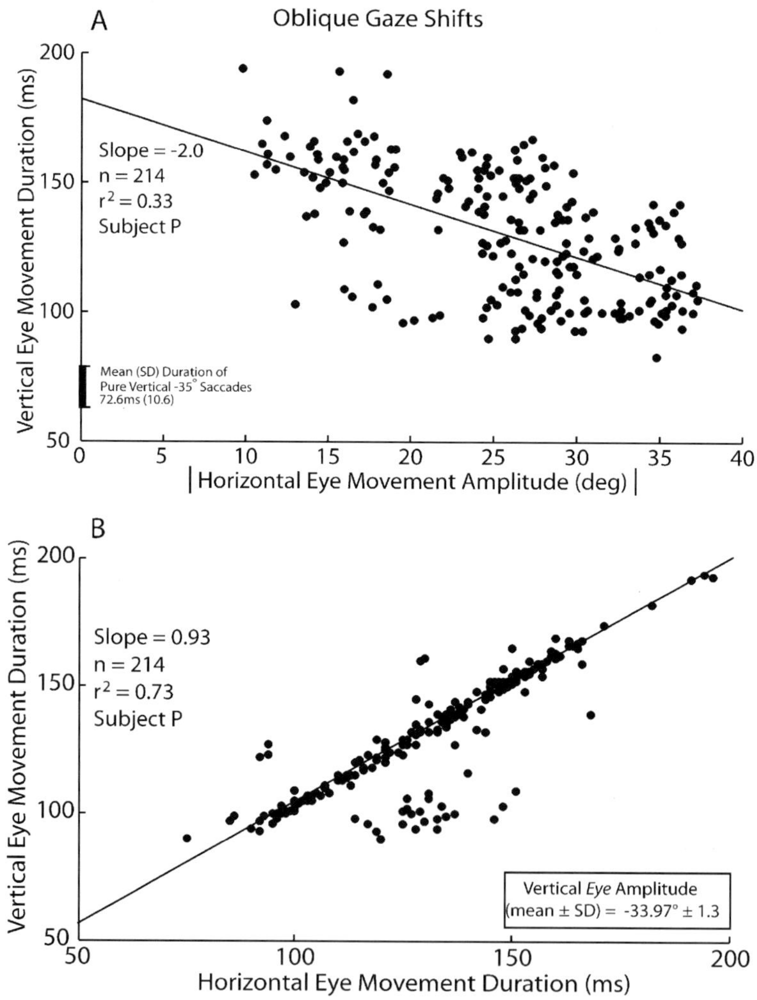

Fig. 3. Summary data for 214 oblique gaze shifts (35° left, 35° down). (A) Vertical saccade duration plotted as a function of the absolute value of horizontal saccade amplitude. The slope of the line of best fit was negative (−2.0), and horizontal saccade amplitude accounted for only 33% of the variance in vertical saccade duration. (B) For the same set of 214 movements, vertical saccade duration is plotted as a function of horizontal saccade duration. The line of best fit had a positive slope (0.93) and horizontal saccade duration accounted for 73% of the variance in vertical saccade duration. For comparison, the mean (SD) duration of 35° movements directed along the vertical meridian is shown along the ordinate.

component amplitude (Fig. 3A). As illustrated the slope of the line of best fit for these data was negative. When the vertical component was three times larger than the associated horizontal component, the duration of the vertical component was increased by a factor of ~2.5 compared to movements of the same amplitude directed along the vertical meridian (mean and SD indicated along the ordinate of Fig. 3). Vertical component duration was also increased when horizontal saccade amplitudes were about one-third the amplitude of the vertical component, compared to other oblique movements when the horizontal component was approximately the same amplitude as the vertical. This stretching of the vertical component results in the major saccade component being stretched in time to match the duration of the minor component. In Fig. 3B, the duration of the vertical component is plotted as a function of the horizontal component, and as shown, the horizontal component duration accounts for 73% of the variance in vertical duration.

Discussion

Several mechanisms have been proposed to account for the coupling of horizontal and vertical components of oblique saccades (vanGisbergen et al., 1985; Grossman and Robinson, 1988; Becker and Jürgens, 1990; vanOpstal and Kappen, 1993; Quaia and Optican, 1997). With the exception of the model proposed by Grossman and Robinson (1988), each of these hypotheses scales the velocity of horizontal and vertical components as a function of the relative component amplitudes (or direction) of the saccade. In the absence of head movements, when oblique saccades are made and the amplitude of one of the components (e.g., the vertical) remains constant while the other (e.g., the horizontal component) varies, the duration of the vertical is positively correlated with the amplitude of the horizontal. Because the amplitude and duration of the horizontal component (in this example) are tightly linked and positively correlated, the duration of the vertical component is also correlated with the duration of the horizontal component. Nonetheless, nearly all extant models of oblique saccades use information about the relative amplitudes of the two saccade components to appropriately scale component velocities.

The assumption that component coupling is determined by the relative amplitudes (and not duration) of the two components is not readily testable when the head is restrained because of the seemingly immutable duration–amplitude relationship observed under these conditions. However, when the head is free to move and under specific conditions, the duration–amplitude relationship for saccadic eye movements can be reversed: as saccade amplitude increases, duration decreases. This separation of saccade duration and amplitude allows for reconsideration of the assumption that relative amplitudes of saccade components govern component coupling.

During repeated gaze shifts to the same target location, when the eyes begin in different orbital positions, eye and head movement amplitudes vary inversely as a function of the initial orbital positions of the eyes (Delreux et al., 1991; Volle and Guitton, 1993; Freedman and Sparks, 1997a, b). In addition, as head movement amplitude and velocity increase, the velocity of concurrent saccades decreases (Freedman and Sparks, 2000; Freedman, 2005). A direct consequence of these two observations is that under the conditions stated above, as saccade amplitude decreases, duration increases (Fig. 1). When repeated oblique gaze shifts to the same target location are made with different initial horizontal positions of the eyes, the duration–amplitude relationship for the horizontal saccade component has a negative slope. The duration of the constant amplitude vertical component varies under these conditions, but variations in vertical component duration are not correlated with the amplitude of the horizontal saccade; in fact horizontal saccade amplitude and vertical saccade duration are anti-correlated (Fig. 3A). The duration of the constant amplitude vertical saccade component is, however, positively correlated with the duration of the horizontal saccade component (Fig. 3B).

Although preliminary, the data presented in this report suggest that the hypothesized mechanisms resulting in coupling of horizontal and vertical saccade components may need to be reconsidered. It appears that the duration of the constant amplitude component is stretched in time when associated with a long duration orthogonal component regardless of the amplitude of that component. For instance, even when the vertical component is three times larger than the associated horizontal component, vertical saccade duration is significantly prolonged. It is difficult to reconcile existing coupling hypotheses with these data. Understanding the ways in which the eyes and head are coordinated during gaze shifts continues to alter our thinking about the saccadic system and the neural control of saccades. The data presented here constrain existing and future hypotheses attempting to account for component coupling during oblique saccades, and may force a reconsideration of the primacy of relative component amplitudes in governing the scaling of component velocities.

Acknowledgements

This work is supported in part by NEI RO1-EY13239. The authors thank Gil Rivlis for developing data acquisition and control software, and G. Parker for technical assistance. Drs. Stephan Quessy, Mark Walton, and Julie Quinet contributed valuable comments and discussions.

References

Bahill, A.T., Clark, M.R. and Stark, L. (1975) The main sequence, a tool for studying human eye movements. Math. Biosci., 24: 191–204.

Baloh, R.W., Konrad, H.R., Sills, A.W. and Honrubia, V. (1975) The saccade velocity test. J. Neurol., 25: 1071–1076.

Becker, W. and Jürgens, R. (1990) Human oblique saccades: quantitative analysis of the relation between horizontal and vertical components. Vision Res., 30: 893–920.

Collewijn, H. (1977) Eye- and head movements in freely moving rabbits. J. Physiol., 266: 471–498.

Collewijn, H., Erkelens, C.J. and Steinman, R.M. (1988a) Binocular co-ordination of human horizontal saccadic eye movements. J. Physiol., 404: 157–182.

Collewijn, H., Erkelens, C.J. and Steinman, R.M. (1988b) Binocular co-ordination of human vertical saccadic eye movements. J. Physiol., 404: 183–197.

Delreux, V., Abeele, S.V., Lefevre, P. and Roucoux, A. (1991) Influences of eye position on the control of head movement amplitude. In: Paillard J. (Ed.), Brain and Space. Oxford University Press, Oxford, pp. 38–48.

Evinger, C., Kaneko, C.R.S. and Fuchs, A.F. (1981) Oblique saccadic eye movements of the cat. Exp. Brain. Res., 41: 370–379.

Freedman, E.G. (2001) Interactions between eye and head control signals can account for movement kinematics. Biol. Cybern., 84: 453–462.

Freedman, E.G. (2005) Head-eye interactions during vertical gaze shifts made by rhesus monkeys. Exp. Brain Res., 1–14.

Freedman, E.G. and Sparks, D.L. (1997a) Activity of cells in the deeper layers of the superior colliculus of the rhesus monkey: evidence for a gaze displacement command. J. Neurophysiol., 78: 1669–1690.

Freedman, E.G. and Sparks, D.L. (1997b) Eye-head coordination during head-unrestrained gaze shifts in rhesus monkeys. J. Neurophysiol., 77: 2328–2348.

Freedman, E.G. and Sparks, D.L. (2000) Coordination of the eyes and head: movement kinematics. Exp. Brain Res., 131: 22–32.

Grossman, G.E. and Robinson, D.A. (1988) Ambivalence in modelling oblique saccades. Biol. Cybern., 58: 13–18.

Guitton, D. and Mandl, G. (1980) Oblique saccades of the cat: a comparison between the durations of horizontal and vertical components. Vision Res., 20: 875–881.

Judge, S.J., Richmond, B.J. and Chu, F.C. (1980) Implantation of magnetic search coils for measurement of eye position: an improved method. Vision Res., 20: 535–538.

King, W.M., Lisberger, S.G. and Fuchs, A.F. (1986) Oblique saccadic eye movements of primates. J. Neurophysiol., 56: 769–784.

Nichols, M.J. and Sparks, D.L. (1996) Component stretching during oblique stimulation-evoked saccade: the role of the superior colliculus. J. Neurophysiol., 76: 582–600.

Phillips, J.O., Ling, L., Fuchs, A.F., Seibold, C. and Plorde, J.J. (1995) Rapid horizontal gaze movement in the monkey. J. Neurophysiol., 73: 1632–1652.

Quaia, C. and Optican, L.M. (1997) Model with distributed vectorial premotor bursters accounts for the component stretching of oblique saccades. J. Neurophysiol., 78: 1120–1134.

Smit, A.C., vanOpstal, A.J. and vanGisbergen, J.A.M. (1990) Component stretching in fast and slow oblique saccades in the human. Exp. Brain Res., 81: 325–334.

Stahl, J.S. (2001) Eye-head coordination and the variation of eye-movement accuracy with orbital eccentricity. Exp. Brain Res., 136: 200–210.

Tomlinson, R.D. and Bahra, P.S. (1986) Combined eye-head gaze shifts in the primate I. Metrics. J. Neurophysiol., 56: 1542–1557.

Tweed, D., Glenn, B. and Vilis, T. (1995) Eye–head coordination during large gaze shifts. J. Neurophysiol., 73: 766–799.

vanGisbergen, J.A.M., vanOpstal, A.J. and Schoenmakers, J.J.M. (1985) Experimental test of two models for the generation of oblique saccades. Exp. Brain Res., 57: 321–336.

vanOpstal, A.J. and Kappen, H. (1993) A two-dimensional ensemble coding model for spatial temporal transformation for saccades in monkey superior colliculus. Network, 4: 19–38.

Volle, M. and Guitton, D. (1993) Human gaze shifts in which the head and eyes are not initially aligned. Exp. Brain Res., 94: 463–470.

Yee, R.D., Schiller, V.L., Lim, V., Baloh, F.G., Baloh, R.W. and Honrubia, V. (1985) Velocities of vertical saccades with different eye movement recording methods. Invest Ophthalmol Vis. Sci., 26: 938–944.

C. Kennard & R.J. Leigh (Eds.)
Progress in Brain Research, Vol. 171
ISSN 0079-6123

CHAPTER 4.16

Head movement control during head-free gaze shifts

Nadine Lehnen*, Ulrich Büttner and Stefan Glasauer

Department of Neurology, Bernstein Center for Computational Neuroscience, Ludwig-Maximilian-University, Munich

Abstract: Humans mostly use combined eye-head movements to shift gaze to a new target. Here, we discuss head motor control during human gaze shifts. We demonstrate that vestibular feedback monitors and guides the head movement on-line. Our results further suggest that cerebellar signals play a role in this on-line control, possibly by contributing to detect vestibular input indicating deviations from the intended movement trajectory.

Keywords: motor control; vestibular system; feedback; cerebellum; head moment of inertia; human; principle of reafference

Head movements during gaze shifts

Most goal-directed gaze shifts larger than 20° are combined eye and head movements. They start with both the eyes and the head moving toward the target (Fig. 1). Once gaze has reached the target, it is stabilized at this position while the head continues to move. This stabilization, which is due to a counter-rotation of the eyes, is mediated by the vestibulo-ocular reflex (VOR). Whereas neural control of gaze shifts has been studied in detail, much less attention has been paid to the head movement component of gaze shifts. Accurate head movements were not considered critical for gaze shift performance because up to the oculomotor range of ±55° the eyes could compensate for any shortcomings in head motor control. In this respect, the requirements for active head movements in humans, which help to bring the eyes to the target, differ from those of many animals that need to bring the head to the target (when, e.g., a lion catches its prey). While predatory animals thus certainly require refined on-line head control strategies, the question remains whether similar mechanisms are used during human gaze shifts.

Challenging the head motor control system by increasing head inertia

We assessed head motor control by changing the properties of the head, i.e., by increasing the head moment of inertia (HMI). If the head motor command is not adjusted to match the new head plant properties, this increase should provoke head oscillations (Peng et al., 1996; Tangorra et al., 2003). Figure 2A illustrates this situation. When an unchanged motor command drives a simulated weighted head plant the predicted head velocity trajectory shows oscillatory undershoots.

We used a helmet with eccentrically attached masses to increase the HMI in healthy subjects while they performed horizontal gaze shifts toward briefly flashed targets in darkness (target step amplitudes were ±75° and ±80°; for details on the

*Corresponding author. Tel.: +49 89 7095 4833; Fax: +49 89 7095 4801; E-mail: NLehnen@nefo.med.uni-muenchen.de

DOI: 10.1016/S0079-6123(08)00648-1

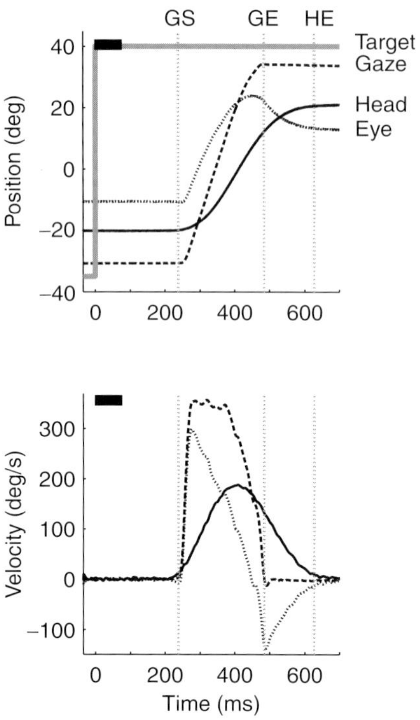

Fig. 1. Position (top) and velocity (bottom) traces of a human gaze shift evoked by a 75° target step (grey, the thick black bars indicate when the target is visible). The gaze shift (dashed, GS marks gaze shift start) is the sum of a movement of the eye in head ("eye," dotted) and of the head in space ("head," solid line). After gaze shift end (GE), the head is still moving in the direction of the gaze shift (HE marks head movement end), while the eyes are moving in the opposite direction under the influence of the vestibulo-ocular reflex (VOR), which stabilizes gaze.

methods see Lehnen, 2006). Figure 2B shows head velocity traces with and without increased head inertia from one subject (individual head velocity traces as part of 75° gaze shifts are grey, means are black). Unweighted, head movements are smooth. When the HMI is increased, the head only slightly oscillates, much less than predicted if no neural control was present (compare to Fig. 2A, bottom). This is true starting from the first movement with increased inertia indicating on-line sensory feedback that guides head movements during gaze shifts and helps to decrease unwanted oscillations.

Vestibular feedback guides head movements on-line

Input to the feedback loop can come from visual, proprioceptive, and vestibular sensors. Flashing the target excludes visual feedback in this study. Vestibular signals are well suited for on-line head control as they provide short-latency information about head velocity in space, whereas proprioceptive input only gives relative positions. We therefore increased the HMI in patients who had undergone complete bilateral vestibulectomy for acoustic neuroma several years prior to the study (vestibular patients; age-matched to the healthy controls; same experimental protocol as above). Figure 2C shows head velocity traces from one vestibular patient with and without increased head inertia (same depiction as Fig. 2B). Whereas head movements do not oscillate in the unweighted situation, there are pronounced oscillations with increased inertia. Importantly, oscillations are the same as in healthy subjects in the unweighted condition, but are much greater when the HMI is increased (compare to Fig. 2B). This indicates that vestibular feedback guides active head movements during gaze shifts.

Cerebellar signals are involved in the on-line control of the head

For accurate on-line guidance the vestibular system must be able to differentiate between input reflecting the intended movement and input showing that the movement has left its intended path. It is currently thought that an internal prediction of the sensory consequences of movements is compared to the actual input and the self-generated part of vestibular activation is cancelled (Cullen, 2004). It has been suggested that the cerebellum is involved in signalling the differences between predicted and real sensory consequences of movements (Blakemore et al., 2001). To investigate whether this cerebellar signalling helps to control active head movements on-line, we increased the HMI in patients with cerebellar atrophy of unknown aetiology (cerebellar patients). All cerebellar patients had intact peripheral vestibular function. They were age-matched to the healthy controls and the vestibular patients. Figure 2D shows head velocity traces from one cerebellar patient. Whereas the head does not oscillate in the unweighted condition, there are marked oscillatory undershoots in the

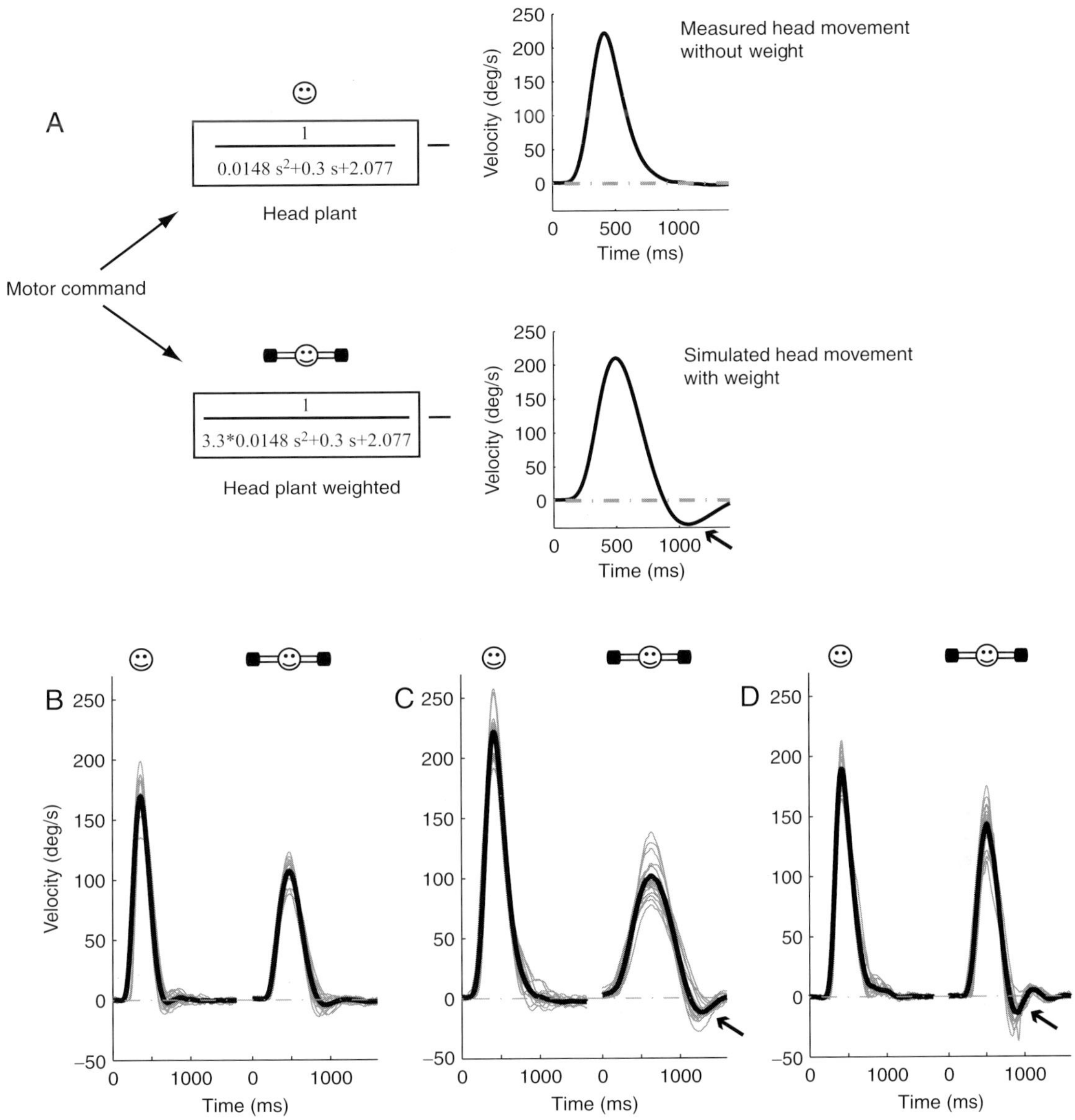

Fig. 2. (A) Velocity profiles of a natural head movement (top) and of a simulated head movement with the head moment of inertia 3.3-fold increased (weighted, bottom), assuming the same motor command. Natural head movements are free of oscillations (top). If the motor command is not adjusted to the new head plant properties, an increase in the head moment of inertia leads to head oscillations (arrow; head plant properties from Peng et al., 1996 and Tangorra et al., 2003). (B) Head velocity traces from a healthy subject with (right) and without (left) increased head inertia. Unweighted (left), the head movement is smooth. No oscillations occur (individual head velocity traces as part of 75° gaze shifts are grey, means are black). With the head moment of inertia increased (right), the healthy subject shows only minor head oscillations, much less than simulated if no neural control was present (compare to Fig. 2A, bottom). (C) Head velocity traces from a patient with chronic bilateral loss of vestibular function with (right) and without (left) increased head inertia. Whereas head movements are smooth without weight, the vestibular patient's head markedly oscillates with the head moment of inertia increased (arrow, same depiction as Fig. 2B). (D) Head velocity traces from a patient with cerebellar atrophy with (right) and without (left) increased head inertia. Head movement behaviour with and without increased head inertia is similar in the cerebellar and the vestibular patient (compare to Fig. 2C, same depiction). Increasing the head moment of inertia leads to head oscillations (arrow).

weighted condition. The amount of oscillations is comparable in cerebellar and vestibular patients (cf. Fig. 2C). This indicates that the cerebellum plays a role in the on-line control of head movements during gaze shifts, probably by predicting expected vestibular input from an efference copy of motor commands.

Summary

The head movement component of gaze shifts is well controlled. A vestibular feedback loop which involves the cerebellum guides active head movements during gaze shifts. In our study, it helps to reduce unwanted head oscillations due to an increase in head inertia. What is the reason for this refined head control? Oscillations occur when gaze has reached its target and could easily be compensated for by the VOR without affecting gaze shift performance. One hypothesis could be that head control during gaze shifts is a phylogenetic remnant of species in which the head is used as a grasping instrument similar to the human hand. Alternatively, the optimal performance of a head-borne sense such as hearing might depend on oscillation-free head movements.

Abbreviations

VOR	vestibulo-ocular reflex
HMI	head moment of inertia

Acknowledgements

We thank Thomas Eggert and Thomas Fritz for advice, and Katie Ogston for copy editing the manuscript. This work was supported by the BMBF (BCCN Munich 01GQ0440).

References

Blakemore, S.J., Frith, C.D. and Wolpert, D.M. (2001) The cerebellum is involved in predicting the sensory consequences of action. Neuroreport, 12: 1879–1884.

Cullen, K.E. (2004) Sensory signals during active versus passive movement. Curr. Opin. Neurobiol., 14: 698–706.

Lehnen, N. (2006) The effect of increased head inertia on eye-head control in human gaze shifts: analysis and mathematical modeling of a motor system. Thesis. Faculty of Medicine. Ludwig-Maximilian-University, Munich.

Peng, G.C., Hain, T.C. and Peterson, B.W. (1996) A dynamical model for reflex activated head movements in the horizontal plane. Biol. Cybern., 75: 309–319.

Tangorra, J.L., Jones, L.A. and Hunter, I.W. (2003) Dynamics of the human head-neck system in the horizontal plane: joint properties with respect to a static torque. Ann. Biomed. Eng., 31: 606–620.

C. Kennard & R.J. Leigh (Eds.)
Progress in Brain Research, Vol. 171
ISSN 0079-6123

CHAPTER 4.17

Postural changes during eye–head movements

J. Fukushima[1,*], T. Asaka[1] and K. Fukushima[2]

[1]*Department of Health Sciences, Hokkaido University School of Medicine, Sapporo, Japan*
[2]*Department of Physiology, Hokkaido University School of Medicine, Sapporo, Japan*

Abstract: We investigated postural sway in young subjects during eye–head movements. Subjects stood on the force plate wearing a helmet on which a laser spot was attached. Visual targets were presented stepwise at 60° to right and left or 50° to upward and downward of the straight-ahead position. Subjects were asked to look at the targets quickly by directing the laser spot to the target. The anterior–posterior component of centre of pressures (COPs) exhibited the largest changes and shortest latencies during downward movements. Most of the subjects showed COP changes preceding the onset of downward head movements, and these changes were associated with EMG activity of biceps femoris and gastrocnemius. When subjects were asked to look at the target only by eyes, no consistent change was detected in ground reaction forces. These results indicate that the eye–head movements elicited the COP changes that preceded the onset of head movements.

Keywords: postural control; eye–head movements

Introduction

Maintenance of postural balance is important to insure clear vision. Voluntary movements are associated with anticipatory postural adjustments (APAs) that occur prior to the movement itself. Their role is to minimize perturbations that would be induced by the movement. Previous studies reported that head extension or flexion increased velocity and trajectory of centre of pressures (COPs) in young subjects (Buckley et al., 2005; Vuillerme and Rougier, 2005). Strupp et al. (2003) reported that only eye movements affected the posture. Although they confirmed COP changes during voluntary head movements, they did not compare the onsets of postural changes with those of the head movements. If eye and head movements work as a perturbation to posture, there is a possibility that APAs may be observed preceding the movements. We, therefore, examined whether the onset of COP change preceded the eye and head movements.

Methods

Subjects stood on the force plate in a magnetic field wearing a helmet which had a laser spot attached (Fukushima et al., 2007). A search coil was also put on the helmet to record head movements. Visual targets were located 60° to right and left and 50° up and down of the subjects (Fig. 1A, B). Subjects were asked to look at the targets quickly by directing the laser spot to the target. We examined the COPs, when subjects were asked to look at the target placed at 15° from the centre without any head movements. Infrared oculography (DC to

*Corresponding author. Tel.: +81 11 706 3329;
Fax: +81 11 706 3329; E-mail: jf002@pop.cme.hokudai.ac.jp

DOI: 10.1016/S0079-6123(08)00649-3

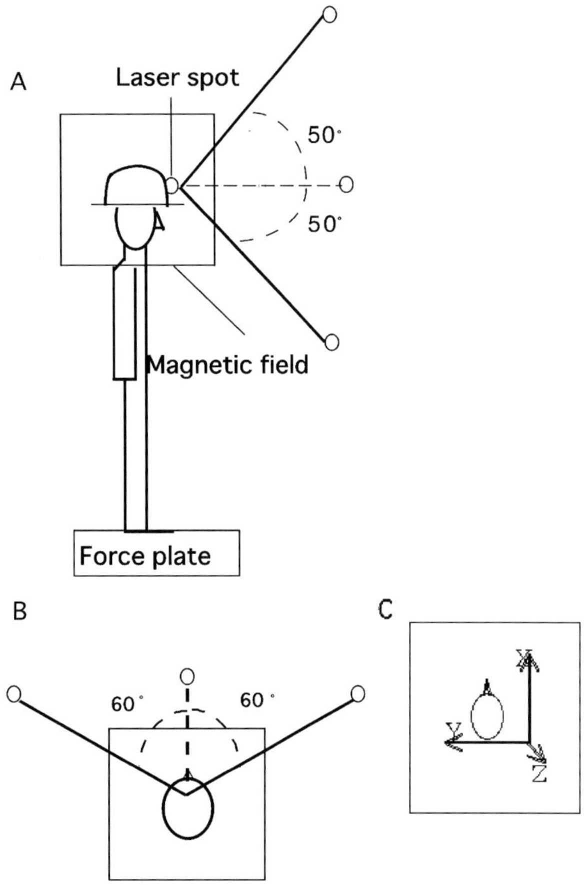

Fig. 1. (A) Experimental set up. Subject were standing on the force plate, wearing a helmet. A search coil was attached to the helmet to detect head movements in a magnetic field. They were instructed to look at the targets 50° upward and downward, adjusting the light on the helmet to the targets. (B) Horizontal targets were located 60° right or left to the subject. (C) *X*, *Y*, and *Z* represent the directions of ground reaction force. Fx: anterior to posterior component of ground reaction force; Fy: medial to lateral component of ground reaction force; Fz: vertical component of ground reaction force. Adapted with permission from Fukushima et al. (2007).

50 Hz, −30 dB/octave) was used to record eye movements. Disposable self-adhesive electrodes were used to record the surface EMG of the following muscles from the right side of the body: trapezius, sternocleidomastoideus, biceps femoris, rectus femoris, long head of the hamstrings, tibialis anterior, and lateral head of gastrocnemius muscles. The vertical and horizontal components of head position, three components of ground reaction forces (Fx, Fy, and Fz in Fig. 1C), and EMG were stored on a data recorder for off-line analysis. All data were digitized at 1 kHz. The trajectory of COPx (anterior–posterior component of COP) was calculated with our interactive computer programme (Fukushima et al., 2004). All data that consisted of head position, head velocity, reaction force (Fx, Fy, Fz, and COPx), acceleration of the trunk and EMG were aligned at the onset of head movement. To find out the onset of head movement, we drew a tangential line along the initial slope of the head position traces. The onset was determined as the time at which the straight line along the head position traces intersected with the base line. The onset of EMG activities was determined at a point beyond the mean + 2 SD of the baseline. Statistical significance was evaluated by paired *t*-test with a significant level of $p < 0.05$.

Results

Each subject showed a consistent change in Fx, Fy, Fz, and COPx associated with head movements. Inset of Fig. 2 shows a typical example of a subject whose head moved towards a downward target. Fx preceded downward head movement in 11 of the 13 subjects. Although force changes preceded the head movement in several subjects, the mean latency in the other movements lagged head movement onset (Fig. 2). Peak distances of COPx were calculated during downward and upward head movement in all subjects. The mean ± SDs of the COP changes in all subjects were 4.0 ± 0.9 cm for downward movement and 1.6 ± 0.7 cm for upward movement. The difference was significant ($t = 5.08$, $p = 0.003$). The averaged latencies of COPx initiation for downward and upward head movements were −17.2 ms and 55.4 ms, respectively, which was also significantly different ($t = 4.04$, $p = 0.001$). However, no consistent change was observed in reaction forces or COP during horizontal or vertical eye movement without head movement. In two subjects, EMG responses of biceps femoris preceded the onset of downward head movements by 22 ms and 54 ms, respectively. However, gastrocnemius EMG did not exhibit consistent activity before and after the initiation of head movements.

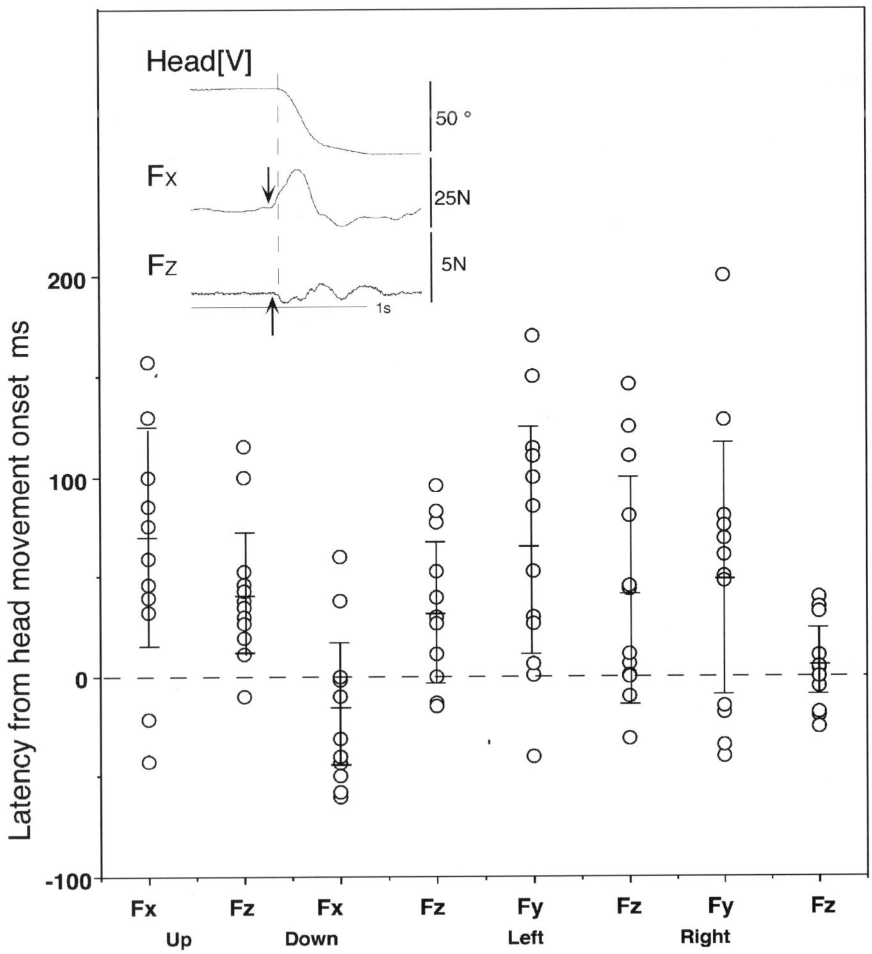

Fig. 2. An open circle shows mean latencies for an individual subject. Lines indicate mean ± SD in each component of reaction forces for all the subjects. Inset: A typical example of a subject. An arrow indicates the onset of Fx. The amplitude of head was the angle measured by neck bending. Fx, Fy, and Fz are the same as those in Fig. 1. Adapted with permission from Fukushima et al. (2007).

Discussion

We observed larger COPx change in downward than that in upward head movement, consistent with Buckley's findings (Buckley et al., 2005). Why did only downward head movements cause larger and preceding COPx changes? During horizontal head rotation, the centre of mass (COM) position of the head remains near the mastoid bone. Therefore, it is unlikely to show large changes in ground reaction force in medial to lateral direction (Fy). However, vertical head rotation changes the COM position of the head, because rotation axis (near C1) is distant from head COM. In our study, the amplitude of actual vertical head movements was not 50°, because the rotational axis for the head movements was at the joint of the occipital bone and the atlas (C1). We measured the angles of neck bending with a goniometer during upward and downward head movements. Actual angles of neck bending were 10°~20° and 25°~45°, respectively. Larger downward head movement must have resulted in the largest perturbing effects on COM.

Unlike the results by Strupp et al. (2003), we could not detect force change during eye movement without head movements. This discrepancy may have been due to the difference in the experimental conditions. Whereas they used the target moving sinusoidally to follow, our task required saccadic eye movements. In the present study, activities preceding the head movements were observed only in the biceps femoris, but not in other muscles such as gastrocnemius. Donna-Dos-Santos et al. (2007) also reported activities in biceps femoris preceding the head movements elicited by hand releasing from the loading. Our findings are consistent with the previous reports (Horak et al., 1994; Anson and Kasai, 1995), although their tasks were different from ours. These studies suggest the possibility that the activities in lower leg muscles such as gastrocnemius and soleus were suppressed before and after the head movements in the present study, although we could not exclude the possibility that we were unable to detect activity in these muscles. In conclusion, the COPx, Fx, and the muscle activities preceding head movement seen in this study may be preprogrammed movements related to APAs.

Acknowledgement

This study was supported by Grant-in-Aid for Scientific Research from the Ministry of Education, Culture, Sports, Science and Technology of Japan (18023004 and 18500394).

References

Anson, J.G. and Kasai, T. (1995) Evidence for soleus-reflex depression induced by ballistic head rotation. Neurosci. Lett., 95: 17–20.

Buckley, J.G., Anand, V., Scally, A. and Elliott, D.B. (2005) Does extension and flexion increase postural instability in elderly subjects when visual information is kept constant? Gait Posture, 21: 59–64.

Donna-Dos-Santos, A., Degani, A.M. and Latash, M.L. (2007) Anticipatory control of head posture. Clin. Neurophysiol., 118: 1802–1814.

Fukushima, J., Akao, T., Takeichi, N., et al. (2004) Pursuit-related neurons in the supplementary eye fields: discharge during pursuit and passive whole body rotation. J. Neurophysiol., 91: 2809–2825.

Fukushima, J., Asaka, T., Ikeda, N., et al. (2007) Postural control during downward head movements in young subjects. J. Phys. Ther. Sci., 19: 205–212.

Horak, F.B., Shupert, C.L., Dietz, V., et al. (1994) Vestibular and somatosensory contributions to responses to head and body displacements in stance. Exp. Brain Res., 100: 93–106.

Strupp, M., Glasauer, S., Jahn, K., et al. (2003) Eye movement and balance. Ann. N.Y. Acad. Sci., 1004: 352–358.

Vuillerme, N. and Rougier, P. (2005) Effects of head extension on undisturbed upright stance control in humans. Gait Posture, 21: 318–325.

C. Kennard & R.J. Leigh (Eds.)
Progress in Brain Research, Vol. 171
ISSN 0079-6123

CHAPTER 4.18

Cortical processing in vestibular navigation

Barry M. Seemungal[1,*], Vincenzo Rizzo[2], Michael A. Gresty[1], John C. Rothwell[2] and Adolfo M. Bronstein[1]

[1]*Department of Clinical Neuroscience, Charing Cross Hospital, Imperial College, London, UK*
[2]*Sobell Department of Motor Neuroscience and Movement Disorders, ION, UCL, London, UK*

Abstract: Visual and vestibular perceptual processes are intimately related. Previous data suggest a reciprocal visual-vestibular inhibition with regard to head motion (velocity) perception with each sensory modality mediated by distinct cerebral cortical loci. The relationship between visual and vestibular perceptual processes with regard to displacement perception is unknown. In a path-reversal vestibular navigation task, we investigated the effects of early visual deprivation on vestibular perception by comparing performance and strategies employed (i.e., displacement vs. velocity matching) by a group of congenitally blind subjects to that of a sighted control group. In a second experiment in a sighted group, we attempted to disrupt firstly displacement encoding and then velocity encoding, during a path-reversal vestibular navigation task, by applying repetitive transcranial magnetic stimulation (rTMS) to the right posterior parietal cortex (PPC), vs. a left motor cortex control. Our data show that for a path-reversal vestibular navigation task, when given free choice, sighted and blind subjects can utilize both displacement- and velocity-matching strategies, and overall, both groups display equivalent performance in performing the path-reversal task. In sighted subjects, when rTMS was applied during encoding in the path-reversal task, displacement but not velocity encoding was disrupted for leftward rotations. In summary, our data suggest that early visual deprivation does not degrade the perception of vestibular signals of head velocity or of derived percepts of angular displacement. The rTMS results in the sighted group show that leftward vestibular-derived displacement perception is encoded in the right PPC, an area prominent in visuo-spatial perception. Thus, in contrast to current theories of vestibular head velocity perception which involve reciprocal inhibition between separate and competing visual and vestibular cortical areas, we suggest that at least for displacement perception, visual and vestibular-derived signals are encoded in a common cortical locus.

Keywords: vestibular cortex; rTMS; congenital blindness; vestibular navigation; vestibular perception

Introduction

During everyday locomotion, multi-sensory inputs generate perceptions of *motion* and *position-in-space*, and in the dark, these perceptions rely upon vestibular input (Glasauer et al., 2002) to enable accurate locomotor turns. Sighted humans can orient in the dark using only vestibular signals (Metcalfe and Gresty, 1992; Seemungal et al., 2007). Although blind subjects can also accurately locomote in the dark (Loomis et al., 2001), it is unknown if they do so using vestibular signals. A second question is whether the neuro-anatomical substrates subserving vestibular perceptions of motion and derived angular displacement are co-localized.

*Corresponding author. Tel.: +44(0)2088467523;
Fax: +44(0)2088467577; E-mail: b.seemungal@imperial.ac.uk

DOI: 10.1016/S0079-6123(08)00650-X

Since vestibular and haptic signals are not isolated during walking, locomotor tasks cannot reliably inform regarding the vestibular contribution locomotor turning orientation. To answer the first question, (i.e., do the congenital blind utilize vestibular signals during angular orientation?) we used a 'vestibular navigation' task that isolates vestibular input (Metcalfe and Gresty, 1992; Seemungal et al., 2007). Subjects are passively rotated in yaw-axis to right or left from a fixed origin (in the dark, with white noise sound masking) on a motorized but vibrationless 'Bárány chair' ('*self-rotation test*'; Fig. 1A). Subjects return themselves to start using a chair-mounted joystick. A given joystick deflection provides a voltage to the chair velocity servo giving a proportionate constant chair angular velocity. In this first experiment, we did not constrain subjects' strategies so that we could compare and contrast performance and strategy between sighted and blind subjects.

Human and animal studies have demonstrated multiple cerebral cortical areas subserving vestibular processing (Guldin and Grüsser, 1998), although cortical areas specifically mediating vestibular-derived displacement perception are unknown. In contrast, regions mediating whole-body motion perception in humans have been found via electrical cortical stimulation (ECS) in several areas including posterior superior temporal gyrus, temporoparietal junction (TMJ), and posterior parietal cortex (PPC) (Foerster, 1936; Blanke et al., 2000; Kahane et al., 2003). Functional imaging data during vertiginous stimulation also demonstrated temporal and parietal cortex activation (Bense et al., 2001; Suzuki et al., 2001; Fasold et al., 2002; Dieterich et al., 2003). The use of vertigo-inducing stimuli in functional imaging studies complicates distinguishing cortical areas subserving vestibular perception of velocity vs. derived displacement. Although use of ECS or repetitive transcranial magnetic stimulation (rTMS) to disrupt focal cortical function during an imagined self-rotation task offer advantages above correlational techniques like functional imaging (Harris and Miniussi, 2003; Zacks et al., 2003), their relevance to vestibular navigation is unclear (viz., imagined self-rotation vs. updating of one's position during/following real rotations). Furthermore, the mental rotation literature is discordant, i.e., mental self-rotation is mediated by left PPC (viz., Zacks et al., 2003) or right TMJ (viz., Blanke et al., 2004). Vestibular perceptual studies in cortical lesion patients are complicated by the multi-focal vestibular-cortical representation (Israël et al., 1995; Farrell and Robertson, 2000; Philbeck et al., 2006; Ventre-Dominey and Vallee, 2007).

To overcome the drawbacks inherent to previous methodologies, we used rTMS to transiently disrupt focal cortical function during the 'self-rotation test' (Seemungal et al., 2008). We hypothesized that right PPC is specifically concerned with (lateralized) vestibular-derived displacement perception and tested this by disrupting right PPC function by applying rTMS *during the encoding phase* of the vestibular navigation task. Primary motor cortex served as a control rTMS site. Vestibular projections to motor cortical areas are important in gait and posture (Guldin and Grüsser, 1998); however, there is no data implicating motor cortex in spatial orientation perception.

Experimental procedures

Experiment 1: self-rotation test performance: sighted vs. blind groups

We tested 6 congenitally blind ('B1–B6': mean age 38, SD 8 years) and 12 sighted subjects ('S1–S12': mean age 30, SD 10 years). The subjects sat on a vibration-free rotating chair in the dark with white noise sound masking via earphones (Fig. 1A). The chair could be rotated by external computer control ('stimulus') or the subject could actively rotate himself ('response') by manipulating a directionally congruent joystick, which provided a velocity demand to the servo-motor of maximum angular velocity set at 140° for Experiment 1. In Experiment 1, chair rotational stimuli were raised cosines of 1, 2, 3, 4, 5, and 6 s duration (i.e., 1–0.33 Hz) at peak angular velocities 30, 60, 90, and 120°/s providing stimulus angles of 15°–360°. The large number of stimulus-velocity profiles presented in random order obviated "counting" strategies in estimating travelled angle. Subjects were instructed to return actively to the start position as accurately as possible, by moving in the opposite direction to the stimulus movement. The task can be performed

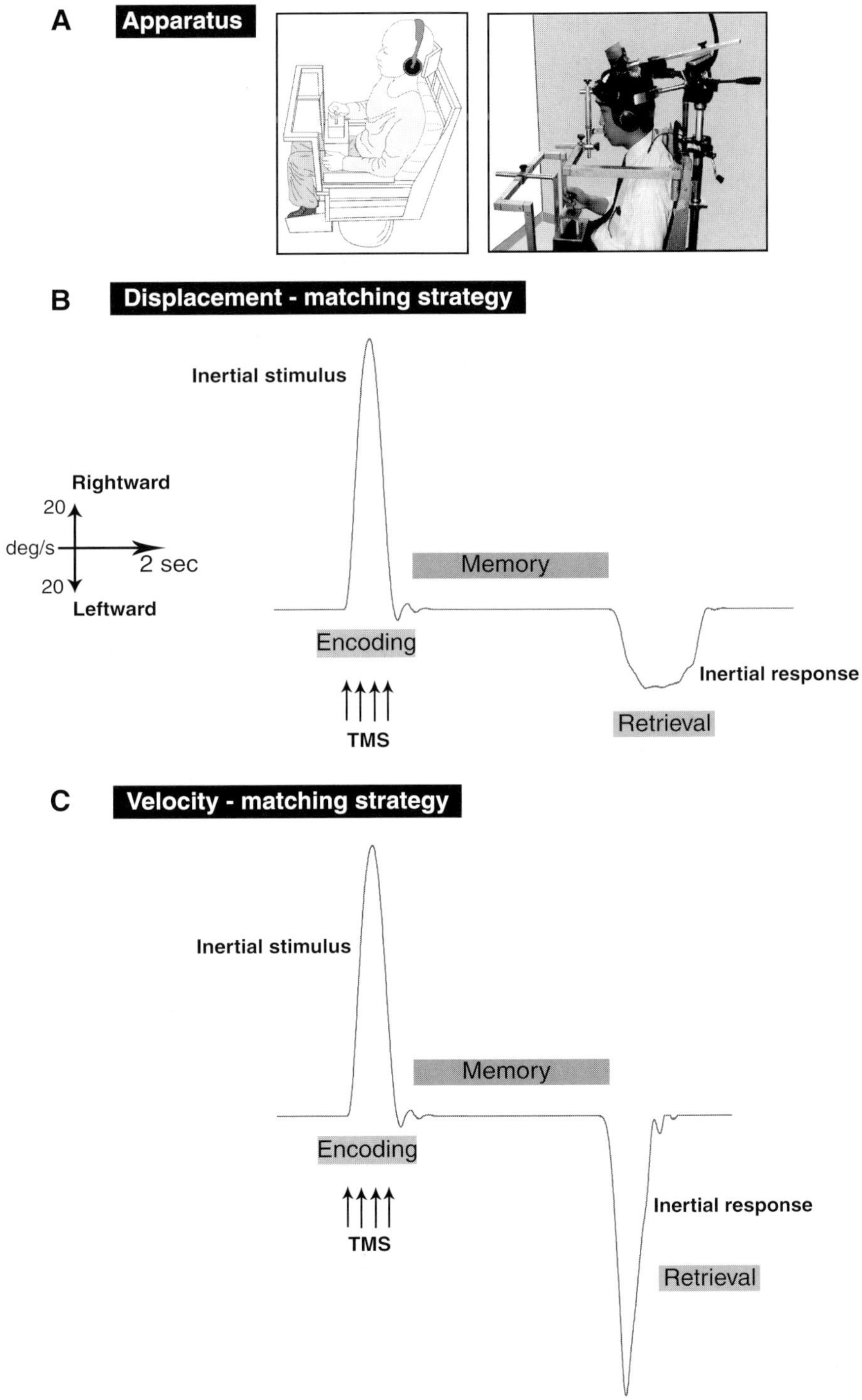

Fig. 1. Apparatus and representative angular chair velocity recordings. (A) Top left panel: The motorized rotating chair with joystick control and headphones with white-noise masking. Top right panel: TMS coil mounting. (B) Representative chair angular velocity trace — rightward stimulus rotation (upward on trace) of 'raised cosine' waveform followed by leftward rotation response of trapezoidal shape. Trapezoidal responses signify displacement-matching responses (required in Experiment 2). (C) Representative velocity-matching trace showing similarity in shape between stimulus and response velocity profiles. Velocity matching was required in Experiment 3.

by different strategies of the type shown for Experiment 2 (Fig. 1B: displacement-matching) or Experiment 3 (Fig. 1C: velocity-matching). In Experiment 1, subject strategy (unlike that for Experiments 2 and 3) was not constrained; subjects benefited from 15 min of free practice and a practice experiment without vision or audition.

Experiment 2: right PPC vs. left motor cortex rTMS during displacement matching

We tested five right-handed normal subjects (average age 34 years, range 25–50 years; four male) who gave their written consent. Subject strategy was constrained to a displacement-matching strategy (Fig. 1B) by explicitly telling subjects to return to the start position by imagining the stimulus angle travelled but also by limiting the maximum joystick-driven velocity to 60°/s (which precluded a velocity-matching strategy). Stimulus rotations were identical to Experiment 1 except that only rotations of 1, 2, and 3 s duration were used to avoid prolonged periods of rTMS. Apart from the constraints on subject strategy and rotation durations, conditions were identical to Experiment 1, including a period of free practice and a practice experiment without TMS.

Experiment 3: right PPC vs. left motor cortex rTMS during velocity matching

The same five subjects from Experiment 2 were told to explicitly recapitulate the kinetics of inertial stimuli (Fig. 1C), and a peak joystick-driven chair velocity of 140°/s enabled this strategy. We have shown equivalent spatial performance when subjects reproduce displacement or kinetic parameters in the self-rotation test (Seemungal et al., 2007). Apart from the required strategy, conditions were identical to those in Experiment 2.

TMS

rTMS was applied during the encoding phase of the vestibular navigation task as shown in Fig. 1B (Experiment 2) and 1C (Experiment 3). The order in which subjects received parietal vs. motor cortex stimulation was alternated between subjects. The handle of a figure-of-eight coil (outer windings 9 cm) was fixed to the chair via an adjustable mount while the coil was located over the scalp stimulation site. The subject's head was stabilized as in Fig. 1A. The best position for stimulating the motor representation of the first dorsal interroseous in the left motor cortex was located by observing the amplitude of TMS-induced muscle contractions while subjects were at rest. The motor threshold intensity at this optimal site was defined as the intensity that produced a visible hand movement in three or more trials out of six. The TMS parietal scalp sites (P3 and P4) were located according to the 10–20 international EEG electrode system (Klem et al., 1999). P3 and P4 are known to overly the parietal cortex close to the intraparietal sulcus (Hilgetag et al., 2001). We stimulated motor and parietal cortex at 10% below motor threshold at 10 Hz. The number of TMS pulses applied per rotation duration was as follows: 1 s: 4 pulses, 2 s: 8 pulses, 3 s: 12 pulses.

Data analysis

Experiment 1

Displacement performance

Quantitative analysis of displacement performance (for group and individuals) was obtained by linear regression between response and stimulus displacements. Regression slopes of sighted and blind groups' displacement performances were compared using a 2-tailed *t*-test, and correlation coefficients were compared using Fisher's *r*-to-*z* transformation.

Navigational strategy

Multiple regression analysis was used to quantify subjects' strategy in vestibular navigational task by identifying the stimulus parameters that most predicted response displacement.

Experiments 2 and 3

Performance was ascertained for each angular response via absolute percentage error. Displacement performance was analysed via repeated-measures two-way ANOVA with factors cortical locus (motor vs. parietal) and inertial stimulus direction (rightward vs. leftward). Absolute percentage errors made in peak velocity reproduction in Experiment 3 were similarly analysed. Responses to 15° stimuli were excluded from all analyses, since a sub-analysis of previous data (Seemungal et al., 2007) showed that this was the only stimulus angle which elicited a significantly different absolute percentage error of angular response (larger) compared to the average response for all stimulus angles combined (119.6% vs. 43.2%; $P = 0.0007$; *t*-test).

Results

Experiment 1

Self-rotation test: Sighted subjects' displacement performance and strategy

Sighted subjects were highly consistent in reproducing the angular displacement with a cumulative regression analysis between response (*R*) and stimulus (*S*) displacements of the non-averaged data ($n = 561$) yielding $R = 0.76S + 31°$ and an r^2 of 0.80. Figure 2 shows the averaged angular responses for each stimulus angle for the sighted group. Individual performances (Table 1), '*Displacement performance*', for subjects S1–S12 were similarly consistent with r^2 for linear regressions between response and stimulus displacements range 0.75–0.94 ($P < 0.0001$ for all). Individuals' strategy is summarized in Table 1, right-hand side, '*Predictors of response angle*,' for subjects S1–S12.

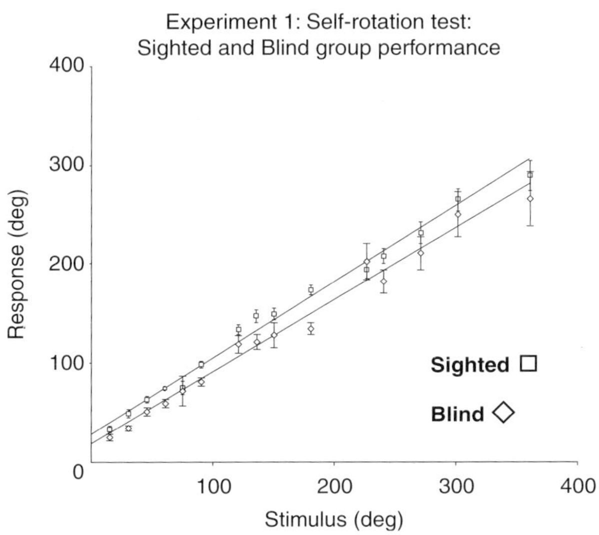

Fig. 2. Experiment 1 results: Group performance in the self-rotation task (±1 standard error).

Self-rotation test: Blind subjects' displacement performance and strategy

The Blind group's self-rotation test performance ($R = 0.74S + 17°$; $r^2 = 0.77$; $n = 287$) was not significantly different ($P > 0.05$) from the sighted for slope (*t*-test) and r^2 values (Fisher's *r* transformation) for their respective regressions of the non-averaged data (Fig. 2). Blind individuals' strategy (Table 1) was predominantly displacement matching, although one subject (B2) utilized a predominantly velocity-matching strategy.

Experiment 2

Angular displacement encoding: right PPC vs. left motor cortex rTMS

Right parietal rTMS (Fig. 3A) elicited an asymmetrical displacement response (2-tailed, paired *t*-test: $P < 0.008$). Control motor cortex rTMS elicited a symmetrical displacement response (*t*-test: $P = 0.56$). A repeated-measures ANOVA with factors, cortical locus of rTMS (parietal vs. motor), and inertial stimulus direction (right vs. leftward) showed no main effect of inertial stimulus direction [$F(1,109) = 2.35$; $P = 0.13$] or cortical locus of rTMS [$F(1,109) = 0.452$; $P = 0.50$] but there was a significant interaction between cortical location of rTMS and stimulus direction [$F(1,109) = 5.541$; $P = 0.020$]. The observed significant interaction implies that right PPC encodes leftward vestibular-derived displacement perception. There were no significant effects of rTMS on response percentage error of peak angular velocity for both rTMS locations (repeated-measures ANOVA).

Table 1. Individual performance and strategy during 'self-rotation' test

Subject (S = sighted; B = blind)	Displacement performance (linear regression response vs. stimulus angle)		Predictors of response angle via multi-regression fit (response angle vs. stimulus angle, velocity, time, acceleration)					
	Linear regression		Multi-regression fit		Angle	Velocity	Time	Acceleration
	r^2	B	N	r^2	β	β	β	β
S1	0.90	1.0	48	0.81	0.57		0.34	
S2	0.83	0.76	48	0.88		0.72	0.38	−0.39
S3	0.92	0.83	48	0.93	0.68	0.29		
S4	0.82	0.61	48	0.83	0.49	0.40		
S5	0.80	0.63	47	0.82	0.70			
S6	0.94	1.0	43	0.94	1.00			
S7	0.89	0.83	48	0.90	0.69			−0.23
S8	0.74	0.78	48	0.78	0.60			−0.36
S9	0.90	0.74	47	0.90	0.78			
S10	0.79	0.55	48	0.81		0.48		
S11	0.75	0.58	48	0.84		0.71		−0.32
S12	0.90	0.74	48	0.90	0.86			
B1	0.73	0.90	48	0.74	1.2			
B2	0.74	0.63	48	0.79		0.73		−0.44
B3	0.71	0.47	47	0.73	**			
B4	0.88	0.81	48	0.90	0.68	0.39		−0.30
B5	0.61	0.71	48	0.63	**			
B6	0.93	1.0	48	0.95	1.2		−0.26	

Linear and multiple regressions for individual data, viz., correlation coefficients r^2, and slopes, both standardized (b) and unstandardized (B) (non-significant values, i.e., PW0.05, are omitted). Strategy, which was unconstrained, is shown by the reported b values under the respective columns, e.g., a spatial strategy was signified by high b values in the 'Angle' column as in S6.

Notes: B3 and B5 respective strategies during the self-rotation test were best described as displacement matching since, although multiple regression analyses did not yield any significant β values, those for 'Angle' approached significance, i.e., $P<0.08$ (denoted by **).

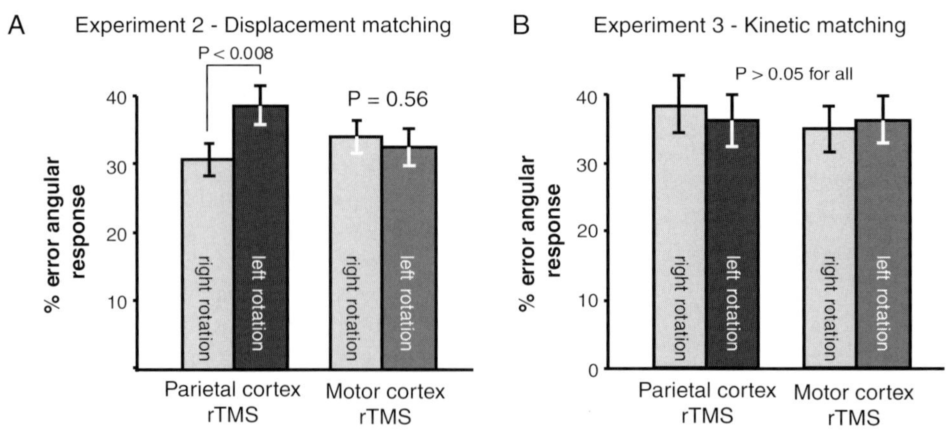

Fig. 3. The main results for Experiments 2 and 3. (A) Experiment 2: Absolute percentage error of response displacement for rightward or leftward inertial stimuli for displacement-matching task. (B) Experiment 3: Absolute percentage error of response displacement divided according to rightward or leftward inertial stimulus direction for velocity-matching task.

Experiment 3

Angular velocity encoding: right PPC vs. left motor cortex rTMS

Repetitive TMS had no effect on displacement (Fig. 3B) or velocity response symmetry (paired *t*-tests and ANOVA: $P > 0.05$ for all).

Discussion

Our findings indicate that angular path reversal under vestibular guidance can be successfully performed via two main strategies, i.e., displacement or velocity matching. The equivalence in performance between blind and sighted groups combined with the fact that blind subjects used either strategy in task performance indicate that the perception of raw and derived vestibular signals are entirely independent of visual mechanisms. Interestingly however, previous data has shown that blind subjects as a group appear to be less able than the sighted at inferential vestibular navigation, i.e., navigation tasks requiring mental manipulation, as opposed to a simple reproduction of the stimulus angle, for their successful completion (Seemungal et al., 2007).

We found that when sighted subjects employed a displacement-matching strategy during the self-rotation task, rTMS to the right PPC disrupted encoding of vestibular-derived displacement perception. In contrast, when subjects employed a velocity-matching strategy, right parietal rTMS had no effect on encoding displacement or velocity perception. Although the use of rTMS to disrupt vestibular perception is novel, rTMS at parietal sites P3 and P4, at similar TMS intensities (10% below motor threshold), has been shown to interfere with sensory perception (Oliveri et al., 2000), mental rotation (Bestmann et al., 2002), and working memory (Mottaghy et al., 2002).

The main human vestibular cortex is thought to be TMJ and contiguous posterior superior temporal gyrus (Foerster, 1936; Penfield, 1957; Friberg et al., 1985; Israël et al., 1995; Blanke et al., 2000, 2004; Bense et al., 2001; Suzuki et al., 2001; Fasold et al., 2002; Dieterich et al., 2003; Kahane et al., 2003). Concomitantly, animal data show primarily a head-velocity-processing role in this region (Guldin and Grüsser, 1998). Vestibular percepts of self-motion have also been demonstrated by electrical cortical activation in the area to that which we applied rTMS; that is, in close proximity to the intraparietal sulcus (Foerster, 1936; Penfield, 1957; Blanke et al., 2000) although, in support of our findings that this area primarily mediates vestibular-derived displacement perception, Snyder et al. (1998) recorded vestibular-derived position signals in primates at this locus.

Lesion studies, although inconsistent in result, support the finding of a right PPC encoding of vestibular-derived displacement encoding (Israël et al., 1995; Farrell and Robertson, 2000; Philbeck et al., 2006; Ventre-Dominey and Vallee, 2007). One problem with lesion studies, however, is that deficits in reference frame transformations rather than deficits in perceptual encoding may complicate data interpretation; for example, different reference frames and neuronal substrates are engaged during reaching under visual versus proprioceptive guidance (Battaglia-Mayer and Caminiti, 2002). Similarly, transformation of vestibular cues into other reference frames required in previous lesion studies (e.g., manual pointing, Philbeck et al., 2006; or saccades, Ventre-Dominey and Vallee, 2007) contrast with our use of vestibular guidance in both stimulus and response phases.

Previous human data suggest that the posterior parietal cortex mediates egocentric visuo-spatial orientation (Spiers and Maguire, 2007). This data combined with our own findings suggests that the PPC may represent a common locus mediating egocentric spatial orientation perception for both visual and vestibular input. This contrasts with theories of visual and vestibular velocity perception that invoke a reciprocal inhibition between visual and vestibular cortical areas (Brandt et al., 2002).

In summary, the right PPC encodes vestibular-derived signals of leftward angular displacement but not angular velocity. Although visual experience does not affect vestibular perception, in sighted subjects, the PPC may represent a common locus for the convergence of visual and vestibular position signals in providing an accurate percept of egocentric spatial orientation.

References

Battaglia-Mayer, A. and Caminiti, R. (2002) Optic ataxia as a result of the breakdown of the global tuning fields of parietal neurones. Brain, 125: 225–237.

Bense, S., Stephan, T., Yousry, T.A., Brandt, T. and Dieterich, M. (2001) Multisensory cortical signal increases and decreases during vestibular galvanic stimulation (fMRI). J. Neurophysiol., 85: 886–899.

Bestmann, S., Thilo, K.V., Sauner, D., Siebner, H.R. and Rothwell, J.C. (2002) Parietal magnetic stimulation delays visuomotor mental rotation at increased processing demands. Neuroimage, 17: 1512–1520.

Blanke, O., Landis, T., Spinelli, L. and Seeck, M. (2004) Out-of-body experience and autoscopy of neurological origin. Brain, 127: 243–258.

Blanke, O., Perrig, S., Thut, G., Landis, T. and Seeck, M. (2000) Simple and complex vestibular responses induced by electrical cortical stimulation of the parietal cortex in humans. J. Neurol. Neurosurg. Psychiatr., 69: 553–556.

Brandt, T., Glasauer, S., Stephan, T., Bense, S., Yousry, T.A., Deutschlander, A. and Dieterich, M. (2002) Visual-vestibular and visuovisual cortical interaction: new insights from fMRI and pet. Ann. N.Y. Acad. Sci., 956: 230–241.

Dieterich, M., Bense, S., Lutz, S., Drzezga, A., Stephan, T., Bartenstein, P. and Brandt, T. (2003) Dominance for vestibular cortical function in the non-dominant hemisphere. Cereb. Cortex, 13: 994–1007.

Farrell, M.J. and Robertson, I.H. (2000) The automatic updating of egocentric spatial relationships and its impairment due to right posterior cortical lesions. Neuropsychologia, 38: 585–595.

Fasold, O., von Brevern, M., Kuhberg, M., Ploner, C.J., Villringer, A., Lempert, T. and Wenzel, R. (2002) Human vestibular cortex as identified with caloric stimulation in functional magnetic resonance imaging. Neuroimage, 17: 1384–1393.

Foerster, O. (1936) Motorische Felder und Bahnen. In: Bumke O. and Foerster O. (Eds.), Hanbuch der Neurologie. Springer, Berlin, pp. 386–387.

Friberg, L., Olsen, T.S., Roland, P.E., Paulson, O.B. and Lassen, N.A. (1985) Focal increase of blood flow in the cerebral cortex of man during vestibular stimulation. Brain, 108(Pt 3): 609–623.

Glasauer, S., Amorim, M.A., Viaud-Delmon, I. and Berthoz, A. (2002) Differential effects of labyrinthine dysfunction on distance and direction during blindfolded walking of a triangular path. Exp. Brain Res., 145: 489–497.

Guldin, W.O. and Grüsser, O.J. (1998) Is there a vestibular cortex? Trends Neurosci., 21: 254–259.

Harris, I.M. and Miniussi, C. (2003) Parietal lobe contribution to mental rotation demonstrated with rTMS. J. Cogn. Neurosci., 15: 315–323.

Hilgetag, C.C., Theoret, H. and Pascual-Leone, A. (2001) Enhanced visual spatial attention ipsilateral to rTMS-induced 'virtual lesions' of human parietal cortex. Nat. Neurosci., 4: 953–957.

Israël, I., Rivaud, S., Gaymard, B., Berthoz, A. and Pierrot-Deseilligny, C. (1995) Cortical control of vestibular-guided saccades in man. Brain, 118(5): 1169–1183.

Kahane, P., Hoffmann, D., Minotti, L. and Berthoz, A. (2003) Reappraisal of the human vestibular cortex by cortical electrical stimulation study. Ann. Neurol., 54: 615–624.

Klem, G.H., Luders, H.O., Jasper, H.H. and Elger, C. (1999) The 10–20 electrode system. The International Federation of Clinical Neurophysiology. Electroencephalogr. Clin. Neurophysiol., 52(Suppl.): 3–6.

Loomis, J.M., Klatzky, R.L. and Golledge, R.G. (2001) Navigating without vision: basic and applied research. Optom. Vis. Sci., 78: 282–289.

Metcalfe, T. and Gresty, M. (1992) Self-controlled reorienting movements in response to rotational displacements in normal subjects and patients with labyrinthine disease. Ann. NYAS, 656: 695–698.

Mottaghy, F.M., Doring, T., Muller-Gartner, H.W., Topper, R. and Krause, B.J. (2002) Bilateral parieto-frontal network for verbal working memory: an interference approach using repetitive transcranial magnetic stimulation (rTMS). Eur. J. Neurosci., 16: 1627–1632.

Oliveri, M., Rossini, P.M., Filippi, M.M., Traversa, R., Cicinelli, P., Palmieri, M.G., Pasqualetti, P. and Caltagirone, C. (2000) Time-dependent activation of parieto-frontal networks for directing attention to tactile space. A study with paired transcranial magnetic stimulation pulses in right-brain-damaged patients with extinction. Brain, 123(9): 1939–1947.

Penfield, W. (1957) Vestibular sensation and the cerebral cortex. Ann. Otol. Rhinol. Laryngol., 66: 691–698.

Philbeck, J.W., Behrmann, M., Biega, T. and Levy, L. (2006) Asymmetrical perception of body rotation after unilateral injury to human vestibular cortex. Neuropsychologia, 44: 1878–1890.

Seemungal, B.M., Glasauer, S., Gresty, A.M. and Bronstein, A.M. (2007) Vestibular perception and navigation in the congenitally blind. J. Neurophysiol., 97: 4341–4356.

Seemungal, B.M., Rizzo, V., Gresty, A.M., Rothwell, J.C. and Brostien, A.M. (2008) Posterior parietal rTMS disrupts human Path Integration during vestibular navigation task. Neurosci. Lett., published online, March 28th.

Snyder, L.H., Grieve, K.L., Brotchie, P. and Andersen, R.A. (1998) Separate body- and world-referenced representations of visual space in parietal cortex. Nature, 394: 887–891.

Spiers, H.J. and Maguire, E.A. (2007) A navigational guidance system in the human brain. Hippocampus, 17(8): 618–626.

Suzuki, M., Kitano, H., Ito, R., Kitanishi, T., Yazawa, Y., Ogawa, T., Shiino, A. and Kitajima, K. (2001) Cortical and subcortical vestibular response to caloric stimulation detected by functional magnetic resonance imaging. Brain Res. Cogn. Brain Res., 12: 441–449.

Ventre-Dominey, J. and Vallee, B. (2007) Vestibular integration in human cerebral cortex contributes to spatial remapping. Neuropsychologia, 45: 435–439.

Zacks, J.M., Gilliam, F. and Ojemann, J.G. (2003) Selective disturbance of mental rotation by cortical stimulation. Neuropsychologia, 41: 1659–1667.

C. Kennard & R.J. Leigh (Eds.)
Progress in Brain Research, Vol. 171
ISSN 0079-6123

CHAPTER 4.19

Foot rotation contribution to trunk and gaze stability during whole-body mediated gaze shifts: a principal component analysis study

S. Sklavos[1], D. Anastasopoulos[1,*], N. Ziavra[2], M.A. Hollands[3] and A.M. Bronstein[2]

[1]*Department of Physiology, School of Nursing, University of Athens, Goudi, Athens, Greece*
[2]*Academic Department of Neuro-otology, Charing Cross Hospital, Imperial College London, London, UK*
[3]*School of Exercise and Sport Sciences, University of Birmingham, Birmingham, UK*

Abstract: Large gaze displacements are mediated by combined motion of the eye, head, trunk, and foot. We applied principal component analysis (PCA) to examine the degree of variability and linearity in the angular velocity pattern of the various segments involved that participate in this task. Ten normal subjects stood up and had to visually fixate and realign their bodies with LED targets separated 45° apart, ranging from ±45 to 360°. The outbound movement in this paradigm is unpredictable whereas the return (inbound) movement occurs under spatially predictable conditions. Under such predictable conditions, subjects generate in approximately 15% of the trials gaze shifts, with periods of fairly constant high gaze velocity (single-step gaze shifts). PCA showed that gaze velocity variability did not change if the feet were rotating or not. Foot velocity was variable and showed additional PCs suggestive of non-linear motion components. Trunk and head-in-space velocity showed intermediate levels of variability but its variability decreased during the foot stepping movements. The results suggest that the feet, trunk, and head are less tightly controlled by the central nervous system than gaze velocity. Movements of the feet seem to aid trunk stability and motion rather than gaze control.

Keywords: gaze; posture; coordination; principal component analysis

Introduction

Voluntary whole-body turning on the spot represents a common behaviour in everyday life (Land, 2004; Solomon et al., 2006). However, little is known about the coordination pattern among the various body segments and specifically about the kinematic relationship between the rotations of the lower extremities and those of the eye and head (Grasso et al., 1998; Hollands et al., 2004). This relationship has considerable significance from a clinical point of view, as falls in the elderly and in patients with extrapyramidal disorders occur more frequently when turning than during straight ahead locomotion. In this study we have investigated patterns of coordination in yaw during such pivot turns by means of principal component analysis (PCA). This technique has already been implemented in oculomotor plant analysis, to infer the dimensionality of a set of data belonging to the

*Corresponding author. Tel.: +30-210-7489306;
Fax: +30-210-7713366; E-mail: danastas@nurs.uoa.gr

DOI: 10.1016/S0079-6123(08)00651-1

same variable, as they vary from trial to trial (Sklavos et al., 2005).

Materials and methods

Ten healthy adults (three females), mean age 52 ± 2.6 years, volunteered for the study as approved by the Imperial College Riverside Ethics Committee. Participants stood in the centre of a circular array of eight LEDs, spaced 45° apart at eye level in the dark, as in a previous experiment (Hollands et al., 2004). At the beginning of each trial, subjects (Ss) aligned eyes and body with the central LED (Fig. 1). After 10 s the central target was extinguished thus indicating that another LED in one of the seven eccentric locations (±45, 90, 135, and 180°) had appeared. The subject had to fixate the lit LED and realign his whole body with it (outbound trials). After 15 s the eccentric LED was turned off thus cueing subjects to return to the initial, central position (this protocol eliminated changes in initial eye position in the orbit as an additional source of variability). Subjects performed four trials to each LED location, spatially randomized. Head in space, upper trunk, and feet horizontal (yaw plane) movements were recorded using an electromagnetic motion analysis system (Polhemus Fastrak; 30 Hz; latency 4 ms, accuracy 0.15° RMS), with markers placed on the head, the seventh cervical vertebra, and on each foot. Horizontal eye movements were recorded with bi-temporal DC electro-oculography (EOG).

On–off LED signals, EOG, and body position markers were recorded at 240 Hz. Movement onset/offset for the various segments was calculated from the time derivative (angular velocity). Gaze (eye position in space) was obtained by adding EOG + head in space signals. Subtracting the trunk from the head signal resulted in an estimate of head-on-trunk movement.

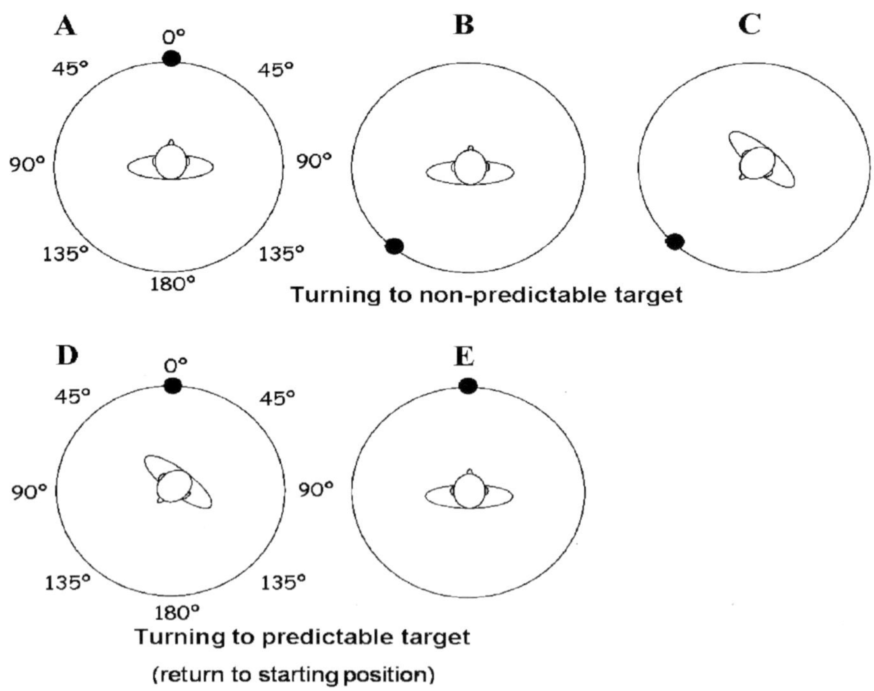

Fig. 1. Experimental protocol. The subject is standing at the centre of a circular array of targets. Extinction of the centrally located target signals the subject that a second LED has been turned on (B, here at 135° on the left). He has to locate, fixate, and rotate his whole body towards this new target (C). After a few seconds this eccentric target is switched off and the subject has to return back to the initial (primary) position (D–E).

Principal component analysis

We used PCA to examine how the velocity of the participating body segments varied during gaze re-orientations. Generally, PCA explores whether a set of variable data (e.g., eye or head-in-space velocity) during repetition of an event ('trial'), can be thought to represent variations of an underlying function ('system') and whether this system is linear or not. In this case, PCA examines the relationship among a set of n correlated traces $X_1, \ldots, X_n$ that refer to the same variable. In general, this method derives n uncorrelated principal components (PCs) $Z_1, \ldots, Z_n$ in decreasing order of the original data variance percentage. Percentages are the $\lambda_1 > \lambda_2 > \cdots > \lambda_n$ eigenvalues ($\sum_{i=1}^{n} \lambda_1 = 1$) of the covariance matrix $\sum X^T \cdot X(X = [X_1, \ldots, X_n]$, T for transpose) with corresponding eigenvectors $(v_i)_{i=1,\ldots,n}$: $(\Sigma - \lambda_i \cdot I)v_i = 0$ and $|\sum - \lambda_i \cdot I| = 0$, where I is the $n \times n$ unique matrix. The more variance the 1st PC explains, the lesser the variability among the set of the original traces is. PCA currently defines 'large' PCs as those accounting for more than 1% of the variation of the original variable ($\lambda_i \geq 0.01$). An increase in the number of 'large' PCs presumes a bias effect of a non-linear mechanism on to the original variable (Chatfield and Collins, 1980).

We applied PCA for each original variable separately, i.e., gaze velocity ($\dot{G}$), eye velocity in the orbit ($\dot{E}$), trunk velocity ($\dot{T}$), head on the trunk velocity ($\dot{H}_T$), left ($\dot{L}_F$) and right ($\dot{R}_F$) foot velocity. Only 27 trials in which the visual target was acquired with a single-step gaze shift (i.e., gaze transference achieved in one large gaze saccade) were included for PCA, with data segments of 420 ms (100 data points at 240 Hz sampling frequency). PCA was run using Matlab©. Since in this paper we concentrate on the influence of foot motion on performance, each variable was studied separately for both directions of rotational movement, leftwards ($n = 13$) or rightwards ($n = 14$) and separately for two periods, i.e. (a) between the saccadic initiation and the onset of foot rotation (0–170 ms) and (b) for an interval of 250 ms thereafter.

Results

Elsewhere we have described various gaze transference patterns observed during these experiments (manuscript submitted). In a small proportion of those trials (about 15%; particularly during return, predictable trials) subjects produced a large, single-step gaze shift, which took the eyes onto or very near the target. Those large gaze steps were accompanied by relatively long periods (ca. 500 ms) of a stable gaze velocity (eye velocity in space, i.e., eye plus head plus trunk) with values of 300–450°/s (submitted). In the current paper, we concentrate on the effect of foot rotations on these periods of fairly constant gaze velocity during single-step gaze transference (Fig. 2).

Gaze velocity showed quite low variability among different trials over the whole 420 ms periods, as the 1st PC explained 99% of its variance (Table 1). Despite the high variability of foot rotational velocity over the later part of the movement (1st PC of foot velocity was less than 81%), the 1st PC extracted from the gaze velocity traces was not reduced. In contrast to gaze velocity which had a single PC, foot rotations displayed three or more 'large' PCs (between 22.8% and 1.4%), however this may be partly influenced by the fact that subjects sometimes started the stepping movements with the ipsidirectional foot and sometimes with the contradirectional foot. Rotational velocity of the foot contralateral to the movement direction was found to be less variable than that of the ipsilateral foot (Table 1). Eye velocity variability increased during the time interval before target acquisition as compared with the period just before foot movement initiation (variance explained by its 1st PC fell from 98.3% to 87.7% for leftwards and from 96.8% to 93.6% for rightwards). Both trunk and head-on-trunk velocity variability correspondingly decreased during the last 250 ms of coordinated movement. Such an increase of the 1st PC was accompanied by a corresponding decrease of the 2nd or 3rd PC amount where in some cases it fell under the limit of 1% (trunk and head on the trunk velocity, rightward movements). The 1st PC for each velocity variable is shown in Fig. 2.

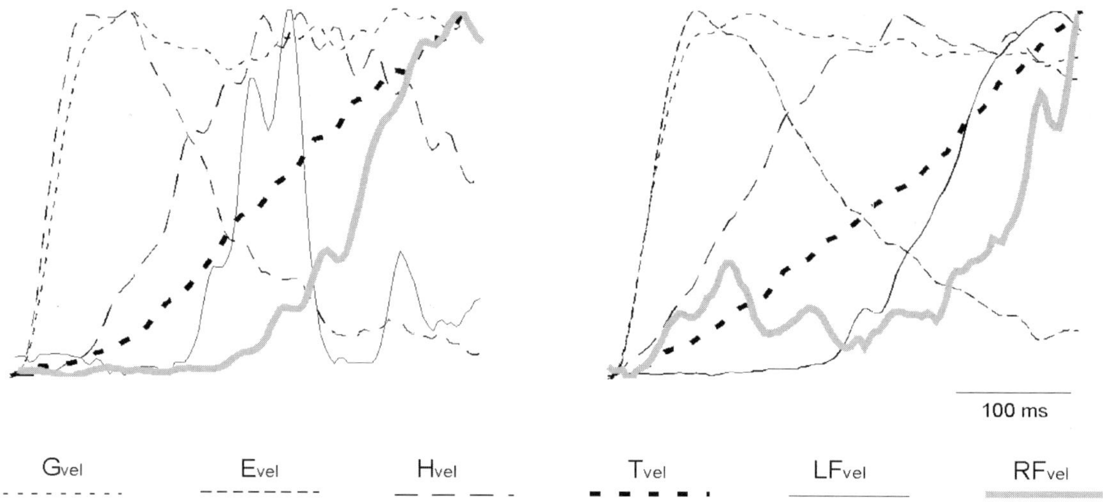

Fig. 2. Normalized 1st PC for each variable (G_{vel} for gaze velocity; E_{vel} for eye velocity; H_{vel} for head-on-trunk velocity; T_{vel} for trunk velocity; LF_{vel} for left foot velocity; RF_{vel} for right foot velocity) over the whole analysis period of 420 ms for leftward (left panel) and rightward (right panel) rotational movements.

Table 1. Principal components (PCs) with greater than 1% of variance explained, in decreasing order

Time interval	Leftward movements		Rightward movements	
	0–170 ms	170–420 ms	0–170 ms	170–420 ms
Variable (%)				
Gaze velocity	99	99.2	98.7	99.1
Eye-in-orbit velocity	98.3	87.7, 8, 1.2, 1	96.8, 2.3	93.6, 3.6
Trunk velocity	88.8, 9.7	97.8, 1.7	91.9, 6.2	98
Head-on-trunk velocity	89.4, 6.9	95.2, 2.8	93.5, 3.5, 1.7	95.5, 2.5
Left foot velocity		73.8, 22.8, 1.4		77.7, 17, 3.4
Right foot velocity		80.3, 16.1, 2.3		70.7, 12.9, 8.9, 3.4, 2.1

Discussion

The purpose of this study was to apply PCA in order to understand the control of various body segments during a specific type of gaze transference (large single-step gaze shifts). By describing the level of variability in the various segments, and in different periods of the motion pattern (namely before and after foot movements), we expected that PCA would provide insight into which of the participating motion segments may be more tightly (and more loosely) controlled by the CNS.

PCA indicated that gaze velocity consistently displayed the lowest levels of variability during the period studied, both during intervals with and without foot motion and during rightwards and leftwards rotations. At the other end of the spectrum, foot velocities showed the highest levels of variability and, according to the presence of additional PCs, probably included non-linear components in the motion pattern. This finding suggests that gaze velocity may be a main parameter under tight central control.

Note, however, that the period when foot movements occur is associated with improved (reduced) variability in trunk and head-on-trunk velocity. These findings would suggest that although foot motion does not seem to contribute significantly to gaze control, they do add to a reduction of trunk motion variability. Since the trunk, due to its mass, is crucial for the overall control of postural stability, it is possible that foot rotations are primarily brought into play for postural rather than gaze control.

References

Chatfield, C. and Collins, A.J. (1980) Introduction to Multivariate Analysis. Chapman & Hall/CRC, USA

Grasso, R., Prevost, P., Ivanenko, Y.P. and Berthoz, A. (1998) Eye-head coordination for the steering of locomotion in humans: an anticipatory synergy. Neurosci. Lett., 253: 115–118.

Hollands, M., Ziavra, N. and Bronstein, A.M. (2004) A new paradigm to investigate the roles of head and eye movements in the coordination of whole-body movements. Exp. Brain Res., 154: 261–266.

Land, M.F. (2004) The coordination of rotations of the eyes, head and trunk in saccadic turns produced in natural situations. Exp. Brain Res., 159(2): 151–160.

Sklavos, S., Porrill, J., Kaneko, C.R. and Dean, P. (2005) Evidence for wide range of time scales in oculomotor plant dynamics: implications for models of eye-movement control. Vision Res., 45(12): 1525–1542.

Solomon, D., Kumar, V. and Jenkins, R.A. (2006) Head control strategies during whole-body turns. Exp. Brain Res., 173: 475–486.

C. Kennard & R.J. Leigh (Eds.)
Progress in Brain Research, Vol. 171
ISSN 0079-6123

CHAPTER 4.20

Supraspinal locomotor control in quadrupeds and humans

Klaus Jahn, Angela Deutschländer, Thomas Stephan, Roger Kalla, Katharina Hüfner, Judith Wagner, Michael Strupp and Thomas Brandt*

Department of Neurology, University of Munich, Klinikum Grosshadern, Marchioninistrasse 15, 81377 Munich, Germany

Abstract: Locomotion in humans and other vertebrates is based on spinal pattern generators, which are regulated by supraspinal control. Most of our knowledge about the hierarchical network of supraspinal locomotion centres derives from animal experiments, mainly in the cat. Here we summarize evidence that the supraspinal network of quadrupeds is conserved in humans despite their transition to bipedalism. By use of mental imagery of locomotion in fMRI we found (1), locomotion modulates sensory systems and is itself modulated by sensory signals. During automated locomotion in healthy subjects cortical sensory inhibition occurs in vestibular and somatosensory areas; this inhibition is cancelled in the congenitally blind; (2), we delineated separate and distinct areas in the brainstem and cerebellum which are remarkably similar to the feline locomotor network. The activations found here include homologues to the pacemakers for gait initiation and speed regulation in the interfastigial cerebellum and bilateral midbrain tegmentum (cerebellar and mesencephalic locomotor regions), their descending target regions in the pontine reticular formation, and the rhythm generators in the cerebellar vermis and paravermal cerebellar cortex. This conservation of the basic organization of supraspinal locomotor control during vertebrate phylogeny opens new perspectives for both, the diagnosis and treatment of common gait disorders. It is conceivable that electrical stimulation of locomotor brain stem centres may initiate and improve gait in selected patients suffering from Parkinson's disease or progressive supranuclear palsy.

Keywords: locomotion; motor imagery; fMRI; multisensory control; blind

Introduction

Quadrupeds and bipedal humans actively explore their environment by locomotion. Most of our knowledge about the sensorimotor control of posture and gait has been gained from different preparations of the intact, decorticate, and decerebrate cat. In all vertebrates this coordination is provided by "central pattern generators" (CPGs), segmentally organized groups of interneurons in the spinal cord (Grillner, 1981, 2006). The CPGs are controlled by descending input from specific locomotor command regions in the brainstem and cerebellum. Several distinct regions that initiate and modulate spinal stepping have been identified by electrical or chemical stimulation in the cat (Shik and Orlovsky, 1976; Armstrong, 1988; Mori

*Corresponding author. Tel.: +498970952570;
Fax: +498970955584;
E-mail: thomas.brandt@med.uni-muenchen.de

DOI: 10.1016/S0079-6123(08)00652-3

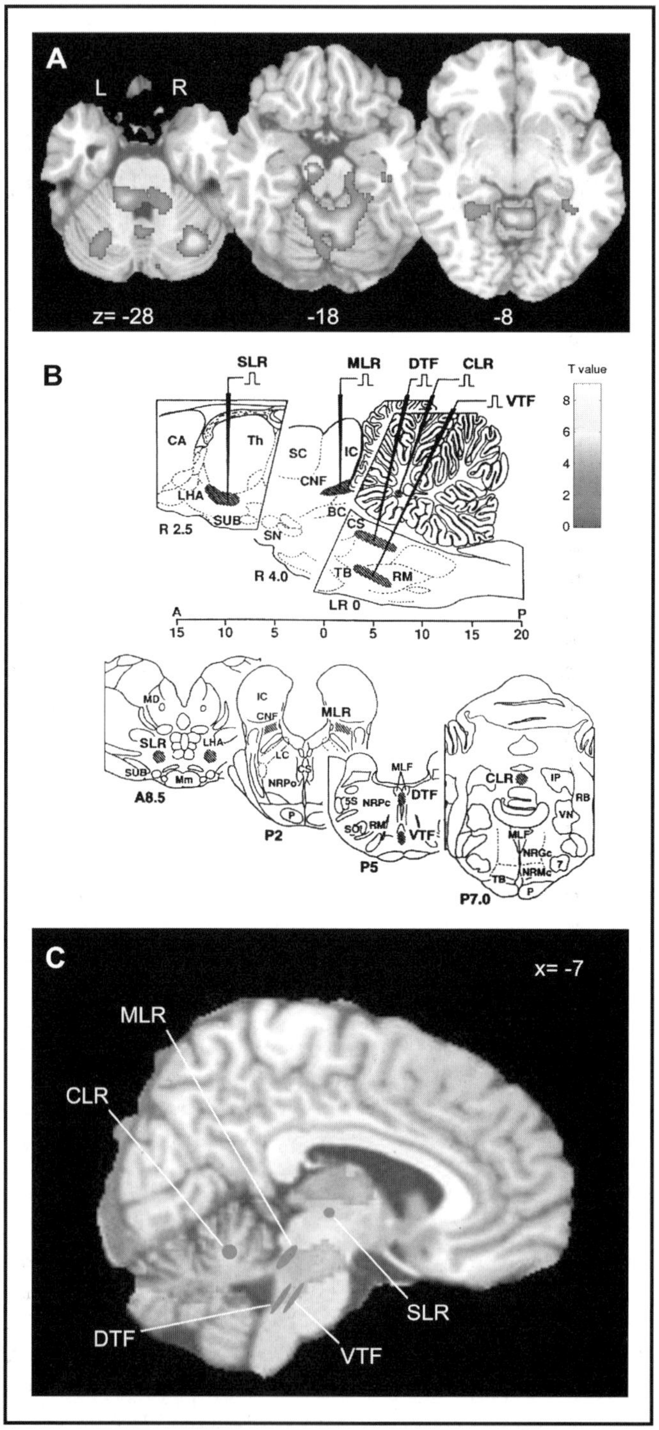
A
L
R
z= -28
-18
-8
B
SLR
MLR
DTF
CLR
VTF
T value
8
6
4
2
0
CA
Th
SC
IC
CNF
LHA
SUB
R 2.5
BC
CS
SN
R 4.0
TB
RM
LR 0
A
P
15
10
5
0
5
10
15
20
MD
SLR
LHA
SUB
Mm
A8.5
IC
CNF
MLR
LC
NRPo
P
P2
MLF
5S
NRPc
DTF
SO
RM
VTF
P5
CLR
IP
RB
VN
MLF
NRGc
7
NRMc
TB
P
P7.0
C
x= -7
MLR
CLR
SLR
DTF
VTF

et al., 2001; Fig. 1B). Proceeding rostrocaudally, the most important regions are the subthalamic locomotor region (SLR) in the lateral hypothalamic area, the mesencephalic locomotor region (MLR), corresponding to the cuneiform and pedunculopontine nuclei in the dorsal midbrain, the cerebellar locomotor region (CLR) located close to the fastigial nuclei in the cerebellar midline, and the pontine locomotor region in the pontomedullary reticular formation (PMRF). The locomotor regions in the cat have been associated with different functions in the network for supraspinal locomotor control (Armstrong, 1988; Whelan, 1996; Mori et al., 2001; Grillner, 2006). For simplification we can assume that the SLR and MLR initiate locomotion when disinhibited from tonic basal ganglia control. The CLR receives rhythmic input from the vermis and paravermal cerebellar cortex to control speed. The CLR output converges with descending MLR projections in the PMRF, where locomotor signals are transmitted to the spinal cord CPGs.

There are several lines of evidence that the organization of supraspinal locomotor control in humans was conserved during their transition from quadrupedal to bipedal locomotion. First, electrophysiological studies have demonstrated that the quadrupedal interlimb coordination remains conserved during bipedal walking (Dietz et al., 2001; Wannier et al., 2001; Dietz, 2002; Balter and Zehr, 2007). Second, anecdotal reports on single patients with vascular midbrain lesions support the view that the MLR also mediates gait initiation in humans (Masdeu et al., 1994; Hathout and Bhidayasiri, 2005). Third, a few brain activation studies have identified homologues to cat locomotor regions during actual and imagined gait (Hanakawa et al., 1999; Jahn et al., 2004, 2008). Surprisingly, the clinical implications of these centres for common central disorders of gait and posture like Parkinson's disease, progressive supranuclear palsy (PSP), and cerebellar ataxia have so far been largely ignored. Postmortem histological studies, however, revealed that Parkinson's disease and PSP are associated with reduced cell counts in the pedunculopontine nucleus (Zweig et al., 1989; Pahapill and Lozano, 2000). Moreover, stimulation of this area with low frequency electrical stimulation improved the gait of patients with advanced Parkinson's disease (Plaha and Gill, 2005; Stefani et al., 2007).

The locomotor network in all mammals is modulated by cortical control and sensory feedback on multiple levels. This feedback originates from muscles and skin afferents as well as from the visual and vestibular senses. It modulates transmission in locomotor pathways in a state- and phase-dependent manner (Rossignol et al., 2006). Besides modulating the phases of the gait cycle, certain aspects of sensory control seem to be in general suppressed during locomotion to prevent adverse interactions with an optimized sensorimotor pattern. In humans, it was shown that the H-reflex excitability is lower during locomotion (Capaday et al., 1986). Changes in the gain of reflexes were explained by a task-dependent increase of tonic presynaptic inhibition in group Ia afferents (Dietz, 1992; Faist et al., 1996). For

Fig. 1. Infratentorial BOLD-signal increases during human locomotion and locomotor regions in the cat. (A) Statistical parametric maps are shown for the contrast locomotion (mental imagery of walking and running) vs. lying in 26 subjects ($p<0.01$, FDR). The slices selected show bilateral activity in the dorsal brainstem, in the cerebellar vermis and hemispheres, as well as in the parahippocampal gyrus ($z=-8$). (B) The anatomical locations of the subthalamic locomotor region (SLR), the mesencephalic locomotor region (MLR), the cerebellar locomotor region (CLR), the dorsal tegmental field (DTF), and the ventral tegmental field (VTF) in the cat are shown. The horizontal axis represents the rostrocaudal Horsley–Clarke coordinates. 5S, spinal trigeminal nucleus; 7, facial nucleus; BC, brachium conjunctivum; CA, caudate nucleus; CNF, cuneiform nucleus; CS, nucleus centralis superior; IC, inferior colliculus; IP, nucleus interpositus; LHA, lateral hypothalamic area; LC, nucleus locus caeruleus; MD, mediodorsal nucleus of the thalamus; MLF, medial longitudinal fasciculus; Mm, mammillary body; NRGc, nucleus reticularis gigantocellularis; NRMc, nucleus reticularis magnocellularis; NRPc, nucleus reticularis pontis caudalis; NRPo, nucleus reticularis pontis oralis; P, pyramidal tract; RB, restiform body; RM, nucleus raphe magnus; SUB, subthalamic nucleus; SC, superior colliculus; SO, superior olive; SN, substantia nigra; TB, trapezoid body; Th, thalamus; VN, vestibular nuclei. (C) Sagittal section of the human brain with transparent overlay of BOLD-signal increases during locomotion ($p<0.05$, FDR). The anatomical locations of the cat locomotor regions are superimposed. B adapted from Mori et al. (2001). (See Color Plate 4.20.1 in color plate section.)

the vestibular system it was shown in the cat that locomotion decreased the responses of neurons in the lateral vestibular nucleus to tilt stimulation (Orlovsky, 1972). In humans, galvanic vestibular stimulation was used to demonstrate reduced vestibular control during running compared to walking by measuring gait deviations (Jahn et al., 2000). Further, slow phase velocity of spontaneous nystagmus in patients with acute unilateral vestibular failure decreases during locomotion (Jahn et al., 2002). Functional magnetic resonance imaging (fMRI) data show reduced activity of multisensory cortical areas during walking and even more so during running (Jahn et al., 2003).

Here we present an overview of human supraspinal control of locomotion. Drawing on additional analyses of data from a series of experiments using mental imagery of locomotion in fMRI, we correlate the locations and functions of locomotor regions in the human brainstem and cerebellum with those of the cat. These experiments focused on (1) the demonstration of brainstem and cerebellar locomotor regions in humans (Jahn et al., 2008); (2) the investigation of task-dependent suppression of cortical sensory control during locomotion (Jahn et al., 2003, 2004); and (3) the changes of multisensory supraspinal control in the congenitally blind (Deutschländer et al., 2008).

Methods

A total of 45 subjects were examined during three different studies using mental imagery of locomotion in fMRI in sighted healthy adults ($n = 38$, Jahn et al., 2004, 2008) and congenitally blind subjects ($n = 7$, Deutschländer et al., 2008). Subjects were trained to perform (eyes open) and then to imagine (eyes closed) these four conditions: lying (rest condition), standing, walking, and running. The experimental procedure is described in detail in Jahn et al. (2008). Functional imaging was done immediately after the training using a standard clinical scanner (1.5 T Siemens Magnetom Vision, Erlangen, Germany). A T2*-weighted gradient echo multislice sequence (EPI, TR 4500 ms, TE 60 ms, voxel size $3.75 \times 3.75 \times 3.75\,mm^3$, matrix 64×64) was used to acquire 34 slices covering the whole brain and the cerebellum. Data processing was done using statistical parametric mapping software (SPM2; http://www.fil.ion.ucl.ac.uk/spm). All stereotactic coordinates given in this paper refer to the MNI coordinate system.

Results

Mental imagery of locomotion in fMRI caused BOLD-signal increases in frontal and precentral gyri with left hemispheric dominance, bilaterally in the precuneus, the parahippocampal gyrus, the thalamus, and basal ganglia. A large cluster of activation was found in the brainstem and cerebellum, including the anterior vermis and paravermal cerebellar cortices, and extending bilaterally to the midbrain tegmentum and pontine reticular formation ($n = 26$, Table 1).

The infratentorial activations are shown in Fig. 1 in their relation to the anatomical locations of locomotor regions defined in the cat. Significant activity clusters cover the CLR, the MLR, and the PMRF. This distribution of activity implies that there is a widespread infratentorial network associated with locomotion.

The differential activity for standing, walking, and running is shown in Fig. 2. Under all conditions a large cluster of activation occurs in the left anterior insula. Walking and running, but not standing, are associated with BOLD-signal changes in the anterior vermis, brainstem, and parahippocampal gyrus. The activity in the infratentorial locomotor network is most pronounced during running.

Cortical areas associated with sensory functions showed decreased BOLD signals in healthy subjects during locomotion. Figure 3 illustrates deactivations in multisensory areas which can be attributed to the vestibular (supramarginal gyrus, superior temporal gyrus), somatosensory (postcentral gyrus), and visual (middle occipital gyrus) systems ($n = 26$). In contrast, multisensory cortical areas for the vestibular and somatosensory systems showed increased BOLD signals during locomotion in the congenitally blind ($n = 7$), indicating that blind subjects are more dependent on these sensory systems.

Table 1. BOLD-signal changes during mental imagery of locomotion (walking and running vs. lying)

	BA	Cluster	T	P	x	y	z
Cerebral hemispheres							
L/R Frontal mid + inf gyri	48	1309	5.22	0.026	44	34	22
	48	40	3.73	0.040	−54	18	8
L Precentral gyrus	6	153	4.23	0.034	−30	−4	38
	6	69	3.91	0.037	−22	8	64
L Insula	48	319	4.24	0.034	−32	26	8
L/R Parahippocampal gyrus	37	127	5.00	0.026	−32	−42	−6
	37	80	4.40	0.032	34	−42	−6
L/R Precuneus	29	89	4.70	0.027	−18	−50	14
	30	104	5.76	0.026	20	−46	16
L/R Thalamus/Pallidum		539	4.76	0.026	−4	−22	14
		19	3.89	0.038	−16	−6	−4
Brainstem and cerebellum							
L/R Midbrain + Vermis + Cerbl. 4–5		1554	5.96	0.026	−6	−48	−14
L/R Cerebellum Crus1		417	5.62	0.026	34	−64	−30
		125	4.15	0.034	−32	−58	−34

Significance level $p<0.05$, FDR ($n=26$). R, right; L, left; BA, Brodmann area; Cluster, cluster size in voxels; T, T-value; P, p-value FDR-corrected; x,y,z coordinates in the MNI space.

Discussion

The fMRI data presented here strongly support the assumption that the organization of supraspinal locomotor control was conserved during vertebrate phylogeny, despite the transition from quadrupedal to bipedal human locomotion.

Supraspinal motor control of locomotion

We delineated regions in the dorsal brainstem and midline cerebellum which correspond to the anatomical location of the MLR, the CLR, and the pontine reticular formation. We did not detect the SLR in this study. The MLR lies just ventral to the inferior colliculus and corresponds to the cuneiform and pedunculopontine nuclei, bordering on the locus coeruleus, brachium conjunctivum, and mesencephalic trigeminal nucleus (Shik et al., 1966; Pahapill and Lozano, 2000). Electrical stimulation of the MLR in cats initiates locomotion and dictates the level of force developed during stepping (Drew et al., 2004). The MLR receives afferent projections from the basal ganglia (including the hypothalamic region and subthalamic nucleus), the sensorimotor cortex, and the limbic system. Its descending projections reach the spinal cord via the PMRF. The rhythmic activity that can be recorded from MLR cells in cats is derived from cerebellar input from the vermis and paravermal cerebellar cortex (Armstrong, 1988). Surgical removal of the cerebellum transforms the rhythmic pattern into a non-rhythmic one. By analogy, midline cerebellar injuries and atrophies in humans mainly produce instabilities in posture and locomotion (truncal and gait ataxia). Conversely, the cluster of cerebellar activations in our study was largest during running, thus agreeing with the hypothesis of a cerebellar pacemaker that increases activity with increasing speed.

Further, Mori and co-workers described a locomotion-inducing site at the midline of the cerebellar white matter, rostral to the rostral-most pole of the fastigial nucleus (Mori et al., 1999). The vermis also integrates proprioceptive, vestibular, and visual afferent information into the locomotor program (Mori et al., 2001). The BOLD-signal increase in the dorsal tegmentum at the caudal pontine level during standing (Fig. 2) corresponds to the dorsal tegmental field (DTF), a locomotion suppression site delineated in cats. During MLR-evoked locomotion, the concomitant stimulation of the DTF in the caudal pons inhibits locomotion (Mori et al., 2001). Ventral tegmental field (VTF) and DTF regulate postural muscle tonus by differential interaction.

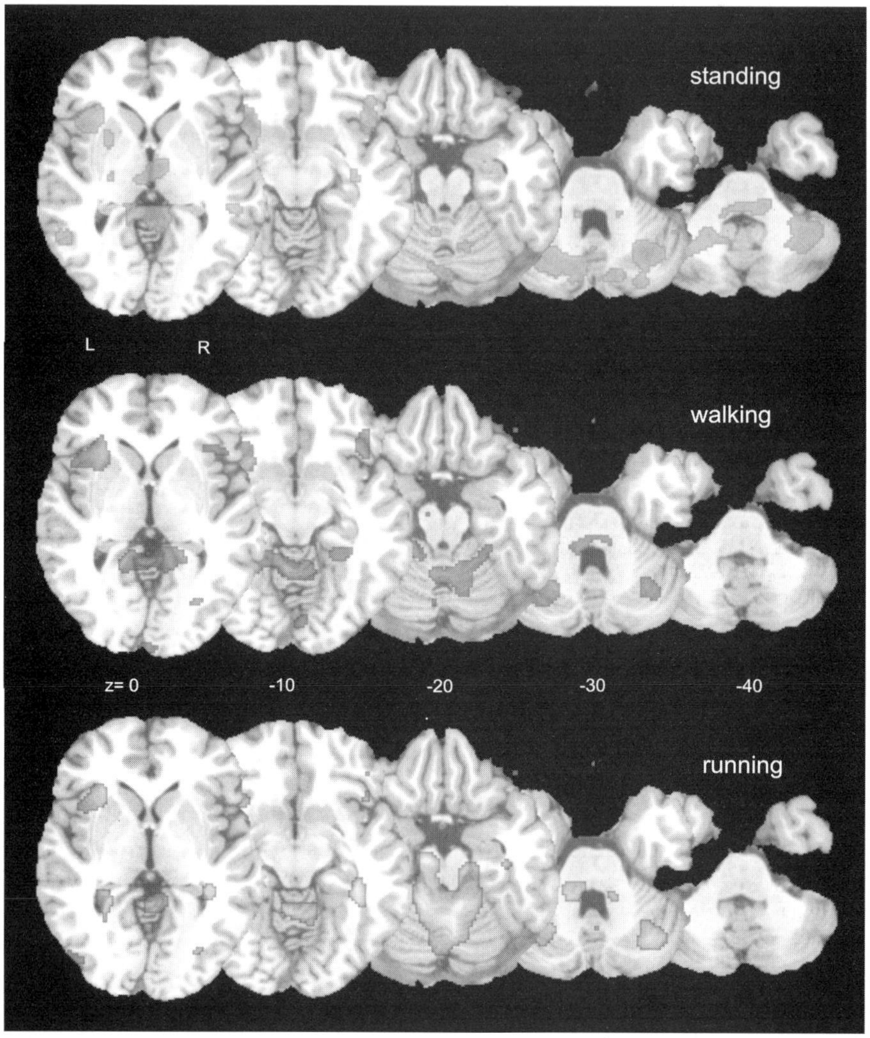

Fig. 2. BOLD-signal increases during mental imagery of standing, walking, and running ($p<0.05$, FDR). During standing, activations were prominent in the frontal gyri, precuneus, thalamus, and cerebellar hemispheres. Note the activation in the dorsal pons ($z=-40$, DTF). During walking, activations were found in the frontal gyri, parahippocampal gyri, midline cerebellum, and in the pontine reticular formation. During running, a large cluster of activation was found in the cerebellum, extending bilaterally to the midbrain. There was also activation in parahippocampal gyri ($z=0$ and -10). (See Color Plate 4.20.2 in color plate section.)

The activations found in the cortex and in the basal ganglia also fit well to the models of input to the brainstem network established on the basis of animal data (Whelan, 1996; Grillner, 2006). It has been postulated anatomically and physiologically that the basal ganglia output nuclei (homologues to the pallidum) have mutual connections to the MLR (Garcia-Rill and Skinner, 1986; Inglis and Winn, 1995; Hashimoto, 2006). The cerebral cortex, mainly area 6, sends projections to the brainstem, forming the cortico-reticulo-spinal system for automated and voluntary control of posture and gait (Mori et al., 2001). The frontal cortices have mutual connections with the cerebellum through the thalamus and pontine nuclei and with the basal ganglia through the basal ganglia-thalamocortical circuit (e.g., Hashimoto, 2006). To initiate and terminate locomotor movements in the biped and quadruped, a

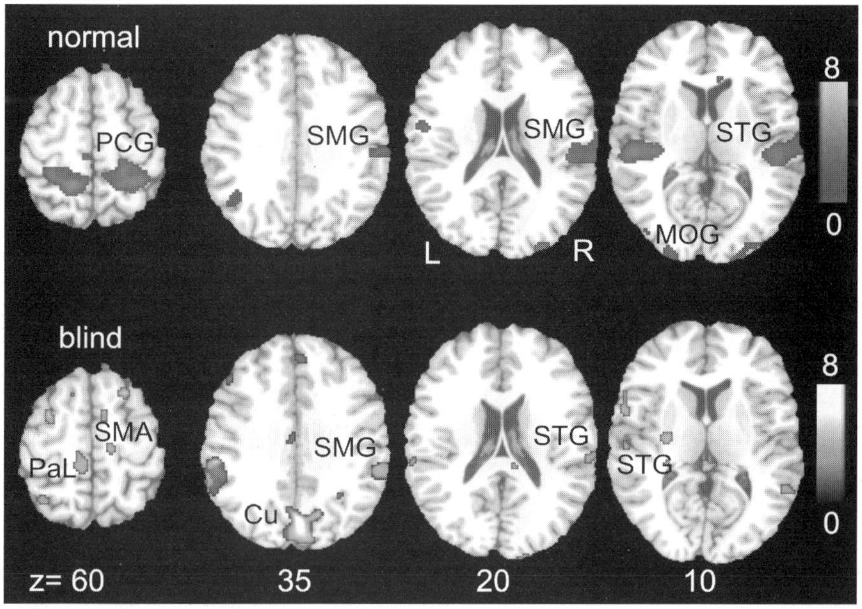

Fig. 3. Multisensory cortical areas during mental imagery of running in sighted and blind subjects ($p < 0.01$, uncorrected). The upper panel shows BOLD-signal decreases in 26 healthy subjects. Deactivations were detected bilaterally in the post-central (PCG), supramarginal (SMG), superior temporal (STG), and middle occipital gyri (MOG). These cortical areas can be attributed to somatosensory (PCG), vestibular (SMG, STG), and visual (MOG) functions. The lower panel shows BOLD-signal increases in seven congenitally blind subjects. Note that SMG and STG show increased signals. Deactivations were not detected in this group. SMA, supplementary motor area; PaL, paracentrale lobule; Cu, cuneus; R, right; L, left. *T*-value colour coding is indicated on the right side. (See Color Plate 4.20.3 in color plate section.)

smooth transition is required both from and to a standing upright posture. This transition appears to be accomplished by two different control mechanisms: an automatic, lower-order control process and a volitional, higher-order process (Mori et al., 2001). The lower-order process is controlled by brainstem motor programs, which permit the decerebrate cat to walk on a treadmill in a straight line. Natural locomotion involves the higher-order process and thereby the cerebellum, basal ganglia, and cerebral cortex (Armstrong, 1988).

A comparison of neural mechanisms related to postural and locomotor control in primates and cats has suggested significant analogies (Mori et al., 1996). Primates have a spinal rhythm generator (Fedirchuk et al. 1998), and locomotion can be elicited by stimulation in the SLR and MLR (Eidelberg et al., 1981). Human locomotion is assumed to be under more supraspinal control than animal locomotion. For example, it is difficult for human patients to achieve spinal stepping after traumatic spinal transection; it is achieved only when body weight is partially unloaded (Dietz et al., 1994; Dietz, 2003). As mentioned above, the conservation of quadrupedal limb-coordination in humans, the gait disturbances in patients with restricted lesions in brainstem and cerebellum, and the fMRI findings reported here suggest more similarities than differences between human and animal locomotion (Fig. 4). Clinical testing of these qualities during the evaluation of patients with gait disorders such as Parkinson's disease might disclose dysfunction in supraspinal locomotor control. Such an approach may prove beneficial for patients with gait disturbances as it provides the foundations for developing alternative therapeutic strategies.

Cortical sensory activity during locomotion

The locomotor neural network is modulated by sensory input from a variety of sources. Peripheral,

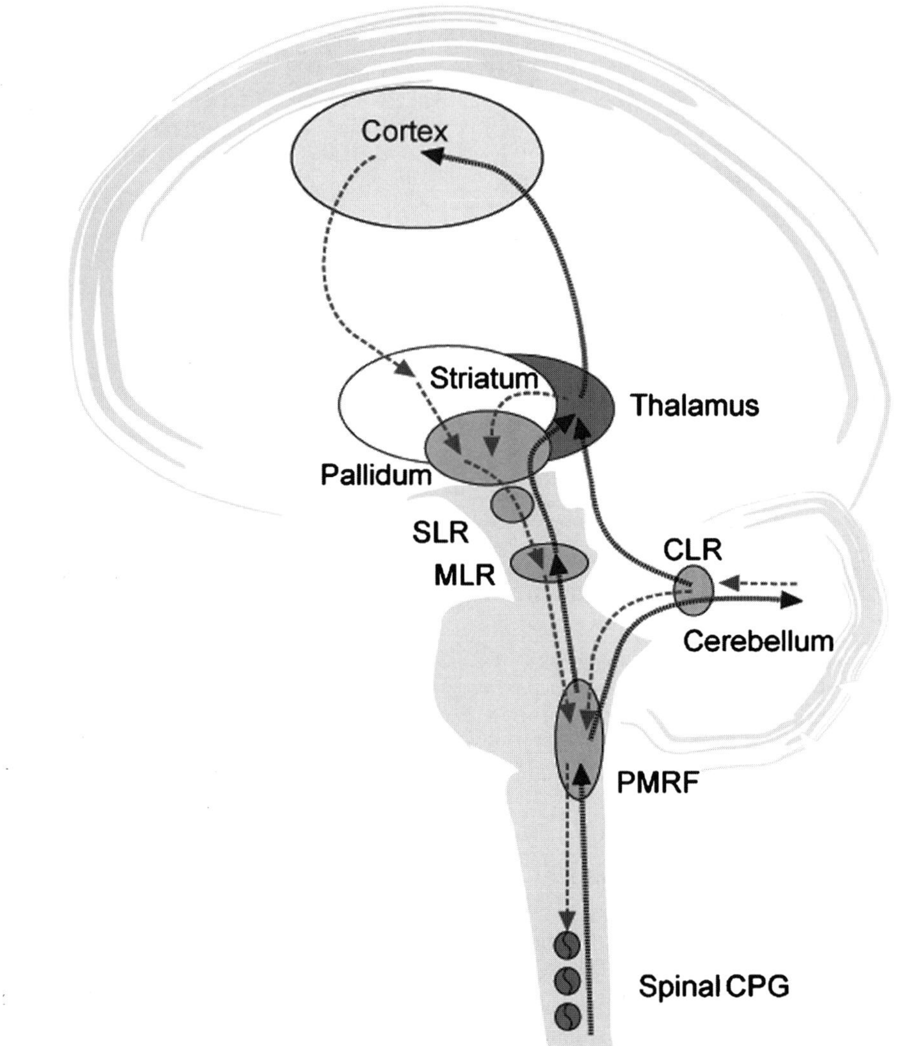

Fig. 4. Schematic drawing of the hierarchical locomotor control from cortex to spinal cord. The drawing shows brainstem and cerebellar locomotor regions (CLR, cerebellar locomotor region; MLR, mesencephalic locomotor region; PMRF, ponto-medullary reticular formation; SLR, subthalamic locomotor region), the spinal central pattern generators (CPG), basal ganglia, thalamus, cerebellum, and cortex. Descending pathways are drawn in dotted line, ascending pathways in solid line. Cortical signals project to the brainstem locomotor regions via the striatum and pallidum. The locomotor command conveys a message from the pallidum via the SLR to the MLR, which is further transmitted to the PMRF, where it converges with cerebellar signals from the CLR. The CLR also projects to the MLR via the thalamus and basal ganglia. It receives input from the vermal and paravermal cerebellar cortex. The PMRF is a major site of interaction between ascending and descending pathways. Cortical signals are modulated via the thalamo-cortical-basal ganglia circuit (Inglis and Winn, 1995; Whelan, 1996; Pahapill and Lozano, 2000; Burke, 2001; Mori et al., 2001; Grillner, 2006).

spinal, and supraspinal mechanisms are involved in such sensory control, which can be modulated according to environmental demands (perturbations, unfamiliar tasks) and the stride cycle (Dietz, 1992; Schubert et al., 1997; Rossignol et al., 2006). It has been shown for the somatosensory system that Ia sensitivity and dynamic gamma-drive are low during routine movements (walking), but they dramatically increase during motor tasks that are either difficult or unfamiliar (Hulliger, 1993). The inhibition of sensory signals during locomotion is not limited to peripheral and spinal mechanisms; it also includes cortical representations of the somatosensory and the vestibular systems. In our fMRI studies we were able to visualize deactivations of the vestibular and somatosensory cortex during mental imagery of locomotion. These deactivations were not found during mental imagery of standing. The cluster sizes for deactivations of the vestibular and somatosensory cortex areas are larger during imagined running than during imagined walking. This fits earlier findings on the inhibitory interaction between locomotion and vestibular function (Brandt et al., 1999). The major deactivations found in the retroinsular areas (supramarginal gyrus) and superior temporal gyrus correspond best to the parieto-insular vestibular cortex (PIVC), which was delineated electrophysiologically in monkeys (Grüsser et al., 1990). Brain activation studies in humans showed that cortical areas corresponding to the PIVC are activated by caloric and galvanic vestibular stimulation (Brandt and Dieterich, 1999). Electrical stimulation of these areas also induced vestibular sensations in humans (Kahane et al., 2003). The fMRI deactivations were predominant in the right hemisphere, thus confirming that the non-dominant hemisphere is dominant for the vestibular system (Bense et al., 2001; Dieterich et al., 2003). Blind subjects show activation rather than deactivation of the PIVC, which indicates cross-modal substitution between sensory systems.

Acknowledgements

We thank Judy Benson for copy-editing the manuscript. This work was supported by the Deutsche Forschungsgemeinschaft (JA 1087/1-1).

References

Armstrong, D.M. (1988) The supraspinal control of mammalian locomotion. J. Physiol., 405: 1–37.

Balter, J.E. and Zehr, E.P. (2007) Neural coupling between the arms and legs during rhythmic locomotor-like cycling movement. J. Neurophysiol., 97: 1809–1818.

Bense, S., Stephan, T., Yousry, T.A., Brandt, T. and Dieterich, M. (2001) Multisensory cortical signal increases and decreases during vestibular galvanic stimulation (fMRI). J. Neurophysiol., 85: 886–899.

Brandt, T. and Dieterich, M. (1999) The vestibular cortex. Its locations, functions, and disorders. Ann. N.Y. Acad. Sci., 871: 293–312.

Brandt, T., Strupp, M. and Benson, J. (1999) You are better off running than walking with acute vestibulopathy. Lancet, 355: p. 233.

Burke, R.E. (2001) The central pattern generator for locomotion in mammals. In: Ruzicka E., Hallet M. and Jankovic J. (Eds.), Gait disorders, Advances in Neurology, Vol. 87. Lippincott Williams & Wilkins, Philadelphia, pp. 11–24.

Capaday, C., Cody, F.W. and Stein, R.B. (1986) Amplitude modulation of the soleus H-reflex in the human during walking and standing. J. Neurosci., 6: 1308–1313.

Deutschländer, A., Stephan, T., Hüfner, K., Wagner, J., Strupp, M., Brandt, T. and Jahn, K. (2008) Brain activation patterns during imagined stance and locomotion in blind subjects, *submitted*.

Dieterich, M., Bense, S., Lutz, S., Drzezga, A., Stephan, T., Bartenstein, P. and Brandt, T. (2003) Dominance for vestibular cortical function in the non-dominant hemisphere. Cereb. Cortex, 13: 994–1007.

Dietz, V. (1992) Human neuronal control of automatic functional movements: interaction between central programs and afferent input. Physiol. Rev., 72: 33–69.

Dietz, V. (2002) Do human bipeds use quadrupedal coordination? Trends Neurosci., 25: 462–467.

Dietz, V. (2003) Spinal cord pattern generators for locomotion. Clin. Neurophysiol., 114: 1379–1389.

Dietz, V., Colombo, G. and Jensen, L. (1994) Locomotor activity in spinal man. Lancet, 344: 1260–1263.

Dietz, V., Fouad, K. and Bastiaanse, C.M. (2001) Neuronal coordination of arm and leg movements during human locomotion. Eur. J. Neurosci., 14: 1906–1914.

Drew, T., Prentice, S. and Schepens, B. (2004) Cortical and brainstem control of locomotion. Prog. Brain Res., 143: 251–261.

Eidelberg, E., Walden, J.G. and Nguyen, L.H. (1981) Locomotor control in Macaque monkeys. Brain, 104: 647–663.

Faist, M., Dietz, V. and Pierrot-Deeillingny, E. (1996) Modulation, probably presynaptic in origin, of monosynaptic Ia excitation during human gait. Exp. Brain Res., 109: 441–449.

Fedirchuk, B., Nielsen, J., Petersen, N. and Hultborn, H. (1998) Pharmacologically evoked fictive motor patterns in the acutely spinalized marmoset monkey (Callithrix jacchus). Exp. Brain Res., 122: 351–361.

Garcia-Rill, E. and Skinner, R.D. (1986) The basal ganglia and the mesencephalic locomotor region. In: Grillner S., Stein P.S.G., Stuart D.G., Forssberg H. and Hermann R.M. (Eds.), Neurobiology of vertebrate locomotion. Macmillan Press, London, pp. 77–104.

Grillner, S. (1981) Control of locomotion in bipeds, tetrapods and fish. In: Brooks, V.B. (Ed.), Handbook of Physiology, Sect. 1. The nervous system II. Motor control. Am. Physiol. Soc., Waverly Press, Maryland, pp. 1179–1236.

Grillner, S. (2006) Biological pattern generation: the cellular and computational logic of networks in motion. Neuron, 52: 751–766.

Grüsser, O.J., Pause, M. and Schreiter, U. (1990) Localisation and responses of neurons in the parieto-insular vestibular cortex of awake monkeys (Macaca fascicularis). J. Physiol., 430: 537–557.

Hanakawa, T., Katsumi, Y., Fukuyama, H., Honda, M., Hayashi, T., Kimura, J. and Shibasaki, H. (1999) Mechanisms underlying gait disturbance in Parkinson's disease. A single photon emission computed tomography study. Brain, 122: 1271–1282.

Hashimoto, T. (2006) Speculation on the responsible sites and pathophysiology of freezing of gait. Parkinsonism Relat. Disord, 12: S55–S62.

Hathout, G.M. and Bhidayasiri, R. (2005) Midbrain ataxia: an introduction to the mesencephalic locomotor region and the pedunculopontine nucleus. Am. J. Radiol., 184: 953–956.

Hulliger, M. (1993) Fusimotor control of proprioceptive feedback during locomotion and balancing: can simple lessons be learned for artificial control of gait? Prog. Brain Res., 97: 173–180.

Inglis, W.L. and Winn, P. (1995) The pedunculopontine tegmental nucleus: where the striatum meets the reticular formation. Prog. Neurobiol., 47: 1–29.

Jahn, K., Deutschländer, A., Stephan, T., Brückmann, H., Strupp, M. and Brandt, T. (2003) An fMRI study of vestibular and somatosensory cortex deactivation during imagined locomotion. Ann. N.Y. Acad. Sci., 1004: 469–472.

Jahn, K., Deutschländer, A., Stephan, T., Kalla, R., Wiesmann, M., Strupp, M. and Brandt, T. (2008) Imaging human supraspinal locomotor centers in brainstem and cerebellum. Neuroimage, 39: 786–792.

Jahn, K., Deutschländer, A., Stephan, T., Strupp, M., Wiesmann, M. and Brandt, T. (2004) Brain activation patterns during imagined stance and locomotion in fMRI. Neuroimage, 22: 1722–1731.

Jahn, K., Strupp, M. and Brandt, T. (2002) Both actual and imagined locomotion suppress spontaneous vestibular nystagmus. NeuroReport, 13: 2125–2128.

Jahn, K., Strupp, M., Schneider, E., Dieterich, M. and Brandt, T. (2000) Differential effects of vestibular stimulation on walking and running. NeuroReport, 11: 1745–1748.

Kahane, P., Hoffmann, D., Minotti, L. and Berthoz, A. (2003) Reappraisal of the human vestibular cortex by cortical electrical stimulation study. Ann. Neurol., 54: 615–624.

Masdeu, J.C., Alampur, U., Cavaliere, R. and Tavoulareas, G. (1994) Astasia and gait failure with damage of the pontomesencephalic locomotor region. Ann. Neurol., 35: 619–621.

Mori, S., Matsui, T., Kuze, B., Asanome, M., Nakajima, K. and Matsuyama, K. (1999) Stimulation of a restricted region in the midline cerebellar white matter evokes coordinated quadrupedal locomotion in the decerebrate cat. J. Neurophysiol., 82: 290–300.

Mori, S., Matsuyama, K., Miyashita, E. and Nakajima, K. (1996) Basic neurophysiology of primate locomotion. Folia Primatol., 66: 192–203.

Mori, S., Matsuyama, K., Mori, F. and Nakajima, K. (2001) Supraspinal sites that induce locomotion in the vertebrate central nervous system. In: Ruzicka E., Hallet M. and Jankovic J. (Eds.), Gait Disorders, Advances in Neurology, Vol. 87. Lippincott Williams & Wilkins, Philadelphia, pp. 25–39.

Orlovsky, G.N. (1972) Activity of vestibulospinal neurons during locomotion. Brain Res., 46: 85–98.

Pahapill, P.A. and Lozano, A.M. (2000) The pedunculopontine nucleus and Parkinson's disease. Brain, 123: 1767–1783.

Plaha, P. and Gill, S.S. (2005) Bilateral deep brain stimulation of the pedunculopontine nucleus for Parkinson's disease. NeuroReport, 16: 1883–1887.

Rossignol, S., Dubuc, R. and Gossard, J.P. (2006) Dynamic sensorimotor interactions during locomotion. Physiol. Rev., 86: 89–154.

Schubert, M., Curt, A., Jensen, L. and Dietz, V. (1997) Corticospinal input in human gait: modulation of magnetically evoked motor responses. Exp. Brain Res., 115: 234–246.

Shik, M.L. and Orlovsky, G.N. (1976) Neurophysiology of locomotor automatism. Physiol. Rev., 56: 465–501.

Shik, M.L., Severin, F.V. and Orlovsky, G.N. (1966) Control of walking and running by means of stimulation of the midbrain. Biophysics, 11: 756–765.

Stefani, A., Lozano, A.M., Peppe, A., Stanzione, P., Galati, S., Tropepi, D., Pierantozzi, M., Brusa, L., Scarnati, E. and Mazzone, P. (2007) Bilateral deep brain stimulation of the pedunculopontine and subthalamic nuclei in severe Parkinson's disease. Brain, 130(6): 1596–1607.

Wannier, T., Bastiaanse, C., Colombo, G. and Dietz, V. (2001) Arm to leg coordination in humans during walking, creeping and swimming activities. Exp. Brain Res., 141: 375–379.

Whelan, P.J. (1996) Control of locomotion in the decerebrate cat. Prog. Neurobiol., 49: 481–515.

Zweig, R.M., Jankel, W.R., Hedreen, J.C., Mayeux, R. and Price, D.L. (1989) The peduncolopontine nucleus in Parkinson's disease. Ann. Neurol., 26: 41–46.

C. Kennard & R.J. Leigh (Eds.)
Progress in Brain Research, Vol. 171
ISSN 0079-6123

CHAPTER 4.21

Private lines of cortical visual information to the nucleus of the optic tract and dorsolateral pontine nucleus

Claudia Distler and Klaus-Peter Hoffmann*

Allgemeine Zoologie & Neurobiologie, Ruhr-Universität Bochum, Bochum, Germany

Abstract: The subcortical nucleus of the optic tract and dorsal terminal nucleus of the accessory optic system (NOT-DTN), along with the dorsolateral pontine nucleus (DLPN), has been shown to play a pivotal role in controlling slow eye movements. Both nuclei are known to receive cortical input from striate and extrastriate cortex. To determine to what degree this cortical input arises from the same areas, and potentially from the same individual neurons, in one set of experiments we placed different retrograde tracers into the NOT-DTN and the DLPN. In the ipsilateral cortical hemisphere the two projections mainly overlapped in the middle temporal (MT) area, the middle superior temporal (MST) area, and the visual area in the fundus of the STS (FST) and the surrounding cortex. In these areas, neurons projecting to the NOT-DTN or the DLPN were closely intermingled. Nevertheless, only 3–11% of the labelled neurons in MT and MST were double-labelled in our various cases. In a second set of experiments, we identified neurons in areas MT and MST projecting to the DLPN and/or to the NOT-DTN by antidromic electrical stimulation. Again, neurons projecting to either target were located in close proximity to each other and in all subregions of MT and MST sampled. Only a small percentage of the antidromically identified projection neurons (4.4%) sent branches to both the NOT-DTN and the DLPN. On the population level, only neurons activated from the NOT-DTN had a clear preference for ipsiversive stimulus movement whereas the neurons activated from the DLPN, and neurons not antidromically activated from either target, had no common directional preference. These results indicate that the cortical input to the NOT-DTN and DLPN arises from largely separate neuronal subpopulations in the motion sensitive areas in the posterior STS. Only a small percentage of the projection neurons bifurcate to supply both targets. These findings are discussed in relation to the effects of cortical lesions on the optokinetic and smooth pursuit system.

Keywords: optokinetic; smooth pursuit; slow eye movements; monkey; MT/MST

Neuronal substrate of eye movement control

Eye movement control is widely distributed over cortical and subcortical structures. There is a wealth of literature correlating neuronal response properties in certain cortical, pretectal, tectal, and brainstem areas with different kinds of eye

*Corresponding author. Tel.: +49-234-3224363; Fax: +49-234-3214185; E-mail: kph@neurobiologie.rub.de

DOI: 10.1016/S0079-6123(08)00653-5

movements or phases thereof. Assuming that such a widespread network exists, a crucial question is how information is exchanged between related areas and how specificity is generated and maintained.

Here we want to discuss the connectivity between cortical areas: middle temporal (MT) and middle superior temporal (MST) with the pretectal nucleus of the optic tract and the dorsolateral pontine nucleus (DLPN) subserving slow eye-movements during optokinetic reactions, ocular following, and smooth pursuit. Slow eye movements are crucial for stabilizing the image of the world on the retina and thus for compensation of self-motion or externally generated movement of the visual world. For many years, the nucleus of the optic tract and the dorsal terminal nucleus of the accessory optic system (NOT-DTN) were regarded as the key visuomotor interface for the optokinetic reflex that stabilizes large field retinal image motion or slip on the retina. The DLPN has been largely implicated in slow eye movements during smooth pursuit of small stimuli and in the ocular following response (OFR).

Probably the most significant and directly eye-movement-related cell type, both in the NOT and the DTN, is retinal slip neurons that are directionally selective for ipsiversive image motion and project to the dorsal cap of the inferior olive and nucleus praepositus hypoglossi (NPP) in all mammals investigated so far. In monkeys, in addition to the retinal slip neurons (Hoffmann et al., 1988; Mustari and Fuchs, 1990), omnidirectional pause neurons, anatomically clearly offset with respect to the retinal slip neurons, have been described in the NOT (Mustari et al., 1997) but not the DTN (Mustari and Fuchs, 1990). Based on the homogeneous response properties, the orthodromic and antidromic latencies from their input and output structures, and morphological distribution of retrogradely labelled retinal slip cells after injection of the inferior olive, we consider the NOT and DTN as a functional unit (Hoffmann et al., 1988; Hoffmann and Fischer, 2001; for further discussion see Distler et al., 2002).

Retinal slip neurons in the monkey NOT-DTN prefer large field ipsiversive visual motion (Hoffmann et al., 1988; Mustari and Fuchs, 1990). Some NOT-DTN neurons also respond to small spots of light moving ipsiversively (Hoffmann and Distler, 1989; Mustari and Fuchs, 1990; Ilg and Hoffmann, 1991), a stimulus that elicits smooth eye movements in awake animals. Electrical stimulation in the NOT-DTN provokes optokinetic eye movements with the slow phase towards the stimulated side (Schiff et al., 1988; Mustari and Fuchs, 1990; Taylor et al., 2000; Hoffmann and Fischer, 2001). Similarly, lesion or inactivation of the NOT-DTN leads to deficits of optokinetic eye movements during visual stimulation towards the lesion (Kato et al., 1988; Schiff et al., 1990; Ilg et al., 1993; Inoue et al., 2000; Hoffmann and Fischer, 2001). Interestingly, lesions of the NOT-DTN also impair smooth pursuit and ocular following (Ilg et al., 1993; Inoue et al., 2000; Yakushin et al., 2000b). Thus, the NOT-DTN in the monkey not only plays an important role for optokinetic nystagmus (OKN), but for all slow eye movements.

In DLPN, the visual and visual-pursuit neurons respond direction-selectively to moving large area random dot patterns, and, in part, to moving single spots of light. In contrast to NOT-DTN, these neurons as a population do not code for a common direction of movement (Suzuki and Keller, 1984; Mustari et al., 1988; Thier et al., 1988; Suzuki et al., 1990). Especially the eye movement-related and visual-pursuit neurons of the DLPN are supposedly part of the smooth pursuit pathway. Lesion of the DLPN causes impairment of smooth pursuit and of the initial phase of optokinetic eye movements whereas the steady-state optokinetic response is unaffected (May et al., 1988; Thier et al., 1991). In addition, DLPN contains neurons that fire before eye movements, and whose response properties implicate them in the OFR (Kawano et al., 1992, 1996).

Within the cerebral cortex pursuit-related neurons have been identified in several cortical areas including the frontal eye fields, the lateral intraparietal area, and the superior temporal areas MT and MST. In MT and MST, these direction selective neurons include the representation of the fovea in their receptive fields but differ in their preferred stimulus and receptive field size, with MT neurons having smaller receptive fields and preferring small instead of large area visual stimuli

(Komatsu and Wurtz, 1988). Lesions in MT produce a retinotopic deficit in smooth pursuit whereas MST lesions lead to retinotopic as well as directional deficits in smooth pursuit and the slow build up of OKN (Newsome et al., 1985; Duersteler and Wurtz, 1988). Area MST also contains neurons related to ocular following (Kawano et al., 1994). Both MT and MST project to the DLPN (Maunsell and Van Essen, 1983; Ungerleider et al., 1984; Boussaoud et al., 1992). The main cortical input to the NOT-DTN comes from MT (Distler and Hoffmann, 2001).

Anatomical overlap but cellular specificity of cortical input to NOT-DTN and DLPN

To examine the relative strength of cortical projections to the NOT-DTN and DLPN, respectively, we performed dual retrograde tracer injections into these nuclei and analysed which cortical areas and to what extent these areas contained retrogradely labelled cells (Distler et al., 2002). The cortical input to the NOT-DTN turned out to originate mainly from motion sensitive areas MT, MST, fundus of the STS (FST), and visual areas in the anterior part of the STS, followed by V1, V2, and V3. By contrast, injections into the DLPN yielded retrogradely labelled cells mainly in MST, followed by MT and FST with some sparse labelling in intraparietal and frontal cortex. Significantly, the only region of substantial overlap of neurons labelled from the NOT-DTN and the DLPN was the posterior part of the STS with the motion sensitive areas MT, MST, and FST. In these areas, neurons labelled from both targets were closely intermingled, i.e., they were lying very close to each other. However, double-labelled cells were rather rare indicating that only few cortical neurons simultaneously project both to the NOT-DTN and the DLPN. Depending on the size of the injection, 0–3.4% of the NOT-DTN projecting cells in MT also projected to the DLPN, 0–10.4% of the DLPN-projecting cells in MT also projected to the NOT-DTN. For area MST, 0–9.6% of the NOT-DTN projecting neurons and 0–11.1% of the DLPN projecting population simultaneously also projected to the other target. Thus, the quantitatively strongest input to the NOT-DTN originates from area MT, and to the DLPN from area MST. The specificity within each projection is very high. Even though in both cortical areas neuronal discharges related to optokinetic responses and smooth pursuit have been reported, and lesions of these areas affect eye movements in both behavioural contexts, only about 4% of the subcortically projecting cortical neurons (averaged over all cases and over both MT and MST) project both to the NOT-DTN and the DLPN thus distributing their information to both targets (Distler et al., 2002).

Cortical information transmitted to the NOT-DTN and the DLPN

In order to specify the kind of information that is transmitted from motion sensitive areas in the STS to the NOT-DTN and DLPN, respectively, we employed electrical stimulation in these subcortical nuclei while simultaneously analysing the response properties of antidromically activated cortical neurons in anaesthetized and paralysed monkeys. Similarly to our anatomical results, neurons in MT and MST that could be activated from the NOT-DTN, or the DLPN, and neurons that could not be activated from either target were found next to each other and were distributed over the whole extent of MT and MST sampled in our experiments. Only about 4% of the cortical projection neurons could be antidromically activated from both the NOT-DTN and the DLPN thus confirming our anatomical results. Antidromic latencies did not differ after stimulation of the NOT-DTN or the DLPN suggesting that the conduction velocity and therefore axon calibre of the cortical neurons projecting to either target were in the same range.

There was one response parameter that clearly differed between cortical neurons projecting to the NOT-DTN and to the DLPN: the preferred direction of stimulus motion. As expected, the vast majority of cortical neurons recorded in MT and MST were direction selective, regardless whether they projected to the NOT-DTN, the DLPN, or to none of these targets. As reported earlier (Ilg and

Hoffmann, 1993; Hoffmann et al., 2002), cortical neurons projecting to the NOT-DTN as a population strongly prefer ipsiversive stimulus movement thus matching the direction preference of their target neurons, the retinal slip cells in the NOT-DTN. By contrast, neither cortical neurons projecting to the DLPN nor neurons not projecting to either target showed a common directional preference. Again, in the case of the DLPN this matches the directional properties of the target neurons in this nucleus because direction selective neurons in the DLPN do not exhibit a common preferred direction.

Private lines?

Our data on the cortical projections to subcortical centres involved in eye movement control, i.e., the NOT-DTN and the DLPN, clearly indicate that indeed there are private lines between cortical areas MT and MST and their target nuclei NOT-DTN and DLPN.

These results seem surprising. Although information from almost completely separate populations of neurons in MT and MST reaches the NOT-DTN and DLPN, lesions especially of the NOT-DTN lead to a deterioration of slow eye movements, not only in OKN but also in ocular following and smooth pursuit. Lesions in the DLPN spare most of the slow phases of OKN. Nevertheless the different subcortical routes for slow eye movements, despite their private lines of cortical input seem to be highly interdependent. Lesions in NOT-DTN could create a tonus asymmetry almost like a lesion in the vestibular nuclei. Such a lesion leads to spontaneous nystagmus that would interfere with other slow eye movements. Even after spontaneous nystagmus ceases this asymmetry in the signal from the left and right NOT-DTN which is transformed into eye velocity at later stages would be a handicap for normal gain in smooth pursuit and ocular following. The ipsiversive deficits after lesions of MT and MST in OKN, ocular following, and smooth pursuit might all be attributed to the altered cortical input to the NOT-DTN with its projection to the NPP and from here directly to the vestibular nuclei. The projection to the DLPN can reach the brainstem oculomotor circuitry only via the cerebellum and may be much more important for the predictive properties, like keeping a selected object within the fovea, of smooth pursuit. A comparison of the cortical neurons projecting to NOT-DTN and DLPN in awake, trained monkeys will have to give us the answer to these questions.

In many cases, projections from a given cortical area to different cortical or subcortical areas are not identical. Interestingly, most studies showing evidence for functionally distinct projections concern brain regions that have been very well characterized, e.g., V1, MT, LIP, FEF (Finlay et al., 1976; Segraves, 1992; Movshon and Newsome, 1996; Paré and Wurtz, 1997; Everling and Munoz, 2000; Sommer and Wurtz, 2000, 2001; Hoffmann et al., 2002; Ferraina et al., 2002). It may be a prerequisite for recognizing functionally distinct subpopulations that the diversity of response properties in a brain area be fully understood. This hypothesis, however, does not make room for exceptions like the findings of Churchland and Lisberger (2005) in MST. Nevertheless, more importantly strong similarities in the response properties of the projection neurons and the target area can always be found. This raises an important issue for developmental neurobiology. How are these distinct projections established? Clearly, in addition to molecular guidance cues, activity-dependent matching mechanisms (Hebb, 1948) have to play an important role to set up these parallel cortical and corticofugal processing streams.

References

Boussaoud, D., Desimon, R. and Ungerleider, L.G. (1992) Subcortical connections of visual areas MST and FST in macaques. Vis. Neurosci., 9: 291–302.

Churchland, A.K. and Lisberger, S.G. (2005) Discharge properties of MST neurons that project to the frontal pursuit area in macaque monkeys. J. Neurophysiol., 94: 1084–1090.

Distler, C. and Hoffmann, K.-P. (2001) Cortical input to the nucleus of the optic tract and dorsal terminal nucleus (NOT-DTN) in macaques: a retrograde tracing study. Cereb. Cortex, 11: 572–580.

Distler, C., Mustari, M.J. and Hoffmann, K.-P. (2002) Cortical projections to the nucleus of the optic tract and dorsal terminal nucleus and to the dorsolateral pontine nucleus in macaques: a dual retrograde tracing study. J. Comp. Neurol., 444: 144–158.

Duersteler, M.R. and Wurtz, R.H. (1988) Pursuit and optokinetic deficits following chemical lesions of cortical areas MT and MST. J. Neurophysiol., 60: 940–965.

Everling, S. and Munoz, D.P. (2000) Neuronal correlates for preparatory set associated with pro-saccades and anti-saccades in the primate frontal eye field. J. Neurosci., 20: 387–400.

Ferraina, S., Paré, M. and Wurtz, R.H. (2002) Comparison of cortico-cortical and cortico-collicular signals for the generation of saccadic eye movements. J. Neurophysiol., 87: 845–858.

Finlay, B.L., Schiller, P.H. and Volman, S.F. (1976) Quantitative studies of single-cell properties in monkey striate cortex. IV. Corticotectal cells. J. Neurophysiol., 39: 1352–1361.

Hebb, D.O. (1948) Organization of Behavior. Wiley, New York.

Hoffmann, K.-P. and Distler, C. (1989) Quantitative analysis of visual receptive fields of neurons in the nucleus of the optic tract and dorsal terminal nucleus of the accessory optic tract in macaque monkey. J. Neurophysiol., 62: 416–428.

Hoffmann, K.-P., Distler, C., Bremmer, F. and Thiele, A. (2002) Directional asymmetry of neurons in cortical areas MT and MST projecting to the nucleus of the optic tract and dorsal terminal nucleus (NOT-DTN) in macaques. J. Neurophysiol., 87: 2113–2123.

Hoffmann, K.-P., Distler, C., Erickson, R. and Mader, W. (1988) Physiological and anatomical identification of the nucleus of the optic tract and dorsal terminal nucleus of the accessory optic tract in monkeys. Exp. Brain Res., 69: 635–644.

Hoffmann, K.-P. and Fischer, W.H. (2001) Directional effect of inactivation of the nucleus of the optic tract on optokinetic nystagmus in the cat. Vision Res., 41: 3389–3398.

Ilg, U.J., Bremmer, F. and Hoffmann, K.-P. (1993) Optokinetic and pursuit system: a case report. Behav. Brain Res., 57: 21–29.

Ilg, U.J. and Hoffmann, K.-P. (1991) Responses of monkey nucleus of the optic tract neurons during pursuit and fixation. Neurosci. Res., 12: 101–110.

Ilg, U.J. and Hoffmann, K.-P. (1993) Functional grouping of the cortico-pretectal projection. J. Neurophysiol., 70: 867–869.

Inoue, Y., Takemura, A., Kawano, K. and Mustar, M.J. (2000) Role of the pretectal nucleus of the optic tract in short-latency ocular following responses in monkeys. Exp. Brain Res., 131: 269–281.

Kato, I., Harada, K., Hasekawa, K. and Koike, Y. (1988) Role of the nucleus of the optic tract in monkeys in optokinetic nystagmus and optokinetic after nystagmus. Brain Res., 474: 16–26.

Kawano, K., Shidara, M., Watanabe, Y. and Yamane, S. (1994) Neural activity in cortical area MST of alert monkey during ocular following responses. J. Neurophysiol., 71: 2305–2324.

Kawano, K., Shidara, M. and Yamane, S. (1992) Neural activity in dorsolateral pontine nucleus of alert monkey during ocular following responses. J. Neurophysiol., 67: 680–703.

Kawano, K., Takemura, A., Inoue, Y., Kitama, T., Kobayashi, Y. and Mustari, M.J. (1996) Visual inputs to cerebellar ventral paraflocculus during ocular following responses. Prog. Brain Res., 112: 415–422.

Komatsu, H. and Wurtz, R.H. (1988) Relation of cortical areas MT and MST to pursuit eye movements. I. Localization and visual properties of neurons. J. Neurophysiol., 60: 580–603.

Maunsell, J.H.R. and Van Essen, D.C. (1983) The connections of the middle temporal visual area (MT) and their relationship to a cortical hierarchy in the macaque monkey. J. Neurosci., 3: 2563–2586.

May, J.G., Keller, E.L. and Suzuki, D.A. (1988) Smooth pursuit eye movement deficits with chemical lesions in the dorsolateral pontine nucleus of the monkey. J. Neurophysiol., 59: 952–977.

Movshon, J.A. and Newsome, W.T. (1996) Visual response properties of striate cortical neurons projecting to area MT in macaque monkeys. J. Neurosci., 16: 7733–7741.

Mustari, M.J. and Fuchs, A.F. (1990) Discharge patterns of neurons in the pretectal nucleus of the optic tract (NOT) in the behaving primate. J. Neurophysiol., 64: 77–90.

Mustari, M.J., Fuchs, A.F. and Pong, M. (1997) The response properties of pretectal omnidirectional pause neurons in the behaving primate. J. Neurophysiol., 77: 116–125.

Mustari, M.J., Fuchs, A.F. and Wallman, J. (1988) Response properties of dorsolateral pontine units during smooth pursuit in the rhesus macaque. J. Neurophysiol., 60: 664–686.

Newsome, W.T., Wurtz, R.H., Duersteler, M.R. and Mikami, A. (1985) Deficits in visual motion processing following ibotenic acid lesions of the middle temporal visual area of the macaque monkey. J. Neurosci., 5: 825–840.

Paré, M. and Wurtz, R.H. (1997) Monkey posterior parietal cortex neurons antidromically activated from superior colliculus. J. Neurophysiol., 78: 3493–3497.

Schiff, D., Cohen, B., Büttner-Ennever, J. and Matsuo, V. (1990) Effects of lesions of the nucleus of the optic tract on optokinetic nystagmus and after-nystagmus in the monkey. Exp. Brain Res., 79: 225–239.

Schiff, D., Cohen, B. and Raphan, T. (1988) Nystagmus induced by stimulation of the nucleus of the optic tract in the monkey. Exp. Brain Res., 70: 1–14.

Segraves, M.A. (1992) Activity of monkey frontal eye field neurons projecting to oculomotor regions of the pons. J. Neurophysiol., 68: 1967–1985.

Sommer, M.A. and Wurtz, R.H. (2000) Composition and topographic organization of signals sent from the frontal eye field to the superior colliculus. J. Neurophysiol., 83: 1979–2001.

Sommer, M.A. and Wurtz, R.H. (2001) Frontal eye field sends delay activity related to movement, memory, and vision to the superior colliculus. J. Neurophysiol., 85: 1673–1685.

Suzuki, D.A. and Keller, E.L. (1984) Visual signals in the dorsolateral pontine nucleus of the alert monkey: their relationship to smooth-pursuit eye movements. Exp. Brain Res., 47: 145–147.

Suzuki, D.A., May, J.G., Keller, E.L. and Yee, R.D. (1990) Visual motion response properties of neurons in dorsolateral pontine nucleus of alert monkey. J. Neurophysiol., 63: 37–59.

Taylor, R.B., Wennberg, R.A., Lozano, A.M. and Sharpe, J.A. (2000) Central nystagmus induced by deep-brain stimulation for epilepsy. Epilepsia, 41: 1637–1641.

Thier, P., Bachor, A., Faiss, J., Dichgans, J. and Koenig, E. (1991) Selective impairment of smooth pursuit eye movements due to an ischemic lesion of the basal pons. Ann. Neurol., 29: 443–448.

Thier, P., Koehler, W. and Buettner, U.W. (1988) Neuronal activity in the dorsolateral pontine nucleus of the alert monkey modulated by visual stimuli and eye movements. Exp. Brain Res., 70: 496–512.

Ungerleider, L.G., Desimone, R., Galkin, T.W. and Mishkin, M. (1984) Subcortical projections of area MT in the macaque. J. Comp. Neurol., 223: 368–386.

Yakushin, S.B., Gizzi, M., Reisine, H., Raphan, T., Buettner-Ennever, J. and Cohen, B. (2000) Functions of the nucleus of the optic tract (NOT). II. Control of ocular pursuit. Exp. Brain Res., 131: 433–447.

C. Kennard & R.J. Leigh (Eds.)
Progress in Brain Research, Vol. 171
ISSN 0079-6123

CHAPTER 4.22

Gravity perception in cerebellar patients

A.A. Tarnutzer*, S. Marti and D. Straumann

Department of Neurology, Zurich University Hospital, Zurich, Switzerland

Abstract: Patients with midline cerebellar atrophy typically suffer from impaired postural balance. We asked whether this deficit is, in part, caused by a deficient perception of body position relative to gravity, and therefore measured the subjective visual vertical (SVV) in patients with degeneration predominantly of the midline cerebellar structures ($n=11$) and age-matched healthy human subjects ($n=9$). On a motorized turntable, subjects were placed in different roll positions [0, 75° right ear down (RED), 75° left ear down (LED)] and had to align a luminous arrow with the perceived earth-vertical. Both SVV deviations and intra-individual SD of SVV adjustments were not significantly different between patients and age-matched controls. Our findings suggest that the perception of verticality in patients with cerebellar ataxia may only deteriorate in a more advanced stage of the disease. To which degree the deterioration of perceived vertical is caused by a collateral vestibular impairment, awaits further clarification.

Keywords: subjective visual vertical; vestibular; roll; a-effect; otolithic organs; posture; verticality perception

Introduction

Head and body orientation in space relative to the gravity vector is continuously assessed by several sensory systems (mainly by the semicircular canals, the otolith organs and the somatosensory and visual systems), integrated within the central graviceptive pathways (CGP) and appropriately adapted by the motor systems to maintain postural stability. CGP consist of several brainstem areas (Brandt and Dieterich, 1987; Halmagyi et al., 1990; Dieterich and Brandt, 1992), posterolateral thalamic areas (Dieterich and Brandt, 1993), and cortical areas, e.g., within the posterior insula (Brandt et al., 1994; Yelnik et al., 2002). It is likely that certain cerebellar structures are also part of the CGP (Mossman and Halmagyi, 1997). Lesions within the CGP result in misperception of the gravity vector, and therefore lead to specific changes in gravity-dependent reflexive responses, both within the ocular motor system (resulting in skew deviation and ocular torsion) and posture (resulting in head tilt and lateropulsion). However, not only brainstem reflexes may be altered by lesions, but also the perceived gravity vector, as assessed by the subjective visual vertical (SVV), may be systematically shifted in these patients using psychophysical tasks as the SVV (for review, see Howard, 1982). Whereas shifts in perceived vertical have been systematically studied in patients with brainstem lesions (Dieterich and Brandt, 1992), corresponding studies in cerebellar patients are rare, most likely because cerebellar structures have only recently been considered to belong to the CGP. A precise estimate of self-orientation relative to the gravity vector is crucial

*Corresponding author. Tel.: +41 44 255 11 11;
Fax: +41 44 255 43 80;
E-mail: alexander.tarnutzer@access.unizh.ch

DOI: 10.1016/S0079-6123(08)00654-7

for balance control. In patients with midline cerebellar pathologies, typically suffering from impaired postural control, besides gait ataxia and downbeat nystagmus (for review, see Leigh and Zee, 2006), an impairment of the CGP could result in erroneous tilts of the perceived gravity vector. In two patients with acute unilateral vascular lesions of cerebellar structures and postural imbalance contraversive tilts of SVV together with ocular torsion have been reported by Mossman and Halmagyi (1997). Here we asked whether impaired balance in cerebellar patients is, in part, caused by a deficient perception of body position relative to gravity.

Material and methods

We included 9 age-matched healthy human subjects (3 females, 6 males; aged 66 ± 6 years, average ± 1 SD) and 11 patients with cerebellar degeneration predominantly of the midline cerebellar structures (4 females, 7 males; aged 67 ± 9 years). Subjects were placed upright in a turntable with three servo-controlled motor-driven axes (Acutronic, Jona, Switzerland). A 4-point safety belt was used to secure subjects. The head was restrained in the straight-ahead position with a thermoplastic mask (Sinmed BV, Reeuwijk, The Netherlands). Experiments were performed in otherwise total darkness. Subjects were asked to rapidly (<6 s) adjust the orientation of an arrow (9.5° long) to earth-vertical by turning a knob. The starting roll orientation of the arrow was offset clockwise (CW) or counterclockwise (CCW) by 40° and 60°, respectively. Adjustments of the arrow were tested in three different whole-body roll orientations in pseudo-randomized order [upright, 75° right ear down (RED); 75° left ear down (LED)]. The presentation of the arrow started 10 s after the turntable came to a full stop. Seventy-two trials (24 trials in each roll position) were recorded per subject and sorted according to whole-body roll orientation and direction of arrow rotation. We used the right-hand rule for the definition of torsional directions (CW as seen by the subject: positive).

Results

Since the direction of arrow rotation did not significantly affect the final arrow roll orientation, data from CW and CCW trials were pooled. In upright position, SVV deviated only slightly from earth-vertical both in patients ($-0.9 \pm 4.1°$; average ± 1 SD) and controls ($0.8 \pm 2.5°$), as shown in Fig. 1.

In roll-tilted positions, deviations were markedly increased and directed towards the side of body roll, consistent with roll underestimation (A-effect; for review, see Howard 1982). This was true both for patients (75° RED: $12.9 \pm 13.5°$; 75° LED: $-12.9 \pm 12.3°$) and controls (75° RED: $9.7 \pm 9.2°$; 75° LED: $-11.2 \pm 9.4°$). No significant differences of SVV deviations between the two groups were found (non-parametric Wilcoxon rank sum test, $p > 0.05$). Inter-individual variability was larger for the patients than for the controls, underlining the heterogeneity of the patients' group. When comparing the intra-individual variability, which is a measure of precision of SVV adjustments, both patients ($4.9 \pm 1.8°$ and $4.8 \pm 1.4°$; 75° RED and 75° LED) and controls ($4.2 \pm 2.4°$ and $4.7 \pm 1.3°$; 75° RED and 75° LED) showed increased variabilities in roll-tilted orientations relative to upright ($2.8 \pm 3.3°$ vs. $1.5 \pm 0.5°$; patients vs. controls), as depicted in Fig. 2. Comparing intra-individual variabilities of both patients and controls, no significant differences were found (Wilcoxon rank sum test, $p > 0.05$).

Discussion

Our findings suggest that the perception of verticality, as measured with the SVV, in patients with degeneration predominantly of the midline cerebellar structures may only deteriorate in a more advanced stage of the disease. On the other hand, postural imbalance is often an early sign of disease in these patients. Balance therefore seems to be impaired in these patients despite a functionally intact perception of the gravity vector, which suggests that the postural ataxia is due to disordered sensorimotor transformation of graviceptive information or predominantly motor deficits.

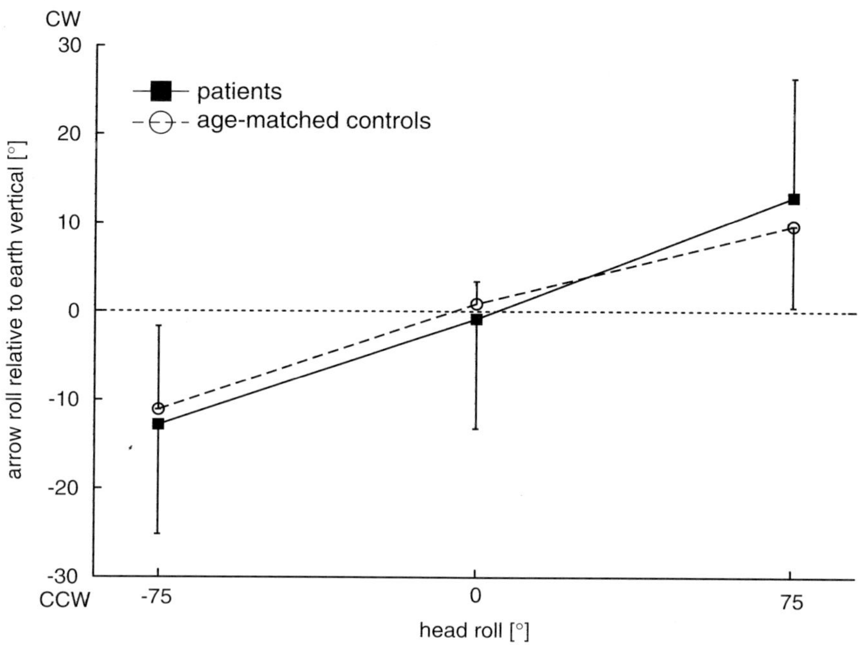

Fig. 1. Average SVV deviations relative to earth-vertical ± 1 SD (error bars) plotted against head roll are shown for patients (filled squares) and age-matched controls (empty circles).

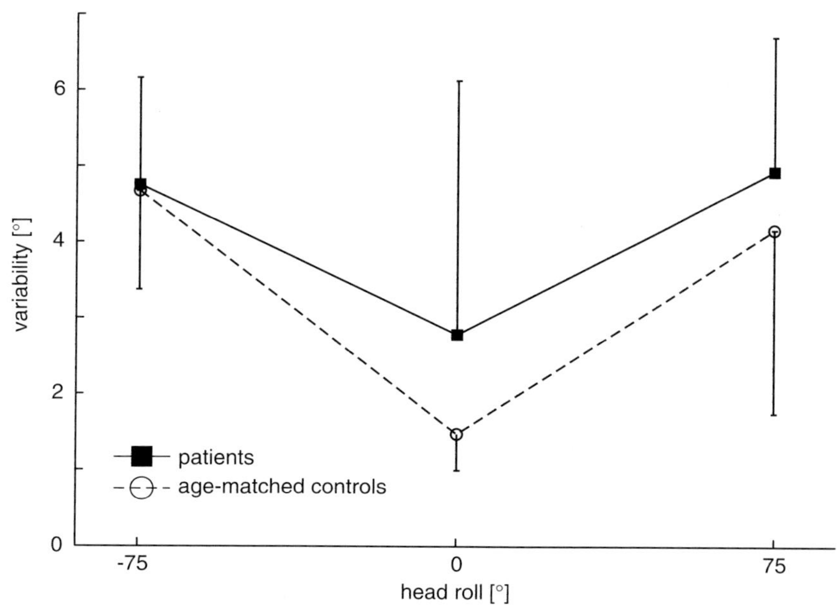

Fig. 2. Average intra-individual variabilities ± 1 SD (error bars) are plotted against head roll, both for patients (filled squares) and age-matched controls (empty circles).

Acknowledgements

This study was financially supported by the Swiss National Science Foundation, the Betty and David Koetser Foundation for Brain Research, Zurich, Switzerland, and the Center of Integrative Human Physiology, University of Zurich, Switzerland, and Bonizzi-Theler Foundation, Zurich, Switzerland.

References

Brandt, T. and Dieterich, M. (1987) Pathological eye–head coordination in roll: tonic ocular tilt reaction in mesencephalic and medullary lesions. Brain, 110: 649–666.

Brandt, T., Dieterich, M. and Danek, A. (1994) Vestibular cortex lesions affect the perception of verticality. Ann. Neurol., 35(4): 403–412.

Dieterich, M. and Brandt, T. (1992) Wallenberg's syndrome: lateropulsion, cyclorotation and subjective visual vertical in thirty-six patients. Ann. Neurol., 31(4): 399–408.

Dieterich, M. and Brandt, T. (1993) Thalamic infarctions: differential effects on vestibular function in the roll plane (35 patients). Neurology, 43(9): 1732–1740.

Halmagyi, G.M., Brandt, T., Dieterich, M., Curthoys, I.S., Stark, R.J. and Hoyt, W.F. (1990) Tonic contraversive ocular tilt reaction due to unilateral meso-diencephalic lesion. Neurology, 40(10): 1503–1510.

Howard, I.P. (1982) Human Visual Orientation. Wiley, New York.

Leigh, R.J. and Zee, D.S. (2006) The Neurology of Eye Movements (4th ed.). Oxford University Press, New York.

Mossman, S. and Halmagyi, G.M. (1997) Partial ocular tilt reaction due to unilateral cerebellar lesion. Neurology, 49(2): 491–493.

Yelnik, A.P., Lebreton, F.O., Bonan, I.V., Colle, F.M., Meurin, F.A., Guichard, J.P. and Vicaut, E. (2002) Perception of verticality after recent cerebral hemispheric stroke. Stroke, (33): 2247–2253.

SECTION 5

Using Eye Movements as a Probe of Cognition, Memory, and Prediction

C. Kennard & R.J. Leigh (Eds.)
Progress in Brain Research, Vol. 171
ISSN 0079-6123

CHAPTER 5.1

Brain mechanisms for switching from automatic to controlled eye movements

Okihide Hikosaka[1,*] and Masaki Isoda[2,**]

[1]*Laboratory of Sensorimotor Research, National Eye Institute, National Institutes of Health, Bethesda, MD, USA*
[2]*Laboratory for Symbolic Cognitive Development, RIKEN Brain Science Institute, Wako, Saitama, Japan*

Abstract: Human behaviour is mostly composed of habitual actions that require little conscious control. Such actions may become invalid if the environment changes, at which point we need to switch behaviour by overcoming habitual actions that are otherwise triggered automatically. It is unclear how the brain controls this type of behavioural switching. Here we show that the presupplementary motor area (pre-SMA) in the medial frontal cortex has a function in switching from automatic to volitionally controlled action. This was demonstrated using colour-matching saccade tasks performed by rhesus monkeys. We found that a group of pre-SMA neurons was selectively activated when subjects successfully switched from a habitual saccade to a controlled alternative saccade. Electrical stimulation in the pre-SMA replaced automatic incorrect saccades with slower correct saccades. A further test suggested that the pre-SMA enabled switching by first suppressing an automatic unwanted saccade and then boosting a controlled desired saccade. Our data suggest that the pre-SMA resolves response conflict so that the desired action can be selected. Possible neuronal circuits through which the pre-SMA might exert its switching functions will be discussed.

Keywords: presupplementary motor area; medial frontal cortex; subthalamic nucleus; substantia nigra pars reticulata; basal ganglia; monkeys; saccadic eye movement; habitual action; conscious control; decision-making

Parallel neural network for controlling behaviour

Our daily behaviour consists mostly of habitual actions. You get up in the morning, take a shower, eat breakfast, drive your car to workplace, and start your computer to check emails. You may pay little attention to each of these steps and may not remember where you parked your car. One day, however, you may find that your usual parking lot is under construction. Then, you decide not to go into the parking lot and think hard where to go.

Habitual actions make our life efficient and easy. They can be performed smoothly and quickly with little mental efforts (owing to learning based on everyday practice) (Anderson, 1982). This is because, as previous research suggests (Hikosaka et al., 1999), such learned procedures or habits are controlled by the motor part of the brain (such as the motor cortex, anterior part of the cerebellum, posterior part of the basal ganglia) whereas the cognitive part of the brain (such as the prefrontal and parietal association cortices, posterior part of the cerebellum, anterior part of the basal ganglia) is at rest.

*Corresponding author. Tel.: +301-402-7959;
Fax: +301-402-0511; E-mail: oh@lsr.nei.nih.gov
**Corresponding author. Tel.: +81-48-467-9611;
Fax: +81-48-467-9645; E-mail: isodam@brain.riken.jp

DOI: 10.1016/S0079-6123(08)00655-9

However, such habitual actions may suddenly become inappropriate if something changes in our environment (Monsell, 2003). Then, the ongoing routine processes in the motor part of the brain must be stopped and instead the cognitive part of the brain must kick in swiftly. This is an important moment of decision-making.

Several lines of research have suggested the presupplementary motor area (pre-SMA) plays a key role in such a behavioural switching process (Shima et al., 1996; Nakamura et al., 1998; Ullsperger and von Cramon, 2001; Rushworth et al., 2002; Nachev et al., 2005; Crone et al., 2006; Woodward et al., 2006). However, the underlying mechanism is still unclear.

Switching from automatic to controlled behaviour

Suppose, in context alpha, action A is appropriate (Fig. 1A). By repeating action A in this context, it becomes habitual or automatic. At some point the context may change, say, to beta, and action A is no longer appropriate and instead action B becomes appropriate. In other words, you have to switch from action A to action B. To enable the switching, you have to do two things: (1) suppress action A which has been automatic or habitual, and (2) boost action B which has been suppressed.

An important step in our research was to create this sort of switching in a behavioural task for monkeys. After many attempts we came up with

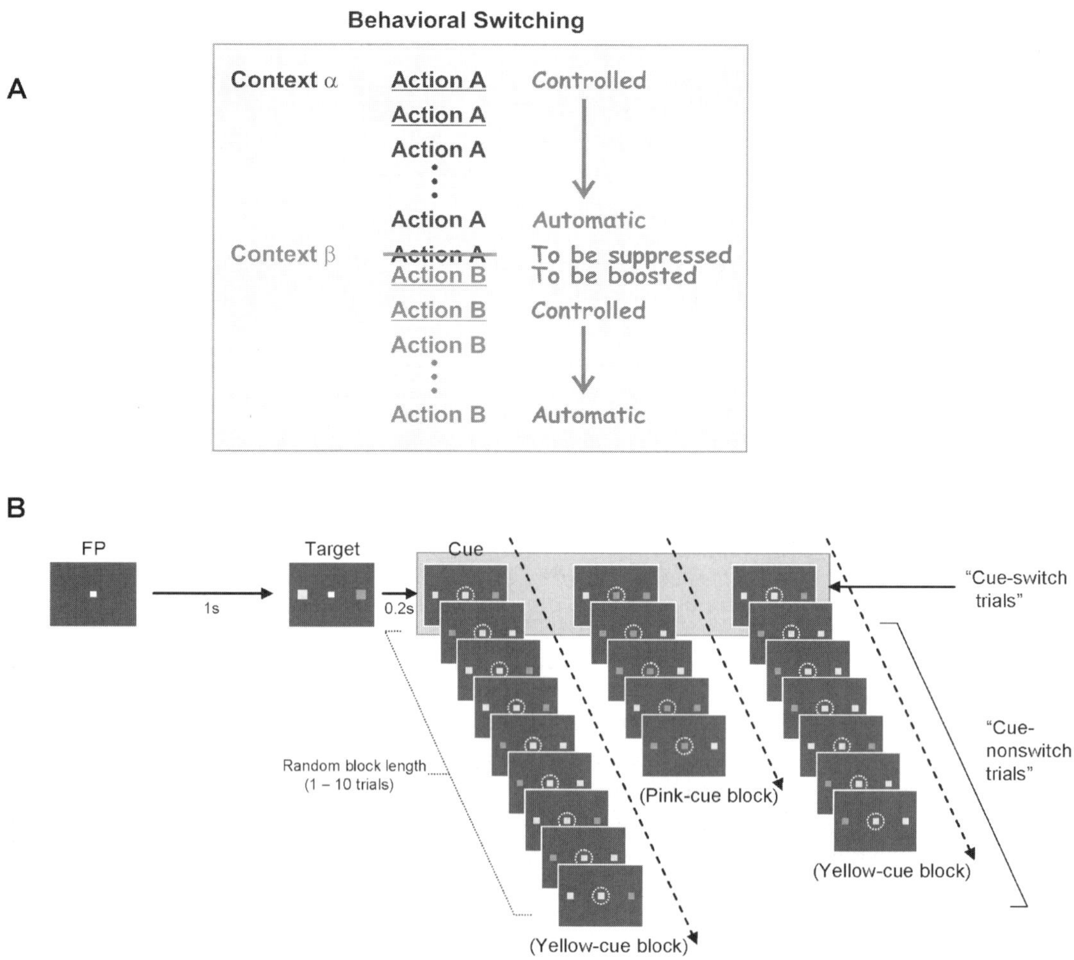

Fig. 1. (A) Scheme of switching from automatic to controlled action. (B) Saccade overriding task — a behavioural task to examine switch-selective neuronal activity. (See Color Plate 5.1.1 in color plate section.)

a task which we call 'saccade overriding task' (Fig. 1B) (Isoda and Hikosaka, 2007). It is basically a colour-matching task. First, a central fixation point appeared and the monkey had to fixate it. After 1 s, two stimuli appeared on the left and right in two different colours, for example, pink and yellow. The positions of the pink and yellow stimuli were randomized. After a short period, the fixation point turned its colour to either pink or yellow as a cue. The monkey's task was to make a saccade to the stimulus whose colour was the same as the central cue.

A critical feature of this task is that the cue colour remained the same in a block of 1–10 trials. It is thus likely that the colour of the upcoming cue will be the same as that in the preceding trial, and it seems natural for the monkey to start preparing for a saccade to the same colour stimulus by assuming that the cue colour will not change.

However, our main interest was in the trial when the cue colour changed. Let's call it 'switch trial,' as opposed to 'nonswitch trial.' In this switch trial, the habitual action is to make a saccade to the stimulus with the same colour, but it would lead to an error. Therefore, the monkey has to suppress the habitual action and instead make a saccade following the cue instruction. Indeed, the monkey's saccade behaviour was different between the switch trials and the nonswitch trials. The saccade reaction times (RTs), measured from the change in the colour of the fixation point, were much longer and the error rate was higher in the switch trial than in the nonswitch trials. In the switch trials, most errors occurred when the saccade RTs were too short; later saccades were usually successful (Fig. 3D). This is because the saccades with short RTs were based on the preceding colour rule and therefore quick, and therefore were bound to be wrong in the current colour rule. In the nonswitch trials, habitual saccades with short RTs were correct because the colour rule remained the same.

We obtained another important piece of information from the RT data in the switch trials. The monkey's performance became above the chance level at 213 ms from cue onset or later (Fig. 3C, D). We call it *behavioural differentiation time*. Suppose there are neurons somewhere in the brain that enable the switch. Their activity must be differentiated between the switch and nonswitch trials before the behavioural differentiation time.

Switch-selective neurons in the pre-SMA

Figure 2 shows the macaque brain (left) and the human brain (right), viewed from the mesial side (Picard and Strick, 2001). The pre-SMA is located in the medial frontal cortex anterior to the SMA (supplementary motor area). The pre-SMA

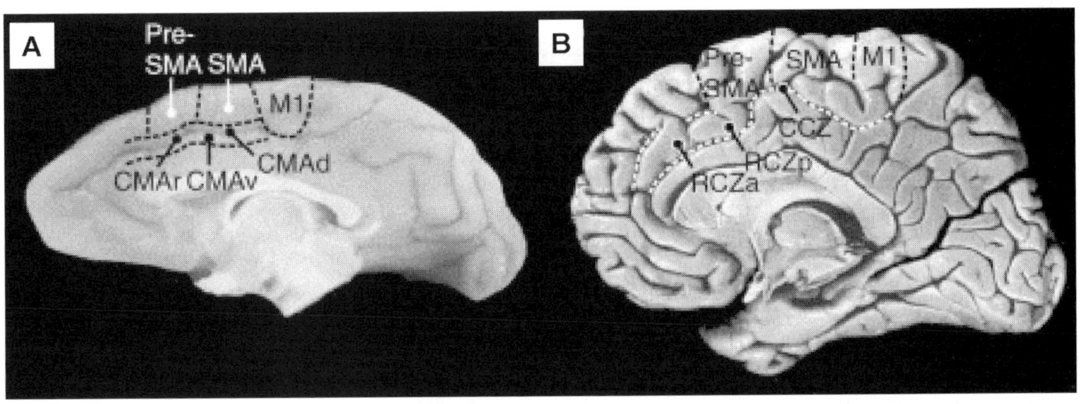

Fig. 2. The brains of the macaque monkey (A) and the human (B) viewed from the mesial side. CCZ: caudal cingulate zone, CMAd: dorsal cingulate motor area, CMAr: rostral cingulate motor area, CMAv: ventral cingulate motor area, M1: primary motor cortex, Pre-SMA: pre-supplementary motor area, RCZ: rostral cingulate zone, RCZa: anterior rostral cingulate zone, RCZp: posterior rostral cingulate zone, SMA: supplementary motor area. Adapted with permission from Picard and Strick (2001).

was originally characterized and localized in the macaque monkey (Matelli et al., 1991; Matsuzaka et al., 1992), but human brain imaging studies have supported the existence of the pre-SMA in the human brain as well (Picard and Strick, 2001). Human brain imaging studies have shown that the pre-SMA is activated in many cognitive tasks rather than motor tasks (Picard and Strick, 1996).

We found many neurons in the pre-SMA that became active selectively in switch trials in the saccade overriding task. Interestingly, the switch-selective activity was usually direction-selective. Some ($n=26$) were active only when the saccade was switched to the contralateral side. The others ($n=15$) were active only when the saccade was switched to the ipsilateral side. A smaller number of neurons ($n=9$) were active in either direction of switch.

Let's consider the switch from the yellow block to the pink block (Fig. 3A, B). It means that the yellow stimulus has been the target to choose in the last several trials. Therefore, when the two

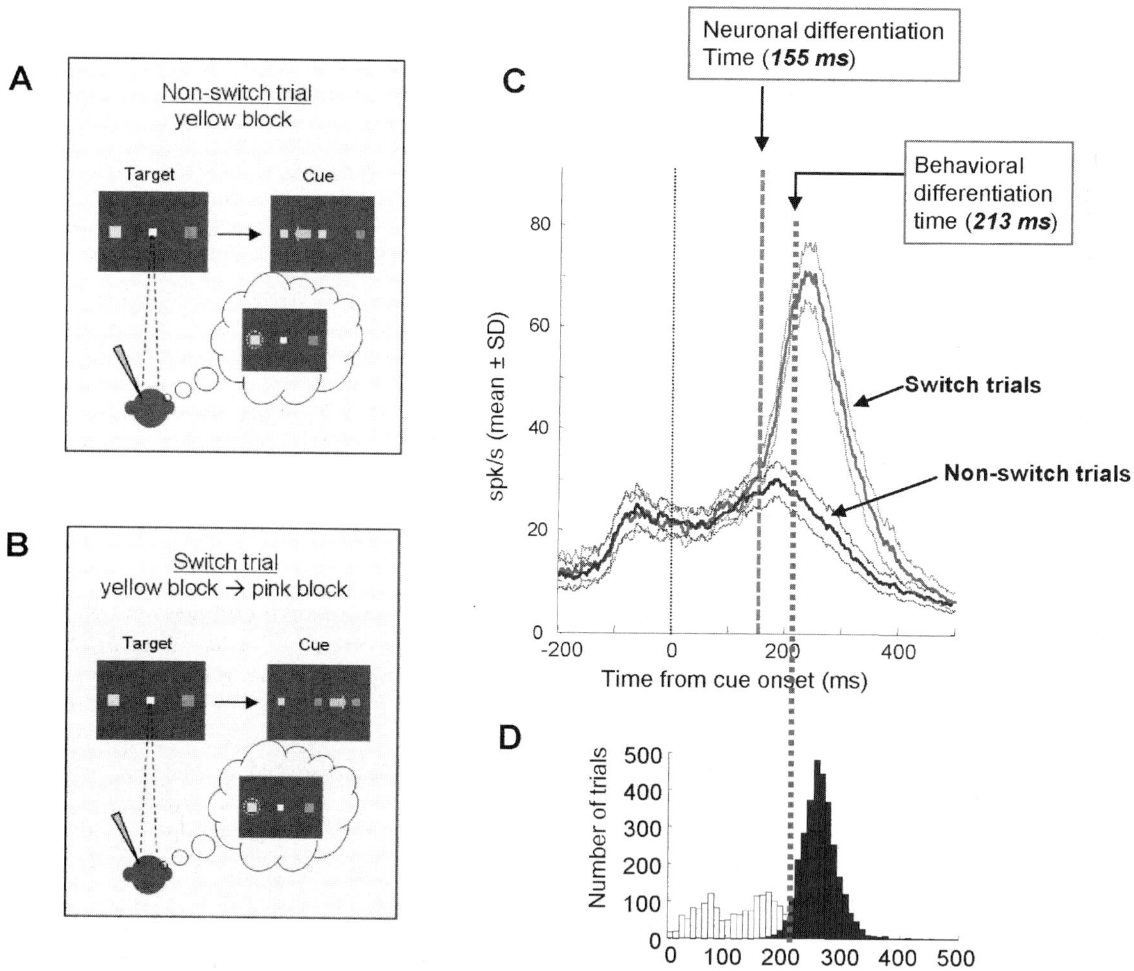

Fig. 3. Switch-selective neuronal activity in the monkey pre-SMA. (A–B) In the saccade overriding task the monkey prepares for a habitual saccade to the stimulus whose colour is the same as the target colour in preceding trials (i.e., yellow). This preparation is valid in nonswitch trials (A), but not in switch trials (B). (C) The population activity of switch-selective pre-SMA neurons for nonswitch trials (blue) and switch trials (red). (D) The distribution of saccade reaction times in switch trials, shown separately for incorrect saccades (light blue) and correct saccades (dark blue). (See Color Plate 5.1.3 in color plate section.)

stimuli come on, the monkey is likely to attend to the yellow stimulus and may start preparing for a saccade to the stimulus. The saccade will be successful in no-switch trials (Fig. 3A). In the switch trial, however, the fixation point turns into pink, and the saccade must be made to the pink stimulus (Fig. 3B). This is the time when pre-SMA neurons fire. Is the firing early enough?

In Fig. 3C we align the population activity of switch-selective pre-SMA neurons on the time of cue onset, and compared their activity between the nonswitch and switch trials. The two sets of activity, which were almost identical initially, diverged at 155 ms after the cue onset. We call it *neuronal differentiation time*. Importantly, the *neuronal differentiation time* was significantly earlier than the *behavioural differentiation time*, 213 ms (Fig. 3D). To summarize, the switch-selective activity of pre-SMA neurons appears to be early enough to enable the switch.

To test whether the pre-SMA activity is actually used for the switch, we applied electrical stimulation in the pre-SMA only in half of the switch trials, and compared the monkey's performance between the switch trials without stimulation and the switch trials with stimulation. We found that the correct rate increased considerably with pre-SMA stimulation. This was true for both contralateral and ipsilateral switches, indicating that the stimulation effect was not due to directional bias. As we indicated before, the erroneous saccades occurred too early based on the previous cue colour. One effect of pre-SMA stimulation was to eliminate such premature and habitual saccades.

Mechanisms of pre-SMA-induced behavioural switching

The above findings may be interpreted in two different ways: (1) the pre-SMA suppresses the unwanted saccade that is triggered by the habitual process, or (2) the pre-SMA facilitates the desired saccade based on the controlled process. To examine these possibilities, we characterized the activity of the switch-selective pre-SMA neurons using a saccade-Go/Nogo task. The saccade-Go/Nogo task was similar to the saccade overriding task. A critical difference was that there was only one target. It appeared on the left or right randomly, and in yellow or pink randomly. If the cue colour was the same as the target colour, the monkey had to make a saccade to the target. If their colours were different, the monkey had to withhold a saccade and continue to fixate. If a neuron becomes active in Go trials, its action is likely to facilitate the saccade. If the neuron becomes active in Nogo trials, its action is likely to suppress the saccade. Furthermore, we should be able to tell whether the neuron influences leftward saccades or rightward saccades.

Let's consider a neuron in the left pre-SMA that becomes active only when the saccade is switched from left to right (as in Fig. 3B). Since the cue colour has been yellow in the preceding trials, the monkey is ready for a saccade to the yellow target on the left side, which was ipsilateral to the pre-SMA neuron. In the switch trial, the cue colour turns out to be pink, and the monkey has to make a saccade to the pink target on the right side, contralateral to the neuron. This is the time when the neuron is selectively active. This switch requires two actions. First, the habitual saccade to the ipsilateral side must be suppressed. Second, the saccade to the contralateral side must be boosted. Does this pre-SMA neuron perform such actions? If so, which action does it perform?

Using the Go-Nogo task, we found that there are three groups of switch-selective neurons in the pre-SMA: suppression only, facilitation only, and both suppression and facilitation. Importantly, these actions were usually consistent with the direction of switching. For example, the switch-selective neuron considered in relation to Fig. 3B may be active only in left Nogo trials suggesting that it suppresses leftward (ipsilateral) saccades, or may be active in right Go trials suggesting that it facilitates rightward (contralateral) saccades, or may be active in both cases. We also found that the Nogo-type neurons tended to be active earlier than the Go-type neurons.

In conclusion, the pre-SMA enables switching from a habitual action to an alternative action, first by suppressing the habitual action and then boosting the alternative action.

Discussion

At the beginning of this article, we presented a scheme of behavioural switching (Fig. 1). In Fig. 4A, we present a circuit diagram that would realize these operations. There are an automatic or habitual process and a controlled process, either of which can trigger a motor execution mechanism. For an appropriate behavioural switch, we need a switch mechanism. This mechanism must be able to suppress the automatic process (or its output) and boost the output of the controlled process. The switch mechanism should be deployed only when the context is changed.

Our study has suggested that the pre-SMA contains such a switch mechanism. Switch-selective pre-SMA neurons are usually not active, but become active when the environment or context

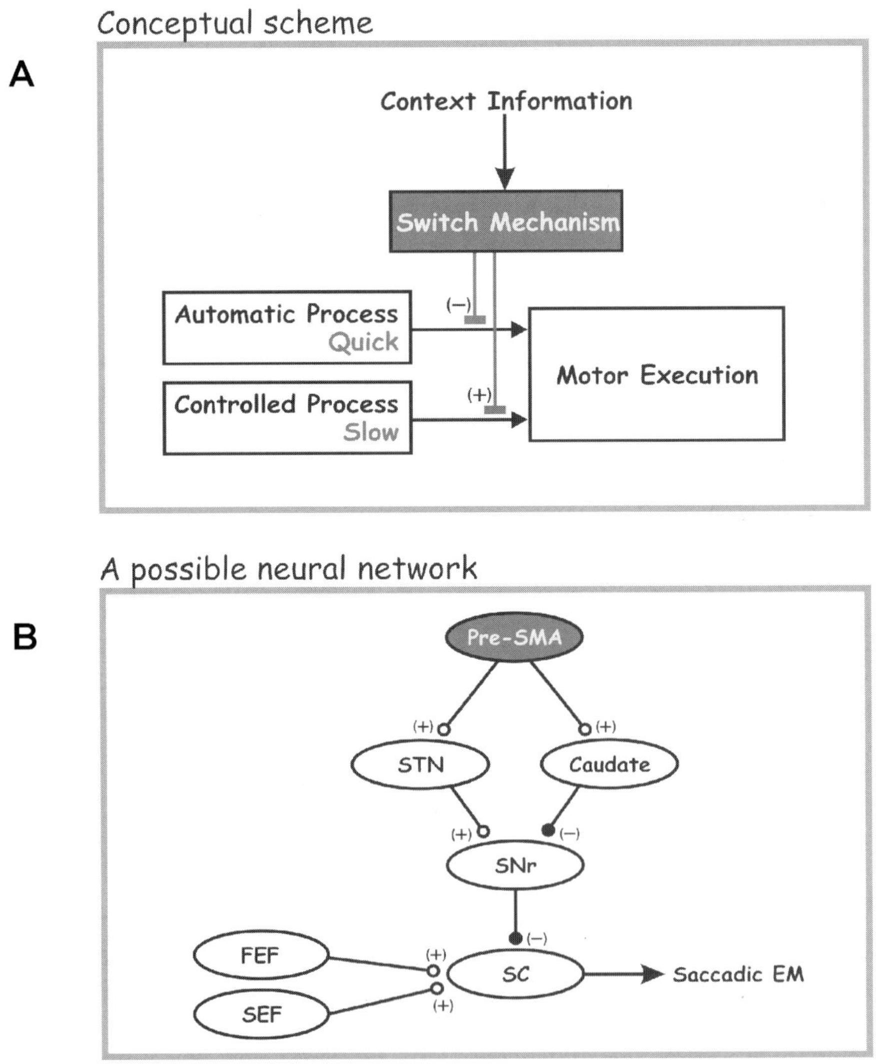

Fig. 4. Hypothetical schemes for switching from automatic to controlled action. (A) Conceptual scheme. (B) Neural network scheme for switching of saccadic eye movements. FEF: frontal eye field, SC: superior colliculus, SEF: supplementary eye field, STN: subthalamic nucleus. (+) and (−) indicate excitatory and inhibitory connections, respectively.

changes. The switch-selective activation occurs well before the monkey switches saccade behaviour. Electrical stimulation of the pre-SMA increases the likelihood of successful switches.

We also showed that the switch-selective activity starts at different timings depending on the neuron's action. It is well known that automatic or habitual actions are very quick while cognitively controlled actions are slow. In other words, the automatic process produces outputs more quickly than the controlled process. Therefore, in order for the switch mechanism to work efficiently, the inhibition of the automatic process should occur first to prevent the execution of the habitual response. The facilitation of the controlled process may occur later. This is actually what we have seen in the pre-SMA: The switch neurons with Nogo action become active earlier than those with Go action.

The next question is: How does the pre-SMA exert its Nogo and Go effect on saccade outputs? Figure 4B shows our hypothesis. The main circuit for saccade initiation is the excitatory pathways from the cortical eye fields to the superior colliculus. The Nogo action requires a powerful inhibition, which we assigned to the inhibition from the substantia nigra pars reticulata (SNr). The pre-SMA, among other frontal cortical area, is known to project to the subthalamic nucleus (STN) (Inase et al., 1999) which in turn project to the SNr (Kita and Kitai, 1987). Since these connections are excitatory, the output of pre-SMA neurons would lead to an inhibition of saccadic neurons in the superior colliculus. This might represent the Nogo or inhibitory action of the switch mechanism. The Go action may be accomplished with the signal originating from the pre-SMA going through the caudate nucleus (Inase et al., 1999). Since the caudate nucleus has direct inhibitory connections to the SNr, the net effect of the pre-SMA signal would be facilitatory (Hikosaka et al., 2000).

We have performed similar experiments on STN neurons, and found similar switch-selective neurons. Importantly, their switch-selective activity of STN neurons starts slightly later, on the average, than that of pre-SMA neurons, but before the behavioural differentiation time. This result is consistent with the hypothetical circuit diagram in Fig. 3B. Human brain imaging studies have also shown the role of the STN in behavioural switching (Aron and Poldrack, 2006; Monchi et al., 2006). Unexpectedly, however, we found not only Nogo neurons (expected from the diagram) but also Go neurons in the STN. One way to explain this result may be to postulate indirect connections from the STN to the SNr through the globus pallidus external segment (GPe). Since GPe neurons are known to be inhibitory, the net effect of the STN output would be reversed from Nogo to Go. This suggests that there are many combinations of neural connectivity between the STN and the GPe (Nambu et al., 2000), which may allow STN-projecting cortical areas to control body movements in many different ways.

Currently we have no data on the possible Go mechanism involving the caudate nucleus. There is some evidence from human brain imaging studies indicating that the caudate nucleus becomes active in switch trials (Monchi et al., 2006). One small problem of this pathway is that the conduction time of caudate neurons is quite long (almost 20 ms) (Hikosaka et al., 1993), so that it may not be suitable for quick switching actions. It may still be functional, however, since the Go action can be delayed compared with the Nogo action. Although not shown in this diagram, cortico-cortical connections from the pre-SMA to the prefrontal areas including the supplementary eye field may serve as the Go mechanism (Husain et al., 2003; Nachev et al., 2005; Stuphorn and Schall, 2006). These possibilities are testable using our experimental paradigm.

References

Anderson, J.R. (1982) Acquisition of cognitive skill. Psychol. Rev., 89: 369–406.

Aron, A.R. and Poldrack, R.A. (2006) Cortical and subcortical contributions to stop signal response inhibition: role of the subthalamic nucleus. J. Neurosci., 26: 2424–2433.

Crone, E.A., Wendelken, C., Donohue, S.E. and Bunge, S.A. (2006) Neural evidence for dissociable components of task-switching. Cereb. Cortex, 16: 475–486.

Hikosaka, O., Nakahara, H., Rand, M.K., Sakai, K., Lu, X., Nakamura, K., Miyachi, S. and Doya, K. (1999) Parallel neural networks for learning sequential procedures. Trends Neurosci., 22: 464–471.

Hikosaka, O., Sakamoto, M. and Miyashita, N. (1993) Effects of caudate nucleus stimulation on substantia nigra cell activity in monkey. Exp. Brain Res., 95: 457–472.

Hikosaka, O., Takikawa, Y. and Kawagoe, R. (2000) Role of the basal ganglia in the control of purposive saccadic eye movements. Physiol. Rev., 80: 953–978.

Husain, M., Parton, A., Hodgson, T.L., Mort, D. and Rees, G. (2003) Self-control during response conflict by human supplementary eye field. Nat. Neurosci., 6: 117–118.

Inase, M., Tokuno, H., Nambu, A., Akazawa, T. and Takada, M. (1999) Corticostriatal and corticosubthalamic input zones from the presupplementary motor area in the macaque monkey: comparison with the input zones from the supplementary motor area. Brain Res., 833: 191–201.

Isoda, M. and Hikosaka, O. (2007) Switching from automatic to controlled action by monkey medial frontal cortex. Nat. Neurosci., 10: 240–248.

Kita, H. and Kitai, S.T. (1987) Efferent projections of the subthalamic nucleus in the rat: light and electron microscopic analysis with the PHA-L method. J. Comp. Neurol., 260: 435–452.

Matelli, M., Luppino, G. and Rizzolatti, G. (1991) Architecture of superior and mesial area 6 and the adjancent cingulate cortex in the macaque monkey. J. Comp. Neurol., 311: 445–462.

Matsuzaka, Y., Aizawa, H. and Tanji, J. (1992) A motor area rostral to the supplementary motor area (presupplementary motor area) in the monkey: neuronal activity during a learned motor task. J. Neurophysiol., 68: 653–662.

Monchi, O., Petrides, M., Strafella, A.P., Worsley, K.J. and Doyon, J. (2006) Functional role of the basal ganglia in the planning and execution of actions. Ann. Neurol., 59: 257–264.

Monsell, S. (2003) Task switching. Trends Cogn. Sci., 7: 134–140.

Nachev, P., Rees, G., Parton, A., Kennard, C. and Husain, M. (2005) Volition and conflict in human medial frontal cortex. Curr. Biol., 15: 122–128.

Nakamura, K., Sakai, K. and Hikosaka, O. (1998) Neuronal activity in medial frontal cortex during learning of sequential procedures. J. Neuro-physiol., 80: 2671–2687.

Nambu, A., Tokuno, H., Hamada, I., Kita, H., Imanishi, M., Akazawa, T., Ikeuchi, Y. and Hasegawa, N. (2000) Excitatory cortical inputs to pallidal neurons via the subthalamic nucleus in the monkey. J. Neuro-physiol., 84: 289–300.

Picard, N. and Strick, P.L. (1996) Motor areas of the medial wall: a review of their location and functional activation. Cereb. Cortex, 6: 342–353.

Picard, N. and Strick, P.L. (2001) Imaging the premotor areas. Curr. Opin. Neurobiol., 11: 663–672.

Rushworth, M.F., Hadland, K.A., Paus, T. and Sipila, P.K. (2002) Role of the human medial frontal cortex in task switching: a combined fMRI and TMS study. J. Neurophysiol., 87: 2577–2592.

Shima, K., Mushiake, H., Saito, N. and Tanji, J. (1996) Role for cells in the presupplementary motor area in updating motor plans. Proc. Natl. Acad. Sci. U.S.A., 93.

Stuphorn, V. and Schall, J.D. (2006) Executive control of countermanding sccades by the supplementary eye field. Nat. Neurosci., 9: 925–931.

Ullsperger, M. and von Cramon, D.Y. (2001) Subprocesses of performance monitoring: a dissociation of error processing and response competition revealed by event-related fMRI and ERPs. Neuroimage, 14: 1387–1401.

Woodward, T.S., Ruff, C.C. and Ngan, E.T. (2006) Short- and long-term changes in anterior cingulate activation during resolution of task-set competition. Brain Res., 1068: 161–169.

C. Kennard & R.J. Leigh (Eds.)
Progress in Brain Research, Vol. 171
ISSN 0079-6123

CHAPTER 5.2

The frontal eye field as a prediction map

Trinity B. Crapse* and Marc A. Sommer

Department of Neuroscience and the Center for the Neural Basis of Cognition, University of Pittsburgh, Pittsburgh, PA, USA

Abstract: Predictive processes are widespread in the motor and sensory areas of the primate brain. They enable rapid computations despite processing delays and assist in resolving noisy, ambiguous input. Here we propose that the frontal eye field, a cortical area devoted to sensorimotor aspects of eye movement control, implements a prediction map of the postsaccadic visual scene for the purpose of constructing a stable percept despite saccadic eye movements.

Keywords: Bayesian; forward model; saccadic eye movements; macaca mulatta; corollary discharge; shifting receptive fields; visual stability; feedback connections

Introduction

The brain is an inferential machine. Both its motor areas and sensory networks engage in predictive computations (Miall and Wolpert, 1996; Bullier, 2001; Friston, 2005). Hierarchical models suggest that visual cortical neurons fire predominately to signal deviations from predicted inputs (Rao and Ballard, 1999; Lee and Mumford, 2003). Studies in the motor realm, too, suggest that predictions are used for adaptive control (Miall and Wolpert, 1996; Hwang and Shadmehr, 2005). Here we propose that predictive operations for both the sensory and motor domains find unification in the primate visuosaccadic system for the purpose of constructing a stable transaccadic percept.

The primate visuosaccadic system faces a serious problem (Fig. 1). It must generate a stable percept despite the frequent disruptions in gaze induced by saccadic eye movements. To appreciate this, consider the saccade as an event comprised of three constituent parts: presaccadically, a target is selected; intrasaccadically, the world as sensed actually shifts; and postsaccadically, the target is foveated. The intrasaccadic component is the problem. Were the animal to perceive the world exactly as sensed during the intrasaccadic epoch, the visual scene would seem to leap from place-to-place dozens of times per minute. Yet it is not perceived as such, and the brain can even distinguish what aspects of the jumpy visual inflow are artefactual (due to saccades) as opposed to real (due to changes in the world). How does it do it?

An important cortical gaze control component is the frontal eye field (FEF) (Fig. 2; Sommer and Wurtz, 2008). Located within the anterior bank of the arcuate sulcus, the FEF contains topographically arranged neurons with response properties that span the continuum from purely visual to purely movement. The FEF receives input from many cortical and subcortical areas including the superior colliculus (SC) via thalamic relay neurons. We recently demonstrated that

*Corresponding author. Tel.: +1 412 268 7229; Fax: +1 412 268 5060; E-mail: tbc6@pitt.edu

DOI: 10.1016/S0079-6123(08)00656-0

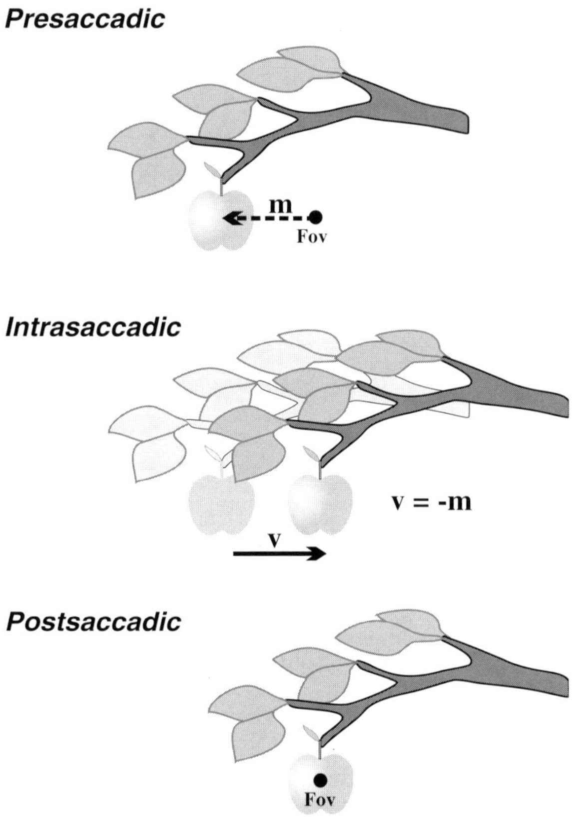

Fig. 1. The saccadic eye movement as a visual problem. Each saccade is composed of three elements: a presaccadic component when a target is selected and the saccade initiated, an intrasaccadic component during which time the world as sensed actually shifts, and a postsaccadic component when the target is reached. The intrasaccadic component is the most disruptive element for the visual system. The fact that the world shifts with each saccade, and yet is not perceived to do so, implies a corrective mechanism within the visual system. This mechanism is rooted in predictive operations. Fov, fovea; **m**, movement vector; **v**, retinal vector emphasizing direction and magnitude of the shift.

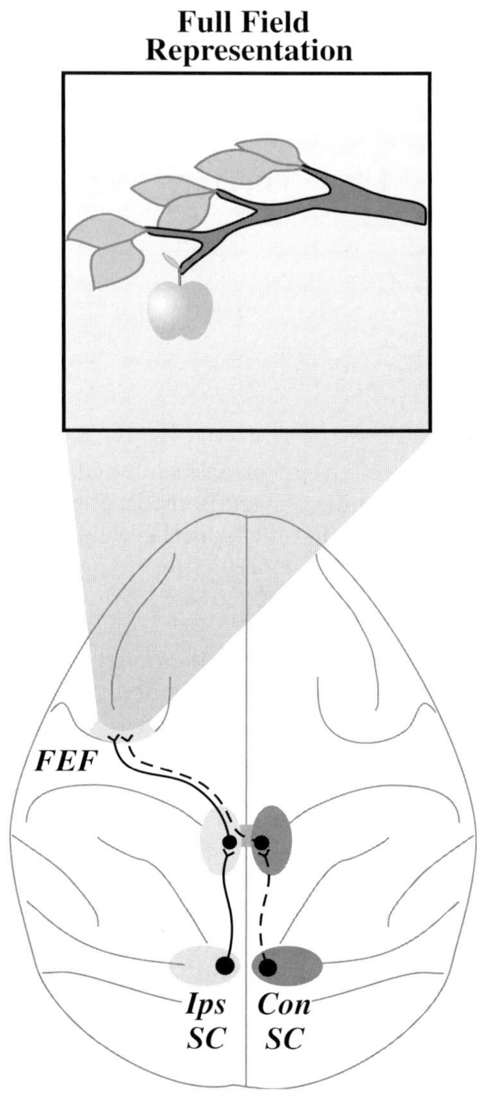

Fig. 2. Full-field representation in the frontal eye field (FEF). The FEF has information about all of visual space and all eye movements via pathways from ipsilateral (Ips) and contralateral (Con) superior colliculus (SC).

convergent inputs from both SCs provide each FEF with a full-field representation of all saccades and all of visual space (Crapse and Sommer, 2007).

Like several other primate areas, the FEF contains neurons that shift their response fields (RFs) before eye movements (a "shifting RF") (Umeno and Goldberg, 1997). While a typical RF is firmly retinotopic and samples a new part of the visual field (the new RF) only after the eye moves, a shifting RF is dynamic and starts sampling the new RF location even before a saccade. Such neurons depend on corollary discharge (CD) from the midbrain to trigger the shift and are thought to contribute to a percept of visual stability (Sommer and Wurtz, 2006, 2008). How might these neurons influence the rest of the brain?

A prediction map in primate frontal eye field

We propose that shifting neurons of the FEF are components of a larger FEF inferential architecture that engages in predictive coding. That is, the FEF is a prediction map. This scheme assigns a causal role to the FEF in the construction of a stable visual percept despite saccadic interruptions. According to this conception, predictions of the future scene (postsaccadic) are generated based on extrapolations from the current scene (presaccadic). Neurons with shifting RFs, informed spatiotemporally of the imminent change in gaze by CD, collect data about what will fall in their RF after the saccade. These data are convolved with estimates of an internal model of the visual world. What follows is a prediction of the visual structure of the postsaccadic scene.

How would the prediction be used? We hypothesize that around the time of the saccade, the predictive signals are sent from the FEF to the posterior lobes. The visual lobes are primed with activity-constraining expectations. Deviations from the predicted inputs (prediction errors) are reported back to the FEF. The prediction errors therefore are ultimately manifest in the visual responses of FEF neurons. This residual alerts the FEF to unpredicted events. This may entail calibration/updating if the error is related to a miscalculation/noise or it may provide useful information about the environment that was unexpected, i.e., perhaps something moved or suddenly appeared. Either way, the final result of the iterative signalling would be the perception of a stable world despite the change in gaze. In the rest of this article we will formalize these ideas.

Computational basis of the prediction map

The prediction map may be grounded in an inferential process based on empirical Bayesian principles. Neurons throughout the brain seem to engage in probabilistic and inferential computations (Gold and Shadlen, 2007). The probabilistic aspect seems necessary because of the noise and ambiguity intrinsic to neural computation (Knill and Pouget, 2004). Bayesian inference would allow prior and conditional probability distributions to be utilized to generate a posterior distribution, i.e., the outcome expected on the balance of known probabilities and other inputs (in other words, a prediction; Fig. 3). For the case of a single saccade, the proposed Bayesian computation would express the conditional expectation or probability of observing the postsaccadic scene (p) given the relative probability of two independent pieces of evidence: the visual structure of the current scene (c), and various saccadic parameters (s), written:

$$\mathrm{P(p|c,s)} = \frac{\mathrm{P(c|p)P(s|p)P(p)}}{\mathrm{P(c)P(s)}} = \lambda^{\rho}$$

Bayesian priors about the structure of the visual scene P(c) and how the world behaves during a saccade P(s) could be learned and updated through experience and stored in FEF networks. The priors would permit calculation of local conditional probabilities relating the postsaccadic scene to the current scene P(c|p) and the postsaccadic scene to the saccade P(s|p). Ultimately, these computations would result in a global conditional expectation λ^{ρ} that would be convolved with a generative model Ψ to generate a prediction of the postsaccadic scene $\Psi(\lambda^{\rho})$. The prediction would be compared with the actual input λ^{α} to yield a prediction error $\varepsilon = \lambda^{\alpha} - \Psi(\lambda^{\rho})$.

Physiological mechanism of the prediction map

Mechanistically, frontal modulatory control of the posterior lobes could be implemented through imposed patterns of synchronization mediated by cortico-cortical connections (Womelsdorf et al., 2007). Visually evoked activity of single neurons is surprisingly quite deterministic (Arieli et al., 1996). The oft-encountered variability in single neuron responses seems to emerge from the dynamics of ongoing network activity. This fact could be exploited by the prediction map for purposes of ensuring transaccadic perception. The imposed predictions could alter the spatiotemporal properties of visual cortical network activity to influence the specific visually evoked responses that are ultimately integrated into the network and expressed as spikes. The FEF

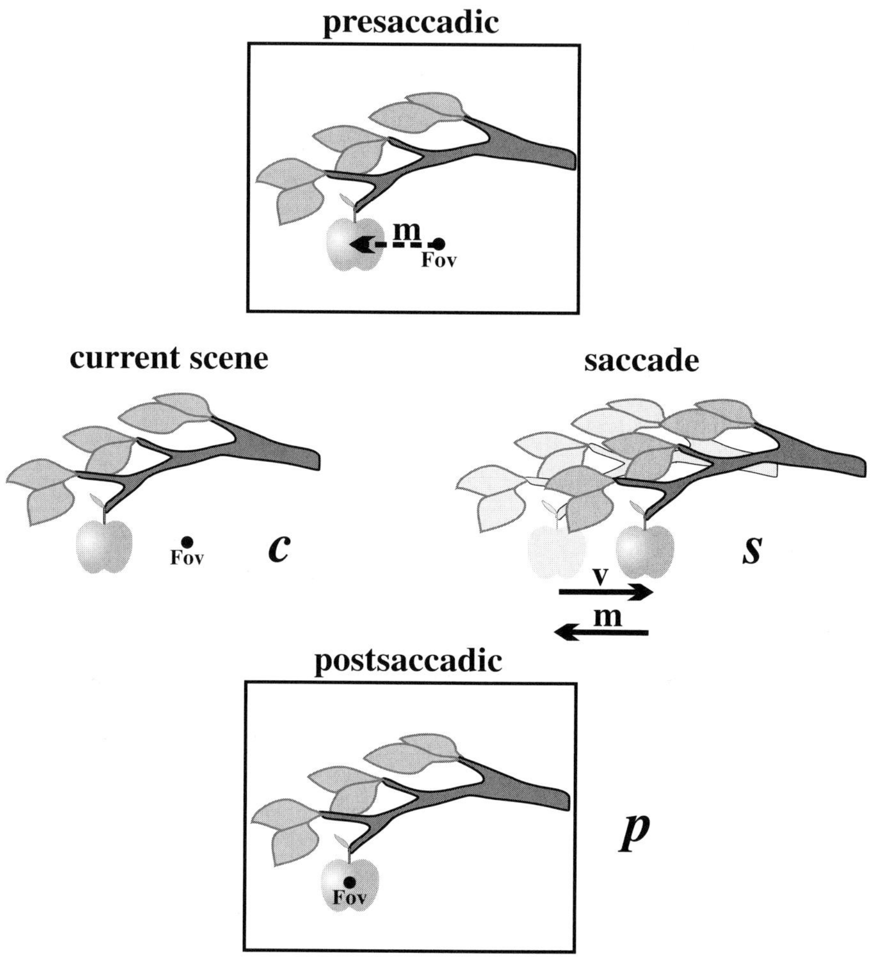

$$P(p|c,s) = \frac{P(c|p)P(s|p)P(p)}{P(c)P(s)} = \lambda^{\rho}$$ conditional expectation

$$\Psi(\lambda^{\rho})$$ prediction

$$\lambda^{\alpha}$$ actual input

$$\varepsilon = \lambda^{\alpha} - \Psi(\lambda^{\rho})$$ prediction error

Fig. 3. Computational basis of the prediction map. A prediction map could be based on Bayesian computations. See text for details.

would initiate a specific synchronization pattern among the relevant visual neurons (those activated by features of the current scene) and this would serve to amplify behaviourally relevant signals in the cortex.

The sequence of events

A detailed account of what we propose occurs during each voluntary saccade is as follows (Fig. 4). The imminent saccade would initiate an iteration

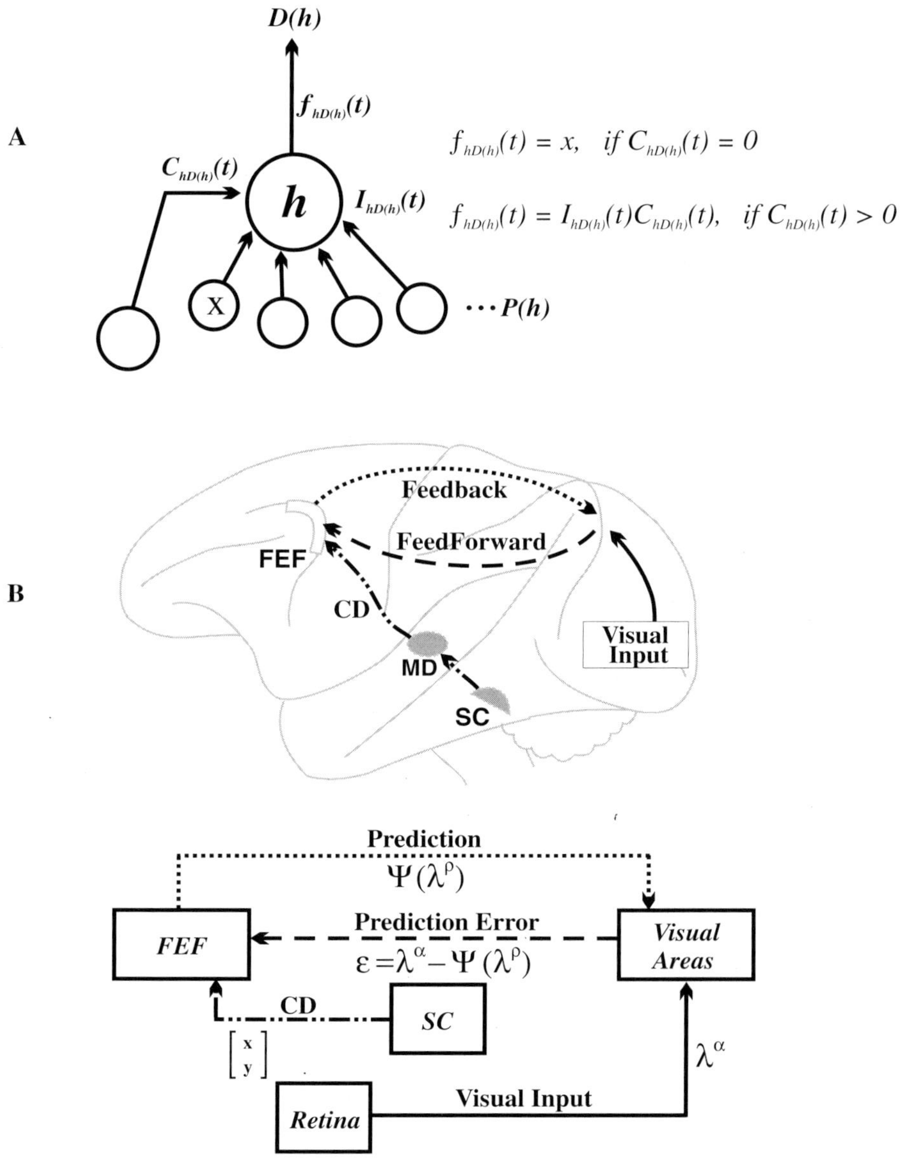

Fig. 4. Mechanics of the prediction map. (A) A centrepiece of the prediction map would be neurons with shifting receptive fields. These would serve as network hubs (h) that selectively route visual information $f_{hD(h)}(t)$ through the network based on the information $I_{hD(h)}(t)$ provided by upstream neurons P(h) and CD from the midbrain $C_{hD(h)}(t)$. This information would then be forwarded to neurons downstream D(h) that perform additional computations for the sake of prediction. Equations are explained in text. (B) Predictions $\Psi(\lambda^{\rho})$ would exit the FEF as feedback that targets various areas of the visual brain. Errors $\varepsilon = \lambda^{\alpha} - \Psi(\lambda^{\rho})$ between the predicted visual input $\Psi(\lambda^{\rho})$ and actual visual input λ^{α} would be then routed back to the FEF.

of recurrent processing between the FEF and an assortment of visual cortical areas, beginning with receipt in the FEF of CD (leftmost line) from the SC. This information represents the when and where (a vector quantity) of the imminent change in gaze and induces a transient alteration in local functional topology of the FEF network. At the centre of the alteration would be FEF network hubs (h) (Sporns et al., 2007). FEF network hubs have access to information about the entire visual field ($I_{hD(h)}(t)$) and are considered to be neurons with shifting RFs. The CD serves a gating function ($C_{hD(h)}(t)$) and routes the appropriate visual information through the hub ($f_{hD(h)}(t)$) to downstream neurons D(h). [If CD is not present ($C_{hD(h)}(t) = 0$), the hub simply passes on the visual information provided to it from node x.] These relationships may be written as:

$$f_{hD(h)}(t) = I_{hD(h)}(t)C_{hD(h)}(t), \quad \text{if } C_{hD(h)}(t) > 0$$
$$\text{Otherwise}: \quad f_{hD(h)}(t) = x, \quad \text{if } C_{hD(h)}(t) = 0$$

The RF is said to shift and corresponds to a collection of visual information about the portions of space that the RF will encompass after the saccade. This information then would be combined with the causal estimates of the generative model embedded within the FEF network. A prediction would be generated and exported, manifested as a round of perisaccadic synchronized activity between the frontal and posterior lobes. The synchronized activity would alter the visual cortical network dynamics and effect how the visual stimulus-driven activity is integrated into the network. The synchronized activity is equivalent to a prediction, the output of the generative model. If the actual visual input does not match this prediction, then residuals (prediction errors) are routed downstream back to the FEF and iterative processes are continued.

Site of prediction error calculation

The prediction errors could be calculated at any number of visual cortical depots. Virtually every portion of the cortical mantle exhibits some degree of saccadic modulation (Baker et al., 2006). Dorsal stream components seem most likely for two reasons. First, the temporal structure of information flow through the primate visual system points to a dorsal stream speed advantage over the ventral stream (Bullier, 2001; Bar, 2007). Dorsal stream components exhibit activation latencies that often trail V1 responses by a meagre 10 ms. Some even activate before V1 (Schmolesky et al., 1998; Bullier et al., 2001). This speed advantage would be optimal for the rapid detection and routing of prediction errors back to the FEF. Second, the dorsal stream is known as the where pathway (Ungerleider and Mishkin, 1982). Components are concerned with motion and the physical location of objects in the visual scene. Since the primary concern of the visual system during a saccade is to ensure a stable world, the dorsal stream would seem to be particularly important in controlling the illusory percept of motion concomitant to saccade generation. Any errors related to the prediction could be a consequence of an object that actually moved on its own accord and thus was not predicted. The object would be with high probability worthy of further inspection. The FEF would then direct the allocation of attentional and targeting mechanisms to these objects via retrojections of activity to stations of the ventral stream such as V4 and IT (Moore and Armstrong, 2003; Hamker, 2005).

Relation to previous FEF studies

Previous studies have uncovered a number of FEF response properties consistent with the notion of a prediction map. One implication of prediction error in general is that single neurons should not respond if the stimulus falling in its RF is predicted. Burman and Segraves (1994) found that when monkeys rescanned a previously scanned image, visual activity was virtually unaffected by the contents of the image that fell within the RF. In contrast, these same neurons fired vigorously during the initial scan and when a target light suddenly appeared in the RF. After the initial scanning, a memory trace of the image was likely formed, and only deviations from the predicted image components were signalled as visual bursts.

This suggests that the visual response of FEF neurons signals prediction error, the unexpected.

An additional implication of a prediction map applies to the movement-related aspects of the FEF, generally. If a function of the FEF is to generate predictions, then movement responses, too, may be viewed probabilistically. In fact, some pure movement cells have been found that generate vigorous bursts even when the monkey errs and does not perform the saccade (Bruce and Goldberg, 1985). Other movement neurons fail to fire even when a saccade is launched into its RF. This is evidence that the movement signal in FEF is closely akin to a prediction or probability estimate of a saccade being generated into its movement field. It bears little relationship to saccade dynamics (Segraves and Park, 1993). We suggest that the forced manner in which FEF movement signals have been studied classically, counting spikes while monkeys make required eye movements into a neuron's RF, may mask this probabilistic aspect of FEF function and redirect focus on alleged deterministic aspects of movement generation.

Other frontal lobe functions

The prediction map is consistent with a number of other phenomena involving the frontal lobes and visual function. Among other things, the frontal lobes are thought to play a role in resolving visual ambiguity. Visual scenes are often ambiguous; the sensory data are consistent with multiple interpretations. This often results in illusions and multistable percepts, i.e., depth reversals, binocular rivalry, ambiguous figures, etc. A host of studies point to a role of the frontal lobes, FEF included, in resolving visual ambiguity (Lumer et al., 1998). Some contend that the frontal lobes bias or control the posterior lobes in an attempt to resolve the ambiguity and arrive at a stable percept (Miller and Cohen, 2001). We submit that the same frontal circuits that are responsible for resolving visual ambiguity and arriving at a stable percept of the present scene are part of the prediction map and are utilized by the oculomotor system for transsaccadic stability.

In a more speculative sense, the frontal lobes are implicated in processes related to mental imagery (Goebel et al., 1998). When a subject is asked to "imagine" a particular image the frontal lobes invariably come on line. Feedback connections from the frontal lobes to the posterior lobes are thought to activate the same circuits that would be activated if the actual image had been presented to the retinas (Kosslyn et al., 1995). Furthermore, the eye movements of subjects asked to imagine a particular scene reflect the content and spatial arrangement of objects populating the scene (Mast and Kosslyn, 2002). Again this may involve the same circuits, as ancestral primates were likely making saccades and requiring mechanisms of transsaccadic stability long before they were engaging in mental imagery and imaginative musings.

References

Arieli, A., Sterkin, A., Grinvald, A. and Aertsen, A. (1996) Dynamics of ongoing activity: explanation of the large variability in evoked cortical responses. Science, 273: 1868–1871.

Baker, J.T., Patel, G.H., Corbetta, M. and Snyder, L.H. (2006) Distribution of activity across the monkey cerebral cortical surface, thalamus and midbrain during rapid, visually guided saccades. Cereb. Cortex, 16: 447–459.

Bar, M. (2007) The proactive brain: using analogies and associations to generate predictions. Trends Cogn. Sci., 11: 280–289.

Bruce, C.J. and Goldberg, M.E. (1985) Primate frontal eye fields. I. Single neurons discharging before saccades. J. Neurophysiol., 53: 603–635.

Bullier, J. (2001) Integrated model of visual processing. Brain Res., 36: 96–107.

Bullier, J., Hupe, J.M., James, A.C. and Girard, P. (2001) The role of feedback connections in shaping the responses of visual cortical neurons. Prog. Brain Res., 134: 193–204.

Burman, D.D. and Segraves, M.A. (1994) Primate frontal eye field activity during natural scanning eye movements. J. Neurophysiol., 71: 1266–1271.

Crapse, T.B. and Sommer, M.A. (2007) Frontal eye field neurons receiving input from both superior colliculi: are their receptive fields tuned or untuned? Program No. 19.10. 2007 Neuroscience Meeting Planner, San Diego, CA, Society for Neuroscience (online).

Friston, K. (2005) A theory of cortical responses. Philos. Trans. R. Soc. Lond., 360: 815–836.

Goebel, R., Khorram-Sefat, D., Muckli, L., Hacker, H. and Singer, W. (1998) The constructive nature of vision: direct evidence from functional magnetic resonance imaging studies of apparent motion and motion imagery. Eur. J. Neurosci., 10: 1563–1573.

Gold, J.I. and Shadlen, M.N. (2007) The neural basis of decision making. Annu. Rev. Neurosci., 30: 535–574.

Hamker, F.H. (2005) The reentry hypothesis: the putative interaction of the frontal eye field, ventrolateral prefrontal cortex, and areas V4, IT for attention and eye movement. Cereb. Cortex, 15: 431–447.

Hwang, E.J. and Shadmehr, R. (2005) Internal models of limb dynamics and the encoding of limb state. J. Neural Eng., 2: S266–S278.

Knill, D.C. and Pouget, A. (2004) The Bayesian brain: the role of uncertainty in neural coding and computation. Trends Neurosci., 27: 712–719.

Kosslyn, S.M., Thompson, W.L., Kim, I.J. and Alpert, N.M. (1995) Topographical representations of mental images in primary visual cortex. Nature, 378: 496–498.

Lee, T.S. and Mumford, D. (2003) Hierarchical Bayesian inference in the visual cortex. J. Opt. Soc. Am., 20: 1434–1448.

Lumer, E.D., Friston, K.J. and Rees, G. (1998) Neural correlates of perceptual rivalry in the human brain. Science, 280: 1930–1934.

Mast, F.W. and Kosslyn, S.M. (2002) Eye movements during visual mental imagery. Trends Cogn. Sci., 6: 271–272.

Miall, R.C. and Wolpert, D.M. (1996) Forward models for physiological motor control. Neural Netw., 9: 1265–1279.

Miller, E.K. and Cohen, J.D. (2001) An integrative theory of prefrontal cortex function. Annu. Rev. Neurosci., 24: 167–202.

Moore, T. and Armstrong, K.M. (2003) Selective gating of visual signals by microstimulation of frontal cortex. Nature, 421: 370–373.

Rao, R.P. and Ballard, D.H. (1999) Predictive coding in the visual cortex: a functional interpretation of some extra-classical receptive-field effects. Nat. Neurosci., 2: 79–87.

Schmolesky, M.T., Wang, Y., Hanes, D.P., Thompson, K.G., Leutgeb, S., Schall, J.D. and Leventhal, A.G. (1998) Signal timing across the macaque visual system. J. Neurophysiol., 79: 3272–3278.

Segraves, M.A. and Park, K. (1993) The relationship of monkey frontal eye field activity to saccade dynamics. J. Neurophysiol., 69: 1880–1889.

Sommer, M.A. and Wurtz, R.H. (2006) Influence of the thalamus on spatial visual processing in frontal cortex. Nature, 444: 374–377.

Sommer, M.A. and Wurtz, R.H. (2008) Brain circuits for the internal monitoring of movements. Annu. Rev. Neurosci., 31: 317–338.

Sporns, O., Honey, C.J. and Kotter, R. (2007) Identification and classification of hubs in brain networks. PLoS ONE, 2: p. e1049.

Umeno, M.M. and Goldberg, M.E. (1997) Spatial processing in the monkey frontal eye field. I. Predictive visual responses. J. Neurophysiol., 78: 1373–1383.

Ungerleider, L.G. and Mishkin, M. (1982) Two cortical visual systems. In: Ingle D.J., Goodale M.A. and Mansfield R.J.W. (Eds.), Analysis of Visual Behavior. The MIT Press, Cambridge, MA.

Womelsdorf, T., Schoffelen, J.M., Oostenveld, R., Singer, W., Desimone, R., Engel, A.K. and Fries, P. (2007) Modulation of neuronal interactions through neuronal synchronization. Science, 316: 1609–1612.

C. Kennard & R.J. Leigh (Eds.)
Progress in Brain Research, Vol. 171
ISSN 0079-6123

CHAPTER 5.3

Volition and eye movements

Parashkev Nachev*, Masud Husain and Christopher Kennard

Department of Clinical Neuroscience, Imperial College London, and Institute of Cognitive Neuroscience, UCL, London, UK

Abstract: Although the conceptual distinction between voluntary and automatic acts seems intuitively obvious, its neural basis remains opaque. Assigning volition — or some paraphrase such as action selection — to discrete parts of the brain arguably tells us nothing about what volition actually is in neural terms. Equally, exploring the relative sensitivity of discrete brain areas to manipulations of action choice, including its asymptote — free choice — would only be informative if voluntary processes could thereby be reliably isolated. Unfortunately, such manipulations are subject to ineliminable confounds, such as the complexity of the underlying condition–action associations. Here we propose an adaptation of a classic oculomotor task — saccadic choice with asynchronous targets — where the processes engaged in free choice manifest as interference in the performance of an automatic task, thereby circumventing the difficulties in parameterising volition. We suggest that this task may be useful in probing deficits in voluntary action in pathological states.

Keywords: volition; eye movements; saccadic choice; executive control; intention

Introduction

To arrive at a satisfactory account of the neural basis of voluntary action is arguably to have found the Holy Grail of movement neuroscience. If there is little doubt about the value of the aim, there is, or at least ought to be, a great deal of disagreement about the nature of the quest. Here we identify some of the major conceptual pitfalls and show how using the oculomotor system as our empirical test bed allows us to circumvent some of them.

Will, volition, and voluntariness

We have an intuitive understanding of what actions are voluntary and what actions are not; giving a formal definition, however, is much harder. The following points seem uncontroversial: for both neuroscientists and philosophers (Passingham, 1995; Bennett and Hacker, 2003).

An action cannot be voluntary unless it is contingent. If a subject never fails to perform an action whenever a set of circumstances obtains then we would not call that action voluntary but an automatic reflex of some sort.

Second, an action cannot be voluntary unless it can be said to have been *chosen*. Thus a sneeze is not a voluntary act because one could never choose to sneeze, only to simulate or withhold a sneeze.

The clarity and simplicity of these conceptual criteria make the empirical investigation of the distinctive capacities on which voluntary action depends seem straightforward. It is tempting to believe that the conceptual divisions are paralleled by analogous neural divisions, and that will,

*Corresponding author. Tel.: +44 208 846 7502;
Fax: +44 208 846 7715; E-mail: p.nachev@imperial.ac.uk

DOI: 10.1016/S0079-6123(08)00657-2

volition, or voluntariness — it does not seem to matter much what we call these capacities — are readily isolable empirically by the right kind of contrast. Here we examine the most influential of such contrasts.

Internally versus externally guided action

Our criteria for voluntary action lead to an obvious paradox. To the extent to which an action is independent of the circumstances that trigger it, it must be dependent on something else: either chance, or information — in the broadest possible sense of the term — from elsewhere. But what could that other source of information be? Unless we posit a homunculus, any additional information is merely part of the circumstances. All action would therefore seem to be either automatic or else dependent on a mysterious "ghost in the machine."

One solution seems to be to speak of the conditions for an action being *external* — referring to the outside world — or *internal* — referring to internal states such as expected rewards — and to identify the first with automatic and second with voluntary actions (Deiber et al., 1991; Frith et al., 1991; Jahanshahi et al., 1995; Jenkins et al., 2000). Actions may thus be considered more or less voluntary depending on the extent to which they are specified by the external world.

Although superficially attractive, this simple dichotomy is flawed for at least two reasons. First, however potently the external circumstances may specify an action such as a pro-saccade, one may still *choose* not to execute it. Conversely, a yawn seems as internal an action as any, and yet it can hardly be called voluntary. Furthermore, consider one's response to hearing someone say, "You need new conceptual glasses!": would the response be any less voluntary than if one did not hear the question from someone else but spontaneously said it to oneself, unprompted? Crucially, in every situation both internal and external conditions will inevitably impinge on any action: if they do so to differing degrees we would have to explain *who* or *what* determines their relative contribution.

Secondly, whereas the external conditions of an action are easy to manipulate — for example, by providing or withholding information about the location of a saccade — the internal conditions are largely opaque. It is not at all clear how we parameterise a choice between two identical targets where the subject is asked to choose "freely" between them (Nachev et al., 2005). The choice here clearly cannot be criterionless — the decision has to come from somewhere — moreover, being unconstrained, the number of variables that may be critical to the decision is likely to be much larger. Manipulations of freedom of choice are therefore also inevitably manipulations of conditional complexity: internally guided actions are thus unquantifiably more complex in their specification than externally guided ones. Our dichotomy is thus riven by an ineliminable confound making empirical studies based on it difficult to interpret.

Executive control

If voluntary actions may be said to be *chosen*, then there must be circumstances under which an action that might ordinarily be performed for some reason is not; otherwise one could not speak of choice. This has lead to the notion that there may be a distinct process of *inhibiting* prepotent actions, coordinated by some kind of *executive controller* that polices simpler, lower-order condition–action associations (Norman and Shallice, 1980; Logan, 1985; Cohen et al., 1990; Schall et al., 2002).

Indeed, there appears to be broad agreement in the literature that making a response and withholding it are fundamentally asymmetrical (Logan et al., 1984; Schall et al., 2002). When we speak of behaviour and of action we nearly always have something dynamic in mind: a subject reaches towards something, says something, looks in a particular direction, etc. By contrast, not doing something seems to be neither an action nor an entity that can be diversified in any way; not doing something is simply not doing. It is therefore easy to give in to the notion that the inhibition of an action must be a unitary process distinct from the action itself. So conceived, response inhibition may be attributed to a discrete system functionally, and thereby anatomically.

The difficulty here is that the grounds for giving a higher position to response inhibition — at least when identified with withholding a movement — are invalid conceptually and unsubstantiated empirically. There is no reason to think that the choice between making an action and not making it is any different than the choice between two actions. In both situations we have some kind of response under some kind of conditions: that we should identify action with movement and not with the maintenance of a static posture is a matter of convention, nothing more. Empirically, in any particular circumstance, one finds neurones sensitive to the absence of action just as one finds neurones sensitive to an action (Kurata and Tanji, 1985). Where action and inaction are placed in competition — as during saccadic countermanding — the outcome simply depends on the balance of activity between gaze-holding and gaze-shifting neurones: no mysterious executive controller is necessary (Hanes et al., 1998).

Intention

One might be tempted to circumvent these difficulties by a simple expedient: asking the subject! If voluntary acts may be said to be chosen, then the choice has to happen at some point in time *before* the action is executed, or so the argument goes. A convenient time for this might be the time when the subject formed the *intention* to act, as revealed *post hoc* by the subject's report (Lau et al., 2004; Brass and Haggard, 2007). The logic here is derived from William James's flawed ideomotor theory of action (James, 1890) (via Libet et al., 1983) which suggests that voluntary action is defined by being preceded by an intention to act so. Having an intention is, of course, neither necessary nor sufficient for an action to occur (Bennett and Hacker, 2003); indeed to conceive of voluntary action as a *causally-related* consequence of intentions is to make it automatic, like sneezing: something that can only be suppressed but never chosen. It would also mean that anyone who had not grasped the concept of intending — for example because he never acquired a language — may not be said to act voluntarily, which is clearly absurd. In any case, introspecting about one's intentions is no more likely to identify an "intention centre" than introspecting about infidelity is likely to identify an "infidelity centre," or about anger an "anger centre," and so on.

A minimalist model

Instead of attempting to impose psychological concepts to the brain, it may be more profitable to

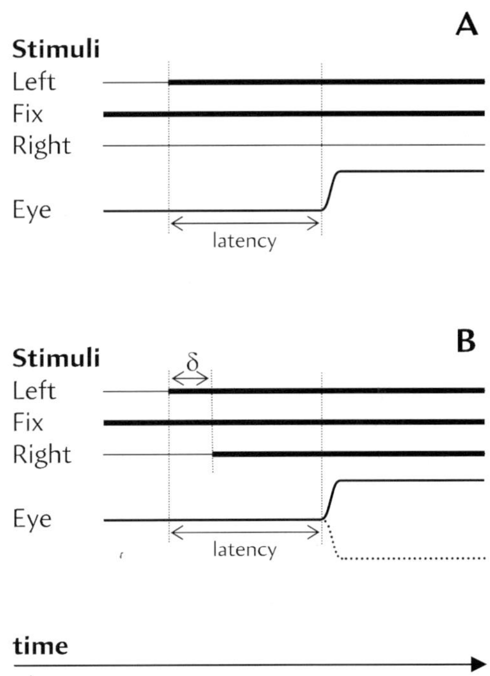

Fig. 1. Diagrammatic sketch of the asynchronous target saccadic choice task. Subjects view a horizontally arranged array of targets with the central target serving as the fixation point. Two types of trial are presented in randomised order. On single target trials (A), the subject fixates the central target and responds to the sudden illumination of either peripheral target with a foveating saccade. On double target trials (B), both peripheral targets are illuminated, but with a slight asynchrony (δ) between them, and the subject is instructed to foveate as rapidly as possible whichever target catches his attention first. The relation between the probability of choosing the first target and the target asynchrony is a monotonically increasing function (Leach and Carpenter, 2001).

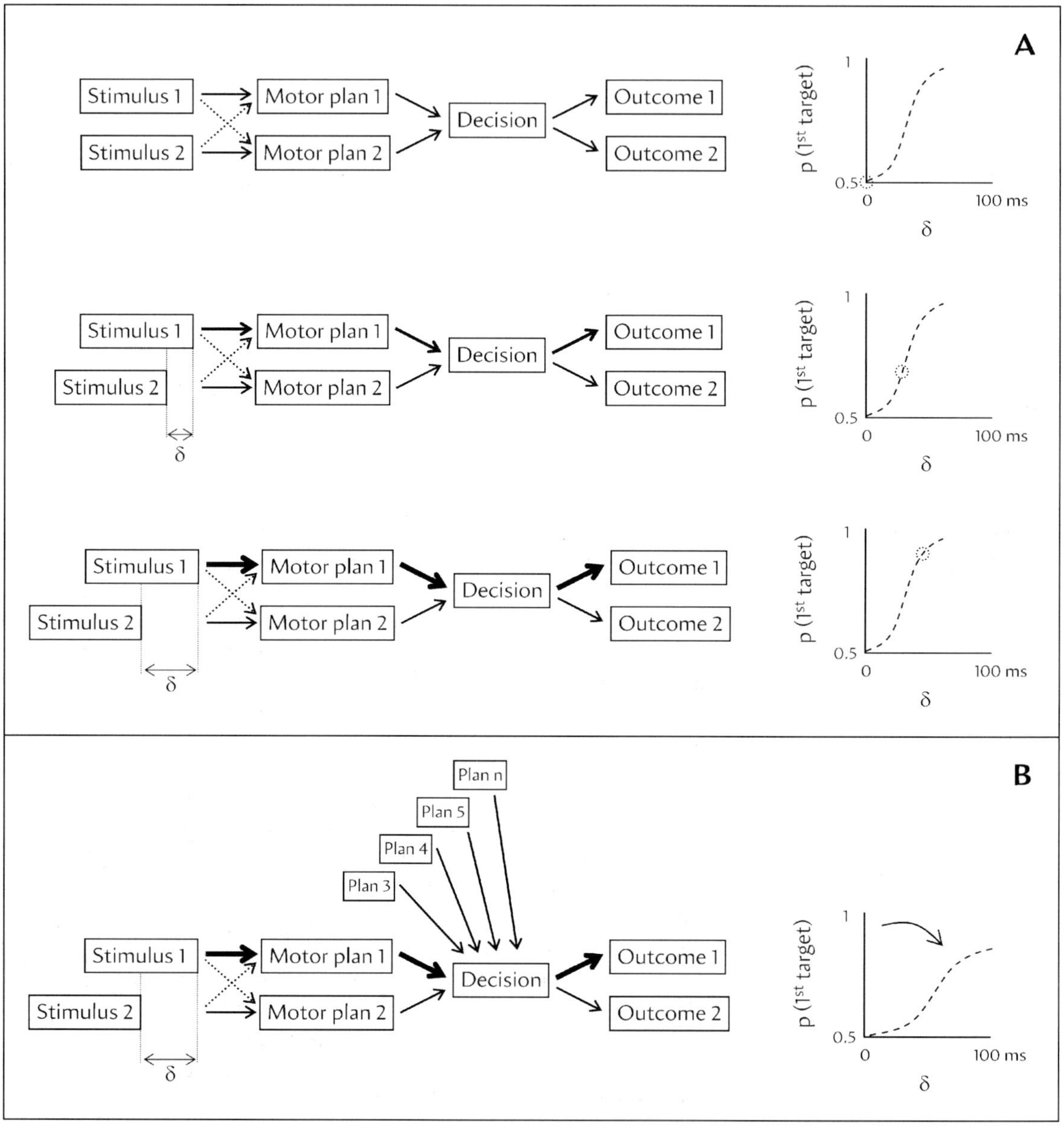

Fig. 2. (A) Simple model of saccadic choice with asynchronous targets. Here the action outcome is simply the result of competition between neuronal pools encoding the two different stimulus–response associations. Where one stimulus is given priority by occurring earlier by an interval δ, the associated motor plan will be more likely to be executed, leading to a bias towards executing the action associated with that stimulus. The size of the bias, expressed as the probability of choosing the first target (p(1st target)) is a monotonic function of the asynchrony between the two targets, δ. (B) A more realistic model. The two stimulus–response associations will not be the only condition–action associations competing for the outcome. A wide range of other, unquantified, probably more complex, associations will impinge on the decision process, tending to flatten the asynchrony function. It is this interference — and its attenuation where the range of alternative condition–action associations is limited by pathological changes — that is the measure of interest here.

examine what *neural* model based on known properties of neurones might allow psychological concepts such as voluntary action to be applied to its owner. Here we can be surprisingly minimalist. At any particular time, the subject's behavioural state — any combination of actions and inactions — may be specified by the outcome of competition between disparate condition–action associations linking different possible actions or inactions to different states of affairs in the external and internal world. Such condition–action associations, although of unspecified complexity, are merely transformations of a set of inputs into a set of outputs: a task for which neural networks appear to be optimised. Decision-making or choice here may be simply explained by the outcome of fair competition between different condition–action associations, all competing for some decision threshold (Carpenter, 1981).

Probing volition

How do we test such a model, if its essential feature is that the aspects giving rise to the kind of behaviour we tend to call voluntary are conditional terms of *unspecified* complexity? Here we tentatively propose a simple solution, based on a familiar oculomotor task: saccadic choice between asynchronous targets (Schiller and Chou, 2000; Leach and Carpenter, 2001).

In this task, subjects are asked on each trial to make rapidly a single saccade directed towards the sudden onset of an eccentrically placed target (Fig. 1). Trials on which there is only one target are randomly interspersed with trials on which there are two identical targets, with the second target displayed at the same eccentricity but the opposite side of the first. On two target trials, there is a small, variable asynchrony between the two peripheral onsets. Subjects are instructed to make a single saccade to whichever peripheral target catches their attention first, but not to choose the target on the basis of the *perceived* difference in onset. There are two principal measures of interest: the relation between the size of the asynchrony and the probability of choosing the first target (the asynchrony–choice function), and the difference in latency between single and double target trials (choice cost).

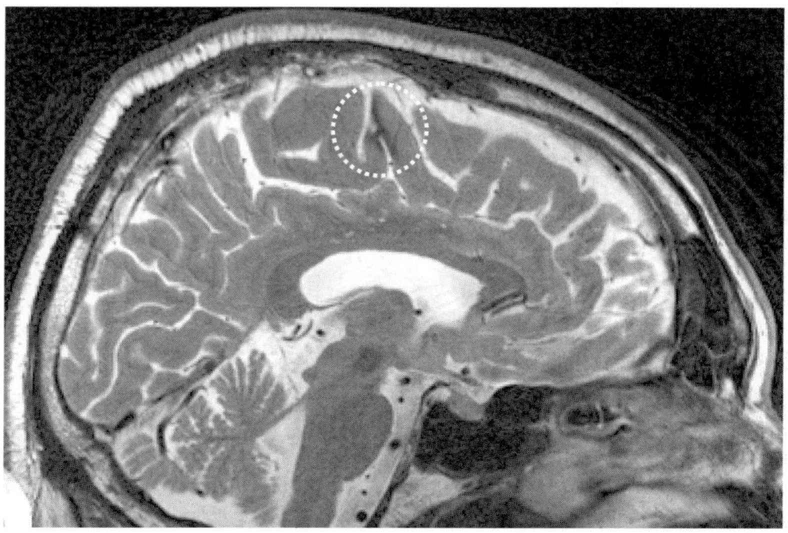

Fig. 3. Mesial frontal lesion of Patient JR. Sagittal T2-weighted scan showing an area of venous infarction on the mesial wall. The damaged area has previously been shown to correspond closely to the left supplementary eye field, as predicted by the location of eye-movement related mesial frontal activity in the unaffected contralateral hemisphere. See Parton et al. (2007) and Sumner et al. (2007) for details.

Critically, when subjects perform this task their choice of saccade is heavily biased by asynchronies as low as 40–50 ms, well below the perceptual threshold for identifying the order of target onset. Such choice as there is here therefore appears wholly automatic, seemingly determined purely by the temporal priority of the stimulus, according to the kind of conceptual framework shown in Fig. 2A.

Naturally, the asynchrony cannot be the only factor that impinges on performance here. Since the subject has no awareness of the choice involved, one may expect various more complex condition–action associations to be automatically activated in this situation, just as they would be if this were a free choice task (Fig. 2B). Such co-activation will thus tend to interfere with the asynchrony bias, raising the asynchrony threshold, flattening the asynchrony function, and increasing the choice cost. A pathological process that disrupts such more complex condition–action associations — but leaves those involved in making reflexive saccades intact — may therefore be expected to result in *less* interference on this task compared with matched controls. Patients so afflicted may thus be expected to have a lower threshold, a steeper function, and a reduced choice cost.

Since the effects we are interested in here arise as interference with the performance of a simple automatic task, it does not matter that we do not know what particular condition–action associations are disrupted in any one case: *all* associations but those describing the asynchrony response will tend, if anything, to degrade performance.

An empirical illustration of this idea is provided by a study of a single patient, JR who has a micro-lesion that falls within the region of the left supplementary eye field (Fig. 3) (Husain et al., 2003; Parton et al., 2007; Sumner et al., 2007). Although asymptomatic, and capable of essentially normal performance on a range of oculomotor tasks including reflexive saccades, centrally guided saccades, and blocked antisaccades, this patient has been shown to have deficits when switching between different condition–action associations. We have therefore hypothesised that the SEF lesion has affected his capacity to perform actions based on complex condition–action associations while preserving simple one such as those describing pro- or antisaccades. In agreement with the model we have proposed here, this patient showed a remarkably steep asynchrony function — without any significant training — in the context of relatively high saccadic latencies (304 ms), suggesting that these results cannot be explained by general expertise in saccadic tasks (Fig. 4). Rather, without testing a specific rule, they suggest an impairment in the ability to perform more complex condition–action associations, such as following a rule to interpret a visual signal differently in the context of

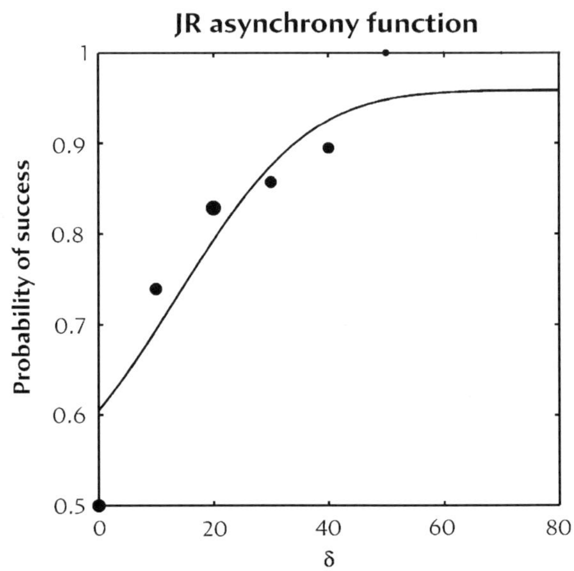

Fig. 4. Asynchrony–choice function for Patient JR. The subject performed a version of the task shown in Fig. 1, with the targets being luminance onsets (squares subtending 0.5°) displayed at 10° eccentricity on a CRT screen refreshed at 100 Hz. Eye movements were monitored at 240 Hz by a remote infra-red, video-based eye-tracker (ASL). The subject performed 320 trials, during which the asynchrony was modulated (in 10 ms steps) by a Bayesian adaptive algorithm designed to maximise the information obtained about the threshold and slope of the underlying psychometric function (Kontsevich and Tyler, 1999). The final best estimate of the function is plotted as a solid line. The points show the raw data for each of the asynchronies used by the algorithm, with the area of the points indexing the number of trials at that asynchrony. Note the surprisingly low threshold and steep slope.

another: the essence of mixed pro- and antisaccades, the change paradigm, and saccadic countermanding, amongst others. These findings would seem to justify exploring the utility of this task in patient groups characterised by impairments in voluntary actions such as patients with Huntington's Disease.

Conclusion

The neural mechanisms subserving voluntary action are rather harder to isolate — both conceptually and empirically — than may at first seem. Conventional approaches such as the exploration of free choice and situations said to require executive control are arguably either conceptually incoherent or admit of too many alternative explanations to be useful empirically. Guided by a simple theoretical framework — describing action in terms of the conditional complexity of the rule that may be said to describe it at any one point — we have suggested that the established paradigm of asynchronous saccadic choice may be used to probe the mechanisms critical to voluntary action. Preliminary data from a patient with a highly localised lesion of the SEF indicate the potential value of the paradigm in studying deficits in voluntary action.

Acknowledgement

This work is funded by the Wellcome Trust.

References

Bennett, M.R. and Hacker, P.M.S. (2003) Philosophical Foundations of Neuroscience. Blackwell Publishing, Oxford.

Brass, M. and Haggard, P. (2007) To do or not to do: the neural signature of self-control. J. Neurosci., 27: p. 9141.

Carpenter, R.H.S. (1981) Oculomotor procrastination. In: Fisher D.F., Monty R.A. and Senders J.W. (Eds.), Eye Movements: Cognition and Visual Perception. Lawrence Erlbaum, Hillsdale, NJ, pp. 237–246.

Cohen, J.D., Dunbar, K. and McClelland, J.L. (1990) On the control of automatic processes: a parallel distributed processing account of the Stroop effect. Psychol. Rev., 97: 332–361.

Deiber, M.P., Passingham, R.E., Colebatch, J.G., Friston, K.J., Nixon, P.D. and Frackowiak, R.S. (1991) Cortical areas and the selection of movement: a study with positron emission tomography. Exp. Brain Res., 84: 393–402.

Frith, C.D., Friston, K., Liddle, P.F. and Frackowiak, R.S. (1991) Willed action and the prefrontal cortex in man: a study with PET. Proc. R. Soc. Lond. B Biol. Sci., 244: 241–246.

Hanes, D.P., Patterson, W.F., II and Schall, J.D. (1998) Role of frontal eye fields in countermanding saccades: visual, movement, and fixation activity. J. Neurophysiol., 79: 817–834.

Husain, M., Parton, A., Hodgson, T.L., Mort, D. and Rees, G. (2003) Self-control during response conflict by human supplementary eye field. Nat. Neurosci., 6: 117–118.

Jahanshahi, M., Jenkins, I.H., Brown, R.G., Marsden, C.D., Passingham, R.E. and Brooks, D.J. (1995) Self-initiated versus externally triggered movements. I. An investigation using measurement of regional cerebral blood flow with PET and movement-related potentials in normal and Parkinson's disease subjects. Brain, 118(Pt 4): 913–933.

James, W. (1890) The Principles of Psychology. Henry Holt and Co., NJ.

Jenkins, I.H., Jahanshahi, M., Jueptner, M., Passingham, R.E. and Brooks, D.J. (2000) Self-initiated versus externally triggered movements. II. The effect of movement predictability on regional cerebral blood flow. Brain, 123(Pt 6): 1216–1228.

Kontsevich, L.L. and Tyler, C.W. (1999) Bayesian adaptive estimation of psychometric slope and threshold. Vision Res., 39: 2729–2737.

Kurata, K. and Tanji, J. (1985) Contrasting neuronal activity in supplementary and precentral motor cortex of monkeys. II. Responses to movement triggering vs. nontriggering sensory signals. J. Neurophysiol., 53: 142–152.

Lau, H.C., Rogers, R.D., Haggard, P. and Passingham, R.E. (2004) Attention to intention. Science, 303: 1208–1210.

Leach, J.C. and Carpenter, R.H. (2001) Saccadic choice with asynchronous targets: evidence for independent randomisation. Vision Res., 41: 3437–3445.

Libet, B., Gleason, C.A., Wright, E.W. and Pearl, D.K. (1983) Time of conscious intention to act in relation to onset of cerebral activity (readiness-potential). The unconscious initiation of a freely voluntary act. Brain, 106(Pt 3): 623–642.

Logan, G.D. (1985) Executive control of thought and action. Acta Psychol., 60: 193–210.

Logan, G.D., Cowan, W.B. and Davis, K.A. (1984) On the ability to inhibit simple and choice reaction time responses: a model and a method. J. Exp. Psychol. Hum. Percept Perform., 10: 276–291.

Nachev, P., Rees, G., Parton, A., Kennard, C. and Husain, M. (2005) Volition and conflict in human medial frontal cortex. Curr. Biol., 15: 122–128.

Norman, D.A. and Shallice, T. (1980) Attention to Action: Willed and Automatic Control of Behavior. Center for Human Information Processing, University of California, San Diego, CA.

Parton, A., Nachev, P., Hodgson, T.L., Mort, D., Thomas, D., Ordidge, R., Morgan, P.S., Jackson, S., Rees, G. and Husain, M. (2007) Role of the human supplementary eye field in the control of saccadic eye movements. Neuropsychologia, 45: 997–1008.

Passingham, R.E. (1995) The frontal lobes and voluntary action. Oxford University Press, Oxford.
Schall, J.D., Stuphorn, V. and Brown, J.W. (2002) Monitoring and control of action by the frontal lobes. Neuron, 36: 309–322.
Schiller, P.H. and Chou, I. (2000) The effects of anterior arcuate and dorsomedial frontal cortex lesions on visually guided eye movements: 2. Paired and multiple targets. Vision Res., 40: 1627–1638.
Sumner, P., Nachev, P., Morris, P., Peters, A.M., Jackson, S.R., Kennard, C. and Husain, M. (2007) Human medial frontal cortex mediates unconscious inhibition of voluntary action. Neuron, 54: 697–711.

C. Kennard & R.J. Leigh (Eds.)
Progress in Brain Research, Vol. 171
ISSN 0079-6123

CHAPTER 5.4

Negative motivational control of saccadic eye movement by the lateral habenula

Masayuki Matsumoto and Okihide Hikosaka*

Laboratory of Sensorimotor Research, National Eye Institute, National Institutes of Health, Bethesda, MD, USA

Abstract: Reward is crucial for survival of animals and influences animal behaviours. For example, an approaching behaviour to reward is more frequently and quickly elicited when a big reward is expected than when a small reward is expected. Midbrain dopamine neurons are thought to be crucial for such reward-based control of motor behaviour. Indeed, dopamine neurons are excited by cues predicting reward and inhibited by cues predicting no-reward. These excitatory and inhibitory signals would then be used for enhancing and depressing sensorimotor processing, respectively, in the brain areas targeted by dopamine neurons (e.g., striatum). However, it was unknown which parts of the brain provide dopamine neurons with reward-related signals necessary for their responses. We recently showed evidence that the lateral habenula transmits reward-related signals to dopamine neurons, especially to inhibit dopamine neurons. This recent study suggested that the lateral habenula suppresses less rewarding saccadic eye movements by inhibiting dopamine neurons. In the present review, we first summarize anatomical and functional aspects of the lateral habenula. We will then describe our own study. Finally, we will discuss how the lateral habenula, as well as dopamine neurons, contributes to the reward-based control of saccadic eye movements.

Keywords: reward; lateral habenula; dopamine neuron; saccade; monkey

What is the lateral habenula?

The lateral habenula, part of the structure called the epithalamus, is a good candidate for a source of reward-related signals in dopamine neurons. It is considered to carry information from the forebrain limbic system down to midbrain structures (Fig. 1A). The origins of its afferents are the medial prefrontal cortex, basal forebrain regions, lateral hypothalamus, and globus pallidus internal segment (Herkenham and Nauta, 1977; Lecourtier and Kelly, 2007). Its main targets are the ventral tegmental area and substantia nigra pars compacta containing dopamine neurons, and raphe nuclei containing serotonin neurons (Herkenham and Nauta, 1979; Lecourtier and Kelly, 2007). Thus, the lateral habenula is in a good position to regulate the brain's monoaminergic (dopaminergic and serotonergic) systems. Indeed, electrical stimulation of the lateral habenula inhibits dopamine neurons (Christoph et al., 1986) and serotonin neurons (Wang and Aghajanian, 1977).

Consistent with the view from the anatomy, the lateral habenula has been implicated in many emotional and cognitive functions. These include anxiety, stress, pain, learning, attention, and reward processing (Sutherland, 1982; Lecourtier and Kelly, 2007). However, the physiological properties of the lateral

*Corresponding author. Tel.: +1 (301) 402 7959;
Fax: +1 (301) 402 0511; E-mail: oh@lsr.nei.nih.gov

DOI: 10.1016/S0079-6123(08)00658-4

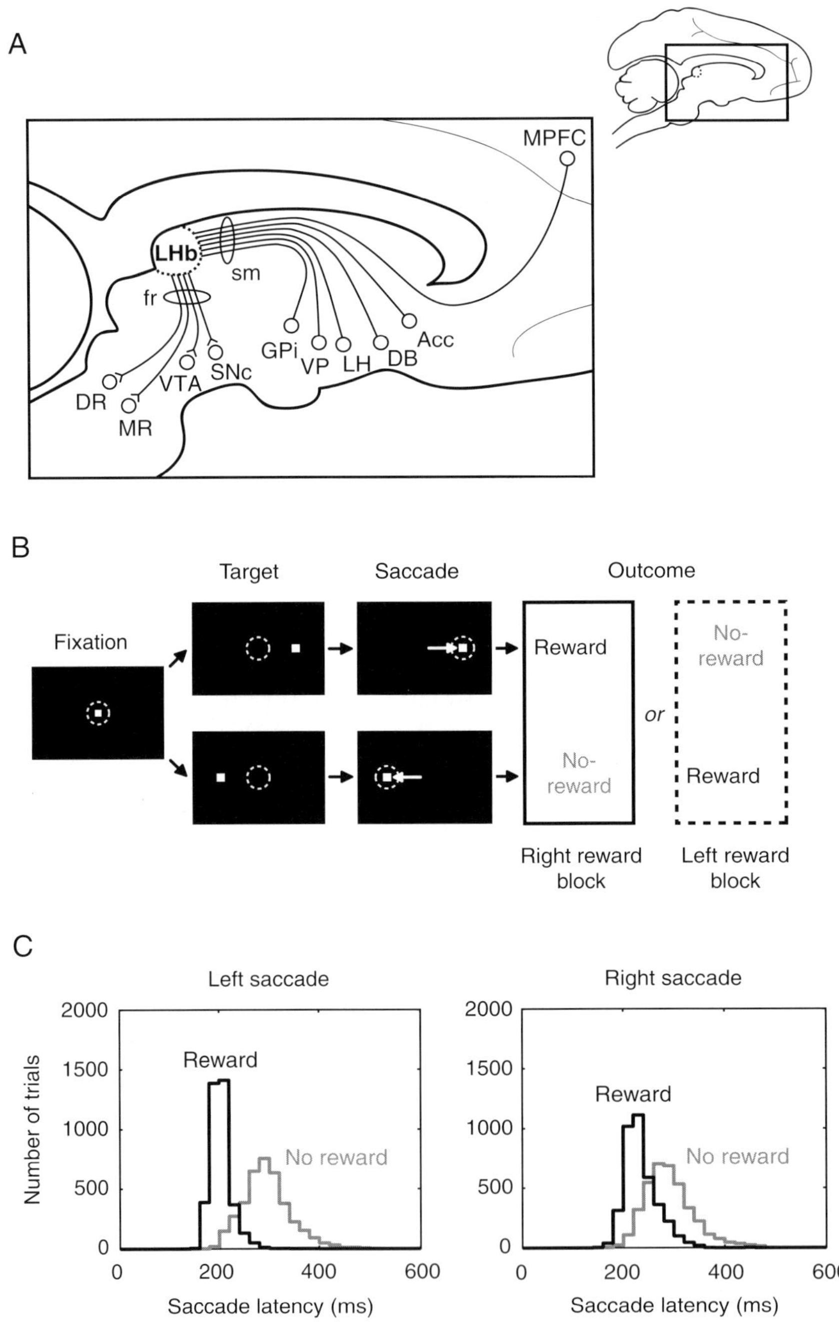

Fig. 1. (A) Scheme showing the location of the lateral habenula and its afferent and efferent connections in a parasagittal section of the macaque brain. Acc, nucleus accumbens; DB, diagonal band of Broca; DR, dorsal raphe; fr, fasciculus retroflexus; GPi, globus pallidus internal segment; LH, lateral hypothalamus; LHb, lateral habenula; MPFC, medial prefrontal cortex; MR, medial raphe; sm, stria medullaris; SNc, substantia nigra pars compacta; VP, ventral pallidum; VTA, ventral tegmental area. (B) Sequence of events in the visually guided saccade task. (C) Distribution of saccade latencies in reward trials (black) and in no-reward trials (grey) in one example monkey. Modified from Nature (Matsumoto and Hikosaka, 2007).

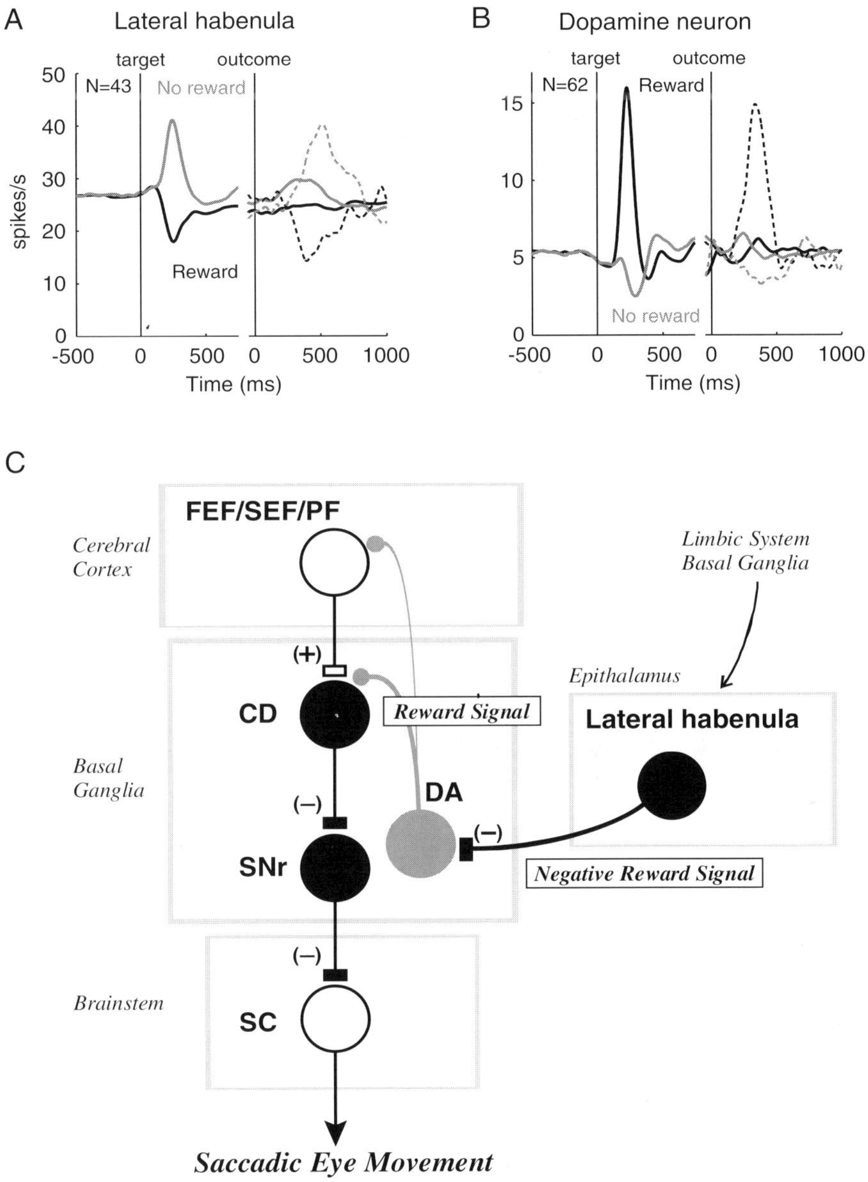

Fig. 2. (A and B) Averaged activity of lateral habenula neurons (A, $n=43$) and dopamine neurons (B, $n=62$) during the visually guided saccade task with positional reward bias. Spike density functions (SDFs) are aligned at the onset of target (left) and at the onset of outcome (right). The SDFs are shown for reward trials (black) and no-reward trials (grey). Continuous curves indicate activity in trials excluding the first trials after the reversal of rewarded direction. Dotted curves indicate activity in the first trials. The data from ipsilateral and contralateral saccades are combined. (C) A hypothetical scheme showing the role of the lateral habenula in reward-based control of saccadic eye movements. CD, caudate nucleus; DA, dopamine neurons; FEF, frontal eye field; PF, prefrontal cortex; SC, superior colliculus; SEF, supplementary eye field; SNr, substantia nigra pars reticulata. Excitatory and inhibitory connections are indicated by (+) and (–), respectively. It is unknown whether lateral habenula neurons themselves are inhibitory. Modified from Nature (Matsumoto and Hikosaka, 2007).

habenula remained poorly understood. In the next section, we summarize our recent study which provided physiological data suggesting that the lateral habenula transmits reward-related signals to dopamine neurons (Matsumoto and Hikosaka, 2007).

Lateral habenula as a source of negative reward signals in dopamine neurons

To test the hypothesis that the lateral habenula transmits reward-related signals to dopamine neurons, we compared the activity of lateral habenula neurons and dopamine neurons while monkeys were performing a visually guided saccade task with positional reward bias (Fig. 1B). The target was presented randomly on the right or left and the monkey had to make a saccade to it immediately. In one block of trials (24 trials) saccades to one fixed direction were rewarded while saccades to the other direction were not rewarded. The rewarded direction was reversed in the next block with no external instruction. Saccade latencies were much shorter when saccades were followed by reward than when they were followed by no-reward (Fig. 1C).

We found that a majority of lateral habenula neurons were excited by targets predicting no-reward and inhibited by targets predicting reward (Fig. 2A). The excitation and inhibition depended on the reward contingency (reward or no-reward), regardless of target position. In contrast, dopamine neurons were excited by targets predicting reward and inhibited by targets predicting no-reward (Fig. 2B). In addition, lateral habenula neurons and dopamine neurons also responded to the delivery and omission of reward in the opposite manner, especially at the first trials after the reversal of rewarded direction (dotted lines in Fig. 2A, B).

In no-reward trials, the excitatory response of lateral habenula neurons started earlier than the inhibitory response of dopamine neurons. In reward trials, however, the inhibitory response of lateral habenula neurons started later than the excitatory response of dopamine neurons. Furthermore, electrical stimulation of the lateral habenula inhibited dopamine neurons. These results suggest that the lateral habenula transmits negative reward signals to dopamine neurons, especially to inhibit dopamine neurons.

Discussion

Recent studies from our laboratory have suggested that the reward-modulated activity of dopamine neurons plays a key role in the reward-based control of saccadic eye movements (Hikosaka et al., 2006). These studies proposed that the efficacy of cortico-caudate synapses carrying visuo-saccadic signals is enhanced or depressed depending on the concurrent increase or decrease, respectively, in dopaminergic inputs. The activity of the superior colliculus would then be enhanced or depressed through the caudate nucleus and substantia nigra pars reticulata (Fig. 2C). Indeed, neurons in the superior colliculus are more excited by saccadic targets predicting reward than targets predicting no-reward (Ikeda and Hikosaka, 2003). By transmitting negative reward signals to dopamine neurons, the lateral habenula would suppress less rewarding saccadic eye movements.

References

Christoph, G.R., Leonzio, R.J. and Wilcox, K.S. (1986) Stimulation of the lateral habenula inhibits dopamine-containing neurons in the substantia nigra and ventral tegmental area of the rat. J. Neurosci., 6: 613–619.

Herkenham, M. and Nauta, W.J. (1977) Afferent connections of the habenular nuclei in the rat. A horseradish peroxidase study, with a note on the fiber-of-passage problem. J. Comp. Neurol., 173: 123–146.

Herkenham, M. and Nauta, W.J. (1979) Efferent connections of the habenular nuclei in the rat. J. Comp. Neurol., 187: 19–47.

Hikosaka, O., Nakamura, K. and Nakahara, H. (2006) Basal ganglia orient eyes to reward. J. Neurophysiol., 95: 567–584.

Ikeda, T. and Hikosaka, O. (2003) Reward-dependent gain and bias of visual responses in primate superior colliculus. Neuron, 39: 693–700.

Lecourtier, L. and Kelly, P.H. (2007) A conductor hidden in the orchestra? Role of the habenular complex in monoamine transmission and cognition. Neurosci. Biobehav. Rev., 31: 658–672.

Matsumoto, M. and Hikosaka, O. (2007) Lateral habenula as a source of negative reward signals in dopamine neurons. Nature, 447: 1111–1115.

Sutherland, R.J. (1982) The dorsal diencephalic conduction system: a review of the anatomy and functions of the habenular complex. Neurosci. Biobehav. Rev., 6: 1–13.

Wang, R.Y. and Aghajanian, G.K. (1977) Physiological evidence for habenula as major link between forebrain and midbrain raphe. Science, 197: 89–91.

C. Kennard & R.J. Leigh (Eds.)
Progress in Brain Research, Vol. 171
ISSN 0079-6123

CHAPTER 5.5

Eye movements as a probe of attention

Albert Hoang Duc, Paul Bays and Masud Husain*

Institute of Neurology & Institute of Cognitive Neuroscience, UCL, London, UK

Abstract: Most studies of visual attention have used indirect manual reaction time measures to make inferences about where processing priority is allocated in the scene. Eye tracking offers an important opportunity to probe more directly where attention is deployed, in both health and disease. Here we provide a selective overview of oculomotor investigations which have demonstrated how attention can be captured "bottom-up" by conspicuous elements in the scene as well as how it can be allocated "top-down" on the basis of task goals and reward outcomes. We consider the concept of *salience maps* within the brain that might be responsible for computing attentional priorities and saccade targets in the visual scene, discussing the evidence for their existence. Finally, we consider how one supposedly crucial role of attention — that of binding features which belong to an object — might operate so that object representations are veridically maintained and remapped across eye movements.

Keywords: visual salience; oculomotor capture; unilateral neglect; object files

Introduction

Many of the pioneering studies in attention research have focused on visual attention. Almost invariably, early studies used manual reaction time as an index of where attention is deployed. But such measures are quite indirect. Thus, Posner's highly influential spatial cuing paradigm makes inferences which depend crucially upon differences in reaction time in response to a stimulus that appears at either a previously cued or uncued location (Posner, 1980). Tracking eye movements has the potential to provide a more direct measure of where attention is deployed since the direction of gaze is generally considered to be tightly coupled to the orienting of attention — at least, under normal circumstances (Hoffman and Subramaniam, 1995; Deubel and Schneider, 1996). Uncoupling of gaze direction and attention can, of course, occur as Posner's task clearly demonstrates. The value of eye tracking is that in natural scene viewing — where the visual environment is complex compared to many simple experimental situations — it should provide a good guide to the locus of attention. In recent years, researchers have capitalized on this possibility, seeking eventually to understand how attention and gaze are deployed to make sense of the visual world.

Oculomotor capture, attention, and reward

Visual scenes typically contain many objects that compete for the control of attention and eye movements. When an object is intentionally chosen to be the focus of attention, selection is said to occur in a voluntary, goal-directed manner. However, when

*Corresponding author. Tel.: +44(0)20 7679 1177;
Fax: +44(0)2079168517; E-mail: m.husain@ion.ucl.ac.uk

DOI: 10.1016/S0079-6123(08)00659-6

specific properties in the visual environment determine what is selected, regardless of the observer's goal or intention, selection is said to occur in an involuntary or stimulus-driven manner. Salient objects that appear suddenly in the visual field, for example, often capture attention even if they are irrelevant for the task at hand (Yantis and Jonides, 1984; Remington and Johnston, 1992); similarly for the oculomotor system.

Theeuwes and his colleagues used a paradigm (Fig. 1) in which observers viewed a display containing six grey circles presented on an imaginary circle around a central fixation point (Theeuwes et al., 1998; Theeuwes and Irwin, 1999). Subsequently all but one of the circles changed to red and participants were explicitly instructed to saccade towards the only grey stimulus remaining. On half the trials, an irrelevant red distractor circle was presented with abrupt onset which observers were told to ignore. When there were no such onsets, participants generated saccades that went directly towards the target. However, on distractor

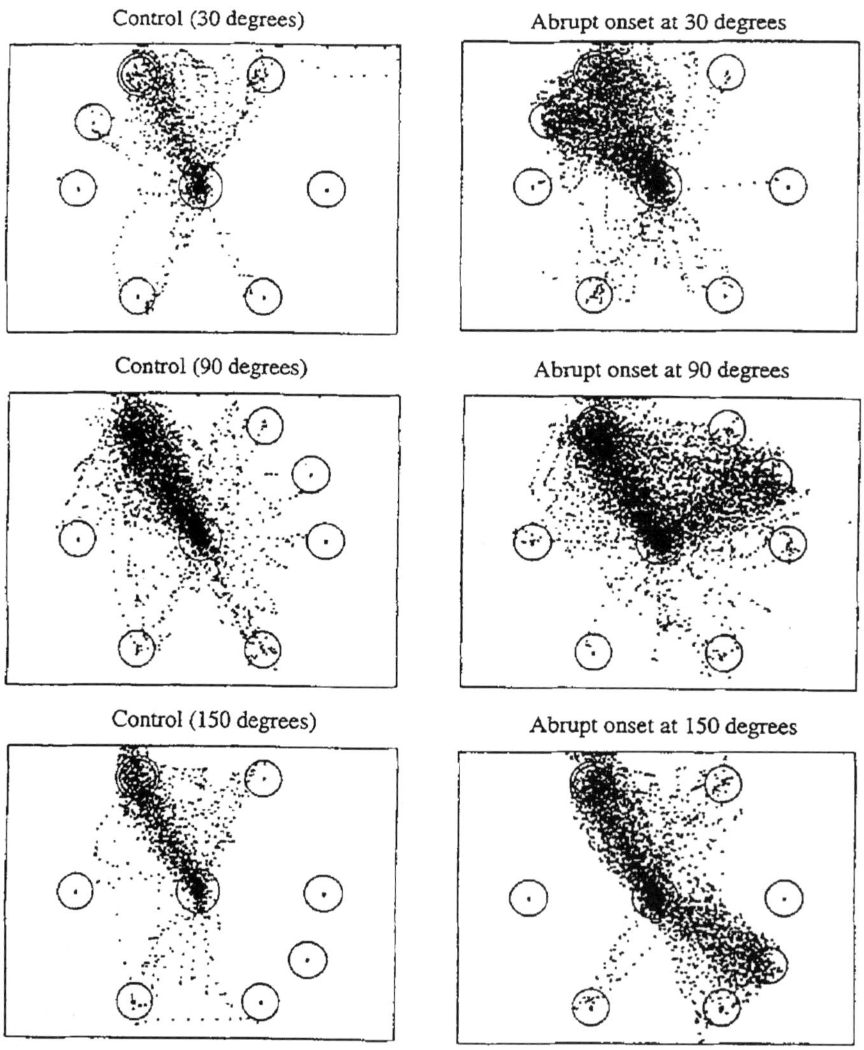

Fig. 1. Oculomotor capture. When an abrupt-onset distractor was presented (right panels), the eyes often went to it first, instead of the target. Left panels depict the control condition without an abrupt-onset distractor. (Adapted from Theeuwes et al., 1998.)

trials, the eyes went in the direction of the irrelevant abrupt-onset stimulus on about 30–40% of these trials, stopped briefly, before proceeding to the target: in other words, gaze was frequently captured by the distractor. When asked, none of the observers was aware that their eye movement behaviour was influenced by the abrupt onset, even though in many cases the eyes went in a direction completely opposite to the target. The presence of such oculomotor capture demonstrates that top-down, or goal-directed control is not always able to override bottom-up capture by abrupt-onset stimuli in the visual scene.

More recently, Milstein and Dorris have used oculomotor capture as a novel behavioural index to investigate the influence of expected value on saccadic control, along with the more conventional measure of saccadic reaction time (Milstein and Dorris, 2007). These authors wanted to investigate whether saccadic preparation toward a particular target was influenced by its expected reward value, defined as the product of reward probability and reward magnitude associated with making a saccade to that particular location. Participants were instructed to fixate centrally and then make a saccade to a red target, presented either to the left or right after a 400 ms warning period. The magnitude of the reward subjects would obtain depended on target location. The influence of expected reward value on saccadic preparation was probed occasionally (30% of trials) by flashing a green distractor during the warning period. Distractors sometimes erroneously triggered saccades to them instead of the target.

The question here was whether the frequency of such oculomotor captures was modulated by the location occupied by the distractor. Would more erroneous saccades be made to a distractor occupying a location associated with a high target value? The investigators found that the greater the reward associated with a target location, the shorter the reaction time for correct saccades. Importantly, distractors presented towards the side of higher-valued targets induced more oculomotor captures than distractors presented toward the side of lower-valued targets. Moreover, the higher the expected value of a target, the higher the proportion of oculomotor captures to distractors flashed nearby those targets. Attention to a spatial location therefore appears to be modulated by the reward associated with that position, as well as its salience.

Salience or priority maps in the brain

What about the allocation of attention in more real-world scenes? In recent years, the concept of a *salience map* has been put forward to account for how visual items are selected for attention (Itti et al., 1998; Itti and Koch, 2001). In such schemes (Koch and Ullman, 1985), visual input is first broken down into a set of topographic maps coding image dimensions such as colour, intensity, or contrast (Fig. 2). Different spatial locations then compete for saliency within each dimension map, such that only locations which locally stand out from their surroundings can persist. The maps generated for each image dimension are then combined to create a single master "saliency map," which codes for local conspicuity over the entire visual scene. Competition among locations in this map gives rise to a single winning location that corresponds to the next saccade target. This location is then suppressed, and a new winner-takes-all process selects the following target, and so on. Thus, the salience map approach can provide an explicit model that generates precise quantitative predictions about the spatial distribution of fixation locations and their sequence.

Parkhurst and his co-workers have examined the degree to which stimulus salience guides the allocation of attention in natural scene viewing (Parkhurst et al., 2002). Human eye movements were recorded while participants viewed a series of images of complex natural and artificial scenes. The correlation between stimulus salience (computed on a similar basis to Fig. 2) and fixation location was found to be significantly greater than that expected by chance alone. Moreover, attention was significantly more stimulus-driven just after stimulus onset when top-down influences are presumably weakest. The results indicate that stimulus-driven mechanisms contribute significantly to attentional guidance under natural

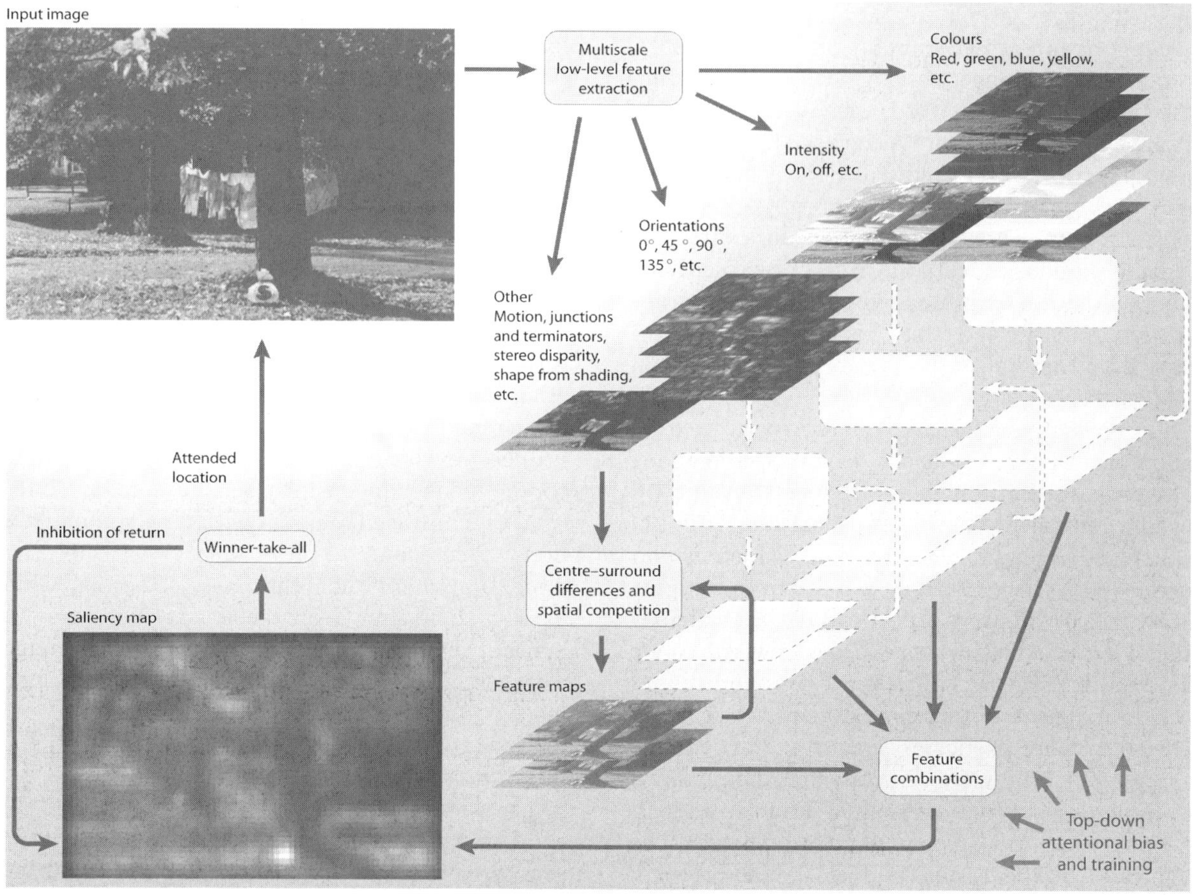

Fig. 2. Modelling the bottom-up control of attention. According to Koch and Ullman (1985), shifts of attention can be predicted by a master "salience map." The input image is decomposed into several pre-attentive feature detection mechanisms, which operate in parallel over the entire visual scene. Neurons in the feature maps then encode for spatial contrast in each of those feature channels. In addition, neurons in each feature map spatially compete for salience. After competition, the feature maps are combined into a unique saliency map, which topographically encodes for salience irrespective of the feature channel in which stimuli appeared salient. Adapted with permission from Itti and Koch (2001).

viewing conditions. However, the focus on free-viewing of static images may underestimate the importance of top-down control on eye movements, and it is unlikely that human gaze-control can be predicted solely on the basis of image characteristics (Henderson, 2007).

Although numerous models of attentional orienting have been proposed, none of these models specified the neural mechanisms or the exact underlying cortical areas responsible for encoding salience. Studies of neglect patients (Rafal, 1994; Friedrich et al., 1998; Mesulam, 1999; Husain and Rorden, 2003) suggest that posterior parietal cortex (PPC) might play a role in the ability to orient visual attention. Single neuron recording studies in monkeys have brought deeper insight into the neurophysiology of this region. Gottlieb and her colleagues hypothesized that the lateral intraparietal area (LIP) could provide a selective spatial representation of objects that are likely to attract attention, either by their salience or task relevance (Gottlieb et al., 1998; Goldberg et al., 2006). LIP is thought to participate in covert orienting of attention and in saccade planning

(Colby and Goldberg, 1999; Snyder et al., 2000; Goldberg et al., 2002) and LIP neurons have been shown to have visual on-responses to abrupt onsets in their receptive fields, as well as sustained activity preceding saccades to the receptive field (Barash et al., 1991). Results from Gottlieb's study showed that the visual representation in LIP is sparse, with only the *most salient* or behaviourally relevant objects represented (Gottlieb et al., 1998).

A different study has explored how the PPC responds to more complex visual displays (Constantinidis and Steinmetz, 2001). The underlying hypothesis was that activity of PPC neurons selectively represents the spatial location of the stimulus that attracts attention. In order to address this issue, single neurons in area 7a of the PPC were recorded in monkeys performing a spatial version of a match-to-sample task. The task required them to locate salient stimuli in multiple-stimulus displays and release a lever after a subsequent stimulus appeared at the same location. Neurons responded preferentially to the appearance of salient stimuli inside their receptive fields. The presence of multiple stimuli did not change the overall activity for the majority of neurons or the population code for the location of the salient stimulus. These results suggest that area 7a neurons represent the location of the stimulus that attracts animal's attention and could provide the spatial information required for directing attention to a salient stimulus in a complex scene. However, several other areas of the brain have also been proposed as the locus of the theoretical salience map, including the frontal eye fields (Thompson and Bichot, 2005), primary visual cortex (Li, 2002), and regions in the ventral visual pathway (Mazer and Gallant, 2003). It may be that the principle of a topographically organized map of stimulus priority is a common mechanism to be found in multiple brain regions.

Breakdown of such priority or salience maps might be one contributory factor to disorders associated with difficulty in directing attention to certain parts of space, such as in the syndrome of unilateral neglect. Patients with this disorder, most prominently associated with right PPC lesions (Mort et al., 2003), often fail to direct their attention and gaze to stimuli to the left (Mannan et al., 2005), although the degree of neglect may be modulated by the attentional demands of the task (Kaplan et al., 1991; Mannan et al., 2005). Thus, the degree of inattention is not simply for one sector of space, regardless of the contents of the visual scene. Neglect is more severe in densely-cluttered environments where distractor stimuli may serve to capture attention more frequently (Kaplan et al., 1991). Moreover, neglect patients often re-fixate items they have already fixated, sometimes erroneously considering them to be novel items that they have not previously inspected (Husain et al., 2001; Mannan et al., 2005).

Attended objects and their representations across saccades

So far we have been interested in how eye movements may be used as a surrogate marker for the locus of attention. In the real world, a key issue is how representations of objects selected for attention are maintained across the potentially disruptive influence of eye movements. Treisman and Gelade (1980) had originally proposed in their highly influential feature integration theory that a key role of spatial attention is to bind features (colour, form, etc.) together. Later, Kahneman, Treisman, and Gibbs developed a theory of *object files* — episodic representations that gather and maintain information about objects they represent *over time* (Kahneman et al., 1992). An object file contains information about features of the object, such as its colour, shape, and luminance. It integrates the features that belong to an object and therefore might be considered as a representation in which features are kept bound. Importantly, it also serves as a representation in which information is updated over time, as objects move or change. Crucially, spatial location is used to address the file (Treisman, 2006). But what happens when the eyes move? How do spatial locations of objects keep updated and how does the brain keep track of object files across saccades?

We decided to investigate this issue using the object reviewing paradigm (Kahneman et al., 1992). In this task subjects view two objects (e.g., squares) each of which has a letter within it. After a short period of viewing (the "previewing" phase) both letters are removed, and the objects move to new positions. A single letter is then presented in one of the two objects, and the subject's task is to name it. Typically, this response is faster if the letter appears in the same object that it had been in, than if it appears in the other object. The benefit in response time is referred to as an *Object-Specific Preview Benefit* (OSPB), because it is derived from previewing a letter within the same object across time.

What are the effects of eye movements on object feature integration or "object files" created by attending to an object? We used a modified version of Kahneman's paradigm to address this question (Kahneman et al., 1992). First, Kahneman and his colleagues, and subsequent studies on object files (Henderson and Anes, 1994; Noles et al., 2005) used letters as stimuli. However, we reasoned that it might be more appropriate to use more complex stimuli to test the fidelity of information stored in object files. Moreover, single feature objects are uncommon in real environments. Consequently, we used a combination of a colour and texture as stimuli. Second, in order to achieve a more robust OSPB, we asked participants to press a key on each trial to indicate that the final stimulus matched one of the initially presented stimuli, and to withhold any key press if not. This response, unlike voice-key naming, requires participants to attend to the initially presented stimuli, but still allows us to measure OSPBs. Third, in addition to the OSPB measure, we used a novel probe of the quality of object file representations by introducing irrelevant feature conjunctions in our identification task. We reasoned that the maintenance of faithful object files should be associated with a low number of false alarms even when irrelevant feature conjunctions are presented.

In our experiment, the initial display consisted of two empty boxes, one centred above fixation and the other below (Fig. 3). Eye position was recorded continuously. After fixation had been maintained for 1 s, a stimulus made of a colour and a texture was presented for 500 ms within each box. There was then a delay of 2 s. During this period the fixation point could either remain on at the centre (*fixation condition*) or shift to the periphery and then back centrally (*saccade condition*), with the subject in this condition having to make an outward saccade to the new peripheral location of the fixation cross and back again to the centre. Finally, a single stimulus was presented in one of the two boxes for 500 ms. The final stimulus could be identical to one of the preview stimuli and be presented in the same box (termed *match congruent* trial) or in the other box (*match incongruent* trial). Alternatively, it could be completely different with no features in common with either preview stimulus (*non-match* trial), or different but share one feature with one of the initial stimuli (*non-match conjunction* trial). The subject's task was to judge whether the final stimulus was the same or different from the initial stimuli by pressing a key as fast as possible.

The OSPB is calculated as the difference in response times between match incongruent and match congruent trials: that is, the reaction time advantage in recognizing a final stimulus if it had been previewed *within the same object*. In this experiment we found that in the fixation condition, the OSPB was 18.9 ms, compared to 28.93 ms in the saccade condition (Fig. 4). A two-way repeated measure ANOVA revealed a main effect of congruency (congruent vs. incongruent; $p < 0.01$) but no interaction with condition (saccade vs. fixation). Thus, the result here would suggest that there is a significant OSPB, which is invariant across eye movements, perhaps indicating that object-specific information is indeed maintained equally well in both fixation and saccade conditions. However, this measure alone does not interrogate the fidelity of feature binding. To do this we also need to examine accuracy across all types of trial.

While subjects made almost no errors on non-match trials, their performance dropped on *non-match conjunction* trials, where the final stimulus shared a single feature with one of the preview stimuli (81.3% accurate, SE = 3.2%). In order to quantify subjects' ability to discriminate between non-match conjunction trials and match trials, we calculated perceptual sensitivity (d') by running

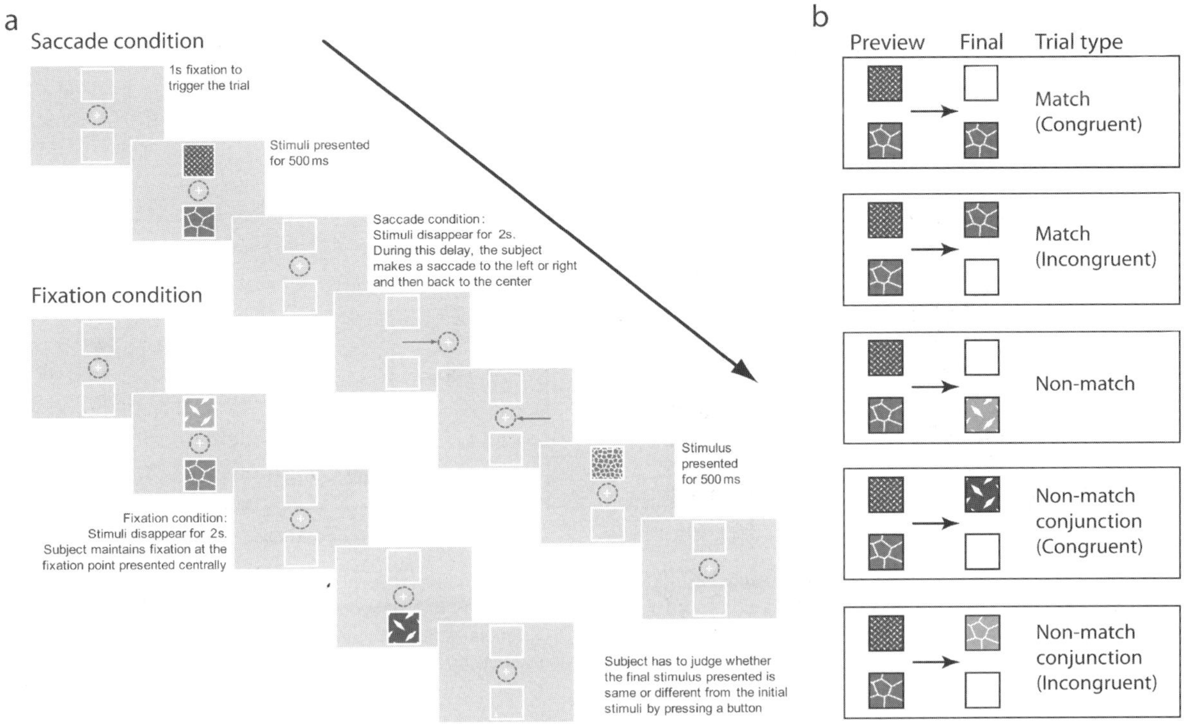

Fig. 3. An experiment to study the fidelity of object feature binding across saccades. (a) The task was to determine whether the final stimulus was same or different from the initial (preview) stimuli by pressing a key as fast as possible. The figure shows both the fixation and saccade condition. Stimuli were composed of a colour and a texture. The dashed circle shows where fixation should be maintained at each time epoch. (b) Examples of the different trial types. Subjects were instructed to press the button only on *match* trials, in which the final stimulus was identical to one of the preview stimuli.

a signal detection analysis (Fig. 4). The results revealed that participants were better in discriminating the final stimulus in the fixation condition ($d' = 2.86$, SE = 0.17) than the saccade condition ($d' = 2.58$, SE = 0.14). A two-way ANOVA showed a significant effect of condition (fixation vs. saccade; $p < 0.02$) but not congruency. Importantly, however, there was a significant interaction between condition and congruency ($p < 0.01$).

In the fixation condition, perceptual sensitivity rose from 2.70 (SE = 0.14) on incongruent trials to 2.93 (SE = 0.18) on the congruent trials. In other words, participants were better at discriminating a final stimulus when it was presented in the same box as the matching preview stimulus — analogous to the reaction time difference indexed by the OSPB. However, in the saccade condition, subjects were no better in discriminating that difference, as revealed by equal d' values for congruent and incongruent trials ($d' = 2.55$, STE = 0.15 and $d' = 2.59$, STE = 0.13, respectively; Fig. 4). Thus, the object-specific advantage of discriminating stimuli in the congruent condition (identical stimulus in same box) compared to the incongruent condition (identical stimulus but in different box) is lost with intervening saccades.

These results suggest that, although an object file is not destroyed by eye movements as demonstrated by a positive OSPB in the saccade condition, the fidelity of information carried over across eye movements is not as well maintained as when the eyes are kept static. In order to rule out the possibility that subjects had a lower perceptual sensitivity in the saccade condition due to a faster reaction time (speed-accuracy trade off), a statistical correlation between perceptual

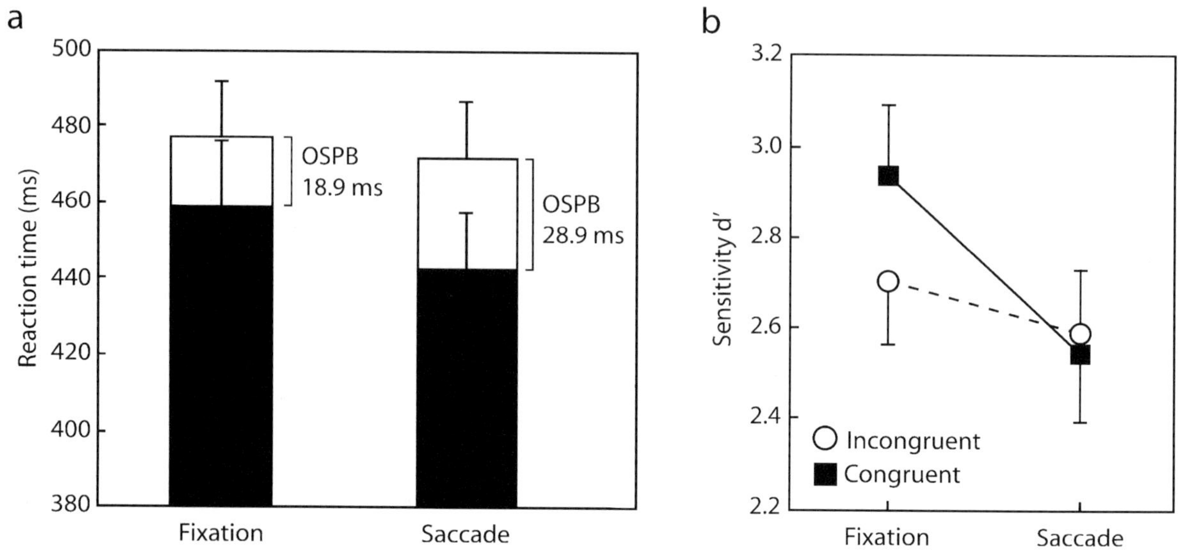

Fig. 4. Performance on congruent and incongruent trials in the fixation and saccade conditions. (a) Median reaction time on congruent (black) and incongruent (white) trials in each condition. (b) Perceptual sensitivity measured by *d*′. A value of 0 indicates inability to judge whether the probe stimulus is same or different from previously seen targets, whereas higher values mean better sensitivity. Note the discrimination advantage for congruent trials is lost when there is an intervening saccade. Error bars indicate ±1 SE.

sensitivity and reaction times was performed. This showed no significant effect ($r = 0.28$, $p > 0.1$). The present findings provide the first evidence that eye movements can interfere with the maintenance of object representations and may therefore constrain theories of feature integration across gaze shifts when viewing natural scenes.

Conclusions

The study of eye movements has much to offer for improving our understanding of visual attention. By measuring eye position we can obtain a far better index of the allocation of attention than more indirect manual reaction time measures. Eye tracking has revealed how attention can be captured by abrupt-onset stimuli or salient items in the visual scene. Such studies have also led to the concept that there might be salience maps in the brain, which guide the deployment of attention. When such representations are damaged, for example, after injury to the PPC, there may be florid inattention of one sector of space, as in the neglect syndrome. Finally, the investigation of eye movements may also constrain models of visual attention. The integration of features into object representations or files appears to be degraded to some extent by gaze shifts and this has implications for the fidelity of such representations when subjects view real-world scenes.

References

Barash, S., Bracewell, R.M., Fogassi, L., Gnadt, J.W. and Andersen, R.A. (1991) Saccade-related activity in the lateral intraparietal area. I. Temporal properties; comparison with area 7a. J. Neurophysiol., 66(3): 1095–1108.

Colby, C. and Goldberg, M.E. (1999) Space and attention in parietal cortex. Annu. Rev. Neurosci., 22: 319–349.

Constantinidis, C. and Steinmetz, M.A. (2001) Neuronal responses in area 7a to multiple-stimulus displays: I. Neurons encode the location of the salient stimulus. Cereb. Cortex, 11: 581–591.

Deubel, H. and Schneider, W.X. (1996) Saccade target selection and object recognition: evidence for a common attentional mechanism. Vision Res., 36(12): 1827–1837.

Friedrich, F.J., Egly, R., Rafal, R.D. and Beck, D. (1998) Spatial attention deficits in humans: a comparison of superior parietal and temporal-parietal junction lesions. Neuropsychology, 12(2): 193–207.

Goldberg, M.E., Bisley, J., Powell, K.D., Gottlieb, J. and Kusunoki, M. (2002) The role of the lateral intraparietal area

of the monkey in the generation of saccades and visuospatial attention. Ann. N.Y. Acad. Sci., 956: 205–215.
Goldberg, M.E., Bisley, J.W., Powell, K.D. and Gottlieb, J. (2006) Saccades, salience and attention: the role of the lateral intraparietal area in visual behavior. Prog. Brain Res., 155: 157–175.
Gottlieb, J.P., Kusunoki, M. and Goldberg, M.E. (1998) The representation of visual salience in monkey parietal cortex. Nature, 391: 481–484.
Henderson, J. (2007) Regarding scenes. Curr. Dir. Psychol. Sci., 16: 219–222.
Henderson, J. and Anes, M. (1994) Roles of object-file review and type priming in visual identification within and across eye fixations. J. Exp. Psychol. Hum. Percept. Perform., 20(4): 826–839.
Hoffman, J.E. and Subramaniam, B. (1995) The role of visual attention in saccadic eye movements. Percept. Psychophys., 57(6): 787–795.
Husain, M., Mannan, S., Hodgson, T., Wojciulik, E., Driver, J. and Kennard, C. (2001) Impaired spatial working memory across saccades contributes to abnormal search in parietal neglect. Brain, 124(Pt 5): 941–952.
Husain, M. and Rorden, C. (2003) Non-spatially lateralized mechanisms in hemispatial neglect. Nat. Rev. Neurosci., 4(1): 26–36.
Itti, L. and Koch, C. (2001) Computational modelling of visual attention. Nat. Rev. Neurosci., 2(3): 194–203.
Itti, L., Koch, C. and Niebur, E. (1998) A model of saliency-based visual attention for rapid scene analysis. IEEE Trans. Pattern Anal. Mach. Intell., 20: 1254–1259.
Kahneman, D., Treisman, A. and Gibbs, B. (1992) The reviewing object files: object-specific integration of information. Cogn. Psychol., 24: 175–219.
Kaplan, R.F., Verfaellie, M., Meadows, M.E., Caplan, L.R., Pessin, M.S. and DeWitt, L.D. (1991) Changing attentional demands in left hemispatial neglect. Arch. Neurol., 48(12): 1263–1266.
Koch, C. and Ullman, S. (1985) Shifts in selective visual attention: towards the underlying neural circuitry. Hum. Neurobiol., 4(4): 219–227.
Li, Z. (2002) A saliency map in primary visual cortex. Trends Cogn. Sci., 6: 9–16.
Mannan, S.K., Mort, D.J., Hodgson, T.L., Driver, J., Kennard, C. and Husain, M. (2005) Revisiting previously searched locations in visual neglect: role of right parietal and frontal lesions in misjudging old locations as new. J. Cogn. Neurosci., 17(2): 340–354.
Mazer, J.A. and Gallant, J.L. (2003) Goal-related activity in V4 during free viewing visual search. Evidence for a ventral stream visual salience map. Neuron, 40(6): 1241–1250.
Mesulam, M.M. (1999) Spatial attention and neglect: parietal, frontal and cingulate contributions to the mental representation and attentional targeting of salient extrapersonal events. Philos. Trans. R. Soc. Lond. B Biol. Sci., 354(1387): 1325–1346.
Milstein, D.M. and Dorris, M.C. (2007) The influence of expected value on saccadic preparation. J. Neurosci., 27(18): 4810–4818.
Mort, D.J., Malhotra, P., Mannan, S.K., Rorden, C., Pambakian, A., Kennard, C., et al. (2003) The anatomy of visual neglect. Brain, 126(Pt 9): 1986–1997.
Noles, N.S., Scholl, B.J. and Mitroff, S.R. (2005) The persistence of object file representations. Percept. Psychophys., 67(2): 324–334.
Parkhurst, D., Law, K. and Niebur, E. (2002) Modeling the role of salience in the allocation of overt visual attention. Vision Res., 42: 107–123.
Posner, M. (1980) Orienting of attention. Q. J. Exp. Psychol., 32: 3–25.
Rafal, R.D. (1994) Neglect. Curr. Opin. Neurobiol., 4: 231–236.
Remington, R.W. and Johnston, J.C. (1992) Involuntary attentional capture by abrupt onsets. Percept. Psychophys., 51(3): 279–290.
Snyder, L.H., Batista, A.P. and Andersen, R.A. (2000) Intention-related activity in the posterior parietal cortex: a review. Vision Res., 40: 1433–1441.
Theeuwes, J. and Irwin, D. (1999) Influence of attentional capture on oculomotor control. J. Exp. Psychol., 25(6): 1595–1608.
Theeuwes, J., Kramer, A., Hahn, S. and Irwin, D. (1998) Exogenous and endogenous control of attention: the effect of visual onsets and offsets. Psychol. Sci., 9(5): 379–486.
Thompson, K.G. and Bichot, N.P. (2005) A visual salience map in the primate frontal eye field. Prog. Brain Res., 147: 251–262.
Treisman, A. (2006) Object tokens, binding and visual memory. In: Zimmer H., Mecklinger A. and Lindenberger U. (Eds.), Handbook of binding and memory. Perspectives from cognitive neuroscience. Oxford University Press, Oxford, pp. 315–338.
Treisman, A. and Gelade, G. (1980) A feature integration theory of attention. Cogn. Psychol., 12: 97–136.
Yantis, S. and Jonides, J. (1984) Abrupt visual onsets and selective attention: evidence from visual search. J. Exp. Psychol. Hum. Percept. Perform., 10: 601–621.

C. Kennard & R.J. Leigh (Eds.)
Progress in Brain Research, Vol. 171
ISSN 0079-6123

CHAPTER 5.6

Using transcranial magnetic stimulation to probe decision-making and memory

René M. Müri* and Thomas Nyffeler

Perception and Eye Movement Laboratory, Departments of Neurology and Clinical Research, Inselspital, Bern University Hospital, and University of Bern, Bern, Switzerland

Abstract: Decision-making and memory are fundamental processes for successful human behaviour. For eye movements, the frontal eye fields (FEF), the supplementary eye fields (SEF), the dorsolateral prefrontal cortex (DLPFC), the ventrolateral frontal cortex and the anterior cingulum are important for these cognitive processes. The online approach of transcranial magnetic stimulation (TMS), i.e., the application of magnetic pulses during planning and performance of saccades, allows interfering specifically with information processing of the stimulated region at a very specific time interval (chronometry of cortical processing). The paper presents studies, which showed the different roles of the FEF and DLPFC in antisaccade control. The critical time interval of DLPFC control seems to be before target onset since TMS significantly increased the percentage of antisaccade errors at that time interval. The FEF seems to be important for the triggering of correct antisaccades. Bilateral stimulation of the DLPFC could demonstrate parallel information-processing transfer in spatial working memory during memory-guided saccades.

Keywords: antisaccades; memory; guided saccades; TMS; Chronometry; FEF; SEF; DLPFC; PPC

Introduction

Decision-making and working memory are two associated functions of the brain important for a successful behaviour in everyday life. Working memory allows holding online information needed for decision-making. Decisions are required for many sensory-motor tasks, and can be thought of as a form of statistical interference between the state of the world and the noisy data provided by the sensory systems (Gold and Shadlen, 2007). Planning an action is characterized by mentally anticipating possible courses of actions, which is critical for decision-making. Eye movements, particularly combined with transcranial magnetic stimulation (TMS), are valuable methodological approaches to test these cognitive functions.

TMS to probe the chronometry of cerebral processes

TMS technique is based on Faraday's principles of electromagnetic induction. The changing magnetic field, generated by a pulse of current flowing through a coil of wires induces a secondary current in the subject's brain, which stimulates the neural tissue. The short interference effect of TMS allows a specific analysis of the region at which time

*Corresponding author. Tel.: +41 31 632 3081;
Fax: +41 31 632 9770; E-mail: rene.mueri@insel.ch

DOI: 10.1016/S0079-6123(08)00660-2

interval it is relevant for the performance of a paradigm. Depending on the stimulation parameters, single and double pulse TMS may have facilitatory or inhibitory effects on oculomotor behaviour. Hence, single pulses appropriately delivered in time and space make it possible to study two aspects of the contribution of a given cortical region to a specific behaviour: "what does it do?" and "when does it do it?" (Pascual-Leone et al., 2000).

Chronometry of antisaccades

Patient studies with lesions of the dorsolateral prefrontal cortex (DLPFC) (Guitton et al., 1985; Pierrot-Deseilligny et al., 1991, 2003; Walker et al., 1998; Ploner et al., 2005; Hodgson et al., 2007) have shown that this region is critically involved in suppressing reflexive saccades during the antisaccade task. Damage to this region leads to an increased percentage of antisaccade errors. The triggering of correct antisaccades seems to be controlled by the frontal eye fields (FEF) (Rivaud et al., 1994; Gaymard et al., 1999).

TMS has been used to examine the role of these two cortical areas during the antisaccade task. Two studies have shown that the FEF is critically implied in triggering of the correct antisaccade: Müri et al. (1991) have applied single pulses over the right FEF and measured the latencies of antisaccades. When stimulation occurred 50–90 ms after the target flashed up, latencies of antisaccades were increased. In another study where TMS was applied over the right FEF, similar results were found (Olk et al., 2006). However, the percentage of antisaccade errors was not increased in both studies.

To define the critical time interval at which the DLPFC inhibits reflexive saccade triggering during the antisaccade task, Nyffeler et al. (2007) used single pulse TMS. Stimulation of the right DLPFC 100 ms before target onset significantly increased the percentage of antisaccade errors to both sides; other stimulation intervals (i.e., at target onset or 100 ms after target onset) had no significant effects. In monkeys, it has been shown that DLPFC neurons projecting to the superior colliculus (SC) show higher activity in the pre-target period for antisaccades than for pro-saccades (Johnston and Everling, 2006). The TMS results support the view that DLPFC sends inhibitory signals to the SC before target onset (Munoz and Everling, 2004).

The results of both studies (Olk et al., 2006; Nyffeler et al., 2007) are presented in Fig. 1. Based on their results, we calculated the percentage of increase or decrease in errors of antisaccades, compared to no stimulation condition.

Chronometry of the organization of spatial working memory

Spatial working memory can be assessed with the memory-guided saccade task. In this task, subjects are instructed to memorize the location of a horizontal flashed lateral target while fixating a central point. After a memorization delay, the central fixation point extinguishes, which is the go-signal for the subject to perform a saccade towards the memorized target. The accuracy of the saccade is considered as a reflection of spatial working memory. Several cortical areas have been shown to be implicated during the performance of the memory-guided saccade task. Studies in humans have shown that lesions of the posterior parietal cortex (PPC) (Pierrot-Deseilligny et al., 1991), FEF (Rivaud et al., 1994; Gaymard et al., 1999), and DLPFC (Pierrot-Deseilligny et al., 1991) result in a delay-dependent increase of memory-guided saccade amplitude error.

TMS studies suggest that the initial stage of visuo-spatial integration, during the first 300 ms following target presentation, is under control of the PPC (Müri et al., 1996) and that subsequent memorization is first controlled by the DLPFC (Müri et al., 1996, 2000). Ploner et al. (1999) have shown that spatial memory at delays longer than 20 s is impaired after lesions of the medial temporal lobe (MTL). Taken together, these studies point to a temporal hierarchy in oculomotor control. However, it is not clear whether the MTL receives visuo-spatial signals directly from the PPC or indirectly via the DLPFC. The latter possibility would argue for an entirely serial, the

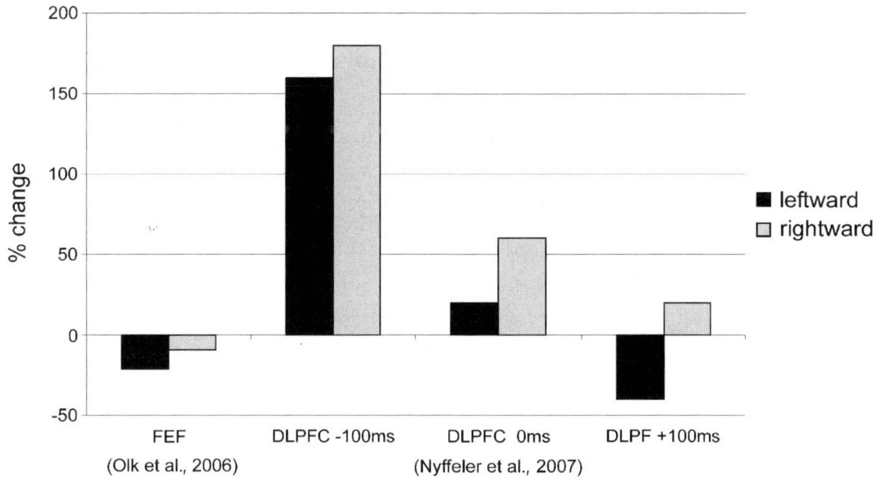

Fig. 1. Effect of TMS on antisaccade errors. The percentage of change in errors was calculated; 0% corresponds to the number of errors without stimulation. Stimulation of the FEF had no effect on antisaccade errors. TMS over the DLPFC significantly increased the percentage of errors when applied 100 ms before target onset. Later stimulation had no significant effect.

former for an additional parallel, information-processing component.

In order to test the two hypotheses, Nyffeler et al. (2002, 2004) examined the influence of TMS on memory-guided saccade amplitude with two different memorization delays of 3 and 30 s. Transient inactivation of the right DLPFC by double pulse TMS 1 s after target presentation significantly impaired the accuracy of contralateral memory-guided saccades, and thus revealed a functional dominance of DLPFC during early memorization. Since DLPFC stimulation reduced saccade accuracy irrespective of the duration of the delay, a serial information processing was suggested. However, evidence for additional parallel information processing was also found: the degrading effect of TMS on saccade accuracy was significantly greater with the short memory delay. This suggested that during longer memorization periods, the MTL obtained access to additional spatial information from other structures, enabling a partial rectification of the simulation-induced saccade inaccuracy. Theoretically, additional information to the MTL could have come from the PPC directly or via the DLPFC of the other hemisphere. In fact, both types of anatomic connections exist between these regions (Goldman-Rakic et al., 1984; Ding et al., 2000). Therefore, in a second study (Nyffeler et al., 2004), bilateral simultaneous transient inhibition of the DLPFC by TMS was performed. Indeed, the degrading effect of bilateral DLPFC stimulation 1 s after target presentation on memory-guided saccade accuracy was again significantly greater in short than in long memory delay, suggesting parallel information-processing transfer to the MTL. Thus, the DLPFC is crucial during the initial phase of working memory for the subsequent long memorization. In contrast, TMS applied late during the long memory delay paradigm (i.e., after 28 s of memorization), had no effect on the accuracy of contralateral memory-guided saccades, confirming the view that the role of DLPFC is limited in time for 'holding' the information in memory. The results of both studies are presented in Fig. 2.

These results are also in agreement with results of fMRI experiments (Leung et al., 2002), where it has been shown that lateral frontal activations decrease during the delay of a spatial working memory task.

Chronometry of prediction

Prediction is of major importance in decision-making. For the oculomotor system, different

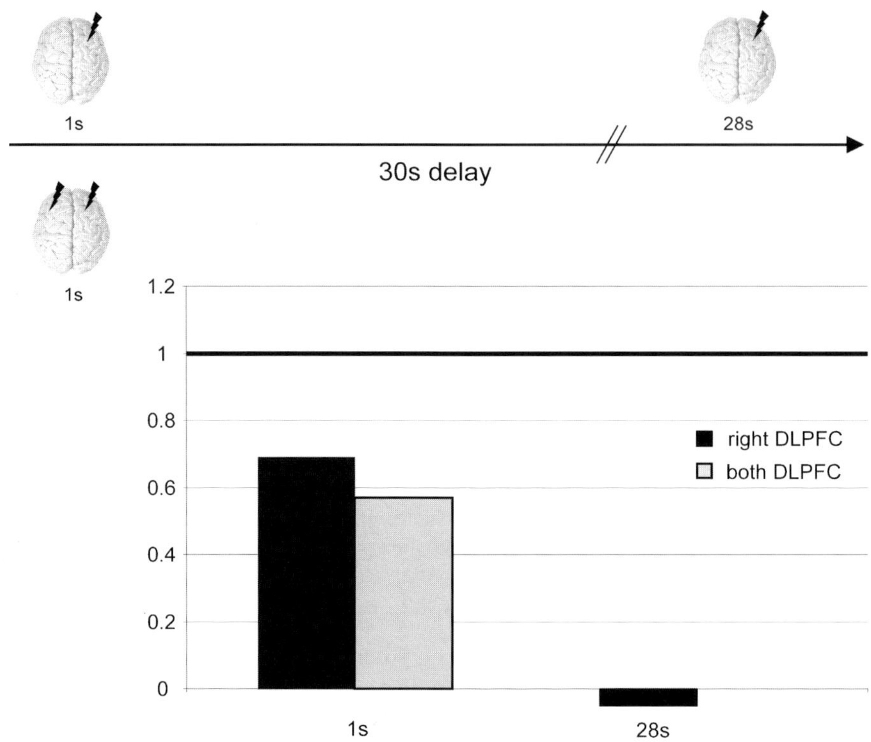

Fig. 2. To illustrate the contribution of parallel processing, the effect of TMS on saccade accuracy during short delay memorization was normalized to 1 (the value 0 corresponds to the accuracy without stimulation). The degrading effect of unilateral stimulation on saccade accuracy was significantly smaller in the long memory delay as shown by the values <1, suggesting that during longer memorization periods the MTL got access to additional spatial information from other structures. Late unilateral stimulation had no significant effect, suggesting that the DLPFC at 28 s is no longer involved in memorization. Bilateral stimulation of the DLPFC was applied 1 s after target presentation and had the same effect as unilateral stimulation, suggesting that information processing is not due to the homologous area.

components of predictive tracking behaviour may be assessed as to whether smooth pursuit or saccades are required (Leigh and Zee, 2006). Under natural conditions, ocular tracking is usually performed as a combination of smooth pursuit and saccades. Recent studies have proposed that these two oculomotor subsystems may share common physiological processes such as the encoding of visual error, attention, or memory (for a review, see Krauzlis, 2005).

It is currently accepted that supplementary eye field (SEF) and FEF are critical for high-level prediction during smooth pursuit (Heinen and Liu, 1997; O'Driscoll et al., 2000). Neurons in the SEF were found to discharge slightly in advance of predictive direction reversal, suggesting a role of this area in prediction during pursuit (Heinen and Liu, 1997). Several lines of evidence suggest that the SEF influences the oculomotor system at a high, cognitive level (e.g., Amador et al., 2000; Nachev et al., 2005). Furthermore, a recent TMS study in humans showed different effects of SEF and FEF stimulation during a periodic pursuit task, the predictive role of the SEF being primarily for direction reversal (Gagnon et al., 2006).

Nyffeler et al. (2007) designed a new paradigm in which predictive pursuit and non-predictable visually guided saccades were associated. This smooth pursuit paradigm required the triggering of unpredictable saccades, performed either during low (ongoing pursuit) or high (pursuit direction reversal) predictive behaviour. Saccade latency was used as a probe to reveal the sharing of prediction information between the saccadic and smooth pursuit subsystem. In healthy subjects, saccade latencies were markedly decreased when triggered around pursuit direction reversal and performed in the direction of the predicted pursuit. If single-pulse TMS over the SEF was applied at a critical time interval, i.e., −75 ms before pursuit direction reversal, decreased saccade latencies were no longer observed, whereas it remained when TMS was applied over the occipital cortex. These results suggest the existence of a common predictive process shared by both oculomotor subsystems. Furthermore, the SEF seems to function as a higher-order structure able to elaborate complex processes, such as prediction, independently of the oculomotor output.

The temporal dynamics of the TMS effect fit nicely with previous electrophysiological and clinical findings during predictable pursuit eye movements. In the monkey, cells in the SEF were found to progressively increase their discharge rate around the time of pursuit direction reversal, with the initial rise being observed 75 ms prior to direction reversal (Heinen and Liu, 1997). Finally, Gagnon et al. (2006) found that TMS over the SEF affected pursuit velocity when applied 75 ms prior to expected target direction reversal but not during ongoing pursuit.

References

Amador, N., Schlag-Rey, M. and Schlag, J. (2000) Reward-predicting and reward-detecting neuronal activity in the primate supplementary eye field. J. Neurophysiol., 84: 2166–2170.

Ding, S.L., Van Hoesen, G. and Rockland, K.S. (2000) Inferior parietal lobule projections to the presubiculum and neighboring ventromedial temporal cortical areas. J. Comp. Neurol., 425: 510–530.

Gagnon, D., Paus, T., Grosbras, M.H., Pike, G.B. and O'Driscoll, G.A. (2006) Transcranial magnetic stimulation of frontal oculomotor regions during smooth pursuit. J. Neurosci., 26: 458–466.

Gaymard, B., Ploner, C.J., Rivaud-Pechoux, S. and Pierrot-Deseilligny, C. (1999) The frontal eye field is involved in spatial short-term memory but not in reflexive saccade inhibition. Exp. Brain Res., 129: 288–301.

Gold, J.I. and Shadlen, M.N. (2007) The neural basis of decision making. Annu. Rev. Neurosci., 30: 535–574.

Goldman-Rakic, P.S., Selemon, L.D. and Schwartz, M.L. (1984) Dual pathways connecting the dorsolateral prefrontal cortex with the hippocampal formation and parahippocampal cortex in the rhesus monkey. Neuroscience, 12: 719–743.

Guitton, D., Buchtel, H.A. and Douglas, R.M. (1985) Frontal lobe lesions in man cause difficulties in suppressing reflexive glances and in generating goal-directed saccades. Exp. Brain Res., 58: 455–472.

Heinen, S.J. and Liu, M. (1997) Single-neuron activity in the dorsomedial frontal cortex during smooth-pursuit eye movements to predictable target motion. Vis. Neurosci., 14: 853–865.

Hodgson, T., Chamberlain, M., Parris, B., James, M., Gutowski, N., Husain, M. and Kennard, C. (2007) The role of the ventrolateral frontal cortex in inhibitory oculomotor control. Brain, 130: 1525–1537.

Johnston, K. and Everling, S. (2006) Monkey dorsolateral prefrontal cortex sends task-selective signals directly to the superior colliculus. J. Neurosci., 26: 12471–12478.

Krauzlis, R.J. (2005) The control of voluntary eye movements: new perspectives. Neuroscientist, 11: 124–137.

Leigh, R.J. and Zee, D.S. (2006) The Neurology of Eye Movements. Oxford University Press, New York.

Leung, H.C., Gore, J.C. and Goldman-Rakic, P.S. (2002) Sustained mnemonic response in the human middle frontal gyrus during online storage of spatial memoranda. J. Cogn. Neurosci., 14: 659–671.

Munoz, D.P. and Everling, S. (2004) Look away: the antisaccade task and the voluntary control of eye movement. Nat. Rev. Neurosci., 5: 218–228.

Müri, R.M., Gaymard, B., Rivaud, S., Vermersch, A.I., Hess, C.W. and Pierrot-Deseilligny, C. (2000) Hemispheric asymmetry in cortical control of memory-guided saccades. A transcranial magnetic stimulation study. Neuropsychologia, 38: 1105–1111.

Müri, R.M., Hess, C.W. and Meienberg, O. (1991) Transcranial stimulation of the human frontal eye field by magnetic pulses. Exp. Brain Res., 86: 219–223.

Müri, R.M., Vermersch, A.I., Rivaud, S., Gaymard, B. and Pierrot-Deseilligny, C. (1996) Effects of single pulse transcranial magnetic stimulation over the prefrontal and posterior parietal cortices during memory-guided saccades in humans. J. Neurophysiol., 76: 2102–2106.

Nachev, P., Rees, G., Parton, A., Kennard, C. and Husain, M. (2005) Volition and conflict in human medial frontal cortex. Curr. Biol., 15: 122–128.

Nyffeler, T., Müri, R.M., Bucher-Ottiger, Y., Pierrot-Deseilligny, C., Gaymard, B. and Rivaud-Pechoux, S. (2007) Inhibitory control of the human dorsolateral prefrontal cortex during the anti-saccade paradigm — a transcranial magnetic stimulation study. Eur. J. Neurosci., 26: 1381–1385.

Nyffeler, T., Pierrot-Deseilligny, C., Felblinger, J., Mosimann, U.P., Hess, C.W. and Müri, R.M. (2002) Time-dependent hierarchical organization of spatial working memory: a transcranial magnetic stimulation study. Eur. J. Neurosci., 16: 1823–1827.

Nyffeler, T., Pierrot-Deseilligny, C., Pflugshaupt, T., von Wartburg, R., Hess, C.W. and Müri, R.M. (2004) Information processing in long delay memory-guided saccades: further insights from TMS. Exp. Brain Res., 154: 109–112.

O'Driscoll, G.A., Wolff, A.L.V., Benkelfat, C., Florencio, P.S., Lal, S. and Evans, A.C. (2000) Functional neuroanatomy of smooth pursuit and predictive saccades. Neuroreport, 6: 1335–1340.

Olk, B., Chang, E., Kingstone, A. and Ro, T. (2006) Modulation of antisaccades by transcranial magnetic stimulation of the human frontal eye field. Cereb. Cortex, 16: 76–82.

Pascual-Leone, A., Walsh, V. and Rothwell, J. (2000) Transcranial magnetic stimulation in cognitive neuroscience — virtual lesion, chronometry, and functional connectivity. Curr. Opin. Neurobiol., 10: 232–237.

Pierrot-Deseilligny, C., Müri, R.M., Ploner, C.J., Gaymard, B., Demeret, S. and Rivaud-Pechoux, S. (2003) Decisional role of the dorsolateral prefrontal cortex in ocular motor behaviour. Brain, 126: 1460–1473.

Pierrot-Deseilligny, C., Rivaud, S., Gaymard, B. and Agid, Y. (1991) Cortical control of reflexive visually guided saccades in man. Brain, 114: 1473–1485.

Ploner, C.J., Gaymard, B.M., Ehrle, N., Rivaud-Pechoux, S., Baulac, M., Brandt, S.A., Clemenceau, S., Samson, S. and Pierrot-Deseilligny, C. (1999) Spatial memory deficits in patients with lesions affecting the medial temporal neocortex. Ann. Neurol., 45: 312–319.

Ploner, C.J., Gaymard, B.M., Rivaud-Pechoux, S. and Pierrot-Deseilligny, C. (2005) The prefrontal substrate of reflexive saccade inhibiton in humans. Biol. Psychiatry, 57: 1159–1165.

Rivaud, S., Müri, R.M., Gaymard, B., Vermersch, A.I. and Pierrot-Deseilligny, C. (1994) Eye movement disorders after frontal eye field lesions in humans. Exp. Brain Res., 102: 110–120.

Walker, R., Husain, M., Hodgson, T.L., Harrison, J. and Kennard, C. (1998) Saccadic eye movement and working memory deficits following damage to human prefrontal cortex. Neuropsychologia, 36: 1141–1159.

C. Kennard & R.J. Leigh (Eds.)
Progress in Brain Research, Vol. 171
ISSN 0079-6123

CHAPTER 5.7

Supplementary eye field contributions to the execution of saccades to remembered target locations

Clive R. Rosenthal[1,*], Timothy L. Hodgson[2], Masud Husain[3] and Christopher Kennard[1]

[1]*Imperial College London, London, UK*
[2]*Department of Psychology, Exeter University, UK*
[3]*University College London, London, UK*

Abstract: Multiple lines of evidence indicate that an anatomically discrete region within the dorsal medial frontal cortex — the supplementary eye field (SEF) — is involved in oculomotor control. To delineate this role further, repetitive transcranial magnetic stimulation (rTMS; 10 Hz, 500 ms) was administered either immediately after the presentation of three-step saccade sequences, or, immediately before response execution of memory-guided saccades. In addition, the effects of changes to visuospatial and temporal order demands were examined by contrasting performance in the presence and the absence of target location information. Results revealed that the SEF supports the processing of spatial information relevant to saccade amplitude. Independently of the time of stimulation, saccade gain was reduced by rTMS applied over the SEF, though only when response execution was performed in the absence of target location information. These results provide evidence of a causal role for the SEF in oculomotor control in the absence of visual feedback.

Keywords: memory-guided saccades; oculomotor control; TMS; eye movements

Introduction

Oculomotor behaviour is supported by a complex interaction between visuospatial, motor, and cognitive processes. Investigation of these processes using functional neuroimaging, neurological lesion, and nonhuman primate neurophysiological methods has led to the identification of a region in the dorsal medial frontal cortex — the supplementary eye field (SEF) — with numerous features of oculomotor control (Tehovnik et al., 2000). Activation of the SEF is observed during tasks that involve eye movements, and appears to be selectively modulated by the operation of higher-order oculomotor processes implicated in the antisaccade task, conditional motor learning, and the learning of new sequences. The SEF has also been shown to have a causal role in supporting the execution of saccades to remembered target locations. In particular, a discrete lesion to this region can result in reduced saccade gain when visual information about target location is unavailable (Parton et al., 2006), just as has been found in patients with Parkinson's disease (Hodgson et al., 1999). We investigated the functional specificity of this role in neurologically intact participants by presenting three-step novel sequences for retention in visuospatial working

*Corresponding author. Tel.: +44 020 8846 7505;
Fax: +44 020 8846 7715; E-mail: c.rosenthal@imperial.ac.uk

DOI: 10.1016/S0079-6123(08)00661-4

memory until a "go-signal" prompted the delayed execution of each sequence (memory-guided saccade (MGS) task).

Repetitive transcranial magnetic stimulation (rTMS; 10 Hz, 500 ms) was administered immediately after the presentation of each sequence and immediately before the "go-signal." On the basis of evidence from nonhuman primate studies, there is reason to assume that the SEF is involved in the planning and control of oculomotor sequences, because neurons in the SEF become active immediately preceding the initiation of internally guided eye movements (Coe et al., 2002), and these neurons also exhibit sequence-specific activity (Isoda and Tanji, 2002). Performance was also assessed in the presence and absence of visual cues to target locations. Removal of the visual cues during response execution changes the demands associated with responding. Critically, in the absence of visual cues, the direction, temporal order, and spatial extent related to each impending saccade are endogenously determined, whereas in the presence of visual cues, only direction and temporal order information are necessary for responding.

Methods

Participants

Twelve volunteers took part ($M \pm SD$ for age: 22.0 ± 0.74 years; 11 males).

Design and materials

The experiment comprised a 2 (site of stimulation: SEF, control [primary sensorimotor cortex]) × 2 (TMS: rTMS, no-rTMS) × 2 (TMS onset: post-presentation, pre-response execution) × 2 (target location: cued, uncued) repeated measures factorial design. Separate experimental and control sessions were conducted to administer the rTMS over the SEF and hand area of primary sensorimotor cortex ($SM1_{HAND}$), respectively; the behavioural protocol was identical in both sessions. Four variants of a basic MGS task were employed across four blocks in order to implement the factorial manipulation of TMS onset and target location (Fig. 1). Each block was comprised of 24 unique three-step saccade sequences, counterbalanced for laterality. Half of the trials within a block were associated with rTMS; trial order was random. Visual stimuli were comprised of red light visual emitting diodes positioned bilaterally on a white screen at discrete 3.75° horizontal increments of the visual angle from central fixation (range: −30° and +30°).

Data analysis

Eye movements were recorded using an infra-red limbus monitoring system (Skalar, IRIS; bandwidth: 100 Hz [−3 dB]), with the digitised output stored for off-line analysis using a proprietary software application developed within the LabView environment (National Instruments, Austin, TX, USA).

TMS protocol

Each event-related train of rTMS comprised a 10 Hz, 5-pulse delivery, with a stimulus intensity of 90% of individually determined motor thresholds. Final sites of stimulation corresponding to the locations on the scalp immediately underlying the area of maximum magnetic field strength were marked using a disc opaque to the signal associated with magnetic resonance imaging.

Results

Analyses focus on saccade gain of each recollected target step in the sequence, because negligible effects were obtained as a function of the rTMS manipulation for final eye position, saccade latency, and sequence accuracy. A 2 × (site: SEF, $SM1_{HAND}$) × 2 (TMS: rTMS, no-rTMS) × 2 cue-type (cued, uncued) × 2 (time of TMS onset: post-presentation; pre-response execution) × 3 (step position: 1, 2, and 3) repeated measures ANOVA revealed significant main effects of TMS, $F(1, 10) = 6.36$, $p < 0.05$, cue-type, $F(1, 10) = 16.39$, $p < 0.01$, and step position, $F(2, 20) = 4.08$, $p < 0.05$, and significant interactions between site and step position, $F(2, 20) = 3.87$, $p < 0.05$, and

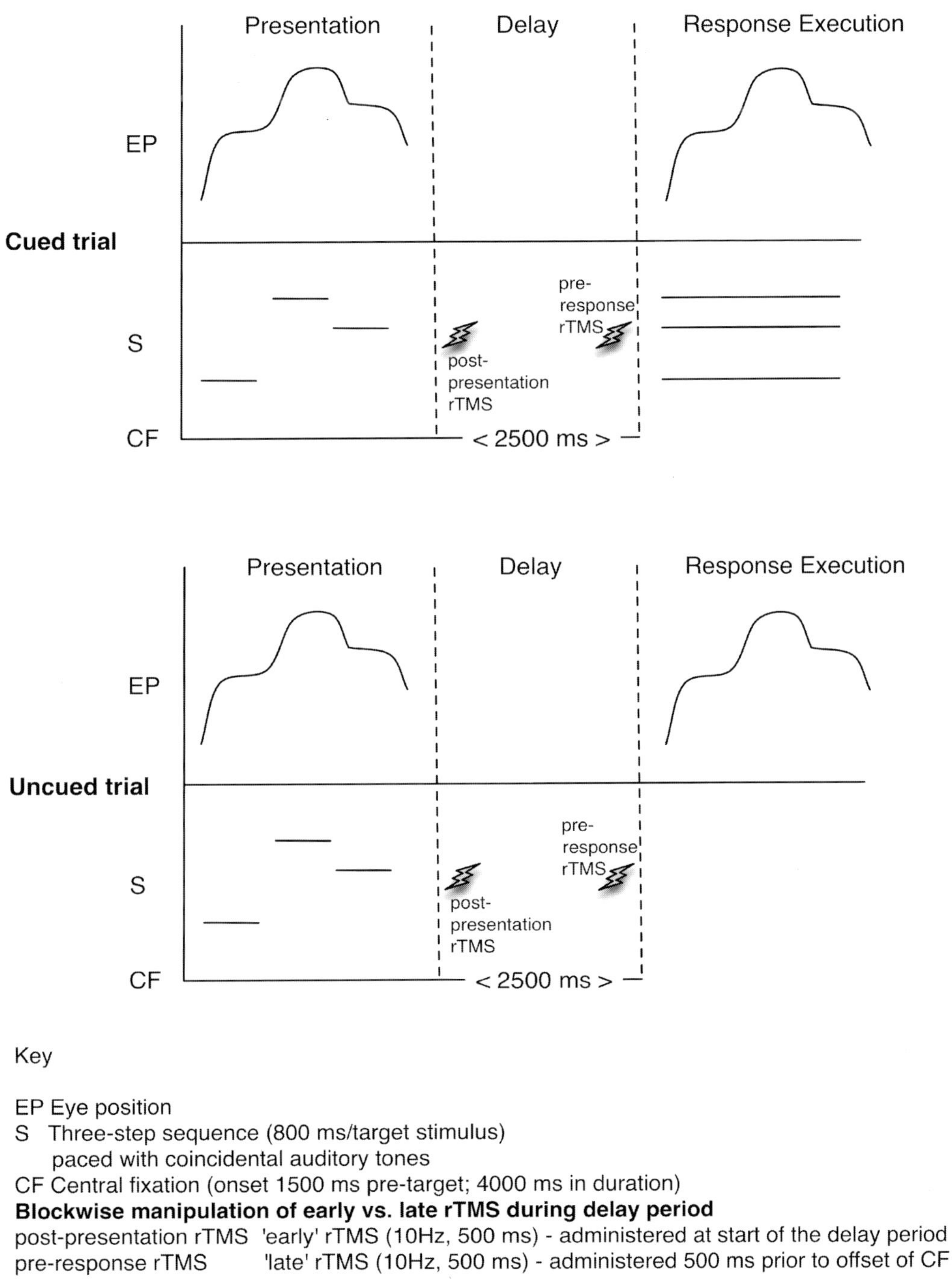

Fig. 1. Schematic of the three discrete phases of a single trial in the memory-guided saccade task performed in a dark visual "Ganzfeld": (1) presentation: three-step sequence of consecutively illuminated LEDs. Each target appeared in an unpredictable direction and amplitude from central fixation (CF); (2) delay: 2500 ms interval. Participants maintained CF during the presentation and delay; and (3) response execution: extinction of the CF served as the "go-signal" to execute three consecutive saccades to the locations of previously presented targets. Presentation and execution phases were paced using three auditory tones. In the two cued blocks, responses were guided by three targets that were simultaneously illuminated 200 ms after extinction of the CF in locations corresponding to the targets shown in the immediately preceding presentation phase. In the two uncued blocks, participants were instructed to execute saccades to the locations of previously presented targets, without any external reference to previous target locations or sequence.

between cue-type and step position, $F(2, 20)=4.44$, $p<0.05$. In addition, the interaction between TMS, cue-type, and step was significant, $F(2, 20)=4.28$, $p<0.05$. The main effect of time of TMS onset was not significant, $F<1$, and time of TMS onset did not interact with any other variable.

Following both post-presentation and pre-response execution stimulation in the uncued condition, a reduction in primary saccade gain was observed in the first step on rTMS trials relative to no-rTMS trials after rTMS was administered over the SEF, $F(1, 10)=9.19$, $p<0.01$, whereas in the cued condition the equivalent contrast was not significant, $F<1$. Correspondingly, a significant reduction in the saccade gain to the third step position in the uncued condition was obtained after rTMS was applied over the SEF, $F(1, 10)=13.87$, $p<0.01$; the equivalent contrast in the cued condition was not significant, $F(1, 10)=2.36$, $p=0.15$. None of these contrasts were significant when rTMS was applied over $SM1_{HAND}$ area. Accuracy of coil positioning was not included as a covariate because the functionally guided coil localisation procedure corresponded closely to anatomical reference landmark of the SEF in all participants.

Discussion

The results demonstrate that independently of the time of stimulation, saccade gain to targets was reduced on trials associated with rTMS applied over the SEF, though, only when MGS sequences were executed in the absence of target location information. The reduction in saccade gain only emerged when stimulation was administered over the SEF, which excludes the possibility that the induced pattern of changes originated from non-specific effects of rTMS. Even though saccade gain reflects a spatial parameter related to the execution of a saccade, it is important to note that dysmetria is not exclusively diagnostic of regions involved in motor preparatory activity, and can also suggest a problem with remembering the target distance or the correct saccade to acquire the target.

The results are in agreement with the data from a patient with a highly discrete lesion of the left SEF: in that case, impairment was restricted to hypometric initial saccades, but only when MGS sequences were performed in the dark (Husain et al., 2003; Parton et al., 2006). Furthermore, results from nonhuman primate neurophysiological studies indicate that spatial location is a particularly important stimulus attribute mediated by the SEF, though again only in the absence of visual cues and for novel sequences (Shima and Tanji, 1998). Arguably similar functional specificity has been reported for the sensorimotor homologue of the SEF, the supplementary motor area (Passingham, 1993).

Several features of the null results obtained on two of the other dependent measures — final eye position and sequence accuracy — warrant consideration. First, impairment in final eye position might still be anticipated under longer delays, because error rapidly increases with delay (White et al., 1994). Second, although the number of elements in a sequence directly influences saccade parameters such as the latency of first saccade, an insufficient sequence length is unlikely to account for the failure to obtain sequence errors (inversions in chronological order, extra or missing saccades, or errors in saccade direction) because such errors have been obtained when using sequences of similar length to those used in the current experiment. A more compelling condition for the induction of sequence errors seems to be supra-threshold TMS — other studies that report sequence errors have stimulated at supra-threshold (Muri et al., 1994). An abiding concern with supra-threshold (vs. sub-/threshold) stimulation is that the effects induced in neural areas proximate and distal to the site of stimulation are amplified. Plausibly, this argument can also be extended to studies in which lesions are not sufficiently selective, because deficits in MGS order occur in patients with gross lesions involving the SEF (Gaymard et al., 1993), but are absent when the lesion is circumscribed to the SEF (Parton et al., 2006); however, dysfunction rendered by diaschisis remains a relevant issue in both such lesion types.

In contrast to the functional and anatomical specificity observed here, the TMS onset manipulation had no discernable temporally specific impact on responding. Summation of the effects of rTMS is unlikely to have led to this outcome because rTMS was administered using a random, rather than blocked, design. It is conceivable that late stimulation did not coincide with the "critical period" that induces maximal interference in saccade initiation — the critical period for anticipatory activity within the SEF appears to be task specific (Coe et al., 2002). Only limited success, however, might be achieved with a modular, serial stage account of saccadic generation.

Taken together, our data are compatible with the proposal that the SEF is involved in the programming of saccade amplitude when there is response uncertainty, as in the case of making MGSs in the absence of visual cues (Parton et al., 2006). The role of the SEF in saccade generation may, in fact, be characterised as one that determines the detailed oculomotor properties of saccadic movements in tasks with cognitive requirements. Future research will need to determine the component processes underlying oculomotor sequence performance by systematically examining the impact of changes in motor sequence learning parameters on the SEF contribution to oculomotor control.

Acknowledgements

This research was supported by a Programme grant from the Wellcome Trust. We thank Sarah Offord for her assistance with the collection of data.

References

Coe, B., Tomihara, K., Matsuzawa, M. and Hikosaka, O. (2002) Visual and anticipatory bias in three cortical eye fields of the monkey during an adaptive decision-making task. J. Neurosci., 22: 5081–5090.

Gaymard, B., Rivaud, S. and Pierrot-Deseilligny, C. (1993) Role of the left and right supplementary motor areas in memory-guided saccade sequences. Ann. Neurol., 34: 404–406.

Hodgson, T.L., Dittrich, W.H., Henderson, L. and Kennard, C. (1999) Eye movements and spatial working memory in Parkinson's disease. Neuropsychologia, 37: 927–938.

Husain, M., Parton, A., Hodgson, T.L., Mort, D. and Rees, G. (2003) Self-control during response conflict by human supplementary eye field. Nat. Neurosci., 6: 117–118.

Isoda, M. and Tanji, J. (2002) Cellular activity in the supplementary eye field during sequential performance of multiple saccades. J. Neurophysiol., 88: 3541–3545.

Muri, R.M., Rosler, K.M. and Hess, C.W. (1994) Influence of transcranial magnetic stimulation on the execution of memorized sequences of saccades in man. Exp. Brain Res., 101: 521–524.

Parton, A., Nachev, P., Hodgson, T.L., Mort, D., Thomase, D., Ordidgee, R., Morgan, P.S., Jackson, S., Rees, G. and Husain, M. (2006) Role of the human supplementary eye field in the control of saccadic eye movements. Neuropsychologia, 45: 997–1008.

Passingham, R.E. (1993) The frontal lobes and voluntary action. Oxford University Press, Oxford.

Shima, K. and Tanji, J. (1998) Both supplementary and pre-supplementary motor areas are crucial for the temporal organisation of multiple movements. J. Neurophysiol., 80: 3247–3260.

Tehovnik, E.J., Somner, M.A., Chou, I.-H., Slocum, W.M. and Schiller, P.H. (2000) Eye fields in the frontal lobes of primates. Brain Res. Rev., 32: 413–448.

White, J.M., Sparks, D.L. and Stanford, T.R. (1994) Saccades to remembered target locations: an analysis of systematic and variable errors. Vision Res., 34: 79–92.

C. Kennard & R.J. Leigh (Eds.)
Progress in Brain Research, Vol. 171
ISSN 0079-6123

CHAPTER 5.8

Multiple memory-guided saccades: movement memory improves the accuracy of memory-guided saccades

Silvia Colnaghi[1,*], Giorgio Beltrami[2], Andrea Cortese[1,3], Wolfgang H. Zangemeister[3], Vittorio Cosi[1] and Maurizio Versino[1]

[1]*Neuro-otology and Neuro-ophthalmology Laboratory, IRCCS "Neurological Institute C.Mondino" Foundation; Department of Neurology, University of Pavia, Pavia, Italy*
[2]*Department of Biomedical Engineering, University of Pavia, Pavia, Italy*
[3]*Department of Neurology, UKE, University of Hamburg, Hamburg, Germany*

Abstract: Memory-guided saccades (MGSs) with 3 s memorization delay were recorded in healthy subjects using four different paradigms: two "regular" MGS paradigms with the peripheral target lit for 0.2 s (MGS2) and for 1.8 s (MGS18); a multiple memory-guided saccade (MMGS) paradigm with the target lit for 1.8 s and the instruction to perform a visually guided saccade (VGS) towards it before the MGS; a trained memory-guided saccades (TMGSs) paradigm where the same target was presented so that the subjects should made 10 VGSs before the MGS. The longer target presentation interval (MGS18 paradigm) did not improve the accuracy of MGS. The execution of the VGSs improved the accuracy of the corrective saccades made after the first MGS to drive the eyes closer to the target, and this improvement was independent from the number of the VGSs (there was no difference between the MMGS and the TMGS paradigms). The VGSs provide a template that improves the capability of the corrective saccades to compensate for the residual position error at the end of the first saccade.

Keywords: memory-guided saccades; movement memory; efference copy

Introduction

Memory-guided saccades (MGSs) (Leigh and Zee, 2006) are coded in a craniotopic reference system and, after the first MGS, subjects make corrective saccades aimed to drive the eyes closer to the memorized target position, even though there is no visual-feed back. We assessed if one ore more visually guided saccades (VGSs) made previously towards the same target location could improve the accuracy of MGSs. In order to enable the subjects to make a reflexive saccade, we prolonged the target presentation as compared to a classical MGS paradigm, and we evaluated in an ad-hoc task the effect of this reinforcement of the visuo-spatial information. Moreover, we assessed if the VGSs made previously could provide not only a movement memory but also a training effect.

*Corresponding author. Tel.: +39 0382 380340; Fax: +39 0382 24714; E-mail: si.co@libero.it

DOI: 10.1016/S0079-6123(08)00662-6

Methods

We used four paradigms:

1. MGS2: Regular MGSs with the target presented for 0.2 s. The subjects tried to memorize the location of the target while staring at a central fixation point that was switched off after a memorization delay of 3 s, which was the go-signal for the subject to perform a saccade towards the memorized location.
2. MGS18: MGSs with a 1.8 s target presentation interval, and a memorization delay of 3 s. An improvement of accuracy as compared to MGS2 would be attributable to an improvement of spatial memory due to a longer availability of the visual information.
3. Multiple memory-guided saccades (MMGS): MGSs with a 1.8 s target presentation interval plus the instruction to make a VGS towards the target, then immediately look back at the central fixation point and wait for the go-signal to execute the MGS, with a memorization delay of 3 s. The VGS could provide the ocular motor system with the memory of the movement, and it could also reinforce spatial memory with a non-retinal input (efference copy or extraocular proprioception) corresponding to the target position. In addition, making a VGS to the same target just prior to the MGS could result in a training effect.
4. Trained memory-guided saccades (TMGS): As in MMGS but the peripheral target was shown 10 times, and the subjects had to make 10 VGSs before making the MGS. Although we expect that a training effect would mainly improve the accuracy of the first MGS both in the MMGS and in the TMGS paradigms, we predicted it to be larger in the TMGS than in the MMGS paradigm. A first set of experiments based on paradigms 1, 2, and 3 was carried out in Pavia by using the scleral search coil technique (Skalar 3020 system) on 12 healthy subjects (8 women, 4 men, mean age: 26.9 years); a second set of experiments based on paradigms 1, 3, and 4 was carried out in Hamburg by using the infrared reflection technique (Ober 2 system) on (8 women, 3 men, mean age: 37.5 years). Eye movements were recorded and calibrated monocularly. Interactive software analysed the saccades off-line by identifying the beginning and the end of each saccade based on threshold velocity criteria. The difference in eye position at these two points corresponded to the pulse amplitude (pulse$A$). One additional mark identified the position the eye reached after all the corrective saccades and before the reappearance of the target (final position). The difference between the starting and final positions corresponded to the final amplitude (final$A$). The target amplitude was identified by target$A$. We computed the amplitude errors (E) as: pulse$E = ln|1-(\text{pulse}A/\text{target}A)|$ and final$E = ln|1-(\text{final}A/\text{target}A)|$, and the amplitude error difference (ED) as $\text{ED} = |1-(\text{final}A/\text{target}A)| - |1-(\text{pulse}A/\text{target}A)|$. We approximated a normal distribution of the values using a natural logarithm transformation, and we used the absolute value to express a scatter of the MGSs endpoints despite the presence of both hypo- and hyper-metric saccades. For each subject, and paradigm, we computed the mean value of pulse$E$, final$E$, and ED. For each set of experiments, we compared the three paradigms by using repeated measure analyses of variance and contrast analyses on all the parameters previously listed.

Results

The first set of experiments (Table 1) showed that the pulse$E$ and the final$E$ mean values did not differ in the three paradigms (MGS2, MGS18, and MMGS), whereas a significant difference was detectable for the ED. The ED was always negative, namely the residual error at the position reached by the pulse was always bigger than the residual error at the final position. The MGS2 paradigm showed the smallest ED (namely the smallest improvement of accuracy obtained by the corrective saccades), and the MMGS paradigm

Table 1. Repeated measure analyses of variance (F and p values) to compare the mean value of pulse amplitude error (pulse$E$), final amplitude error (final$E$), and amplitude error difference (ED) from different memory-guided saccade paradigms (see Methods)

Paradigm	First set of experiments			Second set of experiments		
	pulse$E$	final$E$	ED	pulse$E$	final$E$	ED
MGS2	-2.28 ± 0.61	-2.75 ± 0.54	-0.06 ± 0.07	-1.78 ± 0.35	-1.96 ± 0.28	-0.04 ± 0.05
MGS18	-2.17 ± 0.49	-2.71 ± 0.54	-0.07 ± 0.05			
MMGS	-2.15 ± 0.45	-3.00 ± 0.52	-0.10 ± 0.08	-2.17 ± 0.41	-2.52 ± 0.33	-0.04 ± 0.03
TMGS				-2.98 ± 0.33	-2.48 ± 0.33	-0.10 ± 0.05
F	0.56	1.7	4.44	3.31	21.66	3.01
p	0.59	0.23	0.04	0.99	0.01	0.10

showed the largest, and this difference proved to be significant ($F=9.61$, $p=0.01$); the ED in the MGS18 paradigm was not statistically different from both MGS2 and MMGS. The second set of experiments (Table 1) showed that the pulse$E$ and the ED mean values did not differ in the three paradigms (MGS2, MMGS, and TMGS). The final$E$ mean values were significantly different, since the final$E$ of MGS2 was larger than that of MMGS ($F=21.9$, $p=0.001$) and of that TMGS ($F=38.5$, $p<0.001$), whereas MMGS and TMGS were not statistically different. The ED values in the first and the final$E$ values in the second set of experiments both suggest that corrective saccades are more effective in the MMGS and in the TMGS than in the MGS2 paradigm. In the first set of experiment the final$E$ was not significantly influenced by the different paradigms, but was smaller in the MMGS than in the MGS2 and in the MGS18 paradigms. In the second set of experiments the significant effect was detected for the final$E$ but not for the ED values probably because not only the final error but the pulse$E$ also showed a slight (and not significant) reduction from the MGS2 to the MMGS and the TMGS paradigms.

Discussion

Our results showed that the accuracy of MGSs improves when they follow one or many VGSs made towards the same target location (MGS2 vs. MMGS and TMGS comparisons), but not when the target can be visualized for a longer period of time (MGS2 vs. MGS18 comparison). Accordingly, this improvement cannot be explained by a reinforcement of the visual information. We suggest that the observed effect is due to the memory of the ocular movement made previously; moreover, it cannot be explained by a training effect, since (i) the second set of experiments showed no differences between the MMGS and the TMGS paradigms and (ii) the accuracy improvement did not come from an improvement of the pulse accuracy (we never detected a pulse error reduction in the MMGS and in the TMGS paradigms). The increased effectiveness of the corrective saccades (that led to an ED increase in the MMGS paradigm and to a final$E$ reduction in the TMGS paradigm) suggests that the VGS provided a motor template that was memorized and used to check the position error at the end of the first saccade, and to make corrective saccades more accurately. This motor template can correspond either to a memory of movement or of a proprioceptive coding of target position, both of which can be derived from the efference copy of that VGS.

References

Leigh, R.J. and Zee, D.S. (2006) The neurology of eye movements. Oxford University Press, New York.

C. Kennard & R.J. Leigh (Eds.)
Progress in Brain Research, Vol. 171
ISSN 0079-6123

CHAPTER 5.9

Visual vector inversion during memory antisaccades — a TMS study

Thomas Nyffeler*, Mathias Hartmann, Christian W. Hess and René M. Müri

Perception and Eye Movement Laboratory, Departments of Neurology and Clinical Research, Inselspital, Bern University Hospital, University of Bern, Bern, Switzerland

Abstract: In the memory antisaccade task, subjects are instructed to look at an imaginary point precisely at the opposite side of a peripheral visual stimulus presented short time previously. To perform this task accurately, the visual vector, i.e., the distance between a central fixation point and the peripheral stimulus, must be inverted from one visual hemifield to the other. Recent data in humans and monkeys suggest that the posterior parietal cortex (PPC) might be critically involved in the process of visual vector inversion. In the present study, we investigated the temporal dynamics of visual vector inversion in the human PPC by using transcranial magnetic stimulation (TMS). In six healthy subjects, single pulse TMS was applied over the right PPC during a memory antisaccade task at four different time intervals: 100 ms, 217 ms, 333 ms, or 450 ms after target onset. The results indicate that for rightward antisaccades, i.e., when the visual target was presented in the left screen-half, TMS had a significant effect on saccade gain when applied 100 ms after target onset, but not later. For leftward antisaccades, i.e., when the visual target was presented in the right screen-half, a significant TMS effect on gain was found for the 333 ms and 450 ms conditions, but not for the earlier ones. This double dissociation of saccade gain suggests that the initial process of vector inversion can be disrupted 100 ms after onset of the visual stimulus and that TMS interfered with motor saccade planning based on an inversed vector signal at 333 ms and 450 ms after stimulus onset.

Keywords: sensorimotor transformation; human; posterior parietal cortex

Introduction

In the antisaccade task, two mental processes are required: the inhibition of triggering a reflexive saccade towards the stimulus, and the inversion of the visual vector, i.e., the amplitude of the stimulus from one hemifield to the other. Whereas the dorsolateral prefrontal cortex seems to play the crucial role in inhibiting reflexive saccade triggering (Pierrot-Deseilligny et al., 2003; Munoz and Everling, 2004; Ploner et al., 2005), recent data from human and monkey studies suggest that the posterior parietal cortex (PPC) might be the area involved in the process of visual vector inversion (Everling et al., 1998; Zhang and Barash, 2000; Zhang and Barash, 2004; Medendorp et al., 2005; Moon et al., 2007; Nyffeler et al., 2007). It is assumed that the vector of a visual stimulus is first perceived and inverted in the PPC contralateral to the stimulus and then transferred to the PPC ipsilateral to the stimulus for motor planning of the antisaccade.

The aim of the present study was to investigate the temporal dynamics of visual vector inversion in the human PPC by using an interference

*Corresponding author. Tel.: +41316322111;
Fax: +41316329679; E-mail: thomas.nyffeler@insel.ch

DOI: 10.1016/S0079-6123(08)00663-8

technique. In six healthy subjects, the right PPC was stimulated with single pulse transcranial magnetic stimulation (TMS) at four distinct time intervals during a memory-delayed antisaccade task, and the amplitude of leftward and rightward antisaccades was analysed. We hypothesized that early stimulation during the delay should impair the process of inversion of the visual vector from the left to the right hemifield. Hence, we expected that, after early stimulation, the gain of rightward antisaccades would be reduced. Later stimulation, on the other hand, was expected to impair the inverted visual vector signal, which is stored to plan the antisaccade, resulting in a decreased gain of leftward antisaccades.

Methods

Subjects

Six subjects volunteered for the study (five males and one female). Their mean age was 24 (range 20–31 years), one subject was left-handed. All participants had normal or corrected-to-normal visual acuity and gave their informed consent prior to participation. The study was approved by the ethical committee of the State of Bern and is consistent with the latest Declaration of Helsinki.

Eye movement recording and saccade paradigm

Subjects were seated in total darkness with the head fixed on a chin rest to avoid head movements. Eye movements were measured with an infrared corneal reflection device (Iris Skalar, Delft, Netherlands) with a spatial resolution of 0.1° and sampling rate of 1000 Hz. The digitized signals were stored on the computer for off-line analysis. The recording session began with a calibration procedure, during which the subjects were instructed to look at lateral targets appearing in a staircase pattern to the right and then to left, with amplitudes of 8°, 10°, 12°, 14°, and 16°. Additional calibrations were regularly performed throughout the session. At the outset of each trial, a central fixation point was presented. After pseudo-randomized durations of 1500 ms to 2900 ms, a left or right lateral target was presented for 250 ms, with variable amplitudes (8°, 10°, 12°, 14°, or 16° from the central fixation point). After a delay of 1000 ms, the central fixation point extinguished, which was the "go" signal for the subjects to perform a saccade towards the mirrored location of the peripheral target. Then, after another 1000 ms, the mirror-positioned target was presented to allow for a corrective saccade, if necessary.

TMS procedure

TMS was applied by using a MagStim high-speed stimulator (MagStim Company Limited, Wales, UK) with a round coil (diameter 120 mm). The individual motor threshold was determined by stimulating the right motor cortex with single pulses of increasing intensity, until a slight muscle twitch of the subject's relaxed small hand muscles was detected. Stimulus intensity was then set to 65% of the stimulator output, which was 110–130% of the relaxed individual motor threshold of the small hand muscles. The right PPC was localized as previously described (Müri et al., 1996), using the international 10–20 system of electrode placement (P4). The direction of the inducing current was from posterior to anterior, the handle of the coil was pointed backwards.

Four different TMS conditions were tested: single pulses were applied 100 ms, 217 ms, 333 ms, or 450 ms after target onset. The experiment was performed in two test sessions, separated by one week. In the first session, saccade trials without TMS as well as the 100 ms and 450 ms conditions were conducted. In the second session, trials without TMS and with TMS and the 217 ms and 333 ms conditions were performed. Each session consisted of 23 blocks with eight trials in fixed pseudo-random order. Thus, for each saccade direction and TMS condition, 30 trials were acquired per session.

Data analysis

In a first step, saccades performed in the wrong direction or executed before the "go" signal were excluded from analysis (22%). Then, for each subject, the mean gain of mirror antisaccades (i.e., the ratio of first saccade amplitude to theoretical saccade amplitude) was calculated for each condition (i.e., no stimulation, TMS at 100 ms, TMS at 217 ms, TMS at 333 ms, and TMS at 450 ms) and direction (leftward and rightward saccades). Since two different sessions were performed, the data of the four TMS conditions were normalized in relation to their respective baseline condition (i.e., without stimulation), by setting the no stimulation condition to 100%, while the gain for each stimulation condition was expressed as percentage of it (mean relative saccade gain). For statistical analyses, Mann–Whitney U-Tests were used.

Results

For rightward antisaccades, i.e., when the visual target was presented in the left screen-half, TMS had a significant effect on saccade gain when applied 100 ms after target onset (mean relative gain: 93.45, SD = 3.76; $p < .01$). No significant effect was observed when TMS was applied 217 ms (mean relative gain: 96.63, SD = 9.08), 333 ms (mean relative gain: 97.27, SD = 6.45), and 450 ms (mean relative gain: 99.07, SD = 9.90) after target onset. For leftward antisaccades, i.e., when the visual target was presented in the right screen-half, a significant TMS effect was found for the stimulation time 333 ms (mean relative gain: 94.32, SD = 3.92; $p = 0.002$) and 450 ms (mean relative gain: 92.47, SD = 6.48; $p = 0.002$) but not earlier, i.e., at 100 ms (mean relative gain: 102.37, SD = 4.96) and at 217 ms (mean relative gain: 98.23, SD = 4.06).

The mean relative saccade gain after TMS is shown in Fig. 1.

Discussion

The present study shows that TMS applied over the right PPC significantly interfered with visual vector inversion during a memory-delayed antisaccade paradigm. This interference was time and side specific: stimulation early during the memory delay, i.e., 100 ms after target presentation, caused ipsilateral antisaccade inaccuracy, while later stimulation, i.e., at 333 ms and 450 ms, resulted in contralateral antisaccade inaccuracy. This

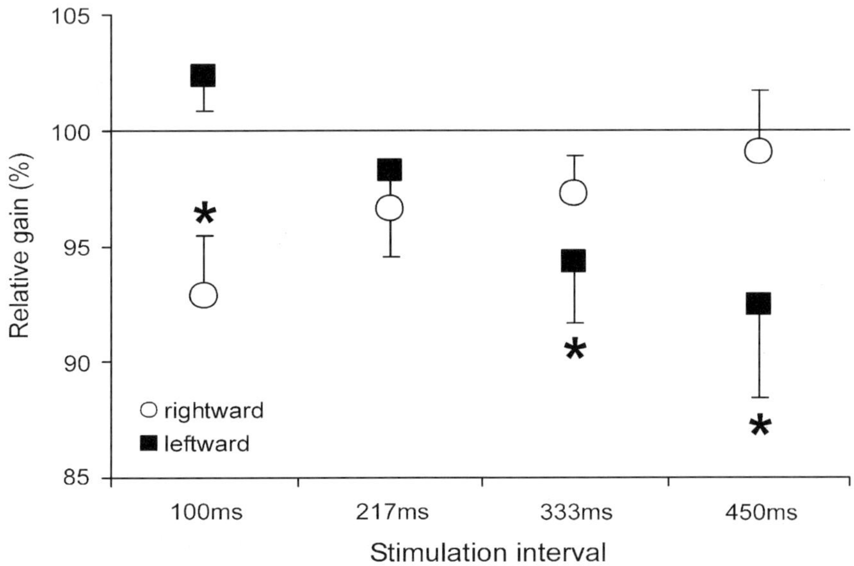

Fig. 1. Mean relative gain of antisaccades (SEM) is shown for the four stimulation conditions. TMS had a significant effect on rightward antisaccade gain when applied 100 ms after target onset ($p = 0.002$), but not for later conditions. For leftward antisaccades, a significant TMS effect was found at 333 ms ($p = 0.002$) and 450 ms ($p = 0.002$), but not for earlier conditions.

double dissociation suggests that TMS disrupted the initial process of vector inversion during the antisaccade task 100 ms after onset of the visual stimulus and interfered with motor saccade planning based on a stored vector signal at 333 ms and 450 ms after stimulus onset.

Until today, the role of the human PPC in visual vector inversion during the antisaccade task has been examined with a lesion approach (Nyffeler et al., 2007) and by using correlational methods (Everling et al., 1998; Medendorp et al., 2005; Moon et al., 2007). Using event-related potential analysis, Everling et al. (1998) showed that a negative potential occurs in the PPC contralateral to the visual stimulus, followed by a second potential in the PPC ipsilateral to the visual stimulus after a delay of 30–90 ms. A similar shift of activity from the PPC of one hemisphere to the other was found in a study using event-related functional magnetic resonance imaging (fMRI) in a paradigm of memory-delayed pro and antisaccades (Medendorp et al., 2005). These results led to the assumption that the vector of a visual stimulus is perceived and inverted in the contralateral PPC. The vector inversion signal is then transferred, probably transcallosally, to the ipsilateral PPC for motor planning of the antisaccade.

The results of the present TMS study show that the process of vector inversion can be critically disrupted 100 ms after onset of the visual stimulus. The relative gain of rightward antisaccades was significantly reduced when the right PPC was stimulated at this time interval, but not later. This finding suggests that the signal of visual vector inversion is rapidly computed and ready to be transferred to the homologue PPC of the other hemisphere where it is used to plan the antisaccade. In fact, the double dissociation found in the relative gain of antisaccades suggests that vector inversion takes place between 100 ms and 333 ms. Our findings, obtained with an interference approach, nicely fit with results of a recent study using a correlational method. To characterize the temporal profile of visual vector inversion during the antisaccade task, Moon et al. (2007) used magnetencephalography. A significant activity was found in the left PPC at 130 ms when a leftward antisaccade had to be made and an activity beginning at 220 ms when a rightward antisaccade was performed. Furthermore, our results are in line with findings from electrophysiological studies in the monkey. Zhang and Barash (2000, 2004) described a subset of neurons in the lateral intraparietal area (LIP) that show an early activity when the visual stimulus matched the contralateral receptive field of the cell and, later, a "paradoxical" activity when the visual target was presented ipsilaterally. A subanalysis revealed that the mean time interval between stimulus onset and this paradoxical activity, i.e., the time it needed to perceive, invert, and transfer the visual vector signal from the opposite hemisphere to the LIP, was 285 ms.

References

Everling, S., Spantekow, A., Krappmann, P. and Flohr, H. (1998) Event-related potentials associated with correct and incorrect responses in a cued antisaccade task. Exp. Brain Res., 118: 27–34.

Medendorp, W.P., Goltz, H.C. and Vilis, T. (2005) Remapping the remembered target location for anti-saccades in human posterior parietal cortex. J. Neurophysiol., 94: 734–740.

Moon, S.Y., Barton, J.J.S., Mikulski, S., Polli, F.E., Cain, M.S., Vangel, M., Hämäläinen, M.S. and Manoach, D.S. (2007) Where left becomes right: a magnetencephalographic study of sensorimotor transformation for antisaccades. Neuroimage, 36: 1313–1323.

Munoz, D.P. and Everling, S. (2004) Look away: the antisaccade task and the voluntary control of eye movement. Nat. Rev. Neurosci., 5: 218–228.

Müri, R.M., Vermersch, A.I., Rivaud, S., Gaymard, B. and Pierrot-Deseilligny, C. (1996) Effects of single-pulse transcranial magnetic stimulation over the prefrontal and posterior parietal cortices during memory-guided saccades in humans. J. Neurophysiol., 76: 2102–2106.

Nyffeler, T., Rivaud-Pechoux, S., Pierrot-Deseilligny, C., Diallo, R. and Gaymard, B. (2007) Visual vector inversion in the posterior parietal cortex. NeuroReport, 18: 917–920.

Pierrot-Deseilligny, C., Müri, R.M., Ploner, C.J., Gaymard, B., Demeret, S. and Rivaud-Pechoux, S. (2003) Decisional role of the dorsolateral prefrontal cortex in ocular motor behaviour. Brain, 126: 1460–1473.

Ploner, C.J., Gaymard, B.M., Rivaud-Pechoux, S. and Pierrot-Deseilligny, C. (2005) The prefrontal substrate of reflexive saccade inhibition in humans. Biol. Psychiatry, 57: 1159–1165.

Zhang, M. and Barash, S. (2000) Neuronal switching of sensorimotor transformations for antisaccades. Nature, 408: 971–975.

Zhang, M. and Barash, S. (2004) Persistent LIP activity in memory antisaccades: working memory for a sensorimotor transformation. J. Neurophysiol., 91: 1424–1441.

C. Kennard & R.J. Leigh (Eds.)
Progress in Brain Research, Vol. 171
ISSN 0079-6123

CHAPTER 5.10

Predictive signals in the pursuit area of the monkey frontal eye fields

K. Fukushima*, T. Akao, N. Shichinohe, T. Nitta, S. Kurkin and J. Fukushima

Department of Physiology, Hokkaido University School of Medicine, Sapporo, Japan

Abstract: In order to pursue a moving target with our eyes, visual motion-signals are converted into eye movement commands. Because of delays in processing visual information, prediction is necessary to compensate for those response-delays and maintain target images on the foveae. Previous studies showed that the majority of FEF pursuit neurons receive visual signals related to actual and predicted target motion. However, in those studies, discharge related to the memory of visual motion could not be separated from that related to prediction. To distinguish the two, while fixating a stationary spot, monkeys were required to memorize the direction of random dot motion (cue-1). After a delay (delay-1), a second cue (cue-2) instructed the monkeys to prepare either pursuit in the memorized direction or to maintain fixation. After a second delay (delay-2), the monkeys selected the correct response. In virtually all tested neurons that showed a visual motion-response to cue-1, the response was not maintained during the delay-1. The majority of responsive neurons were modulated during cue-2 and delay-2. Changing the delay-2 duration also changed the duration of discharge modulation, suggesting that delay-2 modulation was predictive. These results suggest that activity related to visual motion-memory was not conveyed by the discharge of caudal FEF pursuit neurons.

Keywords: frontal eye fields; smooth pursuit; visual motion; memory; prediction; monkey

Introduction

Obtaining accurate visual information about moving objects is essential in an environment in which we are surrounded by moving as well as stationary objects. The smooth pursuit system uses retinal image-slip-velocity information of targets to match eye velocity in space to actual target velocity. To compensate for the delays involved in processing visual-motion information and/or eye-velocity commands, the pursuit system must use prediction and is quite efficient at doing so, especially when target movement is periodic. Prediction occurs not only in the motor commands, but also in the sensory and/or perception circuits. For example, visual responses anticipate the eventually renewed direction and speed of the target movement of a temporarily occluded visual input (Umeno and Goldberg, 1997). Such a mechanism may use memory (see Leigh and Zee, 2006 for a review).

The caudal part of the simian frontal eye fields (FEF) in the fundus of the arcuate sulcus contains pursuit neurons (MacAvoy et al., 1991; Gottlieb et al., 1993, 1994; Tanaka and Fukushima, 1998; Akao et al., 2005, 2007; Kurkin et al., 2007). The majority of FEF pursuit neurons carry visual signals about the direction and velocity of target motion (Fukushima et al., 2000, 2002). These studies also

*Corresponding author. Tel.: +81 11 706-5038; Fax: +81 11 706-5041; E-mail: kikuro@med.hokudai.ac.jp

DOI: 10.1016/S0079-6123(08)00664-X

indicate that caudal FEF pursuit neurons are involved in prediction of target motion (see Fukushima et al., 2006 for a review). As illustrated in Fig. 1A, response delays seen at the onset of abrupt sinusoidal target motion were compensated in the succeeding cycles (arrows). The response delays clearly seen in the first cycle were compensated by the second cycle (Fig. 1B). Moreover, as illustrated in Fig. 1C–E, these neurons also showed predictive "visual" responses to target motion when trained monkeys fixated a stationary spot while a second spot moved in various directions even when it was visible only for half of the trajectory (Fig. 1D). Preferred directions were similar with and without a visible target (blanking vs. non-blanking conditions, Fig. 1E). These results suggest that the predictive responses of pursuit neurons in the caudal FEF contain visual components that reflect direction and speed of invisible target motion (i.e., estimates of target-velocity, Robinson et al., 1986) when its movement is predictable. However, in these studies (Fig. 1A, B), modulation related to the visual motion-memory could not be separated from predictive responses related to pursuit. In the present study, we used an oculomotor task designed to separate the two, and examined discharge characteristics of neurons in the FEF pursuit area.

Methods

Two monkeys (*Macaca fuscata*) were used. All the procedures were evaluated and approved by the Animal Care and Use Committee of the Hokkaido University School of Medicine. Our methods for animal preparation, training, recording, and data analysis are described elsewhere in detail, and summarized here only briefly. Monkeys' heads were firmly restrained in the primate chair in the stereotaxic plane, and they were trained to pursue a target spot that moved in the frontal plane. Vertical and horizontal components of eye movements were recorded using a scleral search coil. Extracellular recordings were made mostly in the fundus of the arcuate sulcus as reported previously, while the monkeys pursued a moving target (Fukushima et al., 2000, 2002).

The task conditions are illustrated in Fig. 2. A red fixation spot appeared and the monkeys were required to fixate it (Fig. 2-1). At cue-1, two random-dot patterns with different colours were shown for 0.5 s (Fig. 2-2); one moved in one of the eight directions separated by 45° at 10°/s, and the other remained stationary. We presented the moving pattern in one of the four directions in a block of trials: horizontal, vertical, or two diagonal directions. In each direction (e.g., horizontal), the frequency of occurrence of the direction of the moving pattern (e.g., right or left) was equal. The monkeys were required to remember which pattern moved in which direction. After a delay (Fig. 2-3, delay-1 of 1 s), a stationary pattern with one of the two previous colours was presented as the second cue (Fig. 2-4, cue-2). The colour of cue-2 instructed the monkeys to prepare to pursue a spot that would move in the direction instructed by cue-1 or to maintain fixation of the stationary spot. The pattern used for cue-2 was different from the cue-1 stationary pattern to make sure that the monkeys did not respond to the pattern itself. After the second delay, which lasted for 1–4 s (Fig. 2-5, delay-2) with a fixed interval (e.g., 2 s) in a block of trials, the monkeys were required to perform the required eye movement by selecting the correct spot (Fig. 2-6, action). For this, the fixation spot, although it remained stationary, was also split into two other identical spots, and one moved in the direction instructed by the cue-1 moving pattern and the other moved in the opposite direction at 10°/s. The monkeys were required to act correctly, either to pursue the correct spot or not to pursue by maintaining fixation of the spot that remained stationary. The frequency of occurrence of the fixation condition was set at 24% of the trials, and in the remaining 76% of the trials the monkeys were required to pursue one of the two moving spots as described above.

The reward circuits compared position signals of the fixation spot (Fig. 2-1–5) and the correct target spot during the action period (Fig. 2-6) with the monkey's eye position signals. If the monkey's gaze was within the error window of $\pm 3°$ while it was on, apple juice was automatically delivered to the animal at the end of each trial (Fig. 2, reward). If the monkey's gaze was outside the error

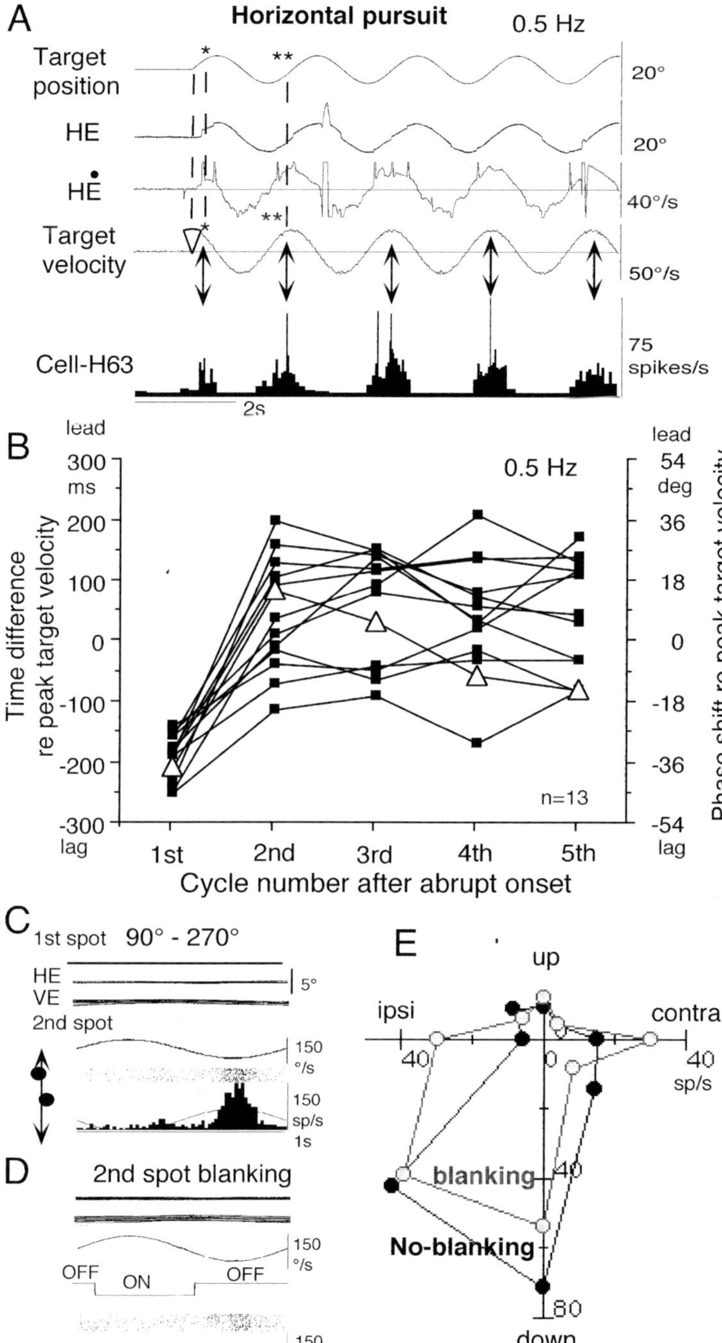

Fig. 1. Predictive visual responses of pursuit neurons in the caudal FEF during abrupt onset, sinusoidal target motion. (A) Discharge of a representative neuron. (B) Plot of the time difference between peak discharge and peak target velocity of 13 neurons for the first 5 cycles of sinusoidal target motion at 0.5 Hz. Abrupt target motion was applied along the preferred direction of each neuron. Responses of each cell are connected by lines. The neuron shown in A is plotted as open triangles in B. HE and HE vel are horizontal eye position and velocity, respectively. (C–E) Visual motion-response of another pursuit neuron. The monkey fixated a stationary spot, while the second test spot moved sinusoidally along four different directions (E, vertical, horizontal, two diagonal directions). In C and D, the second spot moved vertically when it was visible continuously (C) and when it was extinguished for more than half of each cycle as indicated (OFF). Traces in C and D are first spot position horizontal and vertical eye positions (HE, VE), second spot velocity, and rasters and histograms of cell responses. E shows directional tuning of this neuron with (open circles) and without (filled circles) blanking the second spot. Adapted with permission from Fukushima et al. (2002).

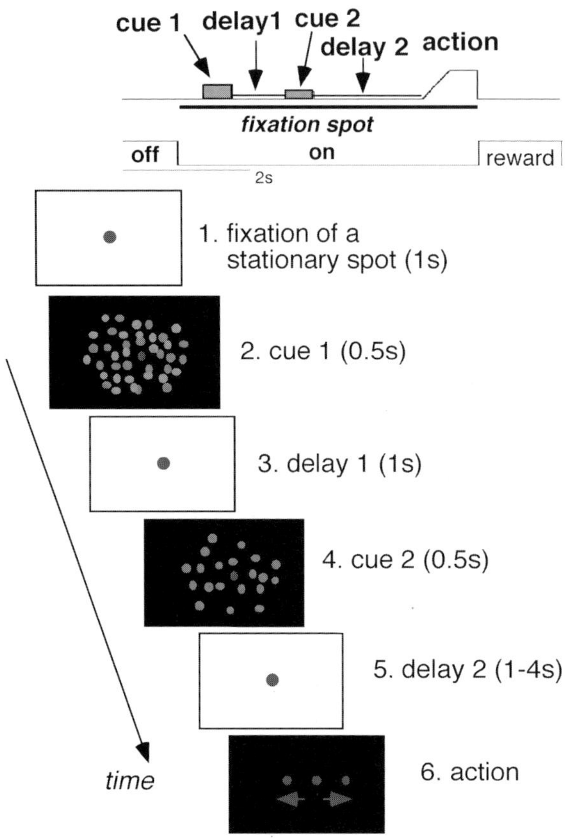

Fig. 2. The task conditions used in the present study. For further explanation, see text.

window, the trial was aborted and then started again. Typically, the monkeys were trained to perform this task over several months. After the error rate became less than 20%, we started recording from the caudal FEF in the fundus of the arcuate sulcus (MacAvoy et al., 1991; Tanaka and Fukushima, 1998; Fukushima et al., 2000, 2002).

Results

Eye-movement responses

Early in their training (typically after 6 months of training), the two monkeys learned the tasks (error rate < 20%) and performed the final action using saccades with latencies typically between 260 ms and 300 ms, followed by smooth-pursuit eye movements (Fig. 3, early training, eye pos and vel). Later (typically after 8 months of training), saccade latency shortened typically to 220 ms. Moreover, preceding the saccades, smooth-pursuit eye movements appeared with latencies typically of 130–150 ms (Fig. 3, eye vel, arrow). The latencies of smooth pursuit were longer than those when a single spot was presented and moved in the identical trajectory along one of the eight directions in darkness, typically 100 ms as reported previously (Leigh and Zee, 2006). Thus, our monkeys performed voluntary smooth-pursuit eye movements with reasonably short latencies in our task conditions.

Discharge of task-related neurons in the caudal FEF

We analysed the activity of a total of 94 neurons that exhibited modulation during our task (Fig. 2) and that were also tested for sinusoidal pursuit. Of these, 63 neurons (67%) also exhibited consistent modulation during sinusoidal pursuit, indicating that they were pursuit neurons (MacAvoy et al., 1991; Fukushima et al., 2000).

As illustrated in Fig. 4A, B, the majority of task-related neurons (64/94 = 68%) responded during cue-2 and delay-2. This example neuron was a pursuit neuron with a rightward-preferred direction. It exhibited a clear response to cue-2 (Fig. 4B, arrow). This response was not simply a visual response because, as shown in Fig. 4C, the identical pattern stimuli elicited minimal modulation when the identical pattern in cue-1 instructed the monkey in cue-2 to prepare leftward pursuit after the delay-2 (Fig. 4C, arrow). This difference in discharge modulation suggests that this neuron responded to the instruction of cue-2 (Fig. 4B, arrow). This was further confirmed by changing the colour of the cue-1 moving patterns. Even with different colours, this neuron started responding in a few trials to the cue-2, similar to the response shown in Fig. 4B, when it required the monkey to prepare for rightward pursuit after the delay-2 (not shown). In about half of neurons that exhibited delay-2 modulation (31/64 = 48%), their discharge

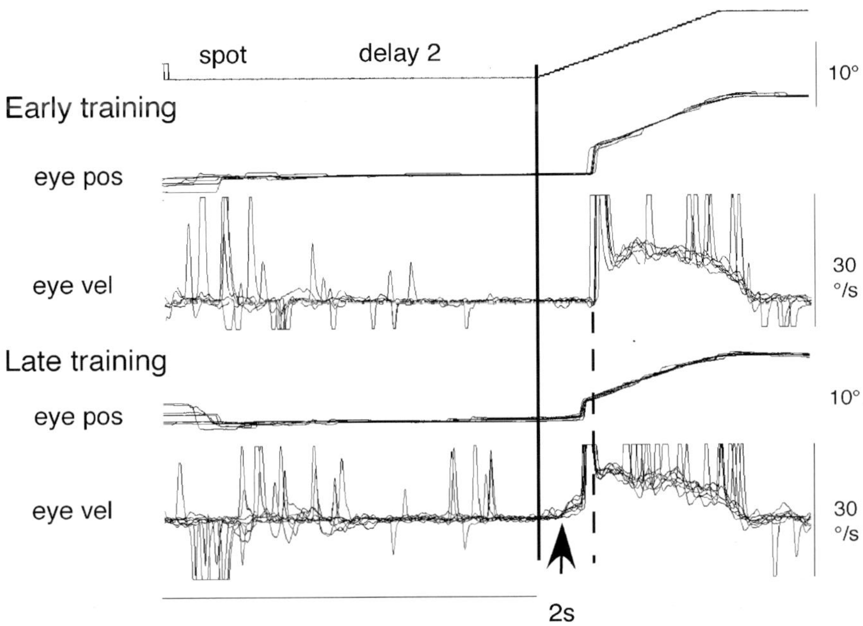

Fig. 3. Representative eye movement responses during early and late training. Eye position (pos) and velocity (vel) traces are shown during the delay-2 and action period. The vertical straight line indicates the onset of spot motion during the action period illustrated in Fig. 2. The dashed line indicates the onset of saccades during the early training. The arrow indicates that smooth pursuit appeared prior to saccades during the late, but not early, training.

was specific to the preferred pursuit directions (Fig. 4B, C).

Changing delay-2 duration also changed the duration of discharge modulation of this neuron (Fig. 4B) during delay-2, if the monkey was required to perform rightward pursuit after delay-2 (not shown), thus confirming a predictive nature of the delay-2 modulation. Similar predictive discharge was observed in both pursuit neurons and non-pursuit neurons in the caudal FEF.

About one-third of the total (34/94 = 36%) exhibited visual motion-responses to cue-1. Visual responses were directional and were more frequently observed in pursuit neurons (29/63 = 46%) than non-pursuit neurons (5/31 = 16%). For pursuit neurons, visual-preferred directions were consistent with pursuit-preferred directions of individual neurons as reported previously (Fukushima et al., 2000, 2002). Representative discharge is illustrated in Fig. 4D. This neuron was also a pursuit neuron with a rightward-preferred direction. It exhibited a visual motion-response when the cue-1 pattern moved rightwards, thus requiring the monkey to perform rightward pursuit during the action period. However, the visual motion-response to cue-1 was not maintained, and the response decreased approximately to the resting level during the delay-1. Discharge rate increased again during delay-2, shortly before the actual eye movements towards right (Fig. 4D). The visual motion-responses were not maintained during the delay-1 in any of the 34 neurons that responded to cue-1.

To rule out the possibility that the lack of maintained discharge modulation during delay-1 was due to our task conditions in which we presented the two patterns during the cue-1 (one stationary and the other moving, see Methods section), we modified the task condition so that only a moving pattern was presented during cue-1. If the colour of the stationary pattern presented as cue-2 was the same as that of the cue-1 moving pattern, the monkeys were required to perform smooth pursuit during the action period (Fig. 2-6). However, if their colours were different, the monkeys were required not to pursue but to

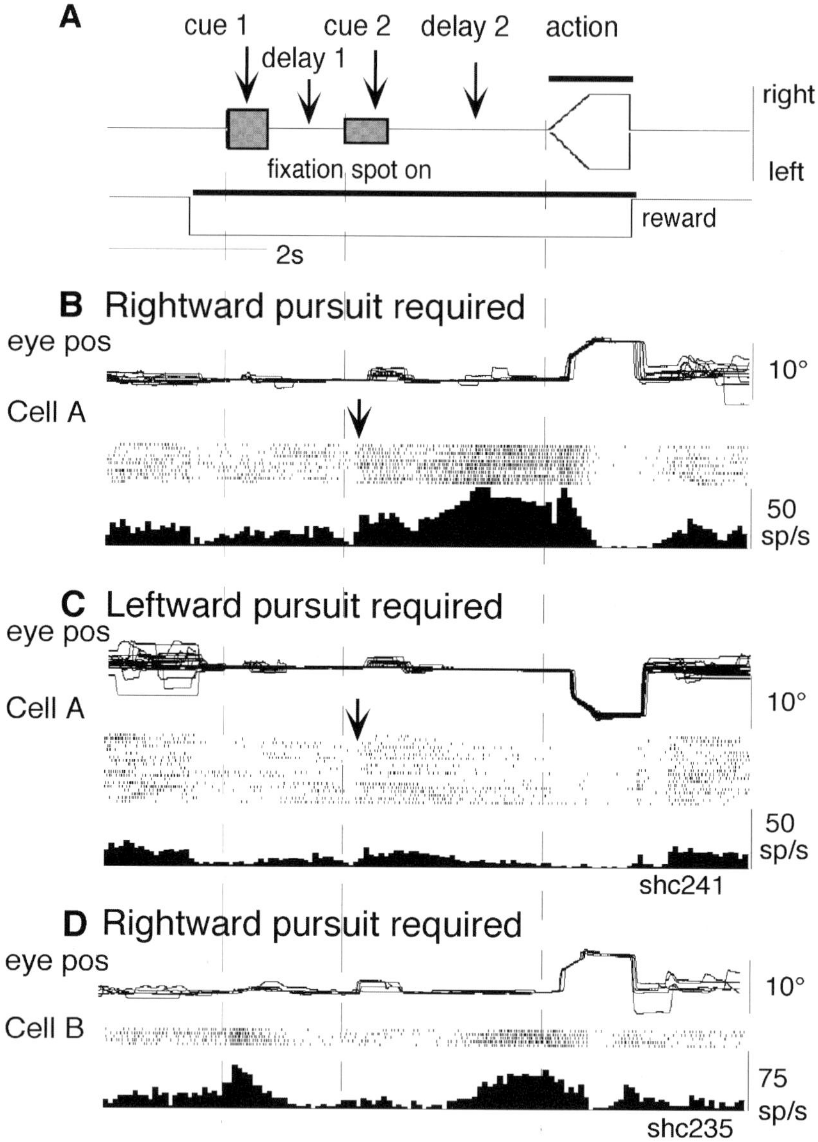

Fig. 4. Discharge of representative pursuit neurons in the caudal FEF. The time scale of the task condition shown in A is common for the two representative neurons (B–C) cell A and (D) cell B as indicated by dashed lines. Arrows in C and D indicate that the identical pattern stimuli (cue-2) induced very different discharge depending on the instruction of cue-2. For further explanation, see text.

maintain fixation of a stationary spot during the action period. Using this revised task condition, we recorded seven pursuit neurons that exhibited visual motion-responses to cue-1. The results were similar to the original cue-1 condition (Fig. 4D), and none of the seven neurons exhibited maintained discharge during the delay-1 period.

Lesion or chemical inactivation of the FEF pursuit area is known to impair smooth pursuit (Lynch, 1987; Keating, 1991, 1993; Shi et al., 1998; Fukushima et al., 1999). Muscimol injection (10 μg) into the FEF pursuit area in this study not only resulted in low gain (~0.5) in smooth-pursuit eye movements, but also impaired delay compensation of pursuit eye movements during sinusoidal pursuit (Fig. 1A). Moreover, the initial smooth eye velocity that appeared prior to the saccades (Fig. 3, late training, eye vel, arrow) was virtually abolished after muscimol infusion into the caudal FEF. These results suggest that this area is necessary for predictive pursuit and response delay compensation during pursuit.

Discussion

For the neurons tested in the caudal FEF that exhibited modulation in our task conditions ($n = 34$, Fig. 2), none showed a visual motion-response to cue-1 that was maintained during delay-1 despite the fact that these neurons increased their discharge again during delay-2 and the action periods (Fig. 4A, D). Because changing the delay 2 duration also changed the duration of the discharge modulation and because in about half of the neurons that exhibited delay-2 modulation their discharge was specific to the preferred pursuit directions (Fig. 4B, C), we conclude that the discharge modulation during the delay-2 period reflects the prediction of the forthcoming target motion and/or pursuit eye movements. These results suggest that the activity coding visual motion-memory is not conveyed directly as the predictive discharge of caudal FEF pursuit neurons. Because our monkeys performed the task appropriately, visual motion-memory must have been stored somewhere outside the caudal FEF. To understand where and how this is done, future studies must compare discharge modulation of pursuit-related cortical areas (i.e., supplementary eye field, SEF and medial superior temporal, MST) using the same task conditions. Such experiments are in progress.

Acknowledgements

We thank Dr. C.R.S. Kaneko of the Washington National Primate Research Center, University of Washington, Seattle, for his valuable comments on the manuscript. This research was supported by Grant-in-Aid for Scientific Research on Priority Areas (System study on higher-order brain functions) (17022001) and (B)(18300130) from the MEXT of Japan.

References

Akao, T., Kurkin, S., Fukushima, J. and Fukushima, K. (2005) Visual and vergence eye movement related responses of pursuit neurons in the caudal frontal eye fields to motion-in-depth stimuli. Exp. Brain Res., 164: 92–108.

Akao, T., Saito, H., Fukushima, J., Kurkin, S. and Fukushima, K. (2007) Latency of vestibular responses of pursuit neurons in the caudal frontal eye fields to whole body rotation. Exp. Brain Res., 177: 400–410.

Fukushima, J., Akao, T., Kurkin, S., Kaneko, C.R.S. and Fukushima, K. (2006) The vestibular-related frontal cortex and its role in smooth-pursuit eye movements and vestibular-pursuit interactions. J. Vestib. Res., 16: 1–22.

Fukushima, K., Sato, T. and Fukushima, J. (1999) Vestibular-pursuit interactions: gaze-velocity and target-velocity signals in the monkey frontal eye fields. Ann. N.Y. Acad. Sci., 871: 248–259.

Fukushima, K., Sato, T., Fukushima, J., Shinmei, Y. and Kaneko, C.R.S. (2000) Activity of smooth pursuit-related neurons in the monkey periarcuate cortex during pursuit and passive whole body rotation. J. Neurophysiol., 83: 563–587.

Fukushima, K., Yamanobe, T., Shinmei, Y. and Fukushima, J. (2002) Predictive responses of peri-arcuate pursuit neurons to visual target motion. Exp. Brain Res., 145: 104–120.

Gottlieb, J.P., MacAvoy, M.G. and Bruce, C.J. (1993) Smooth eye movements elicited by microstimulation in the primate frontal eye field. J. Neurophysiol., 69: 786–799.

Gottlieb, J.P., MacAvoy, M.G. and Bruce, C.J. (1994) Neural responses related to smooth pursuit eye movements and their correspondence with electrically elicited slow eye movements in the primate frontal eye field. J. Neurophysiol., 72: 1634–1653.

Keating, E.G. (1991) Frontal eye field lesions impair predictive and visually-guided pursuit eye movements. Exp. Brain Res., 86: 311–323.

Keating, E.G. (1993) Lesions of the frontal eye field impair pursuit eye movements, but preserve the predictions driving them. Behav. Brain Res., 53: 91–104.

Leigh, R.J. and Zee, D.S. (2006) The Neurology of Eye Movements (4th ed.). Oxford University Press, New York, pp. 188–382.

Lynch, J.C. (1987) Frontal eye field lesions in monkeys disrupt visual pursuit. Exp. Brain Res., 68: 437–441.

Kurkin, S.A., Akao, T., Fukushima, J. and Fukushima, K. (2007) Activity of pursuit neurons in the caudal part of the frontal eye fields during static roll-tilt. Exp. Brain Res., 176: 658–664.

MacAvoy, M.G., Gottlieb, J.P. and Bruce, C.J. (1991) Smooth pursuit eye movement representation in the primate frontal eye field. Cereb. Cortex, 1: 95–102.

Robinson, D.A., Gordon, J.L. and Gordon, S.E. (1986) A model of the smooth pursuit eye movement system. Biol. Cybern., 55: 43–57.

Shi, D., Friedman, H.R. and Bruce, C.J. (1998) Deficits in smooth pursuit eye movements after muscimol inactivation within the primate frontal eye field. J. Neurophysiol., 80: 458–464.

Tanaka, K. and Fukushima, K. (1998) Neuronal responses related to smooth pursuit eye movements in the periarcuate cortical area of monkeys. J. Neurophysiol., 80: 28–47.

Umeno, M.M. and Goldberg, M.E. (1997) Spatial processing in the monkey frontal eye field. I. Predictive visual responses. J. Neurophysiol., 78: 1373–1383.

C. Kennard & R.J. Leigh (Eds.)
Progress in Brain Research, Vol. 171
ISSN 0079-6123

CHAPTER 5.11

Internally generated smooth eye movement: its dynamic characteristics and role in randomised and predictable pursuit

G.R. Barnes* and C.J.S. Collins

Faculty of Life Sciences, University of Manchester, Manchester, UK

Abstract: Although originally examined in the context of prediction, it is now apparent that smooth eye movements generated by internal (extra-retinal) mechanisms play a role in both predictable and randomised pursuit responses. Internally generated responses are characterised by trajectories that begin with an increasing acceleration that develops much more slowly than responses generated through retinal feedback, but which can, nevertheless, reach high velocities. They can be evoked by regularly repeated motion stimuli or by cues that occur at a regular time before target motion onset. Although frequently observed as anticipatory movements, we now provide evidence that such movements also form the basis of the extra-retinal component of a randomised step-ramp response. In such circumstances they also build up slowly in the first second or so of the initial response. They are normally masked in the presence of visual feedback, but can be revealed by prolonged target extinction immediately after response initiation. The key to release internally generated responses in both random and predictable conditions is expectation of future target motion. The key to their functionality is rapid acquisition and storage of velocity and timing information.

Keywords: eye movements; ocular pursuit; anticipation; randomisation; human

Introduction

Although ocular pursuit is clearly dependent on visual input, it is widely accepted that internal (extra-retinal) drive plays a large part in the response, probably through an efference copy, positive-feedback loop. Such an internal loop operates as a form of velocity memory for the perpetuation of pursuit. The fact that smooth eye movement tends to continue during short periods when the moving target disappears, albeit at reduced velocity, provides evidence for this concept (von Noorden and Mackensen, 1962). However, in order to achieve this, the subject must have a strong expectation of target reappearance (Becker and Fuchs, 1985; Bennett and Barnes, 2003), otherwise eye velocity simply decays towards zero with a time constant of ~0.1–0.2 s (Mitrani and Dimitrov, 1978; Pola and Wyatt, 1997).

To date, there have been many experiments investigating the influence of visual input on pursuit response initiation (Carl and Gellman, 1987; Krauzlis and Lisberger, 1994b) but no demonstration of the independent activity of the internal loop in the initial stages of pursuit. On the other hand, it is known that smooth

*Corresponding author. Tel.: +44 161 306 5761;
Fax: +44 161 306 3887; E-mail: g.r.barnes@manchester.ac.uk

DOI: 10.1016/S0079-6123(08)00665-1

eye movements can be generated prior to target movement when a strong expectation of future target is created and reliable timing information is provided (Kowler, 1989; Barnes and Asselman, 1991; Kao and Morrow, 1994). These anticipatory eye movements have much slower onset than the visually driven response, are initiated ~200 ms before target motion onset and are uninfluenced by visual feedback for ~80 ms after target onset. What happens beyond this time has remained controversial, although experiments with stabilised images suggest that the internal drive can be sustained if retinal position and velocity error are eliminated (Barnes et al., 1995).

Although a number of pursuit models based on the efference copy principle have been developed (Robinson et al., 1986; Krauzlis and Lisberger, 1994a), they are unable to account for anticipatory behaviour. On the other hand, the predictions of a model specifically developed to simulate anticipatory behaviour (Barnes and Asselman, 1991), have not been examined in relation to pursuit initiation. A crucial aspect of this model (Fig. 1) is that it relies on sampling and holding visuomotor drive information and relaying it to a short-term memory (MEM) that can retain motion information even after eye movement has ceased. Independent timing of the output from MEM subsequently allows the initiation of appropriately scaled anticipatory eye movements. If the pursuit system does operate by sampling target velocity information in the first moments of target presentation, a very brief presentation of the target should be sufficient to allow eye velocity to continue increasing during a period of extinction. In contrast, a conventional model, with continuous efference copy feedback would, at best, predict that eye velocity is sustained during extinction. In a previous study Churchland et al. (2003) found no evidence of increasing eye velocity, but their extinction period was only 200 ms. In light of other recent experiments (Bennett and Barnes, 2003) we have used much longer extinction periods (up to 600 ms) and do find evidence to support the hypothesis that target velocity information is sampled over the first 150–200 ms of target presentation. We also find evidence that expectation modulates the output of the internal feedback pathway.

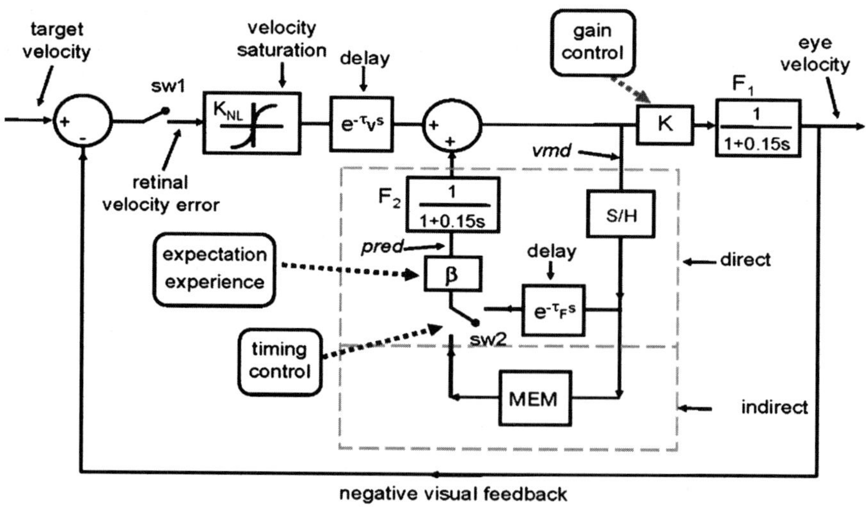

Fig. 1. Model of ocular pursuit. Negative visual feedback is supplemented by direct and indirect internal positive feedback pathways, which have common input through sample and hold system (S/H) that captures data over a period of ~150 ms. Both direct and indirect pathways feed out through an expectation-modulated gain control (β) and filter F_2. In a reactive response S/H output is fed out directly after delay τ_F and is also temporarily stored in MEM. In predictive mode, output of MEM is fed out under timing control to form anticipatory response. K_{NL} = non-linear gain (velocity saturation); K = open-loop gain; τ_V = visuomotor processing delay (~80 ms).

General methods

Two experiments are described; they were conducted with local ethics committee approval and six subjects participated with informed consent. All subjects had normal or corrected-to-normal vision, were healthy and had no relevant medical history. They sat in a dark room with the head fixed by pads to the side and supported on a chin rest. They viewed a flat screen 1.5 m from the head on which a target (diameter 1°) was projected by a purpose-built projector using LEDs as a light source. Eye movements were recorded by infra-red limbus tracking (Skalar Iris) with a resolution of ~5–10 min arc and the signals were sampled at a frequency of 250 Hz. Image motion was controlled by a motor-driven mirror. Eye movements were analysed by methods described previously (Bennett and Barnes, 2003).

Experiment 1: extraction of motion from brief target presentation

In experiment 1, we examined whether velocity information could be reliably extracted within the first 200 ms of motion by presenting step-ramp stimuli in which the initial motion of the target was visible for periods of 50, 100, 150, or 200 ms. In the Mid-ramp Extinction (ME) condition the target was then extinguished for a period of 400 or 600 ms before reappearing and continuing to move along the same trajectory. Subjects thus had a strong expectancy of target reappearance. In the Short Ramp (SR) condition the target did not reappear and subjects thus had no expectation of target reappearance. However, they were still instructed to follow the imagined target motion trajectory. In both conditions the speed (5, 10, 15, and 20°/s), direction (L,R), initial presentation duration (PD), extinction duration (ED), and initial step size of stimuli were fully randomised, so that subjects were unable to predict the upcoming motion. A Control condition without extinction was also examined.

In the ME condition all responses were reactive to target motion onset. Smooth eye velocity responses were characterised by two phases (Fig. 2A–C). There was an initial rapid increase in velocity that started after target extinction for PD < 150 ms and reached a peak well before attaining target velocity. This was followed by a more slowly developing secondary phase that continued throughout extinction. Although Fig. 2 shows means across subjects, similar response patterns were observed in individual subjects. As PD increased from 50 ms to 200 ms, there was a systematic change in response pattern (Fig. 2C). When PD was only 50 ms (Fig. 2A) the initial phase reached a low peak velocity, but then continued to increase throughout extinction. In contrast, when PD was 200 ms (Fig. 2B), the initial phase reached a higher peak and eye velocity then declined to a plateau level that was sustained throughout extinction. The plateau level was notably similar for PD = 150 and 200 ms.

The impression gained from the plots in Fig. 2C is that the response generated during target extinction is composed of two parts, one part that gives rise to the initial rapid increase, but then decays in the manner revealed by the 200 ms initial exposure, and a second component that starts around the same time, but develops more slowly over the period of extinction. Analysis of the magnitude and timing of the initial peak showed that it increased significantly with initial exposure and target velocity (Fig. 3A). Moreover, when this response was compared with the Control condition (Fig. 2C) in which the target remained visible throughout the ramp, they were found to be very similar. This and other evidence suggests that the initial phase is a response to the briefly exposed visual motion stimulus.

In order to analyse the secondary phase of the ME responses we examined eye velocity at the end of extinction for ED = 600 ms. There was a significant increase in end-extinction eye velocity as PD increased from 50 ms to 100 ms, but no further increase as PD increased to 200 ms (Fig. 3B). This analysis reflects mean levels of end-extinction eye velocity averaged across target velocity (see Fig. 3B). When a more detailed examination was made for each level of PD, it was found that the scaling of eye velocity to target velocity gradually improved (i.e., had increasing gain) as initial exposure was increased. In fact, the

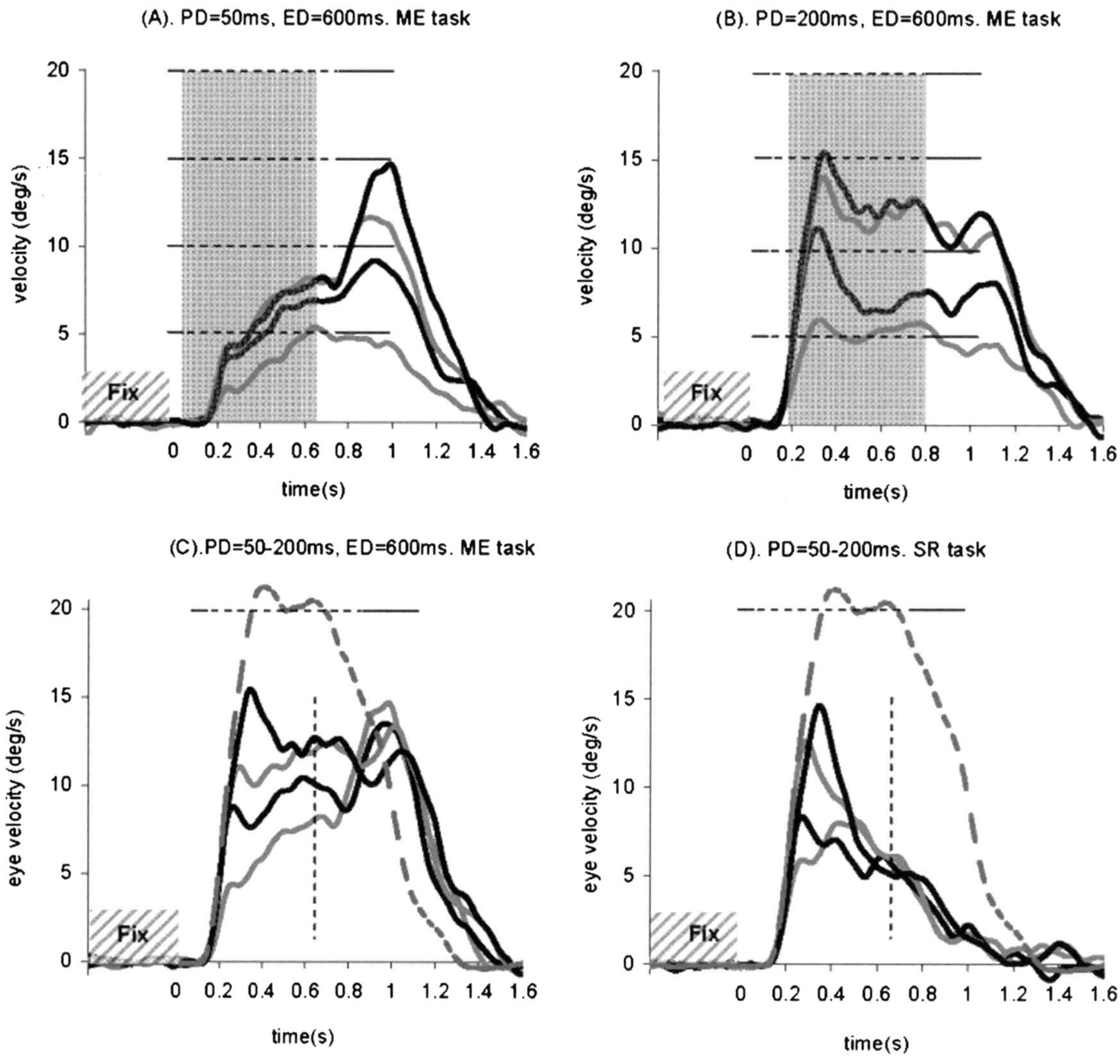

Fig. 2. (A and B) Smooth eye velocity, averaged across six subjects, in Mid-ramp Extinction (ME) condition for each level of target velocity (alternate grey /black traces) and extinction duration (ED = 600 ms). Initial target exposure PD = 50 ms in A, PD = 200 ms in B. Shading indicates extinction period. (C) Average ME responses to 20°/s stimulus as PD increases from 50 ms to 200 ms (alternate grey /black traces), plus 20°/s Control response (dashed line). (D) Average velocity responses to 20°/s stimulus in Short Ramp (SR) condition as PD increases from 50 ms to 200 ms (alternate grey/black traces). Dashed vertical lines mark 650 ms after onset.

relationship between eye velocity and target velocity only attained significance for the 150 ms and 200 ms conditions, suggesting that effective speed discrimination takes at least 150 ms.

The eye movement generated during extinction must predominantly represent the activity of internal drive mechanisms, since it is unrealistic to expect that the visual response would continue to generate an increasing response up to 600 ms after target extinction. However, to say that it is *wholly* dependent on internal drive presumes that the after-effects of the visual stimulus have totally died away. To test this, we also examined responses to the SR condition, in which the same

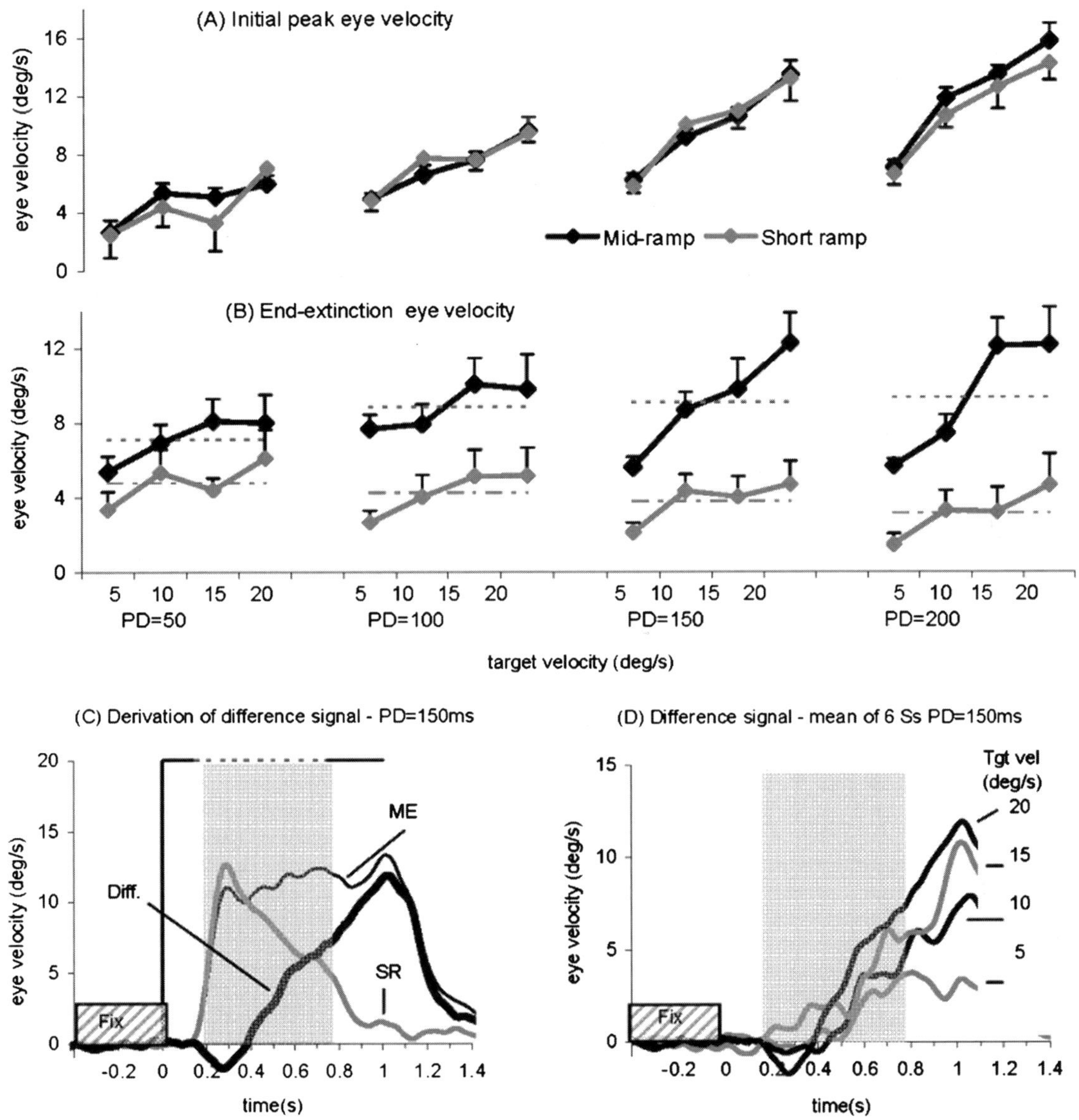

Fig. 3. (A) Initial peak eye velocity and (B) smooth eye velocity (V_{END}) at the time corresponding to the end of extinction for the Mid-ramp Extinction (ME) and Short Ramp (SR) tasks as a function of PD and target velocity. Broken lines indicate mean V_{END} (averaged across target velocity) for each task. Mean of six subjects ±1SE. (C) Derivation of Difference signal by subtraction of SR from ME response. (D) Comparison of Difference signals for PD = 150 ms as target velocity increases (alternate grey/black traces). Mean of six subjects.

initial exposure conditions were presented to the subjects, but in which there was no expectation that the target would reappear. As expected, eye velocity reached an initial peak that was not significantly different from that in the ME condition (Fig. 3A). However, there was no comparable secondary phase of sustained or increasing velocity. Rather, eye velocity decayed to zero in a manner similar to that reported previously (Fig. 2D). This occurred even though the subjects did manage to estimate the position trajectory of the unseen target reasonably well by making a combination of smooth and saccadic eye movements. Although eye velocity initially decayed rapidly in the manner described previously (Mitrani and Dimitrov, 1978; Pola and Wyatt, 1997), once velocity had fallen to $\sim 5^{\circ}/s$, it decayed much more slowly (Fig. 2D). There was thus some residual smooth eye velocity at the end of the extinction period, but its average level did not change with PD (Fig. 3B). Most importantly, it was significantly less than end-extinction velocity in the ME condition (Fig. 3B).

The difference between responses to the ME and SR conditions represents an expectation-dependent component that is responsible for sustaining pursuit in the ME condition. When velocity profiles obtained in the SR condition were subtracted from those in the ME condition, as shown in Fig. 3C, the initial phases cancelled out and the remaining Difference signal exhibited a slowly increasing velocity profile that started ~200–300 ms after target onset. [N.B. Small differences in initial peak velocity led to relatively large errors in determining Difference signal magnitude around the time of initiation, precluding accurate measurement of latency.] Notably, it had a much slower temporal development than the visually driven Control response, but was evidently scaled to target velocity, indicating that it had been based on motion information that was sampled during the initial exposure.

Experiment 2: randomly timed initiation of isolated, internally driven responses

In experiment 2, we attempted to segregate the internal mechanism and test it independently by removing the initial visual input but retaining the randomised timing of stimulus initiation. Since the magnitude of the necessary internal drive could no longer be estimated from target velocity on-the-fly, it was necessary to derive it from another source. One way to accomplish this is to use techniques that have previously been used to evoke anticipatory smooth pursuit. These experiments have shown that when identical motion stimuli are repeated, motion information can be stored and used to scale the magnitude of subsequent anticipatory smooth pursuit (Barnes and Donelan, 1999). Moreover, a recent experiment (Collins and Barnes, 2006) has shown that prolonged periods of anticipatory pursuit can be revealed by extinguishing the target during the initial part of the repeated motion stimulus, thus allowing comparison with the prolonged extinction in the ME task.

In an attempt to show how the internally driven smooth movement might be initiated in a randomised pursuit response, we determined how quickly anticipatory smooth eye movements could be initiated if target motion onset itself formed the relevant timing cue for response initiation. Subjects initially fixated a stationary target for a randomised period of 500–1000 ms. The target was then extinguished for 600 ms but the subject was informed that target extinction indicated that its motion had started. Thus, when it reappeared, the target was already in an eccentric position. We refer to this as the Initial Extinction paradigm. The stimulus was presented in predictable blocks, in which target speed, direction, and ED remained identical for six consecutive presentations. Successive blocks had randomised speed (5, 10, 15, or $20^{\circ}/s$) and direction (L/R). Subjects attempted to follow target motion from its initiation despite being unable to see the initial part of the trajectory. [N.B. This method is similar to the 'gap' technique used by Boman and Hotson (1988), but differs in the important detail that the target was in motion, not stationary, during the 'gap'.]

In the first presentation of the stimulus (Fig. 4A) the subject had no knowledge of target speed, direction, or timing and a reactive response was made ~100 ms after target appearance. In the second presentation (Fig. 4B) the subject began to make anticipatory smooth movement during extinction and, in the third (Fig. 4C), generated a higher velocity response that remained similar for subsequent presentations. Mean velocity profiles were obtained by averaging responses to the final

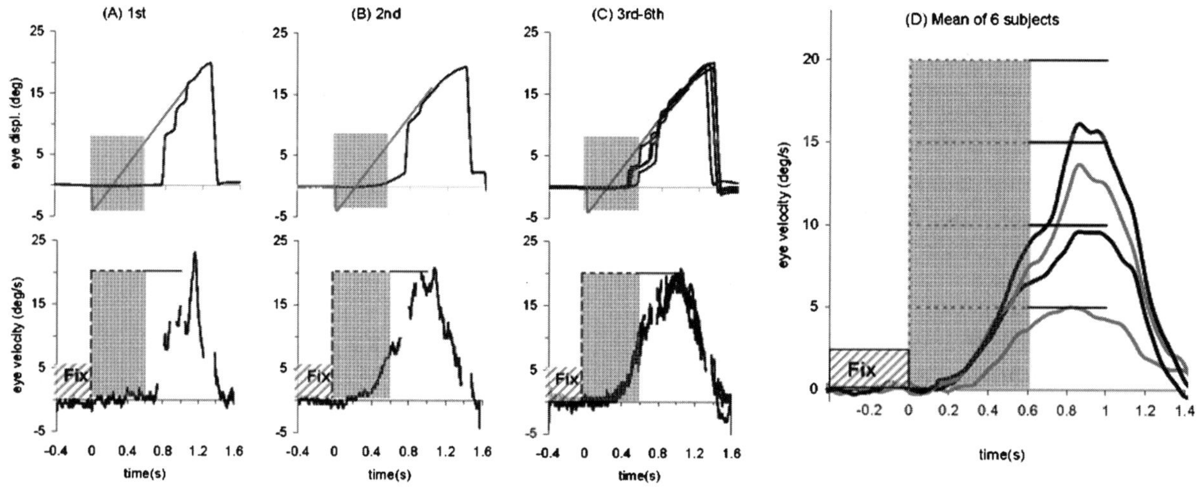

Fig. 4. (A–C) Examples of eye displacement (upper traces) and velocity (lower traces) responses of one subject in Initial Extinction paradigm. Saccades removed without replacement in eye velocity traces. (A) 1st response; (B) 2nd response; (C) 3rd–6th responses overlaid. Subject fixated stationary target during Fix period (actually of randomised duration). (D) Mean smooth eye velocity averaged across six subjects for each level of target velocity (alternate grey/black traces). Shading indicates extinction period.

four presentations (Fig. 4D). All responses were reactive to the offset of fixation, but once initiated, exhibited a relatively slow build up of eye velocity during the remaining extinction interval and increasing eye velocity as target velocity increased. There was no significant difference in latency with speed or direction and the overall mean was 196.8 ms (SD 15.6 ms). Eye velocity at the time of target appearance (V600) increased significantly with target velocity, in line with previous observations relating to anticipatory eye movements (Collins and Barnes, 2006).

Discussion

The objective of these experiments was to investigate the behaviour of internal drive mechanisms in the initial stage of a reactive pursuit response. We assessed this in the ME task by giving a brief presentation of the moving target prior to a long extinction period. There were two parts to the response; an initial rapidly accelerating component that appeared to be dependent on visual input and a secondary component that developed more slowly throughout extinction. When initial exposure was only 50 ms, eye velocity continued to increase during the secondary phase. This finding, of itself, is incompatible with conventional pursuit models in which there is continuous feedback of efference copy (Robinson et al., 1986), since such models would, at best, predict a sustained, not an increasing, velocity. Such behaviour suggests that velocity information (initially of the target) is sampled and held in short-term memory as a reference level in the manner proposed in the model of Fig. 1. If eye velocity has not reached the sampled reference level at the time of extinction (Fig. 2A), eye velocity will simply continue to increase during the extinction period towards the reference level, whereas, if it has exceeded the reference level at extinction (Fig. 2B) it will decrease to that level. Its slow temporal development towards the reference level is dictated by the filter characteristics (F2) embedded in the output of this internal loop. The fact that scaling of end extinction eye velocity to target velocity reaches an optimum for an initial exposure of 150–200 ms implies that this is close to the minimum time required to complete the sampling process.

Crucially, the output of the internal loop is dependent on expectation of future target motion. If that expectation is not present, as in the SR task, the internal drive is weak and results in the eventual decay of eye velocity ($\beta << 1$ in the model). The fact that there is still some residual velocity 600 ms after target extinction implies that there may still be some residual output from this internal loop (i.e., $\beta \neq 0$). Nevertheless, the Difference signal derived by subtraction of SR from ME responses (Fig. 3D) reveals how this internal drive develops over a period when there is no visual input. The increase in the Difference signal with target velocity indicates that its magnitude is dependent on velocity information sampled before target extinction. Its output is delayed by 200–300 ms after target onset, which, in the model, is represented by a further delay (τ_F) within the direct pathway (Fig. 1).

The response in the Initial Extinction paradigm (Fig. 4D) also exhibits a slow development of smooth eye velocity, similar to that of the Difference signal (Fig. 3D) and eye velocity at target onset is scaled to target velocity in line with previous studies of anticipatory pursuit. However, these responses also show that the shortest latency at which this internal drive could be released, even when target velocity is known in advance, is ~200 ms after the cue (fixation offset) that is known to be associated with the start of target motion. In the model (Fig. 1) it is assumed that the anticipatory response is generated through the indirect internal loop, which uses velocity information derived from prior presentations, stored in MEM, as its reference level. Again, its slow development over time is dictated by the filter F2 and the timing of its release is governed by the fixation offset cue. Both the Difference signal (Fig. 3D) and the anticipatory response (Fig. 4D) are very dissimilar to the visually driven control response because the latter does not pass through the additional filter (F2) within the internal feedback loop. Of particular importance is that output from the internal loop is again dependent on expectation of future target motion, a crucial factor in anticipatory response initiation (Kowler, 1989; Barnes et al., 1997).

Acknowledgement

This work was supported by the Medical Research Council.

References

Barnes, G.R. and Asselman, P.T. (1991) The mechanism of prediction in human smooth pursuit eye movements. J. Physiol. (Lond.), 439: 439–461.

Barnes, G.R. and Donelan, A.S. (1999) The remembered pursuit task: evidence for segregation of timing and velocity storage in predictive oculomotor control. Exp. Brain Res., 129: 57–67.

Barnes, G.R., Goodbody, S.J. and Collins, S. (1995) Volitional control of anticipatory ocular pursuit responses under stabilized image conditions in humans. Exp. Brain Res., 106: 301–317.

Barnes, G.R., Grealy, M.A. and Collins, S. (1997) Volitional control of anticipatory ocular smooth pursuit after viewing, but not pursuing, a moving target: evidence for a re-afferent velocity store. Exp. Brain Res., 116: 445–455.

Becker, W. and Fuchs, A.F. (1985) Prediction in the oculomotor system: smooth pursuit during transient disappearance of a visual target. Exp. Brain Res., 57: 562–575.

Bennett, S.J. and Barnes, G.R. (2003) Human ocular pursuit during the transient disappearance of a moving target. J. Neurophysiol., 90: 2504–2520.

Boman, D.K. and Hotson, J.R. (1988) Stimulus conditions that enhance anticipatory slow eye movements. Vision Res., 28: 1157–1165.

Carl, J.R. and Gellman, R.S. (1987) Human smooth pursuit: stimulus-dependent responses. J. Neurophysiol., 57: 1446–1463.

Churchland, M.M., Chou, I.-H. and Lisberger, S.G. (2003) Evidence for object permanence in the smooth-pursuit eye movements of monkeys. J. Neurophysiol., 90: 2205–2218.

Collins, C.J.S. and Barnes, G.R. (2006) The occluded onset pursuit paradigm: prolonging anticipatory smooth pursuit in the absence of visual feedback. Exp. Brain Res., doi:10.1007/s00221-006-0527-2.

Kao, G.W. and Morrow, M.J. (1994) The relationship of anticipatory smooth eye movement to smooth pursuit initiation. Vision Res., 34: 3027–3036.

Kowler, E. (1989) Cognitive expectations, not habits, control anticipatory smooth oculomotor pursuit. Vision Res., 29: 1049–1057.

Krauzlis, R.J. and Lisberger, S.G. (1994a) A model of visually-guided smooth pursuit eye movements based on behavioral observations. J. Comput. Neurosci., 1: 265–283.

Krauzlis, R.J. and Lisberger, S.G. (1994b) Simple spike responses of gaze velocity Purkinje cells in the floccular lobe

of the monkey during the onset and offset of pursuit eye movements. J. Neurophysiol., 72: 2045–2050.
Mitrani, L. and Dimitrov, G. (1978) Pursuit eye movements of a disappearing moving target. Vision Res., 18: 537–539.
Pola, J. and Wyatt, H.J. (1997) Offset dynamics of human smooth pursuit eye movements: effects of target presence and subject attention. Vision Res., 39: 2767–2775.
Robinson, D.A., Gordon, J.L. and Gordon, S.E. (1986) A model of the smooth pursuit eye movement system. Biol. Cybern., 55: 43–57.
von Noorden, G.K. and Mackensen, G. (1962) Pursuit movements of normal and amblyopic eyes: an electro-ophthalmographic study. 1. Physiology of pursuit movements. Am. J. Ophthalmol., 53: 325–336.

C. Kennard & R.J. Leigh (Eds.)
Progress in Brain Research, Vol. 171
ISSN 0079-6123

CHAPTER 5.12

Predictive disjunctive pursuit of virtual images perceived to move in depth

P. Fu[1], R. Chang[1], K. O'Beirne[1] and W.M. King[2,*]

[1]*University of Michigan School of Literature, Science and the Arts, Ann Arbor, MI, USA*
[2]*Department of Otolaryngology and Kresge Hearing Research Institute, Ann Arbor, MI, USA*

Abstract: Human and nonhuman primates predictively use smooth pursuit and saccades to track visual targets that move in a fronto-parallel plane. This behaviour is believed to be facilitated by short-term memory of the target motion and/or an efference copy of the subject's motor effort. Subjects in our experiments tracked dichoptically viewed targets that appeared to move vertically, right or left and towards the subject. The virtual image of the target was tracked using disjunctive smooth pursuit and saccades. To reveal predictive tracking, targets were blanked 100 ms after the onset of motion for intervals of 800 ms. During the blanked interval, subjects initiated pursuit and predictively tracked the unseen virtual image using memory guided eye movements. Our data are consistent with recent electrophysiological studies that describe cells that encode target or eye movements in depth when a target is briefly blanked but pursuit is maintained. However, predictive pursuit of a virtual target with disjunctive eye movements poses a challenge for understanding how a short-term memory store might encode the desired eye movement, its coordinate frame, and how it is transformed into motor commands.

Keywords: humans; smooth pursuit; vergence; anticipatory pursuit; eye movements; sensorimotor

Introduction

Unlike saccades, smooth pursuit eye movements cannot, in general, be produced voluntarily in the absence of a moving, visible target (Heywood and Churcher, 1971). Several studies have shown, however, that once initiated, smooth pursuit eye movements may be continued predictively for periods of up to several seconds when the target is transiently occluded if subjects expect it to reappear (Becker and Fuchs, 1985; Barnes et al., 1987; Bennett and Barnes, 2003, 2004; Churchland et al., 2003; Madelain and Krauzlis, 2003). Furthermore, Barnes and others have shown that under certain circumstances, robust anticipatory pursuit movements may be initiated prior to the onset of visible target motion and the initiation of visually guided smooth pursuit (Boman and Hotson, 1988; Kowler, 1989; Barnes and Asselman, 1991; Barnes and Donelan, 1999; Bennett and Barnes, 2003). In general, these experiments have employed step-ramp target movement (Rashbass, 1961) in a single direction on a tangent screen or monitor placed in front of the viewer. When such targets are transiently blanked or occluded, smooth eye movements are observed to continue, but at a slower

*Corresponding author. Tel.: +01 734 936 9418;
Fax: +01 734 936 9412; E-mail: wmking@umich.edu

DOI: 10.1016/S0079-6123(08)00666-3

speed during the intervals of occlusion. This observation has naturally led to "inertial" models of predictive pursuit wherein pursuit velocity is stored in an integrator that discharges during the occlusion interval (Robinson et al., 1986; Krauzlis and Lisberger, 1994; Churchland et al., 2003). The neural correlate of this process is not known but may be a neural representation of the target's motion or an efference copy of the motor command for the pursuit eye movements (Churchland et al., 2003). Neurons in the frontal eye fields (Fukushima et al., 2002; Barborica and Fererra, 2003; Akao et al., 2005a) and in area MST (Ilg and Thier, 2003; Akao et al., 2005a) have been recorded that discharge during occluded pursuit consistent with this idea. However, this simple model is likely to be insufficient to account for all aspects of predictive pursuit. Barnes and colleagues have argued persuasively for an additional cognitive component underlying anticipatory and predictive pursuit (Bennett and Barnes, 2006; Collins and Barnes, 2006). They employ the term "cognitive expectation," which improves upon repeated exposure to repetitive target presentations. Cognitive expectation is a more complex memory process than implied by simple integration. Their hypothesis was further supported by experiments wherein the target accelerated during an interval of occlusion. Subjects scaled their predictive pursuit responses according to the expected acceleration after a series of blocked trials (Bennett and Barnes, 2006).

To further distinguish reflexive models of pursuit from models employing a cognitive memory element, we studied pursuit during which target trajectory was altered in direction while the target was transiently blanked. In these experiments, target motion shifted abruptly from horizontal to oblique in 2-dimensions, or if viewed dichoptically, appeared to abruptly emerge from the tangent screen and move towards the subject in 3-dimensions. Subjects anticipated and predictively tracked changes in target direction, in 2- and in 3-dimensions. Such behaviour would be unexpected based on "inertial" models of pursuit but could be produced by cognitive models such as those hypothesized by Bennett and Barnes (2006).

Methods

Data were collected from three human subjects recruited for this study. All methods and procedures were approved by the University of Michigan Medical School's Institutional Review Board for the protection of human subjects. Subjects were seated in a dimly illuminated room facing a rear projection screen located in front of them and approximately 1 m from their eyes. Subjects were instructed to fixate and track any target presented on the screen as accurately as possible even if it transiently disappeared.

Target presentation

Targets consisted of small squares projected onto the tangent screen. Targets were presented separately to each eye using paired DLP projectors (PT-D5500, Panasonic, Inc.) and polarizing filters placed in front of the projectors and the subjects' eyes. The targets were created using custom software developed for this project. The software ran on a dedicated PC (2 GHz Reason Computer) with an Nvidia Quadro FX1100 graphics card that operated in stereo mode and drove each projector at a 60 Hz refresh rate. The pixel location of a target, its size, and the eye to which it was visible were specified throughout a trial by the experimenter. A target could be blanked or occluded simply by not specifying its location during the desired time interval. Virtual motion of a target towards the subject was achieved by separating the left and right eye target positions on the screen; if a target seen by the left eye were positioned to the right of a target seen by the right eye (creating crossed disparity), then the fused image was perceived to float in space between the screen and the subject. To assist subjects in fusing the right and left eye images, the size of the target images was increased with their separation as a monocular cue to target distance.

To accurately time target motion and instants when changes in direction occurred, we turned on or off pixels in a corner of the screen out of sight of the subject. The state of these pixels was detected by light sensitive diodes and these signals were sampled at the same time as the eye movement data.

Measurement of eye movements

Eye movements were measured binocularly using a video tracking system (Chronos Vision, Skalar Medical BV). The system consists of paired infrared sensitive CCD cameras attached to a headband that fixes the camera positions relative to the eyes. During fixation, this system can measure eye position with an accuracy of 0.1° (Clarke et al., 2002). At the beginning of each experiment, the eye tracker was calibrated by having the subject fixate targets located at $\pm 10°$ up, down, and left or right of straight ahead. During an experiment, eye movements were sampled at 200 frames per second per eye. Video data and signals from the light sensitive diodes were saved for later off-line analysis.

Experimental procedures

After calibration of the eye tracker, subjects performed 6 blocks of trials, each consisting of 20 trials. The first, third, and fifth blocks were training blocks consisting of 20 identical trials wherein the target was always visible. The second, fourth, and sixth blocks were experimental blocks wherein the first 5 trials were identical to the previous training block but the last 15 trials included a time interval during which no target was visible. In all trials, the target initially moved horizontally to the right at a constant speed of ~5, ~10, or ~15°/s. After ~500 ms, the target abruptly changed direction and appeared to move downward and towards the subject. In order to track the target, subjects were required to generate conjugate horizontal eye movements for 500 ms followed by downward and convergent eye movements. The target positions were specified so that vergence was asymmetric and entirely by the left eye (monocular pursuit). During blanked trials, the target was turned off 100 ms after the onset of motion for ~800 ms. Thus, subjects saw 100 ms of horizontal target motion, 800 ms without any visible target, and then the reappearance of a target moving in a different direction. Data were collected continuously throughout each block of 20 trials.

Data analysis

Using software supplied with the eye tracker (Iris), the video data were analysed to produce records of horizontal and vertical eye movements sampled at 200 Hz. These data and the diode signals were transformed into Spike2 (Cambridge Electronic Design) files for display and analysis. Within the Spike2 environment, the data were re-sampled at 1 kHz and smoothed over a 10 ms interval. This process removed artefacts associated with the video recording without materially affecting the subsequent analysis. Horizontal and vertical eye velocities were computed for each eye and displayed. Saccade parameters were measured and the saccades were deleted from each trial using an interactive routine. After saccade removal, each trial was divided into 75 bins of 25 ms duration and mean eye position and speed were computed within each bin. These data were transferred to Matlab (Mathworks, Inc.) for final analysis and the creation of figures.

Results

Subjects varied in their ability to fuse and track the virtual target motion, but all of them produced segments of anticipatory and predictive pursuit as shown in Fig. 1. Figure 1A shows vertical eye position plotted as a function of horizontal eye position. Initially, the target (heavy black trace) and both eyes moved conjugately to the right. After ~500 ms, the target seen by the left eye abruptly began to move downward and horizontally so that its trajectory was oblique. The left eye followed the target and changed direction after about a 100 ms delay (thin black traces). Simultaneously, the target seen by the right eye also moved downward but stopped moving horizontally. The right eye (thin grey traces) faithfully followed the new target trajectory producing an asymmetric vergence eye movement. After the direction change, subjects perceived the target to move downward and towards themselves. For the

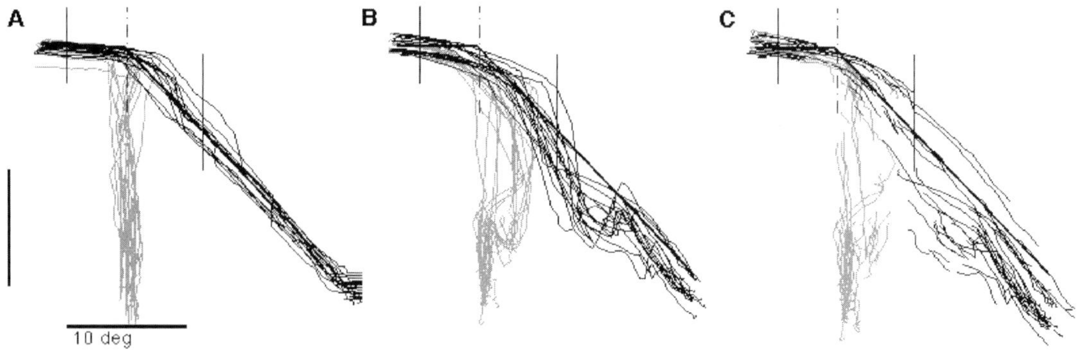

Fig. 1. Three-dimensional smooth pursuit of a virtual target that changes direction. Multiple trial data showing vertical eye position plotted against horizontal eye position. (A) Visible target; (B) target transiently occluded for 800 ms; (C) same as B except saccades removed. In all panels, thin black traces are left eye position, thin grey traces are right eye position, heavy black trace is left eye target position. For the trials shown in panels B and C, solid vertical lines indicate the interval during which the target was blanked. Dashed vertical line is the time at which target direction changes. Calibration bars are 10°.

experiment shown, the target was aligned horizontally with the right eye, so the convergent pursuit was monocular.

After training, subjects generated anticipatory and predictive pursuit of the target before and during the interval of occlusion (Fig. 1B). The solid vertical lines indicate the occlusion interval which began ~100 ms after the onset of the target's horizontal motion. The overlaid traces show that both eyes continued to predictively pursue the target after it was turned off (first solid line) and anticipated the target's downward motion before it began (dashed line). Furthermore, the binocular trajectory of the target was remembered and the eyes converged during the occluded interval in the absence of any visible target. The target was tracked using coordinated disjunctive saccades and smooth pursuit eye movements. To visualize the contribution of smooth pursuit alone, Fig. 1C shows the same data as Fig. 1B except saccades have been removed from the data. This panel shows that disjunctive and even monocular smooth pursuit contributed to tracking performance.

Figure 2 shows averaged eye velocity during the same trials depicted in Fig. 1. Consider the time interval prior to the target's change of direction (dashed line). Whether the target was visible or occluded, the eye velocity profiles in both eyes were conjugate and similar. Figure 2A, B show that when the target was visible, horizontal eye velocity increased rapidly and reached a peak before the direction change occurred. Both eyes appeared to anticipate the change in direction and there was a decline in smooth eye velocity that preceded the direction change. Following the direction change, right eye velocity was reduced to nearly 0°/s regardless of occlusion whilst left eye velocity recovered towards left eye target velocity. If the target was occluded, however, left eye velocity was less than target velocity. Vertical eye velocity was characterized by a conjugate anticipatory component that is evident in the time interval before the direction change (Fig. 2C, D), and which was the same regardless of whether or not the target was occluded during the block of trials. Following the direction change, vertical eye velocity rapidly increased to match target velocity if the target were visible (black traces). During blocks of occluded trials (grey traces), vertical eye velocity also increased, but at the same rate as the prior anticipatory response, suggesting a continuation of that response.

Qualitatively similar results were obtained for the 5 and 10°/s blocked trials except that the anticipatory and predictive eye velocities were scaled according to the speed of the target. The scaling of eye velocity with different target velocities demonstrates that the subjects were not simply producing a default eye movement.

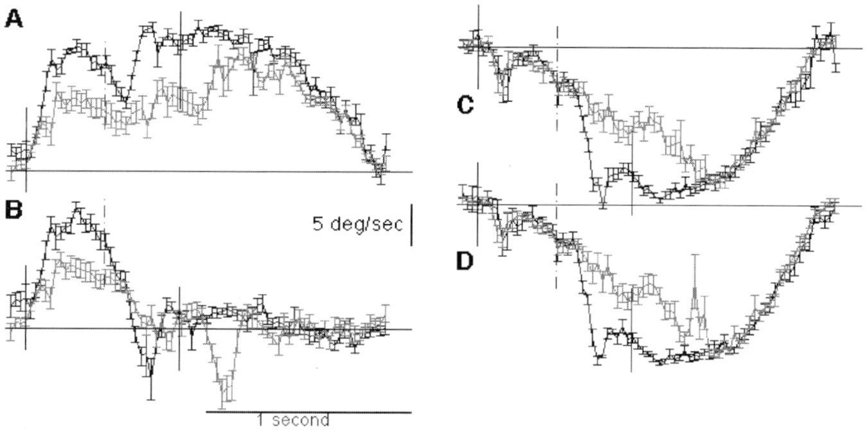

Fig. 2. Averaged eye velocity ±SE during visible (black traces) and occluded trials (grey traces). (A) Horizontal right eye velocity. (B) Horizontal left eye velocity. (C) Vertical right eye velocity. (D) Vertical left eye velocity. Same trials as shown in Fig. 1.

Discussion

Previous studies have demonstrated what have been called anticipatory and predictive pursuit of smoothly moving visual targets (Becker and Fuchs, 1985; Boman and Hotson, 1988; Kowler, 1989; Bennett and Barnes, 2003, 2004; Collins and Barnes, 2006). Predictive or memory guided pursuit is shown to occur when a target is transiently occluded and the pursuit eye movement continues during the interval of occlusion (Becker and Fuchs, 1985; Barnes and Asselman, 1991; Bennett and Barnes, 2003). Anticipatory pursuit is most clearly identified if subjects are asked to pursue a target whose motion onset is cued by an auditory tone (Barnes and Donelan, 1999). Previous studies also showed that anticipation is enhanced after a series of repetitive trials (Becker and Fuchs, 1985; Barnes and Asselman, 1991). The extent to which these two processes might share common processing elements has not been determined.

We have employed a novel stimulus to investigate the nature of predictive smooth pursuit eye movements. Previous studies used blocked trials to investigate the influence of prior experience on anticipatory and predictive pursuit. Bennett and Barnes also used accelerating target trajectories to probe the voluntary or reflexive nature of predictive pursuit. We used an abrupt change of direction in the target trajectory to extend these studies to situations where predictive movements occurred in the absence of prior eye velocity in that direction. For example, prior to the direction change, there was no vertical velocity associated with our target movement (although there was anticipatory vertical pursuit). During occluded trials, the onset of vertical eye velocity occurred at the middle of the occlusion interval. Thus, there was no prior vertical target movement to use as an input to a predictive controller such as a velocity memory element or integrator. However, there were anticipatory vertical eye movements. For example, in the vertical direction, there was substantial vertical eye velocity in the interval prior to the onset of vertical target velocity (Fig. 2C, D). The anticipatory vertical component of eye velocity was the same whether or not the target was occluded up to the moment of direction change. If the target were visible, vertical eye velocity rapidly increased after the direction change to match target velocity. However, if the target were occluded, the anticipatory vertical velocity increased at the same rate as it had before. The simplest interpretation of these data is that the vertical eye velocity is entirely anticipatory and characterized by a single process that is independent of the target's occlusion.

Horizontal eye velocity cannot be characterized in the same way. Left and right eye speeds are identical during the interval prior to the direction change (Fig. 2A, B). The eye movements are conjugate, but eye speed is less than target speed during trials when the target is occluded. This result is similar to the findings of previous studies (Becker and Fuchs, 1985) and could be modelled using a simple predictive controller based on short-term memory or velocity integration (Churchland et al., 2003). However, during blocked trials when the target is occluded (Fig. 2A, B, grey traces), left eye velocity increases after the direction change (Fig. 2A) and right eye velocity decreases (Fig. 2B). This response cannot be easily modelled by a simple velocity integrator unless one imagines a change in time constant linked to the expected time of the direction change and separate integrators assigned to store left and right eye velocity. As an alternative, we propose that these data may also be explained by a "cognitive expectation" model such as proposed by Barnes and his colleagues (Bennett and Barnes, 2006). This model suggests that after several trials, there is a build-up in short-term memory of a model of the target's trajectory in space and time that can be used to generate anticipatory and/or predictive tracking eye movements (Collins and Barnes, 2006).

The virtual target paradigm employed in these experiments raises questions about the coordinate frame within which such a model might be constructed. For example, the 3-dimensional aspect of the target's trajectory is a perceptual construct, illusory and dependent on the subject's ability to fuse the disparate images in both eyes. Is there a short-term memory of the perceived target trajectory or of the different retinal paths taken by the targets seen by both eyes?

A related question is concerned with the coordinate frame of the remembered target trajectory and the coordinate frame of the tracking eye movements. Our data clearly show anticipatory or predictive asymmetric vergence eye movements in the absence of any visual target. The velocity of these movements was scaled to the expected speed of the target during a block of trials. If the target were visible, then the disjunctive eye movements could be accounted for by fusional vergence. But what is the neural basis for the generation of pursuit in depth in the absence of a visual stimulus? An alternative hypothesis is that the 3-dimensional trajectory of the target is transformed into left and right eye motor commands so that each eye is controlled monocularly (King and Zhou, 1995, 2000, 2002; Zhou and King, 1998). This hypothesis is not dependent on visually guided fusional vergence and can account for the disjunctive movements observed during intervals of occlusion; however, it does not address the question of where the coordinate transformation might occur. Studies of neurons in FEF and in MST have identified cells that continue to discharge when nonhuman primates track targets during intervals of occlusion (Ferrera and Lisberger, 1997; Ilg and Thier, 2003; Fukushima et al., 2004; Akao et al., 2005a, b). These cells encode target motion in what appears to be a head centred coordinate frame and not an eye centred frame. Thus, if there is a coordinate transformation, it must occur nearer the motor output, perhaps in the cerebellum or brainstem where neurons encode eye movements in an eye centred frame (Zhou and King, 1998; King and Zhou, 2000).

Acknowledgement

We are grateful to Mo Kakwan who developed the programme used to create the virtual target display.

References

Akao, T., Kurkin, S.A., Fukushima, J. and Fukushima, K. (2005a) Visual and vergence eye movement-related responses of pursuit neurons in the caudal frontal eye fields to motion-in-depth stimuli. Exp. Brain Res., 164: 92–108.

Akao, T., Mustari, M.J., Fukushima, J., Kurkin, S. and Fukushima, K. (2005b) Discharge characteristics of pursuit neurons in MST during vergence eye movements. J. Neurophysiol., 93: 2415–2434.

Barborica, A. and Fererra, V.P. (2003) Estimating invisible target speed from neuronal activity in monkey frontal eye field. Nat. Neurosci., 6: 66–74.

Barnes, G.R. and Asselman, P.T. (1991) The mechanism of prediction in human smooth pursuit eye movements. J. Physiol. (Lond.), 439: 439–461.

Barnes, G.R. and Donelan, A.S. (1999) The remembered pursuit task: evidence for segregation of timing and velocity storage in predictive oculomotor control. Exp. Brain Res., 129: 57–67.

Barnes, G.R., Donnelly, S.F. and Eason, R.D. (1987) Predictive velocity estimation in the pursuit reflex response to pseudo-random and step displacement stimuli in man. J. Physiol., 389: 111–146.

Becker, W. and Fuchs, A.F. (1985) Prediction in the oculomotor system: smooth pursuit during transient disappearance of a visual target. Exp. Brain Res., 57: 562–575.

Bennett, S.J. and Barnes, G.R. (2003) Human ocular pursuit during the transient disappearance of a moving target. J. Neurophysiol., 90: 2504–2520.

Bennett, S.J. and Barnes, G.R. (2004) Predictive smooth ocular pursuit during the transient disappearance of a visual target. J. Neurophysiol., 92: 578–590.

Bennett, S.J. and Barnes, G.R. (2006) Smooth ocular pursuit during the transient disappearance of an accelerating target: the role of reflexive and voluntary control. Exp. Brain Res., 175: 1–10.

Boman, D.K. and Hotson, J.R. (1988) Stimulus conditions that enhance anticipatory slow eye movements. Vision Res., 28: 1157–1165.

Churchland, M.M., Chou, I.-H. and Lisberger, S.G. (2003) Evidence for object permanence in the smooth-pursuit eye movements of monkeys. J. Neurophysiol., 90: 2205–2218.

Clarke, A.H., Ditterich, J., Druen, K., Schonfeld, U. and Steineke, C. (2002) Using high frame rate CMOS sensors for three-dimensional eye tracking. Behav. Res. Methods Instrum. Comput., 34: 549–560.

Collins, C.J.S. and Barnes, G.R. (2006) The occluded onset pursuit paradigm: prolonging anticipatory smooth pursuit in the absence of visual feedback. Exp. Brain Res., 175: 11–20.

Ferrera, V.P. and Lisberger, S.G. (1997) Neuronal responses in visual areas MT and MST during smooth pursuit target selection. J. Neurophysiol., 78: 1433–1446.

Fukushima, J., Akao, T., Takeichi, N., Kurkin, S., Kaneko, C.R. and Fukushima, K. (2004) Pursuit-related neurons in the supplementary eye fields: discharge during pursuit and passive whole body rotation. J. Neurophysiol., 91: 2809–2825.

Fukushima, K., Yamanobe, T., Shinmei, Y., Fukushima, J., Kurkin, S. and Peterson, B.W. (2002) Coding of smooth eye movements in three-dimensional space by frontal cortex. Nature, 419: 157–162.

Heywood, S. and Churcher, J. (1971) Eye movements and the afterimage 1. Tracking the afterimage. Vision Res., 11: 1163–1168.

Ilg, U.J. and Thier, P. (2003) Visual tracking neurons in primate area MST are activated by smooth pursuit eye movements of an "imaginary" target. J. Neurophysiol., 90: 1489–1502.

King, W.M. and Zhou, W. (1995) Initiation of disjunctive smooth pursuit in monkeys: evidence that Hering's law of equal innervation is not obeyed by the smooth pursuit system. Vision Res., 35: 3389–3400.

King, W.M. and Zhou, W. (2000) New ideas about binocular coordination of eye movements: is there a chameleon in the primate family tree? Anat. Rec. New Anat., 261: 153–161.

King, W.M. and Zhou, W. (2002) Neural basis of disjunctive eye movements. Ann. N.Y. Acad. Sci., 956: 273–283.

Kowler, E. (1989) Cognitive expectations, not habits, control anticipatory smooth oculomotor pursuit. Vision Res., 29: 1049–1057.

Krauzlis, R.J. and Lisberger, S.G. (1994) A model of visually-guided smooth pursuit eye movements based on behavioral observations. J. Comput. Neurosci., 1: 265–283.

Madelain, L. and Krauzlis, R.J. (2003) Effects of learning on smooth pursuit during transient disappearance of a visual target. J. Neurophysiol., 90: 972–982.

Rashbass, C. (1961) The relationship between the saccadic and smooth tracking eye movements. J. Physiol. (Lond.), 159: 326–338.

Robinson, D.A., Gordon, J.L. and Gordon, S.E. (1986) A model of the smooth pursuit eye movement system. Biol. Cybern., 55: 43–57.

Zhou, W. and King, W.M. (1998) Premotor commands encode monocular eye movements. Nature, 393: 692–695.

C. Kennard & R.J. Leigh (Eds.)
Progress in Brain Research, Vol. 171
ISSN 0079-6123

CHAPTER 5.13

Tracking in 3-D space under natural viewing condition

H.A. Rambold[1,2,*], T. Sander[1], A. Sprenger[1] and C. Helmchen[1]

[1]*Department of Neurology, University of Lübeck, Lübeck, Germany*
[2]*Laboratory of Sensorimotor Research, National Eye Institute, National Institutes of Health, Bethesda, MD, USA*

Abstract: To track a small visual target in 3-D space, the two eyes have to move in different directions and/or at different velocities. This tracking might be accomplished by a disjunctive pursuit system, which uses separate motion processing of each individual eye but no disparity signal (hypothesis 1), or by the conjugate pursuit and the vergence system (hypothesis 2). To test the validity of the two hypotheses we recorded eye movements in five healthy human subjects with the scleral search-coil method. A small dim laser stimulus was presented on an earth horizontal platform. A position-ramp stimulus was presented in eight different directions: rightward or leftward, convergence or divergence, or a combination of them. We compared a fusible with an un-fused and a monocular viewing condition to assess whether a disparity signal is needed for 3-D tracking. Fusion was prevented by a vertical prism. We compared the monocular with the prism viewing condition to examine the effect of retinal motion signals of either one or both eyes on the tracking performance in the absence of disparity signals. Results revealed severe impairment of tracking in depth, while tracking in pure horizontal directions remained unaffected during the prism and monocular as compared to the binocular viewing condition. These data support hypothesis 2.

Keywords: vergence; eye movements; smooth pursuit

Introduction

Tracking a target in natural space, which includes different depth planes, requires disjunctive eye movements. The two eyes might move in different directions and/or at different velocities. Controversy still surrounds the questions of how these eye movements are generated and what signals are used. Two major hypotheses are put forward: (1) the disjunctive pursuit hypothesis proposes that retinal motion signals of the two eyes are processed independently for the right and the left eye (King and Zhou, 1995). (2) Hering's hypothesis predicts that tracking in space is accomplished by two independent eye movement systems, the conjugate smooth pursuit and the disparity vergence system (Maxwell and King, 1992; Semmlow et al., 1998). The latter hypothesis is supported by a study using large un-fuse images which impaired vergence eye movements (Erkelens and Collewijn, 1985). A study using un-fused slowly moving small targets in 3-D space in humans has not yet been conducted. This study is designed to elucidate whether either hypothesis can account for visual tracking in 3-D space.

*Corresponding author. Tel.: +301 496 9080;
Fax: +301 402 0511; E-mail: ramboldh@nei.nih.gov

DOI: 10.1016/S0079-6123(08)00667-5

Methods

Binocular recordings were obtained from five healthy subjects (five males), aged 28–44 years, in accordance with the Declaration of Helsinki (1994) and the local ethics committee. All subjects had normal vision and stereovision (Stereo Optical Co., Inc., Chicago, OH, USA). Binocular eye movements were recorded with the Scleral Search Coil System (Remmel Labs, MD, USA). The search coils were calibrated using a combined offline in vitro and in vivo calibration, as described elsewhere (Eggert et al., 1999; Rambold et al., 2002). Single laser targets (spot diameter 0.1°, 635 nm, LISA laser products OHG, Katlenberg-Lindau, Germany) were projected on a horizontal plane in an otherwise dark surrounding (Rambold et al., 2006). This plane was located 5 cm below the level of the eyes and was slightly tilted towards the subject by 5° (Rambold et al., 2006). For binocular stimulation the stimulus was aligned as exactly as possible in the mid-sagittal plane in between the eyes.

Ramp stimuli with a constant velocity were used in eight different directions starting from the mid-sagittal plane at an initial vergence angle of 7°. The stimulus moved in the plane of fixation on an isovergence line to the right or to the left (rightward or leftward condition), in a sagittal plane towards or away from the subject's nose (vergence condition) or in combination of both stimulus directions (oblique condition). Stimulus velocity was 10°/s angular velocity for the horizontal (isovergence) directions (calculated for a cyclopean eye) and 3°/s vergence velocity for vergence directions. The directions were randomized, and 20–30 repetitions were performed for each motion direction. In a first experiment, a vertical prism of 6 prism dioptre (pd) base down was mounted in front of one eye to examine retinal slip in the two eyes without horizontal fusion. The stimuli were repeatedly presented in three experimental conditions: prism in front of the right eye, prism in front of the left eye, and viewing with no prism. In a second experiment the eye was covered to serve as control for estimating the residual vergence movements caused by different cues during monocular viewing. Similarly, this stimulus condition was repeated three times, with the right eye or the left eye covered, or with no eye covered.

Horizontal and vertical vergence was calculated as left minus right horizontal eye position. Positive values indicated a rightward, upward, and convergence eye movement direction. All data were low-pass filtered by a median (20 samples) and a Gaussian filter (cut-off frequency at 30 Hz) and further processed as described elsewhere (Rambold et al., 2006). The eye velocity was analysed for each trial. Subsequently vergence velocities were averaged and analysed. Latency and the initial acceleration of the open loop period were quantified according to the algorithm described (Carl and Gellman, 1987). The steady-state velocity was analysed by averaging a 100 ms interval 350 ms after target onset. Statistical differences were significant for $p < 0.05$.

Results

Eye positions

Figure 1A shows representative traces of the vergence (ordinate) and version (abscissa) position, in the binocular viewing condition, for different target directions for one subject (no. 1). During tracking conditions the eye position was variable but generally matched the target trajectory. A vertical prism before one eye (Fig. 1B) or covering one eye (Fig. 1C) impaired tracking in depth, whereas conjugate components were preserved. Data for right and left eye in comparable conditions (covered, uncovered, prism viewing, and non-prism viewing) were not different and the data were pooled and presented as if the right eye was viewing or the prism mounted in front of the right eye.

The effect of fusion: prism versus binocular viewing

Movements in one depth plane (rightward and leftward, isovergence viewing condition)

Figure 1D–F shows averaged eye velocities for the left (dashed) and right eye (black line) for rightward moving targets (black rectangular line) in the binocular, prism, and monocular viewing conditions for one subject (no. 1). The eyes were conjugate and had a latency of on average

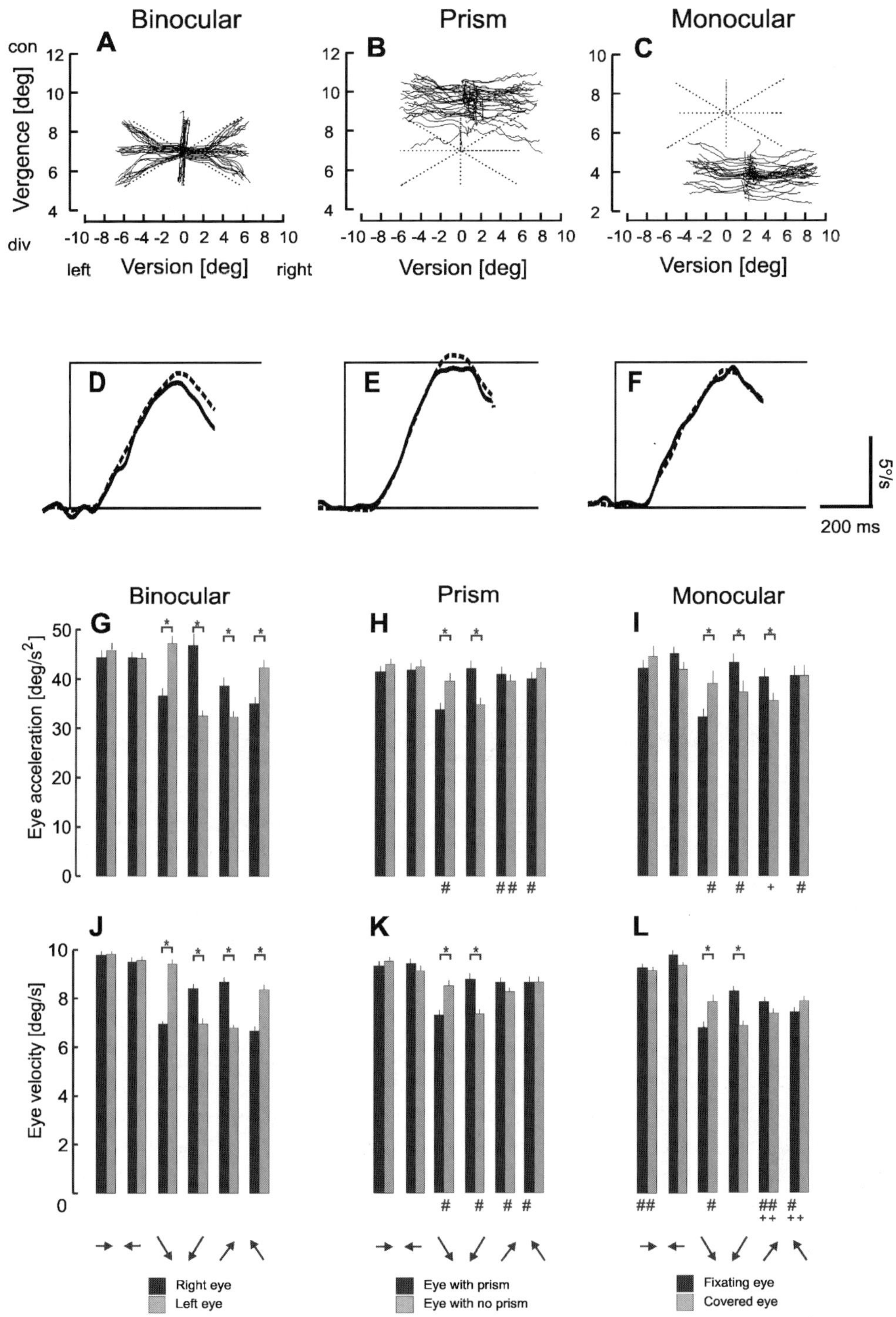

Fig. 1. Version (mean of the two eyes) and vergence (left minus right eye) positions (solid lines) are plotted for one subject (no. 1) in the three different viewing conditions. (A) Binocular viewing; (B) prism; (C) monocular viewing condition. Target trajectories are shown in dotted lines. In the prism (B) and monocular (C) viewing condition there is nearly no depth component and the vergence starting position is shifted (convergence: con; divergence: div). Horizontal right (solid line) and left eye (dashed line) velocities for rightward directions are shown for the binocular viewing condition (D), the prism (E), and the monocular viewing condition (F) in subject no. 1. Additionally, the zero and the target velocities are indicated by horizontal and vertical black lines. The initial eye acceleration (G–I), and steady-state eye velocity (J–L) are shown as group means ± one standard error (error bars) for the different viewing (binocular, prism, and monocular) conditions. The stimulus directions are indicated by arrows below (see also Fig. 2). Statistically significant differences ($p < 0.05$) are indicated by different symbols: differences between the two eyes are indicated by a bracket and (*, above the bars) difference between the monocular or the prism condition and the binocular condition by # (below the bars), and differences between the monocular and the prism condition by (+) (below the bars).

138 ± 32 ms. Generally, there was no difference for individual subjects and the group data for the rightward and leftward stimulus direction with respect to eye latency, initial eye acceleration (open loop), and steady-state eye velocity (closed loop), neither in the binocular, prism, or monocular viewing condition (Fig. 1G–L).

Movements in the depth (vergence condition)

In the prism viewing condition (Fig. 2B) eye velocities were decreased compared to the binocular viewing condition (Fig. 2A). There was no difference in initial acceleration and steady-state velocity between the right and left eye in the prism condition. The steady-state eye velocity, however, was decreased in the prism condition (1.2 ± 0.5°/s) compared to binocular viewing without the prism (1.6 ± 0.3°/s).

Vergence latency was increased (prism: 202 ± 50 ms; binocular: 153 ± 39 ms) and the initial vergence acceleration decreased (prism: 13.1 ± 9.0°/s^2; binocular: 30.0 ± 20.0°/s^2) in the prism condition compared to binocular viewing without prisms. The vergence steady-state velocity was decreased by 61 ± 12% compared to the binocular no-prism condition (Fig. 2D).

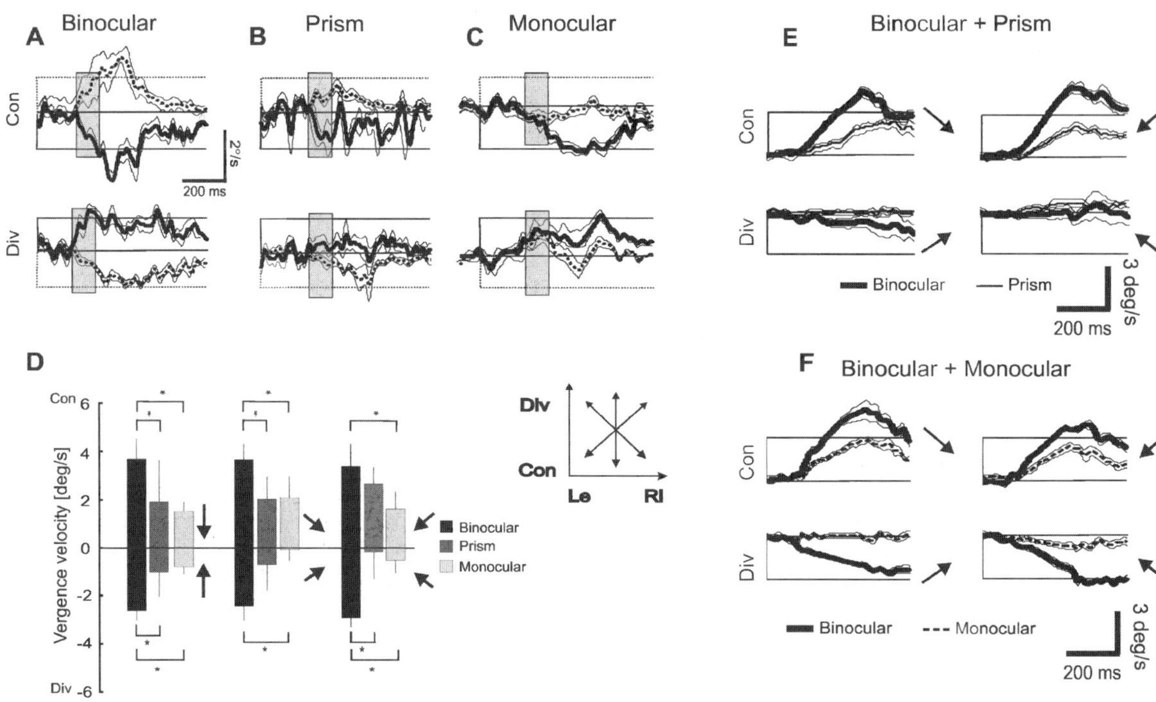

Fig. 2. Horizontal right (RE; solid line) and left eye velocity (LE; dashed line) in the convergence (Con) and divergence (Div) direction are shown for the binocular condition (A), the prism (B), and monocular viewing condition (C). Target velocities are indicated by horizontal black lines for each eye (right eye: solid line; left eye: dashed line). The first 100 ms after eye movement onset (open loop period) are indicated by a grey rectangle. Group means ± one standard deviation (error bars) for vergence steady-state velocity is shown in D. The stimulus directions are indicated by the small arrows next to the bars as defined in the schematic drawing. Significant differences are indicated by asterisks ($p < 0.05$). In the monocular and the prism viewing conditions there is a significant decrease in vergence velocity when compared with the binocular viewing condition (convergence: Con; divergence: Div; rightward: Ri; leftward: Le). Averaged vergence velocity traces (± one standard error, thin lines) are shown for subject no. 1 for the prism (E) and the monocular viewing condition (F) and compared with the binocular condition (thick black trace). Vergence velocity is decreased in the monocular and prism compared to the binocular viewing condition (convergence: Con; divergence: Div).

Movements in depth (oblique condition)

The differences in initial eye acceleration and the steady-state eye velocity between the left and right eye were smaller in the prism and monocular viewing condition as compared to the binocular condition (Fig. 1G–L). The vergence component was nearly absent in the oblique-divergence and reduced in the oblique-convergence condition (Fig. 2E). Vergence latency was increased in the prism viewing condition (prism: 203 ± 43 ms; binocular: 142 ± 41 ms). Vergence acceleration (prism: $11 \pm 10\,°/s^2$; binocular: $28 \pm 15°/s^2$) and vergence steady-state velocity were decreased (by $68 \pm 28\%$) when prism and binocular conditions were compared (Fig. 2D).

The effect of monocular versus binocular retinal image motion

Movements in depth (vergence condition)

When one eye was occluded divergence and convergence eye velocities were decreased when compared to the binocular viewing condition (Fig. 2A, C). However, a small disconjugate component remained. There was no difference in steady-state velocity of the viewing eye between the monocular ($1.2 \pm 0.5°/s$) and prism condition ($1.2 \pm 0.5°/s$). The initial eye acceleration ($8.3 \pm 4.0°/s^2$) in the monocular condition did not differ from the prism condition ($8.3 \pm 3.5°/s^2$). Vergence latencies (monocular: 182 ± 56 ms; prism: 201 ± 45 ms), initial vergence acceleration (monocular: $13.1 \pm 9.3°/s^2$; prism: $13.2 \pm 9.0°/s^2$), and the vergence steady-state velocity were not significantly different in the monocular and prism condition.

Movements in depth (oblique conditions)

In general, there were no differences of the oblique tracking eye movements in the prism and monocular condition with respect to steady-state eye velocity, latency, and initial acceleration, except for small differences in the divergence directions as indicated by (+) in Fig. 1I, L.

Vergence latencies (monocular: 185 ± 98 ms; prism: 203 ± 50 ms), vergence acceleration (monocular: $11.0 \pm 10°/s^2$; prism: $11.0 \pm 10°/s^2$), and steady-state vergence velocity (Fig. 2D) were not different in the monocular and prism condition. There was a consistent small vergence component (Fig. 2F) during monocular tracking in the depth. Our data show that the tracking behaviour was not improved by providing the second eye with a motion signal.

Discussion

Our study shows that disabling the fusional mechanism with a vertical prism impairs tracking of a small slowly moving target in 3-D space in the depth component but not in a conjugate pursuit (rightward, leftward) condition. A similar impairment was found in the monocular viewing condition, e.g., in the presence of only one retinal slip signal.

Hering's law of equal innervation states that eye movements result from a summation of conjugate version and vergence eye movements at the motoneuron level. While Hering's law has been shown to be valid for closed loop tracking eye movements (Rashbass and Westheimer, 1961; Miller et al., 1980; Semmlow et al., 1998; Maxwell and Schor, 2004), a disjunctive pursuit system has been proposed for the open loop period based on a non-human primate study (King and Zhou, 1995). In accord with the latter study eye acceleration (open loop phase) of the viewing eye in the monocular viewing condition of our study was not different from the binocular viewing condition. However, eye velocity trajectories in the open loop phase were different during disjunctive and conjunctive tracking, i.e., vergence velocity was severely decreased in the monocular and prism condition during tracking in depth (vergence) compared to the binocular viewing condition. In contrast, eye velocity during conjugate pursuit (isovergence) was not different from the binocular viewing condition.

In contrast to previous studies (Rashbass and Westheimer, 1961; King and Zhou, 1995) we additionally examined eye movements to slowly moving, small, un-fused targets. Pursuit movements to un-fused targets have previously been studied by Erkelens and Collewijn (1985) by sinusoidally moving half images of a larger random dot stereogram in the closed loop phase. In contrast to them, we analysed the open loop phase of eye movements following small un-fused targets with ramp stimuli: (i) to better dissociate Hering's hypothesis from the disjunctive pursuit model, and (ii) to compare the findings of Erkelens and Collewijn (1985) with stimulation conditions with single spot targets in 3-D space in humans.

We used a vertical prism mounted in front of one eye to disable fusion and to decrease the horizontal retinal disparity signal. This view is in accord with several studies showing that vertical disparity decreases horizontal disparity vergence (Mitchell, 1970; Boman and Kertesz, 1983; Yang and Miles, 2003). Our subjects might have seen two images. One of the two apparent stimuli was tracked with the fovea, while the other was tracked parafoveally which might led to a slight decrease in pursuit gain and initial acceleration (Tychsen and Lisberger, 1986; Carl and Gellman, 1987; Pidcoe and Wetzel, 2006). However, initial eye acceleration and velocity during tracking was not different between the two eyes (eye with vs. without prism) during our prism vergence and pursuit (isovergence) conditions. This indicates that parafoveal tracking probably does not account for the decrease in vergence velocity in the prism paradigm.

Alternatively, a change in motor strategy in the prism condition might take place if, instead of the two, only one target is selected and tracked (Erkelens and Collewijn, 1985; Ferrera and Lisberger, 1995). Some evidence against this hypothesis comes from our data in which vergence velocities gradually decreased with increasing power of the vertical prisms (unpublished data). This gradual decline contradicts an all-or-nothing principle as would be expected from a change in motor strategy and argues in favour of a decrease of disparity vergence.

Since the introduction of a prism severely reduced tracking in depth but not in one depth plane (isovergence) our data, obtained with unfused targets, support Hering's hypothesis. Accordingly, both (disparity) vergence and conjugate pursuit seem to contribute to depth tracking in space. In contrast, according to a disjunctive pursuit model vergence velocity should not have been decreased in the prism condition (retinal slip presented to both retinas).

Our 3-D tracking in the monocular viewing condition shows some residual vergence. This small monocular disjunctive component was delayed compared to the binocular viewing condition and might be caused by different monocular cues: blur driven accommodative vergence (Hung et al., 1983), proximal vergence caused by the proximity of the target, looming, luminance gradient, kinetic depth effect (Erkelens and Regan, 1986; Enright, 1987a, b; McLin and Schor, 1988; Ringach et al., 1996), a learned association of vertical height and depth. However, cues for proximal vergence have little effect on vergence during closed loop accommodation or disparity vergence (Hung et al., 1994). The change in target blur or the change in target luminance which drives accommodative vergence was not large enough to explain the effects on vergence in our setup. There might be a learned association of vertical height and depth due to our target presentation in 3-D space. This factor could not be proved or disproved by our data and is a feasible explanation of our disjunctive vergence eye movements. We conclude that tracking a target in 3-D space can largely be explained by Hering's hypothesis.

References

Boman, D.K. and Kertesz, A.E. (1983) Interaction between horizontal and vertical fusional responses. Percept. Psychophys., 33: 565–570.

Carl, J.R. and Gellman, R.S. (1987) Human smooth pursuit: stimulus-dependent responses. J. Neurophysiol., 57: 1446–1463.

Eggert, T., Mezger, F., Robinson, F. and Straube, A. (1999) Orbital position dependency is different for the gain of externally and internally triggered saccades. Neuroreport, 10: 2665–2670.

Enright, J.T. (1987a) Art and the oculomotor system: perspective illustrations evoke vergence changes. Perception, 16: 731–746.

Enright, J.T. (1987b) Perspective vergence: oculomotor responses to line drawings. Vision Res., 27: 1513–1526.

Erkelens, C.J. and Collewijn, H. (1985) Eye movements and stereopsis during dichoptic viewing of moving random-dot stereograms. Vision Res., 25: 1689–1700.

Erkelens, C.J. and Regan, D. (1986) Human ocular vergence movements induced by changing size and disparity. J. Physiol., 379: 145–169.

Ferrera, V.P. and Lisberger, S.G. (1995) Attention and target selection for smooth pursuit eye movements. J. Neurosci., 15: 7472–7484.

Hung, G.K., Ciuffreda, K.J., Semmlow, J.L. and Horng, J.L. (1994) Vergence eye movements under natural viewing conditions. Invest. Ophthalmol. Vis. Sci., 35: 3486–3492.

Hung, G.K., Semmlow, J.L. and Ciuffreda, K.J. (1983) Identification of accommodative vergence contribution to the near response using response variance. Invest. Ophthalmol. Vis. Sci., 24: 772–777.

King, W.M. and Zhou, W. (1995) Initiation of disjunctive smooth pursuit in monkeys: evidence that Hering's law of equal innervation is not obeyed by the smooth pursuit system. Vision Res., 35: 3389–3400.

Maxwell, J.S. and King, W.M. (1992) Dynamics and efficacy of saccade-facilitated vergence eye movements in monkeys. J. Neurophysiol., 68: 1248–1260.

Maxwell, J.S. and Schor, C.M. (2004) Symmetrical horizontal vergence contributes to the asymmetrical pursuit of targets in depth. Vision Res., 44: 3015–3024.

McLin, L.N., Jr. and Schor, C.M. (1988) Voluntary effort as a stimulus to accommodation and vergence. Invest. Ophthalmol. Vis. Sci., 29: 1739–1746.

Miller, J.M., Ono, H. and Steinbach, M.J. (1980) Additivity of fusional vergence and pursuit eye movements. Vision Res., 20: 43–47.

Mitchell, D.E. (1970) Properties of stimuli eliciting vergence eye movements and stereopsis. Vision Res., 10: 145–162.

Pidcoe, P.E. and Wetzel, P.A. (2006) Oculomotor tracking strategy in normal subjects with and without simulated scotoma. Invest. Ophthalmol. Vis. Sci., 47: 169–178.

Rambold, H., Neumann, G., Sander, T. and Helmchen, C. (2006) Age-related changes of vergence under natural viewing conditions. Biol. Aging, 27: 163–172.

Rambold, H., Sprenger, A. and Helmchen, C. (2002) Effects of voluntary blinks on saccades, vergence eye movements, and saccade-vergence interactions in humans. J. Neurophysiol., 88: 1220–1233.

Rashbass, C. and Westheimer, G. (1961) Independence of conjugate and disjunctive eye movements. J. Physiol., 159: 361–364.

Ringach, D.L., Hawken, M.J. and Shapley, R. (1996) Binocular eye movements caused by the perception of three-dimensional structure from motion. Vision Res., 36: 1479–1492.

Semmlow, J.L., Yuan, W. and Alvarez, T.L. (1998) Evidence for separate control of slow version and vergence eye movements: support for Hering's Law. Vision Res., 38: 1145–1152.

Tychsen, L. and Lisberger, S.G. (1986) Visual motion processing for the initiation of smooth-pursuit eye movements in humans. J. Neurophysiol., 56: 953–968.

Yang, D.S. and Miles, F.A. (2003) Short-latency ocular following in humans is dependent on absolute (rather than relative) binocular disparity. Vision Res., 43: 1387–1396.

C. Kennard & R.J. Leigh (Eds.)
Progress in Brain Research, Vol. 171
ISSN 0079-6123

CHAPTER 5.14

Exploring the pulvinar path to visual cortex

Rebecca A. Berman* and Robert H. Wurtz

Laboratory of Sensorimotor Research, National Eye Institute, National Institutes of Health, Bethesda, MD, USA

Abstract: The primary pathway for visual signals from the retina to cerebral cortex is through the lateral geniculate nucleus of the thalamus to primary visual cortex. A second visual pathway has been postulated to pass through the thalamic pulvinar nucleus and to project to multiple regions of visual cortex. We have explored this second visual pathway using a method that allows us to identify the inputs and outputs of pulvinar neurons. Specifically, we applied microstimulation in the superficial layers of superior colliculus (SC) to test for orthodromic activation of pulvinar neurons receiving input from SC. We also microstimulated the cortical motion area MT and tested for antidromic activation of pulvinar to identify neurons projecting to MT (and to determine the presence of orthodromic input back to pulvinar). In this initial report, we concentrate on two observations. First, we find that there are clusters of neurons in the pulvinar that receive input from SC along with neurons that project to MT or receive input from MT. Second, we find that neurons with input from SC have characteristics of the SC superficial layers: they respond to visual stimuli but do not discharge before saccadic eye movements. Neurons projecting to MT respond similarly to these SC-input neurons, while those receiving input from MT more frequently show directional selectivity as does MT. These findings indicate the visual nature of the signals conveyed in this pathway and shed light on the functional role of the thalamus in a possible second visual pathway.

Keywords: pulvinar; superior colliculus; MT; two visual pathways; monkey visual pathways

Introduction

The classic visual pathway studied in the primate is that from the retina to the lateral geniculate nucleus (LGN) to the primary visual cortex, V1. Diamond and Hall in 1969 (Diamond and Hall, 1969) postulated that a second pathway from retina to visual cortex may also provide significant visual information. This proposed second pathway extends from the retina to the superficial layers of the superior colliculus (SC) and from there to the pulvinar nucleus of the thalamus. Through this second pathway the pulvinar could provide information to multiple cortical visual areas.

The idea of a second visual pathway has had considerable influence since it was first introduced. For example, in cases of damage to the primary visual cortex, the second pathway has been regarded as a possible path for residual visual input, particularly in dorsal stream areas such as the motion area MT (Rodman et al., 1990; Girard et al., 1992). The residual visual abilities of blindsight patients might depend on the input from the pulvinar pathway, particularly for detection of visual motion (Perenin, 1991; Weiskrantz, 2004). Little is known, however, about the existence and functional properties of this

*Corresponding author. Tel.: +1 301 402 4956;
Fax: +1 301 402 0511; E-mail: bermanr@nei.nih.gov

DOI: 10.1016/S0079-6123(08)00668-7

pathway in the primate brain. The exploration of the pulvinar nucleus with single-neuron recording (Bender, 1981, 1988; Petersen et al., 1985; Robinson, 1993) have encountered two major problems. One is the delineation of regions within the pulvinar through which the pathway to cortex passes. This difficulty has been partly alleviated by improved identification of the subdivisions using neurochemical methods (Gray et al., 1999; Adams et al., 2000; Kaas and Lyon, 2007). The second problem is knowing the connectivity of a given neuron recorded in the pulvinar. Even if a neuron can be localized to a neurochemical subdivision of the pulvinar, that alone does not indicate what input it receives or where it projects.

As a first step to understanding this pathway, we took a new approach to characterize pulvinar function in the awake, behaving monkey. We recorded neurons in the thalamus of monkeys trained to do simple visual motor tasks. But we did not sample neurons at random; we studied only those for which we had some information about their input and/or output. To do this we combined microstimulation and single-neuron recording to identify pulvinar neurons that received input from outside the pulvinar and those that projected to visual cortex. To determine input, we placed stimulating electrodes in the lower superficial layers of the SC, the presumed source of input to the pulvinar neurons that we concentrated on, those in the inferior pulvinar. To test for projections to visual cortex, we placed electrodes in the visual motion area, MT, because there is clear evidence that inferior pulvinar projects to MT and that MT also projects back to pulvinar (Adams et al., 2000; Kaas and Lyon, 2007). Our long term goal is to identify the relay neurons that lie in the pathway, that is, those neurons that receive from the SC and project to MT, to find where these neurons lie, and to ascertain what information they convey along this second visual pathway. In this initial report, however, we concentrate on the first steps toward this goal. We address two aims: to determine whether there is a subregion of pulvinar that contains clusters of neurons either receiving input from SC or projecting to MT, and to describe the functional characteristics of neurons lying in these clusters.

Methods

The general techniques used in these experiments, including those for antidromic and orthodromic stimulation, have been described previously (Sommer and Wurtz, 2004). Briefly, the monkey sat in a primate chair and faced a tangent screen on which visual stimuli and saccade targets were projected. The monkey's tasks were either to fixate while the receptive field characteristics of an isolated neuron were examined or to make saccades from one target to another to determine the relation of the neuron's activity to saccades.

Our recordings were concentrated in the visual pulvinar, spanning the inferior and lateral divisions. The data described here were obtained in a single hemisphere from more than 60 recording sites in a zone that extended up to 7 mm medial-lateral and up to 10 mm anterior–posterior. Within this zone, penetrations were typically made at 1 mm spacing, and at 0.5 mm in areas of interest. Once we isolated a single pulvinar neuron, we used microstimulation to determine whether it was connected to either MT or SC. For example, if microstimulation in MT evokes a spike from the pulvinar neuron under study, it indicates that the pulvinar neuron is connected to MT. The nature of the connection is indicated by the kind of activation, *antidromic* (backward-going) or *orthodromic* (forward-going). If microstimulation in SC or MT antidromically activates or backfires the pulvinar cell, this tells us that the pulvinar cell sends output to that structure. By contrast, if microstimulation orthodromically activates or synaptically drives the pulvinar cell, this tells us that the pulvinar cell receives input from that structure. Three features of the evoked spike distinguish between antidromic and orthodromic activation: the reliability of evoking the spike, the electrical latency, and whether it passes the collision test (Lemon, 1984). See Fig. 1 for an outline of the procedure.

Results

In this initial survey, we identified a total of 101 pulvinar neurons with connections to either SC or MT. These neurons fell into three categories: (1) neurons

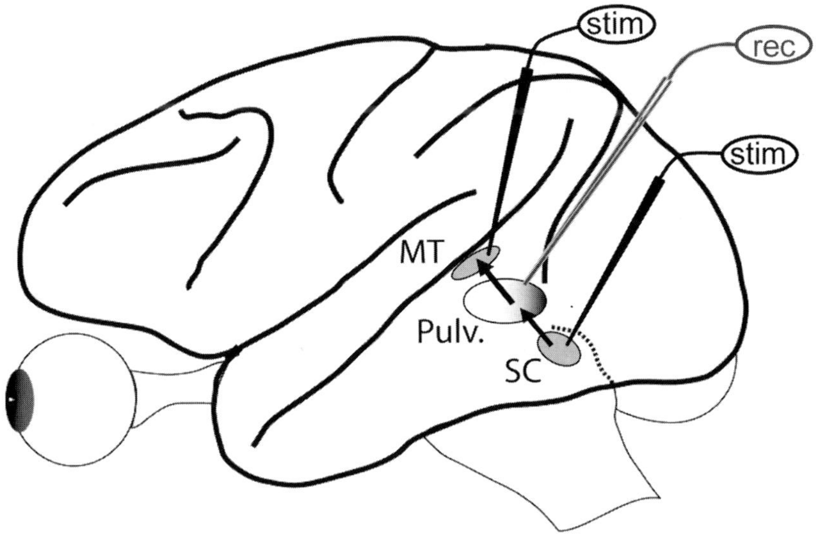

Fig. 1. Studying the second visual pathway by identifying connected neurons in the pulvinar. A sagittal view of the macaque brain shows the configuration of recording and stimulating electrodes used to identify pulvinar neurons with connections to superior colliculus (SC) or area MT. A recording microelectrode is lowered into the pulvinar nucleus (Pulv). Once a single neuron is isolated, microstimulation is used to determine whether the cell receives input or sends output to either SC or MT (see Methods).

with input from the SC ($n = 56$), (2) neurons with output to area MT ($n = 21$), (3) neurons with input from area MT ($n = 24$). A fourth possible group, neurons with output to SC, were exceedingly rare ($n = 2$) and will not be considered further. We first describe the localization of these connected neurons within the pulvinar, and then turn to the characterization of their functional properties.

Location

Connected neurons were located within a zone of the pulvinar just adjacent to the LGN. The posterior pole of the LGN, where neurons represent parafoveal locations, served as our primary landmark. We found connected pulvinar neurons immediately medial and posterior to this part of LGN, in a zone that extended approximately 4 mm medial and 3 mm posterior. A schematic of the area just posterior to the LGN is shown in Fig. 2. We found the greatest concentration of connected pulvinar neurons in a smaller zone centred about 3 mm medial and 2 mm posterior to the LGN landmark. Throughout the general zone in which connected neurons were located, we encountered them at depths spanning up to 6 mm; the majority were located at a depth comparable or dorsal to the LGN landmark, though some were located more ventrally. We did not find strong evidence for a segregation of the three classes of connected neurons in any dimension; rather the three classes of pulvinar neurons were largely intermingled. The localization of these connected cells indicates that there exists a region of the pulvinar in which there may be a convergence of signals travelling from SC to MT, as well as signals returning from MT.

Neuronal responses of pulvinar neurons in the pathway

Our examination of neural activity focused on two questions regarding the salient properties of pulvinar neurons that received input from SC and those that projected to or received input from MT. We first asked: do pulvinar neurons with input from the SC have functional properties similar to neurons in the superficial layers of SC, which have clear visual responses to spots of light over a region of the contralateral visual field? Or do they have properties similar to neurons in the intermediate layers, which have visual responses but also have activity before the onset of the saccade? Our stimulating

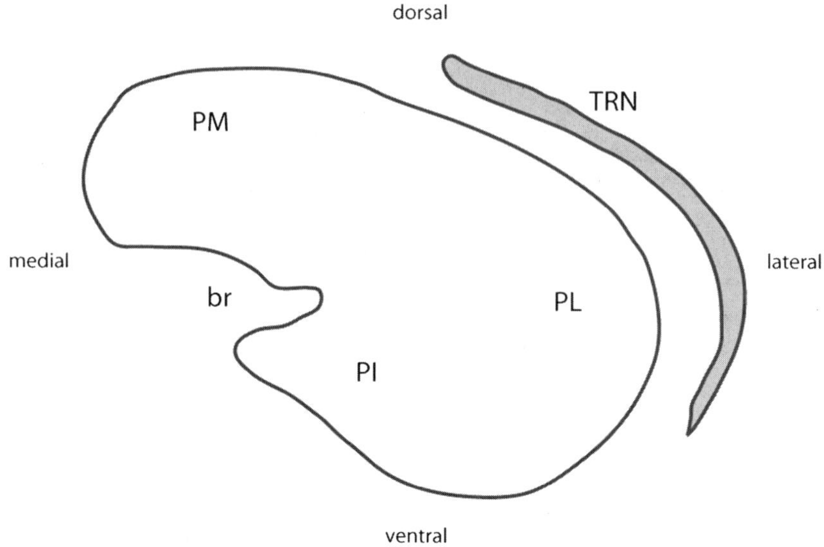

Fig. 2. Regions of the pulvinar with neurons receiving input from SC and neurons projecting to MT. Schematic shows a coronal cross-section through the pulvinar at the level where single-unit recording was concentrated, just posterior to the LGN. The visual subdivisions, PI and PL, are of greatest interest for connected cells. PM = medial pulvinar; PL = lateral pulvinar, PI = inferior pulvinar, br = brachium of the superior colliculus, TRN = thalamic reticular nucleus.

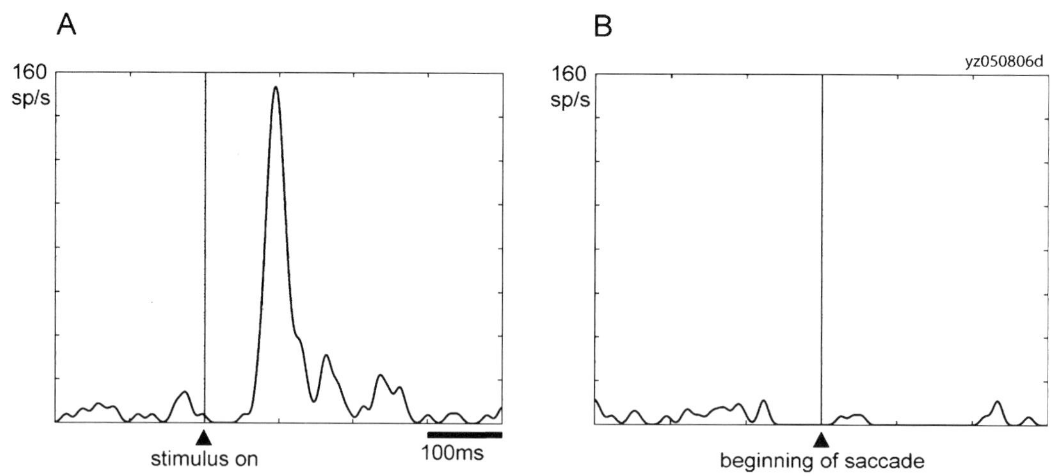

Fig. 3. Pulvinar neurons with input from SC have visual but not presaccadic activity. The activity of a single example neuron with input from SC during the memory-guided saccade task. (A) Activity is aligned on the appearance of the stimulus. The pulvinar neuron gives a brisk visual response. (B) Activity is aligned on the beginning of the eye movement. The neuron has no significant presaccadic build-up or burst activity. In each panel, the spike density function is smoothed with an 8 ms Gaussian.

electrodes targeted the lower superficial layers, but we considered the possibility that the intermediate layers might also be activated by microstimulation. We found that pulvinar neurons with input from the SC had activity profiles that resembled those of neurons in the superficial layers of SC: they had visual responses but no presaccadic activity. These properties are illustrated for an example pulvinar neuron with SC input in Fig. 3. This pulvinar neuron had a restricted receptive field located parafoveally in the contralateral visual field, at an eccentricity of 3.5°. When a small target (1°) appeared in the receptive field, the neuron gave a brisk response. When the monkey made a memory-guided saccade to the same target location, the neuron's activity was not modulated (Fig. 3B). We observed the same pattern in the vast majority of the pulvinar neurons with input from SC. The only deviations from this pattern were neurons for which we were unable to identify a clear visual receptive field; we have not observed presaccadic activity in any of the sampled cells. These findings demonstrate that the properties of pulvinar cells in this pathway reflect input from the superficial but not the intermediate layers of SC.

Second, we asked: do pulvinar neurons with input from or output to MT show evidence of the directional tuning observed in MT itself? Here we found that the functional properties of the pulvinar neurons reflected their relationship to MT. Specifically, directional tuning was prominent in pulvinar neurons with input from MT but not in those that projected to MT. An example of a directionally tuned neuron with *input* from MT is shown in Fig. 4A. Its activity is shown for eight directions of random dot motion, which moved at an optimal speed of 40°/s at 100% coherence in a 10° diameter window centred on its receptive field (40° eccentricity). The neuron fired strongly when motion was up and to the right (preferred direction), but gave little response to motion in the opposite (null) direction. As a measure of its directional selectivity, we computed a directional index (DI) using the formula where $DI = 1-null/pref$, where *null* and *pref* are the average firing rates in the null and preferred conditions, respectively, after subtraction of spontaneous activity. DI values greater than 1 indicate directional selectivity, and values greater than 0.6 indicate directional bias. The DI for this pulvinar neuron was 1.07, comparable to values observed in area MT. An example pulvinar neuron with *output* to MT is shown in Fig. 4B. This neuron gave a brisk transient response to the random dot motion in its receptive field, irrespective of the direction of motion (motion parameters as above; 30° eccentricity). Its activity was not strongly tuned for direction, as indicated by its DI of 0.08. We found a similar trend in the population of pulvinar neurons connected to area MT in which we were able to obtain data for the directional series ($n = 22$). Neurons that received input from MT were likely to be directionally tuned (6/8) and typically had weak responses to stationary stimuli. Neurons that projected to MT were less likely to be directionally tuned, although we did observe directional selectivity or bias in some of these neurons as well (6/14). Overall, these neurons with projections to MT tended to resemble the pulvinar cells with input from SC, in that they often had a brisk visual response to stationary stimuli.

Discussion and conclusion

We draw two conclusions from these initial exploratory experiments. First, the use of orthodromic and antidromic stimulation demonstrates the clear existence of a subregion in pulvinar in which there are neurons that receive input from the SC, neurons that project from pulvinar to MT, and neurons that receive input from MT. The second conclusion is that these neurons are conveying visual information to and from cortex, not saccadic premotor activity. This is reasonable given previous anatomical evidence that the ascending pathway through the inferior pulvinar originates primarily in the superficial layers (Benevento and Fallon, 1975). It is also intriguing to see direct evidence that there are neurons in this subregion of the pulvinar that project to MT, as well as those that receive input from MT. This observation is of particular interest given the argument that communication between cortical areas benefits from pathways through higher-order thalamic structures as well as the direct connections from one cortical area to the next (Sherman and Guillery, 2002).

The next steps in these investigations require the positive identification of relay neurons (those that receive from SC and project to MT) in order to demonstrate unequivocally that a second visual pathway does indeed pass through pulvinar, despite doubts based on neuroanatomical studies (Stepniewska et al., 2000). More important will be to determine the contribution of this pathway to visual processing in cortex and to such functions as

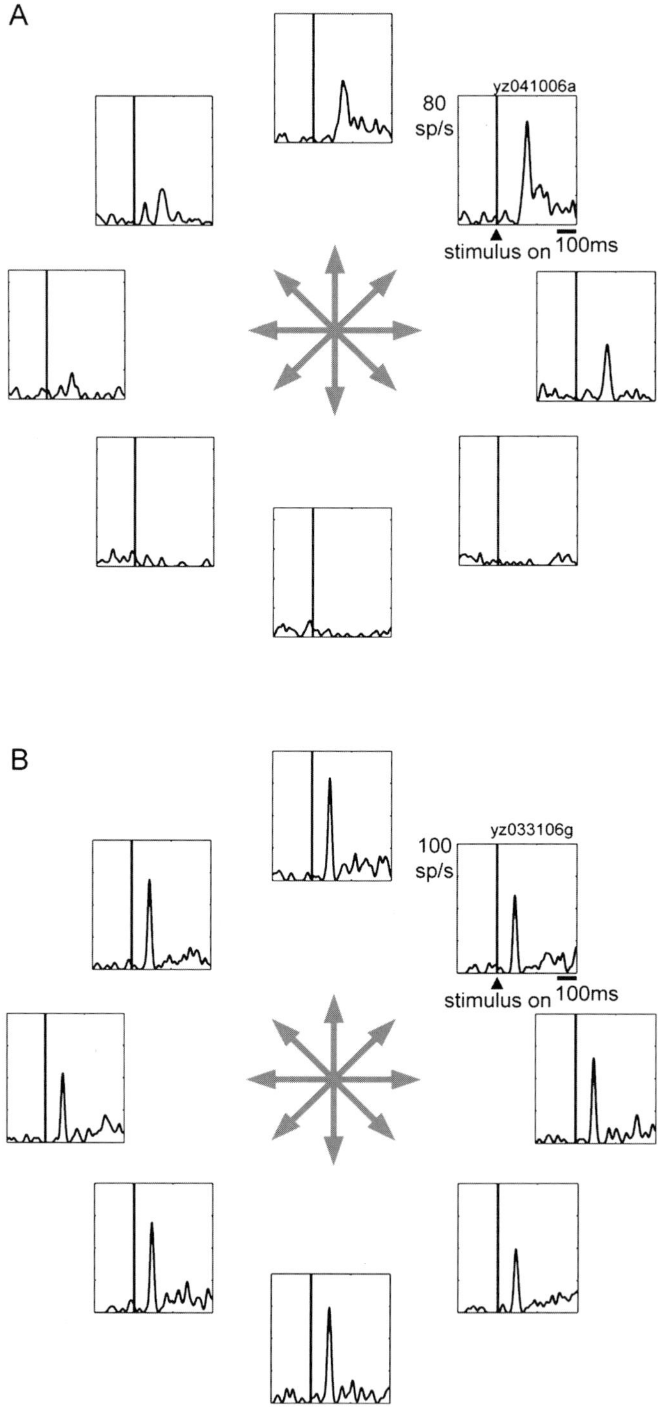

Fig. 4. Pulvinar neurons with input from MT are more likely to be directionally selective than are pulvinar neurons with output to MT. The visual responses to eight different directions of motion are shown for (A) an example pulvinar neuron receiving input from MT and (B) an example pulvinar neuron sending output to MT. Activity is aligned on the appearance of the stimulus.

visual attention already demonstrated in the pulvinar (Petersen et al., 1985, 1987).

References

Adams, M.M., Hof, P.R., Gattass, R., Webster, M.J. and Ungerleider, L.G. (2000) Visual cortical projections and chemoarchitecture of macaque monkey pulvinar. J. Comp. Neurol., 419: 377–393.

Bender, D.B. (1981) Retinotopic organization of macaque pulvinar. J. Neurophysiol., 46: 672–693.

Bender, D.B. (1988) Electrophysiological and behavioral experiments on the primate pulvinar. Prog. Brain Res., 75: 55–65.

Benevento, L.A. and Fallon, J.H. (1975) The ascending projections of the superior colliculus in the rhesus monkey (*Macaca mulatta*). J. Comp. Neurol., 160: 339–361.

Diamond, I.T. and Hall, W.C. (1969) Evolution of neocortex. Science, 164: 251–262.

Girard, P., Salin, P.A. and Bullier, J. (1992) Response selectivity of neurons in area MT of the macaque monkey during reversible inactivation of area V1. J. Neurophysiol., 67: 1437–1446.

Gray, D., Gutierrez, C. and Cusick, C.G. (1999) Neurochemical organization of inferior pulvinar complex in squirrel monkeys and macaques revealed by acetylcholinesterase histochemistry, calbindin and Cat-301 immunostaining, and Wisteria floribunda agglutinin binding. J. Comp. Neurol., 409: 452–468.

Kaas, J.H. and Lyon, D.C. (2007) Pulvinar contributions to the dorsal and ventral streams of visual processing in primates. Brain Res. Rev., 55: 285–296.

Lemon, R. (1984). Methods for neuronal recording in conscious animals. In: Smith A.D. (Ed.), IBRO Handbook Series: Methods in the Neurosciences. Vol. 4. J. Wiley & Sons, New York, pp. 95–102.

Perenin, M.T. (1991) Discrimination of motion direction in perimetrically blind fields. NeuroReport, 2: 397–400.

Petersen, S.E., Robinson, D.L. and Keys, W. (1985) Pulvinar nuclei of the behaving rhesus monkey: visual responses and their modulation. J. Neurophysiol., 54: 867–886.

Petersen, S.E., Robinson, D.L. and Morris, J.D. (1987) Contributions of the pulvinar to visual spatial attention. Neuropsychologia, 25: 97–105.

Robinson, D.L. (1993) Functional contributions of the primate pulvinar. Prog. Brain Res., 95: 371–380.

Rodman, H.R., Gross, C.G. and Albright, T.D. (1990) Afferent basis of visual response properties in area MT of the macaque. II. Effects of superior colliculus removal. J. Neurosci., 10: 1154–1164.

Sherman, S.M. and Guillery, R.W. (2002) The role of the thalamus in the flow of information to the cortex. Philos. Trans. R. Soc. Lond. B Biol. Sci., 357: 1695–1708.

Sommer, M.A. and Wurtz, R.H. (2004) What the brain stem tells the frontal cortex. I. Oculomotor signals sent from superior colliculus to frontal eye field via mediodorsal thalamus. J. Neurophysiol., 91: 1381–1402.

Stepniewska, I., Qi, H.X. and Kaas, J.H. (2000) Projections of the superior colliculus to subdivisions of the inferior pulvinar in New World and Old World monkeys. Vis. Neurosci., 17: 529–549.

Weiskrantz, L. (2004) Roots of blindsight. Prog. Brain Res., 144: 229–241.

C. Kennard & R.J. Leigh (Eds.)
Progress in Brain Research, Vol. 171
ISSN 0079-6123

CHAPTER 5.15

The role of the human pulvinar in visual attention and action: evidence from temporal-order judgment, saccade decision, and antisaccade tasks

Isabel Arend[1], Liana Machado[2], Robert Ward[1], Michelle McGrath[1], Tony Ro[3] and Robert D. Rafal[1,*]

[1]*Wolfson Center for Clinical and Cognitive Neuroscience, Bangor University, Bangor, UK*
[2]*University of Otago, Dunedin, New Zealand*
[3]*Rice University, Huston, TX, USA*

Abstract: The pulvinar nucleus of the thalamus has been considered as a key structure for visual attention functions (Grieve, K.L. et al. (2000). Trends Neurosci., 23: 35–39; Shipp, S. (2003). Philos. Trans. R. Soc. Lond. B Biol. Sci., 358(1438): 1605–1624). During the past several years, we have studied the role of the human pulvinar in visual attention and oculomotor behaviour by testing a small group of patients with unilateral pulvinar lesions. Here we summarize some of these findings, and present new evidence for the role of this structure in both eye movements and visual attention through two versions of a temporal-order judgment task and an antisaccade task. Pulvinar damage induces an ipsilesional bias in perceptual temporal-order judgments and in saccadic decision, and also increases the latency of antisaccades away from contralesional targets. The demonstration that pulvinar damage affects both attention and oculomotor behaviour highlights the role of this structure in the integration of visual and oculomotor signals and, more generally, its role in flexibly linking visual stimuli with context-specific motor responses.

Keywords: pulvinar; thalamus; human; attention; saccade decision; antisaccade; temporal order judgment; visuomotor transformation; saccade; eye movements

Introduction

Converging evidence from humans and experimental animals has implicated the central and lateral thalamus in oculomotor control. Saccades can be elicited by electrical stimulation of these nuclei (Schlag and Schlag-Rey, 1971; Crommelinck et al., 1977), and single units are active in relation to saccades (Crommelinck et al., 1977; Schlag-Rey and Schlag, 1977, 1984). Humans with lesions of central thalamus are impaired in using corollary discharge to remap the location of the second target in a double-step saccade paradigm (Bellebaum et al., 2005).

The role of the pulvinar nucleus in eye-movement control, which is the focus of the current research, remains less explored. This largely visual nucleus has direct connections with extrastriate visual cortex and with the superior colliculus and oculomotor cortex of frontal and parietal lobes (Romanski et al., 1997; Grieve et al., 2000; Shipp, 2003). These

*Corresponding author. Tel.: 01248 383885; Fax: 01248 382599; E-mail: r.rafal@bangor.ac.uk

DOI: 10.1016/S0079-6123(08)00669-9

connections position it to play a vital role in controlling visual attention and eye movements.

The involvement of the pulvinar in visual attention has been demonstrated by electrophysiological and lesion studies in monkeys (Chalupa et al., 1976; Bender, 1981; Beneveto and Miller, 1981; Petersen et al., 1985, 1987), and in humans (LaBerge and Buchsbaum, 1990; Ward et al., 2002; Danziger et al., 2004; Michael and Desmedt, 2004; Ward and Arend, 2007; Ward et al., 2007). Recently, we have also found that lesions to the spatial maps of the pulvinar affect object-based location coding (Ward and Arend, 2007). This complex spatial coding within the pulvinar may reflect its role in integrating and maintaining different frames of reference computed in different cortical areas, and could represent a computation supporting visually guided action including eye movements.

Some lesion studies in monkeys reporting prolonged fixation durations and increases in saccade latency were confounded by lesion extension into the brachium of the superior colliculus (Ungerleider and Christensen, 1977, 1979). No impairments in single- and dual-step saccade tasks were observed in monkeys with restricted pulvinar lesions (Bender and Butter, 1987; Bender and Baizer, 1990), leading Bender and Baizer (1990) to suggest that the pulvinar is more likely to be involved in the integration of saccadic eye movements with visual processing, than in saccade generation.

Nevertheless, single-cell recordings in the pulvinar showed response modulation during eye movements (Petersen et al., 1985; Robinson et al., 1986). Robinson et al. (1990) tested the excitability of pulvinar neurons during eye movements and observed modulation in response to a visual stimulus as a function of the position of the eye in the orbit. These findings resemble those found in lateral intraparietal cortex (Andersen et al., 1990).

In humans, focal lesions to the pulvinar are relatively rare, and some studies that have examined the effects of pulvinar lesions on eye movements have been confounded by the fact that the patients also had cortical lesions (Zihl and von Cramon, 1979; Ogren et al., 1984), or had hemispatial neglect (Watson and Heilman 1979; Brigell et al., 1984; Hirose et al., 1985). Rafal et al. (2004) measured saccade latencies for voluntary saccades (initiated by verbal instructions) and visually summoned saccades (toward peripheral targets) in three patients with chronic focal lesions of posterior thalamus (SM, TN, and GJ, all of whom also participated in the current research; see Fig. 2 in the Methods section). They reported that the presence of a fixation point had the normal effect of prolonging the latencies for voluntary saccades, but did not affect latencies for visually summoned saccades. Also, as shown in Fig. 1, the patients had longer latencies to initiate contralesional saccades for both voluntary and visually summoned saccades.

Here, we report the effects of pulvinar lesions on both perceptual judgments and oculomotor performance in a group of patients with chronic, unilateral pulvinar lesions. Three patients were tested in an antisaccade task, and five patients were tested in two temporal-order judgment tasks: a perceptual decision task and a saccade decision task.

Methods

Patients

All patients had chronic (10–24 months post ictus), unilateral lesions affecting the posterior thalamus (Fig. 2) due to hypertensive intracerebral haematoma, except for CR who had sustained a haemorrhagic contusion avulsing the pole of the pulvinar from closed head injury. In one patient (a woman), the lesion was in the right hemisphere and in the other four (all men) it was on the left. Lesion analyses for these patients have been reported previously (Ward et al., 2002; Danziger et al., 2004; Ward and Arend, 2007). Except CR, all patients had contralesional motor and/or sensory deficits. None had any evidence of cognitive impairment and all were living independently. They all had intact visual fields and normal contrast sensitivity, and had no signs of visual neglect or visual extinction. All five patients were tested in the saccade decision and perceptual temporal-order judgment tasks. SM, TN, and GJ were tested in the antisaccade task.

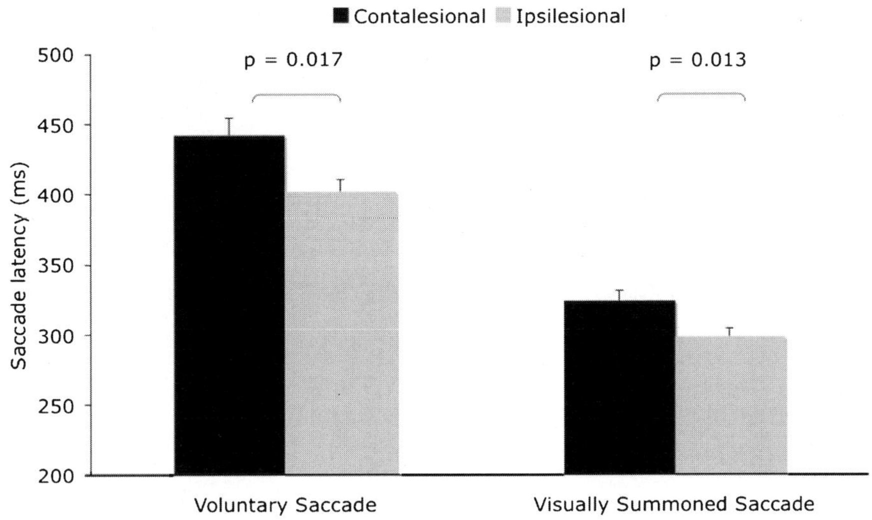

Fig. 1. Mean saccade latency (error bars denote standard error of the mean) for voluntary saccades and visually summoned saccades towards ipsilesional and contralesional fields.

Temporal-order judgment tasks

Patients were instructed to fixate on a 1° dark filled circle at the centre of a white monitor screen for 2 s before the onset of the left and right targets. The targets consisted of grey 1° filled circles that appeared at 10° to the left and to the right of the fixation point on every trial. There were three different stimulus onset asynchronies (SOAs) between presentation of the first target on one side of the screen and the second target on the opposite side of the screen: simultaneously, left or right target first at 17 ms SOA, and left or right target first at 150 ms SOA. The first stimulus event was presented for 1000 ms and the second for 1000 ms minus the asynchrony.

Saccade and perceptual decision tasks were tested in separate blocks. In the saccade task, patients were instructed to make an eye movement towards the target that appeared first. In the perceptual decision task, patients were instructed to maintain fixation and to indicate which target appeared first by pressing the left or right button on the keypad of a Gravis joystick using the index and middle fingers of the ipsilesional hand. Each block consisted of 150 trials taking about 10 min. CR and DG each completed four blocks of both tasks. GJ and SM each completed four blocks of the saccade decision task and three blocks of the perceptual decision task. TN completed three blocks of each task.

Antisaccade task

After an inter-trial interval of 2500 ms, each trial began with a display consisting of a black background on which there were a white fixation point and two unfilled 2.4° white squares each located 10° to the right and left, respectively. Patients maintained their gaze on the fixation point (a white filled circle measuring 0.7°) for a variable interval ranging between 700–1300 ms, after which a target (a white 1.8° asterisk that appeared randomly in the left or right box and remained visible until response) was presented. The instructions were to move the eyes away from the target towards the box in the opposite visual field. In half the trials, the fixation point offset simultaneously with target presentation, and in the other random half of the trials, the fixation point remained visible. Experimental sessions consisted of one block of 80 trials. Patient SM completed seven blocks and the other two patients completed five blocks each.

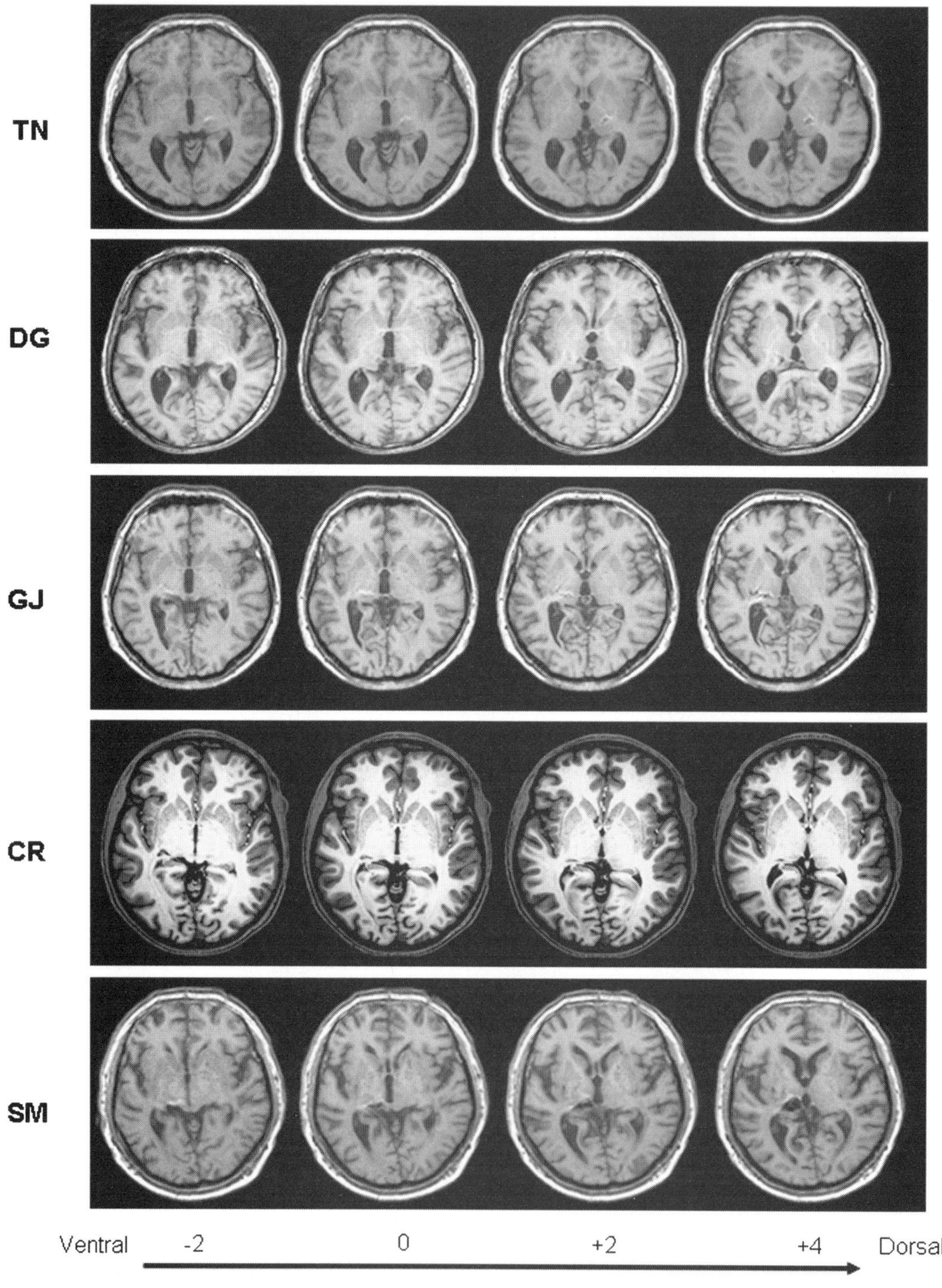

Fig. 2. Reconstructed T1-weighted axial MRI images for each patient. Anatomical convention: slices running from ventral to dorsal and showing millimetres above or below the anterior–posterior commissural (AC–PC) plane.

Results

Saccadic decision task

For the 17 ms condition, the mean proportion of correct ipsilesional and contralesional saccades responses was computed for each patient for both the ipsilesional target first and contralesional target first conditions. A one-way ANOVA showed a main effect of field, $F(1,4)=9.40$, $p=0.037$, confirming a bias to saccade towards ipsilesional targets in preference to contralesional ones (Fig. 3, top.) This was the case for all patients except CR, who showed no difference between fields. For the 0 ms SOA, where ipsilesional and contralesional targets appeared simultaneously, mean proportion of contralesional choices for each patient was computed. All patients, except DG, made more saccades to their ipsilesional than to their contralesional field when both targets were presented simultaneously. Although the proportion of contralesional choices were less than ipsilesional ones (Fig. 3, top), this difference did not reach significance $t(4)=1.307$, $p=0.26$.

Perceptual decision task

In contrast to the bias observed at the 17 ms SOA in the saccade decision task, Fig. 3 (bottom) shows that patients were very accurate in correctly reporting which target appeared first, 97% correct or better in all cases. For the simultaneous (0 ms SOA) condition, mean proportion of contralesional choices for each patient was computed. A one-sample *t*-test showed that patients were more likely to judge the ipsilesional target as appearing first $t(4)=4.32$, $p=0.012$ (Fig. 3, bottom). All five patients had an ipsilesional bias in this task.

Antisaccade task

Error rates were very low (mean$<$4%). No patient made more than 7% errors in any condition, and errors were not further analysed. After excluding trials with latencies more than four standard deviations from the conditional mean, individual *t*-tests for each patient were conducted, with trial as the unit of analysis. GJ and TN showed longer mean latencies to initiate antisaccades away from contralesional targets than ipsilesional targets: GJ — contra 442.4 ms (SEM = 7.5), ipsi 353.1 ms (SEM = 4.6), $t(339)=10.1$, $p<0.0005$; TN — contra 344.0 ms (SEM = 4.9), ipsi 329.5 ms (SEM = 3.5), $t(316)=2.35$, $p=0.020$. SM, in whom the lesion extended into dorsomedial thalamus, did not have an asymmetry of antisaccade latencies in this task: contra 427.3 (SEM = 13.5), ipsi 425.8 (SEM = 9.5), $t(263)=0.03$, $p=0.925$.

Discussion

Results reported here are in accord with previous research in these patients showing deficits of attention in the field contralateral to the pulvinar lesion (Rafal and Posner, 1987; Ward et al., 2002; Danziger et al., 2004; Michael and Desmedt, 2004; Ward and Arend, 2007; Ward et al., 2007). In a temporal-order judgment task, attention to one of the two simultaneous events results in its 'prior entry' into awareness. (Zackon et al., 1999; Shore et al., 2001; Spence et al., 2001; Vibell et al., 2007). All five patients showed an ipsilesional bias in the perceptual decision task. When the two targets were presented simultaneously, the patients were more likely to perceive the ipsilesional target as appearing first.

A similar bias was observed for simultaneous targets in the saccade decision task (although this difference did not reach statistical significance). This performance does not, by itself, indicate a specific oculomotor function of the pulvinar. The attentional deficit identified in the perceptual decision task would be expected to affect saccade responses no less than it did for manual responses. Importantly, however, the patients also showed a bias against making a saccade towards a contralesional target in the 17 ms SOA condition. In the perceptual judgment task, all patients performed almost flawlessly in this condition. At this temporal interval between the two targets, they had no difficulty in reliably judging which appeared first. Thus, in the saccade task, the patients showed a bias against making a saccade towards the contralesional target

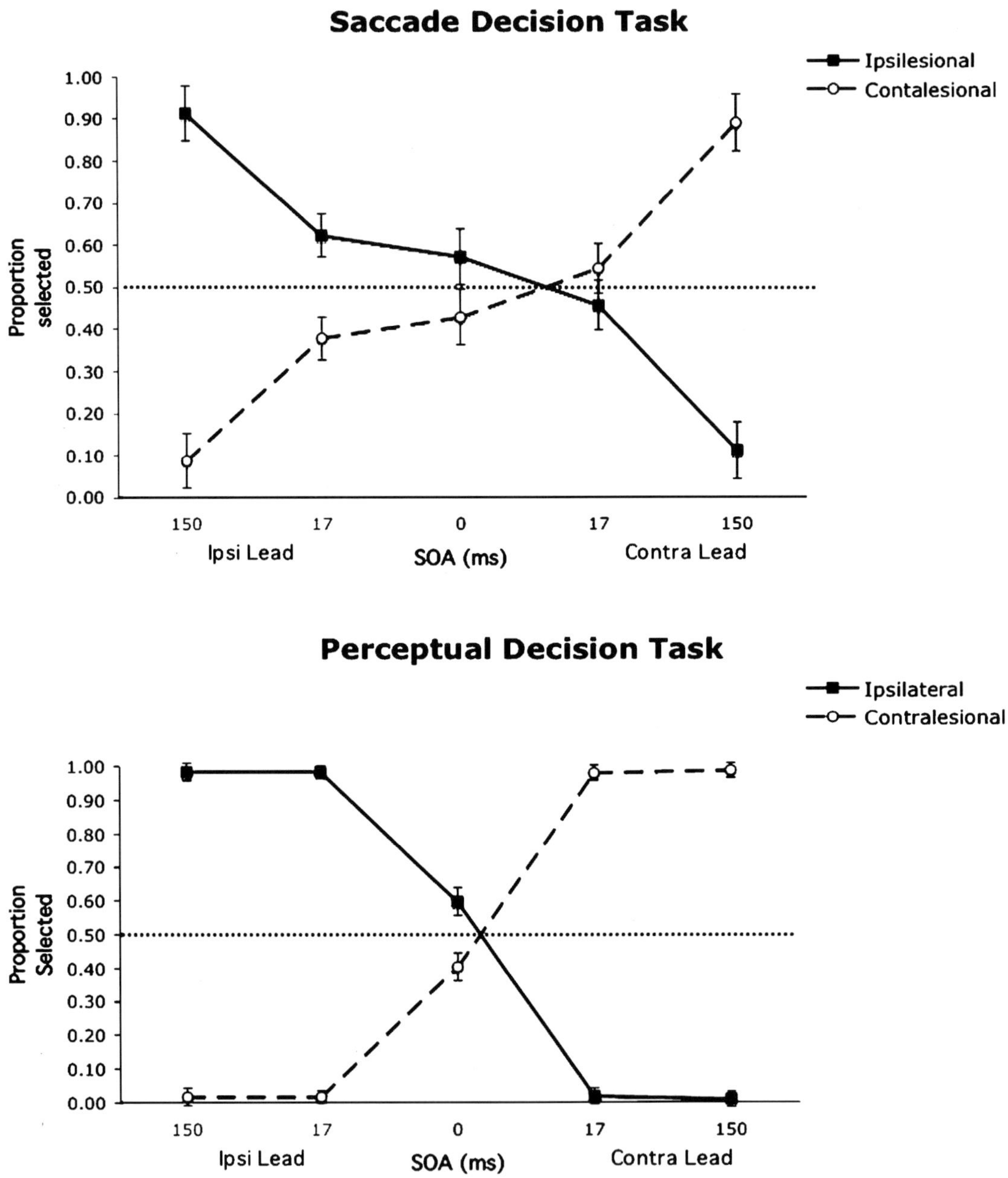

Fig. 3. Mean proportion of contralesional and ipsilesional responses for each target. Target stimulus onset asynchrony (SOA) in the saccade decision task (top) and perceptual decision task (bottom). Error bars display standard errors of the mean.

even under conditions in which they presumably correctly judged that the contralesional target had appeared first. These patients with chronic lesions of the pulvinar show a saccade decision bias similar to patients with chronic lesions of intraparietal cortex (Ro et al., 2001).

Like patients with chronic lesions of intraparietal cortex (Machado and Rafal, 2004), two patients tested in the antisaccade task had longer latencies to initiate saccades away from contralesional targets. One patient, SM, in whom the lesion extended into the dorsomedial thalamus did not show this effect. On the basis of physiological observations in antisaccade and memory-antisaccade tasks, Zhang and Barash (2000) have postulated a specific role of intraparietal cortex in computing visuo-motor transformations. Given its extensive and reciprocal connections not only with IPL, but also with frontal areas (Grieve et al., 2000; Shipp, 2003), the pulvinar may also be involved in such transformations.

This function is consistent with previous research suggesting that the pulvinar is involved more generally in visuo-motor transformations. We have previously found that spatial coding within the pulvinar is defined by multiple frames of reference, including retinotopic and object-based codes most likely generated within the cortex (Ward and Arend, 2007). In conjunction with the present results, it appears that the cortex and pulvinar may be jointly involved in a variety of spatial transformations, presumably to facilitate different kinds of action. More generally, however, as we discuss below, our results support a conceptual framework that considers the pulvinar as critical in flexibly linking visual stimuli with context-specific motor responses.

Sherman (2007) has recently invited us to consider the possibility that "all direct corticocortical pathways are modulatory [rather than 'driver'], which would mean that information routes between cortical areas depend on higher order thalamic relays. This would imply that all information reaching a cortical area, whether originating in the periphery (e.g. retina) or another cortical area, must pass through the thalamus. In other words, just as retinal information is relayed by thalamus, so is corticocortical information."

This hypothesis is consistent with the conceptual framework outlined here that the pulvinar is critical in integrating visual information with action systems. Specifically, we suggest that the pulvinar coordinates activities of visual and motor systems by enabling direct stimulus–action linkages that can be rapidly implemented in a specified context.

Using a flanker task with manual responses, we have shown that pulvinar lesions eliminate automatic response channel activation by contralesional visual stimuli (Danziger et al., 2004). A similar effect is seen in patients with chronic lesions of lateral prefrontal cortex (Rafal et al., 1996). In the flanker task, a specific visual stimulus becomes associated with a specific response that is automatically activated whenever the stimulus is presented (e.g., red square — push a button with index finger), resulting in interference by distracters coding for a response incompatible with the target stimulus. Similar to the case of the antisaccade task, pulvinar lesions do not prevent patients from making appropriate key-press responses to targets in their contralesional field but, when the stimuli are presented there as distractors, they do not automatically activate their associated response code and, hence, produce less interference with responses to a target presented at fixation.

Having implicated the pulvinar in attention and visually guided behaviour across a range of studies in our laboratories, we must also note that these patients generally do not complain about their vision, and do not report impairment of visually guided behaviours in their everyday life. That is, attentional impairments in patients with pulvinar lesions, although pervasive and consistent, are nevertheless 'sub-clinical.' Presumably corticocortical connections can mediate successful, if perhaps somewhat less efficient, perceptual processing and visually guided behaviour, even in the absence of the quick and efficient cortico-thalamo-cortical circuits. Even if cortico-cortical pathways do not normally function as 'drivers,' they could be recruited to do so in a reorganized brain during recovery from brain damage.

Acknowledgements

We want to thank all patients for their kind collaboration in this research programme and to acknowledge support from BBSRC project grant C501417 and Wellcome Trust programme grant 071924.

References

Andersen, R.A., Bracewell, R.M., Barash, S., Gnadt, J.W. and Fogassi, L. (1990) Eye position effects on visual, memory, and

saccade-related activity in areas LIP and 7a of macaque. J. Neurosci., 10: 1176–1196.
Bellebaum, C., Daum, I., Koch, B., Schwarz, M. and Hoffmann, K.P. (2005) The role of the human thalamus in processing corollary discharge. Brain, 128: 1139–1154.
Bender, D.B. (1981) Retinotopic organization of the macaque pulvinar. J. Neurophysiol., 46(3): 672–693.
Bender, D.B. and Baizer, J.S. (1990) Saccadic eye movements following kainic acid lesions of the pulvinar in monkeys. Exp. Brain Res., 79: 467–478.
Bender, D.B. and Butter, C. (1987) Comparison of the effects of the superior culliculus and pulvinar lesions on visual search and tachistoscopic pattern discriminations in monkeys. Exp. Brain Res., 69: 140–154.
Beneveto, L.A. and Miller, J. (1981) Visual responses of single neurons in the caudal lateral pulvinar of the macaque monkey. J. Neurosci., 11: 1268–1278.
Brigell, M., Babikian, V. and Goodwin, J.A. (1984) Hypometric saccades and low-gain pursuit resulting from a thalamic. Ann. Neurol., 15(4): 374–378.
Chalupa, L.M., Coyle, R.S. and Lindsley, D.B. (1976) Effects of pulvinar lesions on visual pattern discrimination in monkeys. J. Neurophysiol., 39: 354–369.
Crommelinck, M., Roucoux, A. and Meulders, M. (1977) Eye movements evoked by stimulation of lateral posterior nucleus and pulvinar in the alert cat. Brain Res., 124: 361–366.
Danziger, S., Ward, R., Owen, V. and Rafal, R. (2004) Contributions of the human pulvinar to linking vision and action. Cogn. Affect. Behav. Neurosci., 4(1): 89–99.
Grieve, K.L., Acuna, C. and Cudeiro, J. (2000) The primate pulvinar nuclei: vision and action. Trends Neurosci., 23(1): 35–39.
Hirose, G., Kosoegawa, H., Saeki, M., Kitagawa, Y., Oda, R., Kanda, S., et al. (1985) The syndrome of posterior thalamic hemorrhage. Neurology, 35: 998–1002.
LaBerge, D. and Buchsbaum, M.S. (1990) Positron emission tomographic measurements of pulvinar activity during an attention task. J. Neurosci., 10: 613–619.
Machado, L. and Rafal, R. (2004) Control of fixation and saccades during an anti-saccade task: an investigation in humans with chronic lesions of oculomotor cortex. Exp. Brain Res., 156: 55–63.
Michael, G. and Desmedt, S. (2004) The human pulvinar and attentional processing of visual distractors. Neurosci. Lett., 362: 176–181.
Ogren, M.P., Mateer, C.A. and Wyler, A.R. (1984) Alterations in visually related eye movements following left pulvinar damage in man. Neuropsychologia, 22: 187–189.
Petersen, S.E., Robinson, D.L. and Keys, W. (1985) Pulvinar nuclei of the behaving rhesus monkey: visual properties and their modulation. J. Neurophysiol., 54(4): 867–886.
Petersen, S.E., Robinson, D.L. and Morris, J.D. (1987) Contributions of the pulvinar to visual spatial attention. Neuropsychologia, 25: 95–106.
Rafal, R., McGrath, M., Machado, L. and Hindle, J. (2004) Effects of lesions of the human thalamus on ocular fixation during voluntary and visually triggered saccades. J. Neurol. Neurosurg. Psychiatr., 75: 1602–1606.
Rafal, R.D., Gershberg, R.E., Ivry, R., Kingstone, A. and Ro, T. (1996) Response channel activation and the lateral prefrontal cortex. Neuropsychologia, 34: 1197–1202.
Rafal, R.D. and Posner, M.I. (1987) Deficits in human visual spatial attention following thalamic lesions. Proc. Natl. Acad. Sci., 84: 7349–7353.
Ro, T., Rorden, C., Driver, J. and Rafal, R. (2001) Ipsilesional biases in saccades but not perception after lesions of the human inferior parietal lobule. J. Cogn. Neurosci., 13: 920–929.
Robinson, DL., McClurkin, J.W. and Kertzman, C. (1990) Orbital position and eye movement influences on visual responses in the pulvinar nuclei of the behaving macaque. Exp. Brain Res., 82: 235–246.
Robinson, D.L., Petersen, S.E. and Keys, W. (1986) Saccade related and visual activities in the pulvinar nuclei of the behaving rhesus monkey. Exp. Brain Res., 62: 625–634.
Romanski, L.M., Giguere, M., Bates, J.F. and Goldman-Rakic, P.S. (1997) Topographic organization of medial pulvinar connections with the prefrontal cortex in the rhesus monkey. J. Comp. Neurol., 379(3): 313–332.
Schlag, J. and Schlag-Rey, M. (1971) Induction of oculomotor responses from thalamic internal medullary lamina in the cat. Exp. Neurol., 33: 498–508.
Schlag-Rey, M. and Schlag, J. (1977) Visual and presaccadic neuronal activity in thalamic internal medullary lamina of cat: a study of targeting. J. Neurophysiol., 40(1): 156–173.
Schlag-Rey, M. and Schlag, J. (1984) Visuomotor functions of central thalamus in monkey. I. Unit activity related to spontaneous eye movements. J. Neurophysiol., 51: 1149–1174.
Sherman, S.M. (2007) The thalamus is more than just a relay. Curr. Opin. Neurobiol., 17: 417–422.
Shipp, S. (2003) The functional logic of cortico-pulvinar connections. Philos. Trans. R. Soc. Lond. B Biol. Sci., 358(1438): 1605–1624.
Shore, D.I., Spence, C. and Klein, R.M. (2001) Visual prior entry. Psychol. Sci., 12(3): 205–212.
Spence, C., Kingstone, A., Shore, D.I. and Gazzaniga, M.S. (2001) Representation of visuotactile space in the split brain. Psychol. Sci., 12(1): 90–93.
Ungerleider, L.G. and Christensen, C.A. (1977) Pulvinar lesions in monkeys produce abnormal eye movements during visual discrimination training. Brain Res., 136: 189–196.
Ungerleider, L.G. and Christensen, C.A. (1979) Pulvinar lesions in monkeys produce abnormal scanning of the complex visual array. Neuropsychologia, 17: 493–501.
Vibell, J., Klinge, C., Zampini, M., Spence, C. and Nobre, A.C. (2007) Temporal order is coded temporally in the brain: early event-related potential latency shifts underlying prior entry in a cross-modal temporal order judgment task. J. Cogn. Neurosci., 19(1): 109–120.
Ward, R. and Arend, I. (2007) An object-based frame of reference within the human pulvinar. Brain, 130: 2462–2469.
Ward, R., Calder, A., Parker, M. and Arend, I. (2007) Emotion recognition following human pulvinar damage. Neuropsycologia, 45: 1973–1978.

Ward, R., Danziger, S., Owen, V. and Rafal, R. (2002) Deficits in spatial coding and feature binding following damage to spatiotopic maps in the human pulvinar. Nat. Neurosci., 5(2): 99–100.

Watson, R.T. and Heilman, K.M. (1979) Thalamic neglect. Neurology, 29: 690–694.

Zackon, D.H., Casson, E.J., Zafar, A., Stelmach, L. and Racette, L. (1999) The temporal order judgment paradigm: subcortical attentional contribution under exogenous and endogenous cueing conditions. Neuropsychologia, 37(5): 511–520.

Zhang, M. and Barash, S. (2000) Neuronal switching of sensorimotor transformations for antisaccades. Nature, 408(6815): 971–975.

Zihl, J. and von Cramon, D. (1979) The contribution of the second visual system to directed visual attention in man. Brain, 102: 835–856.

SECTION 6

Abnormal Eye Movements: Mechanisms and Treatment Strategies

C. Kennard & R.J. Leigh (Eds.)
Progress in Brain Research, Vol. 171
ISSN 0079-6123

CHAPTER 6.1

How disturbed visual processing early in life leads to disorders of gaze-holding and smooth pursuit

Michael J. Mustari[1,2,*], Seiji Ono[1,2] and Katia C. Vitorello[1]

[1]*Division of Sensory-Motor Systems, Yerkes National Primate Research Center, Emory University, Atlanta, GA, USA*
[2]*Department of Neurology, Emory University, Atlanta, GA, USA*

Abstract: When synergistic interactions between visual and oculomotor systems are impaired early in life, strabismus, gaze instability, smooth pursuit asymmetry, and loss of visual function are likely to occur. These disorders are relatively common, permanent, and difficult to treat. We have developed effective animal models for infantile strabismus by raising infant monkeys (*Macaca mulatta*) with restricted binocular visual experience. We have found that the specific oculomotor disorders that occur with early onset strabismus depend on the type of early visual experience. Our approach allows us to examine the neural substrate associated with different components of infantile strabismus including latent nystagmus (LN) and smooth pursuit asymmetry. For example, we have found LN is most associated with loss of binocular visual sensitivity normally present in neurons of pretectal nucleus of the optic tract (NOT). In contrast, nasalward bias in smooth pursuit of strabismic monkeys could be associated with loss of binocular visual and eye movement sensitivity of neurons in medial superior temporal (MST) area.

Keywords: latent nystagmus; strabismus; smooth pursuit; gaze-holding

Introduction

Different eye movement subsystems including vestibular ocular, optokinetic, saccadic, smooth pursuit, vergence, and fixation work together to maintain the image of an object of interest virtually stable on the fovea during self or object motion (see Leigh and Zee, 2006, for review). These oculomotor circuits require postnatal visual and oculomotor experience for final and maintained calibration.

It is well known that disruption of coordinated binocular vision early in life leads to strabismus and varying degrees of amblyopia. This is due, in part, to loss of functional connections in striate cortex (see Boothe and Fulton, 2000). The degree to which visual function is compromised by abnormal early visual experience depends on the onset time, duration, and type of visual deprivation.

The macaque has proven to be a valuable model for studying the postnatal development of the visual system because both monkey and human follow a similar developmental sequence. The first 6 weeks of life comprises the most sensitive part of this early critical period for visual acuity development (Boothe and Fulton, 2000; Harwerth et al., 1990). Other capabilities such as visual motion processing and dependent eye movement systems such as smooth pursuit may have extended critical periods (Wong et al., 2003; Kiorpes and Movshon 2004; Kourtzi et al., 2006).

*Corresponding author. Tel.: +404-727-9194;
Fax: +404-727-9294; E-mail: mjmustar@rmy.emory.edu

DOI: 10.1016/S0079-6123(08)00670-5

Disruption of V1 binocular visual sensitivity is thought to part of a cascade of events preventing sensory and motor fusion, resulting in infantile strabismus (see Tychsen, 1993, for review). Cortical areas outside of V1 are likely to play a role in pathological eye movements in strabismic subjects. Kiorpes et al. (1996) have shown that neurons in middle temporal (MT) cortex of strabismic monkeys have reduced binocular sensitivity but visual direction and speed sensitivity of these neurons appeared to be in a normal range suggesting involvement of distal sites. Our studies are designed to test whether neurons in medial superior temporal (MST) cortex of strabismic monkeys have properties consistent with their playing a role in abnormal smooth pursuit.

Methods

Special rearing

Our infant monkeys (*Macaca mulatta*) were born in captivity at Yerkes National Primate Research Center and raised in a dedicated nursery facility on-site. Here we show preliminary results from a monkey with strabismus, created by recession/resection surgery of the lateral and medial rectus muscles of one eye to produce esotropia. Our eye movement and single unit recording studies were conducted when the monkeys were at least 2 years old.

Surgical procedures

A detailed description of our surgical procedures for scleral search coil and recording chamber implantation can be found in our earlier publications (Mustari et al., 2001; Ono and Mustari, 2006). Briefly, surgical procedures were carried out under aseptic conditions in a dedicated surgical suite using isoflurane anaesthesia (1.25–2.5%) and post-surgical analgesia (Buprenorphine 0.01 mg/kg)/anti-inflammatory (Banamine 1.0 mg/kg) treatment. Eye movement and single-unit recording studies were conducted in juvenile normal or strabismic monkeys (*M. mulatta*), weighing 4–5 kg. All surgical and experimental procedures were performed in strict compliance with NIH guidelines and the protocols were reviewed and approved by the Institutional Animal Care and Use Committee at Emory University.

Behavioural paradigms

A detailed description of our behaviour and neurophysiological methods can be found in earlier publications. Testing was conducted in a sound attenuated and light tight room with visual stimuli rear projected on a tangent screen at 57 cm viewing distance. Eye position data from each eye were calibrated by requiring the monkey to fixate a small diameter (0.2°) target at known horizontal and vertical eccentricities during monocular viewing.

Results

We are able to produce animals with the essential components found in infantile strabismus. In Fig. 1 we show eye alignment and gaze-holding data from a 2-year old esotropic monkey produced by resection/recession surgery (at 6 days of age) of medial and lateral rectus muscles of the right eye. The left panel shows a photograph of an esotropic monkey with the corneal light reflex located at the lower margin of the pupils (illumination from below the line of sight). When the photograph was taken the left eye was attending a target on the midline while the right eye was turned inward (light reflex of the right eye is shifted laterally). We found variable amounts (7°–25°) of esotropic strabismus, depending on the viewing eye and behavioural context. For the eye movement data in Fig. 1, the monkey was viewing a stationary target centred on a stationary patterned background. During monocular viewing, gaze-instability consistent with latent nystagmus (LN) was present. Nasalward LN slow phases were low speed ($<1^\circ$/s left eye viewing; $<5^\circ$/s right eye viewing).

Asymmetric smooth pursuit is an important component of infantile strabismus. In Fig. 2, we document smooth pursuit and vestibular ocular reflex (VOR) performance for an esotropic

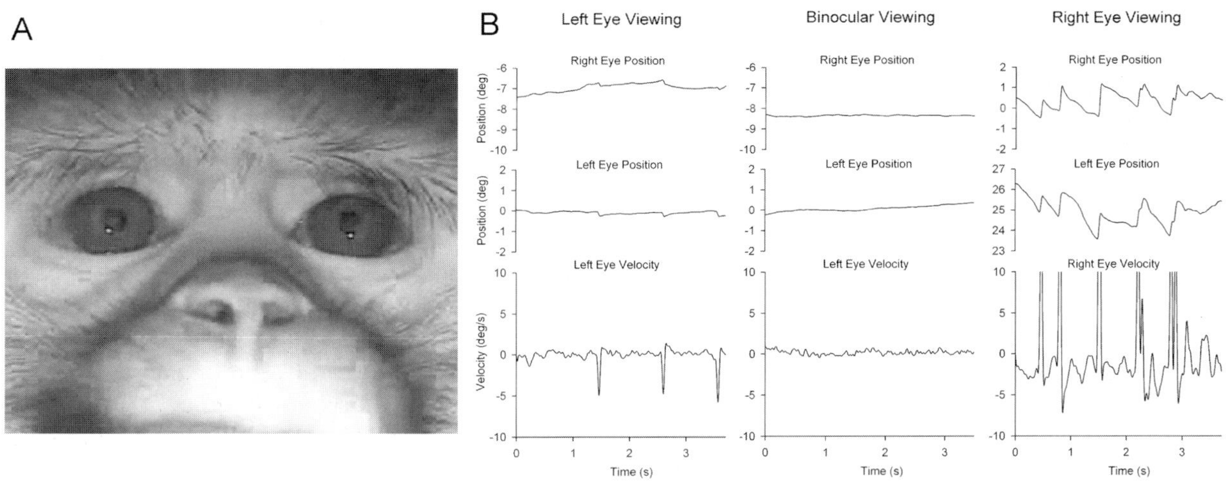

Fig. 1. Examples of latent nystagmus in experimental infantile strabismus in a monkey. Photograph of esotropic strabismic monkey produced by resection (medial rectus) and recession (lateral rectus) surgery of right eye muscles before 1 week of age. Illumination from below produced a corneal light reflex below the pupils (left eye is fixating a midline target; right eye turned inward). Gaze-holding during fixation of a stationary target on a patterned background is defective. LN is evident as the direction of slow-phases remains nasalward with respect to the viewing eye. Eye movement recordings obtained with dual scleral search coils.

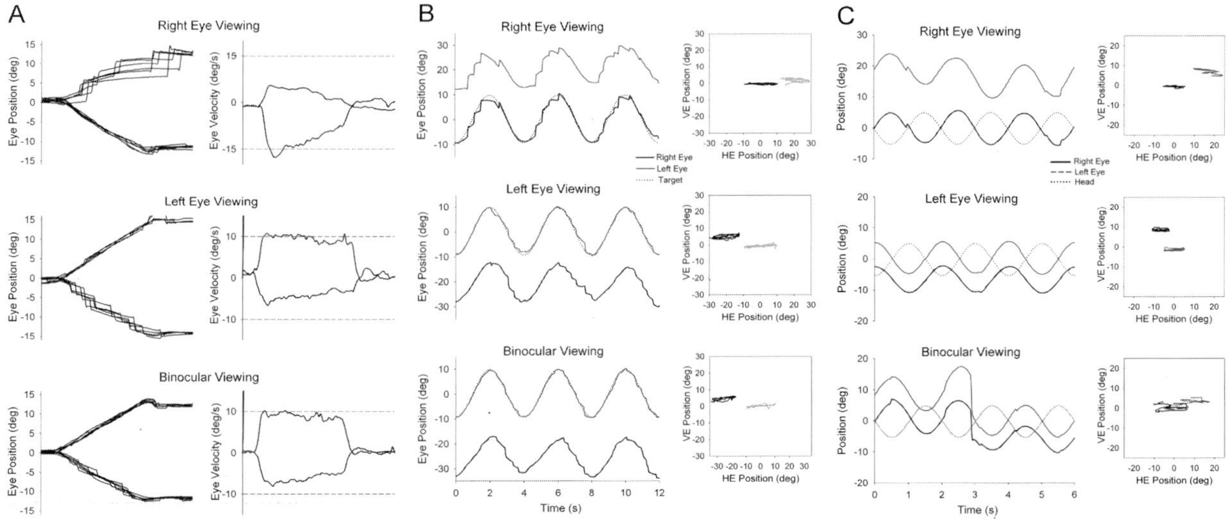

Fig. 2. Smooth pursuit and VOR testing in a strabismic monkey. Horizontal eye movements are shown for horizontal step-ramp SP (left panel), sinusoidal SP (middle panel), and horizontal VOR (chair motion, $\pm 5^{\circ}$; 0.25 Hz). SP evinced a strong nasalward bias during monocular viewing conditions (top panels). In contrast, the VOR of the attending eye remained symmetric during all viewing conditions. Note, alternating fixation during VOR (bottom panel). Right eye (black traces), left eye position (grey traces), and target or chair position (dashed lines). The x,y positions of the eyes are shown in the inset plots.

monkey, during monocular and binocular viewing conditions. The main goal of this testing was to compare and contrast horizontal eye movements of volitional smooth pursuit and reflex driven, non-visual origins. The left and middle columns of Fig. 2 show eye movement recordings taken during smooth pursuit (step-ramp and sinusoidal) of a small diameter (0.1°) target spot moving over a dark background. Eye position and velocity signals from the right and left eyes are shown. The left column of Fig. 2 shows smooth pursuit performance during step-ramp tracking. In this panel, we only show eye position and eye velocity for the viewing eye (left or right) in monocular conditions or the left eye in binocular condition (bottom panel). Smooth pursuit during monocular viewing (top and middle panels) was asymmetric with much higher gain (eye velocity/target velocity) for eye motion directed towards the nose (nasalward). In contrast, smooth pursuit was mostly symmetric during binocular viewing conditions. The animal attempts to perform smooth pursuit in any direction but tracking away from the nose (temporalward) during monocular viewing is interrupted by "catch-up" saccades. These intrusive saccades have been removed from the eye velocity records using standard methods (Mustari et al., 2001). Examination of the eye velocity records (left panel) prior to the onset of smooth pursuit shows that LN was minimal as shown in Fig. 1. Therefore, the observed nasalward bias and significant deficit in temporalward smooth pursuit is not likely to be an artefact of LN.

In the middle column of Fig. 2, we show examples of the monkey performing smooth pursuit of a target moving sinusoidally on the tangent screen. Here, we show left and right eye position recorded during three standard viewing conditions. The angle of strabismus, which averages 20° (esotropic), is easy to see in these records. There is a clear nasalward bias in smooth pursuit during monocular viewing conditions. Temporalward tracking has lower gain and contains catch-up saccades. Notice that the movements of both eyes are mostly conjugate. We also show plots of horizontal eye position against vertical eye position (middle and right columns). For example, when the monkey views the target with his left eye (middle panel) the right eye position traces are shifted across the vertical meridian (note small and variable vertical deviation in the non-viewing eye). Therefore, even though resection/recession surgery was performed only on the right eye, both eyes show similar pathological smooth pursuit performance. We suggest that this binocular smooth pursuit disorder is due to a specific disruption of visual-oculomotor development and not peripheral pathology associated with early surgery.

To test this idea further, we conducted experiments where smooth eye movements were generated by the VOR rather than by volitional smooth pursuit (Fig. 2, right column). We examined VOR eye movements over the same amplitude and frequency ranges as tested during volitional smooth pursuit. We produced a horizontal VOR by delivering whole-body stimulation with a 60 ft-lb torque motor that rotated the monkey in the horizontal plane in a periodic fashion ($\pm 5°$; 0.25 Hz). During VOR testing, we provided the monkey with the same small diameter target used during smooth pursuit testing (Ono and Mustari, 2006). However, the target was maintained stationary on the tangent screen while the monkey was oscillated in front of that screen. This "VOR in light" condition produces an eye movement equal in amplitude and opposite in direction to the motion of the head. VOR driven eye movements in our strabismic monkeys were mostly saccade free except during binocular viewing when monkeys freely alternate their fixating eye (Fig. 2, bottom right). Such alternation also occurs during ongoing sinusoidal smooth pursuit and saccadic tracking (not illustrated). Importantly, the smooth eye movements produced by the VOR in the strabismic monkey are symmetric and unity gain (eye velocity/head velocity). The symmetric VOR during monocular viewing indicates that the abnormal smooth pursuit in our strabismic monkeys represents a specific disorder of visual-oculomotor processing. We suggest the disorder responsible for asymmetric smooth pursuit in some strabismic subjects may be due to a problem in transforming visual information into commands for eye movements at cerebral cortical levels.

Visual sensitivity of MST neurons strabismic monkey

We examined both the visual and eye movement sensitivity of MST neurons in strabismic monkeys. In normal monkeys all neurons in MST evince binocular sensitivity. In contrast, strabismic monkeys have decreased binocular visual sensitivity in MST. For this testing the monkey fixated a stationary target at the centre of the tangent screen while a large-field visual stimulus was moved in the preferred and anti-preferred directions (Mustari et al., 2001; Price et al., 2005). In Fig. 3 we present single-unit data aligned on the start of background motion (Fig. 3, dashed line) without shifting the rasters or spike density functions by visual latency. Visual latency was taken into account in determining direction and speed preference for each neuron. We determined a visual ocular dominance index (ODI) for each neuron using the following equation.

$$\mathrm{ODI} = \frac{A_{\mathrm{contra}}}{(A_{\mathrm{contra}} + A_{\mathrm{ipsi}})}$$

where A, is the average (at least 10 cycles) firing rate for each half-cycle. We used a five category classification ranging from contralateral eye only (group 1) to ipsilateral eye only (group 5) with neurons in group 3 having equal drive from both eyes and the other groups dominated by either the contralateral (group 2) or ipsilateral eye (group 4). Figure 3 shows the response of three typical neurons (groups 1, 3, and 5) during visual testing. We found a breakdown in the binocular visual sensitivity of MST neurons similar to that reported for MT (Kiorpes et al., 1996). We do not find monocular MST neurons in normal monkeys. We found that a full range of visual motion directions and speeds was represented in MST of strabismic monkeys (not illustrated). It is therefore possible that higher cortical areas receive signals with defective binocular information, which could impair generation of smooth pursuit commands thought to begin, at least in part in area MST.

Eye motion sensitivity of MSTd neurons

In normal monkeys smooth pursuit neurons in MST have equal modulation regardless of the viewing eye. We found a profound alteration in the smooth pursuit related (eye motion) responses in some MST neurons in the strabismic monkey. We showed that our MST neurons carried extraretinal signals by blinking the target off briefly (e.g., 200 ms) during smooth pursuit. Neurons with extraretinal signals continue their response in absence of visual motion as long as pursuit continues. Figure 4 shows examples of MST neurons from the strabismic monkey during sinusoidal smooth pursuit of a small target moving over a dark background. We used the best direction and speed to test whether there was a preference for eye motion of either eye. We found examples of neurons that responded preferentially during smooth pursuit when either the left (Fig. 4, top panel) or right eye (Fig. 4, second panel) was viewing. We calculated an ocular preference index for the smooth pursuit related response of our MST neurons in a manner similar to that used for visual motion sensitivity (above). We found clear examples of MST neurons that responded only during left or right eye viewing, i.e., their response was eye specific and related to volitional SP eye motion. Such uniocular smooth pursuit neurons are never found in MST cortex of normal monkeys. We also found a preponderance of left eye dominated neurons, which might be related to the eye preference of this monkey. When we examined smooth pursuit related direction selective neurons further, we found that some eye specific MST neurons have a nasalward smooth pursuit preference (Fig. 4, bottom panel), which could be a source of the nasalward biased smooth pursuit behaviour of strabismic monkeys. For example, the left eye neuron in Fig. 4 (top) responds during nasalward (rightward) tracking. When tracking with the right eye in the same direction (rightward) this neuron was not well-modulated (Fig. 4, top right). The right eye neuron in Fig. 4 (second panel) also shows a nasalward direction preference. Therefore, the ocular and directions preferences of some MST neurons are compatible with their playing a role in asymmetric smooth pursuit. We do not know if the smooth pursuit response of these neurons represents a directional smooth pursuit command originating in MST cortex or whether it might be an efference copy of a pursuit command from distal centres (see Discussion).

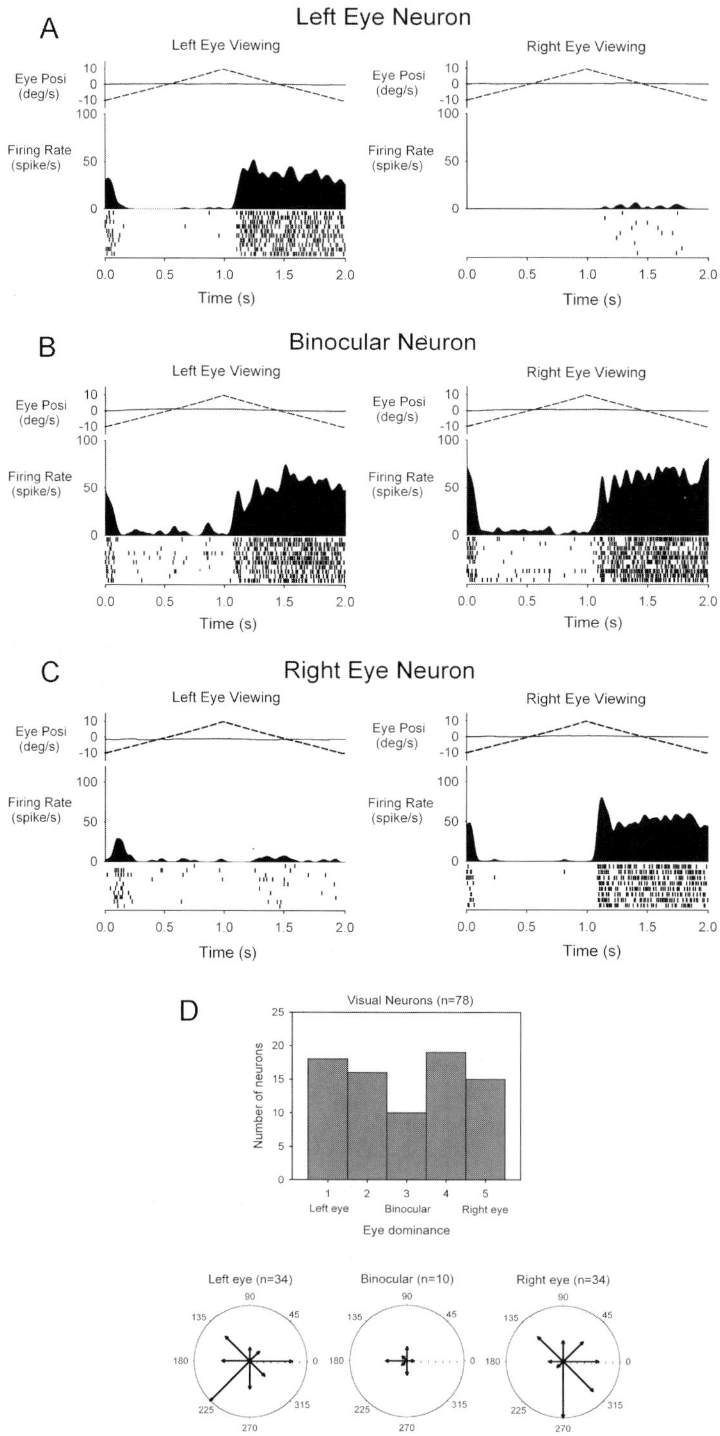

Fig. 3. Visual sensitivity of MST neurons from strabismic monkey. Neurons were driven by visual stimulation delivered to either the left or right eye. Binocular (group 3) neurons were proportionately reduced compared to normal monkeys. Data aligned on start of visual motion with spike density functions and rasters not shifted by visual latency of the neuron. All directions of visual motion were represented with no strong T-N bias.

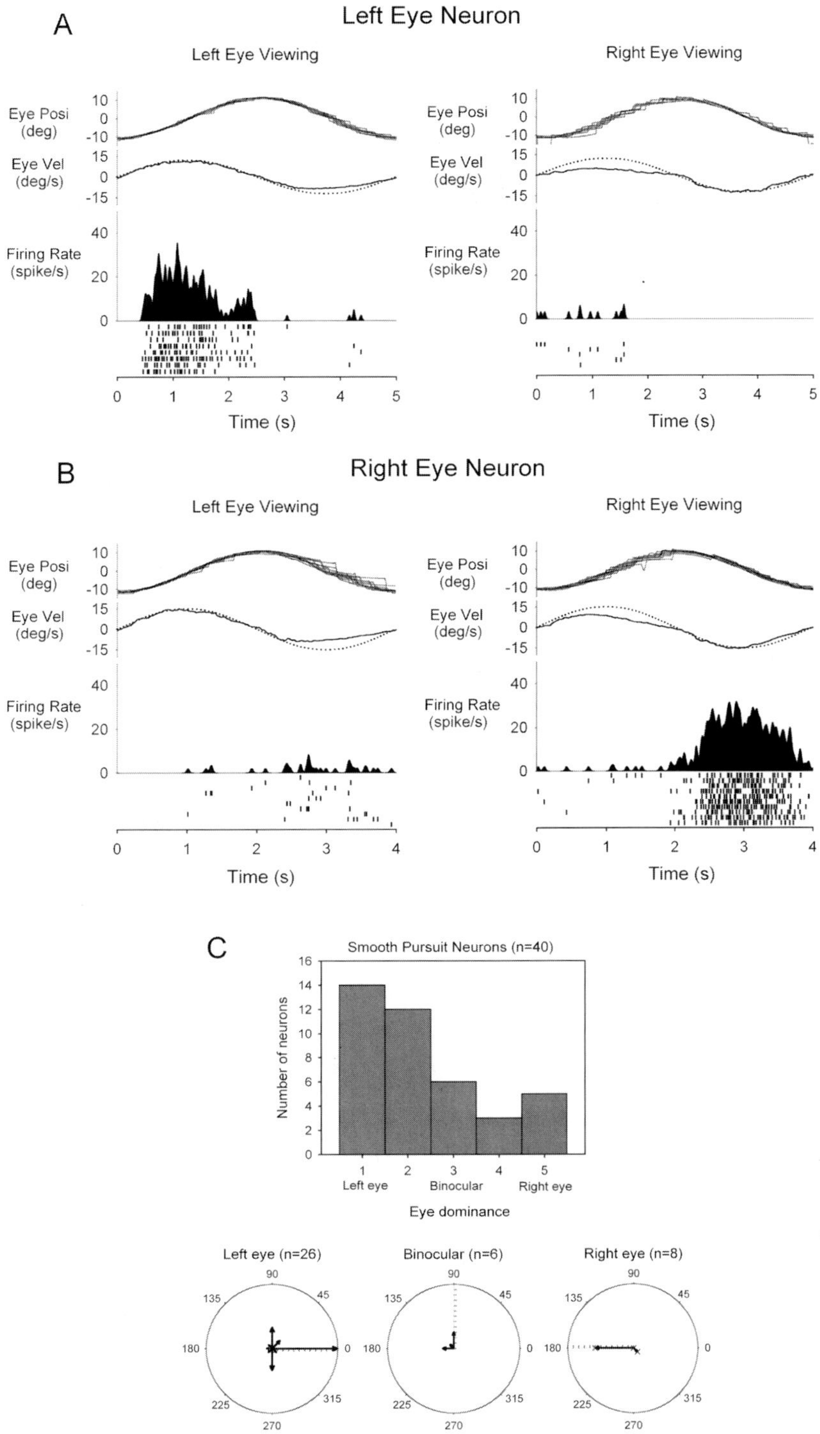

Fig. 4. Smooth pursuit eye motion sensitivity of MST neurons from strabismic monkey. Examples of ocular specific SP related neurons in MST cortex of strabismic monkey. Each neuron has a nasalward bias. The response of these neurons was related mostly to eye motion with only small contributions from visual motion per se. Ocular preference histogram and associated direction preferences are shown. Top panel, neuron responding during nasalward pursuit during left eye viewing but unresponsive during right eye viewing. Bottom panel, neuron responding preferentially during nasalward pursuit during right eye viewing.

Discussion

Infantile strabismus is a common problem affecting about 3% of children. In addition to eye misalignment, infantile strabismus is associated with disorders of gaze-holding, OKN, and smooth pursuit. The neural mechanisms associated with different components of this syndrome have not been fully worked out. Our approach involves production of appropriate animal models to permit studying neural mechanisms associated with specific problems in development of visual-oculomotor function.

Latent nystagmus

LN is commonly associated with some forms of strabismus, especially those with early onset in infancy. The severity of LN in experimental strabismus is related to the onset, duration and type of impoverished binocular visual experience (Tusa et al., 2002). For example, low speed LN has been reported for monkeys made esotropic by surgery (Kiorpes et al., 1996) or prism rearing (Wong et al., 2003). We recently reported that LN was strongly associated with loss of binocular sensitivity and presence of contralateral eye dominance of neurons in the pretectal nucleus of the optic tract (NOT). Forms of deprivation that create strong competitive interactions between inputs from the left and right eyes might be important for altering NOT binocular visual sensitivity (Mustari et al., 2001).

The cortical pursuit system of normal monkeys

Our understanding of neural mechanisms associated with different aspects of smooth pursuit has advanced significantly in the last decade (see Krauzlis, 2004, for review). Briefly, retinal input to the accessory optic system (e.g., NOT) and geniculo-striate dorsal stream pathways provide signals for the afferent limb of smooth pursuit. Especially important are visual motion sensitive areas in the dorsal stream areas including MT and MST. Lesions placed in areas MT and MST produced retinotopic and directional deficits in smooth pursuit, respectively (Newsome et al., 1985; Dürsteler et al., 1987). Both MT and MST have reciprocal connection with the FEF region (see Lynch and Tian, 2006, for review). Lesions of FEF are associated with defective predictive and visually guided smooth pursuit (see Fukushima, 2003, for review). The majority of FEF neurons begin their response before the onset of pursuit, and they contribute to the initiation of pursuit, which is characterized by high retinal slip and eye acceleration (see Fukushima, 2003, for review). All of these cortical areas have direct projections to brainstem areas (e.g., rNRTP, DLPN, NOT, and SC) critical for the generation of smooth pursuit. These brainstem centres project to different regions of the cerebellum (flocculus, paraflooculus, and vermis), which play complimentary roles in smooth pursuit (see Leigh and Zee, 2006, for reviews).

The cortical smooth pursuit system of strabismic monkeys

Although smooth pursuit depends on visual motion processing and generation of early commands for eye movements in the areas MT and MST of the cortical pursuit system (see Krauzlis, 2004, for review), there have been few single-unit recording studies in any of these areas in specially reared monkeys. Kiorpes et al. (1996) tested the neural responses of MT neurons from monkeys with early onset (<11 days old) artificial strabismus. These monkeys had asymmetric monocular smooth pursuit with a nasalward bias and LN. Single-unit recording studies in paralysed and anaesthetized preparations of these monkeys revealed a loss of binocularly sensitive neurons in MT. However, the motion sensitivity of MT neurons was indistinguishable from that found in MT neurons of normal monkeys. Therefore, Kiorpes et al. (1996) suggested that the nasalward smooth pursuit bias of strabismic monkeys could depend on structures "downstream" from MT. Area MST and FEF are two such cortical areas. Our preliminary results indicate that neurons in area MST of strabismic monkeys have altered ocular and directional sensitivity that could play a role in abnormal smooth pursuit. Further studies are required to determine the potential involvement of other parts of the cortical smooth pursuit system in asymmetric smooth pursuit eye movements.

Acknowledgements

We acknowledge the expert technical assistance provided by Mrs. Tracey Fountain and Mr. Anthony Gazy. This work is supported by grants from the National Institutes of Health Grants (NEI EY06069; RR00165).

References

Boothe, R.G. and Fulton, A.B. (2000) Amblyopia. In: Albert D.M. and Jakobiec F.A. (Eds.), Principles and Practice of Ophthalmology, 2nd ed., Vol. 5, Section XVI, Pediatric Ophthalmology. W. B. Saunders Company, Philadelphia, PA, pp. 4340–4357.

Dürsteler, M.R., Wurtz, R.H. and Newsome, W.T. (1987) Directional pursuit deficits following lesions of the foveal representation within the superior temporal sulcus of the macaque monkey. J. Neurophysiol., 57: 1262–1287.

Fukushima, K. (2003) Frontal cortical control of smooth-pursuit. Curr. Opin. Neurobiol., 13: 647–654.

Harwerth, R.S., Smith, E.L., III, Duncan, G.C., Crawford, M.L.J. and Von Noorden, G.K. (1990) Behavioral studies of the sensitive periods of development of visual functions in monkeys. Behav. Brain Res., 41: 179–198.

Kiorpes, L. and Movshon, J.A. (2004) Development of sensitivity to visual motion in macaque monkeys. Vis. Neurosci., 21: 851–859.

Kiorpes, L., Walton, P.J., O'Keefe, L.P., Movshon, J.A. and Lisberger, S.G. (1996) Effects of early onset artificial strabismus on pursuit eye movements and on neuronal responses in area MT of macaque monkeys. J. Neurosci., 16: 6537–6553.

Kourtzi, Z., Augath, M., Logathetis, N.K., Movshon, J.A. and Kiorpes, L. (2006) Development of visual evoked activity in infant macaque monkeys studied longitudinally with fMRI. Magn. Reson. Imaging, 24: 359–366.

Krauzlis, R.J. (2004) Recasting the smooth pursuit eye movement system. J. Neurophysiol., 91: 591–603.

Leigh, R.J. and Zee, D.S. (2006) The neurology of eye movements (4th ed.). Oxford Press, New York. Contemporary Neurology Series.

Lynch, J.C. and Tian, J.R. (2006) Cortico-cortical networks and cortico-subcortical loops for the higher control of eye movements. In: Büttner-Ennever J. (Ed.), Neuroanatomy of the Oculomotor System, Progress in Brain Research, Vol. 151. Elsevier, Amsterdam, pp. 461–502.

Mustari, M.J., Tusa, R.J., Burrows, A.F., Fuchs, A.F. and Livingston, C.A. (2001) Gaze-holding deficits and latent nystagmus in monkeys with early-onset visual deprivation. Role of the pretectal NOT. J. Neurophysiol., 86: 662–675.

Newsome, W.T., Wurtz, R.H., Dursteler, M.R. and Mikami, A. (1985) Deficits in visual motion processing following ibotenic acid lesions of the middle temporal visual area of the macaque monkey. J. Neurosci., 5: 825–840.

Ono, S. and Mustari, M.J. (2006) Extraretinal signals in MSTd related to volitional smooth pursuit. J. Neurophysiol., 96: 2819–2825.

Price, N.S., Ono, S., Mustari, M. and Ibbotson, M.R. (2005) Comparing acceleration and speed tuning in macaque MT: physiology and modeling. J. Neurophysiol., 94: 3451–3464.

Tusa, R.J., Mustari, M.J., Das, V.E. and Boothe, R.G. (2002) Animal models for visual deprivation-induced strabismus and nystagmus. Ann. N.Y. Acad. Sci., 956: 346–360.

Tychsen, L. (1993) Motion sensitivity and the origins of infantile strabismus. In: Simmons K. (Ed.), Early Visual Development, Normal and Abnormal. Oxford, New York, pp. 364–390.

Wong, A.M.F., Foeller, P., Bradley, D., Burkhalter, A. and Tychsen, L. (2003) Early versus delayed repair of strabismus n macaque monkeys: I. Oculomotor effects. J. AAPOS, 7: 200–209.

C. Kennard & R.J. Leigh (Eds.)
Progress in Brain Research, Vol. 171
ISSN 0079-6123

CHAPTER 6.2

Manifest latent nystagmus: a case of sensori-motor switching

Richard Abadi[1,*], Richard Clement[2], Theofana Theodorou[3] and Columba Scallan[1]

[1]*University of Manchester, Faculty of Life Sciences, Manchester, UK*
[2]*Institute of Child Health, University College London, London, UK*
[3]*Hallamshire Hospital, Sheffield, UK*

Abstract: Latent nystagmus (LN) and manifest latent nystagmus (MLN) are closely associated with early visual deprivation and strabismus. In both cases, the eyes oscillate horizontally in an involuntary manner and the fast phases always beat towards the attending or fixing eye. By simultaneously recording eye movements during the dichoptic viewing of dissimilar visual stimuli we present evidence that MLN offers a unique opportunity to examine the nature of sensori-motor switching. In particular, we show how the nystagmus beat direction is strongly influenced by endogenous and exogenous attention. A model describing the possible mechanisms underpinning the sensori-motor switching is proposed.

Keywords: manifest latent nystagmus; sensori-motor switching; binocular rivalry; attention; dichoptic viewing; pause cells

Introduction

Sensory and motor correspondence underpins single binocular vision. On the occasions when greatly dissimilar images fall on the foveae, fusion is compromised, leading to diplopia, rivalry, or suppression. In early infancy a sustained, abnormal visual experience can lead to a unilateral strabismus. Here, the fixing eye is dominant and the nonfixing eye exhibits a lower visual acuity (amblyopia) and sensory suppression. On occasions the strabismus may alternate, with either eye taking up fixation. Clinically, the use of differently orientated Bagollini striated lenses placed over each eye indicates which is the fixing eye. Recording the dynamics of this perceptual switching requires the full co-operation of the subject to report which of the two stimuli are being perceived at any point in time.

In this paper, we report how a particular form of alternating fixation, manifest latent nystagmus (MLN) associated with early visual deprivation can assist our understanding of sensori-motor switching. Specifically, by monitoring the beat direction of the nystagmus, we can ascertain the temporal dynamics of fixation without the need to ask what the subject is seeing. Finally, we propose a neural model underpinning this sensori-motor switching.

Manifest latent nystagmus

Infantile nystagmus is an involuntary, bilateral, conjugate oscillation of the eyes, which is present

*Corresponding author. Tel.: +44 161 306-3875; Fax: +44 161 306-3887; E-mail: richard.abadi@manchester.ac.uk

DOI: 10.1016/S0079-6123(08)00671-7

at birth or develops within the first few months. There are three major manifestations of infantile nystagmus: congenital nystagmus (CN), latent nystagmus (LN), and MLN (Dell'Osso, 1985; Abadi and Dickinson, 1986; Harris, 1997; Abadi and Bjerre, 2002; Leigh and Zee, 2006). Their principal oculomotor differences lie in the form of their slow phases (Dell'Osso et al., 1979; Dell'Osso, 1985; Abadi and Bjerre, 2002). Whereas CN has a slow phase that has an increasing exponential velocity, LN and MLN have slow phases that are invariably decreasing or linear. Additional distinguishing features are found in the nature of the fast phase or the beat direction. In LN and MLN the eyes always beat towards the attending eye (Dell'Osso et al., 1979; Abadi, 1980; Dell'Osso, 1985; Abadi and Scallan, 2000).

It is generally agreed that whilst LN is a rare occurrence, MLN is not uncommon (Dell'Osso, 1985; Abadi and Bjerre, 2002). MLN seen in adults does not appear to be associated with either any one-classification group such as idiopath or afferent visual defect (Abadi and Dickinson, 1986; Abadi and Bjerre, 2002), although primates and human neonates who have experienced early visual deprivation often exhibit an MLN (Tusa et al., 1991; Abadi et al., 2006). Abadi and Scallan (2000) have suggested that instead of treating LN and MLN as separate entities, they may be considered as part of a continuum. Their differences lie in the nystagmus behaviour seen during binocular viewing. Thus, a Type 1 MLN is stable apart from the occasional saccadic intrusion (Abadi and Gowen, 2004). This is essentially an LN. A Type 2 MLN has frequent saccadic intrusions, whilst Types 3 and 4 exhibit a nystagmus that is torsional in Type 3 and horizontal in Type 4. During monocular viewing the nystagmus is always bilateral, conjugate, and horizontal for each of the four MLN types.

MLN is a visually driven ocular motor disorder in which the direction of the fast phase indicates the viewing eye. Thus a right or left beating MLN indicates that the viewing eyes are the right and the left respectively. It has also been shown that removing vision (no fixation target in darkness) or opening the loop reduces the number of fast phases and extends the period of the slow phase (Abadi and Scallan, 2000; Gradstein et al., 2004).

Binocular vision and perceptual multistability

Binocular vision breaks down when the monocular inputs are different. This may be simulated in the laboratory. For example, by presenting a horizontal grating to one eye while the other eye views a vertical grating, one of two possibilities may arise — monocular suppression or binocular rivalry (Levelt, 1965). In the former case, one eye is dominant although the dominance may alternate. In the latter case, a mosaic of vertical and horizontal patterns is seen and they too can alternate. These perceptual situations are often referred to as bistable states (Leopold and Logothetis, 1999). The periods of dominance or eye preference can be manipulated by varying the spatial and temporal properties of the stimulus and has been associated with aspects of visual attention (Lack, 1978).

Dissimilar stimuli have also been employed to study the nature of suppression in strabismus (Schor, 1977; Smith et al., 1985) as well as in LN (Abadi, 1980). Measures of dominance were gained by recording the intra-ocular differences in stimuli size, contrast, and orientation that were able to induce a switch in eye preference.

Experiment 1: perceptual reversals with MLN

Three individuals with MLN dichoptically viewed orthogonally orientated sinusoidal grating stimuli presented on each of the two monochromatic Clinton Monoray DS2000 monitors subtending 37° (horizontally) and 29° (vertically) (Fig. 1a). The mean background luminance of the two 5° displays was 113 cd/m^2 and the ambient illumination was 22 lux. The orthogonal 4c/deg sinusoidal gratings were generated by VSG 2/5 graphics cards (CRS, Rochester, England) and subjects used a response box to record their perceptual experience (Fig. 1b). Binocular horizontal eye movements were simultaneously monitored using a head-mounted infra-red limbal eye tracker (IRIS 6500: Skalar Medical, Delft, The Netherlands).

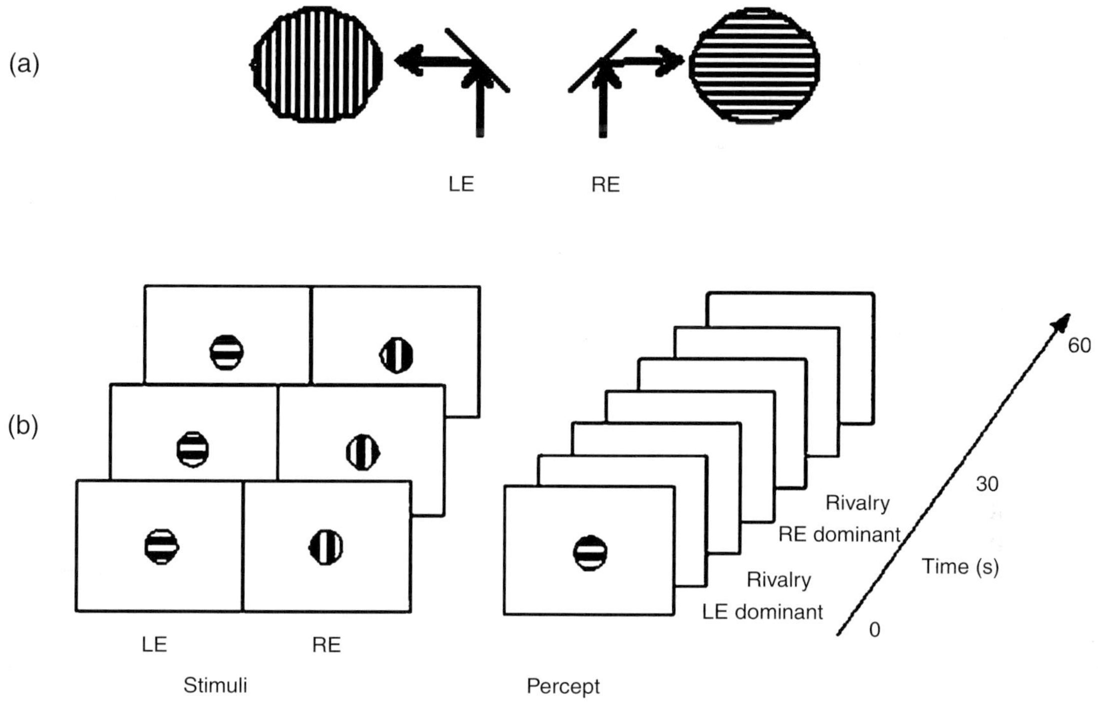

Fig. 1. (a) A schematic illustration of the dichoptic display. (b) A schematic illustration of part of the 60 s random visual presentation trial.

It was critical that, before each experimental trial, the central fixation crosses on the two displays were correctly aligned. This was achieved by subjects aligning the half crosses displayed on each monitor until a single cross was perceived. Each of the monocular orthogonal stimuli was matched for contrast and spatial frequency. Trials lasted a maximum of 60 s. Subjects were instructed to gaze into the centre of the displays and to indicate which of three perceptions they experienced: vertical gratings, horizontal gratings or a mosaic of the two individual patterns. On occasions one of the screens would randomly revert to a blank equivalent luminance field. Each blank presentation had a 300 ms gradual lead-in and a 400 ms fade-out, with the blank duration periods between 1 and 4 s. A minimum period of 4 s elapsed between each blank presentation.

Upon replacing the stimuli viewed by the fixing eye with a blank field the beat direction of the fast phase of the MLN changed to indicate that the fellow eye had taken up fixation (Fig. 2a, b). Figure 3 illustrates schematically how the latency and strength of this response depended on the duration of the blanking stimulus.

Experiment 2: visual attention and sensori-motor switching

The ability to engage in flexible strategies for attending to different images presented to the two eyes is generally driven by stimulus saliency. Thus the attributes of the physical stimulus can manipulate the onset and offset of ocular dominance. This behaviour falls within the category of exogenous attention (Ooi and He, 1999; Mitchell et al., 2004). Attention can also be driven endogenously, where there are no changes in the physical state of the stimulus. These two cognitive behaviours form part of a spectrum of attention but there is strong evidence to suggest that exogenous and endogenous attention arise from different mechanisms (Yantis, 1998; Gowen et al., 2007). There are also recognised strong links between binocular rivalry and attention, in that

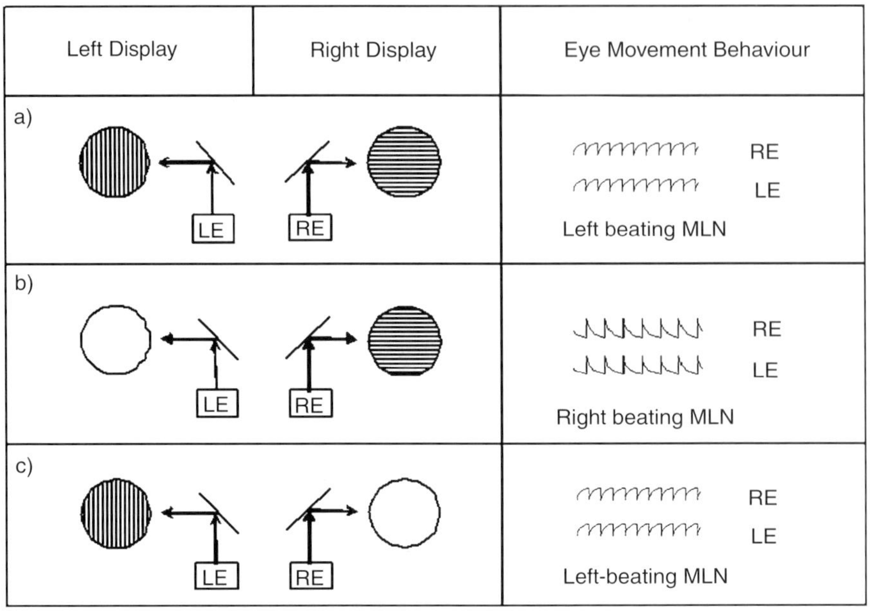

Fig. 2. Exogenous switching of the fast phase of manifest latent nystagmus (MLN). (a) The dichoptic display presented vertically orientated gratings to the left eye and horizontally orientated gratings to the right eye respectively. Here, the subject only perceives the vertical patterns and exhibits a left beating MLN. (b) The vertical grating pattern is replaced by a blank field and the MLN switches to a right beating MLN to reflect that only the right eye is receiving a visual input. (c) The right visual stimulus is now blanked off together with the simultaneous return of the left visual stimulus. A left beating MLN is monitored.

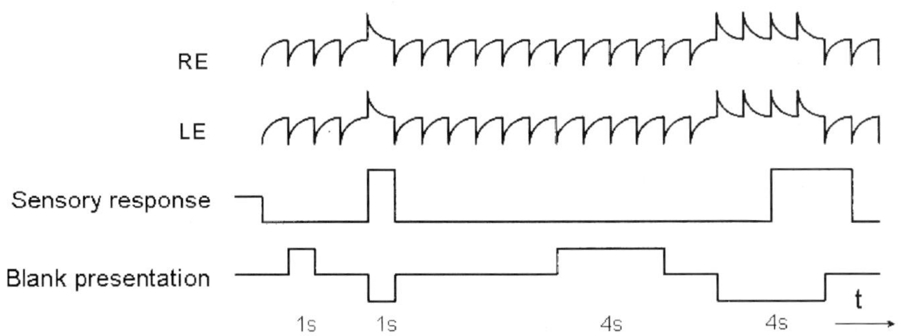

Fig. 3. A schematic illustration of the effect of the monocular random blanking of the left visual field whilst an MLN subject dichoptically views orthogonal grating patterns. Top trace: right eye position; 2nd trace: left eye position; 3rd trace: perceptual response as recorded with the subject's response box; 4th trace: the timing of the monocular blanking. Note that on blanking the left visual stimulus for 4 s, there is a short latency before the subject is perceptually aware of change in the stimulus conditions.

attention can influence the rate of alternation during rivalry (Meridith and Meridith, 1962; Lack, 1978; Chong and Blake, 2006).

In the previous experiment we investigated exogenous attention; here, we examine the role of endogenous attention on the MLN beat direction. In the first instance, six individuals with MLN binocularly viewed a projected cross target on a large screen ($104 \times 41^{\circ}$) and were then requested to imagine viewing the target first with one eye and

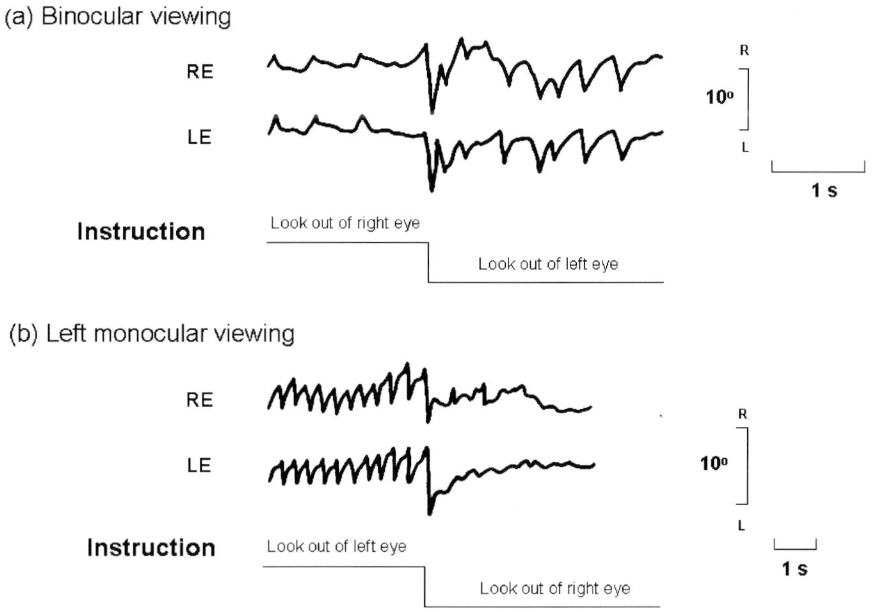

Fig. 4. The effects of endogenous shifts of attention on the fast phase of MLN. (a) MLN subject binocularly viewing a target and being requested to first view the target through the right eye and then the left eye. (b) The same subject monocularly viewing the target through the left eye. The arrow indicates when the subject was instructed to imagine viewing through the occluded eye.

then the other. Clear endogenously driven switching of the beat direction was regularly recorded in accordance with the commands given (Fig. 4a). The same procedure was then repeated whilst viewing the screen monocularly, with the left eye occluded. In this case, Fig. 4b illustrates the manner in which the initial left beating MLN gives way to an indeterminate and irregular waveform with occasional right and left beating oscillations.

Modelling sensori-motor switching in MLN

Previous research has shown that normal fixation control is determined by the fixation cells found in the rostral superior colliculus (in the region corresponding to the foveal portion of the retinotopic map) and the brainstem omni-pause neurones (Munoz and Wurtz, 1993a, b; Gandhi and Keller, 1999; Munoz et al., 2000). Thus steady fixation requires the fixation cells to prevent saccades by inhibiting the caudal collicular saccadic cells and the brainstem burst neurones whilst at the same time exciting the brainstem omni-pause neurones. It is conjectured that during early development the brainstem pause cells are trained to fire when the image is centred on the fovea (a calibration process). In the case of any misalignment of the visual axis (e.g., squint) two classes of pause cells could presumably develop: one which fires whenever the right eye is fixating and one when the left eye is fixating — each exhibiting a different offset. The effect of incorporating pause cells with offsets into models such as those presented by Broomhead et al. (2000), Akman et al. (2005), Laptev et al. (2006), and Clement et al. (2008), is to generate MLN type waveforms which switch beat direction according to which class of pause cell are activated.

Although this could explain the sensori-motor behaviour during exogenous orienting (Fig. 4a), endogenous orienting (Fig. 4b) requires additional top–down mechanisms originating from the dorsolateral frontal and parietal cortex and the frontal eye fields. This pathway is most likely to be involved in manipulating the level of endogenous attention during instruction change (e.g., look vs.

hold eyes steady) or task demand (e.g., passive vs. active/respond to change fixation).

Conclusions

By using incompatible stimuli viewed dichoptically, we have been able to investigate a bistable oculomotor behaviour — MLN. Each perceptual switch was mirrored by a change in the beat direction and a possible role for the pause cells has been proposed as the sensori-motor switching hub. Finally, the accessibility of simultaneously monitoring sensory and motor behaviour would seem to make it an additional powerful tool for studying the distributed neural processes involved in attention and visual awareness.

References

Abadi, R.V. (1980) Visual performance with contact lenses and congenital idiopathic nystagmus. Br. J. Physiol. Opt., 33: 32–37.

Abadi, R.V. and Bjerre, A. (2002) Motor and sensory characteristics of infantile nystagmus. Br. J. Ophthalmol., 86: 1152–1160.

Abadi, R.V. and Dickinson, C.M. (1986) Waveform characteristics in congenital nystagmus. Doc. Ophthalmol., 64: 153–167.

Abadi, R.V., Forster, J.E. and Lloyd, I.C. (2006) Ocular motor outcomes after bilateral and unilateral infantile cataracts. Vision Res., 46: 940–952.

Abadi, R.V. and Gowen, E. (2004) Characteristics of saccadic intrusions. Vision Res., 44: 2675–2690.

Abadi, R.V. and Scallan, C.J. (2000) Waveform characteristics of manifest latent nystagmus. Invest. Ophthalmol. Vis. Sci., 41: 3805–3817.

Akman, O.E., Broomhead, D.S., Abadi, R.V. and Clement, R.A. (2005) Eye movement instabilities and nystagmus can be predicted by a nonlinear dynamics model of the saccadic system. J. Math. Biol., 51: 661–694.

Broomhead, D.S., Clement, R.A., Muldoon, M.R., Whittle, J.P., Scallan, C. and Abadi, R.V. (2000) Modelling of congenital nystagmus waveforms produced by saccadic system abnormalities. Biol. Cybern., 82: 391–399.

Chong, S.C. and Blake, R. (2006) Exogenous attention and endogenous attention influence initial dominance in binocular rivalry. Vision Res., 46: 1794–1803.

Clement, R.A., Broomhead, D.S. and Akman, O. (2008) Dynamics of saccadic oscillations. Prog. Brain Res. (this volume).

Dell'Osso, L.F. (1985) Congenital, latent and manifest latent nystagmus: similarities, differences and relation to strabismus. Jpn. J. Ophthalmol., 29: 351–363.

Dell'Osso, L.F., Schmidt, D. and Daroff, R.B. (1979) Latent, manifest latent and congenital nystagmus. Arch. Ophthalmol., 97: 1877–1884.

Gandhi, N.J. and Keller, E.L. (1999) Comparison of saccades perturbed by stimulation of the rostral superior colliculus, the caudal superior colliculus and the omni-pause neurone region. J. Neurophysiol., 82: 3236–3253.

Gowen, E., Abadi, R.V., Poliakoff, E., Hanson, P.C. and Miall, R.C. (2007) Modulation of saccadic intrusions by exogenous and endogenous attention. Brain Res., 1141: 154–167.

Gradstein, L., Goldstein, H.P., Wizov, S.S. and Reinecke, R.D. (2004) Extended slow phase in latent/manifest latent nystagmus. Invest. Ophthalmol. Vis. Sci., 45: 1139–1148.

Harris, L. (1997) Nystagmus and eye movement disorders. In: Taylor D. (Ed.), Paediatric Ophthalmology. Blackwell, Oxford, pp. 869–896.

Lack, L. (1978) Selective Attention and the Control of Binocular RSivalry. Moulton, The Hague.

Laptev, D., Akman, O.E. and Clement, R.A. (2006) Stability of the saccadic system. Biol. Cybern., 95: 281–287.

Leigh, R.J. and Zee, D.S. (2006) The Neurology of Eye Movements. Oxford University Press, Oxford.

Leopold, D.A. and Logothetis, N.K. (1999) Multistable phenomena: changing views in perception. Trends Cogn. Sci., 3: 254–264.

Levelt, W.J.M. (1965) On Binocular Rivalry. Moulton, The Hague.

Meridith, G.M. and Meridith, C.G.W. (1962) Effects of instructional conditions on rate of binocular rivalry. Percept. Mot. Skill, 15: 655–664.

Mitchell, J.F., Stoner, G.R. and Reynolds, J.H. (2004) Object-based attention determines dominance in binocular rivalry. Nature, 429(6990): 410–413.

Munoz, D.P., Dorris, M.C., Paré, M. and Everling, S. (2000) On your mark, get set: brainstem circuitry underlying saccadic initiation. Can. J. Phys. Pharm., 78: 934–944.

Munoz, D.P. and Wurtz, R.H. (1993a) Fixation cells in monkey superior colliculus: characteristics of cell discharge. J. Neurophysiol., 70: 559–575.

Munoz, D.P. and Wurtz, R.H. (1993b) Fixation cells in monkey superior colliculus: reversible activation and deactivation. J. Neurophysiol., 70: 576–589.

Ooi, T.L. and He, Z.J. (1999) Binocular rivalry and visual awareness: the role of attention. Perception, 28: 551–574.

Schor, C.M. (1977) Visual stimuli for strabismic suppression. Perception, 6: 583–593.

Smith, E.L., III, Levi, D.M., Manny, R.E., Harwerth, R.S. and White, J.M. (1985) The relationship between binocular rivalry and strabismic suppression. Invest. Ophthalmol. Vis. Sci., 26: 80–87.

Tusa, R.J., Repko, M.X., Smith, C.B. and Herdman, S.J. (1991) Early visual deprivation results in persistent strabismus and nystagmus in monkeys. Invest. Ophthalmol. Vis. Sci., 32: 134–141.

Yantis, S. (1998) The control of visual attention. In: Pashler H. (Ed.), Attention. University College London Press, London, pp. 223–256.

C. Kennard & R.J. Leigh (Eds.)
Progress in Brain Research, Vol. 171
ISSN 0079-6123

CHAPTER 6.3

Eye hyperdeviation in mouse cerebellar mutants is comparable to the gravity-dependent component of human downbeat nystagmus

John S. Stahl[1,2,*] and Brian S. Oommen[1,2]

[1]*Department of Neurology, Case Western Reserve University School of Medicine, OH, USA*
[2]*Dell'Osso-Daroff Ocular Motility Laboratory, Veterans Affairs Medical Center, Cleveland, OH, USA*

Abstract: Humans with cerebellar degeneration commonly exhibit downbeat nystagmus (DBN). DBN has gravity-independent and -dependent components, and the latter has been proposed to reflect hyperactive tilt maculo-ocular reflexes (tilt-MOR). Mice with genetically determined cerebellar ataxia do not exhibit DBN, but they do exhibit tonic hyperdeviation of the eyes, which we have proposed to be the DBN equivalent. As such, the tilt-MOR might be predicted to be hyperactive in these mutant mice. We measured the tilt-MOR in 10 normal C57BL/6 mice and in 6 *tottering*, a mutant exhibiting ataxia and ocular motor abnormalities due to mutation of the P/Q calcium channel. Awake mice were placed in body orientations spanning 360° about the pitch axis. The absolute, equilibrium vertical angular deviations of one eye were measured using infrared videooculography. In both strains, eye elevation varied quasi-sinusoidally with tilt angle in the range of 90° nose-up to 90° nose-down. Beyond this range the eye returned to a neutral position. Deviation over ±30° of tilt was an approximately linear function of the projection of the gravity vector into the animal's horizontal plane, and can thus be summarized by its slope (sensitivity). Sensitivity measured 14.9°/g for C57BL/6 and 20.3°/g for *tottering*, a statistically significant difference. Thus the pitch otolithic reflex of the ataxic mutants is hyperactive relative to controls and could explain tonic hyperdeviation of the eyes, consistent with the idea that the tonic hyperdeviation is analogous to DBN.

Keywords: downbeat nystagmus; calcium; channelopathy; cacna1a; otolith; maculo-ocular

Introduction

Downbeat nystagmus (DBN) is a common ocular motor sign of cerebellar disease, and may cause visual symptoms ranging from subtly blurred vision to frank oscillopsia. The pathophysiology of DBN has been debated extensively. That debate is complicated by there being multiple components to the ocular instability, including gravity-dependent and -independent components. The gravity-dependent component manifests as an increase in nystagmus slow-phase velocities as the head assumes progressively greater nose-down pitch orientations. It has been attributed to over-active maculo-ocular reflexes (Halmagyi et al., 1983; Marti et al., 2002). Theories advanced to explain the gravity-independent component(s) have included imbalances of upward and downward

*Corresponding author. Tel.: +216 844 3170;
Fax: +216 844 5066; E-mail: jss6@case.edu

DOI: 10.1016/S0079-6123(08)00672-9

circuits of the angular VOR (Baloh and Spooner, 1981; Gresty et al., 1986; Baloh and Yee, 1989; Bohmer and Straumann, 1998; Walker and Zee, 2005), impairment of the vertical neural integrator (Zee et al., 1980), mismatches between the desired eye positions specified by the gaze-holding neural integrator and the mechanisms that maintain Listing's law during saccades (Straumann et al., 2000; Glasauer et al., 2003), and imbalances within the vertical smooth pursuit pathways (Zee et al., 1974; Glasauer et al., 2005; Marti et al., 2005; Kalla et al., 2006). These explanations are not mutually exclusive, and it is likely that these pathological processes are present in varying combinations across the spectrum of patients with DBN (Leigh and Zee, 2006).

Studies of eye movements in mice may provide insight into DBN for multiple reasons (Stahl, 2004b). First, since the ocular motor repertoire of mice, like other "afoveate" mammals, is restricted to gaze-stabilizing movements, mice present an opportunity to investigate contributions to ocular motor abnormalities from systems other than those supporting gaze-shifting movements such as smooth pursuit. Second, there are many mouse strains with cerebellar dysfunction arising from a broad variety of molecular or developmental defects, providing an opportunity to dissect the specific cerebellar derangements responsible for a given ocular motor abnormality. Third, mice are a convenient animal model in which to explore the efficacy and neurophysiological effects of potential medications for cerebellar dysfunction.

We have focused our efforts on mouse strains whose cerebellar dysfunction stems from mutations of genes encoding the P/Q voltage-activated calcium channel (Stahl, 2002, 2004a; Stahl et al., 2006). P/Q channel mutations are now recognized as causal in the human cerebellar diseases episodic ataxia type 2 and spinocerebellar ataxia type 6 (Ophoff et al., 1996; Zhuchenko et al., 1997), as well as several naturally arising, induced, or engineered mutant mouse strains characterized by various combinations of cerebellar ataxia, seizures, and episodic dystonia. Among the mouse mutants, *rocker* and *tottering* are particularly interesting because they are reported to have histologically normal or near-normal cerebella (Noebels and Sidman, 1979; Isaacs and Abbott, 1995; Zwingman et al., 2001), with the important implication that their behavioural abnormalities may arise from *alterations* in signal processing, as opposed to the simple absence of neurons. Thus, they are an avenue by which to study the electrophysiological and structural determinants of normal cerebellar function. *Rocker* and *tottering* exhibit a number of ocular motor abnormalities in common, including reduced vestibulo-ocular reflex (VOR) gain, increased VOR phase leads, reduced horizontal optokinetic reflex (OKR) gain, decreased neural integrator time constants, decreased VOR gain adaptation, and an abnormally elevated resting position of the eyes (Stahl, 2004a; Stahl et al., 2006). DBN seems to be absent in these mutants. However, it may be that the upward bias in resting *position* is aetiologically related to the upward *velocity* bias that underlies DBN in humans. If the elevated resting position is indeed the murine homologue of DBN, then the mutant mice might exhibit hyperactivity of their tilt maculo-ocular reflex (tilt-MOR) in the pitch plane; just as DBN is accentuated in forward-pitched positions, the mouse hyperdeviation should become more pronounced with forward tilt. We tested this hypothesis by comparing the pitch tilt-MOR in *tottering* mutants and normal C57BL/6 controls. Data from the control animals has previously been published (Oommen and Stahl, 2008).

Methods

Animal experiments were approved by the Institutional Animal Care and Use Committee at Case Western Reserve University. We studied six locally bred homozygous *tottering* mutants and ten C57BL/6J controls, aged 5–11 months. We prepared each animal for eye movement recording by surgically implanting an acrylic head fixation pedestal in such a manner as to render the pedestal's top surface parallel to the animal's lambda-bregma (L-B) axis (Stahl et al., 2006). Eye positions were recorded using 2D video oculography as previously described (Stahl et al., 2000; Stahl, 2004a). Output of the eye tracker was converted to vertical and horizontal eye angles by a trigonometric formula. Vertical positions were further converted to *absolute* elevation of the pupil above the earth-horizontal plane (Stahl, 2004a).

Mice were mounted in a two-axis gimbal allowing for tilting about the animals' pitch or roll axes. During testing, each mouse was placed in pitch orientations spanning 360° in 15° steps. All pitch tilts were done from a 0° roll orientation (inter-aural axis earth-horizontal). The sequence of tilts was randomized. Each position was achieved by rotating the animal in darkness at a slow rate so as to minimize semicircular canal activation. After eye position stabilized, pupil position was recorded for 12 s in the light and 12 s in the dark. Complete sets of tilt responses were obtained for each mouse on two different days and the results averaged. For the purposes of conducting the experiment, the pitch angle of the L-B axis was assumed to be identical to that of the top surface of the pedestal, i.e., we assumed that we were fully successful in constructing pedestals to parallel the L-B pitch plane. After the recordings were completed, the pitch angles were adjusted for each animal to account for any deviation from this assumption. The correction (the angle between the pedestal surface and the L-B pitch plane) was obtained post-mortem in each animal from lateral photographs of the defleshed skulls (Oommen and Stahl, 2008).

Results

Extreme downward tilt angles or inverted positions sometimes engendered transient runs of DBN in the mutants. However, the appearance of this nystagmus was inconsistent. It often appeared only for the first few times the animal was placed in a position, never appeared in some mutants, and occasionally appeared in a control animal. In cases where nystagmus did appear, data was only taken after the nystagmus abated or, if it persisted, during brief plateaus in eye position that usually preceded a fast phase. Figure 1 shows the vertical position of the eye as a function of L-B pitch, recorded in the light (A, B) and dark (C, D). In both control and mutant animals there was a robust tilt-MOR, with the eye moving upward and laterally (data not shown) as the head was pitched downward. As the animals assumed inverted positions, the eye tended to return to the position assumed at 0° pitch. Eye positions tended to be more erratic (animal-to-animal and moment-to-moment within any one animal) in the extreme tilt positions, as evidenced by the larger error bars.

Within the central regions of the curve (approximately −105° to +105°), eye elevations were consistently greater in the mutants. Fig. 1b, d re-plot eye positions versus the component of the gravity vector projecting into the horizontal plane (in units of g) for the central −90° to 90° of tilt, a transformation that linearizes the tilt-eye elevation relationship and allows one to determine a sensitivity (slope) value for the reflex (Maruta et al., 2001; Andreescu et al., 2005). The sensitivity in the light over the central ±30° ($g = \pm 0.5$) averaged 20.3°/g for *tottering* and 14.9°/g for controls and differed significantly (2-tailed *t*-test, $p = 0.001$), consistent with hyperactivity of the pitch tilt-MOR about the central, head-upright position. Sensitivities in the dark also differed significantly (19.7°/g for *tottering* and 14.8°/g for controls, $p = 0.012$).

Discussion

As in a previous assessment of ocular motor behaviour in *tottering*, the mutant exhibited consistently higher resting eye positions (Stahl et al., 2006). The sensitivity of the pitch tilt-MOR was significantly greater in the mutants, indicating that, as in humans with DBN, there is a hyperactivity of otolith-related reflexes. Furthermore, since the average eye elevation of mutants and controls converged as the head was pitched upwards, the greater sensitivity (i.e., steeper slope) of the pitch tilt-MOR could account for most of the difference in elevation between mutants and controls observed at the head "upright" position.

A key question remains, if P/Q mutants suffer some of the same ocular motor abnormalities (hyperactive tilt-MOR, floccular dysfunction leading to a relative reduction in the ability to follow downward-directed visual stimuli) found in

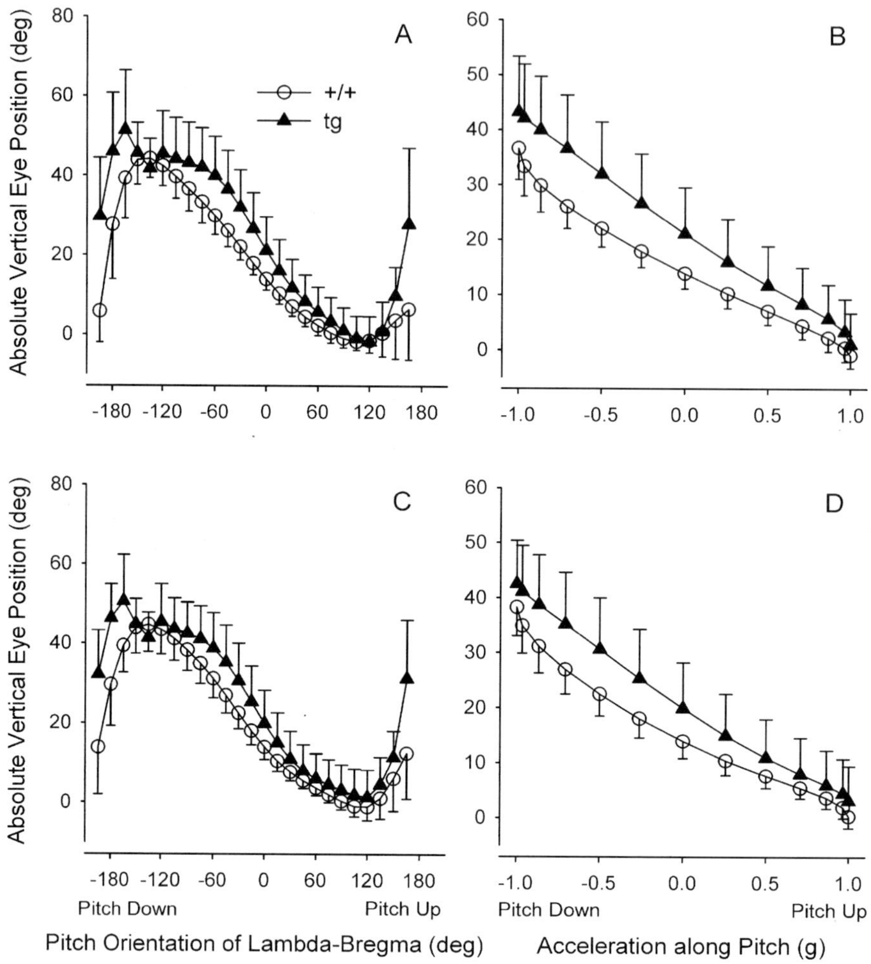

Fig. 1. Effect of pitch tilts on absolute vertical position of the eye. Top panels (A, B) are for eye positions recorded in the light. Bottom panels (C, D) plot data recorded in darkness. Left panels (A, C) plot eye position versus tilt angle of L-B axis. Right panels (B, D) plot eye position versus the fraction of the gravity vector projecting into the horizontal plane, for the central $\pm 90°$ of tilt. Error bars are 1 SD.

patients with DBN, why do these abnormalities manifest as an upward bias in eye position, as opposed to an upward bias in eye velocity? Several factors may contribute to the difference. First, since mice lack a fovea (or equivalent central retinal structure specialized for fixation), they may exert lesser control over absolute position of the retina with respect to the visual world. Thus they would not trigger nystagmus fast phases on the basis of small position errors, and the eyes would be free to move to an equilibrium position. In contrast, in foveate mammals the re-centring fast phases would prevent equilibrium being reached, thereby perpetuating the slow phase drifts. Second, the neural integrator is quite weak in mice and more so in P/Q mutants (Stahl et al., 2006). The centripetal drifts related to leakiness of the neural integrator may counteract the centrifugal forces generated by hyperactive otolithic circuits, resulting in an equilibrium eye position that is not so extreme as to trigger fast phases on the basis of eye-in-orbit eccentricity (Stahl, 2004a). Third, all mice may have a relatively low tendency to initiate fast phases on the basis of eccentricity. For instance, when mice are exposed to prolonged unidirectional optokinetic stimulation, an initial optokinetic response may be followed by the eyes coming to rest at an eccentric position, due to a decline in the occurrence of nystagmus fast phases (Stahl, 2008). Constant gravitational stimulation may, like constant optokinetic stimulation, produce eccentric stable eye positions through this failure to sustain fast phase generation. Fourth and finally, the tilt-MOR in afoveate mammals is commonly regarded as acting to bring the retinas into some favourable orientation within the overall visual world (Collewijn et al., 1985; Maruta et al., 2001), and thus it actually seems more likely that an over-active tilt-MOR would lead to an abnormal eye position, rather than an abnormal eye velocity. Tilt-MOR hyperactivity in humans could manifest as a velocity bias because the primitive eye-orienting effects of the tilt-MOR are masked by the influences of the fovea-directing machinery (e.g., mechanisms supporting fixation or smooth pursuit), leaving the hyperactivity of the tilt-MOR to manifest through some indirect fashion, such as by modulating ocular motor subsystems (such as the VOR) that regulate eye velocity.

DBN is a general sign of vestibulocerebellar dysfunction of any aetiology. Likewise, static ocular hyperdeviation has been demonstrated in every ataxic mouse strain in which absolute vertical eye position has been assessed (Stahl et al., 2006). Therefore the study of the neurophysiological underpinnings or pharmacological responsiveness of ocular hyperdeviation in *tottering* should provide information relevant to understanding and treating DBN in humans.

Abbreviations

DBN	downbeat nystagmus
L-B	lambda-bregma
OKR	optokinetic reflex
tilt-MOR	tilt maculo-ocular reflex
VOR	vestibulo-ocular reflex

Acknowledgements

B.S. Oommen was supported by a Crile Summer Research Fellowship. J.S. Stahl was supported by EY13370 and the Department of Veterans Affairs.

References

Andreescu, C.E., De Ruiter, M.M., De Zeeuw, C.I. and De Jeu, M.T. (2005) Otolith deprivation induces optokinetic compensation. J. Neurophysiol., 94: 3487–3496.

Baloh, R.W. and Spooner, J.W. (1981) Downbeat nystagmus: a type of central vestibular nystagmus. Neurology, 31: 304–310.

Baloh, R.W. and Yee, R.D. (1989) Spontaneous vertical nystagmus. Rev. Neurol. (Paris), 145: 527–532.

Bohmer, A. and Straumann, D. (1998) Pathomechanism of mammalian downbeat nystagmus due to cerebellar lesion: a simple hypothesis. Neurosci. Lett., 250: 127–130.

Collewijn, H., Van der Steen, J., Ferman, L. and Jansen, T.C. (1985) Human ocular counterroll: assessment of static and dynamic properties from electromagnetic scleral coil recordings. Exp. Brain Res., 59: 185–196.

Glasauer, S., Hoshi, M. and Buttner, U. (2005) Smooth pursuit in patients with downbeat nystagmus. Ann. N.Y. Acad. Sci., 1039: 532–535.

Glasauer, S., Hoshi, M., Kempermann, U., Eggert, T. and Buttner, U. (2003) Three-dimensional eye position and slow phase velocity in humans with downbeat nystagmus. J. Neurophysiol., 89: 338–354.

Gresty, M.A., Barratt, H., Rudge, P. and Page, N. (1986) Analysis of downbeat nystagmus: otolithic vs. semicircular canal influences. Arch. Neurol., 43: 52–55.

Halmagyi, G.M., Rudge, P., Gresty, M.A. and Sanders, M.D. (1983) Downbeating nystagmus: a review of 62 cases. Arch. Neurol., 40: 777–784.

Isaacs, K.R. and Abbott, L.C. (1995) Cerebellar volume decreases in the tottering mouse are specific to the molecular layer. Brain Res. Bull., 36: 309–314.

Kalla, R., Deutschlander, A., Hufner, K., Stephan, T., Jahn, K., Glasauer, S., Brandt, T. and Strupp, M. (2006) Detection of floccular hypometabolism in downbeat nystagmus by fMRI. Neurology, 66: 281–283.

Leigh, R.J. and Zee, D.S. (2006) The Neurology of Eye Movements. Oxford University Press, New York.

Marti, S., Palla, A. and Straumann, D. (2002) Gravity dependence of ocular drift in patients with cerebellar downbeat nystagmus. Ann. Neurol., 52: 712–721.

Marti, S., Straumann, D. and Glasauer, S. (2005) The origin of downbeat nystagmus: an asymmetry in the distribution of on-directions of vertical gaze-velocity Purkinje cells. Ann. N.Y. Acad. Sci., 1039: 548–553.

Maruta, J., Simpson, J.I., Raphan, T. and Cohen, B. (2001) Orienting otolith-ocular reflexes in the rabbit during static and dynamic tilts and off-vertical axis rotation. Vision Res., 41: 3255–3270.

Noebels, J.L. and Sidman, R.L. (1979) Inherited epilepsy: spike-wave and focal motor seizures in the mutant mouse tottering. Science, 204: 1334–1336.

Oommen, B.S. and Stahl, J.S. (2008) Eye orientation during static tilts and its relationship to spontaneous head pitch in the laboratory mouse. Brain Res., 1193: 57–66.

Ophoff, R.A., Terwindt, G.M., Vergouwe, M.N., van Eijk, R., Oefner, P.J., Hoffman, S.M.G., Lamerdin, J.E., Mohrenweiser,

H.W., Bulman, D.E., Ferrari, M., Haan, J., Lindhout, D., van Hommen, G.-J., Hofker, M.H., Ferrari, M.D. and Frants, R.R. (1996) Familial hemiplegic migraine and episodic ataxia type-2 are caused by mutations in the Ca(2+) channel gene CACNL1A4. Cell, 87: 543–552.

Stahl, J.S. (2002) Calcium channelopathy mutants and their role in ocular motor research. Ann. N.Y. Acad. Sci., 956: 64–74.

Stahl, J.S. (2004a) Eye movements of the murine P/Q calcium channel mutant rocker, and the impact of aging. J. Neurophysiol., 91: 2066–2078.

Stahl, J.S. (2004b) Using eye movements to assess brain function in mice. Vision Res., 44: 3401–3410.

Stahl, J.S. (2008) Characteristics and applications of mouse eye movements. In: Chalupa L.M. and Williams R.W. (Eds.), Eye, Retina, and the Visual Systems of the Mouse. The MIT Press, Boston, MA.

Stahl, J.S., James, R.A., Oommen, B.S., Hoebeek, F.E. and De Zeeuw, C.I. (2006) Eye movements of the murine p/q calcium channel mutant tottering, and the impact of aging. J. Neurophysiol., 95: 1588–1607.

Stahl, J.S., van Alphen, A.M. and De Zeeuw, C.I. (2000) A comparison of video and magnetic search coil recordings of mouse eye movements. J. Neurosci. Methods, 99: 101–110.

Straumann, D., Zee, D.S. and Solomon, D. (2000) Three-dimensional kinematics of ocular drift in humans with cerebellar atrophy. J. Neurophysiol., 83: 1125–1140.

Walker, M.F. and Zee, D.S. (2005) Asymmetry of the pitch vestibulo-ocular reflex in patients with cerebellar disease. Ann. N.Y. Acad. Sci., 1039: 349–358.

Zee, D.S., Friendlich, A.R. and Robinson, D. (1974) The mechanism of downbeat nystagmus. Arch. Neurol., 30: 227–237.

Zee, D.S., Leigh, R.J. and Mathieu-Millaire, F. (1980) Cerebellar control of ocular gaze stability. Ann. Neurol., 7: 37–40.

Zhuchenko, O., Bailey, J., Bonnen, P., Ashizawa, T., Stockton, D.W., Amos, C., Dobyns, W.B., Subramony, S.H., Zoghbi, H.Y. and Lee, C. (1997) Autosomal dominant cerebellar ataxia (SCA6) associated with small polyglutamine expansions in the alpha1A-voltage-dependent calcium channel. Nat. Genet., 15: 62–69.

Zwingman, T.A., Neumann, P.E., Noebels, J.L. and Herrup, K. (2001) Rocker is a new variant of the voltage dependent calcium channel gene Cacna1a. J. Neurosci., 21: 1169–1178.

C. Kennard & R.J. Leigh (Eds.)
Progress in Brain Research, Vol. 171
ISSN 0079-6123

CHAPTER 6.4

New insights into the upward vestibulo-oculomotor pathways in the human brainstem

Charles Pierrot-Deseilligny[1,*] and Caroline Tilikete[2]

[1]*Service de Neurologie 1, AP-HP, INSERM 679, Hôpital de la Salpêtrière, Paris, France*
[2]*Unité de Neuro-ophtalmologie, Hôpital Neurologique, Hospices Civils de Lyon, INSERM UMR-S864, Université de Lyon-I, Lyon, France*

Abstract: The brainstem vestibulo-oculomotor pathways are not yet fully known. Three different excitatory tracts could be involved in the transmission of upward vestibular eye movement (VEM) signals and upward eye position (EP) signals to the oculomotor nucleus (III): the medial longitudinal fasciculus (MLF), the brachium conjunctivum (BC), and the crossing ventral tegmental tract (CVTT). The involvement of the MLF pathway originating in the medial vestibular nucleus (MVN) in this transmission is experimentally and clinically well established whereas a role of the BC appears to be questionable. Furthermore, there is now accumulating evidence that the CVTT pathway emerging from the superior vestibular nucleus (SVN) also plays an important role in the mediation of excitatory upward EP and VEM signals to the III. This duplication of pathways (MVN-MLF and SVN-CVTT) could be explained by a supplementary and relatively specific function performed by the SVN-CVTT pathway to counteract the gravity pull in the upward eye movement system. Various arguments in support of this hypothesis are reviewed.

Keywords: crossing ventral tegmental tract; downbeat nystagmus; gravity; medial longitudinal fasciculus; otolith pathways; upbeat nystagmus; vestibular eye movements; vestibular pathways; vertical nystagmus

Introduction

The brainstem pathways that connect the vestibular nuclei to the oculomotor nuclei (III) and control the vestibular eye movements (VEM) and eye position (EP) in the vertical plane remain partly unclear. Several excitatory vestibulo-oculomotor tracts — passing both through the medial longitudinal fasciculus (MLF) and outside the MLF — appear to be involved in upward VEM and upward EP, whereas only one excitatory tract (passing through the MLF) is known for the downward system (Leigh and Zee, 2006). We will first review the main findings supporting the roles of the MLF and extra-MLF tracts in the mediation of upward signals, and then try to explain why a duplication of the brainstem excitatory tracts is specifically required for the upward eye movement system.

The MLF tracts

Experimental data

In the cat and the monkey, different excitatory and inhibitory tracts transmitting vertical VEM and

*Corresponding author. Tel.: +33 142161828;
Fax: +33 144245247; E-mail: cp.deseilligny@psl.aphp.fr

DOI: 10.1016/S0079-6123(08)00673-0

EP signals to the III pass through the MLF (Graf and Ezure, 1986; McCrea et al., 1987; Büttner-Ennever, 1992). An excitatory tract involved in upward signals originates in the medial vestibular nucleus (MVN), immediately crosses the midline, passes through the contralateral MLF and projects to the subdivisions of the III controlling the elevator muscle motoneurons. The MLF also contains the only known excitatory tract involved in downward signals: this tract originates in the MVN, passes through the contralateral MLF, and projects to the trochlear nucleus (IV) and the subdivision of the III controlling the inferior rectus motoneurons. The inhibitory upward and downward signals are mediated by two other tracts, both originating in the superior vestibular nucleus (SVN), passing through the ipsilateral MLF before reaching the IV and the different subdivisions of the III. Therefore, the MLF contains four separate tracts which control the activation of the agonist muscle motoneurons and the inhibition of the antagonist muscle motoneurons during upward and downward VEM, constituting a circuitry apparently well balanced to compensate for all types of vertical rotational, or translational head movements.

Human data

In humans, indirect evidence for the existence of this circuitry is provided by the results observed in internuclear ophthalmoplegia (INO), in which the MLF tracts are damaged. As expected, the upward and downward VEM gains in patients with bilateral INO were severely impaired and were reduced to a third of their normal values (Ranalli and Sharpe, 1988a). In INO, relatively balanced impairments in the transmission of upward and downward VEM and EP signals may explain why primary position vertical nystagmus is not observed, with at most in this syndrome the existence of a gaze-evoked nystagmus in the extreme upgaze and/or downgaze (Ranalli and Sharpe, 1988a, and, for a review, Pierrot-Deseilligny and Milea, 2005). Furthermore, in a patient with unilateral INO, performing head rotations with high accelerations, the upward VEM gain was also impaired but less severely than the downward VEM gain, which is an indication that the transmission of the upward VEM signals could be made both through the MLF tract and extra-MLF tracts (Cremer et al., 1999).

The extra-MLF tracts

Brachium conjunctivum

The brachium conjunctivum (BC), i.e., the superior cerebellar peduncle, could contain a tract involved in upward VEM, as suggested by a few, relatively old experimental data in the rabbit (see Yamamoto et al., 1978). It has more recently been argued (Sato and Kawasaki, 1987; Uchino et al., 1994) that the BC could have been confounded in these former studies with the crossing ventral tegmental tract (CVTT) (see below), since these tracts are very close to each other in the lower pontine tegmentum, the CVTT being located just at the anterior limit of the BC (Fig. 1). In humans, the same argument may be used for the few reported clinical cases with primary position upbeat nystagmus (PPUN) attributed to BC lesions (Nakada and Remler, 1981; Benjamin et al., 1986; Kattah and Dagi, 1990). Indeed, in all these cases, the lesions were large bilateral tumours or haematomas, damaging and/or compressing the whole dorsal tegmentum in the lower pons, thus likely impairing both the CVTT and the BC. Furthermore, lesions affecting the BC at the level of its decussation, in the lower midbrain, do not result in PPUN (Mossuto-Agatiello, 2006). Finally, the main argument against the involvement of the BC in upward VEM in humans has been provided by two reports of several clinical cases with small brainstem lesions restricted to the BC, in which either a gaze-evoked downbeat nystagmus (Frohman et al., 2001) or a positional downbeat (not upbeat) nystagmus (Anagnostou et al., 2006, 2 cases) were observed. Accordingly, it now appears unlikely that the human BC is involved in the transmission of excitatory upward VEM signals to the III.

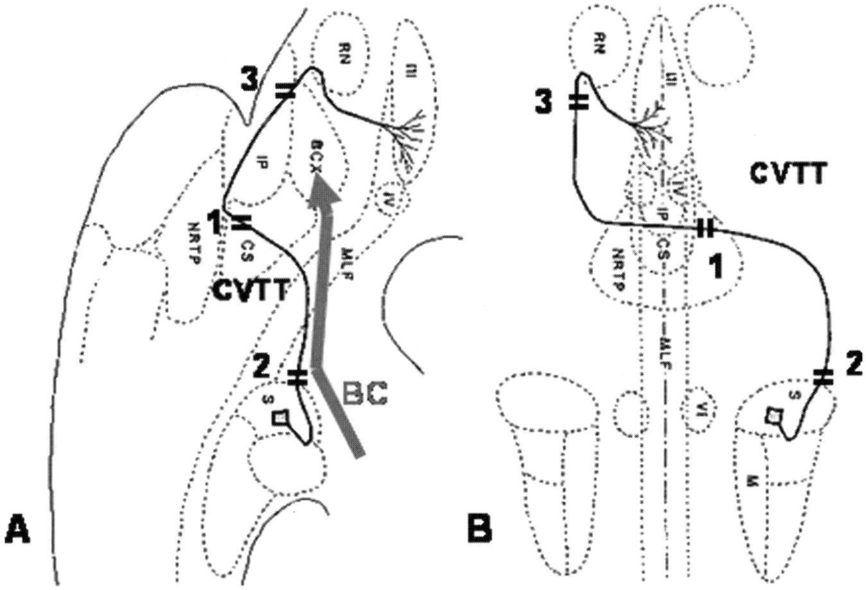

Fig. 1. Crossing ventral tegmental tract in the cat. Sagittal view (A) and coronal view (B) of the cat brainstem (Adapted with permission from Uchino et al., 1994, modified). Approximate locations in this cat brainstem of three small human lesions affecting different portions of the CVTT: (1) lesion in the middle portion of the CVTT, near its decussation, in the anterior part of the midpons (Pierrot-Deseilligny et al., 2005); (2) lesion in the initial portion of the CVTT, in the postero-lateral part of the lower pontine tegmentum (Tilikete et al., 2008); (3) lesion in the final portion of the CVTT, in the antero-medial part of the midbrain tegmentum (Yura et al., 1988). BC: brachium conjunctivum; BCX: decussation of the BC; CS: superior central nucleus of raphé; CVTT: crossing ventral tegmental tract; IP: interpeduncular complex; M: medial vestibular nucleus; MLF: medial longitudinal fasciculus; NRTP: nucleus reticularis tegmenti pontis; RN: red nucleus; S: superior vestibular nucleus; III: oculomotor nucleus; IV: trochlear nucleus; VI: abducens nucleus.

Crossing ventral tegmental tract

The CVTT is involved in upward EP and upward VEM, as reported in detail in the cat (Carpenter and Cowie, 1985; Sato and Kawasaki, 1987; Uchino et al., 1994) and suggested in the monkey (Lang et al., 1979; Sato and Kawasaki, 1991) and humans (Ranalli and Sharpe, 1988b). The CVTT connects the SVN, receiving the excitatory anterior canal inputs generating upward VEM, to the superior rectus and inferior oblique motoneurons in the III (Sato and Kawasaki, 1991). The course of the CVTT in the cat brainstem has been clearly described by Uchino et al. (1994) (Fig. 1): the CVTT leaves the upper pole of the SVN and first ascends rostrally in the postero-lateral part of the lower pontine tegmentum, where it is located just at the anterior limit of the BC; near the midpons, the CVTT arches medially and anteriorly to reach with an almost horizontal course the antero-medial part of the pontine tegmentum, decussating near the upper pole of the nucleus reticularis tegmenti pontis, just above the midpons level; then, the CVTT arches a second time and ascends again rostrally in the contralateral anterior and medial part of the upper pontine tegmentum and lower midbrain tegmentum, passing at this level just laterally to the BC decussation; lastly, the CVTT arches a third time near the inferior pole of the red nucleus, reaching with a final medial and almost horizontal course the superior rectus and inferior oblique motoneurons, bilaterally. In the monkey, the brainstem course of the CVTT could be similar to that of the cat (Lang et al., 1979, Fig. 4, tract 4), with a collateral branch terminating in the thalamus.

What is the evidence in humans for the existence of a tract analogous to the CVTT of the cat and the monkey? Several cases of PPUN due to relatively large and bilateral lesions affecting the anterior pontine tegmentum and the adjacent posterior basis pontis in the midpons and upper pons — without, at such a level, associated damage to the MLF or the BC — suggest the existence of a human CVTT, whose mid course could pass through the very central part (in 3D) of the pons (Fisher et al., 1983; Hankey et al., 1987; Hirose et al., 1991). In a more recent case of PPUN, the lesion was small and unilateral in the central part of the pons (Fig. 2A) and probably located very close to the CVTT decussation (Fig. 1), just above the midpons level, suggesting furthermore that the human decussation lies at the posterior limit of the basis pontis rather than in the anterior pontine tegmentum (Pierrot-Deseilligny et al., 2005). In the same patient, the gain of the upward VEM (tested using the vertical oculocephalic reflex) was decreased to about half that of the downward VEM, suggesting that the CVTT is involved not only in the transmission of upward EP signals but also in that of upward VEM signals. In another patient with PPUN, the lesion could affect the superior pole of the SVN and the initial portion of the CVTT (Figs. 1 and 2B), in the postero-lateral part of the lower pontine tegmentum (Tilikete et al., 2008). Lastly, a case of PPUN due to a lesion affecting the anterior midbrain tegmentum suggests that the final portion of the human CVTT could also be similar to that of the cat (Yura et al., 1988) (Fig. 1). Thus, there is now accumulating evidence for the existence of a human CVTT playing an important role in the transmission of upward excitatory EP and VEM signals, with a brainstem course that appears to be analogous to that of the cat or the monkey CVTT. However, this SVN-CVTT pathway seems to duplicate functionally the excitatory MVN-MLF pathway involved in the transmission of upward signals.

Why two brainstem excitatory pathways for upward vestibular eye movements?

The theory

The asymmetry in the brainstem vertical VEM circuitry, with two excitatory pathways existing for upward movements and only one for downward movements, seems illogical except if one considers that a particular need exists for the upward system. This hypothetical vertical VEM organization has previously been discussed elsewhere (Pierrot-Deseilligny and Milea, 2005) and will be briefly reviewed here. A specific and supplementary need for the upward system, compared to the downward system, could result from the effect of gravity, since the centre of mass of the eyeball is probably located more anteriorly than its rotation centre, at least in the cat (Harris et al., 1993). Thus, in the most physiological, upright position of the head, the gravitational pull could facilitate downward VEM and restrain upward VEM, creating a clear and maximum imbalance between the two vertical VEM systems. This assumes that all other vestibular conditions related to EP and head movements in reference to the orbit are otherwise equal and accurately controlled by the different excitatory and inhibitory MLF tracts. Furthermore, this gravitational imbalance is continuously varying with the static position of the head: in the prone (or supine) position, the gravity vector becomes nil for upward and downward eye movement systems, whereas, in the upside-down position, the gravity effect is reversed compared to the upright position. We suggest that the CVTT is the final tract of a circuitry developed in order to counteract the gravitational pull according to the instantaneous static position of the head, permanently providing the elevator muscle motoneurons with the exact supplement of tonic activity required for rebalancing the vertical eyeball inertial asymmetry. Besides the CVTT, which is the efferent tract of the system, and the otolith afferent pathways, giving information on the current gravity vector via the static position of the head, the core of this anti-gravitational circuitry could comprise (Fig. 3): (1) the SVN, receiving otolithic projections and containing the CVTT neurons; (2) a structure in the caudal medulla, receiving a collateral branch of the CVTT and projecting to the flocculus, probably as feedback control; (3) the flocculus, receiving multiple afferent inputs (from the caudal medulla, the otoliths, and vision), projecting to the SVN and playing an important regulating role in this

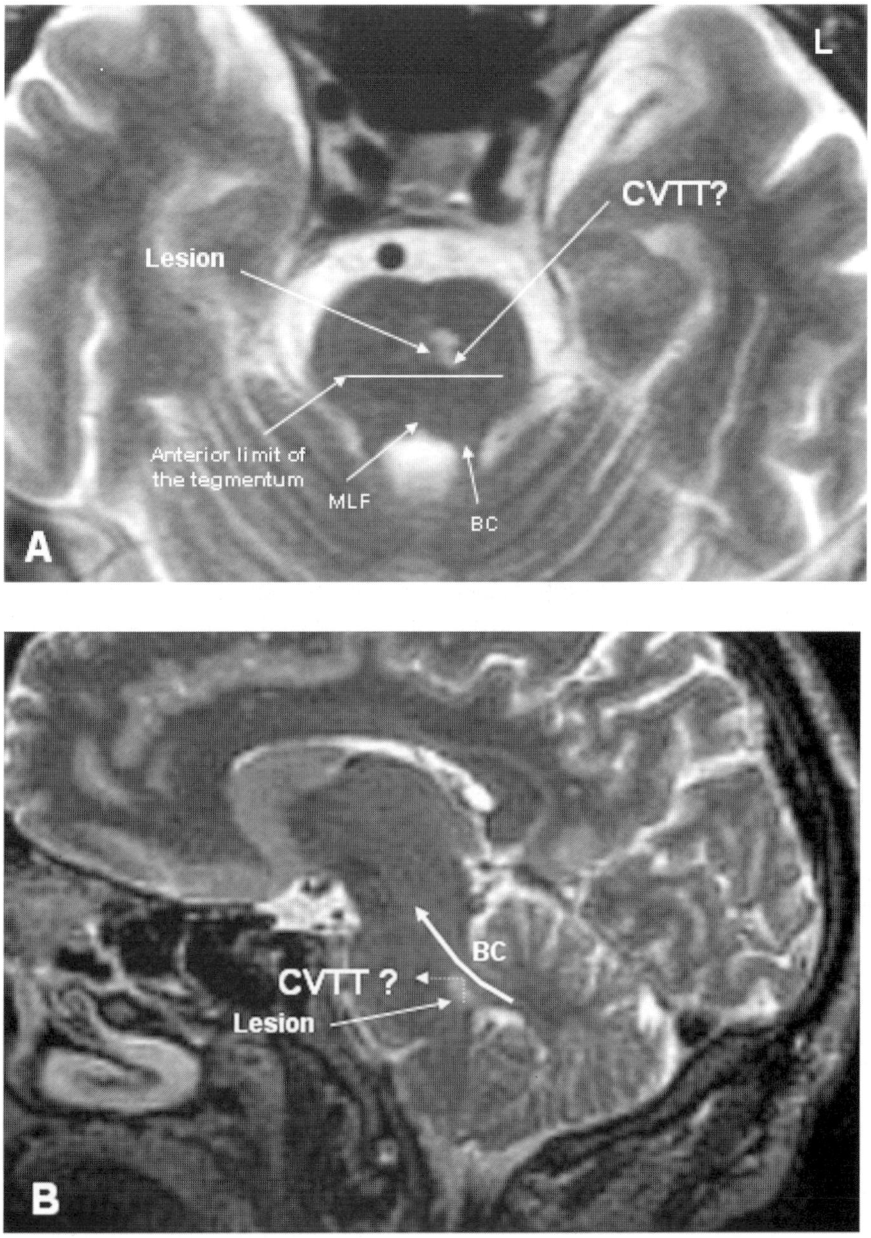

Fig. 2. Small pontine lesions resulting in primary position upbeat nystagmus. (A) Horizontal section (MRI, T2) showing a unilateral lesion located just above the midpons level, in the left postero-paramedian part of the basis pontis (Pierrot-Deseilligny et al., 2005). (B) Sagittal section (MRI, T2) showing a unilateral lesion affecting the postero-lateral part of the right lower pontine tegmentum, sparing the BC (Tilikete et al., 2008). BC: brachium conjunctivum; CVTT: crossing ventral tegmental tract; L: left; MLF: medial longitudinal fasciculus.

circuitry (for details, see Pierrot-Deseilligny and Milea, 2005). Such an organization may explain the existence of PPUN observed after caudal medullary lesions, but also the primary position downbeat nystagmus (PPDN) resulting from floccular lesions. For the latter, since the floccular projection to the SVN is inhibitory, PPDN could result from disinhibition of the SVN with,

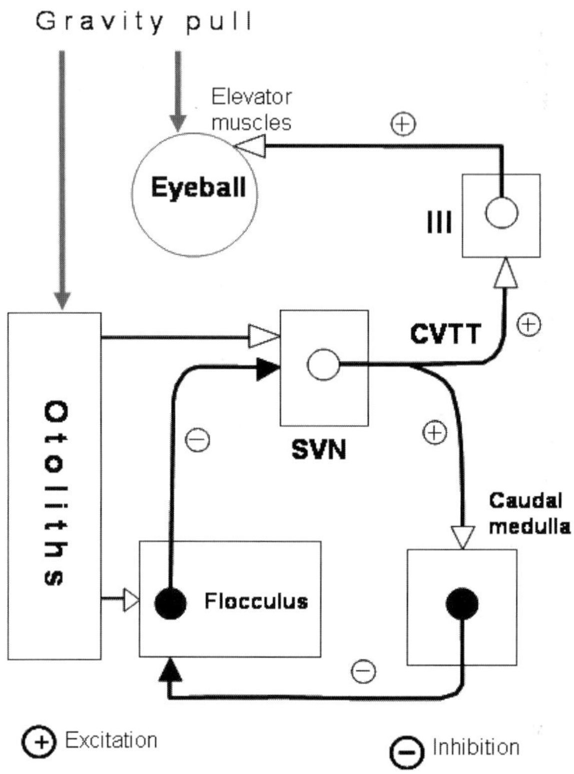

Fig. 3. Hypothetical anti-gravitational circuitry projecting to the elevator muscle motoneurons. CVTT: crossing ventral tegmental tract; SVN: superior vestibular nucleus; III: oculomotor nucleus.

consequently, a permanently increased activity of the CVTT compared to the normal state.

In brief, we suggest that, in the normal state, the CVTT tonic activity is permanently modulated according to the gravity vector and the static position of the head, being maximum in the upright position of the head, decreasing proportionally in all the positions comprised between the upright position and the prone position, becoming almost nil in the prone position, and probably remaining nil in all hanging positions of the head. Incidentally, it is possibly because the latter positions are not very physiological that no equivalent of the CVTT circuitry exists for supplementing the downward eye movement system in such positions. Lastly, in certain pathological states with primary position vertical nystagmus, the CVTT activity would be either no longer modulated or be insufficiently modulated by the head positions, being permanently close to zero after brainstem lesions affecting the CVTT axons (in the case of pontine or midbrain damage) or its main collateral branch (in the case of caudal medullary damage), which result in PPUN, or permanently increased after floccular lesions, which result in PPDN. Accordingly, two different vestibular functions, controlled by separate pathways and two brainstem excitatory tracts, would be exerted in parallel in the upward VEM system: (1) the first calculating the EP and velocity relative to the orbit in response to rotational and/or translational head movements — via the semicircular canals, the otoliths, the MVN, and the MLF; (2) the second (Fig. 3) adjusting these parameters according to the instantaneous gravity vector, i.e., to the position of the head in space — via the otoliths, the flocculus, the SVN, and the CVTT — which could form an otolithic anti-gravitational pathway, specific for the upward gaze. However, these two vestibular functions are probably partly mixed at the vestibular nucleus level and partly distributed in both final efferent tracts (MLF and CVTT), as suggested by the effects of damage to these tracts (see above). Moreover, this partial mixing of functions may explain adaptation mechanisms between the two pathways and the disappearance (after a few weeks or months) of PPUN due to CVTT lesions (Pierrot-Deseilligny and Milea, 2005).

Applications of the theory

Taking into account the main points of this theory, the typical reactions of the ocular motor system according to the different static positions of the head (upright, prone and upside-down positions) may be understood by comparing in each situation the instantaneous effect of gravity exerted on the upward eye movement system and the assumed corresponding CVTT activity (Table 1). Note that the supine position, which has been less studied and raises specific problems, will not be dealt with here in detail. In this comparison, when the CVTT activity exactly compensates for the instantaneous effect of gravity there is no vertical slow eye deviation or nystagmus, whereas when the CVTT activity is no longer adapted, in pathological but also some physiological states, a vertical slow eye deviation in either vertical direction (with the corresponding vertical nystagmus) is observed. In normal subjects in the upright position, the CVTT activity is assumed to be maximal to counteract the maximum effect of gravity being exerted on the upward eye movement system, whereas, in the prone (and supine) position, the CVTT activity could be minimal, as is then the effect of gravity on the upward system, with, consequently, in these two conditions a balance in the upward and downward systems and the absence of spontaneous vertical nystagmus (Table 1). By contrast, in the upside-down position, in which the CVTT activity could remain nil, the reversed effect of the gravity vector would not be compensated for in the absence of a circuitry analogous to that of the CVTT for supplementing the downward eye movement system, with, as a frequent result, the occurrence of an upward slow phase and chin-beating nystagmus, i.e., PPDN (Bisdorff et al., 2000; Kim et al., 2000; Leigh et al., 2002) (Table 1). If the gravity itself rapidly changes for a while, as for example during parabolic flights, with an alternation for about 10 s or 20 s between sudden hypogravity and hypergravity, vertical gaze shifts are observed (see below), which at first would appear to be a strong argument for the existence of a significant permanent influence of the gravity pull on vertical eye movements. Furthermore, during these unusual rapid gravity changes, with a subject remaining in an upright-sitting position, it may be assumed that the CVTT activity cannot immediately be adapted. The result would be that the unchanged CVTT activity becomes (at least transitorily) relatively too high in sudden hypogravity, whereas the same activity is relatively too low in hypergravity: this could account for the upward slow phase (with PPDN) observed in sudden hypogravity and the downward slow phase (with PPUN) observed in hypergravity (Von Baumgarten et al., 1980; Clément et al., 1989; Cheung et al., 1994) (Table 1). By contrast, if, during such a parabolic flight, the subject is sitting in the prone position, i.e., with only a slight impact of gravity changes on the vertical eye movements and a CVTT activity that is nil (as is usual in this position) — neither substantial slow eye deviation nor nystagmus should occur in hypo- or hypergravity: the absence of nystagmus was indeed noted when subjects were in this position during both abnormal gravity states (Von Baumgarten et al., 1980).

In patients with PPUN due to a recent brainstem lesion affecting the CVTT circuitry, with as yet no significant adaptation, the assumed absence of CVTT activity in the upright position does not compensate for the effect of the gravity pull, with, consequently, the existence of a downward slow eye deviation with PPUN (Table 1). In the prone and supine positions, the disappearance of the gravity effect on both vertical VEM systems associated with the permanent absence of CVTT activity, which leads to a situation close to that of the normal state, should theoretically suppress or improve the nystagmus, a finding that was almost always observed when these positions of the head were tested (Fisher et al., 1983, in five patients out of seven; Janssen et al., 1998;

Table 1. Effect of gravity on physiological and pathological primary position vertical nystagmus

		Upright position of the head					Prone position of the head					Upside-down position of the head				
		Effect of gravity[a]		CVTT activity[b]		Nystagmus slow phase[c]	Effect of gravity[a]		CVTT activity[b]		Nystagmus slow phase[c]	Effect of gravity[a]		CVTT activity[b]		Nystagmus slow phase[c]
Normal subjects	Normal gravity	↓	&	↑	=	0	0	&	0	=	0	↑	&	0	=	↑
	Hypo-gravity	0	&	↑	=	↑	0	&	0	=	0	Not tested				
	Hyper-gravity	↓↓	&	↑	=	↓	0	&	0	=	0	Not tested				
Patients with primary position vertical nystagmus	PPUN	↓	&	0	=	↓	0	&	0	=	0	↑	&	0	=	↑
	PPDN	↓	&	↑↑	=	↑	0	&	↑↑	=	↑↑	↑	&	↑↑	=	↑↑↑

[a]The effect of gravity is that exerted on the upward eye movement system in the upright position (↓: normal effect, restraining this system), in the prone position (0: zero effect), and in the upside-down position (↑: reversed effect, facilitating this system); the gravity itself may be nil (0) in hypogravity or increased (↓↓) in hypergravity.

[b]The CVTT activity is assumed to be normally maximum (↑) in the upright position and nil (0) in the prone and upside-down positions; furthermore, this activity could be permanently close to zero in PPUN and permanently increased (↑↑) in PPDN.

[c]The typical ocular motor response would be the result of the comparison of variables [a] and [b], with either a balance in upward and downward eye movement systems and no eye deviation (0: no nystagmus) or an imbalance and the occurrence of a primary position vertical nystagmus (with ↑ or ↓ indicating the direction of the slow phase in relation to the forehead, whatever the head position); this nystagmus may be worsened (↑↑ or ↑↑↑) compared to the initial nystagmus in the upright position. For details, see text. CVTT: crossing ventral tegmental tract; PPDN and PPUN: primary position downbeat and upbeat nystagmus, respectively.

Pierrot-Deseilligny et al., 2007) (Table 1). In the upside-down position, the conditions are again theoretically analogous to those existing in the normal state, with therefore the possible occurrence of an upward slow eye deviation and PPDN, which was indeed observed (Hirose et al., 1991; Minagar et al., 2001). Lastly, in patients with PPDN, the assumed permanently increased CVTT activity is already relatively too high in the upright position (compared to that existing in the normal state), with consequently the occurrence of an upward slow eye deviation with PPDN (Table 1). This situation should theoretically be worsened in the prone position, with the disappearance of any assistance from gravity, and in the upside-down position, with a gravitational effect exerted in the same direction as that of the permanent upward slow phase: a worsening of the PPDN (with increased slow phase velocity) is indeed often observed in such positions of the head (Baloh and Spooner, 1981; Halmagyi et al., 1983; Marti et al., 2002) (Table 1). Of course, exceptions to these typical, pathological ocular motor results exist, but it should be considered that they might then be the result of multiple, complex, appropriate, or inappropriate, sometimes overactive, adaptation mechanisms.

Conclusions

Two different pathways comprising two final brainstem excitatory tracts (MVN-MLF and SVN-CVTT) appear to function in parallel in the upward VEM system, with complementary actions probably combining partly at the vestibular nucleus level and partly at the III level. The reason for this duplication of pathways could mainly result from the existence of a gravity effect on eye movements and the need to counteract this effect specifically in the upward eye movement system. Such an organization could account for the existence in pathology of the main types of primary position vertical nystagmus and explain the influence of head positions on this nystagmus. However, further research is required to determine more precisely the specific roles of these two upward VEM pathways.

Abbreviations

BC	brachium conjunctivum
CVTT	crossing ventral tegmental tract
EP	eye position
III	oculomotor nucleus
INO	internuclear ophthalmoplegia
IV	trochlear nucleus
MLF	medial longitudinal fasciculus
MVN	medial vestibular nucleus
PPDN	primary position downbeat nystagmus
PPUN	primary position upbeat nystagmus
SVN	superior vestibular nucleus
VEM	vestibular eye movements

References

Anagnostou, E., Mandllos, D., Limbitaki, G., Papadimitriou, A. and Anastasopoulos, D. (2006) Positional nystagmus and vertigo due to a solitary brachium conjunctivum plaque. J. Neurol. Neurosurg. Psychiatry, 77: 790–792.

Baloh, R.W. and Spooner, J.W. (1981) Downbeat nystagmus: a type of central vestibular nystagmus. Neurology, 31: 304–310.

Benjamin, E.E., Zimmerman, C.F. and Troost, B.T. (1986) Lateropulsion and upbeat nystagmus are manifestations of central vestibular nystagmus. Arch. Neurol., 43: 962–964.

Bisdorff, A.R., Sancovic, S. and Debatisse, D. (2000) Positional nystagmus in the dark in normal subjects. Neuro-ophthalmology, 24: 283–290.

Büttner-Ennever, J.A. (1992) Patterns of connectivity in the vestibular nuclei. Ann. N.Y. Acad. Sci., 656: 363–378.

Carpenter, M.B. and Cowie, R.G. (1985) Connections and oculomotor projections of the superior vestibular nucleus and cell group 'y'. Brain Res., 336: 265–287.

Cheung, B.S., Money, K.E. and Howard, I. (1994) Human gaze instability during brief exposure to reduced gravity. J. Vestib. Res., 4: 17–27.

Clément, G., Andre-Deshays, C. and Lathan, C.E. (1989) Effects of gravitoinertial force variations on vertical gaze direction. Aviat. Space Environ. Med., 60: 1194–1198.

Cremer, P.D., Migliacio, A.A., Halmagyi, G.M. and Curthoys, I.S. (1999) Vestibulo-ocular reflex pathways in internuclear ophthalmoplegia. Ann. Neurol., 45: 529–533.

Fisher, A., Gresty, M., Chambers, B. and Rudge, P. (1983) Primary position upbeating nystagmus. A variety of central positional nystagmus. Brain, 106: 949–964.

Frohman, E.M., Frohman, T.C., Fleckenstein, J., Racke, M.K., Hawker, K. and Kramer, P.D. (2001) Ocular contrapulsion in multiple sclerosis: clinical features and pathophysiological mechanisms. J. Neurol. Neurosurg. Psychiatry, 70: 688–690.

Graf, W. and Ezure, K. (1986) Morphology of vertical canal related second order vestibular neurons in the cat. Exp. Brain Res., 63: 45–48.

Halmagyi, G.H., Rudge, P., Gresty, M.A. and Sanders, M.B. (1983) Downbeating nystagmus. A review of 62 cases. Arch. Neurol., 40: 777–784.

Hankey, G.J., Silbert, P.L. and Edis, R.H. (1987) Localising value of primary position upbeating nystagmus. Aust. N.Z. J. Med., 17: 333–335.

Harris, L.R., Goltz, H.C. and Steinbach, M.J. (1993) The effect of gravity on the resting position of the cat's eye. Exp. Brain Res., 91: 107–116.

Hirose, G., Kawada, J., Tsukada, K., Yoshioka, A. and Sharpe, J.A. (1991) Upbeat nystagmus: clinico-pathological and pathophysiological considerations. J. Neurol. Sci., 105: 159–167.

Janssen, J.C., Larner, A.J., Morris, H., Bronstein, A.M. and Farmer, S.F. (1998) Upbeat nystagmus: clinicopathological correlation. J. Neurol. Neurosurg. Psychiatry, 65: 380–381.

Kattah, J.C. and Dagi, D.F. (1990) Compensatory head tilt in upbeating nystagmus. J. Clin. Neuro-ophthalmol., 10: 27–31.

Kim, J.I., Somers, J.T., Sthal, J.S., Bhidayasiri, R. and Leigh, R.J. (2000) Vertical nystagmus in normal subjects: effects of head position, nicotine and scopolamine. J. Vestibul. Res., 10: 291–300.

Lang, W., Büttner-Ennever, J.A. and Büttner, U. (1979) Vestibular projections to the monkey thalamus: an autoradiographic study. Brain Res., 177: 3–17.

Leigh, R.J., Das, V.E. and Seidmann, S.H. (2002) A neurobiological approach to acquired nystagmus. Ann. N.Y. Acad. Sci., 956: 380–390.

Leigh, R.H. and Zee, D.S. (2006) The Neurology of Eye Movements (4th ed.). Oxford University Press, Oxford.

Marti, S., Palla, A. and Straumann, D. (2002) Gravity dependence of ocular drift in patients with cerebellar downbeat nystagmus. Ann. Neurol., 52: 712–721.

McCrea, R.A., Strassman, A., May, E. and Highstein, S.M. (1987) Anatomical and physiological characteristics of vestibular neurons mediating the vertical vestibulo-ocular reflexes in the squirrel monkey. J. Comp. Neurol., 264: 571–594.

Minagar, A., Sheremata, W.A. and Tusa, R.J. (2001) Perverted head-shaking nystagmus: a possible mechanism. Neurology, 57: 887–889.

Mossuto-Agatiello, L. (2006) Caudal paramedian midbrain syndrome. Neurology, 66: 1668–1671.

Nakada, T. and Remler, M.P. (1981) Primary position upbeat nystagmus: another central vestibular nystagmus. J. Clin. Neuro-ophthalmol., 1: 181–185.

Pierrot-Deseilligny, C. and Milea, D. (2005) Vertical nystagmus: clinical facts and hypotheses. Brain, 128: 1237–1246.

Pierrot-Deseilligny, C., Milea, D., Sirmai, J., Papeix, C. and Rivaud-Péchoux, S. (2005) Upbeat nystagmus due to a small pontine lesion : evidence for the existence of a crossing ventral tegmental tract. Eur. Neurol., 154: 186–190.

Pierrot-Deseilligny, C., Richeh, W. and Bolgert, F. (2007) Upbeat nystagmus due to a caudal medullary lesion and influenced by gravity. J. Neurol., 254: 120–121.

Ranalli, P.J. and Sharpe, J.A. (1988a) Vertical vestibulo-ocular reflex, smooth pursuit and eye-head tracking dysfunction in internuclear ophthalmoplegia. Brain, 111: 1299–1317.

Ranalli, P.J. and Sharpe, J.A. (1988b) Upbeat nystagmus and the ventral tegmental pathway of the upward vestibulo-ocular reflex. Neurology, 38: 1329–1330.

Sato, Y. and Kawasaki, T. (1987) Target neurons of floccular caudal zone inhibition in y-group nucleus of vestibular nuclear complex. J. Neurophysiol., 57: 460–480.

Sato, Y. and Kawasaki, T. (1991) Identification of Purkinje cell/climbing control by the cerebellar flocculus. Brain Res. Rev., 16: 39–64.

Tilikete, C., Milea, D. and Pierrot-Deseilligny, C. (2008) Upbeat nystagmus: a way to delineate the central vestibular pathways. J. Neuro-Ophthalmol, in press.

Uchino, Y., Sasaki, M., Isu, N., Irai, N., Imagawa, M., Endo, K. and Graf, W. (1994) Second-order vestibular neuron morphology of the extra-MLF anterior canal pathway. Exp. Brain Res., 97: 387–391.

Von Baumgarten, R.J., Baldrighi, H., Vogel, H. and Thümler, R. (1980) Physiological response to hyper- and hypogravity during roller-coaster flight. Aviat. Space Environ. Med., 51: 145–154.

Yamamoto, M., Shimoyama, I. and Highstein, S.M. (1978) Vestibular nucleus neurons relaying excitation from the anterior canal to the oculomotor nucleus. Brain Res., 148: 31–42.

Yura, S., Sako, K. and Yonemasu, Y. (1988) Primary midbrain hemorrhage with upbeat nystagmus. Act. Neuro. (Napoli.), 10: 239–245.

C. Kennard & R.J. Leigh (Eds.)
Progress in Brain Research, Vol. 171
ISSN 0079-6123

CHAPTER 6.5

Mechanisms of vestibulo-ocular reflex (VOR) cancellation in spinocerebellar ataxia type 3 (SCA-3) and episodic ataxia type 2 (EA-2)

Carlos R. Gordon[1,2,*], Avi Caspi[3], Ronen Levite[1] and Ari Z. Zivotofsky[3]

[1]*Department of Neurology, Meir Medical Center, Kfar Saba, Israel*
[2]*Sackler Faculty of Medicine, Tel Aviv University, Tel Aviv, Israel*
[3]*Gonda Multidisciplinary Brain Research Center, Bar Ilan University, Ramat Gan, Israel*

Abstract: The vestibulo-ocular reflex (VOR) mechanism triggers eye movements as a result of head motion in order to keep gaze stationary relative to the world. However, in order to shift the direction of the gaze along with head motion, the VOR mechanism must be overridden ("cancelled"). Two mechanisms have been proposed to explain this cancellation: a reduction of the VOR gain or activation of smooth pursuit (SP) eye movements in the opposite direction. We studied VOR cancellation using the magnetic search coil in six spinocerebellar ataxia type 3 (SCA-3) and four episodic ataxia type 2 (EA-2) patients, conditions that are known to have degraded SP but different degrees of VOR impairment. Abnormal VOR was found in two of the four EA-2 patients and all of the SCA-3 patients. All subjects possessed residual VOR, although when tested using head thrusts it was almost negligible in the SCA-3 patients. The EA-2 patients showed essentially no SP and the SCA-3 patients had poor SP. However, for all patients, the gain during VOR cancellation was comparable to normals. These results provide additional evidence that SP cancellation of the VOR cannot be the sole mechanism utilized in overriding the VOR in these patients.

Keywords: smooth pursuit; VOR gain; head thrust; eye-head tracking; saccadic tracking; magnetic search coil

Introduction

When a person's head is stationary, smooth pursuit (SP) eye movements are used to follow an object moving at low frequency, thus allowing for accurate visual information to be perceived. When making rapid head movements, it is mainly the vestibulo-ocular reflex (VOR) mechanism that triggers eye movements in the direction opposite to head motion in order to keep the gaze stationary relative to the world. During everyday motor activities, combined eye and head movements are used for visual tracking. Therefore, a close interaction between the SP and VOR systems is necessary in order to generate accurate eye movements in space (gaze).

In order to shift the direction of the gaze while the head is in motion, the VOR mechanism must be overridden, modulated, or even cancelled. At least two mechanisms have been proposed to explain this cancellation: a reduction of the VOR gain or activation of SP eye movements in the opposite

*Corresponding author. Tel.: +972-9-7472828;
Fax: +972-9-7471317; E-mail: cgordon@post.tau.ac.il

DOI: 10.1016/S0079-6123(08)00674-2

direction of the VOR (Cullen et al., 1991; Huebner et al., 1993; Roy and Cullen, 1998). Evidence for the first mechanism includes the ability to fixate a head-fixed target during head roll by cancelling torsional VOR, yet there is no torsional SP (Leigh et al., 1989). Evidence for the SP cancellation mechanism is provided by the head-brake experiment in which the head is abruptly stopped during combined eye–head tracking and SP is initiated far quicker than its usual response time, implying that the SP signal was present and negating the VOR (Huebner et al., 1993).

Interactions between SP and VOR have been studied in monkeys, normal humans, and patients with abnormal SP and peripheral vestibular function (Grant et al., 1992; Waterston et al., 1992; Cullen and McCrea, 1993; Huebner et al., 1993; Takeichi et al., 2000). These studies conclude that both SP and additional factors play a role in cancelling the VOR during combined eye–head movements. A review of the subject can be found in Leigh and Zee (2006).

In the present study, we investigated VOR cancellation in patients with spinocerebellar ataxia type 3 (SCA-3) and episodic ataxia type 2 (EA-2), conditions that are known to have abnormal SP but very different vestibular and cerebellar signs and symptoms.

SCA-3, also known as Machado-Joseph disease (MJD), is the most common form of autosomal dominant cerebellar ataxia. In Israel, a Yemenite Jewish genetic isolate of SCA-3, characterized by a relatively large number homozygotes for the CAG trinucleotide repeat expansion at the MJD1 gene and having vestibulo-ocular arreflexia, has been described (Lerer et al., 1996; Gordon et al., 2003). Other ocular-motor findings are saccadic SP, gaze evoked nystagmus, and slow saccades. MRI studies reveal diffuse CNS atrophic changes, particularly in the cerebellar vermis, superior cerebellar peduncle, pontine tegmentum, and frontal lobes (Tokumaru et al., 2003).

EA-2 is an uncommon autosomal dominant disorder characterized by bouts of ataxia with minimal interictal cerebellar findings and symptomatic response to acetazolamide and 4-aminopyridine. EA-2 is caused by mutations of calcium channel gene CACNA1A on chromosome 19p13 that encodes the Ca voltage 2.1 subunit of the P/Q-type calcium channel. The P/Q channel is expressed mainly in the cerebellum explaining patients' ataxia. The main findings on neurological examination are impaired SP, gaze-evoked nystagmus, and lack of optokinetic nystagmus (OKN). Normal MRI or mild cerebellar atrophy, especially of the anterior vermis, have been reported. A recent review of the subject can be found in Strupp et al. (2007).

Considering the structural and functional differences of these diseases, mainly the vestibular arreflexia in SCA-3 and the impaired or absent SP and OKN previously described in EA-2, we assume that investigating VOR cancellation in these two groups of patients can broaden our understanding of the mechanisms regulating combined eye–head tracking.

Methods

Subjects

Six patients with SCA-3 from five unrelated families (three males), four patients with EA-2 from two unrelated families (one male), and five normal control subjects (four males) participated in the study. Written informed consent was obtained from all participants for all procedures. All control subjects had no past history of neurological or psychiatric disease nor were they taking medication that potentially affects eye movements.

Eye-movement recordings and data analysis

The experimental paradigm and data collection were under the control of "virtual instruments" written using Labview (National Instrument) on a PC. Horizontal and vertical gaze of the right eye were measured using the magnetic search coil technique (Robinson, 1963) with 6-ft field coils (CNC Engineering, Seattle, WA) and a scleral search coil embedded in a silastic ring (Skalar, Delft, Netherlands). The eye-movement recordings were monocular because clinically all eye movements appeared to be conjugate and there was no clinical complaint of diplopia. The coils were pre-calibrated using a protractor device.

The coil was placed on the eye after application of one or two drops of local anaesthetic (localin). Coil signals were filtered (bandwidth 0–200 Hz) prior to digitization at 500 Hz sampling rate, with 12-bit resolution. The visual stimulus was a dim (3 mW with a 100 × neutral density filter) red laser spot (1.5 mm diameter) rear-projected onto a semi-translucent screen situated 1.1 m in front of the subject. The position of the laser spot was determined by an X-Y galvanometer (General Scanning). The fly-time of the mirrors is 2–3 ms and they settle to their final location within 5–6 ms. The subject sat in front of the semi-translucent screen in a dark room and was given time to adjust to the dark. The subject's head was immobilized with a head restraint device attached to the chair.

SP eye movements were tested by having the subject track a target moving sinusoidally at 0.2 Hz from eccentric (between 7.5° and 15°; maximum velocity of 19°/s) positions to the centre. Visually enhanced VOR (VVOR) was measured by rotating the chair in a sinusoidal manner, back and forth at approximately 0.4 Hz while the patient fixated a stationary target. In addition, the VOR response to head thrusts was measured by asking the patient to fixate his gaze on the laser spot target, while his head was held from behind by one of the investigators who executed sudden, rapid, unpredictable angular head rotations to both sides. These thrusts had an amplitude of approximately 25° and a maximum velocity of 500°/s. VOR cancellation was measured by having the subject wear a headband that held a front-facing laser pointer, and then rotating the subject in sinusoidal manner back and forth at approximately 0.4 Hz while they were instructed to fixate the moving laser target. Thus, perfect VOR cancellation would have a gain of zero.

Analysis of the data was performed using custom-written m-files on Matlab version 6.1 release 12.1. Gains were calculated by fitting the eye-movement velocity to the best fitting curve at the target frequency and dividing this eye-velocity profile by the target velocity.

Results

Table 1 shows the data for VVOR and head-thrust VOR. Abnormal VVOR (i.e., a gain below 0.8) was found in three of the four EA-2 patients and five of the six SCA-3 patients. Overall, the VVOR was fairly similar for the two patients groups. During head thrusts, decreased VOR gain was

Table 1. Visual enhanced VOR (VVOR), VOR during head thrusts, smooth pursuit (SP) gain, and VOR cancellation (VORCan) gain for all subjects

	Subjects	VVOR	VOR	SP	VORCan
Control	1	0.91	1.06	0.81	0.17–0.15
	2	0.85	1.12	0.97	0.14–0.14
	3	0.88	1.00	0.82	0.11–0.16
	4	0.87	1.02	0.84	–
	X	0.88 ± 0.02	1.05 ± 0.07	0.86 ± 0.06	0.15 ± 0.02
SCA-3	1	0.65	0.21	0.43	0.11–0.10
	2	0.54	0.26	0.81	0.11–0.11
	3	0.86	0.57	0.64	0.19–0.19
	4	0.54	0.54	0.44	0.15–0.16
	5	0.62	0.20	0.28	0.21–0.18
	6	0.70	0.46	0.56	–
	X	0.65 ± 0.12	0.39 ± 0.16	0.53 ± 0.19	0.15 ± 0.04
EA-2	1	0.55	0.77	0.26	0.19–0.13
	2	0.37	0.57	0.15	0.14–0.16
	3	0.48	0.91	0.26	0.16–0.18
	4	0.94	0.96	0.17	0.43–0.39
	X	0.53 ± 0.19	0.75 ± 0.18	0.21 ± 0.06	0.22–0.11

found in two of the four EA-2 patients while all the six SCA-3 patients had very low gains. For the EA-2 patients, the average gain during head thrusts was 0.75 ± 0.18, for the SCA-3 0.39 ± 0.16, and for the normals 1.05 ± 0.07. Thus, although all patients possessed residual VOR that is evident during VVOR, and a much smaller but still present VOR during head thrusts, the difference between the two groups is distinct. There is a significant difference in VOR between the two patient groups: the EA-2 group still had a reasonable working VOR, while the SCA-3 patients during head thrusts had only a negligible VOR.

Table 1 also shows the SP data. The EA-2 patients showed little residual SP and the SCA-3 patients, with one exception, had poor SP. The average SP gain for the EA-2 patients was 0.21 ± 0.06, for the SCA-3 0.53 ± 0.19, and for the normals 0.86 ± 0.06. Here too, it is worth stressing the distinct difference between the patient groups: The EA-2 patients have essentially no SP and track a smoothly moving target predominantly with saccades, while the SCA-3 patients, although not having a normal SP, have gains closer to 0.5. Examples of SP in a normal subject and an EA-2 patient can be seen in Fig. 1.

Table 1 shows the VOR cancellation data. With the exception of one EA-2 patient, the gain during VOR cancellation was similar for all patients and normals. Examples of VOR cancellation in a normal subject and an EA-2 patient can be seen in Fig. 2.

Discussion

In the present study, we evaluated VOR cancellation in two groups of patients with different degrees of SP and VOR abnormalities in order to investigate the interaction of these two systems during different head movements and consequently different visual drives.

The head-thrust data confirms what we have previously shown (Gordon et al., 2003) that SCA-3 patients have minimal VOR, at least at high frequency. This is very different from the EA-2 patients where half of them had almost normal VOR at high frequencies. Two points are relevant here: SCA-3 patients do display a greater gain in VVOR than in head thrust, and during VOR cancellation their gain is significantly lower than their VOR gain during head thrust. Thus, even these patients who are usually treated as lacking a VOR, do have a residual VOR that is "cancelled" during a VOR cancellation task, thus raising the question of how they are doing it. Do they use their impaired SP or do they modulate the residual VOR function by other neural mechanisms?

We also found that the EA-2 patients have minimal SP. Many of these patients have a fairly intact VOR, both at low frequency and during the head thrust, yet show a normal VOR cancellation. Here too, the question of how they cancel their VOR arises. Is it via a residual SP signal or by attenuating the VOR drive? The almost total lack of SP makes it likely that they are attenuating the VOR signal by some other means.

Interestingly, the most symptomatic EA-2 patient (patient 4), who has been severely disabled since she was 2 years old, had normal VVOR, normal head-thrust VOR gain, a very poor SP, and was the only patient with abnormal VOR cancellation. Because all of the other EA-2 patients also lack a normal SP by which to cancel the VOR, it cannot be argued that this patient had abnormal VOR cancellation in combined eye–head tracking because of her poor SP. We might suggest that in healthy persons both mechanisms play a role, while in patients like EA-2 the already low VOR gain can be sufficiently down-modulated without the help of an SP signal. Yet, in a patient like patient 4, the ability to modulate such a high VOR is limited; thus, without the assistance of an SP signal, the patient is left with a relatively large VOR gain during the VOR cancellation task.

We can suggest that it may be "tracking" rather than SP that is cancelling the VOR in the EA-2 patients. In other words, when presented with a smoothly moving target to track, they are unable to generate smooth tracking movements but they do stay close to the target using saccadic tracking. It may be that it is those "tracking" movements that cancel or reduce the VOR, and hence their combined eye–head tracking also looks saccadic. Our results are similar to those found by Takeichi et al. (2000) in their study of VOR cancellation in SCA-6 patients.

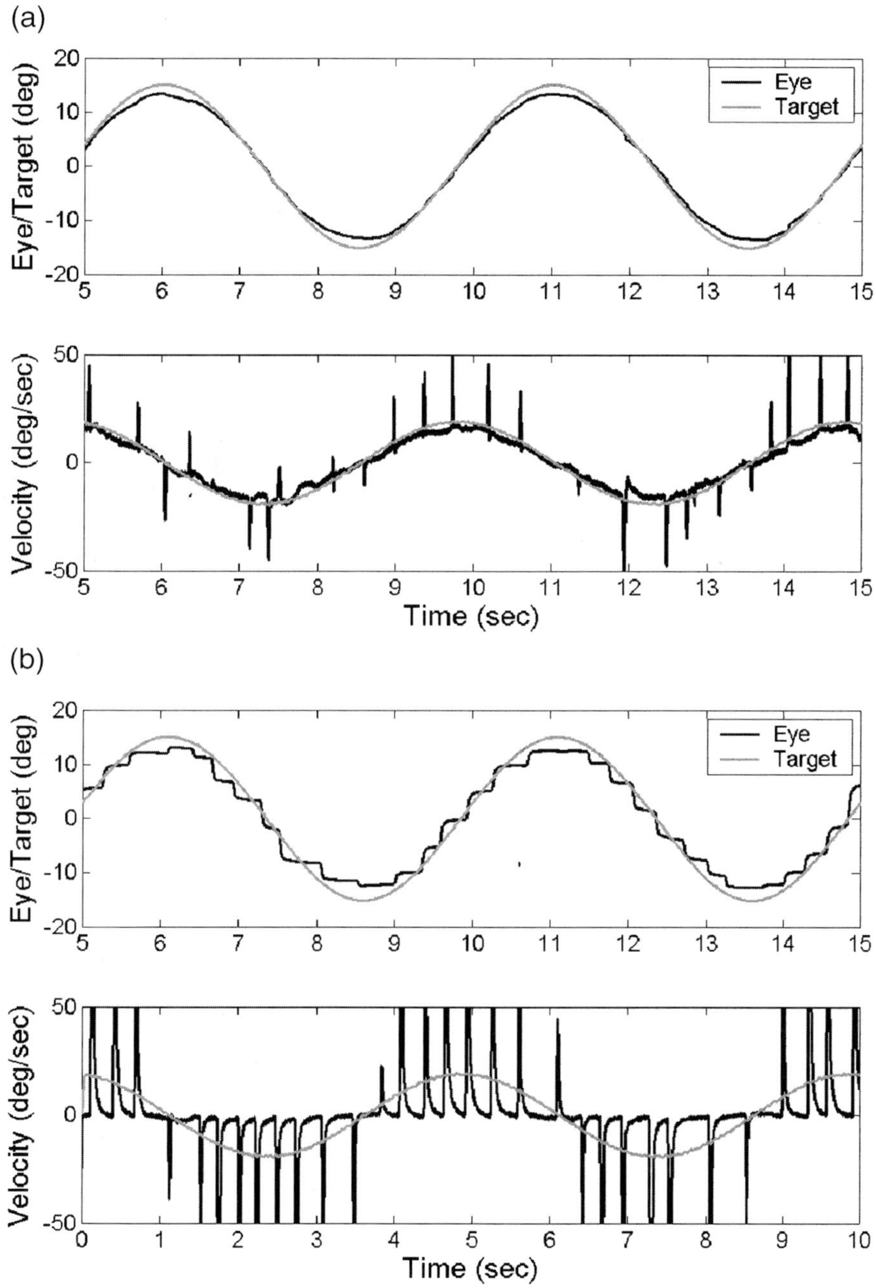

Fig. 1. Examples of smooth pursuit in (a) a normal subject and (b) an EA-2 patient. Traces show eye and target position and velocity.

Electrophysiological single-cell recordings and lesion studies in monkeys have also demonstrated that both the two aforementioned mechanisms, i.e., the reduction of the VOR gain and activation of SP eye movements in the opposite direction, are involved in overriding the VOR during combined eye–head tracking (Cullen and McCrea, 1993; Lisberger, 1990; Belton and McCrea, 1999).

The present data, while not settling the long-standing question of what happens to the VOR

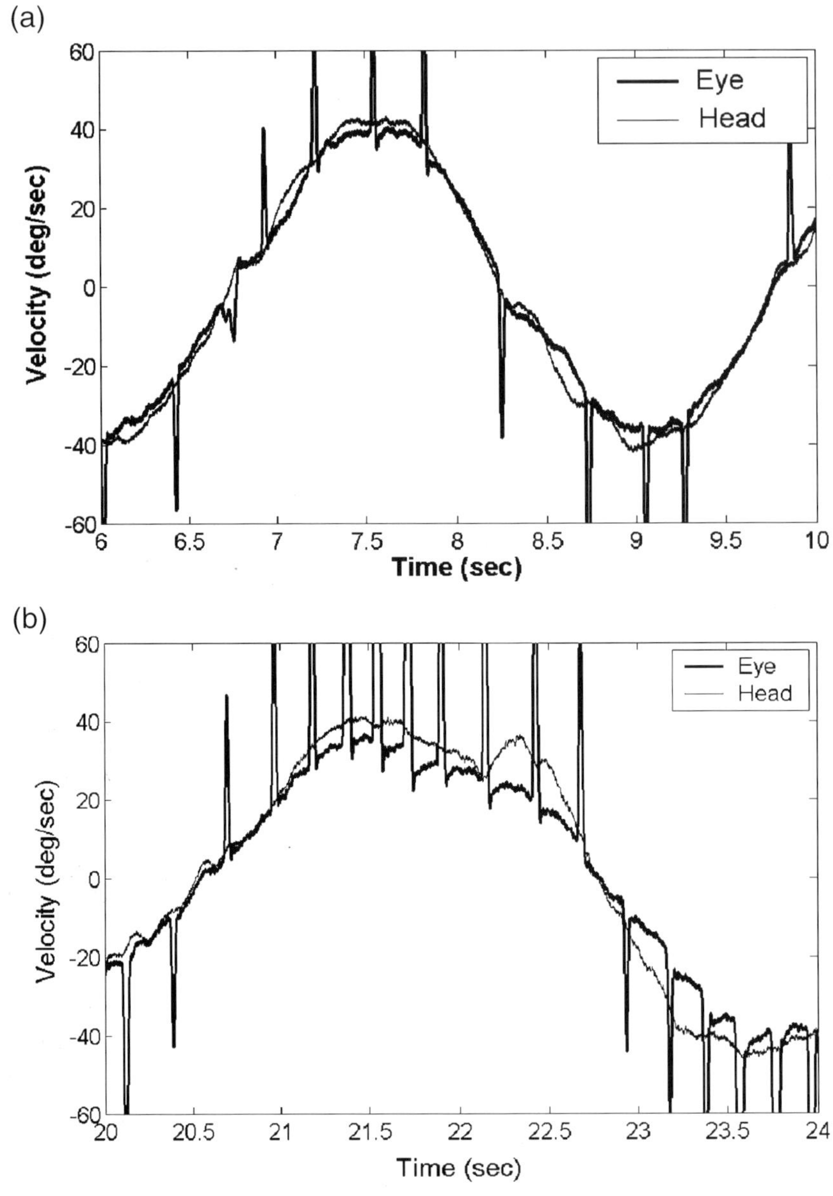

Fig. 2. Examples of VOR cancellation in (a) a normal subject and (b) an EA-2 patient. Traces show head and eye-in-space velocity.

during combined eye–head tracking, do provide additional evidence that SP cancellation of the VOR cannot be the sole mechanism utilized in overriding the VOR. There must be an additional mechanism(s) by which the VOR gain is modulated, and in addition to the usually suggested SP signal, it may be that a saccadic pursuit signal plays a role as well. A future experiment can test this hypothesis by performing the head-brake experiment (Huebner et al., 1993) on subjects such as ours in whom there is limited SP, a reasonable VOR, and significant VOR cancellation, and looking for saccadic tracking to commence with a shorter than normal latency.

Abbreviations

EA-2	episodic ataxia type 2
SCA-3	spinocerebellar ataxia type 3
SP	smooth pursuit
VOR	vestibulo-ocular reflex
VVOR	visual enhanced vestibulo-ocular reflex

Acknowledgement

This study was partially supported by the German-Israel Foundation (GIF) for Scientific Research and Development and the Gonda fund for Brain Research.

References

Belton, T. and McCrea, R.A. (1999) Contribution of the cerebellar flocculus to gaze control during active head movements. J. Neurophysiol., 81(6): 3105–3109.

Cullen, K.E., Belton, T. and McCrea, R.A. (1991) A non-visual mechanism for voluntary cancellation of the vestibulo-ocular reflex. Exp. Brain Res., 83(2): 237–252.

Cullen, K.E. and McCrea, R.A. (1993) Firing behavior of brain stem neurons during voluntary cancellation of the horizontal vestibuloocular reflex. J. Neurophysiol., 70(2): 828–843.

Gordon, C.R., Joffe, V., Vainstein, G., et al. (2003) Vestibulo-ocular arreflexia in families with spinocerebellar ataxia type 3 (Machado-Joseph disease). J. Neurol. Neurosurg. Psychiatr., 74(10): 1403–1406.

Grant, M.P., Leigh, R.J., Seidman, S.H., et al. (1992) Comparison of predictable smooth ocular and combined eye–head tracking behavior in patients with lesions affecting the brainstem and cerebellum. Brain, 115(5): 1323–1342.

Huebner, W.P., Leigh, R.J., Seidman, S.H., et al. (1993) An investigation of horizontal combined eye–head tracking in patients with abnormal vestibular and smooth pursuit eye movements. J. Neurol. Sci., 116(2): 152–164.

Leigh, R.J., Maas, E.F., Grossman, G.E., et al. (1989) Visual cancellation of the torsional vestibulo-ocular reflex in humans. Exp. Brain Res., 75(2): 221–226.

Leigh R.J. and Zee D.S. (Eds.), (2006) The Neurology of Eye Movements.·Oxford University Press, pp. 325–329.

Lerer, I., Merims, D., Abeliovich, D., et al. (1996) Machado-Joseph disease: correlation between clinical features, the CAG repeat length and homozygosity for the mutation. Eur. J. Hum. Genet., 4(1): 3–7.

Lisberger, S.G. (1990) Visual tracking in monkeys: evidence for short-latency suppression of the vestibulo-ocular reflex. J. Neurophysiol., 63(4): 676–688.

Robinson, D.A. (1963) A method of measuring eye movement using a scleral search coil in a magnetic field. IEEE Trans. Biomed. Eng., 10: 137–145.

Roy, J.E. and Cullen, K.E. (1998) A neural correlate for vestibulo-ocular reflex suppression during voluntary eye–head gaze shifts. Nat. Neurosci., 1(5): 404–410.

Strupp, M., Zwergal, A. and Brandt, T. (2007) Episodic ataxia type 2. Neurotherapeutics, 4(2): 267–273.

Takeichi, N., Fukushima, K., Sasaki, H., et al. (2000) Dissociation of smooth pursuit and vestibulo-ocular reflex cancellation in SCA-6. Neurology, 54(4): 860–866.

Tokumaru, A.M., Kamakura, K., Maki, T., et al. (2003) Magnetic resonance imaging findings of Machado-Joseph disease: histopathologic correlation. J. Comput. Assist. Tomogr., 27(2): 241–248.

Waterston, J.A., Barnes, G.R. and Grealy, M.A. (1992) A quantitative study of eye and head movements during smooth pursuit in patients with cerebellar disease. Brain, 115(5): 1343–1358.

C. Kennard & R.J. Leigh (Eds.)
Progress in Brain Research, Vol. 171
ISSN 0079-6123

CHAPTER 6.6

Modelling drug modulation of nystagmus

Stefan Glasauer[1,2,*] and Christian Rössert[1]

[1]*Department of Neurology, Ludwig-Maximilian University, Munich, Germany*
[2]*Bernstein Center for Computational Neuroscience, Munich, Germany*

Abstract: A better understanding of the neural and functional mechanisms underlying drug-induced changes in pathological nystagmus is likely to improve medical treatment. A treatment option for downbeat nystagmus (DBN), a common form of acquired fixation nystagmus that often occurs with cerebellar degeneration, is low doses of the potassium channel blocker 4-aminopyridine (4-AP). The upward ocular drift in DBN has a spontaneous and a vertical gaze-evoked component. Detailed analysis of the effect of 4-AP in patients showed that the drug consistently improved the gaze-evoked component, but had less effect in reducing the spontaneous drift. We show by a combination of computational modelling at the systems level and at the neuronal level how this differential effect can be investigated. We have previously postulated that DBN is caused by damage to the floccular lobe (FL). 4-AP, which has been shown to increase the excitability of Purkinje cells (PCs) in slice experiments, may thus suppress DBN by partly restoring floccular function. We simulated the effect of low concentrations of 4-AP on the cellular level using a multicompartment model of a PC, in which we changed ion channel properties to simulate damage. The transition from the cellular level to the systems level was achieved by constructing a population response. Systems level modelling predicted that the effect of 4-AP on the PCs should reduce DBN, but the predicted effect on the gaze-dependent component was less than is observed in patients. Our results suggest that the beneficial effect of 4-AP on DBN cannot be solely explained by its effect at the neuronal level of PCs, and suggests added effects at the level of the population of neurons.

Keywords: downbeat; nystagmus; cerebellum; neural integrator

Introduction

Understanding the action of drugs on the function of the central nervous system poses a major challenge. While it is quite often possible to determine the effect of a specific drug on healthy nervous tissue in vitro, extrapolating these results to account for the drug's effects on behaviour in animal experiments, and its results in therapeutic trials in patients may be problematic. In vitro experiments provide, e.g., data on how the drug influences a neuron's input–output relation, and thus hints which ion channels are involved or whether synaptic transmission is affected. Animal experiments may provide data on changes in behaviour. However, if no good animal model of the specific disease is available, such changes may be side effects rather than indicating the mechanisms for the beneficial effects of treatment. Patient

*Corresponding author. Tel.: +49 89 7095 4839;
Fax: +49 89 7095 4801;
E-mail: sglasauer@nefo.med.uni-muenchen.de

DOI: 10.1016/S0079-6123(08)00675-4

data of the sensorimotor or cognitive improvements in response to medication are certainly the most relevant for clinical practice and often provide important evidence. However, inferring the underlying mechanisms from the behavioural improvement of the patients is difficult. Evidently, to fully understand the drug's effect on CNS function requires multiple levels of analysis. Subsequently, connecting the different levels of analysis may provide hypotheses of how the drug's effect on the neural level causes the effects seen behaviourally. In the present work, we attempt to show how computational modelling may aid in this task.

One important example of pathological nystagmus is downbeat nystagmus (DBN), a common form of acquired fixation nystagmus. The slow upward drift of the eye in DBN is composed of two components, a gaze-direction independent bias component and a gaze-dependent component (Straumann et al., 2000; Glasauer et al., 2003). The gaze-dependent component usually increases with downward gaze, thus following Alexander's law, which implies that the neural integrator responsible for gaze holding is damaged. Based on anatomical and neurophysiological data, we recently have proposed a hypothesis of the pathogenesis of DBN (Marti et al., 2005). In particular, we proposed that damage to the cerebellar gaze-velocity Purkinje cells (PCs) in the floccular lobe (FL) would not only cause DBN, but also all of its associated oculomotor syndromes. According to our hypothesis, the upward drift is a direct consequence of the asymmetric distribution of on-directions of the FL–PCs: the great majority of these inhibitory cells increase their firing rate for downward visually evoked eye movements (Krauzlis and Lisberger, 1996). Accordingly, damage to the FL would cause a decrease in overall firing rate, disinhibition of floccular target neurons in the vestibular nuclei, and thus an upward drift of the eye corresponding to the gaze-independent part of the nystagmus slow phase velocity. The gaze-dependent drift is a consequence of a decrease in efficacy of the neural integrator function, which critically depends on the FL (Zee et al., 1981), and which normally assures gaze holding independently of visual input.

In previous work, we have shown that the potassium channel blockers 3,4-diaminopyridine and 4-aminopyridine (4-AP) can efficiently be used as treatment for DBN (Strupp et al., 2003; Kalla et al., 2007). Kalla et al. (2007) assessed the effect of 4-AP on DBN in 15 patients by measuring their three-dimensional eye movements in response to target steps. In this study, it was shown that 4-AP has a differential effect on DBN: drift with gaze-straight ahead (i.e., the gaze-independent component) was predominantly reduced in patients with cerebellar atrophy, but less so in the remaining patients; 4-AP on the average improved neural integrator function, i.e., gaze-evoked drift, regardless of aetiology. Based on previous in vitro experiments on the effect of 4-AP on cerebellar PCs (Etzion and Grossman, 2001), we hypothesized that the beneficial effect of 4-AP on DBN was caused by increasing PC excitability in the cerebellar flocculi.

In the present work, we simulated the effect of PC damage and of 4-AP at the cellular level using a detailed multicompartment model (Miyasho et al., 2001). We then translate these results to the population level and utilize them in a systems-level model of the vertical oculomotor system to predict the respective eye movements.

Methods

For the neuronal simulation we modified a multicompartment model of a PC (Miyasho et al., 2001). Dual exponential synapses (De Schutter and Bower, 1994) were added to all spiny dendrites (1002 synapses, time constants 0.5 and 1.2 ms, conductance 0.2 nS) to mimic synaptic excitation. Application of 4-AP was simulated by reduction of effective low-threshold potassium channels, the D-type potassium channels (ID) in the Miyasho model. To model damaged PCs in DBN patients, we assumed that the conductance of the Ca^{2+} P-type channel (CaP) was lowered, mimicking the mutation within the gene encoding the $\alpha_1 2.1$ subunit (Weisz et al., 2005). The input–output relation was obtained by stimulating the model with Poisson distributed spike trains with mean input frequencies between 0 and 200 Hz in steps of

2 Hz. For each input frequency, 100 runs (600 ms duration each) were simulated and the last 500 ms of each run were used to compute the mean output firing rate. The simulations were done using the software package NEURON.

The population response was constructed from the single-cell results by superimposing the firing rate responses of 2000 PCs. In order to simulate the physiological distribution of PC responses, each neuron received a different baseline input (normally distributed with 0 ± 50 spk/s, mean$\pm$SD) and had a different overall synaptic strength (normally distributed 1.0 ± 0.5). For subsequent use in the systems-level model, we normalized the population response of healthy PCs (normal channel conductances) by scaling it to a maximum output of 1 and a maximal slope of 1, and finally shifted the activation function so that the point of maximal slope was at zero input. The same input and scaling parameters were used for the population responses with modified ion channel conductances.

We then used the normalized population response in a simplified version of the systems-level model of the oculomotor circuitry of vertical eye movements described previously (Marti et al., 2005). The present model (Fig. 1, for details see Appendix) could simulate saccadic eye movements and gaze holding only; the vestibulo-ocular reflex and smooth pursuit pathways were removed for simplicity. To assure that this simplification would not lead to major changes in model behaviour, the validity of the results was crosschecked with the original model. The core assumption of the model is that the FL contains a forward model of the eye plant to transform motor commands into an estimate of eye velocity (Glasauer, 2003). This estimate is then compared to desired eye velocity (in the case of gaze holding, the desired velocity is zero) and the resulting error signal is amplified and inhibits its brainstem target neurons. The motor command is sent from the vestibular nuclei directly and via the leaky brainstem integrator to motor neurons, and via the paramedian tract neurons

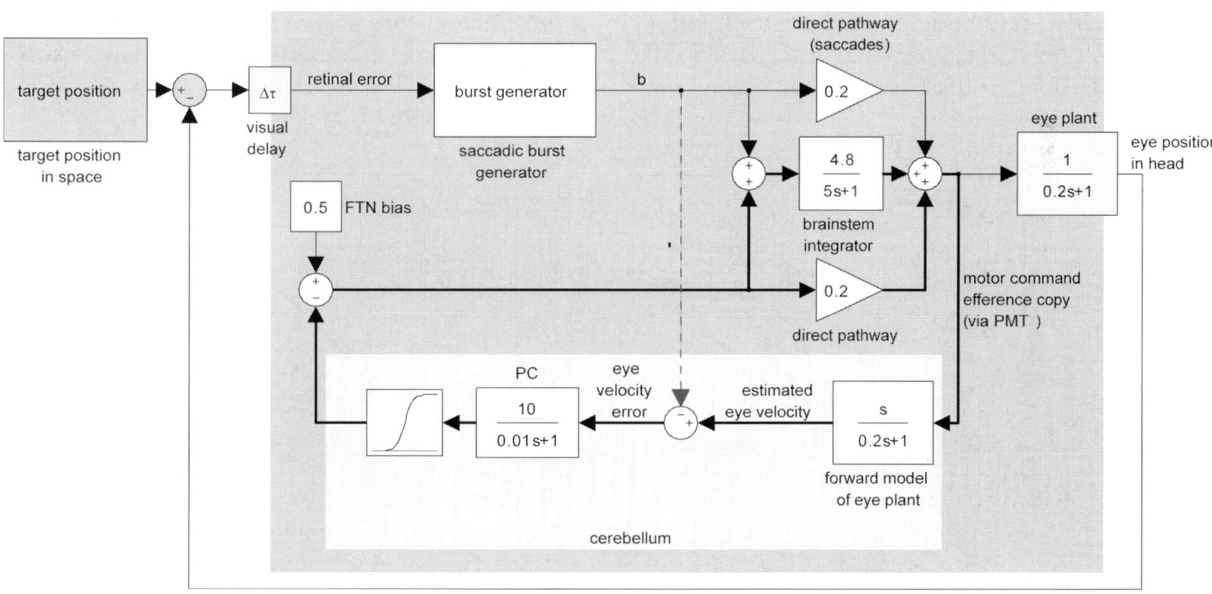

Fig. 1. Outline of the simplified model of the ocular motor system used to simulate DBN and the effect of 4-AP (for details, see Methods and Appendix). The model contains elements necessary for saccades and gaze holding: visual input on the left, a simple saccadic burst generator, direct pathways to the ocular motor nuclei, a leaky brainstem integrator, and the cerebellar feedback loop (thick black) implementing enhancement of integration. The FL is indicated as light grey box. The PC population was modelled using sigmoid nonlinear activation functions. The dashed line carrying the saccadic burst to the FL provides the desired eye velocity signal during saccades and thus silences gaze-velocity PCs.

(Büttner-Ennever and Horn, 1996) back to the FL. The resulting negative velocity feedback loop augments the leaky brainstem integrator and maintains successful gaze holding. The simulations were done using Matlab/Simulink.

Results

Figure 2 (left) shows the firing rate response of the multicompartment model of a single PC (Miyasho et al., 2001) to different levels of mean excitation and reduction of channel conductances. We simulated four cases: (1) a normal PC, (2) reduction of CaP conductance to 80%, (3) reduction of ID conductance to 60% (not shown), and (4) the combination of case 2 and 3. Lowering the CaP conductance to 80% (case 2) leads to a reduction of sensitivity. Reducing the ID conductance to 60% (case 3), mimicking 4-AP application, increases the sensitivity (not shown). Reducing conductance of both channels (case 4) leads to a counteracting of both effects, although the frequency response at higher stimulation frequencies is unaffected by 4-AP, since it depends predominantly on CaP channel conductance. The influence of ID channel conductance is mainly confined to the onset sensitivity.

The resulting scaled population responses are shown in Fig. 2 (right). As expected from the single-cell simulations, the population response of PCs with reduced CaP conductance (case 1, now labelled DBN) is shallower than the normal case, but improves towards normal with 4-AP (case 4, now labelled DBN + 4-AP).

We then used the population response as activation function in the systems-level model. For gaze-straight ahead, i.e., keeping vertical eye position at zero, the model has one fixed point that corresponds to vertical nystagmus slow phase velocity. In Fig. 2 (right), the fixed points of all three cases are shown (black dots). The slope of the activation function at the fixed points determines the integrator time constant, while the x-coordinate of the fixed points is proportional to DBN in gaze-straight ahead (see Appendix). The corresponding simulations of gaze holding in different positions using these population activation functions are shown in Fig. 3. Using the PC activation function derived from single neuron

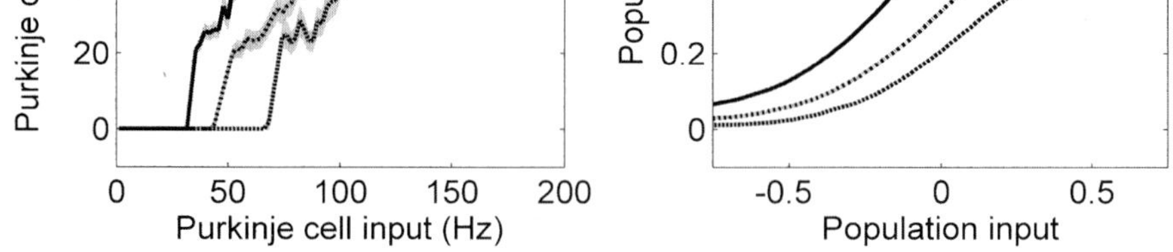

Fig. 2. (Left) Simulation of the input–output relationship of the PC multicompartment model (after Miyasho et al., 2001, see Methods) under synaptic activation with different CaP and ID channel densities. The lines show average firing rate in response to Poisson-distributed spike trains with varying mean frequency. Corresponding 95% confidence intervals are given as grey areas. (Right) PC population response constructed from the single-cell responses shown in the left plot. Intersections of the fine solid line with the activation functions (black dots) give fixed points of the systems model (Fig. 1) for gaze-straight ahead (see Appendix). The x-coordinate of the fixed points is proportional to the nystagmus slow phase velocity in gaze-straight ahead, the corresponding slope of the activation function determines the overall integrator time constant.

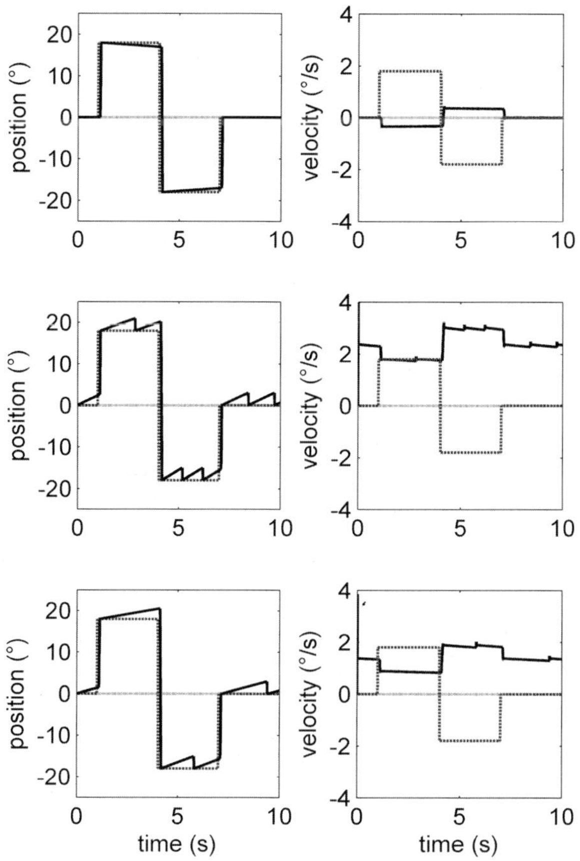

Fig. 3. Simulation of vertical saccades and gaze holding using the activation functions shown in Fig. 2 corresponding to a healthy subject (upper row), a DBN patient (middle row), and the same patient after application of 4-AP. The left column shows target position (grey dashed) and eye position (black solid), and the right column shows slow phase eye velocity (black solid) together with target position (grey dashed, not to scale). Corresponding values: nystagmus slow phase velocity in straight ahead: 0, 2.38, 1.35 °/s; integrator time constant: 55, 30.3, 34.7 s. The gaze-independent nystagmus clearly decreases, while the integrator failure remains almost unchanged.

simulation with reduced CaP conductance shows DBN (2.4 °/s) together with integrator failure (reduction of time constant to 30 s). Subsequent application of 4-AP leads to a decrease of the gaze-independent component of nystagmus (1.4 °/s). However, the concurrent improvement in neural integrator function (from 30 to 35 s) is much smaller than observed experimentally in most patients (Kalla et al., 2007).

Discussion

Our study demonstrates that drug-induced modification of pathological nystagmus can only be understood by including and integrating multiple levels of modelling. Results at the neural level can only be translated to behavioural results, if one takes into account that (1) the effect of the drug has to be simulated for a damaged neuron or neuronal population, (2) the simulated neuron is part of a population, in which each neuron receives different input, and (3) the population is embedded in a network, which, in the present case, is simply a feedback loop. Using this approach, we could partly explain the experimentally observed effect of 4-AP at the behavioural level by the effect of the drug on one single type of neuron, the floccular PC.

The application of low concentrations (1–10 μM) of 4-AP to guinea pig cerebellar PCs markedly shortens the duration of the slowly depolarizing potential, which reduces the latency for Ca^{2+} spikes (Etzion and Grossman, 2001). 4-AP and its structural analogues such as 3,4-DAP are non-selective blockers of the Kv family of the voltage-gated potassium channels. However, it has been shown that 4-AP in low concentration (≤ 1 mM) selectively blocks low-threshold potassium currents (ID) (Storm, 1988). In slice preparations, 4-AP increases the sensitivity of PCs by shortening the slowly depolarizing potential (Etzion and Grossman, 2001). The same effect was found in the multicompartment model after simulating reduction of the ID conductance (not shown), which confirms the model's validity. To model the effect of 4-AP on PCs in DBN patients, we assumed that the CaP conductance was lowered, as may be the case in DBN patients with episodic ataxia type 2 (Strupp et al., 2004). The decrease of CaP conductance used to simulate damaged PCs led to a marked decrease in peak firing and firing onset (Fig. 2, right). In contrast, the effect of 4-AP, a decrease in ID conductance, is confined to firing onset, but has no effect of peak firing rate. The activation functions for the whole PC population (Fig. 2, left) derived from the single-cell simulation mirror these findings. While the decrease in CaP conductance results in a leftward shift and downscaling of the population response, subsequent simulated treatment with 4-AP only leads to a moderate shift back, but does not change the scaling. Consequently, the slope of the activation function at its fixed point, which determines the integrator time constant of the systems model, does not change substantially. The shift back towards normal changed the position of the fixed point, which determines the gaze-independent component of nystagmus slow phase velocity.

One possible solution is that PC damage in DBN also involves changes at the population level, and that 4-AP restores these changes. We simulated these additional effects by assuming a higher variance of baseline inputs leading to a flattening of the population response (Fig. 4), and thus to a larger decrease in integrator function for untreated DBN. This additional simulation better reflects the experimental data (Kalla et al., 2007), but leaves the question how to explain the subsequent improvement after treatment. We hypothesize that 4-AP, in addition to its effect at the single-cell level, may help to recover input specificity of PCs, perhaps via improvement of adaptive mechanisms, or due to improving the intrinsic pacemaking of PC spiking (Walter et al., 2006). Our results thus show that, given the systems level model, the effect of 4-AP on the single neuron level can only partly be reconciled with the data of the patient study (Kalla et al., 2007). Further experiments are needed to investigate whether additional effects, e.g., adaptive changes of parallel fibre synapses caused by increased efficacy of complex spikes, or effects on other parts of the circuit not included in the present model, such as the inferior olive, contribute to the improvement of DBN due to 4-AP observed in patients. Finally, the present neuronal simulation was performed only for a specific case, a decreased conductance of the PC CaP channels. Using our modelling framework, it is now possible to investigate various other deteriorating effects at the single-cell or population level and their effect on system behaviour.

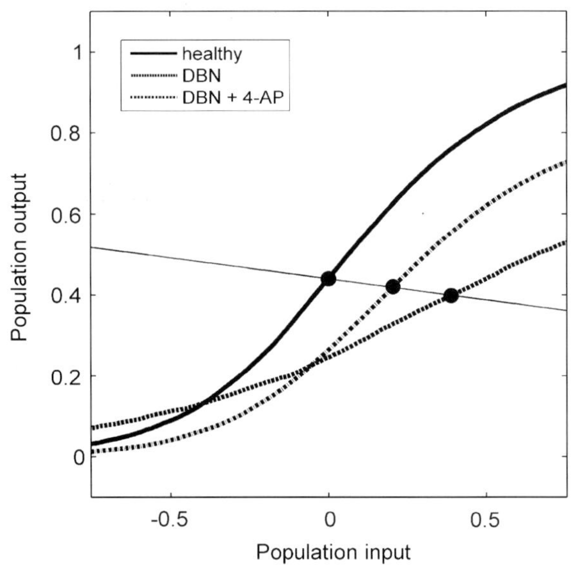

Fig. 4. Example of activation functions of the PC population under the assumption that the neural changes leading to DBN involve not only the CaP reduction, but additional effects at the population level, such as a larger variability of baseline input. If these additional effects are reversed by 4-AP, the simulation approaches experimental values. DBN in straight ahead: healthy: 0 °/s, DBN: 2.22 °/s, DBN + 4-AP: 1.17 °/s; integrator time constant: 55, 22.5, 45.2 s. In this case, the integrator time constant increased substantially due to treatment similarly as observed in patients.

Abbreviations

4-AP	4-aminopyridine
CaP	P-type Ca^{2+} channel
DBN	downbeat nystagmus
FL	floccular lobe
ID	D-type potassium channel
PC	Purkinje cell

Acknowledgements

This work was supported by BCCN Munich, BMBF project number 01GQ0440. CR received a PhD-grant from the Bayerische Forschungsstiftung. We thank U. Büttner for helpful comments, and R.J. Leigh for encouraging us to do this study.

Appendix

Equations for the systems-level model (Fig. 1) are given in Laplace notation (s: complex frequency):

Eye plant output vertical eye position:

$$e_{\mathrm{v}} = m \cdot \frac{1}{1 + \tau_{\mathrm{e}} s} \tag{A.1}$$

Delayed retinal error:

$$r_{\mathrm{e}} = (t - e_{\mathrm{v}}) \cdot e^{-\Delta t \cdot s}$$

Leaky brainstem neural integrator output:

$$e_{\mathrm{i}} = (b - p) \cdot \frac{\tau_{\mathrm{b}} - \tau_{\mathrm{e}}}{1 + \tau_{\mathrm{b}} s} \tag{A.2}$$

Motor command:

$$m = (b - p + c_{\mathrm{ft}}) \cdot \tau_{\mathrm{e}} + e_{\mathrm{i}} \tag{A.3}$$

Internal estimate of eye velocity (Glasauer, 2003):

$$v_{\mathrm{e}} = \frac{s}{1 + \tau_{\mathrm{e}} s} \cdot m \tag{A.4}$$

PC population discharge:

$$p = F\left(g \cdot (v_{\mathrm{e}} - b) \cdot \frac{1}{1 + \tau_{\mathrm{PC}} s}\right) \tag{A.5}$$

with $F(x)$ being the PC population activation function derived from the single-cell model.

Further variables: t: target position; b: saccadic burst. Constants: $\tau_{\mathrm{e}} = 0.2\,\mathrm{s}$: eye plant time constant; $\Delta t = 100\,\mathrm{ms}$: time delay due to visual processing; $\tau_{\mathrm{b}} = 5\,\mathrm{s}$: brainstem integrator time constant; $\tau_{\mathrm{PC}} = 0.01\,\mathrm{s}$: time constant of the PCs; $g = 10$: input gain of the PCs; $c_{\mathrm{ft}} = F(0)$: bias at the level of floccular target neurons compensating for PC resting discharge.

In general, the fixed point of the system for gaze-straight ahead ($e_{\mathrm{v}} = 0$) can be determined as PC input firing rate x_0 being the solution of $F(x) = c_{\mathrm{ft}} - (\tau_{\mathrm{b}}/(g \cdot (\tau_{\mathrm{b}} - \tau_{\mathrm{e}}))) \cdot x$, where x is the PC population input (Fig. 2, right). PC firing at the fixed point $F(x_0)$ is in general not equal to its resting discharge $F(0)$ due to the feedback loop. The eye-position independent slow phase velocity at gaze-straight ahead is x_0/g. For the damaged cases (Fig. 2, right), the fixed point x_0 is shifted away from $x = 0$ towards saturation. Linearizing the PC activation function around the fixed point of the system with gaze-straight ahead allows computing the model's overall transfer function, a leaky integrator. The slope of $F(x_0)$ contributes to the overall gain and thus to time constant of integration with $\tau = \tau_{\mathrm{b}} \cdot (1 + g \cdot (\mathrm{d}F(x))/(\mathrm{d}x)|_{x=x_0})$. For the healthy case, the population response was normalized to have a unity slope at zero input, resulting in a time constant $\tau = \tau_{\mathrm{b}} \cdot (1 + g) = 55$ s.

References

Büttner-Ennever, J.A. and Horn, A.K. (1996) Pathways from cell groups of the paramedian tracts to the floccular region. Ann. N.Y. Acad. Sci., 781: 532–540.

De Schutter, E. and Bower, J.M. (1994) An active membrane model of the cerebellar Purkinje cell II. Simulation of synaptic responses. J. Neurophysiol., 71: 401–419.

Etzion, Y. and Grossman, Y. (2001) Highly 4-aminopyridine sensitive delayed rectifier current modulates the excitability of guinea pig cerebellar Purkinje cells. Exp. Brain Res., 139: 419–425.

Glasauer, S. (2003) Cerebellar contribution to saccades and gaze holding: a modeling approach. Ann. N.Y. Acad. Sci., 1004: 206–219.

Glasauer, S., Hoshi, M., Kempermann, U., Eggert, T. and Büttner, U. (2003) Three-dimensional eye position and slow phase velocity in humans with downbeat nystagmus. J. Neurophysiol., 89: 338–354.

Kalla, R., Glasauer, S., Büttner, U., Brandt, T. and Strupp, M. (2007) 4-aminopyridine restores vertical and horizontal neural integrator function in downbeat nystagmus. Brain, 130: 2441–2451.

Krauzlis, R.J. and Lisberger, S.G. (1996) Directional organization of eye movement and visual signals in the floccular lobe of the monkey cerebellum. Exp. Brain Res., 109: 289–302.

Marti, S., Straumann, D. and Glasauer, S. (2005) The origin of downbeat nystagmus: an asymmetry in the distribution of on-directions of vertical gaze-velocity Purkinje cells. Ann. N.Y. Acad. Sci., 1039: 548–553.

Miyasho, T., Takagi, H., Suzuki, H., Watanabe, S., Inoue, M., Kudo, Y. and Miyakawa, H. (2001) Low-threshold potassium channels and a low-threshold calcium channel regulate Ca^{2+} spike firing in the dendrites of cerebellar Purkinje neurons: a modeling study. Brain Res., 891: 106–115.

Storm, J.F. (1988) Temporal integration by a slowly inactivating K^{+} current in hippocampal neurons. Nature, 336: 379–381.

Straumann, D., Zee, D.S. and Solomon, D. (2000) Three-dimensional kinematics of ocular drift in humans with cerebellar atrophy. J. Neurophysiol., 83: 1125–1140.

Strupp, M., Kalla, R., Dichgans, M., Freilinger, T., Glasauer, S. and Brandt, T. (2004) Treatment of episodic ataxia type 2 with the potassium channel blocker 4-aminopyridine. Neurology, 62: 1623–1625.

Strupp, M., Schüler, O., Krafczyk, S., Jahn, K., Schautzer, F., Büttner, U. and Brandt, T. (2003) Treatment of downbeat nystagmus with 3,4-diaminopyridine: a placebo-controlled study. Neurology, 61: 165–170.

Walter, J.T., Alviña, K., Womack, M.D., Chevez, C. and Khodakhah, K. (2006) Decreases in the precision of Purkinje cell pacemaking cause cerebellar dysfunction and ataxia. Nat. Neurosci., 9: 389–397.

Weisz, C.J.C., Raike, R.S., Soria-Jasso, L.E. and Hess, E.J. (2005) Potassium channel blockers inhibit the triggers of attacks in the calcium channel mouse mutant tottering. J. Neurosci., 25: 4141–4145.

Zee, D.S., Yamazaki, A., Butler, P.H. and Gücer, G. (1981) Effects of ablation of flocculus and paraflocculus of eye movements in primate. J. Neurophysiol., 46: 878–899.

C. Kennard & R.J. Leigh (Eds.)
Progress in Brain Research, Vol. 171
ISSN 0079-6123

CHAPTER 6.7

Aminopyridines for the treatment of cerebellar and ocular motor disorders

Michael Strupp*, Roger Kalla, Stefan Glasauer, Judith Wagner, Katharina Hüfner, Klaus Jahn and Thomas Brandt

Department of Neurology, University of Munich, Klinikum Grosshadern, Munich, Germany

Abstract: Downbeat nystagmus (DBN) is the most frequent form of acquired persisting fixation nystagmus. It is hypothesized to occur when physiological inhibitory cerebellar input, namely of the flocculus, to the vestibular nuclei is inhibited. The second most frequent form of acquired nystagmus is upbeat nystagmus (UBN). UBN is probably caused by an imbalance of vertical vestibulo-ocular reflex tone. GABA-ergic substances like baclofen have been used to treat DBN and UBN, but they have had only moderate success. Animal experiments have shown that aminopyridines [3,4-diaminopyridine (3,4-DAP) and 4-aminopyridine (4-AP)], nonselective blockers of the Kv family of voltage-gated potassium channels, increase Purkinje-cell (PC) excitability. It was assumed that such enhancement of PC activity could restore to normal levels the inhibitory influence of the cerebellar cortex on vertical eye movements. On the basis of these assumptions, we evaluated the efficacy and underlying mechanisms of aminopyridines in DBN and UBN as well as in another cerebellar disorder with an impaired PC function: episodic ataxia type 2 (EA2), which is caused by mutations of the PQ-calcium channel. In a placebo-controlled trial on 17 patients we demonstrated that 3,4-DAP significantly reduces the intensity of DBN. This was confirmed in a recent study with 4-AP, which also showed that 4-AP restores gaze-holding ability independently of fixation in DBN. The efficacy of 4-AP in UBN was demonstrated in single patients. Finally, in an open trial on three patients with EA2 we showed that 4-AP prevents attacks of ataxia. This was also found in an animal model (the tottering mouse) of EA2. The clinical efficacy of 4-AP in EA2 is being further evaluated in an ongoing randomized controlled crossover trial. In conclusion, the use of aminopyridines in DBN, UBN, and EA2 is a new treatment principle for vestibular, cerebellar, and ocular motor disorders.

Keywords: downbeat nystagmus; upbeat nystagmus; episodic ataxia; vertigo; dizziness; aminopyridines; potassium channel blockers

Introduction

On the basis of recent clinical trials, we are now able to medically treat more vestibular, ocular motor, and cerebellar syndromes. This article will focus on the use of aminopyridines in downbeat and upbeat nystagmus as well as episodic ataxia type 2 (EA2). A new treatment principle will also be presented: an increase in the activity and excitability of the cerebellar Purkinje cells (PCs) is achieved by blocking potassium channels with 3,4-diaminopyridine (3,4-DAP) and 4-aminopyridine (4-AP).

*Corresponding author. Tel.: +49 89 7095 3678; Fax: +49 89 7095 6673; E-mail: Michael.Strupp@med.uni-muenchen.de

DOI: 10.1016/S0079-6123(08)00676-6

Downbeat nystagmus

Downbeat nystagmus (DBN) is the most frequent form of acquired persisting fixation nystagmus. It is most often caused by cerebellar degeneration, ischaemia, and cranio-cervical malformations (Wagner et al., 2008). DBN is hypothesized to occur when the physiological inhibitory cerebellar input, namely of the flocculus, to the vestibular nuclei is reduced (Zee et al., 1974; Marti et al., 2005; Leigh and Zee, 2006); this hypothesis is supported by fMRI and voxel-based morphometric studies (Kalla et al., 2006; Hufner et al., 2007) (Figs. 1 and 2). Marti et al. (2005) suggest there is an inherent asymmetry in the distribution of on-directions of vertical gaze velocity PCs, which leads to a spontaneous upward ocular drift and thereby to DBN, when the PC input to the vestibular nuclei is reduced.

Since the inhibitory influence of GABAergic PCs is assumed to be impaired in DBN, several agents that act on this receptor have been investigated. The $GABA_A$ agonist, clonazepam, improved DBN (dosages 0.5 mg tid to 1 mg bid), but these studies were not controlled (Currie et al., 1986; Young and Huang, 2001). The $GABA_B$ agonist, baclofen, is assumed to reduce DBN (Dieterich et al., 1991), but as was shown in a double-blind crossover trial in a few patients, only one out of six responded to baclofen (Averbuch-Heller et al., 1997). Further, the alpha-2-delta calcium channel antagonist gabapentin was assumed to have a positive effect on DBN, but again only one out of six patients responded positively (Averbuch-Heller et al., 1997).

On the basis of the assumed pathomechanism of DBN, the effects of aminopyridines were evaluated in a randomized, controlled, crossover trial involving 17 patients with DBN due to cerebellar atrophy, infarction, Arnold–Chiari malformation, or unknown aetiology (Strupp et al., 2003). Mean peak slow-phase velocity of DBN was measured before and 30 min after randomized ingestion of 20 mg of 3,4-DAP or oral placebo. 3,4-DAP reduced peak slow-phase velocity of DBN from 7.2°/s mean before treatment to 3.1°/s 30 min after ingestion ($p < 0.001$) (Fig. 3). The mean peak slow-phase velocity decreased in 10 of 17 patients by more than 50%. Except for transient perioral or digital paraesthesia (three patients) and nausea and headache (one patient), no other side effects were observed. The authors demonstrated that the single dose of 3,4-DAP significantly improved DBN and visual acuity, and also reduced distressing oscillopsia. From a clinical point of view, it must be kept in mind that only 50% of all patients with DBN respond to this treatment, mainly those without structural lesions of the cerebellum or brainstem. The assumed underlying mechanism is that aminopyridines increase the activity and excitability of the PCs [as was found in animal experiments (Etzion and Grossman, 2001)], thereby augmenting the physiological inhibitory influence of the vestibular cerebellum on the vestibular nuclei. Meanwhile the effect of aminopyridines on the gravity dependence of DBN has also been evaluated (Helmchen et al., 2004), and an improvement of postural imbalance in DBN was demonstrated (Sprenger et al., 2005).

The underlying mechanism of action of 4-AP in DBN was also investigated in two studies with the magnetic search-coil technique (Kalla et al., 2004, 2007). The major findings of these studies were as follows: first, 4-AP improves not only DBN, but also smooth pursuit and the gain of the vertical vestibulo-ocular reflex (Kalla et al., 2004). Second, 4-AP improved fixation by restoring gaze-holding ability and neural integrator function (Fig. 4); further, as regards its aetiology-dependent efficacy in DBN, 4-AP may work best when DBN is associated with cerebellar atrophy (Kalla et al., 2007) (Fig. 4). If DBN is caused by a structural lesion, 4-AP does not improve DBN in most cases.

A PET study showed that 4-AP — in parallel to an improvement of DBN — increases the metabolic activity of the flocculus (Bense et al., 2006). All these studies give additional support both to the above hypothesis about the pathophysiology of DBN and the way that aminopyridines act. On the basis of these findings three trials are now in process. In the first the efficacy of 3,4-DAP is being compared with that of 4-AP, in the second the effects on 4-AP on the patient's quality of life, and in the third the effects of 4-AP on ataxia, postural instability, and quality of life in patients with SCA6 are being evaluated.

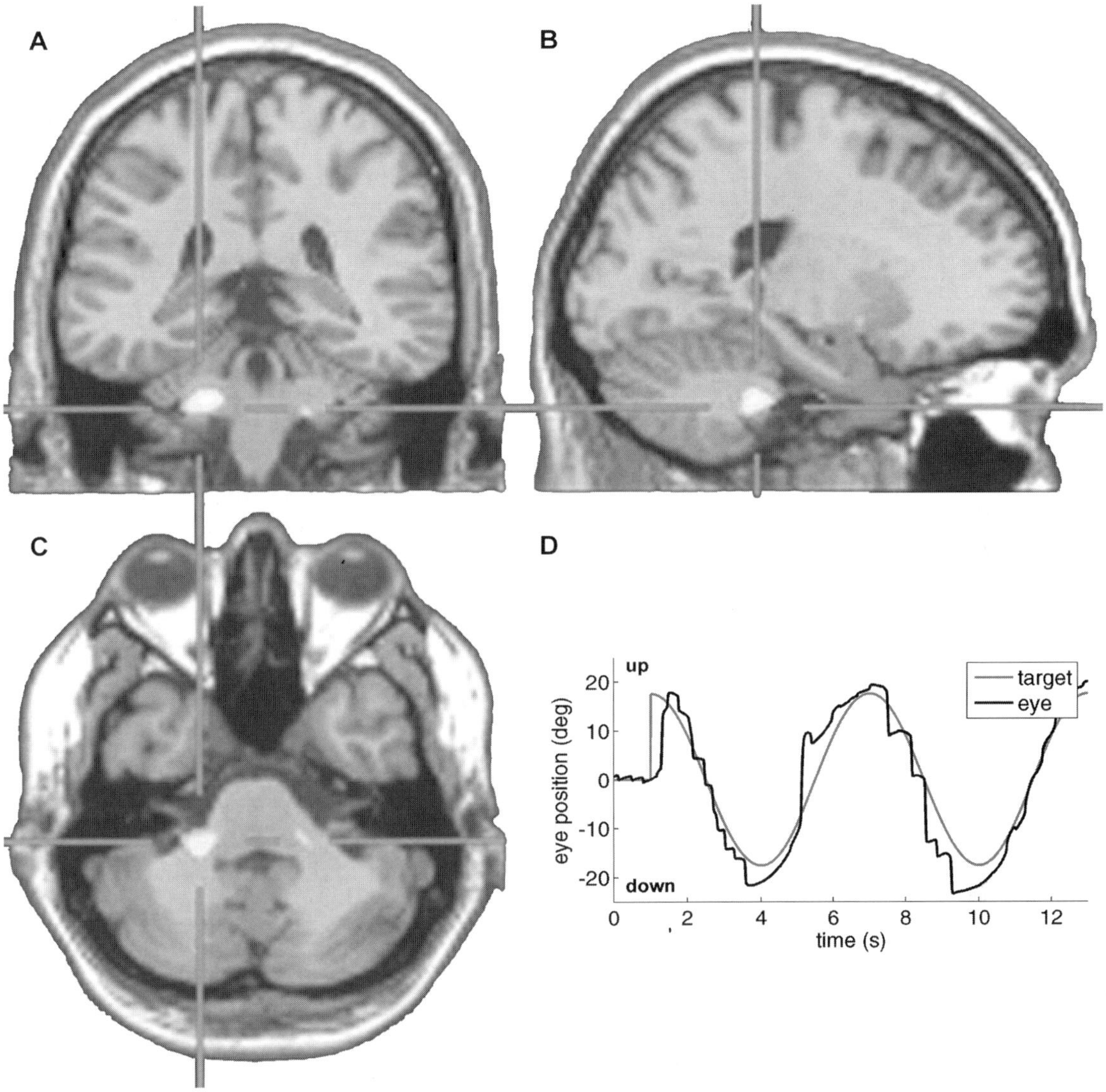

Fig. 1. Activation of the flocculus (red) in controls versus patients for the contrast "smooth pursuit in the downward direction" (SMDOWN) — "fixation of a target in the middle of the display" (FIXMID). Results obtained by region of interest group analysis are superimposed onto orthogonal sections (A: coronal plane, B: sagittal plane, and C: axial plane) at Montreal Neurological Institute coordinates x, y, $z = -20$, -36, -40 through a standard brain template ($p<0.01$). (D) Original recording (search-coil) of vertical pursuit (0.1667 Hz, amplitude $\pm 18°$), which demonstrates normal upward pursuit and impaired downward pursuit in a patient with DBN [adapted with permission from Kalla et al. (2006)].

Upbeat nystagmus

Spontaneous upbeat nystagmus (UBN), i.e., UBN with gaze straight ahead, is an ocular motor disorder that manifests with oscillopsia due to retinal slip of the visual scene and postural instability. UBN usually increases with upgaze. Analogously to DBN, it is associated with impaired upward pursuit. UBN can be caused by lesions in different brainstem and cerebellar regions such as the pontomesencephalic junction, medulla, or cerebellar vermis. Lesions in the

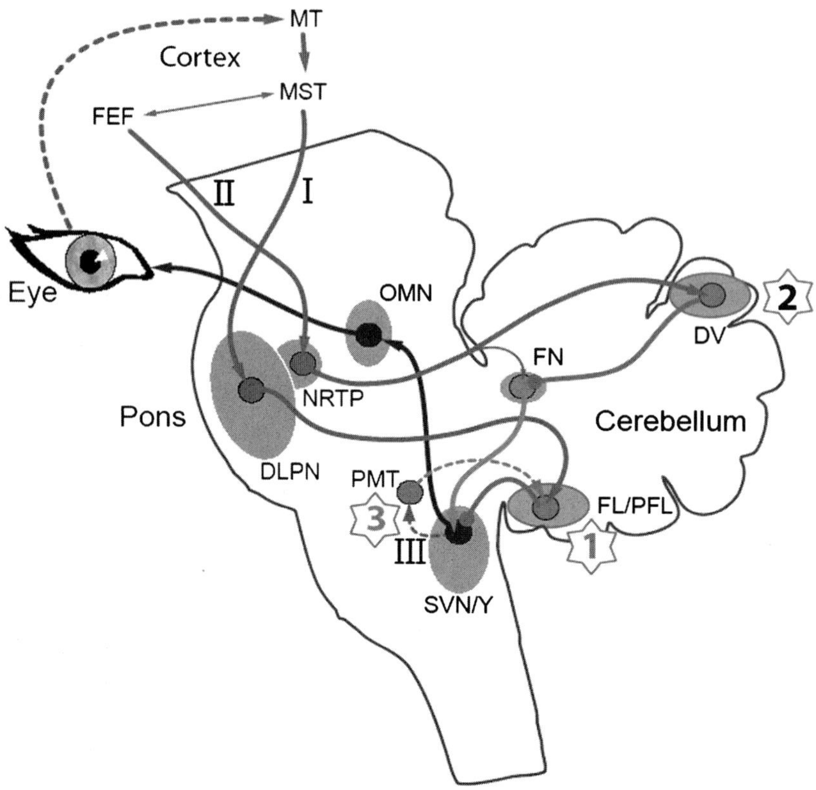

Fig. 2. Schematic representation of a putative model of the pathomechanism of DBN. We propose that all patients with DBN share a final common pathway (disinhibition of the SVN and neurons of the Y group). The ocular motor circuitries involved are the two smooth pursuit eye movement pathways (I, II) and the vertical gaze-holding pathway (III). The different lesion sites that can lead to DBN are shown in red (1–3). See the Discussion for details (FEF = frontal eye field, DLPN = dorsolateral pontine nuclei, DV = dorsal vermis, ocular motor vermis, FN = fastigial nucleus, FL = flocculus, PFL = paraflocculus, MT = middle temporal area, MST = medial superior temporal area, NRTP = nucleus reticularis tegmenti pontis, OMN = ocular motor nuclei, PMT = nucleus of the paramedian tract, SVN = superior vestibular nucleus, Y = neurons of the Y-group) [adapted with permission from Hufner et al. (2007)].

pathways mediating upward eye movements, in particular, from the vestibular nuclei through the brachium conjunctivum to the ocular motor nuclei, might result in slow downward drift of the eyes which is corrected by fast upward movements (Leigh and Zee, 2006). Other hypotheses are that UBN is caused by an imbalance of vertical vestibulo-ocular reflex tone or a mismatch in the neural coordinate systems of saccade generation and neural velocity-to-position integration (see also Pierrot-Deseilligny and Tilikete, this volume).

GABA-ergic substances like baclofen have been used to treat UBN and DBN, but they have had only moderate success. One study demonstrated a beneficial effect of baclofen (5–10 mg tid), but this trial was not controlled (Dieterich et al., 1991). In a single patient with UBN it was shown that 4-AP reduces the peak slow-phase velocity in the light from 8.6°/s to 2.0°/s (Glasauer et al., 2005a). 4-AP did not affect UBN in darkness, but it obviously activated pathways carrying visual information, which could then be used for UBN suppression in the light. Therefore, it was concluded that 4-AP reduces the downward drift in UBN by augmenting smooth pursuit commands. We propose that 4-AP helps to activate parallel pathways that can assume the function of the lesioned structures (Glasauer et al., 2005b). 4-AP may strengthen these parallel pathways by increasing the excitability of cerebellar PCs (Etzion and Grossman,

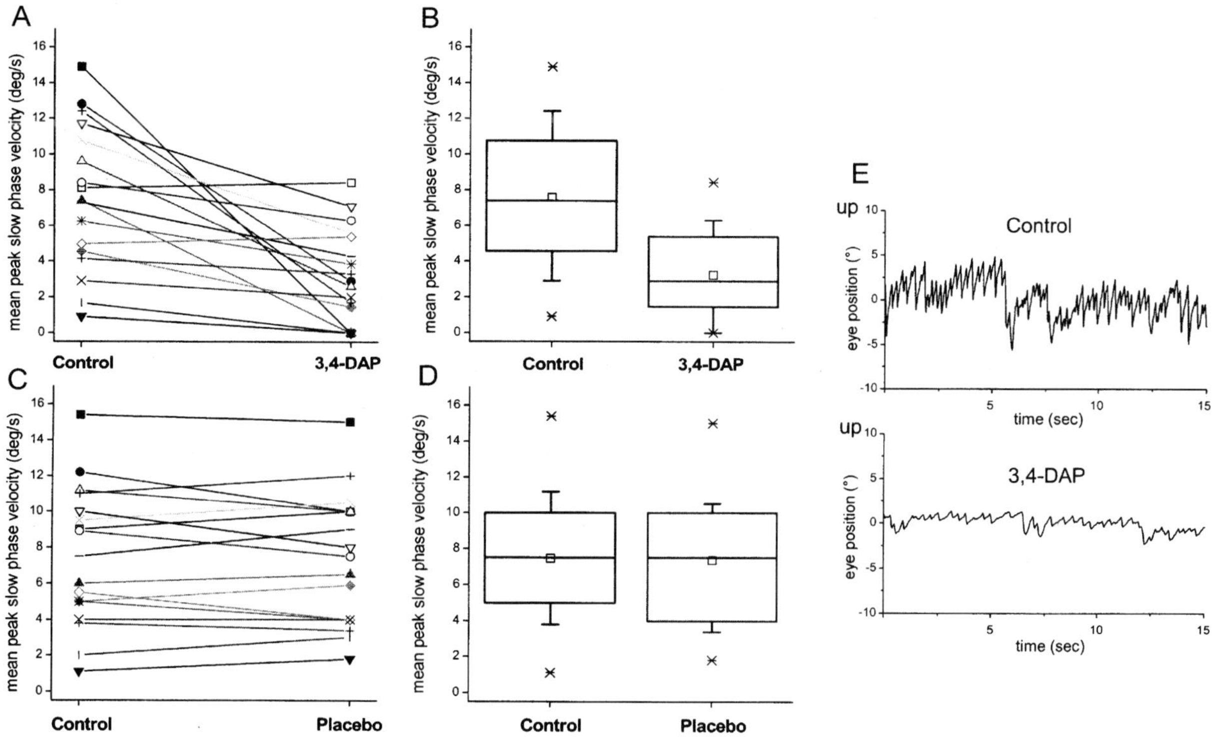

Fig. 3. Mean peak slow phase velocities (PSPV) of DBN measured by 2-D recordings of eye movements. The two graphs on the left show the original data of mean PSVP for each subject: (A) control versus 3,4-diaminopyridine (3,4-DAP), (C) control versus placebo. The two graphs in the middle give the box plot charts with the mean, median, and the 50% percentile as well as the range for control versus 3,4-DAP (B), and control versus placebo (D). 3,4-DAP reduced mean PSPV of DBN from 7.2±4.2°/s (mean±SD) before treatment to 3.1±2.5°/s 30 min after ingestion of the 3,4-DAP ($n = 17$, $p < 0.001$, two-way ANOVA). E shows an original recording of the vertical eye position before (upper trace) and 30 min after ingestion of the drug (lower trace) [adapted with permission from Strupp et al. (2003)].

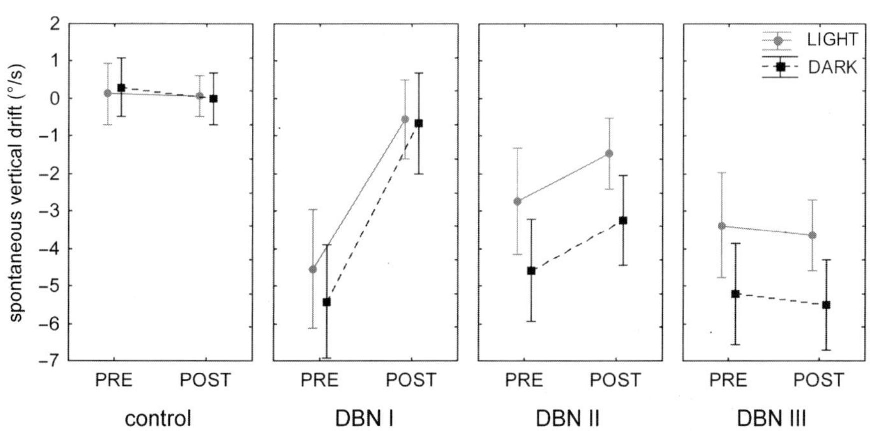

Fig. 4. Spontaneous vertical drift: Vertical drift in control subjects and DBN patients due to cerebellar atrophy (DBN I), unknown aetiology (DBN II), or other aetiologies (DBN III) before (PRE) and after (POST) administration of 4-AP (grey circles: target visible; black squares: target blanked). The pronounced DBN is mainly reduced in DBN I and to a lesser degree in DBN II postmedication. Similar effects are observed while the target is blanked. Error bars indicate 95% confidence intervals [adapted with permission from Kalla et al. (2007)].

2001). It may also evoke complex spikes in PCs similar to those elicited by climbing fibre stimulation (Cavelier et al., 2002).

Episodic ataxia type 2

EA2 is clinically characterized by recurrent attacks of ataxia, provoked by stress or exercise, which last for several hours to days (Griggs and Nutt, 1995; Jen et al., 2004, 2007; Strupp et al., 2007). Associated findings during the non-attack interval include central ocular motor and vestibular dysfunction, mainly DBN. Patients with EA2 can often be successfully treated with acetazolamide (Griggs et al., 1978). Genetically EA2 is an autosomal dominant hereditary disorder caused by mutations of the calcium channel gene CACNA1A (Ophoff et al., 1996), which encodes the Ca_{V21} subunit of the PQ-calcium channel expressed mainly in the PCs. On the basis of the functional changes of the PQ-channel mutation, which leads to a reduced calcium current, it can be assumed that the inhibitory effect of PCs is reduced in EA2 (Kullmann, 2002). This causes the disinhibition of the deep cerebellar nuclei and thus ataxia and DBN. Since aminopyridines were shown to improve DBN most likely by increasing the inhibitory influence of the PCs — a hypothesis supported by animal experiments (Etzion and Grossman, 2001), we evaluated its effects on the occurrence of attacks of EA2 (Strupp et al., 2004). In this pilot study of initially three patients with EA2 (two of whom had proven mutations of the CACNA1A gene; in the meantime six additional patients have joined, five with proven mutations) attacks could be prevented with the potassium channel blocker 4-AP (5 mg tid). The attacks recurred after treatment was stopped. Subsequent treatment alleviated the symptoms (mean follow-up time greater than 12 months). These effects might be due to an improvement of the impaired functioning of the PCs. It must be pointed out that these three patients no longer responded to the standard treatment with acetazolamide. Again on the basis of this open trial a placebo-controlled randomized crossover study is currently in progress; the recruitment of ten patients (seven with genetically proven EA2) has been completed.

These clinical findings are supported by an animal study on the calcium channel mutant tottering mouse. Aminopyridines blocked the attacks characteristic of the tottering mouse via the cerebellar potassium channels by increasing the threshold for attack initiation without mitigating the character of the attack (Weisz et al., 2005).

Other agents also acting on potassium channels were studied in animal models of EA2. It was shown that the precision of PC pacemaking was lost in ataxic P/Q channel mutant mice, which led to a significant degradation of the synaptic information encoded in their activity (Walter et al., 2006). Irregular pacemaking is caused by reduced activation of calcium-activated potassium (K_{Ca}) channels; this was reversed by pharmacologically increasing their activity with 1-ethyl-2-benzimidazolinone (EBIO). Further, in vivo perfusion of EBIO into the cerebellum of ataxic mice significantly improved motor performance. These data support the hypothesis that the precision of intrinsic pacemaking in PCs is essential for motor coordination and suggest that K(Ca) channels may constitute a potential therapeutic target in EA2 (Walter et al., 2006).

These studies show that a new therapeutic principle has been developed: aminopyridines as potassium channel blockers increase the activity and excitability of PCs and may have a beneficial effect on several cerebellar disorders.

Acknowledgement

We thank Judy Benson for copy-editing the manuscript.

References

Averbuch-Heller, L., Tusa, R.J., Fuhry, L., Rottach, K.G., Ganser, G.L., Heide, W., Buttner, U. and Leigh, R.J. (1997) A double-blind controlled study of gabapentin and baclofen as treatment for acquired nystagmus. Ann. Neurol., 41: 818–825.

Bense, S., Best, C., Buchholz, H.G., Wiener, V., Schreckenberger, M., Bartenstein, P. and Dieterich, M. (2006) 18F-fluorodeoxyglucose hypometabolism in cerebellar tonsil and flocculus in downbeat nystagmus. Neuroreport, 17: 599–603.

Cavelier, P., Pouille, F., Desplantez, T., Beekenkamp, H. and Bossu, J.L. (2002) Control of the propagation of dendritic low-threshold Ca(2+) spikes in Purkinje cells from rat cerebellar slice cultures. J. Physiol., 540: 57–72.

Currie, J.N., Matsuo, V., McConnell, H.W., Darlington, C.L., Smith, P.F., Sturge, D.L., Thomson, S.D., Nukada, H., Mair, M.W., Chrousos, G.A., Cowdry, R., Schuelein, M., bdul-Rahim, A.S., Matsuo, V. and Currie, J.N. (1986) The use of clonazepam in the treatment of nystagmus-induced oscillopsia. Ophthalmology, 93: 924–932.

Dieterich, M., Straube, A., Brandt, T., Paulus, W. and Büttner, U. (1991) The effects of baclofen and cholinergic drugs on upbeat and downbeat nystagmus. J. Neurol. Neurosurg. Psychiatry, 54: 627–632.

Etzion, Y. and Grossman, Y. (2001) Highly 4-aminopyridine sensitive delayed rectifier current modulates the excitability of guinea pig cerebellar Purkinje cells. Exp. Brain Res., 139: 419–425.

Glasauer, S., Kalla, R., Buttner, U., Strupp, M. and Brandt, T. (2005a) 4-Aminopyridine restores visual ocular motor function in upbeat nystagmus. J. Neurol. Neurosurg. Psychiatry, 76: 451–453.

Glasauer, S., Strupp, M., Kalla, R., Buttner, U. and Brandt, T. (2005b) Effect of 4-aminopyridine on upbeat and downbeat nystagmus elucidates the mechanism of downbeat nystagmus. Ann. N.Y. Acad. Sci., 1039: 528–531.

Griggs, R.C., Moxley, R.T., Lafrance, R.A. and McQuillen, J. (1978) Hereditary paroxysmal ataxia: response to acetazolamide. Neurology, 28: 1259–1264.

Griggs, R.C. and Nutt, J.G. (1995) Episodic ataxias as channelopathies. Ann. Neurol., 37: 285–287.

Helmchen, C., Sprenger, A., Rambold, H., Sander, T., Kompf, D. and Straumann, D. (2004) Effect of 3,4-diaminopyridine on the gravity dependence of ocular drift in downbeat nystagmus. Neurology, 63: 752–753.

Hufner, K., Stephan, T., Kalla, R., Deutschlander, A., Wagner, J., Holtmannspotter, M., Schulte-Altedorneburg, G., Strupp, M., Brandt, T. and Glasauer, S. (2007) Structural and functional MRIs disclose cerebellar pathologies in idiopathic downbeat nystagmus. Neurology, 69: 1128–1135.

Jen, J., Kim, G.W. and Baloh, R.W. (2004) Clinical spectrum of episodic ataxia type 2. Neurology, 62: 17–22.

Jen, J.C., Graves, T.D., Hess, E.J., Hanna, M.G., Griggs, R.C. and Baloh, R.W. (2007) Primary episodic ataxias: diagnosis, pathogenesis and treatment. Brain, 130: 2484–2493.

Kalla, R., Deutschlander, A., Hufner, K., Stephan, T., Jahn, K., Glasauer, S., Brandt, T. and Strupp, M. (2006) Detection of floccular hypometabolism in downbeat nystagmus by fMRI. Neurology, 66: 281–283.

Kalla, R., Glasauer, S., Buttner, U., Brandt, T. and Strupp, M. (2007) 4-Aminopyridine restores vertical and horizontal neural integrator function in downbeat nystagmus. Brain, 130: 2441–2451.

Kalla, R., Glasauer, S., Schautzer, F., Lehnen, N., Buttner, U., Strupp, M. and Brandt, T. (2004) 4-Aminopyridine improves downbeat nystagmus, smooth pursuit, and VOR gain. Neurology, 62: 1228–1229.

Kullmann, D.M. (2002) The neuronal channelopathies. Brain, 125: 1177–1195.

Leigh, R.J. and Zee, D. (2006) The Neurology of Eye Movements. Oxford University Press, Oxford.

Marti, S., Straumann, D. and Glasauer, S. (2005) The origin of downbeat nystagmus: an asymmetry in the distribution of on-directions of vertical gaze-velocity Purkinje cells. Ann. N.Y. Acad. Sci., 1039: 548–553.

Ophoff, R.A., Terwindt, G.M., Vergouwe, M.N., van-Eijk, R., Oefner, P.J., Hoffman, S.M., Lamerdin, J.E., Mohrenweiser, H.W., Bulman, D.E., Ferrari, M., Haan, J., Lindhout, D., van-Ommen, G.J., Hofker, M.H., Ferrari, M.D. and Frants, R.R. (1996) Familial hemiplegic migraine and episodic ataxia type-2 are caused by mutations in the Ca^{2+} channel gene CACNL1A4. Cell, 87: 543–552.

Sprenger, A., Zils, E., Rambold, H., Sander, T. and Helmchen, C. (2005) Effect of 3,4-diaminopyridine on the postural control in patients with downbeat nystagmus. Ann. N.Y. Acad. Sci., 1039: 395–403.

Strupp, M., Kalla, R., Dichgans, M., Freilinger, T., Glasauer, S. and Brandt, T. (2004) Treatment of episodic ataxia type 2 with the potassium channel blocker 4-aminopyridine. Neurology, 62: 1623–1625.

Strupp, M., Schuler, O., Krafczyk, S., Jahn, K., Schautzer, F., Buttner, U. and Brandt, T. (2003) Treatment of downbeat nystagmus with 3,4-diaminopyridine: a placebo-controlled study. Neurology, 61: 165–170.

Strupp, M., Zwergal, A. and Brandt, T. (2007) Episodic ataxia type 2. Neurotherapeutics, 4: 267–273.

Wagner, J.N., Glaser, M., Brandt, T. and Strupp, M. (2008) Downbeat nystagmus: aetiology and comorbidity in 117 patients. J. Neurol. Neurosurg. Psychiatry. (in press).

Walter, J.T., Alvina, K., Womack, M.D., Chevez, C. and Khodakhah, K. (2006) Decreases in the precision of Purkinje cell pacemaking cause cerebellar dysfunction and ataxia. Nat. Neurosci., 9: 389–397.

Weisz, C.J., Raike, R.S., Soria-Jasso, L.E. and Hess, E.J. (2005) Potassium channel blockers inhibit the triggers of attacks in the calcium channel mouse mutant tottering. J. Neurosci., 25: 4141–4145.

Young, Y.H. and Huang, T.W. (2001) Role of clonazepam in the treatment of idiopathic downbeat nystagmus. Laryngoscope, 111: 1490–1493.

Zee, D.S., Friendlich, A.R. and Robinson, D.A. (1974) The mechanism of downbeat nystagmus. Arch. Neurol., 30: 227–237.

C. Kennard & R.J. Leigh (Eds.)
Progress in Brain Research, Vol. 171
ISSN 0079-6123

CHAPTER 6.8

Baclofen, motion sickness susceptibility and the neural basis for velocity storage

Bernard Cohen[1,*], Mingjia Dai[1], Sergei B. Yakushin[1] and Theodore Raphan[2]

[1]*Department of Neurology, Mount Sinai School of Medicine, New York, NY, USA*
[2]*Computer and Information Science, Brooklyn College, City University of New York, NY, USA*

Abstract: Reduction of the dominant time constant (T_{VOR}) of the angular vestibulo-ocular reflex (aVOR) by habituation is associated with a decrease in motion sickness susceptibility. Baclofen, a $GABA_b$ agonist, reduces the time constant of the velocity storage integrator in the aVOR in a dose-dependent manner. The high frequency aVOR gain is unaltered by baclofen. Here we demonstrate that the reduction in T_{VOR} produced by oral administration of 20 mg of baclofen causes a significant reduction in motion sickness susceptibility, tested with roll while rotating (RWR). These data show that motion sickness susceptibility can be pharmacologically manipulated with a $GABA_b$ agonist and support our conclusion that motion sickness is generated through velocity storage. We also show how baclofen acts on velocity storage at the neural level. A vestibular-plus-saccade (VPS) neuron was recorded in the rostral medial vestibular nucleus (rMVN) of a cynomolgus monkey, an area where we postulate that velocity storage is generated. The cell had a time constant during steps of velocity that was close to that of the T_{VOR}. After parenteral administration of baclofen, there was a similar decrease in the time constants of the VPS neuron and the T_{VOR}. This is the first demonstration of the concurrence of unit and aVOR time constants before and after baclofen. The data support the hypothesis that the velocity storage integrator is generated through activity of vestibular-only (VO) and VPS neurons in rMVN and suggest that $GABA_b$ synapses on VO and VPS neurons are likely to be involved in the baclofen-induced reduction in motion sickness susceptibility.

Keywords: baclofen; vestibulo-ocular reflex (VOR); $GABA_b$; vestibular-plus-saccade (VPS) neuron; roll while rotating (RWR); nodulus/uvula

Introduction

Baclofen, 4-amino-3,4-chlorophenyl butanoic acid (Fig. 1A), is a $GABA_b$ agonist that was initially introduced to reduce muscular spasm. However, baclofen also actively modifies the behaviour of the oculomotor and vestibular systems; it abolishes Periodic Alternating Nystagmus (Halmagyi et al., 1980) by reducing the time constant of the velocity storage integrator in the vestibular nuclei (Leigh et al., 1981; Cohen et al., 1987). In particular, there is a dose-dependent reduction in the time constant of optokinetic after nystagmus (T_{OKAN}) and the angular vestibulo-ocular reflex (T_{VOR}) in monkeys after intramuscular injection of

*Corresponding author. Tel.: +212 241 7068;
Fax: +212 831 1610; E-mail: Bernard.Cohen@mssm.edu

DOI: 10.1016/S0079-6123(08)00677-8

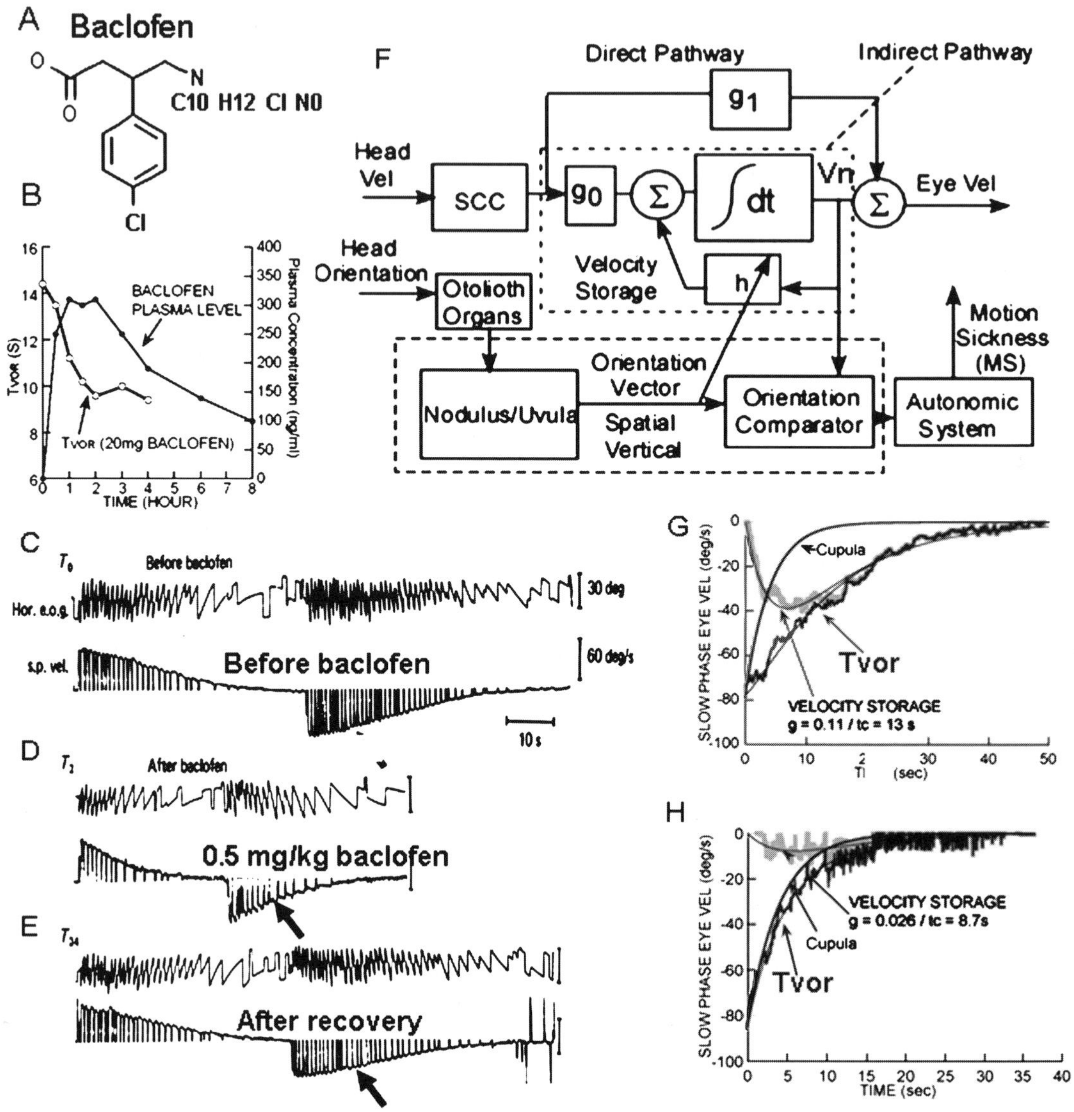

Fig. 1. Baclofen and its effects on the aVOR. (A) Abbreviated chemical structure of baclofen. (B) Reciprocally related plasma levels of baclofen (solid circles) with reduction in human T_{VOR} (open circles) after single doses of baclofen (Adapted with permission from Faigle et al., 1980 and Dai et al., 2006). (C–E) Per- and post-rotatory nystagmus (top traces) and slow phase velocity of a rhesus monkey (bottom trace) induced by a step of velocity of 60°/s in darkness before (C), 2 h after, (D), and 24 h after (E) an injection of 0.6 mg/kg of baclofen (Adapted with permission from Cohen et al., 1987). (F) Model of the aVOR and of control of motion sickness susceptibility. See text for details. (G, H) Model-based per- and post-rotatory responses of human aVOR before (G) and after (H) ingestion of 20 mg of baclofen. The cupula/endolymph and direct pathway response (cupula) were unaffected, but the velocity storage component in G (arrow), was reduced to produce a shortened T_{VOR} (Adapted with permission from Dai et al., 2006).

baclofen (Cohen et al., 1987). The initial jump in eye velocity at the onset and end of rotation was unaffected (Fig. 1C–E). Baclofen similarly affects the human T_{VOR} (Dai et al., 2006).

We have proposed that motion sickness induced by rolling the head while rotating, is produced by a disparity between the yaw axis orientation vector of velocity storage and the vector of the slow phase eye velocity generated through velocity storage (Dai et al., 2003). The orientation vector of velocity storage lies close to the axis of gravity (Dai et al., 1991; Raphan and Sturm, 1991; Raphan and Cohen, 2002), while the slow phase velocity vector can deviate substantially from gravity depending on the stimulus (Dai et al., 2003). If the T_{VOR} is long, then any induced disparity between the eye velocity vector generated through velocity storage and the spatial vertical would persist for a longer time, producing motion sickness more easily. If our formulation is correct, then a reduction in T_{VOR} should produce a reduction in motion sickness susceptibility. This study presents preliminary results that indicate that this hypothesis is correct. We also present findings that support our previous postulate that velocity storage is produced by the action of vestibular-only (VO) and vestibular-plus-saccade (VPS) cells in the rostral medial and superior vestibular nuclei (Reisine and Raphan, 1992; Yokota et al., 1992).

Methods

Limitations of space preclude a full description of the techniques and methods. For a recent review of models of the aVOR, see Dai et al., 1999, Raphan and Cohen, 2002; for eye movement recording in animals and humans, see Yakushin et al., 1995 and Dai et al., 1999, 2003. For effects of baclofen on the time constant of the aVOR in monkeys and humans, see Cohen et al., 1987 and Dai et al., 2006 and for studies of motion sickness, including techniques for producing motion sickness, see Dai et al., 2003. Informed consent was obtained in the human studies, which were approved by the Institutional Review Board of the Mount Sinai School of Medicine. Adequate protection of animals was maintained under the aegis of the Institutional Animal Care and Use Committee. We first present a simplified model of the aVOR to aid in understanding the effect of baclofen ingestion (Fig. 1F). This is followed by a description of the technique of roll while rotating (RWR), which was used to produce motion sickness in this study (Fig. 2). A critical aspect of the RWR paradigm, initially introduced by Graybiel and associates (Miller and Graybiel, 1973), was that the nystagmus induced by head movement was allowed to complete its course before the next head movement began (Dai et al., 2003). This allowed the full participation of velocity storage in producing motion sickness.

Results

Model of velocity storage and determination of its parameters

The aVOR comprises "direct" and "indirect" pathways that process the head velocity signal transduced by the semicircular canals (SCC) (Fig. 1F; Raphan et al., 1979; Raphan and Cohen, 2002). The direct pathway, which is controlled by a gain g_1, has a fast response time that stabilizes the retina in space against rapid angular head perturbations. The dynamics of the direct pathway are dependent on a three-neuron arc (Lorente de Nó, 1932), which has a minimum latency of transmission between the SCC and the eye muscles of about 2.5 ms (Cohen and Suzuki, 1963). Consequently, the eyes can respond to impulses of angular acceleration in 12 ms (Lisberger and Pavelko, 1988).

The indirect pathway is controlled by an integrator whose coupling from the SCC is represented by g_0, and whose dynamics are represented by the feedback, h. The output is realized in velocity storage neurons Vn, which project both to the oculomotor system and the flocculus (Zhang et al., 1993). The feedback loop is controlled by the nodulus and uvula of the vestibulo-cerebellum, which determines the time constant (1/h) and the orientation of velocity storage that tends to align the velocity vector of nystagmus to the spatial vertical (Dai et al., 1991; Raphan and Sturm, 1991; Wearne et al., 1998). The nodulus also controls habituation, which produces a reduction in the time constant of the velocity storage integrator in response to repetitive exposure to rotation (Cohen et al., 1992). A more complete description is presented elsewhere (Raphan et al., 1979; Raphan and Cohen, 2002).

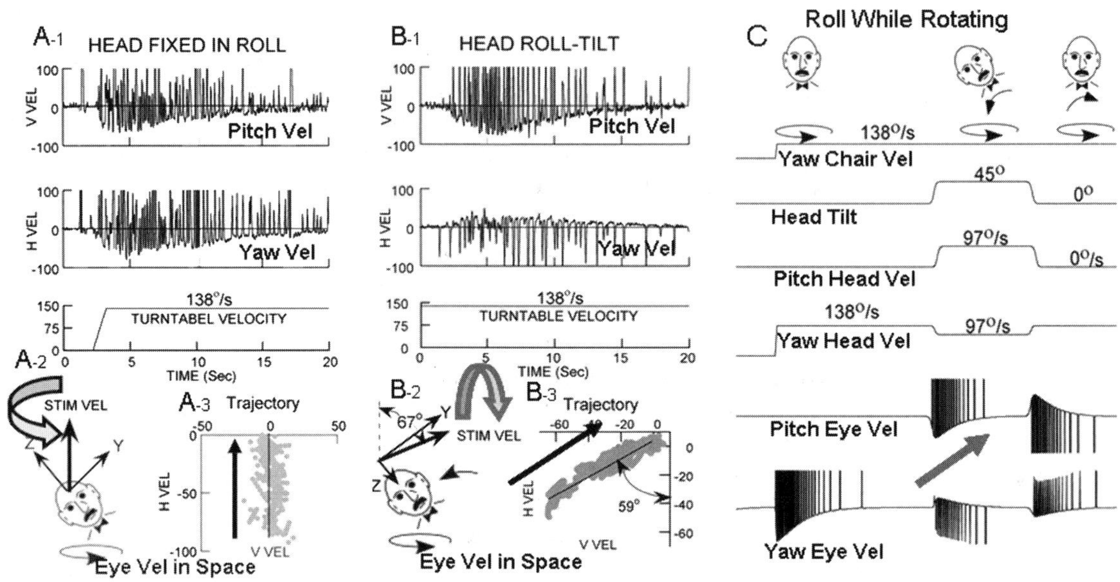

Fig. 2. Roll while rotating (RWR) paradigm used to elicit motion sickness. (A-1) Pitch (top trace) and yaw (second trace) eye velocities during nystagmus induced by rotation at 138°/s (third trace) about a spatial vertical axis with the head statically tilted 45° throughout the rotation (insert below). Velocities down and to the right are positive according to right hand rule. (A-2) Perception during such rotation (circular arrow) was solely that of rotation about a vertical axis; motion sickness was not induced by this stimulus. (A-3) Reconstructed eye velocity vector of the slow phases of pitch and yaw nystagmus. The vector declined toward zero along an axis parallel to gravity (arrow). (B) Scheme as in (A). When the subject tilted his head to the right while rotating to the left, there were upward pitch and left yaw slow phase velocities. This produced a strong sensation of pitch (circular arrow), and an eye velocity vector that was tilted 59° from the spatial horizontal. The eye velocity vector lay close to the stimulus velocity vector (arrow, B-2). This produced a strong sense of nausea, dizziness, and motion sickness. (C) Model-based simulation of the nystagmus induced by pure rotation to the left about a vertical axis (left, insert above), by tilting the head 45° to the right while rotating (middle, insert above), and by bringing the head back to the upright during rotation (right, insert above). Note that the simulated eye velocities in the middle portion of (C) are close to the actual eye velocities, derived experimentally in (B) (Adapted with permission from Dai et al., 2003).

We have proposed that the orientation vector generated by the nodulus/uvula and the eye velocity vector produced by velocity storage is processed by the Orientation Comparator (Fig. 1F). If there is a discrepancy between the two vectors, the autonomic system is activated to produce a buildup of motion sickness (Dai et al., 2003).

For constant velocity rotation about a yaw axis, per- and post-rotatory nystagmus is induced at the onset and end of rotation, shown here for the monkey (Fig. 1C–E; Cohen et al., 1987). The direct pathway response, which is dependent on the dynamics of the cupula/endolymph system, is shown by the solid falling lines labelled "Cupula" in Fig. 1G, H (arrows). The velocity storage or indirect pathway response is shown by the solid lines and arrows labeled "Velocity Storage" in Fig. 1G, H. It has two modes: a rising component due to the input from the SCC and a slower response due to activation of the velocity storage integrator. The induced activity falls with the time constant of the integrator, which is governed by the parameter h, whose value is controlled by the nodulus (Fig. 1F). The signals from the direct and indirect pathways sum in the model at Vn (Fig. 1F) to generate the eye velocity command that produces the slow phase eye velocity and triggers the quick phases of nystagmus. As shown by Fig. 1G, after the first 10 s, there is essentially no input from the labyrinth and the magnitude and decline in the slow phase eye velocity is entirely dependent on the activation of velocity storage. Thus, the time constant of per- and post-rotatory nystagmus in subjects with time constants longer than 8–10 s primarily represents activity in velocity storage (Dai et al., 1999).

Effects of baclofen on the aVOR

Following the injection of baclofen in monkeys, T_{VOR} was reduced but recovered within 24 h (Fig. 1C–E). A similar decline in T_{VOR} occurs in humans after ingestion of 20 mg of baclofen (Figs. 1B and 3A, C). In addition to the changes in velocity storage time constant, there were also changes in the coupling of vestibular input to the velocity storage mechanism in the vestibular nuclei (g_0; Dai et al., 2006).The reductions in T_{VOR} in both humans and monkeys were closely correlated with the rise in the levels of plasma baclofen in humans, determined separately (Fig. 1B; Faigle et al., 1980). Changes in T_{VOR} over 4 h in six subjects after administration of 20 mg and 30 mg of

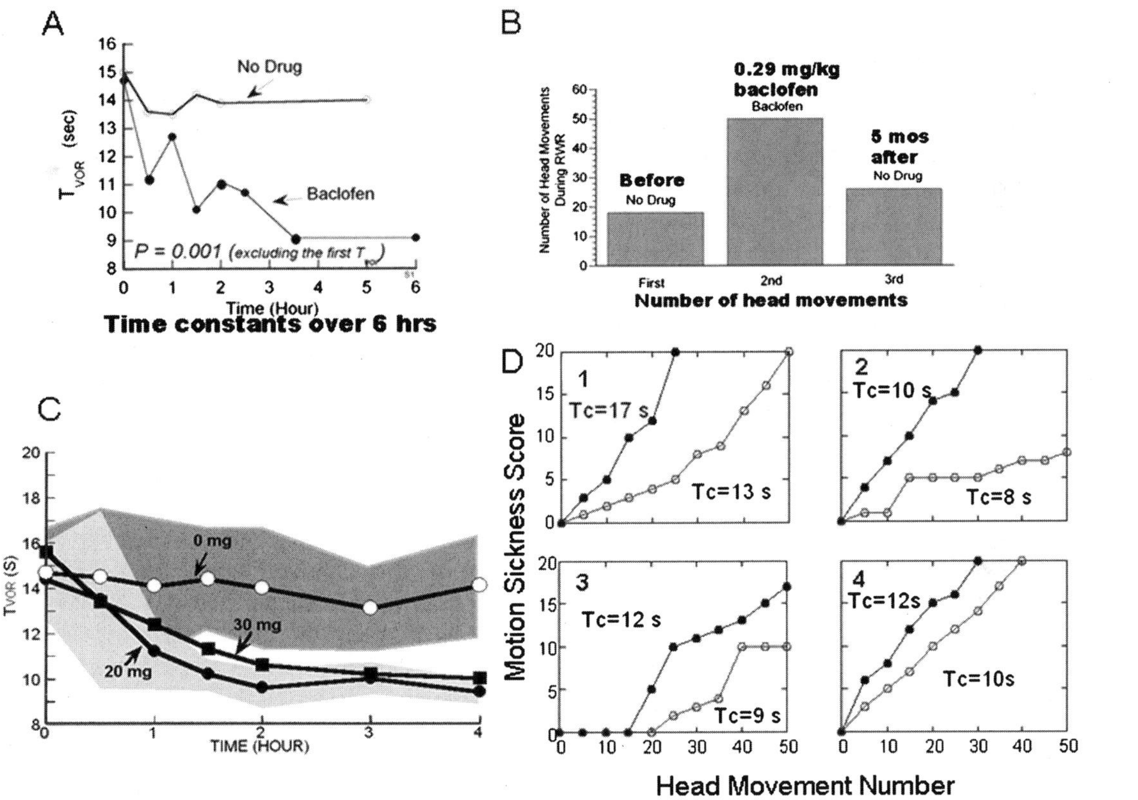

Fig. 3. Changes in motion sickness susceptibility tested with RWR before and after baclofen. (A) Changes in T_{VOR} during testing with 138°/s steps of velocity in dark before (top graph) and after ingestion of 20 mg of baclofen (bottom graph, solid circles). There was little habituation over the 6 h period of testing (top graph), but the subject's T_{VOR} fell from 14 s to 9 s after baclofen over an equivalent period of testing a month later (bottom graph). (B) Number of head movements during RWR before baclofen (left bar), 2 h after baclofen (centre bar), and 5 months after baclofen (right bar). The subject could only initially make 18 head movements without reaching a motion sickness score of 20, but was able to make 50 head movements 2 h after receiving baclofen. Five months later, he was only able to make 23 head movements without baclofen before reaching a motion sickness score of 20 (about to vomit). (C) Comparison of T_{VOR} during rotational testing at 60°/s without medication (open circles), and up to 4 h after receiving 20 mg (filled circles) and 30 mg (filled squares) of baclofen. The grey areas show the 95% confidence intervals. After baclofen, T_{VOR} fell and remained depressed over the 4 h of testing. There was no difference in the reduction in T_{VOR} produced by doses of 20 mg and 30 mg of baclofen. (Adapted with permission from Dai et al., 2006). (D) Increase in motion sickness score, based on steady-state levels of nausea, before (filled circles) and after baclofen (open circles). The ordinate is Head Movement Number, and the abscissa is the Motion Sickness Score. If subjects did not reach a motion sickness score of 20, the test was terminated after 50 head movements. In each subject, the rise in motion sickness was slower after receiving 20 mg of baclofen. The T_{VOR} (Tc's) before and after receiving baclofen are shown adjacent to each of the graphs.

baclofen followed approximately the same time course (Fig. 3C). This indicates that 20 mg is above threshold for inducing the reduction in T_{VOR}.

A consistent finding has been that there is no change in the gain of the aVOR following the administration of baclofen (Fig. 1C–E, G, H; Cohen et al., 1987; Dai et al., 2006). From this, we conclude that baclofen acts primarily on the velocity storage component of the aVOR and has little effect on the direct pathway (Dai et al., 2006). Side effects were minimal at 0.3 mg/kg, being predominantly drowsiness. At considerably higher doses in monkeys ($\approx$ 1–2 mg/kg), the time constant of the final common "neural integrator," which is utilized in producing positions of fixation and in linearizing the slow phases of optokinetic and vestibular nystagmus, also becomes shortened (Yakushin, unpublished data). However, at lower dose levels, baclofen is a convenient way to reduce the time constant of the aVOR, and as we will demonstrate, to reduce motion sickness susceptibility.

Production of motion sickness by roll while rotating (RWR)

Rolling the head while rotating about a spatial vertical axis, activates both the vertical and horizontal SCC. This produces pitch and yaw nystagmus, a sense of pitching forward or backward, and nausea, dizziness, and vertigo. It is a potent stimulus for generating motion sickness (see Dai et al., 2003, for review). Model-generated responses show the direction and magnitude of the expected nystagmus (Fig. 2C). Rotation to the left about a vertical axis induces horizontal nystagmus to the right, which decays to zero as the rotation continues (Fig. 2C, left, Yaw Eye Vel). When the head is rolled to the right (insert above), the vertical canals move into and the lateral canals out of the plane of rotation. This produces upward per-rotatory pitch and leftward post-rotatory yaw eye velocity. The simulated velocity vector of the induced nystagmus is shown by the arrow in the middle portion of Fig. 2C (middle, arrow).

The pitch per- and yaw post-rotatory responses in the vertical and lateral canals produced in the model by head roll, were present experimentally (Fig. 2B, 1st and 2nd traces). The velocity vector of the stimulus in space was 67° (Fig. 2B-2). In response, the eye velocity vector (Fig. 2B-3) declined toward zero along a trajectory that was tilted from the vertical by 59°, similar to that predicted by the model (Fig. 2C, arrow). Concurrently, there is a strong sense of pitching forward (Fig. 2B-2, rotatory arrow) and a sudden sense of disorientation and dizziness. The increase of nausea associated with the roll head movement is momentary, lasting for about 30 s and then declines to a steady state level.

When the head is brought back to the upright (Fig. 2C, right), the direction of the nystagmus reverses, producing a downward pitch post-rotatory and a leftward yaw per-rotatory response. Again there is a transient increase in motion sickness, followed by a decline to a steady state level of nausea that is generally more intense than after the preceding head movement. As the sequence of transient and steady state increments of nausea continue, the level of steady state nausea increases after each head movement until a state of imminent emesis is reached, i.e., a motion sickness score of 20, at which point, the test is terminated. The test is also terminated if a score of 20 is not reached after 50 head movements (A more complete description of the paradigm is given in Dai et al., 2003).

The importance of the disparity between the eye velocity vector and the orientation vector of velocity storage for production of motion sickness is shown in Fig. 2A. If the head is simply held in a tilted position before and during rotation (Fig. 2A-1), horizontal and vertical slow phase velocities are generated. However, in contrast to the situation in Fig. 2B, C, eye velocity decays along a straight line close to the direction of the acceleration of gravity and the orientation vector of velocity storage (Fig. 2A-2; Dai et al., 1991; Raphan et al., 1992). Perceptually, subjects feel as if they are simply rotating around a vertical axis (rotatory arrow, Fig. 2A-2), and such stimulation did not cause motion sickness in any of the six subjects that were repeatedly tested with this paradigm.

Effects of baclofen on motion sickness susceptibility

An important finding from the study of Dai et al. (2003), was that the T_{VOR} declined concurrently with the ability of subjects to make more head movements during repeated testing. That is, the two were reciprocally related. This led to the current hypothesis that motion sickness susceptibility could be reduced if there was a pharmacological reduction in T_{VOR}. Subjects were first tested with RWR to determine their baseline motion sickness susceptibility. A week later, subjects received a single dose of 20 mg of baclofen and were retested with RWR an hour later. Results from one subject are shown in Fig. 3A, B. He made 18 head movements during the pretest before developing severe motion sickness (Fig. 3B, left). Following administration of 20 mg of baclofen, his T_{VOR} dropped from 14 s to 9 s (Fig. 3A), and he was able to make 50 head movements without reaching a motion sickness score of 20 (Fig. 3B, centre). Five months later when effects of baclofen and habituation had disappeared, he could make only 25 head movements during RWR before reaching emesis.

The buildup of the steady state level of motion sickness in four other subjects is shown in Fig. 3D. The T_{VOR}'s of these subjects before and after baclofen are shown by the number adjacent to the individual curves. Two subjects (D-1 and D-4) reached a motion sickness score of 20 (imminent emesis) both with and without baclofen, but the buildup was slower after receiving the drug. One subject (D-3) never reached a score of 20 either with or without baclofen, but the buildup of symptoms was slower after baclofen. A fourth subject (D-2) had a motion sickness score of 20 after 26 head movements before receiving the drug, but was able to make 50 head movements after receiving baclofen and only reached a score of motion sickness score of 6. Thus, 20 mg of baclofen, which shortened the T_{VOR} in all four subjects, was effective in reducing their motion sickness susceptibility.

Neural basis for changes in velocity storage

There is evidence that several of the salient characteristics of velocity storage are embedded in VO and vestibular-plus-saccade (VPS) neurons located in rostral medial vestibular nucleus (rMVN) and SVN (Reisine and Raphan, 1992; Yokota et al., 1992; Holstein et al., 1999). We therefore questioned whether the temporal responses of these types of neurons to optokinetic and vestibular stimuli would be modified by the administration of baclofen. A typical VPS neuron did not respond to optokinetic stimulation at frequencies of 0.1 Hz or higher (Fig. 4A), and had variable responses to a frequency of 0.05 (Fig. 4B). There was a robust response to optokinetic stimulation at 0.02 Hz, however (Fig. 4C). The cell had rapid changes in activity at the onset of constant velocity rotation, followed by slower change in firing rate as the rotation continued (Fig. 4D). This neuron also coded direct pathway activity, and rapid changes in firing rate occurred both at the onset and end of rotation that were approximately linearly related to the rotational velocity (Fig. 4E).

There was a close correspondence between the unit time constants and the T_{VOR} at all tested velocities. The T_{VOR} of the per- and post-rotatory nystagmus shown in Fig. 4D were 20 s and 27 s, respectively (2nd trace), in response to constant velocity rotation at 180°/s (bottom trace). The corresponding time constants of the unit activity associated with this rotation were 23 s and 21 s, respectively (top trace). The close association of the unit and aVOR time constants over trials from 30°/s–180°/s is demonstrated in Fig. 4G by the open circles. After parenteral (IM) administration of 6 mg of baclofen (1.5 mg/kg), the time constants of the per- and post-rotatory responses fell to 9 s and 11 s (Fig. 4F, 2nd trace) for the same 180°/s rotation (bottom trace), while the time constants of the associated unit responses decreased to 8 s and 12 s, respectively (top trace). This parallel reduction in time constants was also present in the repeated trials at the different velocities, as shown by the filled symbols in Fig. 4G.

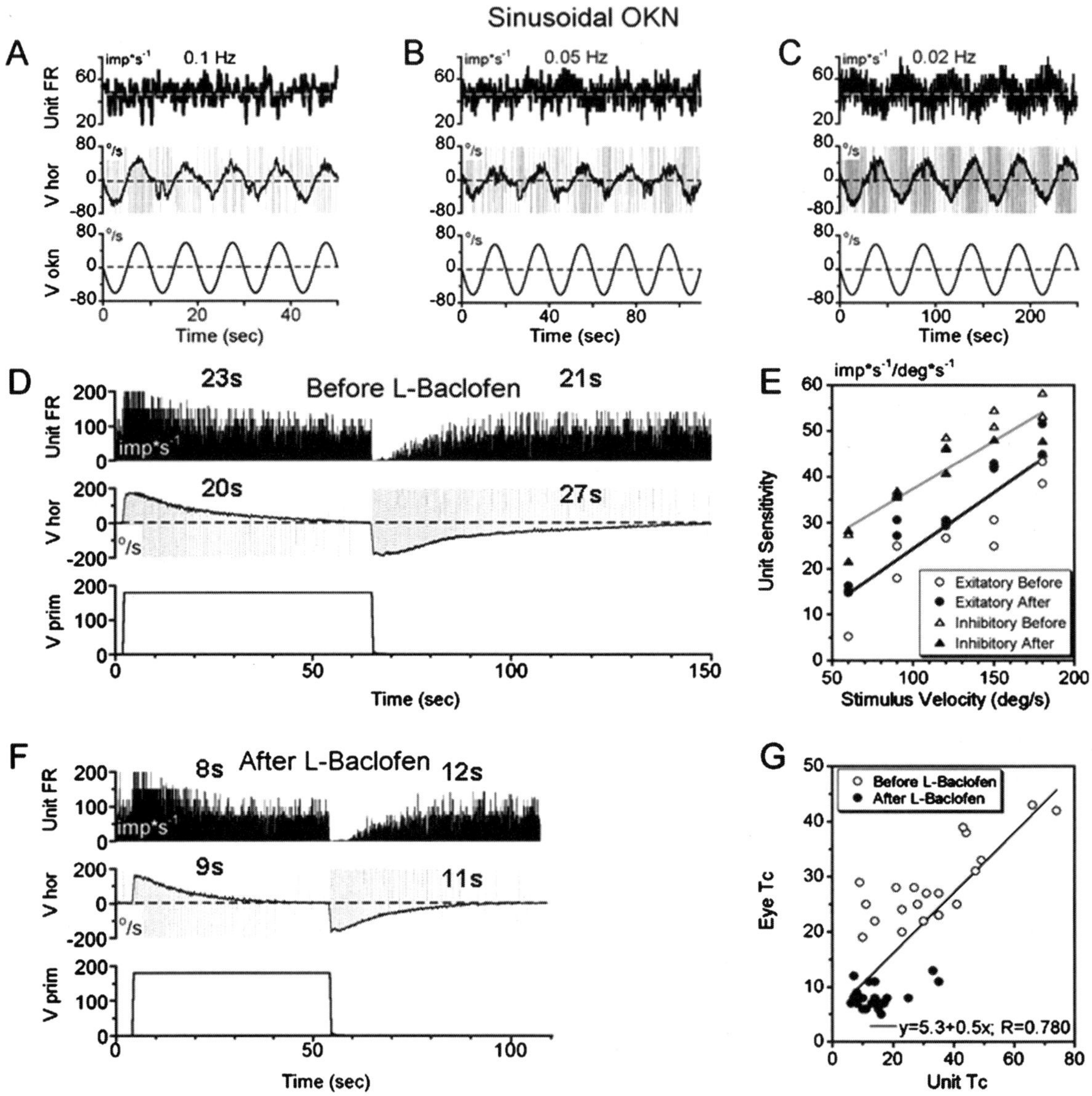

Fig. 4. Graphs of vestibular-plus-saccade neuron and aVOR T_{VOR} before (A-D) and after (F) parenteral injection of 0.2 mg/kg baclofen in a cynomolgus monkey. (A-C) Unit activity (top trace), desaccaded yaw eye velocity (middle trace), and OKN velocity (lower trace). The optokinetic stimulus was oscillated at 0.1 Hz (A), 0.05 Hz (B), and 0.02 Hz (C). There was some modulation in unit firing at 0.05 Hz, but it became robust at 0.02 Hz. (D, F) Comparison of unit activity (top trace) and T_{VOR} (2nd trace) before (D) and after (F) injection of baclofen. The velocity of the rotator is shown in the 3rd trace. The time constants of both the unit activity and the T_{VOR}, were comparable before (D) and after (F) baclofen. (E) Increase and decrease in firing frequency of the unit in response to rotation at velocities from 60°/s to 180°/s in darkness before and after baclofen. Abscissa, stimulus velocity; ordinate, increases (excitatory) and decreases (inhibitory) in firing frequency. There was no effect of baclofen on the initial response to rotation in either the excitatory or inhibitory directions. (G) Comparison of time constants of the aVOR and of unit activity before and after injection of baclofen. There was an increase in the time constants of both the unit activity and eye velocity before baclofen (open circles) in response to increases in stimlulus velocity ($R = 0.780$). After baclofen (solid circles), the time constants were clustered at lower values.

Thus, there was a close correlation of the unit and the nystagmus time constants both before and after administration of baclofen. The changes in firing rate associated with the onset and end of rotation were unaffected by the administration of baclofen (Fig. 4E). This is strong support for the hypothesis that baclofen does not affect the direct pathway response, only the response generated by the velocity storage integrator.

Discussion

We have shown here that oral administration of baclofen, which shortened the aVOR time constant, reduced motion sickness susceptibility. We have further demonstrated that the reduction in T_{VOR} produced by baclofen was closely paralleled by a reduction in the time constant of a VPS neuron lying in rMVN. Consistent with the postulate that baclofen only affects the indirect and not the direct pathway, the high frequency responses of both the aVOR and of the VPS neuron were unaltered. This study also has clinical implications. An average dosage of ≈ 0.3 mg/kg, produced by ingestion of 20 mg of baclofen, which had minimal side effects, was effective in reducing T_{VOR} by 27% on average, in decreasing the coupling gain of efferent vestibular information to velocity storage (g_0) by 24% (Dai et al., 2006), and in reducing motion sickness susceptibility. From the data of Faigle et al. (1980) in human and of Cohen et al. (1987) in monkeys, the reduction in T_{VOR} and presumably in motion sickness susceptibility would last for up to 8 h–10 h. Thus, this or similar $GABA_b$ agonists are likely to be effective agents to reduce motion sickness susceptibility.

Implications of neural activity associated with velocity storage

It has been known since the pioneering studies of Waespe and Henn (1977) that the time constants of VO and VPS neurons reflect the time constants of OKAN and of per- and post-rotatory nystagmus, i.e., of velocity storage (Cohen et al., 1987; Raphan et al., 1979). This is the first demonstration, however, that a VPS neuron changed its time constant in parallel with the changes in T_{VOR} after administration of baclofen. This provides additional evidence that velocity storage is coded in the activity of VPS neurons, and raises the possibility that such neurons could be the neural basis for the effect of baclofen on motion sickness susceptibility. The structural basis for this effect and for the associated changes in g_0 (Dai et al., 2006) may come from the immunocytochemical studies of Holstein et al. (1992a, b) that have demonstrated baclofen-sensitive $GABA_b$ receptors in rMVN. Their ultrastructural studies have further shown that such inhibition is mediated by both axo-dendritic (post-synaptic) and axo-axonic (pre-synaptic) contacts (Holstein et al., 1999). Presumably, activation of axo-dendritic synapses could produce a reduction in membrane resistance that would reduce periods of activation of such neurons, resulting in a reduction of T_{VOR}.

Activity in the direct aVOR pathway is also coded in the firing rates of VO and VPS neurons, but was not affected by baclofen (Fig. 4E). The resting discharge was also not altered. If the reduction in activity related to rotation in the "off-direction" was produced by inhibition, these neurons must have different synaptic inputs and inhibitory transmitters to modulate activity in the direct and indirect pathways. From our results, $GABA_b$ is utilized to inhibit the velocity storage component, but another inhibitory transmitter, possibly $GABA_a$, may be utilized for inhibition of neurons in the direct pathway.

Implications for aVOR structure

The aVOR model of Fig. 1F shows how baclofen could affect the time constant of the aVOR and motion sickness susceptibility. A key element is the control of spatial orientation and temporal properties of velocity storage by the nodulus and uvula, which set the feedback parameter, h, which controls its time constant (Wearne et al., 1998). Whether baclofen also modifies the yaw axis spatial orientation vector of velocity storage is not clear, but it clearly shortens the velocity storage time constant. Thus, any discrepancy between the orientation vector and the actual eye velocity generated by velocity storage is reduced by baclofen, thereby slowing the motion sickness buildup that accompanies each head tilt (Fig. 3D). Thus, more head movements can be made before a critical level of motion sickness buildup is reached.

In the model of Fig. 1F, only the temporal and orientation properties of velocity storage would impact the generation of motion sickness. Yet, VO and VPS neurons reflect both the activity of velocity storage and the direct pathway (Waespe and Henn, 1977; Reisine and Raphan, 1992). This could possibly indicate that the VO and VPS neurons lie beyond the summing junction of the output of the velocity storage integrator, Vn, where the direct and indirect pathways join to form the Eye Velocity Command. It is possible that there are specific neuron classes in the vestibular nuclei that have not been encountered, which only code activity of the velocity storage integrator. If such neurons do not exist, however, it could imply that there is filtering of the signals coming from the vestibular nuclei so that eye velocity commands are generated that only reflect the low frequency velocity storage component. Alternatively, since baclofen does not affect the direct pathway response of the VPS neurons, velocity storage might be an emergent network property of VO and VPS neurons (Reisine and Raphan, 1992) and not a specific property of individual VO or VPS cells. Regardless, the results support the importance of velocity storage as a major factor in the production of motion sickness.

Abbreviations

aVOR	angular vestibulo-ocular reflex
g_0	indirect pathway gain
g_1	direct pathway gain
$GABA_a$, $GABA_b$	gamma amino butyric acid a and b
IM	parenteral (intramuscular) injection
OKAN	optokinetic after nystagmus
rMVN	rostral medial vestibular nucleus
RWR	roll while rotating
T_{OKAN}	time constant of optokinetic after nystagmus
T_{VOR}	time constant of the angular vestibulo-ocular reflex
VO	vestibular-only neuron
VPS	vestibular-plus-saccade neuron

Acknowledgements

Support: DC007847, DC004996; EY011812; EY004148; DC005204, EY001867. We thank Dmitri Ogorodnikov and Sergey Tarasenko for excellent technical support.

References

Cohen, B., Helwig, D. and Raphan, T. (1987) Baclofen and velocity storage: a model of the effects of the drug on the vestibulo-ocular reflex in the rhesus monkey. J. Physiol., 393: 703–725.

Cohen, B. and Suzuki, J.I. (1963) Eye movement induced by ampullary nerve stimulation. Am. J. Physiol., 204: 347–351.

Cohen, H., Cohen, B., Raphan, T. and Waespe, W. (1992) Habituation and adaptation of the vestibuloocular reflex: a model of differential control by the vestibulocerebellum. Exp. Brain Res., 90: 526–538.

Dai, M., Klein, A., Cohen, B. and Raphan, T. (1999) Model-based study of the human cupular time constant. J. Vest. Res., 9: 293–301.

Dai, M., Kunin, M., Raphan, T. and Cohen, B. (2003) The relation of motion sickness to the spatial-temporal properties of velocity storage. Exp. Brain Res., 151: 173–189.

Dai, M., Raphan, T. and Cohen, B. (1991) Spatial orientation of the vestibular system: dependence of optokinetic after-nystagmus (OKAN) on gravity. J. Neurophysiol., 66: 1422–1439.

Dai, M., Raphan, T. and Cohen, B. (2006) Effects of baclofen on the angular vestibulo-ocular reflex. Exp. Brain Res., 171(2): 262–271.

Faigle, J., Keberle, H. and Degen, P. (1980) Chemistry and pharmacokinetics of baclofen. In: Feidman R., Young R. and Koella W. (Eds.), Spaslicity: Disordered Motor Control. Year Book Medical Publishers, Chicago, pp. 461–475.

Halmagyi, G.M., Rudge, P., Gresty, M.A., Leigh, R.J. and Zee, D.S. (1980) Treatment of periodic alternating nystagmus. Ann. Neurology, 8: 609–611.

Holstein, G.R., Martinelli, G.P. and Cohen, B. (1992a) L-baclofen-sensitive $GABA_b$ binding sites in the medial vestibular nucleus localized by immunocytochemistry. Brain Res., 581: 175–180.

Holstein, G.R., Martinelli, G.P. and Cohen, B. (1992b) Immunocytochemical visualization of L-baclofen-sensitive

$GABA_b$ binding sites in the medial vestibular nucleus. Ann. N.Y. Acad. Sci., 656: 933–936.

Holstein, G.R., Martinelli, G.P. and Cohen, B. (1999) The ultrastructure of GABA immunoreactive vestibular commissural neurons related to velocity storage in the monkey. Neurosci., 93: 171–181.

Leigh, R.J., Robinson, D.A. and Zee, D.S. (1981) A hypothetical explanation for periodic alternating nystagmus: instability in the optokinetic vestibular systems. Ann. N.Y. Acad. Sci., 374: 619–635.

Lisberger, S.G. and Pavelko, T.A. (1988) Brain stem neurons in modified pathways for motor learning in the primate vestibulo-ocular reflex. Science, 242: 771–773.

Lorente de Nó, R. (1932) The regulation of eye positions and movements induced by the labyrinth. Laryngoscope, 42: 233–332.

Miller, E. and Graybiel, A. (1973) Experiment M-131-Human vestibular function. Aviat. Space Environ. Med., 44: 593–608.

Raphan, T. and Cohen, B. (2002) The vestibulo-ocular reflex (VOR) in three dimensions. Exp. Brain Res., 145: 1–27.

Raphan, T., Dai, M. and Cohen, B. (1992) Spatial orientation of the vestibular system. Ann. N.Y. Acad. Sci., 22(656): 140–157.

Raphan, T., Matsuo, V. and Cohen, B. (1979) Velocity storage in the vestibulo-ocular reflex arc (VOR). Exp. Brain Res., 35: 229–248.

Raphan, T. and Sturm, D. (1991) Modeling the spatiotemporal organization of velocity storage in the vestibuloocular reflex by optokinetic studies. J. Neurophysiol., 66: 1410–1421.

Reisine, H. and Raphan, T. (1992) Neural basis for eye velocity generation in the vestibular nuclei of alert monkeys during off-vertical axis rotation. Exp. Brain Res., 92(2): 209–226.

Waespe, W. and Henn, V. (1977) Vestibular nuclei activity during optokinetic after-nystagmus (OKAN) in the alert monkey. Exp. Brain Res., 30: 323–330.

Wearne, S., Raphan, T. and Cohen, B. (1998) Control of spatial orientation of the angular vestibulo-ocular reflex by the nodulus and uvula. J. Neurophysiol., 79: 2690–2715.

Yakushin, S., Dai, M., Suzuki, J., Raphan, T. and Cohen, B. (1995) Semicircular canal contributions to the three-dimensional vestibulo-ocular reflex: a model-based approach. J. Neurophysiol., 74: 2722–2738.

Yokota, J., Reisine, H. and Cohen, B. (1992) Nystagmus induced by microstimulation of the vestibular and prepositus hypoglossi nuclei in the monkey. Exp. Brain Res., 92: 123–138.

Zhang, Y., Partsalis, A.M. and Highstein, S.M. (1993) Properties of superior vestibular nucleus neurons projecting to the cerebellar flocculus in the squirrel monkey. J. Neurophysiol., 69(2): 642–645.

C. Kennard & R.J. Leigh (Eds.)
Progress in Brain Research, Vol. 171
ISSN 0079-6123

CHAPTER 6.9

Oculomotor deficits indicate the progression of Huntington's Disease

Stephen L. Hicks[1,*], Matthieu P.A. Robert[1,2], Charlotte V.P. Golding[1], Sarah J. Tabrizi[2] and Christopher Kennard[1]

[1]*Department of Clinical Neuroscience, Imperial College, London, UK*
[2]*Institute of Neurology and National Hospital for Neurology and Neurosurgery, London, UK*

Abstract: The oculomotor deficits associated with Huntington's Disease (HD) are one of the earliest signs of disease onset. They include a marked delay in executing voluntary saccades and a difficulty inhibiting saccades to task-irrelevant stimuli. In addition, HD patients develop a deficit in task-switching, which can be demonstrated by the continued adherence to a rule after it has been recently changed. These deficits are likely to be the result of a progressive neural degeneration of the fronto-striatal system, which is a distinguishing feature of HD neuropathology. It is predicted that as the disease progresses the magnitude of these specific deficits should increase. We tested a cohort of early HD patients and presymptomatic HD gene carriers on a series of oculomotor tasks designed to measure saccade initiation, inhibition and rule switch cost. Saccadic latencies and error rates in early HD patients were found to be systematically higher than controls. Presymptomatic HD subjects showed small increases in saccadic latencies and error rates that were in proportion to the predicted age of disease onset. These results suggest that saccadometry and a cognitively demanding oculomotor task may be useful as an indicator of function in HD.

Keywords: Huntington's Disease; saccades; switch cost; preclinical; biomarker

Introduction

Huntington's Disease (HD) is a neurological disorder caused by an expanded CAG repeat in the Huntingtin gene (The Huntington's Disease Collaborative Research Group, 1993). The disease initially affects GABAergic medium spiny neurons of the caudate nucleus and progresses to affect the entire fronto-striatal network and sensory cortices. HD is characterised by motor, psychiatric, and cognitive symptoms.

Eye movements are impaired at an early stage, often long before other symptoms become clinically relevant. Patients demonstrate a quantifiable delay in executing voluntary saccades (Ali et al., 2006; Blekher et al., 2006; Golding et al., 2006), difficulties inhibiting saccades to irrelevant stimuli, and delays and errors when switching task rules (Aron et al., 2003). The primary aim in this study was to compare the effectiveness of three saccadic paradigms in distinguishing HD patients from presymptomatic carriers of the HD gene mutation. The combination of a cognitively demanding task with an oculomotor response may offer a sensitive functional biomarker of HD.

*Corresponding author. Tel.: +44 208 846 7289; Fax: +44 208 846 7715; E-mail: s.hicks@imperial.ac.uk

DOI: 10.1016/S0079-6123(08)00678-X

Method

Eye-tracking and stimuli presentation were carried out using a Saccadometer Advanced (Ober Consulting); an infra-red system that measures reflectance off the inner canthi of both eyes and averages them for a single value of horizontal eye position. Acquisition rate was 1 KHz at 12 bits and horizontal linearity was assured for $\pm 10^\circ$. Four head mounted lasers projected the stimuli – two red targets at $\pm 10^\circ$ and two green and red central cues – 2 metres onto a blank wall.

Paradigms:

Three oculomotor paradigms were assessed:

(A) Prosaccade — Step task: subject saccades from a central cue to randomly presented stimuli on the left or right – 90 trials.

(B) Centrally cued — Single choice task: subject saccades to left or right target depending on colour of central cue – 120 trials.

(C) Peripheral conflict — Double choice task: central cue informs subject whether to make a pro or antisaccade to a target that is randomly on the left or right – 180 trials.

All paradigms had a cue-target gap of 200 ms and participants were instructed to respond quickly and accurately.

Participants

Twelve early-symptomatic HD patients, 17 pre-symptomatic gene carriers, and 12 controls.

The presymptomatic group was divided into two subgroups according to the probability of developing clinical symptoms within 10 years (based on CAG repeat length and age (Langbehn et al., 2004)): Pre-1, a "far" group, with a probability of less than 50%, and Pre-2, a "close" group, with a probability of greater than 50%.

Results

Error rates

Saccadic error rate increased with HD status. Error rates were highest in paradigm (C). The very

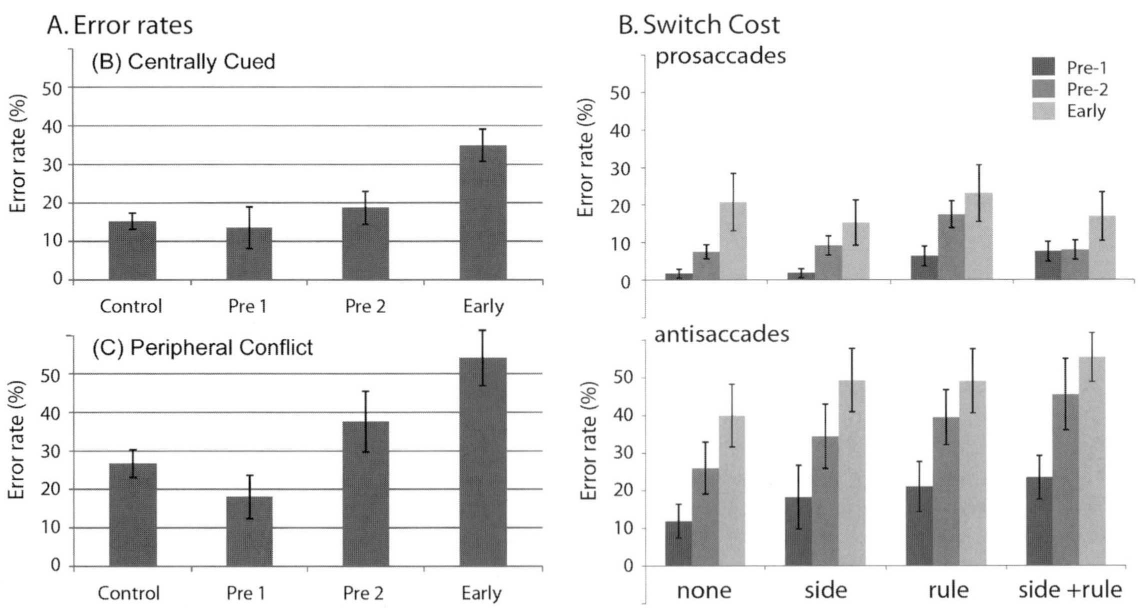

Fig. 1. A. Error rate per group for paradigms (B) and (C). B. Switch cost in increased errors per group for paradigm (C) divided into prosaccade and antisaccade components. Errors increased in proportion to the number of rule switches. Bars indicate SEM.

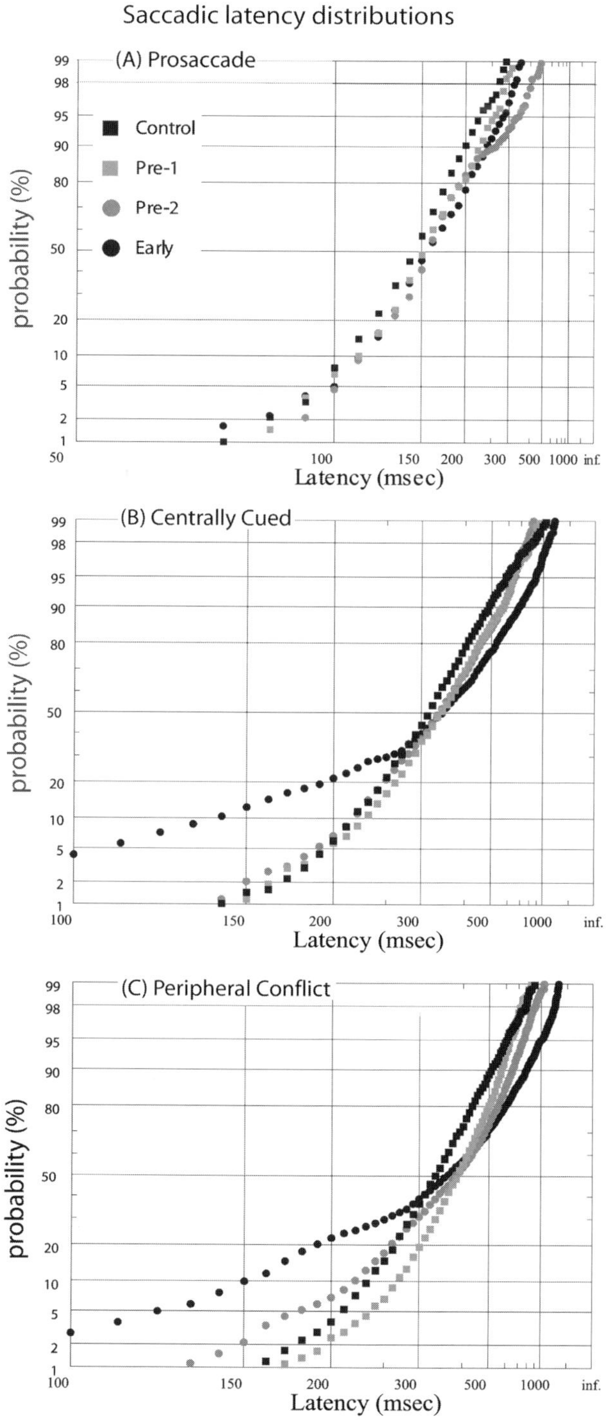

Fig. 2. Grouped LATER plots of the saccade latency distribution for each paradigm. Although the median latencies differ little between groups, the proportion of fast and slow saccades increasingly differentiate the groups in proportion to task difficulty.

high error rate in the Early group is mainly accounted for by errors in the antisaccade condition. Motivation is likely to account for the better-than-control performance of the Pre-1 group (Fig. 1A).

Switch cost

Rule switching increased error rates. Error rates and latency costs for rule switching were derived from paradigm (C) according to the nature of the preceding trial, i.e., same or different target side, same or different rule (central cue). Switch cost was highest for antisaccades, proportional to the number of rules changed and increased with HD progression, subsequently differentiating the two subgroups of presymptomatic participants (Fig. 1B).

Saccadic latencies

Fast saccades separate HD groups. LATER model plots (Carpenter and Williams, 1995) of the distribution of saccadic latencies, pooled by group, demonstrated the discriminating potential of each paradigm. (A) shows little difference between groups. (B) separates the Early HD in respect to a higher proportion of short latency saccades however Pre and controls remain indistinguishable. (C) separates Pre-1 and Pre-2 in both short latency and long latency saccades (Fig. 2).

Discussion

Group differences in saccadic latency profiles suggest that the peripheral conflict task discriminates HD progressing with the highest power. This paradigm appears to be able to differentiate Pre-1 and Pre-2 which is important in considering its use as a marker of preclinical disease status. The error rates are maximally increased for Early and Pre HD participants in the antisaccade component of the peripheral conflict task. This may be indicative of saccadic inhibition deficits, however at this stage it is difficult to rule out task comprehension as a source of errors. Error rates and switch cost are both good indicators of cognitive function, and if combined with saccadic latency measures may produce an index of HD progression. A refined protocol based on the peripheral conflict task is being trialled in the Track-HD study, beginning January 2008.

Acknowledgements

This work is supported by The Wellcome Trust, the HighQ Foundation, and the Fondation pour la Recherche Médicale.

References

Ali, F.R., Michell, A.W., Barker, R.A. and Carpenter, R.H.S. (2006) The use of quantitative oculometry in the assessment of Huntington's disease. Exp. Brain Res., 169(2): 237–245.

Aron, A.R., Watkins, L., Sahakian, B.J., Monsell, S., Barker, R.A. and Robbins, T.W. (2003) Task-set switching deficits in early-stage Huntington's disease: implications for basal ganglia function. J. Cogn. Neurosci., 15(5): 629–642.

Blekher, T., Johnson, S.A., Marshall, J., White, K., Hui, S., Weaver, M., Gray, J., Yee, R., Stout, J.C. and Beristain, X. (2006) Saccades in presymptomatic and early stages of Huntington disease. Neurology, 67(3): p. 394.

Carpenter, R.H. and Williams, M.L. (1995) Neural computation of log likelihood in control of saccadic eye movements. Nature, 377(6544): 59–62.

Golding, C.V.P., Danchaivijitr, C., Hodgson, T.L., Tabrizi, S.J. and Kennard, C. (2006) Identification of an oculomotor biomarker of preclinical Huntington disease. Neurology, 67(3): p. 485.

Langbehn, D.R., Brinkman, R.R., Falush, D., Paulsen, J.S. and Hayden, M.R. (2004) A new model for prediction of the age of onset and penetrance for Huntington's disease based on CAG length. Clin. Genet., 65(4): 267–277.

The Huntington's Disease Collaborative Research Group. (1993) A novel gene containing a trinucleotide repeat that is expanded and unstable on Huntington's disease chromosomes. Cell, 72(6): 971–983.

C. Kennard & R.J. Leigh (Eds.)
Progress in Brain Research, Vol. 171
ISSN 0079-6123

CHAPTER 6.10

Eye movements in visual search indicate impaired saliency processing in Parkinson's disease

Sabira K. Mannan[1,*], Timothy L. Hodgson[2], Masud Husain[3] and Christopher Kennard[1]

[1]*Department of Clinical Neuroscience, Imperial College London, UK*
[2]*Department of Psychology, University of Exeter, UK*
[3]*Institute of Neurology and Cognitive Neuroscience, University College London, UK*

Abstract: Previous studies have produced contradictory evidence on the nature of the visual search impairment in patients with Parkinson's disease (PD). Eye movements were measured during multi-target search in nine individuals with mild-to-moderate PD. Subjects were asked to click on a response button whenever they judged they were fixating a target for the first time. Compared to age-matched healthy volunteers, PD patients were impaired at *efficient* search (detecting "+"s amongst "L"s) but not inefficient search ("T"s amongst "L"s). However, these patients had normal memory for locations as indexed by their rate of re-clicking on previously inspected locations. We suggest that the pattern of gaze for efficient search may reflect impaired saliency processing in PD.

Keywords: visual search; saccades; Parkinson's disease; saliency

Introduction

Idiopathic Parkinson's disease (PD) is a neurodegenerative disorder caused by a loss of dopaminergic innervation of the basal ganglia (BG). Although PD is commonly assumed to be disorder of motor control, recent studies have implicated a range of *cognitive* and *visual* deficits in PD, including deficits in visual search performance. The nature of the visuo-cognitive deficit in visual search in PD, however, remains unclear (Horowitz et al., 2006).

In visual search, participants look for a target item embedded amongst non-target items (distractors). Search for a "+" amongst rotated "L"s, or a search for a "T" amongst rotated "L"s, are considered classic search tasks, the former termed "efficient," and the latter "inefficient." In efficient search, where response times (RT) are independent of the number of distractors, search is commonly assumed to involve stimulus-driven parallel processing in early cortical visual areas. By contrast, inefficient search — where RT increases with distractor number — is often hypothesized to necessitate focal attention and working memory resources for a variety of cognitive operations, e.g., including keeping track of locations already attended.

Several studies have found impairment in efficient, but not inefficient, search in patients with PD using manual RT to index performance (Horowitz et al., 2006). Others have found no

*Corresponding author. Tel.: +44(0)208 846 7712;
Fax: +44(0)208 846 7715; E-mail: s.mannan@imperial.ac.uk

DOI: 10.1016/S0079-6123(08)00679-1

such pattern, but that only a subgroup of PD patients who were classified as "frontally impaired" exhibited impairments of both efficient and inefficient search. Finally, Horowitz et al. (2006) found no impairment on efficient search, but impaired performance in inefficient search when targets were unknown. However, to-date there has been no reports concerning the nature of eye movements during search in patients with PD. Here we measured eye position during efficient and inefficient multi-target search, in order to determine whether differences in gaze pattern between patients and controls resulted in impaired visual search. We also asked the novel question: Do patients with PD fail to remember the locations of items already visited during search? In order to assess this we also asked participants to "click" manually on a response button only when they judged that they were looking at a "new" target, but not to "re-click" on targets they recognized they had found earlier. Clicking does not alter the search display and thus the frequency of re-clicking on a target can index a failure to remember the location of a previously found target across saccades (Husain et al., 2001; Mannan et al., 2005).

Methods

Nine patients with mild-to-moderate, clinically defined idiopathic PD (eight patients at Hoehn and Yahr stage of 1 or 2, and one at stage 3, M = 1.8, SD = 0.6), together with 10 age-matched healthy volunteers. Mean disease duration was 5 years (range 4–7), and mean patient age was 63 years (range 55–70). Control mean age was 66 years (range 61–70). Participants were screened for dementia using a version of the Mini Mental State Examination (MMSE) (all participants scored above 28, with no significant difference between groups). All participants were corrected to 6/12 vision in both eyes. Eight out of nine patients were on levodopa; five were also taking benzhexol while one patient was taking pergolide. Patients were on medication during testing. Eye movements were measured using an infrared video-based eye tracker, the Eyelink I (SR Research Ltd.), whilst patients searched for either (i) 19 "+" targets embedded amongst 44 L's, (ii) 19 "T" targets amongst 44 L's. All participants completed a minimum of nine of each type of search screen (for details see Husain et al., 2001; Mannan et al., 2005). All data was entered into a two-way mixed ANOVA with Group (patients vs. controls) and Search Type (efficient vs. inefficient) as factors. Significance was set at $p<0.05$.

Results

First, although there was no difference in the time taken to complete search between groups, PD patients crucially missed significantly *more* targets than controls in efficient but not inefficient search (Table 1A–B). Thus, in terms of our two outcome measures (time taken to complete search and number of targets found), patients are impaired in *efficient* — but not inefficient — search performance.

Table 1. Comparison of visual search performance in patients with Parkinson's disease and age-matched controls

	Target	Patients	Healthy volunteers
A Response time (s)	+	19.2 (1.4)	18.8 (2.4)
	T	23.5 (1.3)	23.8 (2.5)
B Targets missed	+	2.4 (0.5)	1.9 (0.3)
	T	1.9 (0.5)	1.9 (0.5)
C Re-fixation rate on targets	+	51.2 (6.5)	24.2 (3.4)
	T	65.3 (6.3)	34.0 (5.3)
D Re-click rate on targets	+	5.9 (1.3)	6.3 (0.7)
	T	8.6 (1.8)	8.4 (0.9)
E Median interval between fixation and re-fixation	+	2.8 (0.6)	2.1 (0.5)
	T	2.2 (1.0)	4.4 (1.0)
F Distractors not fixated	+	19.4 (0.8)	31.6 (1.5)
	T	17.1 (0.8)	21.7 (1.4)
G Re-fixation rate on distractors	+	12.1 (1.2)	2.1 (0.4)
	T	22.1 (1.7)	9.6 (1.6)
H Dwell time on targets (ms)	+	319.5 (23)	395.3 (30)
	T	323.2 (22)	356.9 (30)
H Dwell time on distractors (ms)	+	228.2 (17.7)	200.0 (32)
	T	232.0 (15.0)	225.0 (19)

Values represent means and standard deviations are given in parenthesis. Values for efficient search denoted by "+" symbol and for inefficient search by "T" symbol.

Second, the pattern of eye movements for patients differed significantly from that of controls for *both* efficient and inefficient search in several parameters. For example, the refixation rate, defined as the [total number of refixations on targets/total number of targets fixated], was significantly higher for patients in both efficient and inefficient search (Table 1C). The median interval between a fixation on a target, and a subsequent refixation, in terms of the number of intervening fixations, also varied significantly between the two groups (Table 1E). Patients demonstrated significantly *longer* intervals than control subjects for efficient search — but significantly shorter intervals for inefficient search. It appears that patients adopt a different oculomotor strategy to controls during inefficient search, making many, short interval, refixations on a target — despite the fact that overall search performance remains unimpaired. Importantly this high target refixation rate for these patients with mild to moderate PD appears *not* to be due to a spatial working memory deficit for previously fixated targets, as indexed by normal re-click rates (Table 1D). Note also that the distractor refixation rate was also significantly higher for patients on *both* types of search (Table 1G).

Finally, patients also exhibited abnormal gaze patterns selectively for efficient — but not inefficient — search in a number of ways. They fixated more distractors than controls on efficient search (Table 1F). Furthermore, while dwell time on targets (total time spent fixating an item) was significantly shorter for efficient search, dwell times on distractors were longer for patients with PD (Table 1H). Thus, these patients appear to be captured more by distractors while spending less time processing targets than healthy controls.

Discussion

The present study investigated visual search in patients with mild to moderate PD, using a multi-target paradigm. Our results show that patients were impaired on the efficient search task (search for +'s amongst L's), finding fewer targets than age-matched control subjects. By contrast, they were unimpaired searching for T's amongst Ls (inefficient search).

We found no evidence to support the proposal that there might be a spatial working memory deficit for locations visited during search in this group of mildly affected patients. Patient re-click rates did not differ significantly from that of age-matched controls, indicating good memory for locations of targets already identified. Such a finding is consistent with studies suggesting that a spatial working memory deficit in PD does not occur in the early stages of the disease (Kemps et al., 2005).

We interpret our results as follows. Target detection relies both on top-down processing (knowledge of the target properties) and bottom-up (stimulus defined) processing. When the target is visually highly salient from distractors, bottom-up processing pre-dominates, allowing the target to "pop-out" from the background of distractors. A reduction in target saliency will therefore affect efficient, compared to inefficient search, to a greater extent. Thus, we find that our patients, during efficient search, miss more targets, fixate more distractors, and make more refixations on both targets and distractors, exhibiting abnormally short dwell times on targets. Although the pattern of gaze was also abnormal in inefficient search, particularly in terms of refixation rates, patients appear able to compensate for the (putative) saliency deficit in inefficient search. A saliency deficit in PD is consistent with a number of other studies of visual search (see Horowitz et al., 2006, for a review). Dopamine is known to mediate saliency processing within the BG and associated cortico-striatal pathways. The findings presented here would be consistent with a moderate deficit in the representation of visual salience in individuals with PD.

Acknowledgements

This research was support by grants to C. Kennard and M. Husain from the Wellcome Trust (073735).

References

Horowitz, T.S., Choi, W.Y., Horvitz, J.C., Cote, L.J. and Mangels, J.A. (2006) Visual search deficits in Parkinson's disease are attenuated by bottom-up target salience and top-down information. Neuropsychologia, 10: 1962–1977.

Husain, M., Mannan, S., Hodgson, T., Wojcuilik, E., Driver, J. and Kennard, C. (2001) Impaired spatial working memory across saccades contributes to abnormal search in parietal neglect. Brain, 124: 941–952.

Kemps, F., Szmalec, A., Vandierendonck, A. and Crevits, L. (2005) Visuo-spatial processing in Parkinson's disease: evidence for diminished visuo-spatial sketch pad and central executive resources. Parkinsonism Relat. Disord., 11: 181–186.

Mannan, S.K., Mort, D.J., Hodgson, T., Driver, J., Kennard, C. and Husain, M. (2005) Revisiting previously searched locations in visual neglect: role of right parietal and frontal lesions in misjudging old locations as new. J. Cogn. Neurosci., 17: 340–354.

C. Kennard & R.J. Leigh (Eds.)
Progress in Brain Research, Vol. 171
ISSN 0079-6123

CHAPTER 6.11

Ocular motor anatomy in a case of interrupted saccades

Janet C. Rucker[1,*], R. John Leigh[2], Lance M. Optican[3], Edward L. Keller[4] and Jean A. Büttner-Ennever[5]

[1]*Rush Medical Center, Chicago, IL, USA*
[2]*Veterans Affairs Medical Center and Case University, Cleveland, OH, USA*
[3]*National Eye Institute, Bethesda, MD, USA*
[4]*Smith-Kettlewell Eye Institute, San Francisco, CA, USA*
[5]*Ludwig-Maximilian University, Munich, Germany*

Abstract: Saccades normally place the eye on target with one smooth movement. In late-onset Tay–Sachs (LOTS), intrasaccadic transient decelerations occur that may result from (1) premature omnipause neuron (OPN) re-activation due to malfunction of the latch circuit that inhibits OPNs for the duration of the saccade or (2) premature inhibitory burst neuron (IBN) activation due to fastigial nucleus (FN) dysregulation by the dorsal cerebellar vermis. Neuroanatomic analysis of a LOTS brain was performed. Purkinje cells were absent and gliosis of the granular cell layer was present in the dorsal cerebellar vermis. Deep cerebellar nuclei contained large inclusions. IBNs were present with small inclusions. The sample did not contain the complete OPN region; however, neurons in the OPN region contained massive inclusions. Pathologic findings suggest that premature OPN re-activation and/or inappropriate firing of IBNs may be responsible for interrupted saccades in LOTS. Cerebellar clinical dysfunction, lack of saccadic slowing, and significant loss of cerebellar cells suggest that the second cause is more likely.

Keywords: fastigial nucleus; omnipause neurons; burst neurons; latch circuit; brainstem

Introduction

Late-onset Tay–Sachs (LOTS) is a predominantly cerebellar, autosomal recessive disorder of sphingolipid metabolism with intracerebral GM2 ganglioside accumulation (inclusions) caused by deficiency of the enzyme hexosaminidase A. Saccades in LOTS are interrupted by transient decelerations, during which velocity abruptly declines but generally remains greater than 50°/s (Rucker et al., 2004). Normal initial velocity suggests integrity of excitatory burst neurons (EBNs) and motoneurons. Interrupted saccades may be due to (1) premature omnipause neuron (OPN) re-activation due to malfunction of the latch circuit that inhibits OPNs for the saccade duration or (2) premature inhibitory burst neuron (IBN) activation due to fastigial nucleus (FN) dysregulation by the dorsal cerebellar vermis.

*Corresponding author. Tel.: +1 312 942 4500;
Fax: +1 312 942 2380; E-mail: janet_rucker@rush.edu

DOI: 10.1016/S0079-6123(08)00680-8

Methods

Three brainstem and cerebellar tissue blocks in 20% formalin from a patient with LOTS were obtained from the Brain and Tissue Bank for Developmental Disorders (Baltimore, MD) under contracts NO1-HD-3368 and NO1-HD-4-3383. Axial brainstem and parasagittal cerebellar sections were stained with Nissl for highlighting neuronal cytoarchitecture and glia, Luxol Fast Blue (LFB) for cell bodies and fibres, and Holzer for glial fibres and detection of gliosis. Brainstem sections were processed for immunocytochemical detection of parvalbumin (PAV), glycine transporter, and synaptophysin.

Results

There was nearly complete loss of Purkinje cells and severe atrophy of the granular cell layer throughout lobules I–IX, including the dorsal cerebellar vermis (Fig. 1A). Diffuse gliosis was seen in white matter underlying the Purkinje cell layer (Fig. 1B). There was sparing of the cerebellar nodulus and flocculus, with intact Purkinje and granular cell layers without evidence of severe gliosis (Fig. 1C, D). Deep cerebellar nuclei were present but contained large inclusions. Medial deep nuclei appeared to be more severely distorted with inclusions than the dentate nucleus.

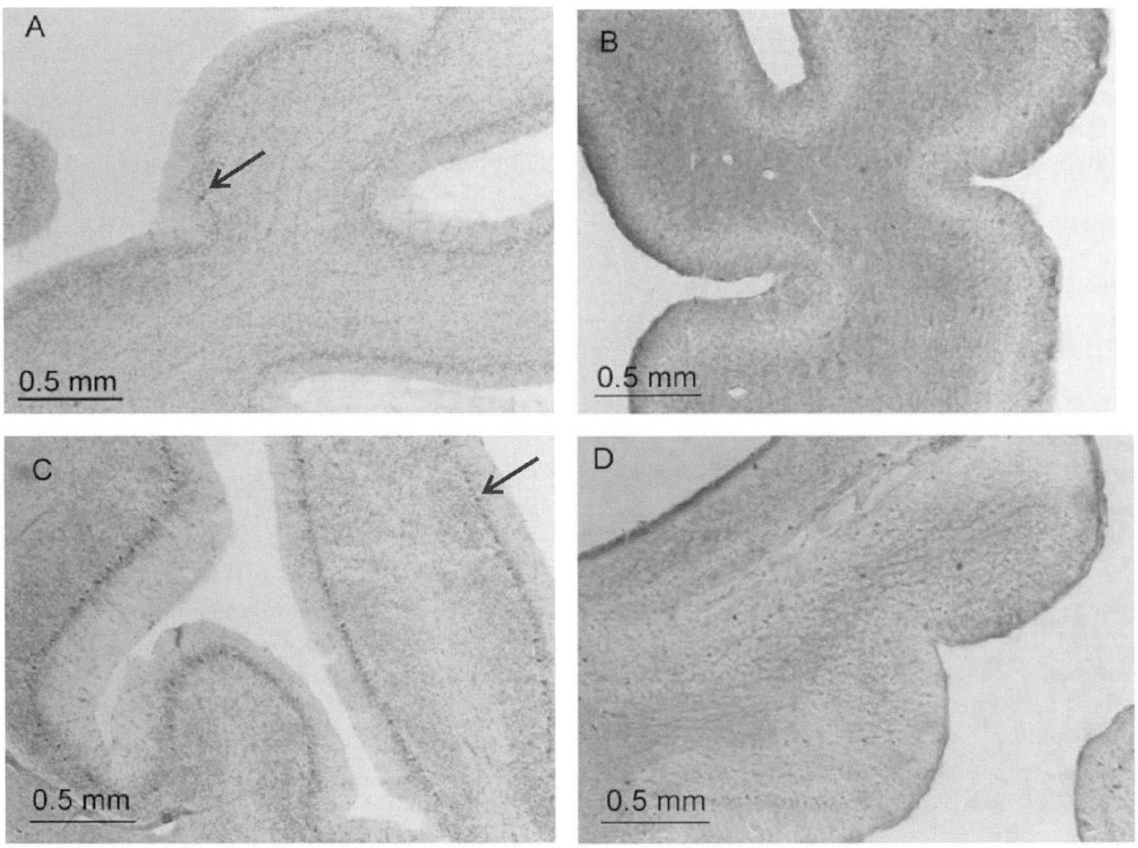

Fig. 1. Cerebellar vermis photomicrographs. (A) Luxol stain through vermis lobules reveals a nearly complete loss of Purkinje cells (one visible at arrow) and atrophy of the granular cell layer. (B) Holzer stain through vermis lobules reveals a thick area of violet staining, consistent with granular cell layer and white matter gliosis. (C) Luxol stain through the nodulus with preservation of Purkinje cells (arrow) and granular cell layer. (D) Holzer stain through the nodulus appears normal without gliosis. (See Color Plate 6.11.1 in color plate section.)

Inclusions and gliosis were present throughout the brainstem. The abducens nucleus appeared normal and there appeared to be cells with PAV-positivity in the abducens nucleus. Putative EBNs with small inclusions were identified in the paramedian pontine reticular formation (PPRF) in the dorsal pons just rostral to the abducens nucleus on account of their medium-size and PAV-positivity. Putative IBNs (medium-sized) were identified caudal and medial to the abducens nucleus and appeared fairly intact despite small inclusions. PAV staining was very light in the IBN region and few PAV-positive cells were identified; however, functional synapses with synaptophysin staining were present on the neurons. Putative-OPNs were found, but the sample did not contain the complete OPN region so that generalized statements on the OPN population are not possible. Neurons in the OPN vicinity and medial pons contained massive inclusions (Fig. 2A–D) and were more severely affected than deep cerebellar nuclei, in general. The OPN region did not stain for PAV

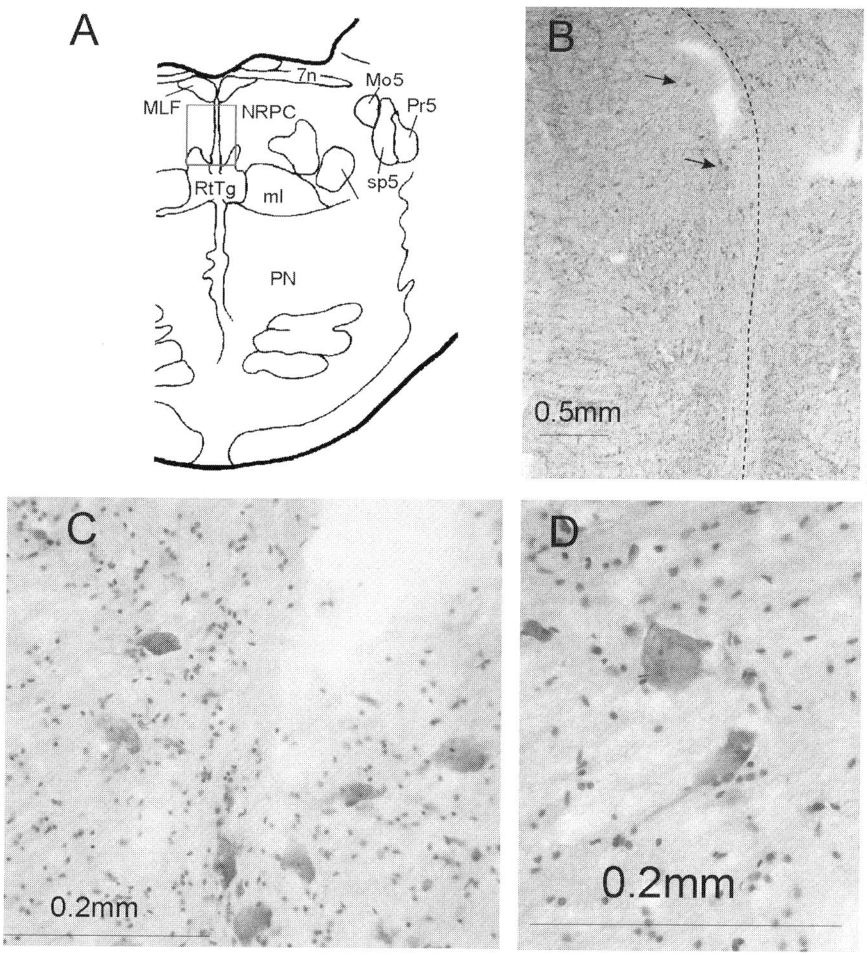

Fig. 2. (A) Drawing of a transverse caudal pontine section demonstrating the region of the nucleus raphe interpositus (RIP) (box), containing OPNs shown in B–D. MLF, medial longitudinal fasciculus; *7n*, seventh nerve; NRPC, nucleus reticularis pontis caudalis; ml, medial lemniscus; PN, pontine nuclei; Mo5, trigeminal motor nucleus; RtTg, reticulotegmental pontine nucleus. (B) Low-power photomicrograph of the Luxol-stained RIP (box in A). Hatched line is pontine midline, with OPNs just off the midline (black arrows). (C and D) High-powered photomicrographs of Luxol-stained OPNs with massive inclusions. (See Color Plate 6.11.2 in color plate section.)

or glycine transporter. The inferior olive and the medial vestibular nuclei appeared normal and the inferior olive stained strongly positive for synaptophysin.

Discussion

The most striking finding was the nearly complete absence of Purkinje cells in the dorsal cerebellar vermis (Fig. 1A, B). As Purkinje cells normally inhibit the FN, their absence suggests that FN is dysregulated. This may cause the cerebellar choke signal on the IBNs to activate prematurely, resulting in intrasaccadic decelerations. Within the brainstem, diffuse neuronal inclusions were seen. As expected, given normal initial velocity of LOTS saccades, EBNs and the abducens nucleus appeared relatively intact with minimal inclusions. Neuronal loss and extensive neuronal inclusions were clearly evident around the pontine raphé (Fig. 2A–D). This, in combination with the lack of staining to PAV and glycine transporter in the OPN region and the presence of degenerating cells nearby, suggests that the OPN region (and possibly the latch circuit, if it is located nearby) is abnormal and may contribute to saccadic interruption, although normal putative-OPN neurons could be found. PAV staining is associated with cells with high oxidative metabolism and fast-firing neurons, including EBNs, IBNs, and OPNs (Celio, 1990; Horn et al., 1995). Brainstem staining was light, raising the possibility of artifactual minimization due to long-standing formalin prefixation of the tissue. However, light PAV staining was identified in the IBN area, but not in the OPN area. In addition, the IBN area stained for synaptophysin. Synaptophysin staining was negative in the OPN, but the reason for this is unclear.

These anatomic results raise an interesting question regarding the correlation between neuronal structure and function. What proportion of a neuronal population must be lost before behavioural changes become manifest? Inclusions were seen in many brainstem neuronal populations but some of these appear to function normally, such as the EBNs, and LOTS patients have a paucity of clinical brainstem involvement. However, it is reasonable to conclude that the more extensive the inclusions, the less well the cell population functions. The most extensive inclusions were seen close to the OPN-area in the pons.

Pathologic findings suggest that interrupted saccades in LOTS may be caused by premature OPN re-activation and/or inappropriate firing of IBNs. Microstimulation in the OPN area mid-saccade in monkeys leads to interrupted saccades similar to those in LOTS (Keller et al., 1996), suggesting that premature OPN re-activation may be responsible. However, cerebellar clinical dysfunction, lack of saccadic slowing, and significant loss of cerebellar cells, suggest that inappropriate IBN firing is more likely.

Acknowledgements

This work was supported by the Office of Research and Development, Medical Research Service, Department of Veterans Affairs; NIH grant EY06717; the Evenor Armington Fund (R.J.L.); and the NEI IRP.

References

Celio, M.R. (1990) Calbindin D-28k and parvalbumin in the rat nervous system. Neuroscience, 35: 375–475.

Horn, A.K., Buttner-Ennever, J.A., Suzuki, Y. and Henn, V. (1995) Histological identification of premotor neurons for horizontal saccades in monkey and man by parvalbumin immunostaining. J. Comp. Neurol., 359: 350–363.

Keller, E.L., Gandhi, N.J. and Shieh, J.M. (1996) Endpoint accuracy in saccades interrupted by stimulation in the omnipause region in monkey. Vis. Neurosci., 13: 1059–1067.

Rucker, J.C., Shapiro, B.E., Han, Y.H., Kumar, A.N., Garbutt, S., Keller, E.L. and Leigh, R.J. (2004) Neuro-ophthalmology of late-onset Tay–Sachs disease (LOTS). Neurology, 63: 1918–1926.

C. Kennard & R.J. Leigh (Eds.)
Progress in Brain Research, Vol. 171
ISSN 0079-6123

CHAPTER 6.12

Mechanism of interrupted saccades in patients with late-onset Tay–Sachs disease

Lance M. Optican[1,*], Janet C. Rucker[2], Edward L. Keller[3] and R. John Leigh[4]

[1]*National Eye Institute, Bethesda, MD, USA*
[2]*Rush Medical College, Chicago, IL, USA*
[3]*Smith-Kettlewell Eye Research Institute, San Francisco, CA, USA*
[4]*Case Western Reserve University, Cleveland, OH, USA*

Abstract: In late-onset Tay–Sachs disease (LOTS), saccades are interrupted by one or more transient decelerations. Some saccades reaccelerate and continue on before eye velocity reaches zero, even in darkness. Intervals between successive decelerations are not regularly spaced. Peak decelerations of horizontal and vertical components of oblique saccades in LOTS is more synchronous than those in control subjects. We hypothesize that these decelerations are caused by dysregulation of the fastigial nuclei (FN) of the cerebellum, which fire brain stem inhibitory burst neurons (IBNs).

Keywords: fastigial nucleus; omnipause neurons; burst neurons; latch circuit

Introduction

Late-onset Tay–Sachs disease (LOTS) is an autosomal recessive disorder of sphingolipid metabolism, caused by deficiency of the enzyme hexosaminidase A, which leads to accumulation of GM2 ganglioside (inclusions) in the nervous system. Patients with LOTS typically make a series of small saccades in response to a target jump, but their peak speeds seem normal (Rucker et al., 2004). LOTS saccades are interrupted by transient decelerations, not always coming to a standstill, after which a new saccade occurs. In a mouse model of LOTS, GM2 storage is prominent in the motor cortex and cerebellum (granular cells and Purkinje cells), along with atrophy in the brainstem (Chern et al., 1976). Recent neuropathological studies of LOTS (Rucker et al., this volume) have demonstrated inclusions and reduction in metabolism in omnipause neurons (OPN), inclusions in fastigial nuclei (FN), and complete loss of Purkinje cells in the dorsal cerebellar vermis. We asked three questions about LOTS saccades to understand how these losses cause the saccade's interruptions: 1. Are saccades truncated if the target is turned off after it jumps? 2. Is there any periodicity to the decelerations? 3. Is the onset of saccade braking different for horizontal and vertical components of oblique saccades?

Methods

We used eye movements recorded with coils in 14 patients diagnosed with LOTS and 10 healthy control subjects (for details, see Rucker et al.,

*Corresponding author. Tel.: +301-496-9375; Fax: +301-402-0511; E-mail: LanceOptican@nih.gov

DOI: 10.1016/S0079-6123(08)00681-X

2004). Horizontal saccades were simulated in Matlab (Mathworks, Natick, MA).

Results

First, we found no difference in any patient between the accuracy of saccades made with the visual stimulus turned off 88 ms after each target jump (complete darkness) versus saccades made with the target continuously visible (Fig. 1A, B). Thus, vision was not needed to restart the saccade.

Second, we found no regularity to the timing of successive dips in peak velocity. This suggests they are not due to oscillations in the closed-loop circuit in the brain stem among inhibitory burst neurons (IBNs) and excitatory burst neurons (EBNs) (cf., Fig. 2A).

Third, we measured the asynchrony between the time of peak deceleration of the horizontal and vertical components of oblique saccades. LOTS patients were less asynchronous than controls, and the difference was significant ($p < 0.001$) for oblique saccades with large vertical and small horizontal components (Fig. 1C, D). Normally, the components of an oblique saccade stop when that component is over. Although there may be some stretching of the duration of the shorter saccade, it is not always enough, and so the two components can stop asynchronously. However, in LOTS, the two components of the decelerations end more synchronously than in controls (e.g., Fig. 1D). If the OPNs or choke (which inhibit both horizontal and vertical bursters) restarted prematurely, both components would decelerate simultaneously.

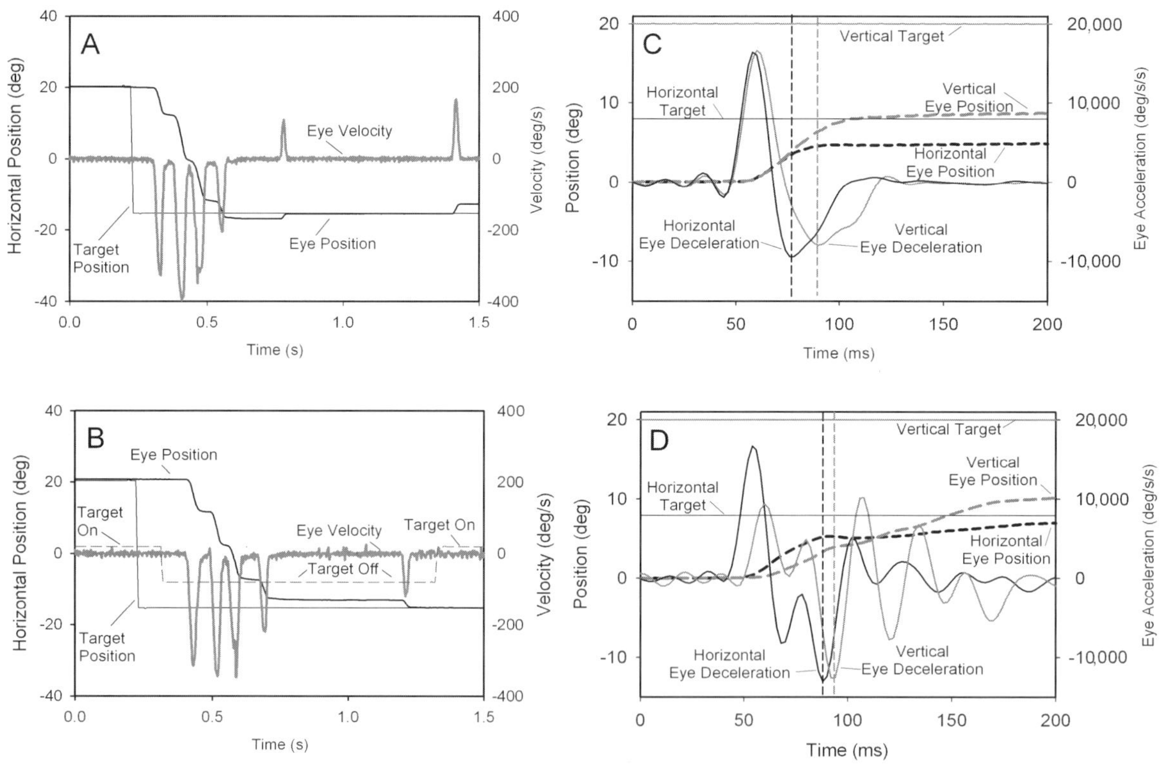

Fig. 1. (A) Horizontal saccades made by a LOTS patient to an illuminated target. (B) Target was turned off for 1 s, 88 ms after its jump. The patient's responses were similar under both conditions. (Note the lack of periodicity of the successive saccadic pulses.) (C) Horizontal and vertical components of a saccade made by a control subject to an oblique target jump (20° up and 8° right) show near-synchronous peak acceleration of both components, but peak decelerations (dashed vertical lines) occur 12 ms apart. (D) Saccade made by a LOTS patient (stimulus as in C) shows asynchronous peak acceleration of both components, but peak decelerations occurred within 5 ms. Note different axes. Upward and rightward movements are positive.

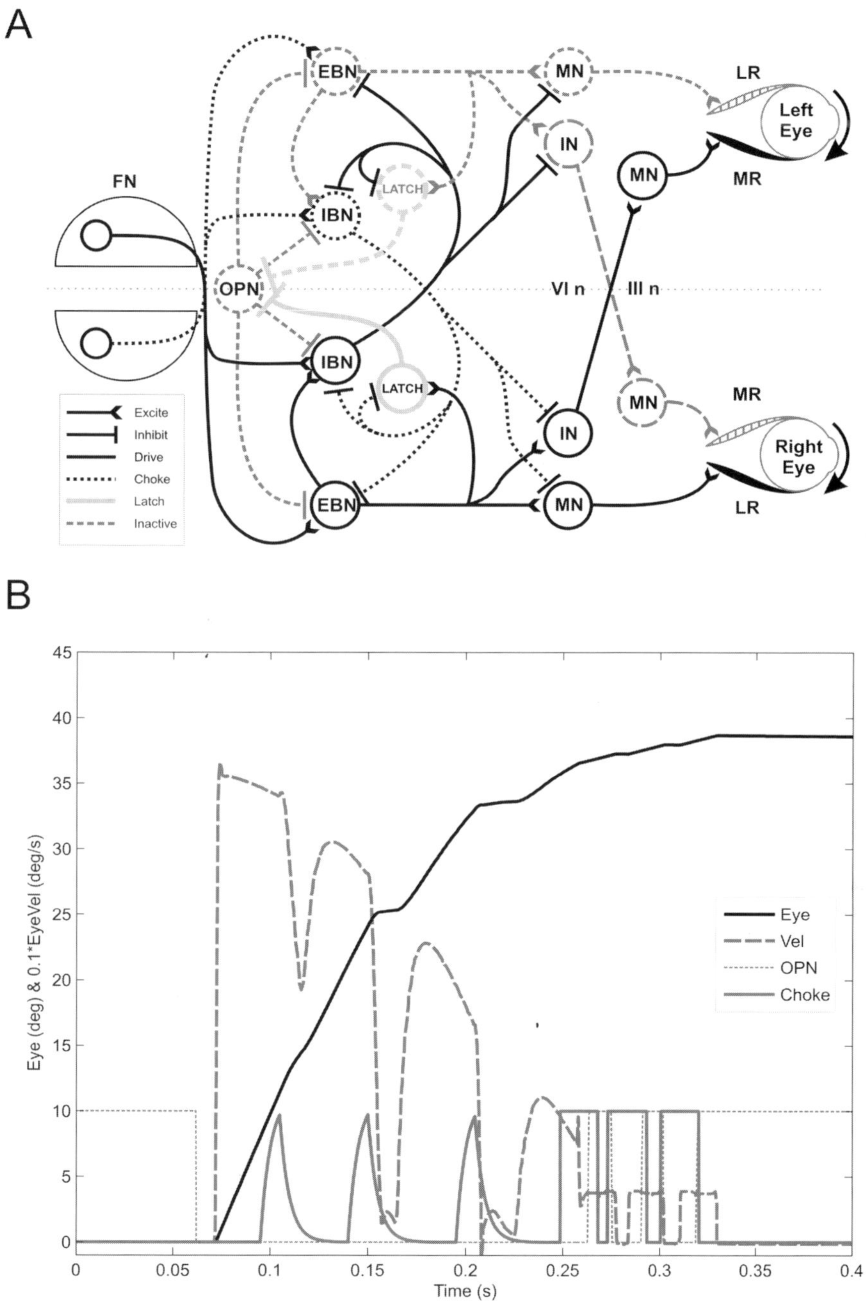

Fig. 2. Simulation of interrupted saccades in LOTS patients. (A) Model. EBNs provide drive for ipsilateral movement. IBNs cross over and prevent activity on the contralateral side. Choke signal on contralateral IBNs (triggered by ipsilateral FN) stops saccade and opens latch. (B) Simulation. OPN turns off, allowing a saccade to the target. EBNs drive Latch neurons, holding off OPN. Eye velocity dips are preceded by bursts in the choke signal. After 0.25 s, the saccade is near its end, and the OPN turns back on. The last two movements are normal corrective saccades (note OPN and choke cycling on and off).

Discussion

Normally, OPNs do not fire during saccades because they are held off by a latch circuit (Fig. 2A) (Keller and Missal, 2003). If EBN firing were greatly reduced during the deceleration because the OPN reactivated, one would expect the latch circuit to open and the saccade to stop. However, even when speed almost reaches zero, the saccade still restarts (Fig. 1A).

A hypothesis more consistent with our data is that contralateral IBNs activate too early, choking off the drive of the ipsilateral EBNs. Premature firing of the ipsilateral FN could generate this choke signal. The key difference between activating IBNs instead of OPNs is that in the former case the ipsilateral EBNs do not have to shut down, as their output can be choked off at the motor neuron (Fig. 2A). Thus, the OPN latch circuit would not be affected and the closed-loop saccadic system would drive the eye to the target.

We modelled lesions in the cerebellar vermis of LOTS patients by reducing their inhibition on FNs several times during a single saccade (Fig. 2B). FNs always receive excitatory inputs from mossy and climbing fibres, and will fire unless inhibited by the vermis. The loss of this inhibition has no effect on the FNs contralateral to the movement, because they normally start firing near the beginning of the movement. However, the ipsilateral FNs normally fire late to stop the saccade (Optican and Quaia, 2002). If they activate prematurely, they will turn on the contralateral IBN in the brain stem, decelerating the saccade.

Acknowledgements

This research was supported by the Office of Research and Development, Medical Research Service, Department of Veterans Affairs; National Eye Institute grants EY06717 and EY08060; the Evenor Armington Fund; and the Intramural Division of the National Eye Institute, NIH, DHHS.

References

Chern, J., Beutler, E., Kuhl, W., Gilbert, F., Mellman, W.J. and Croce, C.M. (1976) Characterization of heteropolymeric hexosaminidase A in human X mouse hybrid cells. Proc. Natl. Acad. Sci. U.S.A., 73(10): 3637–3640.

Keller, E.L. and Missal, M. (2003) Shared brainstem pathways for saccades and smooth-pursuit eye movements. Ann. N.Y. Acad. Sci., 1004: 29–39.

Optican, L.M. and Quaia, C. (2002) Distributed model of collicular and cerebellar function during saccades. Ann. N.Y. Acad. Sci., 956: 164–177.

Rucker, J.C., Shapiro, B.E., Han, Y.H., Kumar, A.N., Garbutt, S., Keller, E.L. and Leigh, R.J. (2004) Neuro-ophthalmology of late-onset Tay–Sachs disease (LOTS). Neurology, 63(10): 1918–1926.

C. Kennard & R.J. Leigh (Eds.)
Progress in Brain Research, Vol. 171
ISSN 0079-6123

CHAPTER 6.13

Conjugacy of horizontal saccades: application of binocular phase planes

Alessandro Serra[1,2,3,*], Ke Liao[1] and R. John Leigh[1]

[1]*Daroff-Dell'Osso Laboratory, Veterans Affairs Medical Center and Case Western Reserve University, Cleveland, OH, USA*
[2]*Institute of Clinical Neurology, University of Sassari, Sassari, Italy*
[3]*OASI Institute for Research and Care on Mental Retardation and Brain Aging, Troina, Italy*

Abstract: The goal of this study was to evaluate the ability of binocular phase planes to detect disconjugacy of horizontal saccades. We recorded horizontal saccades in 22 patients with disease affecting the brainstem reticular formation, medial longitudinal fasciculus (INO), abducens nerve, neuromuscular junction and extraocular muscles, and in 10 age-matched controls. We found that patients with disorders of the brainstem, including INO, or the cranial nerves, showed significant velocity disconjugacy in the first 10% of the displacement. Patients with myasthenia gravis did not show early disconjugacy, but might do so later in the course of the saccade. We conclude that binocular phase planes supplement conventional measures of saccades, such as peak velocity, and are useful to differentiate INO and cranial nerve palsies from myasthenia.

Keywords: internuclear ophthalmoplegia; myasthenia gravis; multiple sclerosis

Introduction

Saccades move the eyes quickly between features of interest in the visual world (Leigh and Zee, 2006). When examining a visual scene at optical infinity, saccades must be conjugate so that the fovea of each eye is aimed at corresponding points. Conjugacy of horizontal saccades depends on the integrity of the brainstem machinery and ocular motor periphery (Fig. 1). Of particular interest to clinicians is the ability to identify disconjugacy of horizontal saccades due to brainstem disease, especially internuclear ophthalmoplegia (INO), which is common in multiple sclerosis (MS) (Leigh and Zee, 2006). A special challenge arises when patients with peripheral ocular motor disorders, such as myasthenia gravis (MG), present with the appearance of INO (pseudo-INO). Prior approaches to identify INO have compared the speed of abducting and adducting eyes measuring, for example, the versional disconjugacy index — the ratio of peak velocity of each eye (Frohman et al., 2002), or the first-pass amplitude — the ratio of the abducting to the adducting eye position, when the abducting eye first reaches an eccentric target of fixation (Frohman et al., 2003). However, these methods measure only one point in the saccade, and do not take into account times of onset and subsequent course of the saccade in each eye, for which normals show differences (Leigh and Zee, 2006). Recently, we showed that patients with

*Corresponding author. Tel.: +(216) 421-3224; Fax: +(216) 231-3461; E-mail: axs505@case.edu

DOI: 10.1016/S0079-6123(08)00682-1

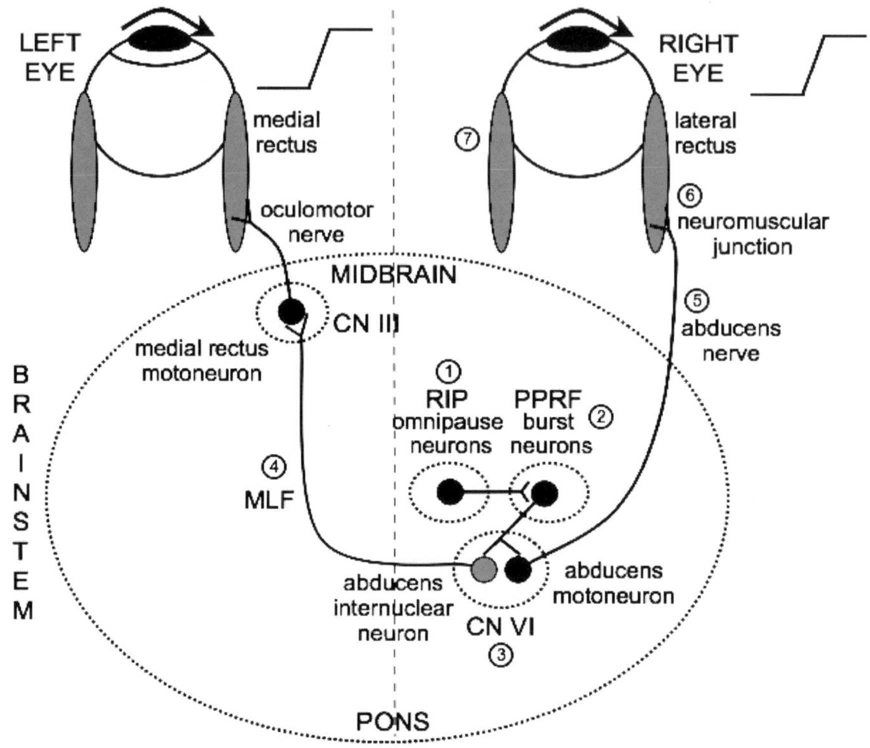

Fig. 1. Schematic of sites of lesions causing disjunctive saccades studied (see text for details).

pseudo-INO due to MG could be differentiated from patients with true INO using binocular phase planes (Khanna et al., 2007). We now extend this approach to address disorders affecting the ocular motor saccadic system at different sites of lesions, from the brainstem to the periphery (Fig. 1).

Subjects and methods

We studied a group of 22 patients (age range 22–70 years, 4 females) with abnormal saccades due to a range of disorders: brainstem stroke (2); saccadic palsy following cardiac surgery (Solomon et al., 2007) (2); putative abducens nucleus lesion (1) (Miller et al., 2002); INO due to MS (7), or brainstem stroke (2), or with episodic ataxia type 2 (EA2 — 1); abducens nerve palsy (2); MG (3); chronic progressive external ophthalmoplegia (CPEO — 2). We also studied 10 age-matched, healthy control subjects (age range 30–60 years, 3 females). All patients and control subjects gave informed written consent, in accordance with the Declaration of Helsinki and the Institutional Review Board of the Cleveland Veterans Affairs Medical Center. Subjects made horizontal saccades in response to 5–40° jumps of a visual target located at 1.2 m on a tangent screen, both under monocular and binocular viewing conditions. We measured horizontal and vertical positions of each eye using the magnetic search coil technique. To compare corresponding velocity profiles of each eye, we used the phase-plane technique, which we have described previously in detail (Khanna et al., 2007). For each patient, the average "velocity disconjugacy" for at least 10 saccades was plotted against the normalized position of the "weak eye" (velocity disconjugacy plots). We defined 95% prediction intervals based on pooled data (1400 saccades) from controls, and determined if patients' data fell outside of those intervals.

Results

Results of binocular phase plane analyses are summarized for representative patients in Fig. 2. Saccades were found to be disconjugate in two patients with post-cardiac surgery saccadic palsy (Fig. 1, site 1), and in one of the two patients with brainstem stroke who had left gaze palsy (Fig. 1, site 2). The patient with abducens nucleus palsy (Fig. 1, site 3) showed conjugate very slow saccades from the intact field of gaze back to centre. All 10 patients with INO (Fig. 1, site 4), and the 2 patients with abducens nerve palsy (Fig. 1, site 5), showed early velocity disconjugacy in the first 10% of the eye displacement, which remained abnormal throughout the whole eye movement. The initial components of horizontal saccades from three patients with MG (Fig. 1, site 6), two with pseudo-INO and one with pseudo-abducens nerve palsy, were found to be similar to controls. These patients only showed horizontal disconjugacy later in the course of the saccades (starting after the first 10% of the eye displacement). Of the two patients with CPEO (Fig. 1, site 7), only the one with advanced disease showed abnormal horizontal saccadic conjugacy.

Discussion

The main finding of this study is that binocular phase-plane analysis of horizontal saccades provides a sensitive means to identify the location of lesions causing disconjugate movements. We attribute the power of this method to two factors: (1) removing time differences between the movements of the two eyes (such as onset) and (2) providing a comparison of the entire course of the saccade for each eye. In comparison, current methods focus on measuring one point in the saccade (e.g., peak velocity). The method was most effective in distinguishing disconjugacy due to INO or cranial

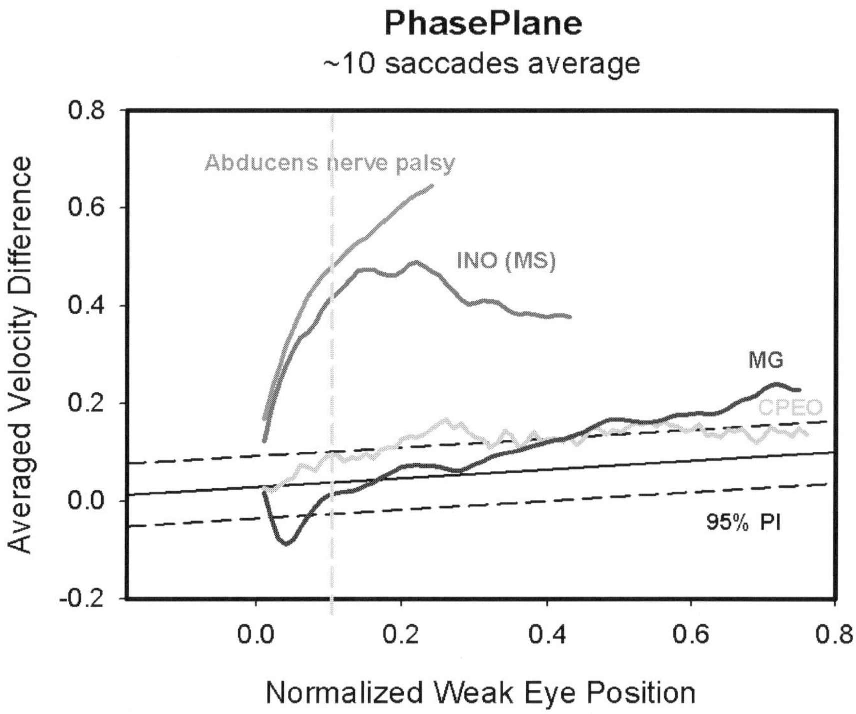

Fig. 2. Data from representative patients showing normalized velocity difference between each eye plotted as a function of normalized eye position. Dotted lines indicate 95% prediction intervals for controls (see text).

nerve palsy from neuromuscular disease (MG). In the case of brainstem gaze palsy, saccades were disconjugate between 20 and 50% of eye movement in 3/5 patients with lesions that may have affected burst neurons in the paramedian pontine reticular formation (PPRF), omnipause neurons in the nucleus raphe interpositus (RIP), or the abducens nucleus (CN VI). In some of these cases, a strategy of using vergence movements to substitute for saccades during gaze shifts might produce disjunctive movements (Solomon et al., 2007). In CPEO, saccades were only disconjugate in the patient with advanced disease.

Current evidence suggests that the pale fibres of the global layer of the extraocular muscles are spared in MG and CPEO, and are therefore able to induce fast saccades, ultimately assuring the two eyes conjugacy at least in the beginning of the eye movement (Khanna et al., 2007). Pale global fibres are singly innervated, have fast-twitch properties, and low fatigue resistance. They are likely to be spared in ocular myasthenia for two reasons: they contract less frequently than the orbital fibres, being active mainly during saccades, and because of the presence of a substantial postjunctional synaptic folding, they have a higher safety factor, when compared to other extraocular fibres (Khanna et al., 2007). These fibres are also probably spared in CPEO since they are relatively poor in mitochondria (Leigh and Zee, 2006). We propose that the initial conjugate component of saccades is generated by the global pale fibres of the extraocular muscles, for which medial and lateral rectus receive similar commands (Hering's law of equal innervation). In summary, binocular phase planes provide a useful way for clinicians to evaluate horizontal saccadic palsies, and for basic scientists to examine mechanisms underlying conjugate saccades.

Acknowledgements

This work was supported by the Department of Veterans Affairs, NIH R01-EY06717, Evenor Armington Fund. We thank Dr. David Zee for making available data due to a putative abducens nucleus lesion.

References

Frohman, E.M., Frohman, T.C., O'Suilleabhain, P., Zhang, H., Hawker, K., Racke, M.K., Frawley, W., Phillips, J.T. and Kramer, P.D. (2002) Quantitative oculographic characterisation of internuclear ophthalmoparesis in multiple sclerosis: the versional dysconjugacy index Z score. J. Neurol. Neurosurg. Psychiatry, 73: 51–55.

Frohman, E.M., O'Suilleabhain, P., Dewey, R.B., Jr., Frohman, T.C. and Kramer, P.D. (2003) A new measure of dysconjugacy in INO: the first-pass amplitude. J. Neurol. Sci., 210: 65–71.

Khanna, S., Liao, K., Kaminski, H.J., Tomsak, R.L., Joshi, A. and Leigh, R.J. (2007) Ocular myasthenia revisited: insights from pseudo-internuclear ophthalmoplegia. J. Neurol., 254(11): 1569–1574.

Leigh, R.J. and Zee, D.S. (2006) The Neurology of Eye Movements (Book/DVD) (4th ed.). Oxford University Press, New York.

Miller, N.R., Biousse, V., Hwang, T., Patel, S., Newman, N.J. and Zee, D.S. (2002) Isolated acquired unilateral horizontal gaze paresis from a putative lesion of the abducens nucleus. J. Neuroophthalmol., 22: 204–207.

Solomon, D., Ramat, S., Tomsak, R.L., Reich, S.G., Shin, R.K., Zee, D.S. and Leigh, R.J. (2007) Saccadic palsy after cardiac surgery: characteristics & pathogenesis. Ann. Neurol., In press.

C. Kennard & R.J. Leigh (Eds.)
Progress in Brain Research, Vol. 171
ISSN 0079-6123

CHAPTER 6.14

The neuroanatomical basis of slow saccades in spinocerebellar ataxia type 2 (Wadia-subtype)

S. Geiner[1], A.K.E. Horn[1], N.H. Wadia[2], H. Sakai[3] and J.A. Büttner-Ennever[1,*]

[1]*Institute of Anatomy, Ludwig-Maximilian University of Munich, Munich, Germany*
[2]*Department of Neurology, Jaslok Hospital and Research Centre, Mumbai, India*
[3]*Department of Anatomy, University of Nagoya, Japan*

Abstract: In a case of spinocerebellar ataxia type 2, Wadia-subtype (SCA2), with slow horizontal saccades, we used parvalbumin immunohistochemistry to identify the omnipause (OPNs) excitatory (EBNs), and inhibitory burst neurons (IBNs) of the saccade generator. Nissl sections was used to measure neuronal diameters, and synaptophysin staining to estimate of synaptic density on the cell somata. Morphometric and synaptic density measurements of the abducens motoneurons were identical in SCA2 and the control. A significant cell loss and reduced synaptic density on somata was found only in the EBN area. We conclude that degeneration of the EBNs is the most likely cause for the slowing of horizontal saccades.

Keywords: horizontal excitatory burst neurons; omnipause neurons; inhibitory burst neurons; parvalbumin; synaptic density; olivopontocerebellar atrophy; abducens motoneurons

Introduction

The classification of hereditary cerebellodegenerative disorders has undergone many alterations in the last 20 years, and in order to interpret the older literature it is important to understand these changes. The cases were usually referred to as olivopontocerebellar atrophies (OPCA) (Duvoisin and Plaitakis, 1984). Harding separated out the disorders into three categories: autosomal dominant cerebellar ataxia type 1 (ADCA1), characterized by degeneration of brainstem nuclei, the spinal cord, and cerebellar afferents; type 2 is similar and includes retinal degeneration; type 3 typically has only cerebellar signs (Harding, 1983). Ten years later molecular genetic studies showed that ADCAs can be caused by an expanded CAG triplet repeat that can occur on different chromosomes (Rosenberg, 1995). The resulting spinocerebellar atrophies (SCA) were then categorized depending on which chromosome the repeat is located. When the CAG repeat locus is on chromosome 6p it results in SCA type 1 (SCA1), on chromosome 12q results in SCA2, on chromosome 14q results in SCA3 (Machado–Joseph disease), and so on. The size of the triplet repeat was found to correlate with the severity of the disease and age of onset (Nance, 1997).

Wadia and Swami (1971) described an unusual type of OPCA in which one of the first clinical signs was the slowing of horizontal saccades. Since this report, similar cases have been recognized all over the world; they belong to the category SCA2

*Corresponding author. Tel.: +0049 89 5160 4851;
Fax: +0049 89 5160 4802;
E-mail: jean.buettner-ennever@med.uni-muenchen.de

DOI: 10.1016/S0079-6123(08)00683-3

(Wadia et al., 1998). In Wadia's SCA2 cases other types of eye movements, such as vertical saccades, the slow compensatory phases of vestibulo-ocular and optokinetic reflexes, convergence and, to a certain extent smooth pursuit, remained initially unaffected. The SCA2 phenotype is thought to be characterized best by the early slowing of horizontal saccades (Dürr et al., 1995; Bürk et al., 1997; Wadia et al., 1998; Klockgether et al., 2000), and this, along with the absence of both square wave jerks and gaze-evoked nystagmus, can be used as a reliable diagnositic feature (Büttner et al., 1998; Bürk et al., 1999).

In SCA2, the cause of the slow saccades was thought to be associated with the marked degeneration of the pons where the saccade generator lies (Wadia et al., 1998; Leigh and Zee, 2006). In this study, we have analysed the excitatory and inhibitory burst neurons (EBNs and IBNs) as well as the omnipause neurons (OPNs) of the saccade generator (Horn et al., 1995; see Büttner-Ennever, this volume) in a case of SCA2 to see if their loss could account for the saccadic slowing. We present here the measurements of cell diameter and the synaptic inputs densities to these cell groups, and show that specific degeneration of the EBNs is the most likely cause for the slowing of horizontal saccades. These results formed the basis of a doctoral thesis (Geiner, 1995).

Case history and methods

At the age of 35 years a female presented with progressive symptoms of SCA2 (case AZ of family AB in Wadia, 1984: same case as Ying et al., this volume). Her oculomotor disorders included slow horizontal saccades, blinking, and head jerks. Vestibular and optokinetic compensatory responses were intact, but fast phases were slow or absent. Smooth pursuit, convergence and pupillary responses could all be elicited. As the disease progressed vertical gaze became affected and the eyes appeared fixed, but could be moved with an effort. She died at the age of 41 years.

The brain was small and weighed 891 g. Paraffin sections (20 μm) were stained with Nissl technique and used for morphometric measurements of cell diameter (Fig. 1.). Parvalbumin (PAV) immunostaining (monoclonal, mouse, Sigma) was used to locate the EBNs, OPNs, and IBN in the caudal pons (Fig. 2) (Horn et al., 1994; Horn et al., 1995). Synaptophysin immunochemistry (monoclonal, Dako Antisynaptophysin SY38) counterstained with haemalaun was used to quantify the density of synapses on neuronal soma and dendrites in the brainstem (Fig. 3). The synaptic count of each cell was made around the perimeter and at one focal setting, only. The measurements were compared with similar sections from control brains with no known neurological deficits.

Results

The brainstem tegmentum in the SCA2 was so shrunken that it was comparable in size to the pyramidal tract, see Fig. 1B. On Nissl sections the cell density was slightly increased and many cells appeared normal; but in PAV sections stained cells were sparse, especially in the EBN area of nucleus reticularis pontis caudalis (NRPC), see arrow in Fig. 1B. In order to compare the general shrinkage parameters of the SCA2 case with the control cases we first compared measurements from the abducens nuclei, where no degenerative changes were expected. The abducens nucleus looked normal and cell diameters were comparable (Normal: $x = 26.18$ μm, SD$\pm$4.32, $N = 72$; SCA2: $x = 26.21$, SD$\pm$4.71, $N = 60$), as were the synaptic densities on the abducens somata (Normal: 0.47 syn/μm, SD$\pm$0.04; SCA2: 0.47 syn/μm, SD$\pm$0.05), see Fig. 3A, B. Since the abducens measurements were so similar we made no further adjustments for shrinkage artefacts.

EBN area

In the PAV sections of the SCA2 there was a significant loss of medium-sized neurons in the EBN area (Fig. 1A and B, arrows). They comprised 12% of the total cell population compared to 30% in normals. In Nissl sections neurons were shrunken (Normal: $x = 28.04$ μm, SD$\pm$5.88, $N = 310$; SCA2: $x = 22.78$, SD$\pm$7.41, $N = 38$) (Fig. 2A, B, arrows) and showed a significant loss of synapses on the soma compared

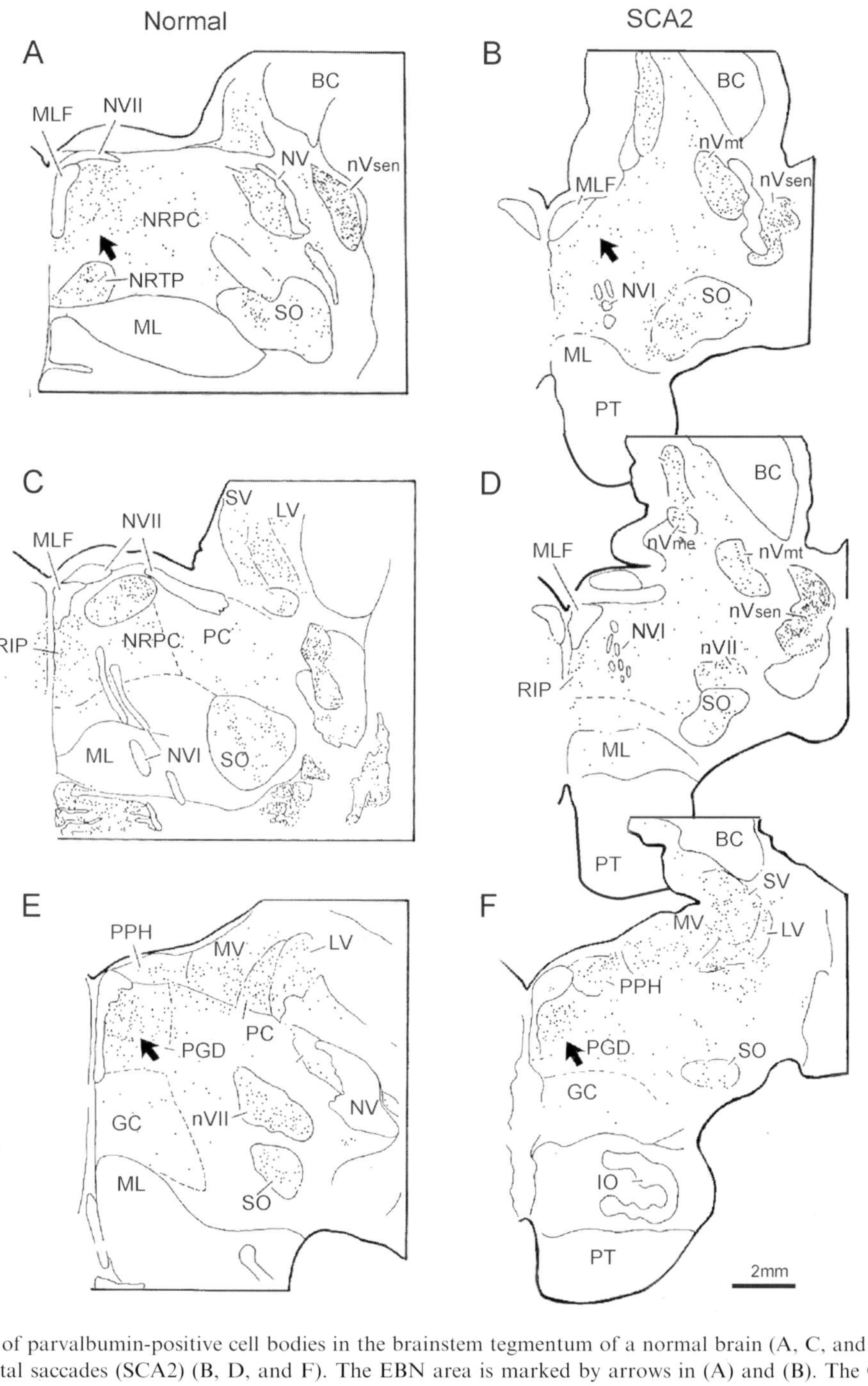

Fig. 1. The location of parvalbumin-positive cell bodies in the brainstem tegmentum of a normal brain (A, C, and E) compared to a case of slow horizontal saccades (SCA2) (B, D, and F). The EBN area is marked by arrows in (A) and (B). The OPNs lie near the midline in RIP in (C) and (D). The IBN area is marked by arrows in (E) and (F). Note the neuronal loss in the EBN area of SCA2. Abbreviations: BC, brachium conjuctivum; GC, nucleus gigantocellularis; IO, inferior olive; LV, lateral vestibular nucleus; ML, medial lemniscus; MLF, medial longitudinal fasciculus; MV, medial vestibular nucleus; nVme, mesencephalic trigeminal nucleus; nVmt, motor trigeminal nucleus; nVsen, sensory trigeminal nucleus; NRPC, nucleus reticularis pontis caudalis; NRTP, nucleus reticularis tegmenti pontis; NV, trigeminal nerve; NVI, abducens nerve; NVII, facial nerve; PC, nucleus reticularis parvocellularis; PGD, nucleus paragigantocellularis dorsalis; PPH, nucleus prepositus; PT, pyramidal tract; RIP, nucleus raphé interpositus; SO, superior olive; SV, superior vestibular nucleus.

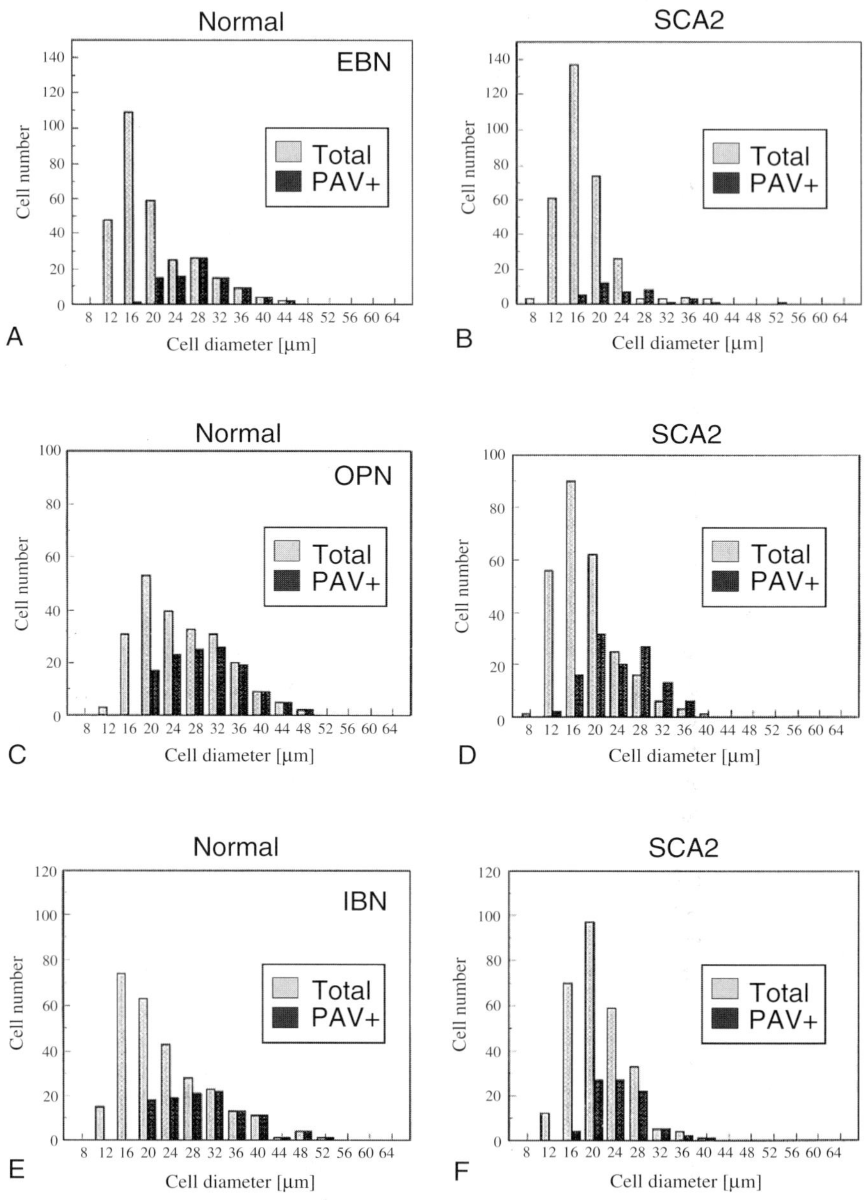

Fig. 2. Measurements of the average neuronal diameter (Dmax + Dmin/2) of normal (A, C, and E) and SCA2 (B, D, and F) brains. Grey bars indicate data taken from Nissl sections to estimate the diameters of the total population in the EBN area (A and B), the OPN area (C and D), and the IBN area (E and F). Black bars indicate the measurements taken from PAV-positive neurons which include EBN, OPN, and IBN neurons (Horn et al., 1995). In the SCA2 EBN, OPN, and IBN cell groups all showed some shrinkage. Note the loss of medium-sized PAV cells in the SCA2 EBN area (B).

with the normals (Normal: 0.48 syn/µm, SD $\pm$0.1; SCA2: 0.13 syn/µm, SD $\pm$0.08), see Fig. 3C, D. Occasionally, the cell dendrites were visible and appeared more fully innervated than the soma.

OPN area

Abundant medium-sized PAV-positive neurons typical of OPNs were found in SCA2 within about

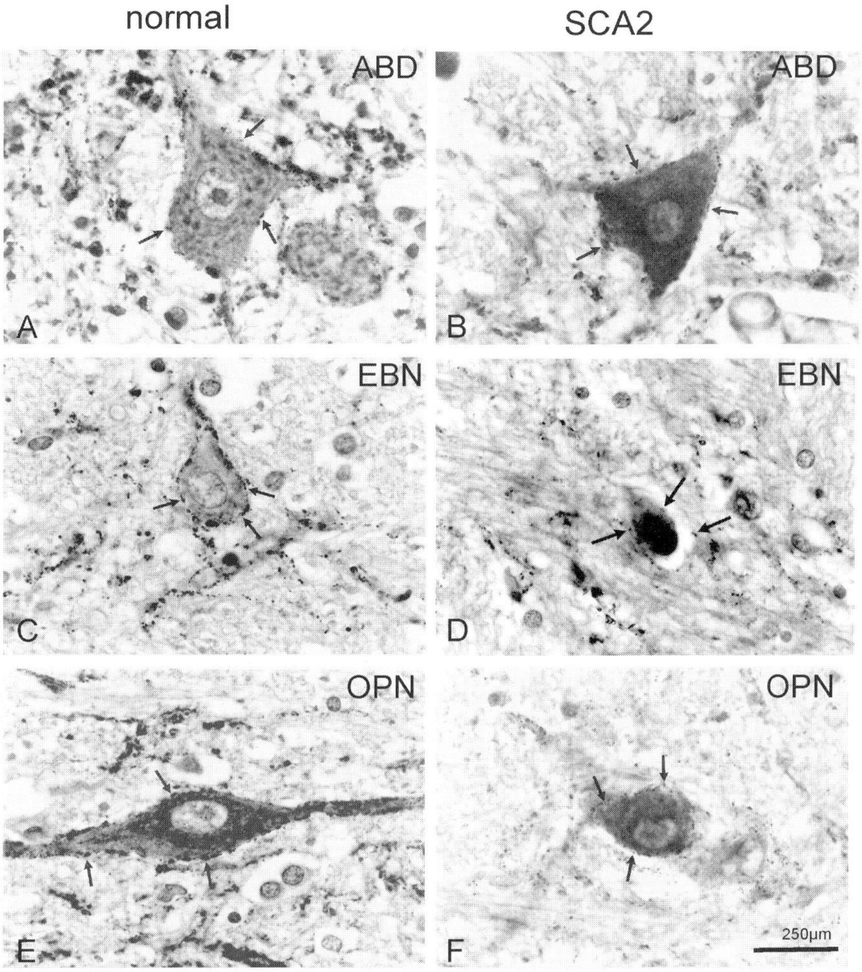

Fig. 3. Photomicrographs of Nissl stained sections from normal and SCA2 material, counter-stained with synaptophysin to label synapse (black dots). (A) and (B) show abducens motoneurons; (C) and (D) EBNs; (E) and (F) OPNS. As evidenced by the increased density of cytoplasm the SCA2 EBN cell has shrunken leaving a halo around it. Synapses can be seen both on the soma and left behind in the neuropil (arrows). In such a case the cell diameter was estimated as the size of the cell + halo. Note the scarcity of synapses in (D).

500 μm of the midline at the level of the abducens rootlets (Fig. 1 in RIP, and Fig. 3E, F). They comprised 45% of the total cell population compared to 61% in normals. In contrast to EBNs in SCA2 the synaptic density on the OPN soma was increased compared with normal (Normal: 0.36 syn/μm, SD ± 0.06; SCA2: 0.50 syn/μm, SD ± 0.06). The synaptic density of the proximal dendrites in the OPN area was not significantly different (Normal: 0.51 syn/μm, SD ± 0.08; SCA2: 0.58 syn/μm, SD ± 0.08). Both the cells and their synaptic boutons in the OPN area were shrunken compared with those of the normal case (Normal: $x = 27.84$ μm, SD ± 6.65, $N = 227$; SCA2: $x = 22.05$, SD ± 5.9, $N = 260$) (Figs. 2C, D and 3C, D).

IBN area

There was less neuronal degeneration in the IBN area than in the EBN area (Fig. 1E, F, arrows), although some shrinkage in overall cell diameter was apparent (Normal: $x = 26.69\,\mu m$, SD$\pm$6.0, $N = 266$; SCA2: $x = 22.47$, SD$\pm$5.44, $N = 283$). In SCA2, the percentage of PAV cells remained high in the IBN region, 27% compared with 31% in normals. In contrast to the EBN area, no reduction in synaptic density on IBN somata was evident, but this was not measured systematically. The vertical burst neurons in the rostral interstitial nucleus of the MLF showed no obvious signs of neuronal degeneration.

Discussion

The main finding in this morphometric study of the saccade generator in SCA2 is the pronounced degeneration of the EBNs compared to the OPNs and IBNs. Not only is NRPC, the region which contains the EBNs, completely shrunken and distorted, but many PAV-positive cells have disappeared from the area. All the remaining cells have a significantly reduced number of synaptic inputs on their soma. In comparison, the OPNs have no reduction in synaptic input, and the OPN and IBN areas show significantly less neuronal degeneration. The PAV population in the lateral brainstem nuclei, like the trigeminal and vestibular nuclei, remain largely intact (Fig. 1). This evidence reveals for the first time in humans that slowed horizontal saccades could be correlated with severe loss of horizontal EBNs, where OPNs were significantly less affected. They also confirm the original hypotheses of Wadia in his pioneering studies of spinocerebellar atrophy with slow saccades (Wadia and Swami, 1971; Wadia et al., 1998). The specificity of the atrophy to one functional cell group of the saccade generator, the EBNs, could explain the early appearance of horizontal slow saccades in this type of SCA. Exactly this cause for the slow saccades was predicted many years ago first by Wadia in his noteworthy studies of spinocerebellar atrophy (Wadia et al., 1971, 1998), and later by a computer simulation study of SCA (Zee et al., 1976; Leigh and Zee, 2006). The reason for the EBN loss in SCA2 is not clear, but hypotheses on the abnormal metabolism of excitatory amino acids, such as glutamate, and excitotoxic effects are supported by several studies (Greenamyre, 1986).

It is now clear that the slowing of horizontal saccades can have several causes (Leigh and Zee, 2006): the loss of OPNs (Kaneko, 1996), the development of a midline scar in the (Büttner-Ennever et al., this volume; Jen et al., this volume), or the loss of EBNs as shown in this study.

Acknowledgements

We are grateful to Ahmed Messoudi for his excellent technical assistance and to Professor P. Mehraein for his continual encouragement. This work was supported by the German Research Council, DFG HO 1639/4-2.

References

Bürk, K., Fetter, M., Abele, M., Laccone, F., Brice, A., Dichgans, J. and Klockgether, T. (1999) Autosomal dominant cerebellar ataxia type I: oculomotor abnormalities in families with SCA1, SCA2, and SCA3. J. Neurol., 246: 789–797.

Bürk, K., Fetter, M., Skalej, M., Laccone, F., Stevanin, G., Dichgans, J. and Klockgether, T. (1997) Saccade velocity in idiopathic and autosomal dominant cerebellar ataxia. J. Neurol. Neurosurg. Psychiat., 62: 662–664.

Buttner, N., Geschwind, D., Jen, J.C., Perlman, S., Pulst, S.M. and Baloh, R.W. (1998) Oculomotor phenotypes in autosomal dominant ataxias. Arch. Neurol., 55: 1353–1357.

Dürr, A., Smadja, D., Cancel, G., Lezin, A., Stevanin, G., Mikol, J., Bellance, R., Buisson, G.G., Chneiweiss, H., Dellanave, J., et al. (1995) Autosomal dominant cerebellar ataxia type I in Martinique (French West Indies)-Clinical and neuropathological analysis of 53 patients from three unrelated SCA2 families. Brain, 118: 101–109.

Duvoisin R.C. and Plaitakis A. (Eds.) (1984). The Olivopontocerebellar Atrophies. Raven press, New York.

Geiner, S. (1995) Bestimmung der synaptischen Dichte an identifizierten praemotorischen Neuronen des okulomotorischen Systems des Menschen und vergleichende Untersuchung eines Falles mit Augenbewegungsstörungen. Doctoral Thesis. Ludwig-Maximilian University, Munich.

Greenamyre, J.T. (1986) The role of glutamate in neurotransmission and in neurologic disease. Arch. Neurol., 43: 1058–1062.

Harding, A.E. (1983) Classification of the hereditary ataxias and paraplegias. The Lancet, 1: 1151–1155.

Horn, A.K.E., Büttner-Ennever, J.A., Suzuki, Y. and Henn, V. (1995) Histological identification of premotor neurons for horizontal saccades in monkey and man by parvalbumin immunostaining. J. Comp. Neurol., 359: 350–363.

Horn, A.K.E., Büttner-Ennever, J.A., Wahle, P. and Reichenberger, I. (1994) Neurotransmitter profile of saccadic omnipause neurons in nucleus raphe interpositus. J. Neurosci., 14: 2032–2046.

Kaneko, C.R.S. (1996) Effect of ibotenic acid lesions of the omnipause neurons on saccadic eye movements in Rhesus macaques. J. Neurophysiol., 75: 2229–2242.

Klockgether, T., Wüllner, U., Spauschus, A. and Evert, B. (2000) The molecular biology of the autosomal-dominant cerebellar ataxias. Mov. Disord., 15: 604–612.

Leigh, R.J. and Zee, D.S. (2006) The Neurology of Eye Movements. Oxford University Press, New York.

Nance, M.A. (1997) Clinical aspects of CAG repeat diseases. Brain Pathol., 7: 881–900.

Rosenberg, R.N. (1995) Autosomal dominant cerebellar phenotypes: the genotype has settled the issue. Neurol., 45: 1–5.

Wadia, N. (1984) A variety of olivopontocerebellar atrophy distinguished by slow eye movements and peripheral neuropathy. In: Duvoisin R.C. and Plaitakis A. (Eds.), The Olivopontocerebellar Atrophies. Raven press, New York, pp. 149–177.

Wadia, N., Pang, J., Desai, J., Mankodi, A., Desai, M. and Chamberlain, S. (1998) A clinicogenetic analysis of six Indian spinocerebellar ataxia (SCA2) pedigrees: the significance of slow saccades in diagnosis. Brain, 121: 2341–2355.

Wadia, N.H. and Swami, R.K. (1971) A new form of heredofamilial spinocerebellar degeneration with slow eye movements (nine families). Brain, 94: 359–374.

Zee, D.S., Optican, L.M., Cook, J.D., Robinson, D.A. and Engel, W.K. (1976) Slow saccades in spinocerebellar degeneration. Arch. Neurol., 33: 243–251.

C. Kennard & R.J. Leigh (Eds.)
Progress in Brain Research, Vol. 171
ISSN 0079-6123

CHAPTER 6.15

Selective, circuit-wide sparing of floccular connections in hereditary olivopontine cerebellar atrophy with slow saccades

Sarah H. Ying[1,*], Anja K.E. Horn[2], Stefan Geiner[2], N.H. Wadia[3] and Jean A. Büttner-Ennever[2]

[1]*Johns Hopkins University School of Medicine, Baltimore, MD, USA*
[2]*Institute of Anatomy, Ludwig-Maximilian University, Munich, Germany*
[3]*Jaslok Hospital and Research Center, Mumbai, India*

Abstract: We present a systems-oriented histopathologic analysis of the ocular motor control circuits in the cerebellum and brainstem from a patient with a hereditary form of olivopontine cerebellar atrophy of the Wadia type, which has a characteristic ocular motor presentation of slow saccades but relative preservation of smooth pursuit and gaze-holding. This differential pattern of clinical involvement is associated with a lobule-specific pattern of cerebellar degeneration. We asked whether these patterns of sparing and degeneration were consistent throughout the associated deep cerebellar and brainstem structures. Specimens were fixed in formalin, embedded in paraffin, and stained for various markers. We found that elements of the floccular and nodular pathways, controlling smooth pursuit and vestibular reflexes, were relatively spared, particularly those structures that are interconnected with the medial regions. Conversely, the elements of the dorsal vermis pathway controlling saccade adaptation were relatively involved. This subregional specificity of degeneration further defines possible areas of investigation for elucidating pathophysiology, testing biomarkers of disease, and developing areas for therapeutic intervention.

Keywords: OPCA; SCA-2; inferior olive; ventrolateral outgrowth; dorsal cap of Kooy

Introduction

We present evidence of clinicopathological correlation for a patient with a hereditary form of olivopontine cerebellar atrophy of the Wadia type (OPCA-W). In this class of disease, which includes spinocerebellar ataxia type 2 (SCA-2), patients show typical ocular motor signs of slow saccades, with relative preservation of smooth pursuit and gaze holding (Wadia et al., 1998). This presentation predicts relative sparing of the floccular system. Although extensive pathological surveys have been conducted in this population (Estrada et al., 1999; Pang et al., 2002; Gierga et al., 2005; Rüb et al., 2005), a systems-based analysis of the floccular circuit (Fig. 1) has not been published. Thus, in addition to an analysis of the integrity of the floccular cerebellar cortex, here we also

*Corresponding author. Tel.: +410-955-3319;
Fax: +410-614-1746; E-mail: sying@dizzy.med.jhu.edu

DOI: 10.1016/S0079-6123(08)00684-5

Circuitry of the flocculus

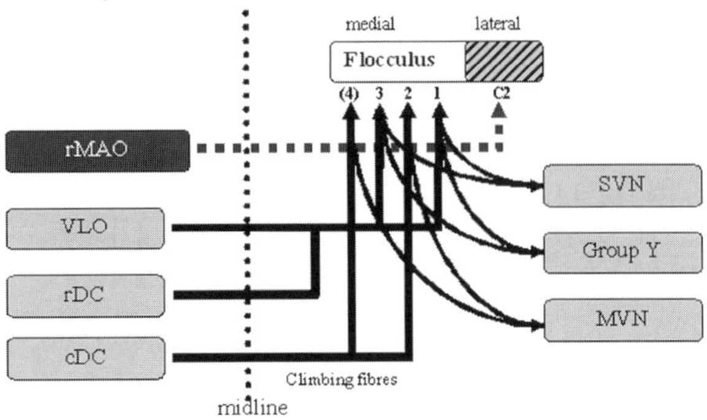

Fig. 1. The climbing fibre inputs to the flocculus show regional specificity. The rMAO, which projects to the C2 region of the lateral flocculus, is severely affected. rMAO, rostral medial accessory olive; VLO, ventrolateral outgrowth; rDC, rostral dorsal cap of Kooy; cDC, caudal dorsal cap of Kooy; SVN, superior vestibular nucleus; MVN, medial vestibular nucleus.

examine the source of olivary climbing fibre inputs (the medial accessory olive, MAO; the ventrolateral outgrowth, VLO; and the dorsal cap of Kooy, DC), the mossy fibre inputs from the paramedian tract nuclei and the raphe nuclei, the nuclear projections (the superior vestibular nuclei, SVN; the medial vestibular nuclei, MVN; and the Y-group), and the subservient white matter region (Buttner-Ennever, 2005).

Patient and methods

The subject was a 41-year-old female with a 16-year progression of imbalance, followed by dysarthria and upper extremity ataxia. Later in the disease course, she was noted to have general wasting, bilateral cataracts, and absent deep tendon reflexes, but with extensor plantar reflexes and intact sensation. Her ocular motor profile was typical of the disease. Range of extraocular movements were full to caloric stimulation or the vestibular ocular reflex, but spontaneous movements were limited. Saccade velocities were slow, and fast phases could not be elicited to either vestibular or optokinetic stimuli. Pursuit movements were normal.

This brain specimen was removed within 24 h postmortem, was placed in 10% formalin, then was transferred to buffered 4% paraformaldehyde. After extended fixation (3 months), the specimen was embedded in paraffin. Serial transverse brainstem sections were cut at 20 μm. These series were stained using a Nissl stain, a Weigert stain, or a "triple stain" consisting of Nissl, luxol fast blue, and silver stains for orientation and distinction of grey and white matter areas. Selected sections were also checked for parvalbumin (PAV, Sigma, St. Louis, MO) and synaptophysin (DAKO SY38, Glostrup, Denmark) immunoreactivity.

Results

We found relative histopathological preservation in the flocculonodular system, including its associated white matter tracts, and associated brainstem nuclei. Results are summarized in Table 1. In the cerebellar cortex, Nissl stain showed relative preservation of the flocculus and the nodulus, whereas the lateral cerebellar hemispheres and the anterior vermis showed significant Purkinje cell dropout, with pallor and thinning of the granular layer. This was corroborated by a corresponding loss of PAV immunoreactivity. The lateral hemisphere showed weak PAV staining, whereas the flocculi were strongly PAV-immunoreactive, with preservation of the dendritic trees. Detailed regional analysis revealed preservation of proximal–medial versus distal–lateral flocculus. This gradient of medial preservation/lateral involvement was also seen in the nodulus. Weigert staining similarly showed that myelinated fibre density was preserved in the flocculus. The arbour vitae subserving the lateral hemispheres only stained weakly.

Beyond the cerebellar cortex, regional differences remained true to type. Circuit-based specificity of the degeneration was illustrated by the pattern of regional degeneration in the inferior olive: there was loss of cellularity in the neighbouring principal olive (PO) and MAO but preservation of the VLO and the DC. Weigert staining of the outflow tract from the inferior olive was weak. In contrast, the superior cerebellar peduncles, the outflow tract from the preserved dentate nucleus, were well preserved. Areas receiving projections

Table 1. Histopathological findings in an OPCA-W brain: elements of the floccular circuit show mild (+), moderate (+ +), or severe (+ + +) involvement

Region of interest	Extent of degeneration
Climbing fibre inputs	
Medial accessory olive	+ + +
Dorsal cap of Kooy	+
Ventrolateral outgrowth	+
Mossy fibre inputs	
Paramedian tract nuclei	+
Raphe nuclei	+
Nucleus prepositus hypoglossi	+
Cerebellar cortex	
Flocculus (medial)	+
Flocculus (lateral)	+ + +
Receive projections	
Superior vestibular nucleus	+
Y-group	+
Medial vestibular nucleus	+
Interposed nuclei	+

from the flocculi, including the SVN, MVN, Y-group, and interposed nuclei, were easily discernible on Nissl stains. A similar examination of the dorsal vermis pathway, a related circuit controlling saccade adaptation, showed that elements were relatively involved.

Discussion

Histopathological analysis of cerebellar lobular involvement shows a pattern of flocculonodular sparing, consistent with prior macroscopic studies and the clinical ocular motor profile of preserved smooth pursuit and vestibular reflexes. This is reminiscent of the pattern of degeneration in pontocerebellar hypoplasia or olivopontocerebellar hypoplasia, in which the flocculonodular lobe is uniquely spared (Barth, 2007). The mechanism of this sparing is unclear. Our study may shed light on the nature of this sparing — beyond the lobules themselves, the associated elements of the floccular and nodular pathways are also relatively spared, particularly structures interconnected with medial regions.

This subregional specificity of degeneration may define possible areas of investigation for elucidating pathophysiology and identifying targets for therapeutic intervention. For example, areas in the flocculus and nodulus connected to the VLO and the DC are uniquely spared. Closer examination of the cellular physiology may help us to understand the mechanism of resistance to neurodegeneration. As distinct from the somatosensory regions of the inferior olives, the oculomotor regions — the VLO and the DC — lack low-threshold calcium currents (provided by T-type calcium channels), hyperpolarization-activated cationic currents (known as the I_h current), and subthreshold oscillations (Urbano et al., 2006).

Furthermore, the circuit-based specificity of preservation or loss suggests that a single element may serve as a biomarker for the entire circuit. This may enable the development of non-invasive biomarker testing, such as magnetic resonance (MR) imaging/spectroscopy.

In summary, this subregional specificity of degeneration further defines possible areas of investigation for elucidating pathophysiology, testing biomarkers of disease, and developing areas for therapeutic intervention.

Acknowledgements

This work has been supported by the Human Frontier Science Program, the Arnold-Chiari Foundation, the Robin Zee Fund, NIH Grant 1 K23 EY015802-01A1, National Alliance for Research on Schizophrenia and Depression, and the Dana Foundation.

References

Barth, P.G. (2007) Pontocerebellar hypoplasia type 2: a neuropathological update. Acta Neuropathol., 114: 373–386.

Buttner-Ennever J.A. (2005). Neuroanatomy of the Oculomotor System. Progress in Brain Research, Vol. 151. Elsevier, Amsterdam.

Estrada, R., Galarraga, J., Orozco, G., Nodarse, A. and Auburger, G. (1999) Spinocerebellar ataxia 2 (SCA2): morphometric analyses in 11 autopsies. Acta Neuropathol. (Berl.), 97(3): 306–310.

Gierga, K., Bürk, K., Bauer, M., Orozco, D.G., Auburger, G., Schultz, C., Vuksic, M., Schöls, L., de Vos, R.A., Braak, H., Deller, T. and Rüb, U. (2005) Involvement of the cranial nerves and their nuclei in spinocerebellar ataxia type 2

(SCA2)Acta Neuropathol. (Berl.), 109(6): 617–631. (Epub May 19, 2005).
Pang, J.T., Giunti, P., Chamberlain, S., An, S.F, Vitaliani, R., Scaravilli, T., Martinian, L., Wood, N.W., Scaravilli, F. and Ansorge, O. (2002) Neuronal intranuclear inclusions in SCA2: a genetic, morphological and immunohistochemical study of two cases. Brain, 125(3): 656–663.
Rüb, U., Gierga, K., Brunt, E.R., de Vos, R.A., Bauer, M., Schöls, L., Bürk, K., Auburger, G., Bohl, J., Schultz, C., Vuksic, M., Burbach, G.J., Braak, H. and Deller, T. (2005) Spinocerebellar ataxias types 2 and 3: degeneration of the precerebellar nuclei isolates the three phylogenetically defined regions of the cerebellum. J. Neural Transm., 112: 1523–1545.
Urbano, F.J., Simpson, J.I. and Llinás, R.R. (2006) Somatomotor and oculomotor inferior olivary neurons have distinct electrophysiological phenotypes. Proc. Natl. Acad. Sci., 103(44): 16550–16555.
Wadia, N., Pang, J, Desai, J., Mankodi, A., Desai, M. and Chamberlain, S. (1998) A clinicogenetic analysis of six Indian spinocerebellar ataxia (SCA2) pedigrees: the significance of slow saccades in diagnosis. Brain, 121: 2341–2355.

C. Kennard & R.J. Leigh (Eds.)
Progress in Brain Research, Vol. 171
ISSN 0079-6123

CHAPTER 6.16

A quick look at slow saccades after cardiac surgery: where is the lesion?

David Solomon[1,*], Stefano Ramat[2], R. John Leigh[3] and David Zee[1]

[1]*Johns Hopkins University, Baltimore, MD, USA*
[2]*Università degli Studi di Pavia, Pavia, Italy*
[3]*Case Western Reserve University, Cleveland, OH, USA*

Abstract: Saccadic palsy is a reported complication of cardiac surgery. One case that came to autopsy showed midline pontine gliosis; however, in most cases, no lesions are evident on neuroimaging. Since the saccadic palsy may range from single large slow saccades to a "staircase" of very small saccades that are normal in speed, it seems plausible that more than one mechanism is possible. Here we postulate that, in those patients who make a staircase of small saccades, loss of cerebellar Purkinje cells could cause fastigial nucleus neurons to fire prematurely, thereby decelerating saccades via inhibitory burst neurons.

Keywords: saccades; saccadic palsy; omnipause neurons; fastigial nucleus; Purkinje cells; saccadic hypometria; excitatory burst neuron; inhibitory burst neuron

Introduction

A range of disorders may impair the ability to generate normal-speed saccades (Leigh and Zee, 2006). We recently reported a series of patients with a selective loss of saccades following cardiac surgery (Solomon et al., 2008). No anatomical lesions were found on imaging studies that could explain this acquired deficit. A previous pathological study of a similar case showed midline pontine gliosis likely involving omnipause neurons (Hanson et al., 1986).

Patients in the study of Solomon et al. (2008) were able to generate normal pursuit and vestibular eye movements, and the head could be moved rapidly when attempting a gaze shift. However, with the head fixed, voluntary gaze shifts to visual targets were either absent or slow; sometimes they could be improved with blinks. When these gaze-shifting movements were measured with precision, different behaviours were observed that were often indistinguishable at the bedside. Thus, some patients generated a single, slow saccade, whereas other patients made saccades with transient decelerations, resulting in dips on the velocity trace. A third behaviour consisted of a staircase of small saccades that eventually carry the eye to the target (Fig. 1A); such small movements were no slower than similar-sized saccades generated by control subjects, although when the patient made larger saccades, they were slow (Fig. 1B). Each of these patterns suggests a different possible problem in saccadic burst generation, with interactions between the several possible mechanisms resulting in the range of responses observed.

*Corresponding author. Tel.: 410-502-2655;
Fax: 410-955-7744; E-mail: dsolomon@dizzy.med.jhu.edu

DOI: 10.1016/S0079-6123(08)00685-7

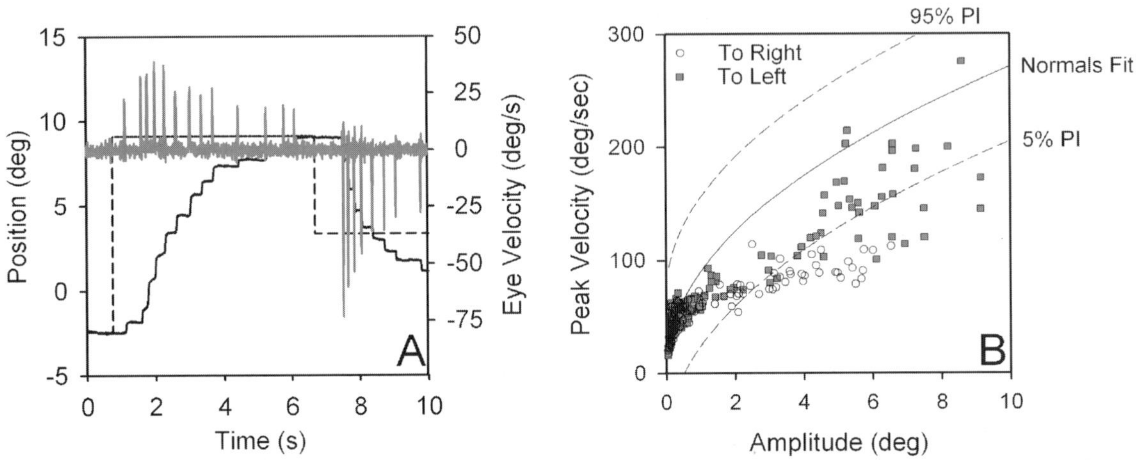

Fig. 1. (A) Attempted head-fixed gaze movements to visual targets made by a patient with acquired saccadic palsy after cardiac surgery. A staircase of hypometric saccades bring the eye to the target, with intersaccadic intervals below 100 ms. (B) Peak velocity–amplitude plot of saccades from same patient. Small saccades are similar to prediction intervals (PI) for control subjects, but larger saccades are slow, more so to the right.

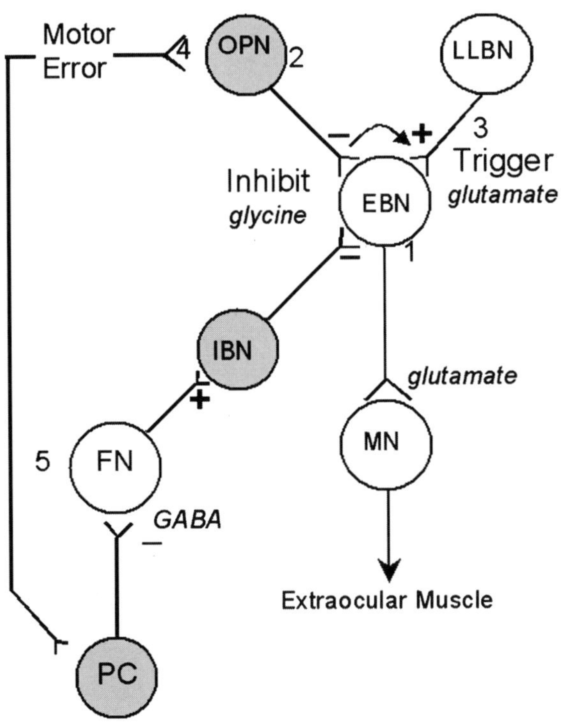

Fig. 2. Summary of hypothetical brainstem sites of lesions that might cause saccades to become slow or small; see text for details.

What are the possible mechanisms that cause saccades movements to become slow and small in these patients? In our prior study (Solomon et al., 2008), we considered several possible mechanisms (Fig. 2). Excitatory burst neurons (EBN) in the paramedian pontine reticular formation (PPRF) drive ocular motoneurons (OMN) during ipsilateral horizontal saccades. EBN receive a trigger signal from the superior colliculus that is relayed by long-lead burst neurons (LLBN), and which uses glutamatergic mechanisms. A second major projection to EBN is from omnipause neurons (OPN), which are tonically active but are inhibited by the superior colliculus when a saccade is to be generated. OPN inhibit burst neurons via glycine. When a saccade is to be triggered, OPN cease discharge, releasing burst neurons from inhibition. The trigger signal is amplified by glycine, which also acts as a neuromodulator at glutamatergic receptors. OPN neurons resume their discharge when a motor error signal falls to near zero, signalling that the saccade is complete. Inhibitory burst neurons (IBN) inhibit EBN during contralateral saccades to impose Sherrington's law of reciprocal innervation (Ramat et al., 2007). Slow saccades could be caused by lesions affecting EBN (site 1); lesions affecting OPN (site 2); or an abnormal trigger signal (site 3). Hypometric saccades might arise if the threshold at which the motor error signal causes OPN to resume discharge is increased (site 4).

Here we consider a further possibility that IBN might inhibit EBN prematurely (Fig. 2, site 5), thereby truncating the planned saccade and causing hypometria, as shown in Fig. 1A. Mossy fibre input to ocular motor vermis from LLBN in the nucleus reticularis tegmenti pontis (Kaneko and Fuchs, 2006) convey a copy of the saccadic motor command, from which the cerebellar cortex constructs the desired amplitude to be compared with the ongoing movement (Optican, 2005). Convergent inhibitory input from Purkinje cells (PC) of the ocular motor vermis control the output of the underlying fastigial neurons (FN). Just prior to, and during, the initial portions of a rightward saccade, the PC in the left vermis are off, so the left FN can fire an early burst to excite the right IBN via projections to the contralateral pontomedullary reticular formation. As the saccade reaches its target, PC on the right ceases; right FN are no longer inhibited and excite the left IBN, choking the drive signal to the right OMN generating the saccade. We hypothesize the following mechanism for the staircase of small saccades we observed: ischaemic injury during surgery results in a selective loss of PC in the ocular motor vermis (Welsh et al., 2002). One consequence of the loss of inhibition of FN could be premature excitation of IBN relaying the choke signal. Thus, for a rightward saccade, the PC on the right are no longer able to inhibit the right FN until the end of the saccade, so the right FN excite the left IBN that in turn inhibit the right OMN well before the end of the desired movement. The right IBN may also inhibit the OPN, perhaps via a latch circuit, so the OPN now can resume firing, bringing eye velocity back to zero after the brief hypometric saccade. Similar closely spaced saccades can be elicited experimentally by maintaining electrical stimulation in the superior colliculus. Recent recordings from OPN during such a staircase of saccades showed that they do pause and resume firing normally with each successive movement (Paul and Gnadt, 2006). Note that transient deceleration of saccades is also observed in late-onset Tay–Sachs disease (Rucker et al., 2004), but in that disorder, the eye may not come to a complete stop before starting another saccade (see papers by Rucker and Optican in this volume).

What drives the subsequent saccades in the series of staircase movements such as shown in Fig. 1A? When saccades are slowed by muscimol injection into the OPN area, saccade-related burst neurons in the superior colliculus continue to fire for the abnormally large duration of the movement. While not coding motor error, this signal could provide the brainstem saccadic generator with the desired target location and drive the sequence of hypometric saccades until the target is reached (Soetedjo et al., 2002). In conclusion, it seems likely that more than one mechanism accounts for the range of forms of saccadic palsy that may follow cardiac surgery. More precise information on the nature of the saccadic disorder, and new approaches such as functional imaging, may elucidate the mechanism in each patient.

Acknowledgement

This work was supported by Christopher Cleveland Memorial Fund, NIH EY06717.

References

Hanson, M.R., Hamid, M.A., Tomsak, R.L., Chou, S.S. and Leigh, R.J. (1986) Selective saccadic palsy caused by pontine lesions: clinical, physiological, and pathological correlations. Ann. Neurol., 20: 209–217.

Kaneko, C.R. and Fuchs, A.F. (2006) Effect of pharmacological inactivation of nucleus reticularis tegmenti pontis on saccadic eye movements in the monkey. J. Neurophysiol., 95: 3698–3711.

Leigh, R.J. and Zee, D.S. (2006) The Neurology of Eye Movements (Book/DVD) (4th ed.). Oxford University Press, New York.

Optican, L.M. (2005) Sensorimotor transformation for visually guided saccades. Ann. N.Y. Acad. Sci., 1039: 132–148.

Paul, K. and Gnadt, J.W. (2006) Activity of omnipause neurons during "staircase saccades" elicited by persistent microstimulation of the superior colliculus. Vision Res., 46: 3430–3442.

Ramat, S., Leigh, R.J., Zee, D.S. and Optican, L.M. (2007) What clinical disorders tell us about the neural control of saccadic eye movements. Brain, 130: 10–35.

Rucker, J.C., Shapiro, B.E., Han, Y.H., Kumar, A.N., Garbutt, S., Keller, E.L. and Leigh, R.J. (2004) Neuro-ophthalmology of late-onset Tay–Sachs disease (LOTS). Neurology, 63: 1918–1926.

Soetedjo, R., Kaneko, C.R. and Fuchs, A.F. (2002) Evidence that the superior colliculus participates in the feedback control of saccadic eye movements. J. Neurophysiol., 87: 679–695.

Solomon, D., Ramat, S., Tomsak, R.L., Reich, S.G., Shin, R.K., Zee, D.S. and Leigh, R.J. (2008) Saccadic palsy after cardiac surgery: characteristics and pathogenesis. Ann. Neurol., 63: 355–365.

Welsh, J.P., Yuen, G., Placantonakis, D.G., Vu, T.Q., Haiss, F., O'Hearn, E., Molliver, M.E. and Aicher, S.A. (2002) Why do Purkinje cells die so easily after global brain ischemia? Aldolase C, EAAT4, and the cerebellar contribution to posthypoxic myoclonus. Adv. Neurol., 89: 331–359.

C. Kennard & R.J. Leigh (Eds.)
Progress in Brain Research, Vol. 171
ISSN 0079-6123

CHAPTER 6.17

Eye and head torsion is affected in patients with midbrain lesions

O. Kremmyda*, S. Glasauer, T. Eggert and U. Büttner

Department of Neurology, Klinikum Grosshadern, Ludwig-Maximilians University, Marchioninistr, Munich, Germany

Abstract: The midbrain has been shown to contain crucial nuclei for the control of vertical and torsional eye movements. Recent studies in monkeys demonstrated that midbrain lesions also affect head movements during gaze saccades, but so far clinical reports on the matter have been missing. We measured 3D eye and head movements in two patients with oculomotor deficits due to unilateral midbrain lesions and in healthy control subjects. Subjects had (1) to perform head-free target directed gaze saccades (head-free task), and (2) to point to the target with a head-laser (head-only task). The patients had vertical eye movement deficits, which were not compensated by the head. Three-dimensional analysis revealed torsional deviations from the normal range of movements for both, the eye and the head movements.

Keywords: midbrain; gaze; head torsion

Introduction

The midbrain is a critical area for the control of combined eye and head (gaze) movements, containing nuclei such as the rostral interstitial nucleus of the medial longitudinal fasciculus (RIMLF), which serves as the vertical-torsional saccade generator and the interstitial nucleus of Cajal (INC), which is regarded as the vertical-torsional neuronal integrator for the eye and head (Klier et al., 2002; Leigh and Zee, 2006). RIMLF and INC lesions lead to torsional nystagmus (Helmchen et al., 2002) and torsional deviations during saccades (Kremmyda et al., 2007). Inactivation of the RIMLF and INC in primates also leads to a contralateral head tilt (Vilis et al., 1989; Klier et al., 2002). Clinically, midbrain lesions can lead to an ocular tilt reaction (OTR), which consists of a contralateral head tilt (with gaze straight ahead), contralateral ocular torsion and skew deviation (ipsilateral higher than the contralateral eye) (Brandt and Dieterich, 1994). Recently, data in non-human primates have shown that INC inactivation affects head posture, not only in the straight-ahead positions, but also after combined eye and head movements in three dimensions (Farshadmanesh et al., 2007). In this study, we describe two patients with acute unilateral midbrain lesions, which showed torsional deviations of the head posture after gaze saccades and head-laser directed movements.

Case reports and methods

Patient 1

A 59-year-old woman woke up with a dysarthric speech, a left head tilt, and a falling tendency to

*Corresponding author. Tel.: +0049-89-70954826; Fax: +0049-89-70954801;
E-mail: okremmyda@nefo.med.uni-muenchen.de

DOI: 10.1016/S0079-6123(08)00686-9

the left. Neurological examination showed slowed vertical saccades with a counterclockwise deviation (upper pole of the eye to the left,) from the pure vertical and a slight skew deviation (right over left eye). MRI showed a right anterior-medial thalamus infarct that included the RIMLF (Kremmyda et al., 2007).

Patient 2

An 80-year-old man experienced a sudden onset of vertigo and falling tendency to the right. Neurological examination showed vertical saccadic palsy and saccadic smooth pursuit. MRI showed a localized left anteromedial midbrain infarct.

3D eye and head movements were recorded in both patients four days after the symptom onset. At this time, both patients did not have any falling tendencies or visible head tilt (patient 1). Seven subjects (age 53–80) without history of eye or head movement pathology were used as controls.

For the eye movement recordings (sampling rate 1 kHz) we used a dual search coil on the left eye (Skalar, Delft, The Netherlands) and for the head movements two coils mounted on a head ring at a 90° angle between them. The patients were seated in complete darkness inside a magnetic field (Remmel Labs, Ashland, MA, USA) and were instructed to follow a laser dot (size 0.1°, distance 145 cm) by gaze movements, i.e., combined eye and head movements (head-free task), or with a head-laser that was mounted on the forehead (head-only task). The laser dot stepped randomly between the centre and eight peripheral positions, which were 28° away from the central target. The dot was only visible for 1 s around the step and then was blanked. Eye and head positions were expressed as quaternions. Further details on the mathematical analysis and the calibration procedure are described elsewhere (Glasauer et al., 2003). The 3D range of eye and head orientation at fixation points (gaze velocity < 10°/s) was described by fitting a second order surface to the acquired head position, expressed by quaternions. The fitted surface was parameterized by:

$$q_1 = a_1 + a_2q_2 + a_3q_3 + a_4q_2q_3 + a_5q_2^2 + a_6q_3^2$$

where q_1, q_2, and q_3 express torsional, horizontal, and vertical quaternion components, respectively (Glenn and Vilis, 1992). This equation gives the torsional position for a given horizontal and vertical position and expresses Donders' law for the eye and head. The parameter a_1 is the torsional offset, the parameters a_2 and a_3 give the tilting of Donders' surface around the horizontal and the vertical axes, respectively, and the parameters a_5 and a_6 determine the curvature of the surface along these two axes. Parameter a_4 is the so-called "twist score." An a_4 value different than zero means that the plotted surface is twisted. All parameters were determined by linear regression using Matlab. For the calculation of the confidence intervals (CI), student *T*-test was used (SPSS, Chicago, IL).

Results

Patient 1

The analysis revealed large (up to 15°) counterclockwise eye torsion with every vertical or oblique gaze movements, as also seen when the patient was performing saccades with the head fixed (Kremmyda et al., 2007). Donders' surface of the head in the head-free task showed that patient 1 tended to rotate the head more counterclockwise when looking up and more clockwise when looking down than normal subjects (Fig. 1A), as expressed by the positive parameter a_2 ($a_2 = 0.218$) (Ceylan et al., 2000), which was outside the 95% CI of the control group (mean $a_2 = 0.12$, CI: −0.044 to 0.15). In the head-only task, the largest tilt of the surface was seen for horizontal movements, where the head was rotating more counterclockwise than in controls when facing left and more clockwise when facing right, as expressed by the negative parameter a_3 (patient $a_3 = -0.29$, control subjects: mean $a_3 = 0.03$, CI: −0.06 to 0.123) (Fig. 1D). The curvature of the surface along the horizontal axis of rotation (vertical movements, parameter a_5) and along the vertical axis of rotation (horizontal movements, parameter a_6) was stronger than in normal subjects (Fig. 2). Both a_5 and a_6 for patient 1 were negative (head-free task: $a_5 = -1.4$, $a_6 = -0.88$, head-only

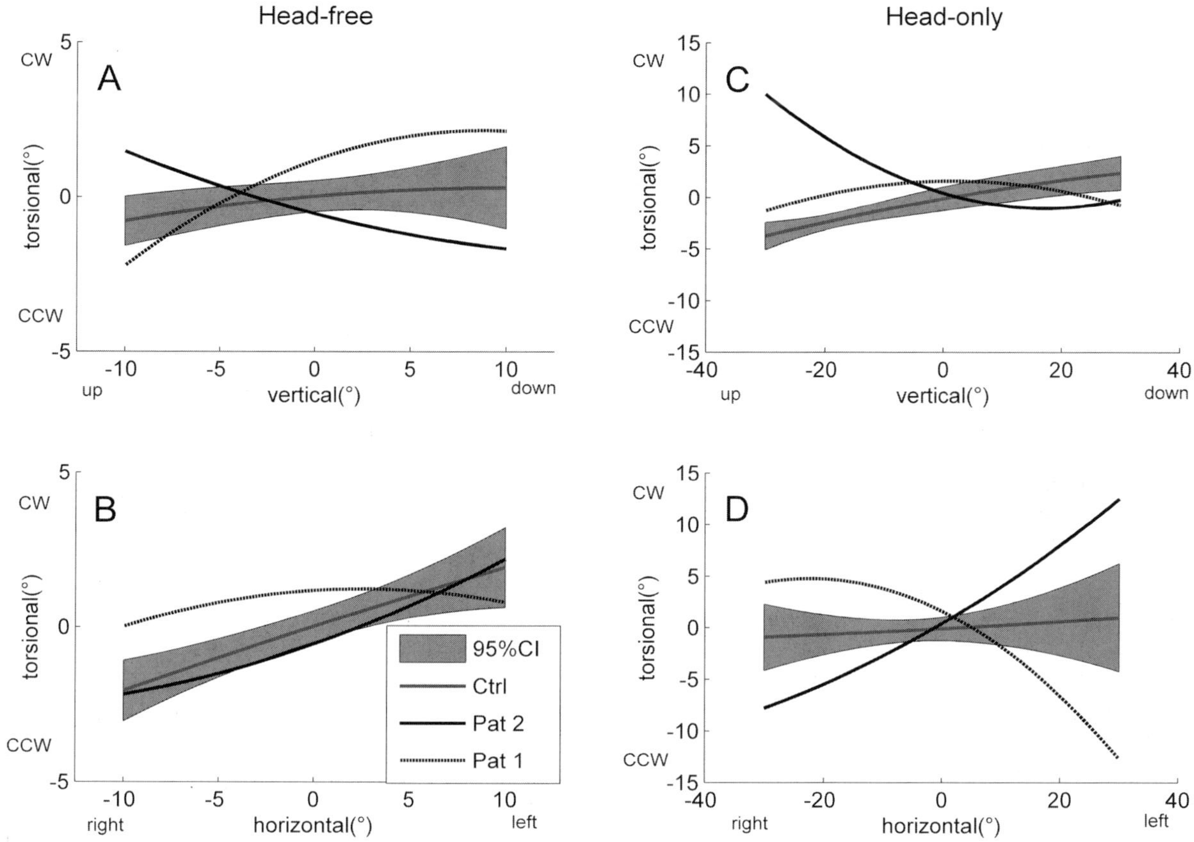

Fig. 1. Second order fit of the head position data during the head-free (A, B) and head-only task (C, D). The plots show a cross-section of the fitted second order surface along the vertical plane at eye level (A, C) or along the horizontal plane (B, D). The grey line shows the mean value of the seven control subjects (shading represents the 95% confidence intervals), the dotted line patient 1 and the solid line patient 2. Positive values represent clockwise, down, and left directions. Different axis scaling in the two tasks represents the actual head movement range during these tasks.

task: $a_5 = -0.33$, $a_6 = -0.73$) and outside the 95% CI of the control group in both tasks except for a_6 of the head-free task (Control group: head-free task: mean $a_5 = -0.296$, CI: -0.90 to 0.031, mean $a_6 = -0.091$, CI: -1.05 to 0.86, head-only task: mean $a_5 = 0.2280$, CI:-0.114 to 0.15, mean $a_6 = -0.16$, CI: -0.57 to 0.25).

Patient 2

Analysis of the eye data of patient 2 revealed a counterclockwise torsional nystagmus (nystagmus beating ipsilateral to the lesion), which implies that the left INC is involved in the lesion (Helmchen et al., 2002). The vertical range of the eye was also limited (up to 8° in both directions), showing a vertical gaze palsy, which was not completely compensated by the head movements (max. 10° in the head-free task), leading to vertical gaze movements up to max 15° (of the requested 28°). In the head-free and the head-only task, the head was rotating much more clockwise when looking up and more counterclockwise when looking down, (head-free: $a_2 = -0.16$, head-only: $a_2 = -0.17$) (Fig. 1A, C). In the head-only task, torsion was also affected for horizontal movements, since the head rotated clockwise when facing left and counterclockwise when facing right ($a_3 = 0.33$), which was outside the CI of the control group and the opposite of

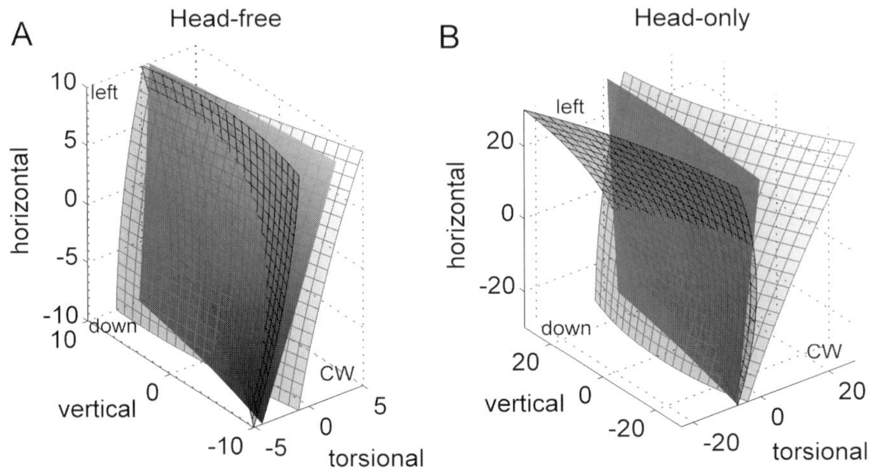

Fig. 2. Donders' 3D surfaces describing head position for the head-free task (A) and for the head-only task (B). The grey surface shows the average plot of all normal subjects, the surface with the black grid patient 1 and the surface with the grey grid patient 2. The increased contralesional curvature of each patient is more obvious in the head-only task where the head movements were larger.

patient 1. During both tasks, the head surface was curved clockwise, in the opposite direction of patient 1, as shown by the positive a_5 and a_6 parameters (head-free task: $a_5 = 0.50$, $a_6 = 0.64$, head-only task: $a_5 = 0.56$, $a_6 = 0.24$, Fig. 2). Parameter a_5 was in both tasks outside the CI of the control group.

Discussion

The neurons in the INC and RIMLF in the primates are considered to be organized such that both sides contain neurons that control up/down directions of the eye movements, but only code ipsitorsional position and velocity signals (Büttner and Büttner-Ennever, 2005). Recent studies have shown that the head movements are similarly organized in the primate INC (Klier et al., 2002; Farshadmanesh et al., 2007). This organization is considered to reflect the direction of the semicircular canals, the extraocular muscles, and the anatomy of the neck (Farshadmanesh et al., 2007). The role of the RIMLF in head control is less investigated, but it is known that RIMLF projects to the neck muscles (Horn, 2005) and that monkeys show contralateral head tilt after RIMLF inactivation (Vilis et al., 1989).

Midbrain lesions in humans, especially those involving the INC, lead to a contralateral head tilt of about 10° (Brandt and Dieterich, 1994), although the head tilt is not as large as in monkeys, in which it can reach more than 40° (Klier et al., 2002). In the present study, we were able to show that midbrain lesions can lead to changes in the 3D head posture during fixation after a gaze or goal-directed movement. These changes were not visible in the straight-ahead position, but were revealed only with eccentric head positions. The head torsion was increased contralateral to the lesion, so that for the two patients with lesions in opposite brainstem sides the 3D surfaces were almost mirror-like. These differences in torsion cannot be attributed only to a compensatory mechanism for the eye movements, but seem to reflect a change in the central head control mechanisms. Our study also shows that the analysis of Donders' surface, a subject so far only poorly investigated in clinical cases, could be a useful tool in understanding the central mechanisms underlying postural pathology of the head in humans.

Acknowledgement

Supported by DFG Research Training Group (GRK) 1091.

References

Brandt, T. and Dieterich, M. (1994) Vestibular syndromes in the roll plane: topographic diagnosis from brainstem to cortex. Ann. Neurol., 36: 337–347.

Büttner, U. and Büttner-Ennever, J.A. (2005) Present concepts of oculomotor organization. Prog. Brain Res., 151: 1–42.

Ceylan, M., Henriques, D.Y., Tweed, D.B. and Crawford, J.D. (2000) Task-dependent constraints in motor control: pinhole goggles make the head move like an eye. J. Neurosci., 20: 2719–2730.

Farshadmanesh, F., Klier, E.M., Chang, P., Wang, H. and Crawford, J.D. (2007) Three-dimensional eye-head coordination after injection of muscimol into the interstitial nucleus of Cajal (INC). J. Neurophysiol., 97: 2322–2338.

Glasauer, S., Hoshi, M., Kempermann, U., Eggert, T. and Büttner, U. (2003) Three-dimensional eye position and slow phase velocity in humans with downbeat nystagmus. J. Neurophysiol., 89: 338–354.

Glenn, B. and Vilis, T. (1992) Violations of Listing's law after large eye and head gaze shifts. J. Neurophysiol., 68: 309–318.

Helmchen, C., Rambold, H., Kempermann, U., Büttner-Ennever, J.A. and Büttner, U. (2002) Localizing value of torsional nystagmus in small midbrain lesions. Neurology, 59: 1956–1964.

Horn, A.K. (2005) The reticular formation. Prog. Brain Res., 151: 127–155.

Klier, E.M., Wang, H., Constantin, A.G. and Crawford, J.D. (2002) Midbrain control of three-dimensional head orientation. Science, 295: 1314–1316.

Kremmyda, O., Büttner-Ennever, J.A., Büttner, U. and Glasauer, S. (2007) Torsional deviations with voluntary saccades caused by a unilateral midbrain lesion. J. Neurol. Neurosurg. Psychiatry, 78: 1155–1157.

Leigh, R.J. and Zee, D.S. (2006) The Neurology of Eye Movements. Oxford University Press, Oxford.

Vilis, T., Hepp, K., Schwarz, U. and Henn, V. (1989) On the generation of vertical and torsional rapid eye movements in the monkey. Exp. Brain Res., 77: 1–11.

C. Kennard & R.J. Leigh (Eds.)
Progress in Brain Research, Vol. 171
ISSN 0079-6123

CHAPTER 6.18

Horizontal saccadic palsy associated with gliosis of the brainstem midline

Jean A. Büttner-Ennever[1,*], Takuya Uemura[2], Yasuko Arai[3] and Jun Tateishi[4]

[1]*Institute of Anatomy, Ludwig-Maximilian University Munich, Munich, Germany*
[2]*Department of Otorhinolaryngology, Kyushu University, Fukuoka, Japan*
[3]*Department of Otorhinolaryngology, Tokyo Women's Medical College, Tokyo, Japan*
[4]*Department of Neuropathology, Kyushu University, Fukuoka, Japan*

Abstract: We studied premotor cell groups involved in the generation of saccades in a patient with a disturbance of voluntary horizontal gaze. The only neurological symptom found was a slowing of horizontal saccades, reported since birth and unaltered over his lifetime. We attribute this disorder, for the first time, to a fibrous gliosis of the brainstem midline, which may disrupt neuronal elements of the horizontal saccade generator crossing the brainstem midline, but it caused no obvious loss of omnipause-, excitatory burst-, and inhibitory burst neurons. No neuronal loss or demyelination, was apparent elsewhere in the brainstem; but there was evidence of an ependymal infection throughout the entire ventricular system. A diagnosis of Gaucher disease was made from the bone marrow of this patient shortly before his death, but for several reasons we considered this complication unlikely to be the cause of the saccadic disorder.

Keywords: slow horizontal saccades; omnipause neurons; excitatory burst neurons; inhibitory burst neurons; astrocytic gliosis; brainstem midline; Gaucher disease

Introduction

Slowed saccadic eye movements can be caused by lesions of the brainstem reticular formation, the location of the saccade generator, but they are more often related to hereditary degenerative disorders such as spinocerebellar atrophy (SCA2), or to metabolic disorders like Gaucher disease (Leigh and Zee, 2006). The neuronal basis for the progressive saccadic slowing in these hereditary diseases has not been identified; although it is best explained by a degeneration of elements within the saccade generator (see Büttner-Ennever Chapter 6.14, this volume; Winkelman et al., 1983). Damage to the excitatory burst neurons (EBNs), the omnipause neurons (OPNs), or the inhibitory burst neurons (IBNs) would all contribute to a slowing of horizontal saccades (Strassman et al., 1986; Kaneko, 1996; Leigh and Zee, 2006). These functional cell groups of the saccade generator have also been localized in humans (Horn et al., 1994, 1995, 1996; Büttner-Ennever and Horn 2004). In an attempt to further elucidate the causes of slow saccades, we have analysed the OPN-, EBN-, and IBN-areas in a patient whose

*Corresponding author. Tel.: (+49) 89 5160 4851 or 4876;
Fax: (+49) 89 5160 4802;
E-mail: jean.buettner-ennever@med.uni-muenchen.de

DOI: 10.1016/S0079-6123(08)00687-0

only neurological symptom was slow horizontal saccades from birth, and whose disorder remained unchanged throughout his lifetime.

Case history and methods

A 15-year-old boy had a history of peculiar head movements from the first year of life, but no other neurological signs. Measurements showed that the patient used head-thrusts to compensate for abnormally slow saccadic eye movements in the horizontal plane (Fig. 1). The quick phases of caloric and optokinetic nystagmus were also abnormally slow. Vertical saccades appeared normal. Smooth pursuit ranged up to 40°/s, and vestibular eye movements evoked by sinusoidal rotation while fixating a target ranged up to 143°/s, exceeding the speed of most ocular saccades. After admission into hospital for cardiac deficiency, a diagnosis of Gaucher disease (type 3) was made from the bone marrow. The patient died approximately 1 month after the examination of head and eye movements (for details see Uemura et al., 1987). Paraffin sections of the formalin fixed brain were stained with haematoxylin–eosin, cresylviolet, periodic acid-Schiff reaction, Klüver-Barrera, Bodian, and, for glial fibrosis, Holzer and GFAP. They were compared with similar sections from control brains with no known neurological deficits.

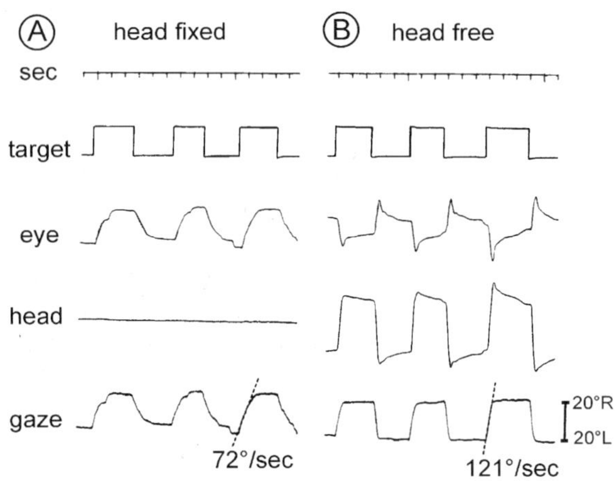

Fig. 1. Measurements of patient's eye movements using DC amplified electro-oculography, and his head movements using a potentiometer coupled to a helmet. Alternate side-to-side gaze test: when the head is fixed (A), the voluntary horizontal saccadic eye movements used to fixate a step-target are slow (72°/s). The patient was able to fixate the target more quickly with his head free (B), using head-thrusts alone (121°/s); note that with the head free, the eyes move with a vestibulo-ocular reflex in the opposite direction to the head and target.

Results

Neuropathological findings

The Holzer sections revealed a feltwork of fine astrocytic fibres on the brainstem midline, which extended from the mesencephalon to the spinal cord (Figs. 2 arrows, 3A and B). Surprisingly it was not visible on Nissl or H and E preparations. Moderate astrocytic gliosis was also evident in the inferior olive and throughout the entire ventricular system under the ependymal layer. There was pronounced granular ependymitis of the ventricle walls throughout the brain — a focal loss of ependymal cells, associated with granular outgrowths of astrocytes caused most likely by a severe infection, such as meningitis (Fig. 3, arrowhead). A fibrous thickening of the leptomeninges was also apparent. Bodian and myelin-stained sections of the caudal pons showed a discrete tangle of axons on the midline, very different from the regular pattern of interlacing fibre bundles in the controls (Fig. 3C, D). There was no obvious neuronal or nerve fibre loss in the brainstem, but there was moderate gliosis accompanied by PAS-granules seen free in the tissue, and within macrophages and nerve cells. The control cases showed no midline gliosis but a weak subependymal gliosis was present in some cases.

Morphometric measurements

No striking changes in the morphology of the EBNs, IBNs, or OPNs could be found to account for the horizontal saccadic palsy.

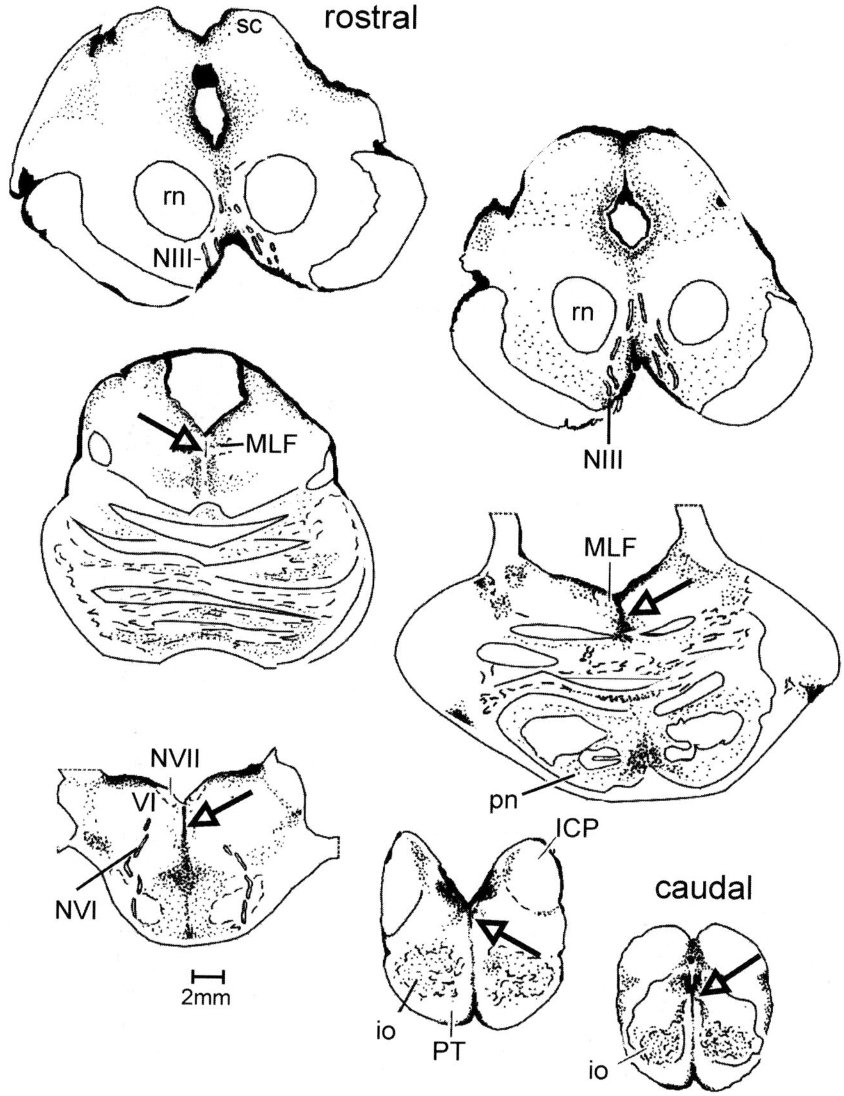

Fig. 2. Drawings of transverse sections (Holzer) through the brainstem to show distribution of gliosis, indicated by black-shading and dots according to its intensity. Note sub-pial, midline, and periventricular gliosis. The fibrous midline plaque in the pons and medulla (arrows) could interfere with the axons or dendrites of the saccade generator crossing the midline. Abbreviations: III, oculomotor nucleus; VI, abducens nucleus; ICP, inferior cerebellar peduncule; io, inferior olive; MLF, medial longitudinal fasciculus; NIII, oculomotor nerve; NVI, abducens nerve; NVII, facial genu; pn, pontine nuclei; PT, pyramidal tract; rn, red nucleus; sc, superior colliculus.

OPN area

Typical OPNs were present in nucleus raphé interpositus (RIP) at the level of the abducens nerve, and no difference in average cross-sectional diameter to normal was found (OPN_{PAT} 23.8 µm SD = 7.2 µm; $N = 250$ compared with OPN_{Norm} 24.1 µm SD = 7.5 µm; $N = 310$); but several were abnormally rounded and shrunken. Their typical horizontal dendrites could not be followed over the midline, probably due to the presence of a dense feltwork of glial fibres.

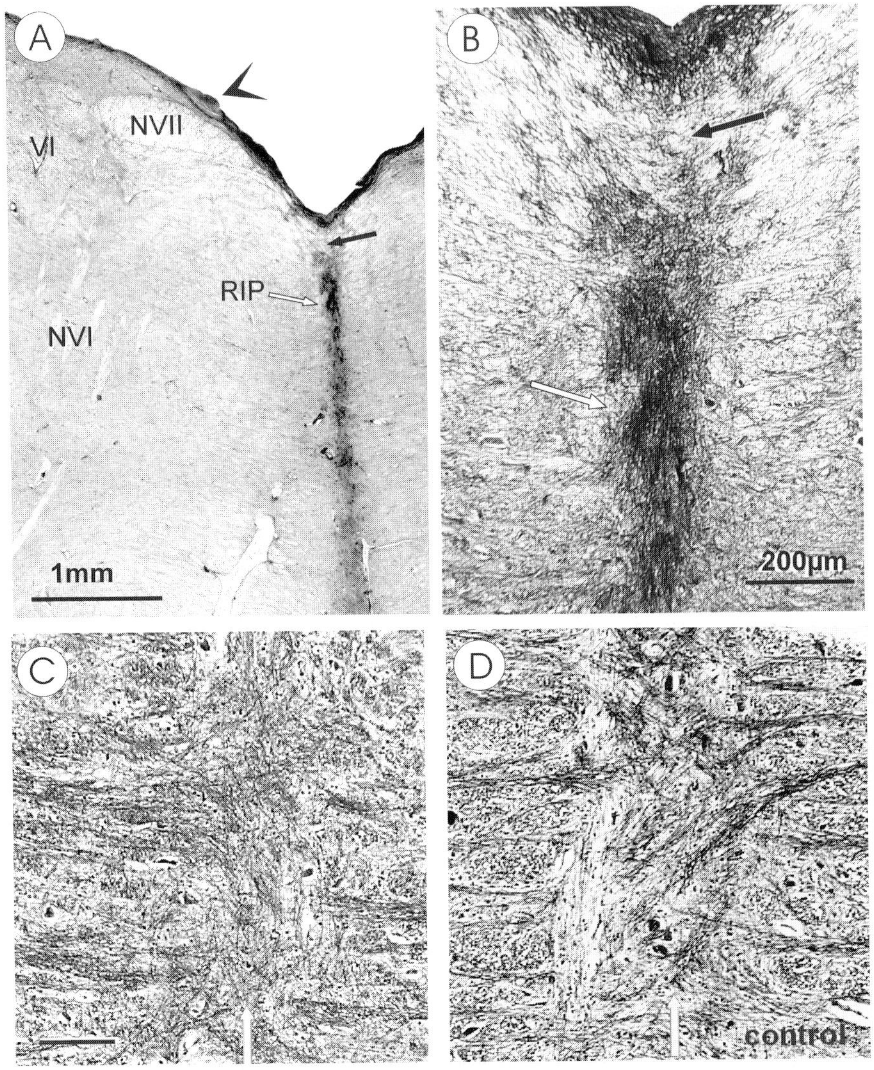

Fig. 3. (A and B) Photomicrographs of Holzer-stained sections of the abducens region in a case with saccadic palsy: glial plaque on the midline is indicated by white arrows. The black arrows indicate a gap in the gliosis just below the fourth ventricle where fibre bundles cross the midline. Note evidence of a granular ependymitis (arrow head) around the ventricle. (C) Myelin stained (Klüver-Barrera) sections of case with saccadic palsy showing axons crossing the midline in a disorderly fashion. (D) Orderly crossing of myelinated axons the midline in caudal pons in normal. Calibration in C and D 100 µm.

EBN area

Medium-sized neurons in the dorsal nucleus reticularis pontis caudalis, the putative EBNs (Horn et al., 1995), were somewhat scarcer than in normals, slightly rounded and shrunken, but measurements showed no significant difference between the average cell diameter in this area to normals (EBN_{PAT} 18.2 µm SD = 6.9 µm; $N = 250$ compared with EBN_{Norm} 20.1 µm SD = 7.2 µm; $N = 310$).

IBN area

In the IBN area, again no statistical difference in the diameter of the total cell population to normals could be seen (IBN_{PAT} 19.7 µm SD = 6.1 µm;

$N = 150$ compared with IBN_{Norm} 25.2 μm SD = 7.8 μm; $N = 266$). However cells were reduced in number and some shrunken.

Discussion

Midsagittal gliosis of the brainstem

The main abnormal features of the brain with horizontal saccadic palsy were a dense plaque of glial fibres on the midline of the dorsal tegmentum and a widespread granular ependymitis. The midline gliosis disrupted, if not prevented, the passage of axons crossing the midline, and was most pronounced at the level of the caudal pons, where OPN and IBN axons decussate and OPN dendrites spread across the midline (Strassman et al., 1986; Ohgaki et al., 1987). It seems likely that it could account for lack of synchrony in the OPNs and EBNs, and result in the slowing of horizontal saccades (Fig. 4) (Leigh and Zee, 2006). The rostral mesencephalon was not affected by the scar, which might explain the preservation of vertical saccades (Solomon et al., 2004).

One weakness of this hypothesis for the slowing of saccades is that if the midline was effectively disrupted by the gliosis then one would expect the patient to have had internuclear ophthalmoplegia, and defective vestibulo-ocular reflexes, which were not seen. A possible explanation is that careful inspection of the sections shows a gap in the scar just below the floor of the 4th ventricle through which some fibres decussate, possibly those of the internuclear and vestibular-ocular pathways (Fig. 3A, B, black arrows) (McCrea et al., 1986, 1987).

The origin of the thick plaque of glial fibres in the midsagittal plane is unclear. It may be associated with the granular ependymitis, which is evidence of a serious illness, probably prenatal meningitis, since his mother noticed no major illness during his lifetime. The midsagittal tegmental region is known to contain specialized glycogen-rich ependyma, extending from the level of the oculomotor complex down through the brainstem to the spinal cord (Sansone 1980; Leonhardt et al., 1987 see Figs. 1–5 in this reference). Furthermore glycogen-rich glial cells are highly sensitive to

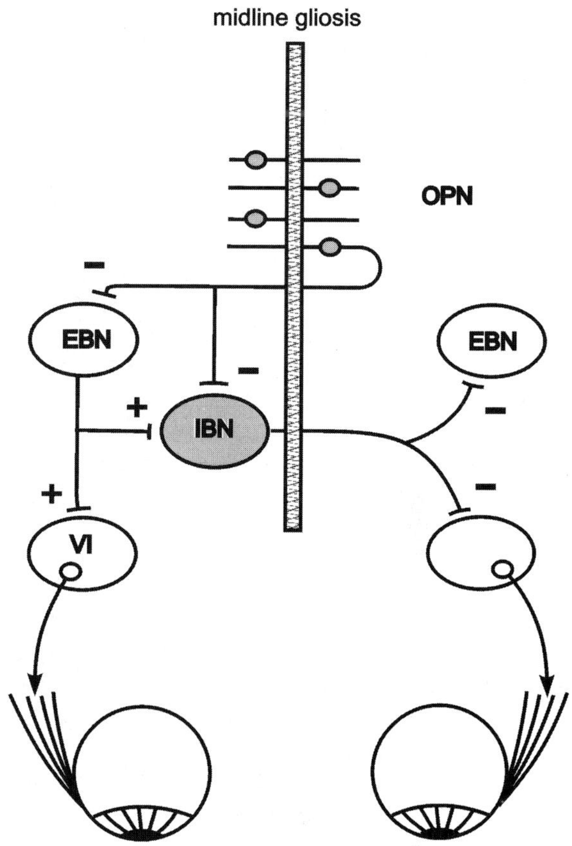

Fig. 4. Simplified diagram of the omnipause neuron (OPN) connections to the excitatory (EBN) and inhibitory burst neurons (IBN) of the horizontal saccade generator. Crossing dendrites and axons of OPNs and IBN may be obstructed at the midline by a glial plaque, and cause slow horizontal saccades.

metabolic disturbances such as ischaemia and hypoxia (Ibrahim, 1972).

Gaucher disease

In our patient the slow saccades were present and unaltered since birth. They are most easily accounted for by the midline gliosis, since no major neuronal abnormalities were found in the saccade generator or elsewhere in the brainstem. However the additional diagnosis of Gaucher's disease in our patient complicates the picture, since it can also cause slowing of horizontal saccades. Gaucher's disease is a disorder of glycolipid metabolism, and is the most common lysosomal storage disease (McGovern, 1998). In type 3 the disease progresses slowly and may include neurological symptoms such as mental retardation, seizure activity, ataxia, and oculomotor disorders (Winkelman et al., 1983; Kaye et al., 1986; Conradi et al., 1991; Solomon et al., 2004; Leigh and Zee, 2006). Initial eye movement deficits are usually horizontal gaze palsy, often sparing vertical saccades, vestibulo-ocular reflex, and smooth pursuit. Our patient had slowed saccades all his life, and did not show the progressive pattern of eye movement disorders typical for Gaucher type 3. For this reason we consider the Gaucher disease unlikely to be the direct cause of the horizontal saccadic palsy.

Personal inspection of brainstem sections from two type 3 Gaucher cases with eye movement deficits (Winkelman et al., 1983, patient HP; Conradi et al., 1991) revealed severe demyelinization and cell loss in the pontine tegmentum of both, where no fibres crossing the midline were found; other regions like the facial nerve or ventral pons were better preserved and stained well. There was no granular ependymitis of the ventricular walls. This alone could account for the oculomotor deficits. In contrast we found no loss of neurons or fibres in our case. This further supports our hypothesis that Gaucher disease is unlikely to be the direct cause of the horizontal saccadic palsy. Interestingly both Gaucher cases showed a midline gliosis of the brainstem, similar but less pronounced than our case. Perhaps the metabolic disturbance in Gaucher also activates the specialized glial cells of the midline, inducing a gliosis, and hence contributes to their eye movement deficits.

In summary, we propose a new mechanism that causes the slowing of horizontal saccades, a gliosis of the brainstem midline ependyma, part of which affects neural processes of the saccade generator crossing the midline. The hypothesis is in accordance with saccadic models (Scudder, 1988; Scudder et al., 1988; Leigh and Zee, 2006), but further lines of supporting evidence are needed to validate it.

Acknowledgements

We are grateful to Professor P. Mehraein for his continual support, to Dr. M. Winkelman, Case Western Reserve University, Cleveland, USA, and Dr. N. Conradi, Gothenburg University, Sweden, for permitting us to inspect sections from their Gaucher cases. This work is supported by the German Research Council DFG HO 1639/4-2.

References

Büttner-Ennever, J.A. and Horn, A.K.E. (2004) Reticular formation: eye movements, gaze and blinks. In: Paxinos G. and Mai J.K. (Eds.), The Human Nervous System. Elsevier Academic Press, Amsterdam, pp. 479–510.

Conradi, N.G., Kyllerman, M., Mansson, J.E., Percy, A.K. and Svennerholm, L. (1991) Late-infantile Gaucher disease in a child with myoclonus and bulbar signs: neuropathological and neurochemical findings. Acta Neuropathol., 82: 152–157.

Horn, A.K.E., Büttner-Ennever, J.A. and Büttner, U. (1996) Saccadic premotor neurons in the brainstem: functional neuroanatomy and clinical implications. Neuroophthalmology, 16: 229–240.

Horn, A.K.E., Büttner-Ennever, J.A., Suzuki, Y. and Henn, V. (1995) Histological identification of premotor neurons for horizontal saccades in monkey and man by parvalbumin immunostaining. J. Comp. Neurol., 359: 350–363.

Horn, A.K.E., Büttner-Ennever, J.A., Wahle, P. and Reichenberger, I. (1994) Neurotransmitter profile of saccadic omnipause neurons in nucleus raphe interpositus. J. Neurosci., 14: 2032–2046.

Ibrahim, M.Z.M. (1972) The response of the brain to hypoxia and ischaemia. J. Neurol. Sci., 17: 271–279.

Kaneko, C.R.S. (1996) Effect of ibotenic acid lesions of the omnipause neurons on saccadic eye movements in Rhesus macaques. J. Neurophysiol., 75: 2229–2242.

Kaye, E.M., Ullman, M.D., Wilson, E.R. and Barranger, J.A. (1986) Type 2 and Type 3 Gaucher disease: a morphological and biochemical study. Ann. Neurol., 20: 223–230.

Leigh, R.J. and Zee, D.S. (2006) The Neurology of Eye Movements. Oxford University Press, New York.

Leonhardt, H., Krisch, B. and Erhardt, H. (1987) Organization of the neuroglia in the midsagittal plane of the central nervous system: a speculative report. In: Scharrer B., Korf H.W. and Hartwig H.-G. (Eds.), Functional Morphology of Neuroendocrine Systems. Springer-Verlag, Berlin, pp. 175–187.

McCrea, R.A., Strassman, A. and Highstein, S.M. (1986) Morphology and physiology of abducens motoneurons and internuclear neurons intracellulary injected with horseradish peroxidase in alert squirrel monkey. J. Comp. Neurol., 243: 291–308.

McCrea, R.A., Strassman, E., May, E. and Highstein, S.M. (1987) Anatomical and physiological characteristics of vestibular neurons mediating the horizontal vestibulo-ocular reflex of the squirrel monkey. J. Comp. Neurol., 264: 547–570.

McGovern, M.M. (1998) Lysosomal storage diseases. In: Fauci A.S., Braunwald E. and Isselbacher K.J. (Eds.), Harrison's Principles of internal medicine (14th edn.). Mc Graw Hill, New York, pp. 2169–2176.

Ohgaki, T., Curthoys, I.S. and Markham, C.H. (1987) Anatomy of physiologically identified eye-movement-related pause neurons in the cat: pontomedullary region. J. Comp. Neurol., 266: 56–72.

Sansone, F.M. (1980) An ultrastructural study of the craniocaudal continuation of the glycogen body. J. Morphol., 163: 45–58.

Scudder, C.A. (1988) A new local feedback model of the saccadic burst generator. J. Neurophysiol., 59: 1455–1475.

Scudder, C.A., Fuchs, A.F. and Langer, T.P. (1988) Characteristics and functional identification of saccadic inhibitory burst neurons in the alert monkey. J. Neurophysiol., 59: 1430–1454.

Solomon, D., Winkelman, A. C., Zee, D. S., Gray, L. and Büttner-Ennever, J. A. (2004) Niemann-Pick Type C disease in two affected sisters: ocular motor recordings and brainstem neuropathology. Ann. N.Y. Acad. Sci; 1039: 436–445.

Strassman, A., Highstein, S.M. and McCrea, R.A. (1986) Anatomy and physiology of saccadic burst neurons in the alert squirrel monkey. II. Inhibitory burst neurons. J. Comp. Neurol., 249: 358–380.

Uemura, T., Tateishi, J. and Kanaseki, T. (1987) Eye-head coordination in a case of congenital oculomotor apraxia: a clinicopathological study. In: Graham M.D. and Kemnik J.L. (Eds.), The Vestibular System: Neurophysiologic and Clinical Research. Raven Press, New York.

Winkelman, M.D., Banker, B.Q., Victor, M. and Moser, H.W. (1983) Non-infantile neuronopathic Gaucher's disease: a clinicopathologic study. Neurology, 33: 994–1008.

Subject Index

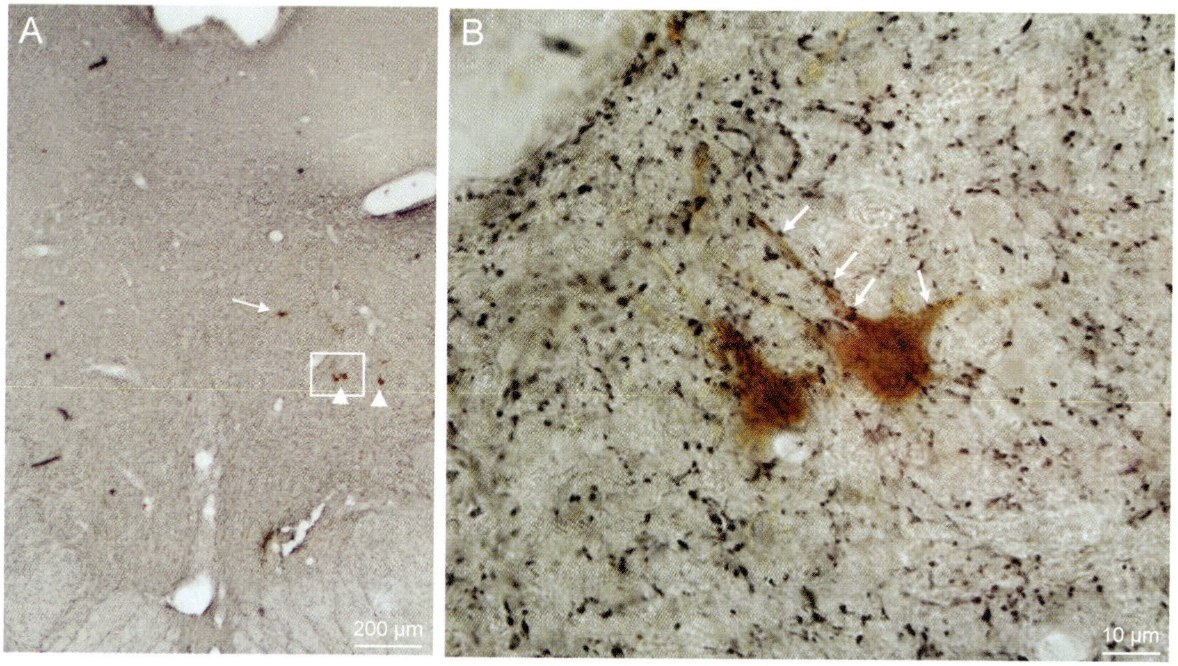

Plate 1.2.1. GAD staining in synaptic boutons (black) was detected adjacent to MIF and SIF motoneurons labelled by cholera toxin (brown). Panel A: Low magnification view of the caudal oculomotor nucleus shows CTB-stained MIF motoneurons (arrow) and SIF motoneurons (arrowheads). Panel B: Higher magnification view of area in white box from panel A shows CTB-stained SIF neurons which are contacted by numerous GAD-positive terminals (arrows).

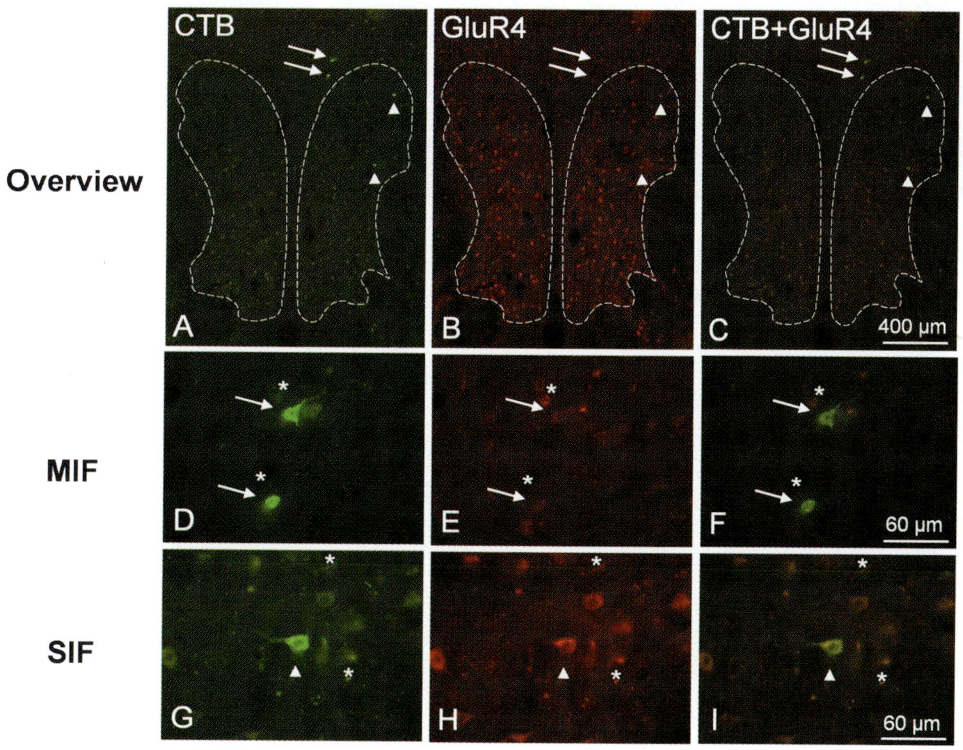

Plate 1.2.2. GluR4 expression (red) was detected in SIF but not MIF motoneurons labelled by cholera toxin (green). Panels A, B, and C show CTB-positive, GluR4-positive, or co-labelled neurons, respectively. The classic oculomotor nucleus is outlined in white. MIF motoneurons are indicated by arrows and SIF motoneurons are indicated by arrow heads. Panels D, E, and F show higher magnification of CTB-stained but GluR4-negative MIF motoneurons. Panels G, H, and I show higher magnification of a CTB-stained and GluR4-positive SIF motoneuron. Blood vessels are indicated by asterisks for reference.

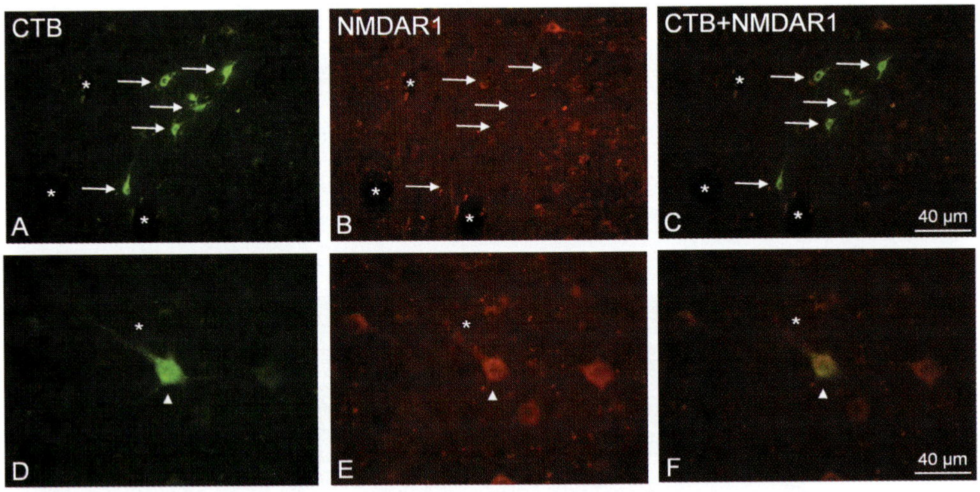

Plate 1.2.3. NMDAR1-immunoreactivity (red) was abundantly detected in SIF but not in MIF motoneurons labelled by CTB (green). Panels A, B, and C show CTB-stained but NMDAR1-negative MIF motoneurons (arrows). Panels D, E, and F show a CTB-stained and NMDAR1-positive SIF motoneuron (arrowhead). Blood vessels are indicated by asterisks for reference.

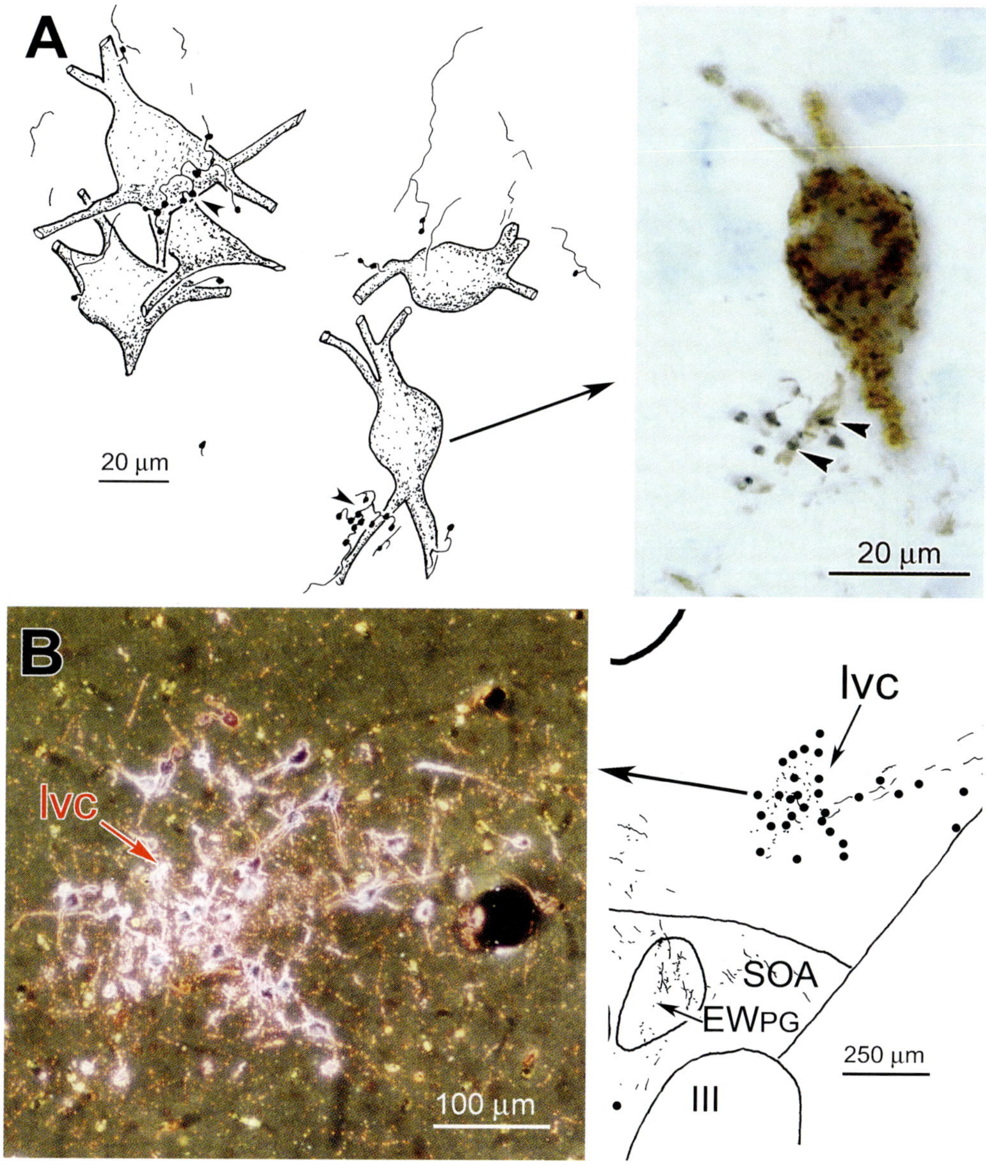

Plate 2.3.3. A: Close associations (arrowheads) between biocytin labelled pretectal axonal boutons and WGA-HRP labelled preganglionic motoneurons. B: WGA-HRP labelled neurons and terminals in the lateral visceral cell column (lvc) following an injection in the pretectum (see Fig. 4). Photomicrograph in B taken with crossed polarizers.

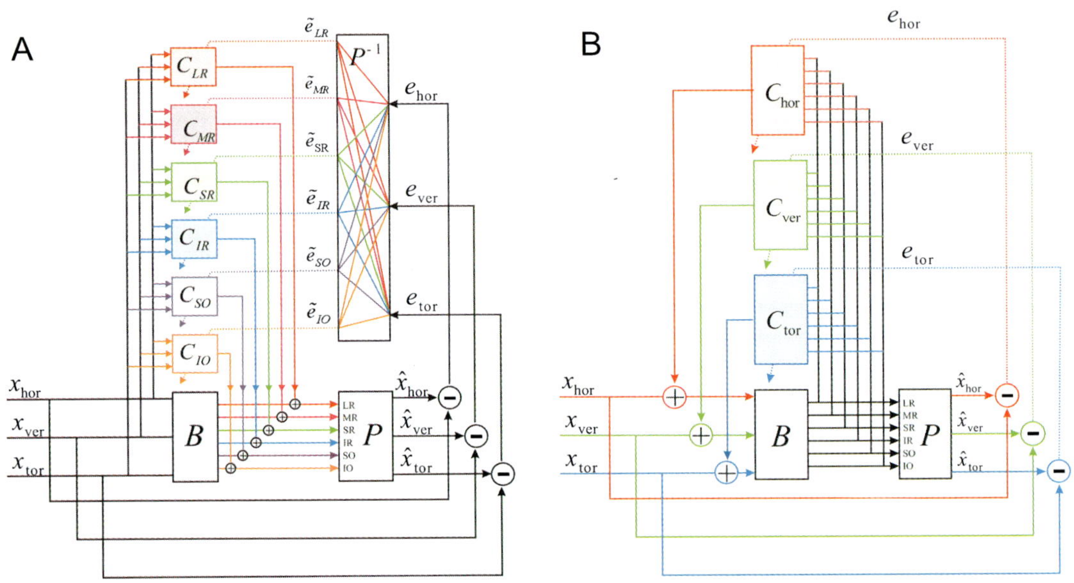

Plate 3.5.3. Alternative architectures applied to simulated VOR in 3D. The vestibular system recovers three components: x_{hor}, x_{ver}, and x_{tor} (horizontal, vertical, and torsional) of head angular velocity. These are processed by the brainstem and cerebellum to produce motor commands to the six extraocular muscles so as to stabilise the eyes rotational position in space. The motor plant is a 3×6 matrix transfer function P and the brainstem contribution to the VOR is a 6×3 matrix transfer function B. (A) Feedback error architecture: the cerebellum takes the three vestibular signals as input and supplies corrections to the six motor commands. Colour is used to highlight the "motor space" modularity. For example, the green component shows those Purkinje cells that contribute to the superior rectus (SR) muscle command, and which require the corresponding motor command error $\tilde{e}_{\mathrm{SR}}$ as training signal on their climbing fibres. This signal must be reconstituted from the horizontal, vertical, and torsional components of retinal slip. (B) Recurrent architecture: the cerebellar takes the six motor commands as input and supplies corrections to the three vestibular signals. Colour is used here to highlight the "task space" modularity. For example, the green component shows Purkinje cells contributing to vertical eye motion; these require the vertical component of optic flow as training signal. (Adapted from Porrill et al., 2004.)

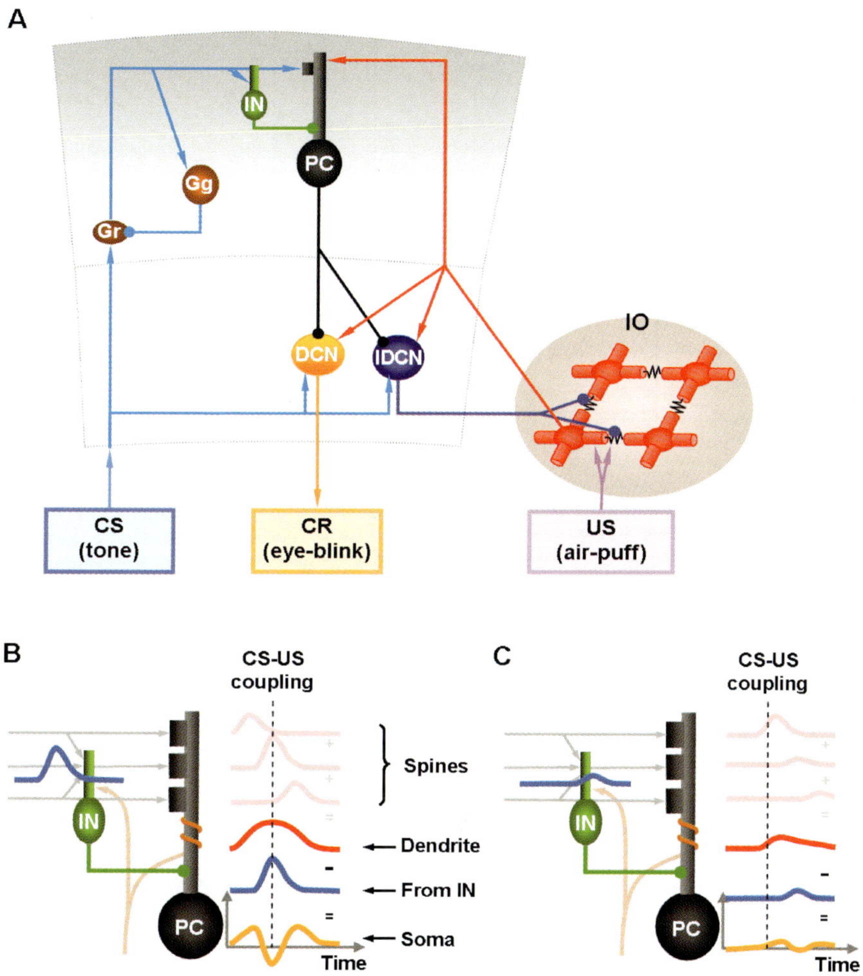

Plate 3.12.1. Schematic of model. (A) Cerebellar modules and IO. See the text for detailed explanations. The zigzags (black) between IO neurons (red) represent gap junctions. Excitatory and inhibitory pathways are represented by triangular heads and round heads, respectively. (B) and (C) PC–IN pairs learn the timing of CS–US coactivation. When the CFs activate after the PF signal with a certain delay, the PC–IN pairs with a matching internal delay timing, such as the one in (B), increase their mGluR-mediated excitability. The "CS–US coupling" with a vertical broken line in (B) and (C) shows the timing. After learning, the increased excitability leads to a big modulation of activity in the PCs of those pairs (the orange wave). This synchronized modulation of a population of PCs leads to the generation of timed CR. For the PC–IN pairs with different internal delay timings, such as the long delay timing in (C), the CS–US coactivation does not change their excitability. Theses pairs remain relatively inactive therefore do not contribute to CR.

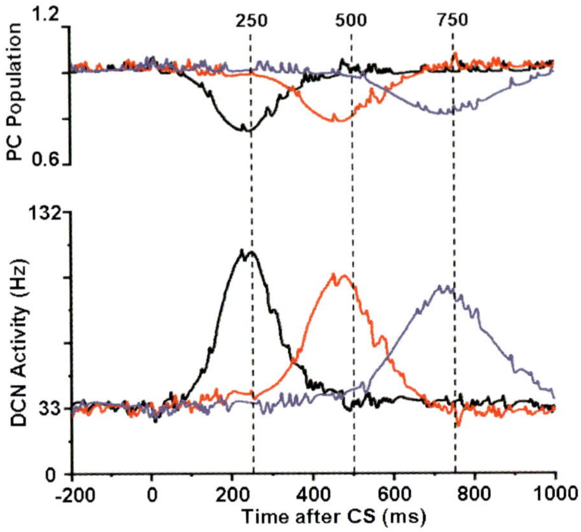

Plate 3.12.2. Simulation of delay eyeblink conditioning. Upper panel: Input given to the simulated DCN neuron from the PC population with different ISIs (black curve: 250 ms, red: 500 ms, blue: 750 ms). Lower panel: DCN activities at different ISIs. PC population reduces the inhibitory output near the arrival of the air-puff US signals (dashed lines). The model DCN peaks near the arrival of the US with appropriate delays because of the reduction of PC inhibition.

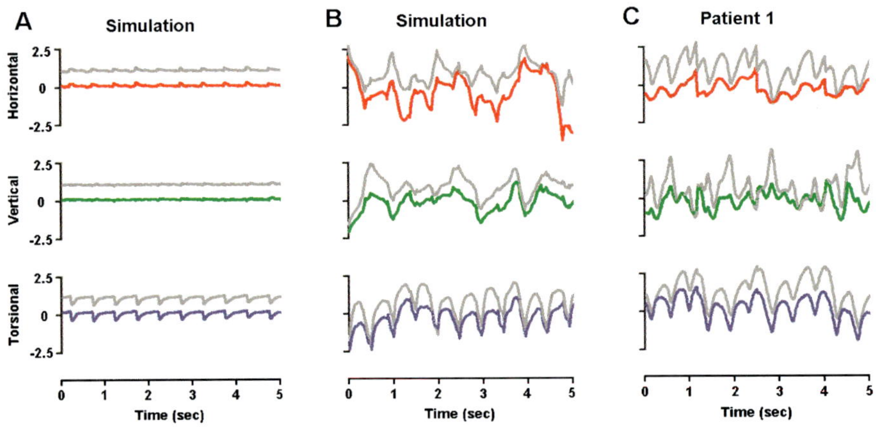

Plate 3.12.4. Three-dimensional eye movement simulations of the OPT without (A) and with (B) the cerebellar cortical learning. (A) The model's simulation with IO hypertrophy but without cerebellar cortical learning shows small, jerky periodic eye movements. (B) When the simulation includes both the IO hypertrophy and the cerebellar cortical learning, the model makes larger, smoother, and more irregular movements, which are qualitatively similar to eye movements recorded in a patient with OPT (C). Gray traces represent right eye, coloured traces represent left eye.

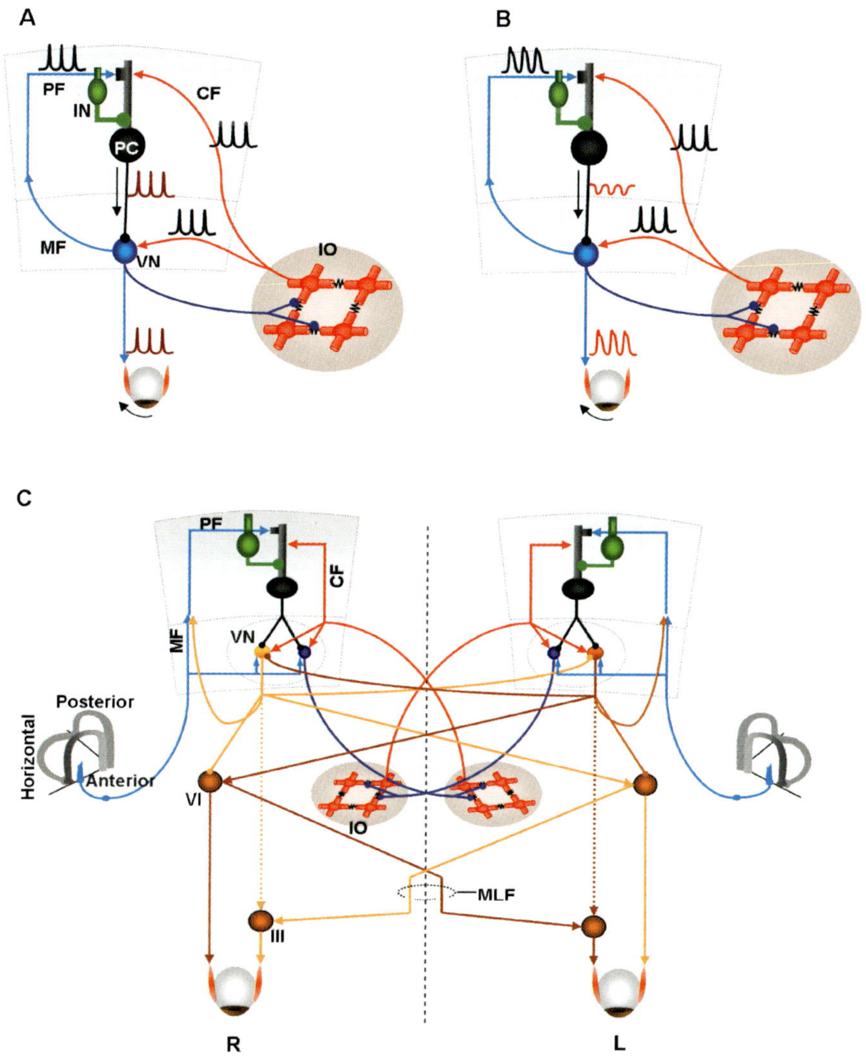

Plate 3.13.1. Summary of the cerebellar circuit that learns the sequences of pulses from the hypertrophied inferior olive. (A) Initial stage of learning. At the beginning of learning, the pulses from the IO (black spiky curves) are simply transmitted to the cerebellar cortex via CFs and PFs. The PC simply reacts to the incoming signals by discharging the same form of spikes as that of the inputs. Note that just one projection from the IO is represented for simplicity. Also for simplicity, just one of three (anterior, posterior, and horizontal) channels is shown. Excitatory and inhibitory projections are indicated by arrow heads and spot heads, respectively. (B) Learned state of the model. The coincidental arrival of PF and CF signals train the cortical PC-IN modules to learn the periodicity of the pulses. After learning, each MF pulse prompts the PC population to pause (the red wavy curve) at the expected time of the next CF signal. This periodic pause of the PC population thereby disinhibits the target vestibular neurons. The vestibular neurons now are driven from both the IO (the black spiky curve) and the PC population (the red wavy curve), and generates a signal (the red curve above the eye ball) that is a mixture of the IO pulse and the PC modulation. (C) A more complete circuit diagram of the model. The diagram has a more detailed representation of the model showing left and right cerebellum and related connections to oculomotor structures. Both excitatory and inhibitory neurons are illustrated in VN. Note that while the inhibitory neuron connects only to the contralateral IO, the excitatory neuron sends its axons to oculomotor neurons and to the cerebellar cortex as a MF. The head rotation triggered phase reset pathway (ear canal → VN → IO) is also illustrated. The indirect projection from VN to the third cranial nucleus (III) via the contralateral sixth cranial nucleus (VI) is assumed to be stronger than the direct projection from the VN to III (dotted line). MLF: medial longitudinal fasciculus.

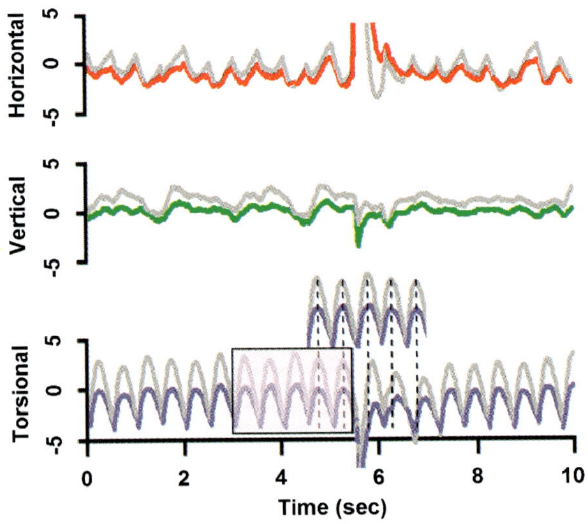

Plate 3.13.2. Simulation of head impulse in OPT. The relatively periodic torsional component of the OPT shifts its phase after the head rotation. For comparison, the five pre-head perturbation wavelets (in box) have been displaced over around the onset of the head perturbation. The simulated head perturbation is a quick horizontal rotation. The overshooting horizontal trace marks the onset of the head rotation. Note that the post-head rotation peaks of the wavelets now do not align with the displaced ones. Grey traces represent right eye, coloured traces represent left eye. Vertical scale: deg/s.

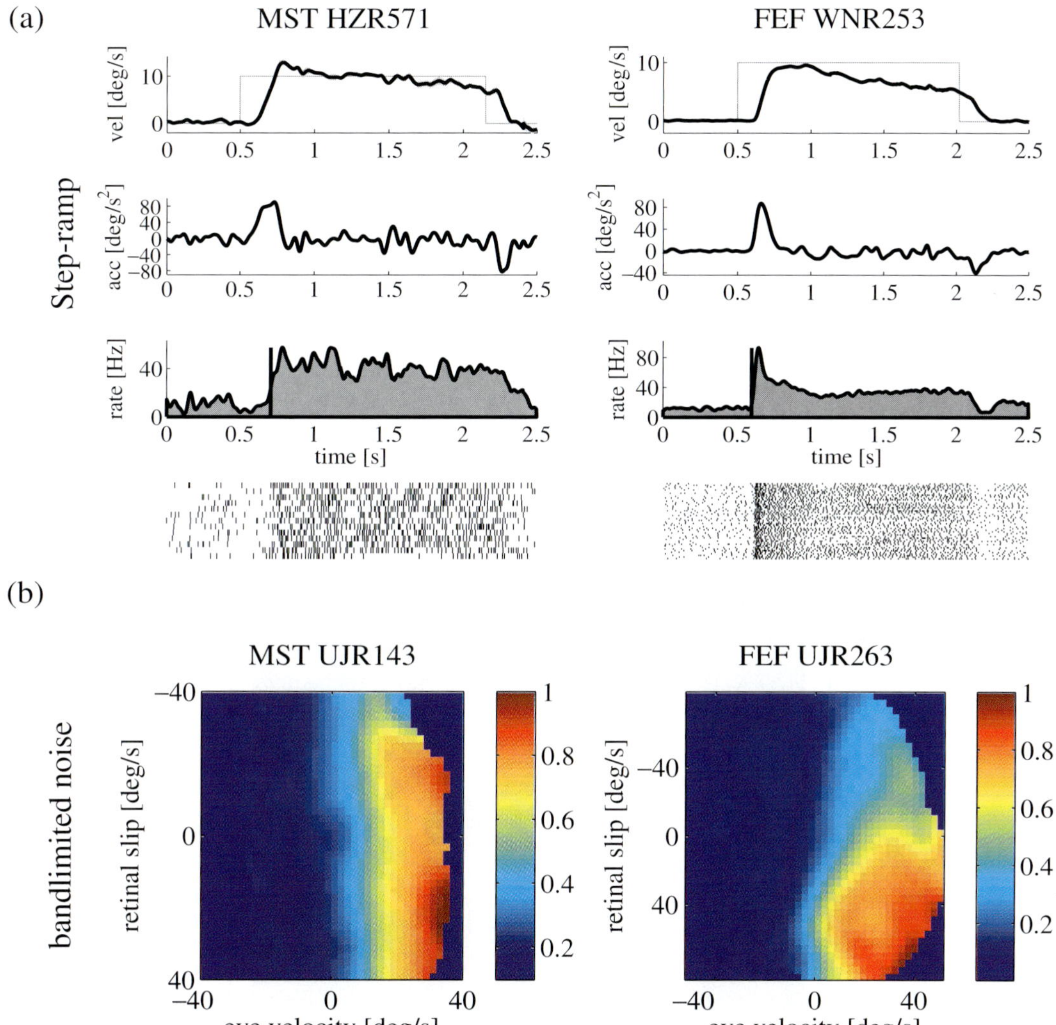

Plate 4.4.2. (a) Typical response behaviour of MST and FEF neurons during step-ramp smooth pursuit. From top to bottom: eye/target velocity, eye acceleration, mean firing rate, spike raster plot. (b) Spike probability (normalized) against eye velocity and retinal slip during pursuit of a bandlimited Gaussian noise stimulus (details see text). Only the FEF neuron shows a curvilinear sensitivity in stimulus space.

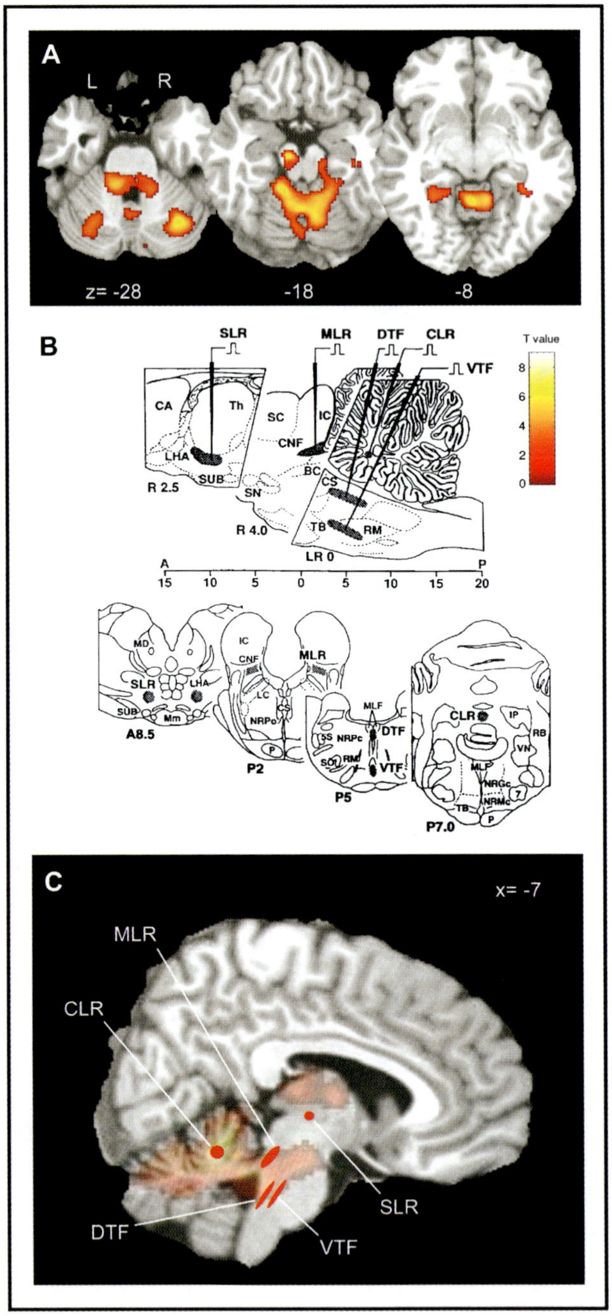

Plate 4.20.1. Infratentorial BOLD-signal increases during human locomotion and locomotor regions in the cat. (A) Statistical parametric maps are shown for the contrast locomotion (mental imagery of walking and running) vs. lying in 26 subjects ($p<0.01$, FDR). The slices selected show bilateral activity in the dorsal brainstem, in the cerebellar vermis and hemispheres, as well as in the parahippocampal gyrus ($z=-8$). (B) The anatomical locations of the subthalamic locomotor region (SLR), the mesencephalic locomotor region (MLR), the cerebellar locomotor region (CLR), the dorsal tegmental field (DTF), and the ventral tegmental field (VTF) in the cat are shown. The horizontal axis represents the rostrocaudal Horsley–Clarke coordinates. 5S, spinal trigeminal nucleus; 7, facial nucleus; BC, brachium conjunctivum; CA, caudate nucleus; CNF, cuneiform nucleus; CS, nucleus centralis superior; IC, inferior colliculus; IP, nucleus interpositus; LHA, lateral hypothalamic area; LC, nucleus locus caeruleus; MD, mediodorsal nucleus of the thalamus; MLF, medial longitudinal fasciculus; Mm, mammillary body; NRGc, nucleus reticularis gigantocellularis; NRMc, nucleus reticularis magnocellularis; NRPc, nucleus reticularis pontis caudalis; NRPo, nucleus reticularis pontis oralis; P, pyramidal tract; RB, restiform body; RM, nucleus raphe magnus; SUB, subthalamic nucleus; SC, superior colliculus; SO, superior olive; SN, substantia nigra; TB, trapezoid body; Th, thalamus; VN, vestibular nuclei. (C) Sagittal section of the human brain with transparent overlay of BOLD-signal increases during locomotion ($p<0.05$, FDR). The anatomical locations of the cat locomotor regions are superimposed. B adapted from Mori et al. (2001).

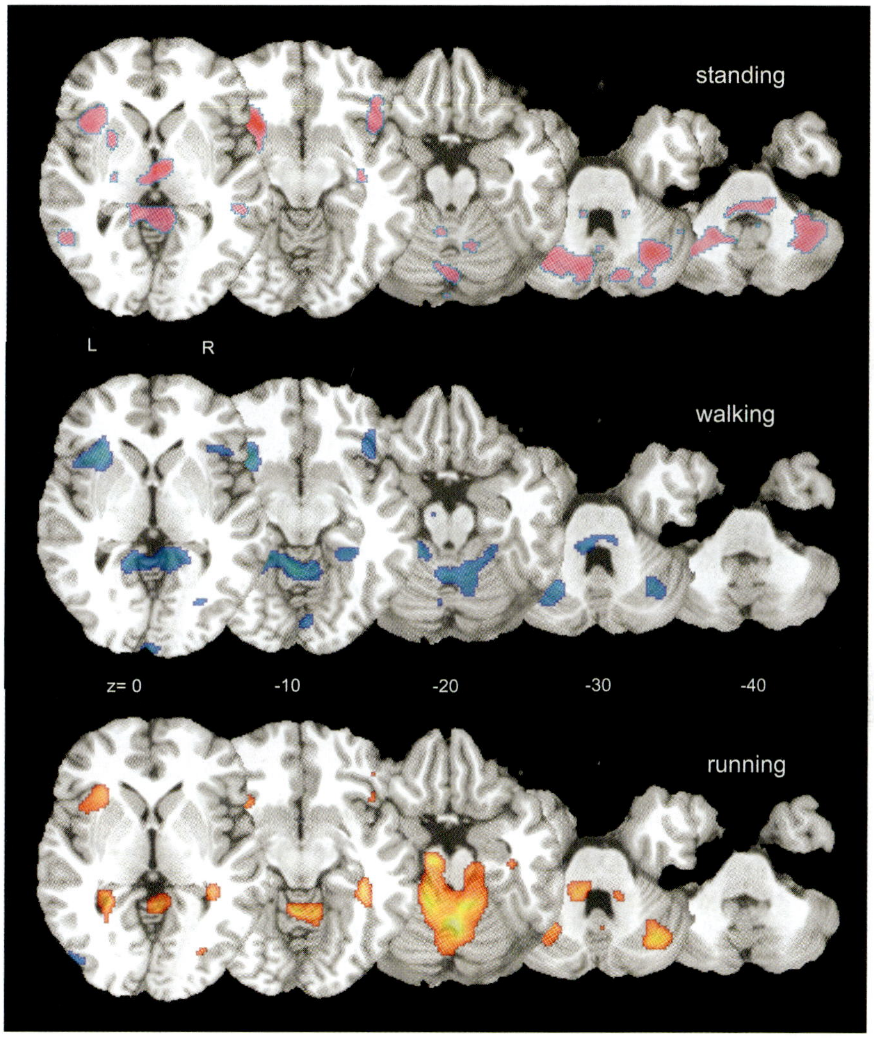

Plate 4.20.2. BOLD-signal increases during mental imagery of standing, walking, and running ($p<0.05$, FDR). During standing, activations were prominent in the frontal gyri, precuneus, thalamus, and cerebellar hemispheres. Note the activation in the dorsal pons ($z=-40$, DTF). During walking, activations were found in the frontal gyri, parahippocampal gyri, midline cerebellum, and in the pontine reticular formation. During running, a large cluster of activation was found in the cerebellum, extending bilaterally to the midbrain. There was also activation in parahippocampal gyri ($z=0$ and -10).

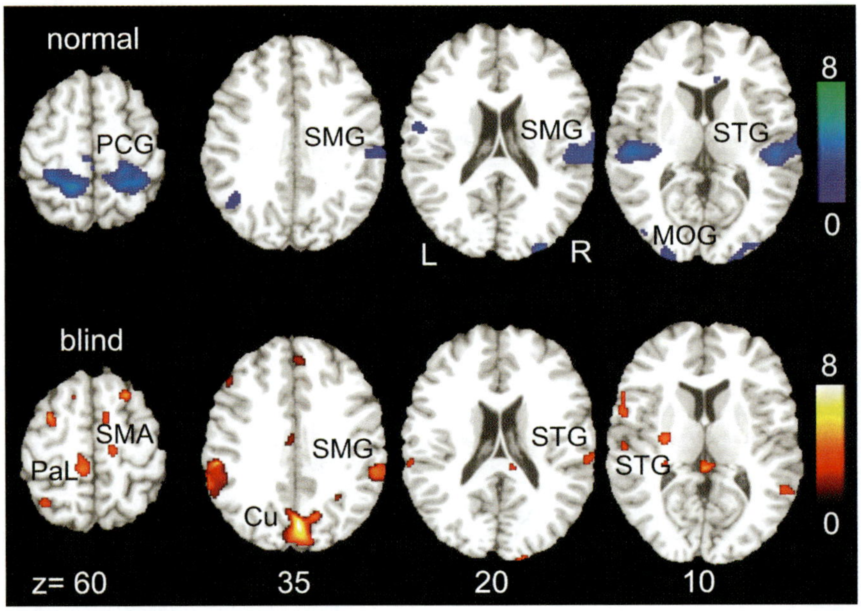

Plate 4.20.3. Multisensory cortical areas during mental imagery of running in sighted and blind subjects ($p<0.01$, uncorrected). The upper panel shows BOLD-signal decreases in 26 healthy subjects. Deactivations were detected bilaterally in the post-central (PCG), supramarginal (SMG), superior temporal (STG), and middle occipital gyri (MOG). These cortical areas can be attributed to somatosensory (PCG), vestibular (SMG, STG), and visual (MOG) functions. The lower panel shows BOLD-signal increases in seven congenitally blind subjects. Note that SMG and STG show increased signals. Deactivations were not detected in this group. SMA, supplementary motor area; PaL, paracentrale lobule; Cu, cuneus; R, right; L, left. *T*-value colour coding is indicated on the right side.

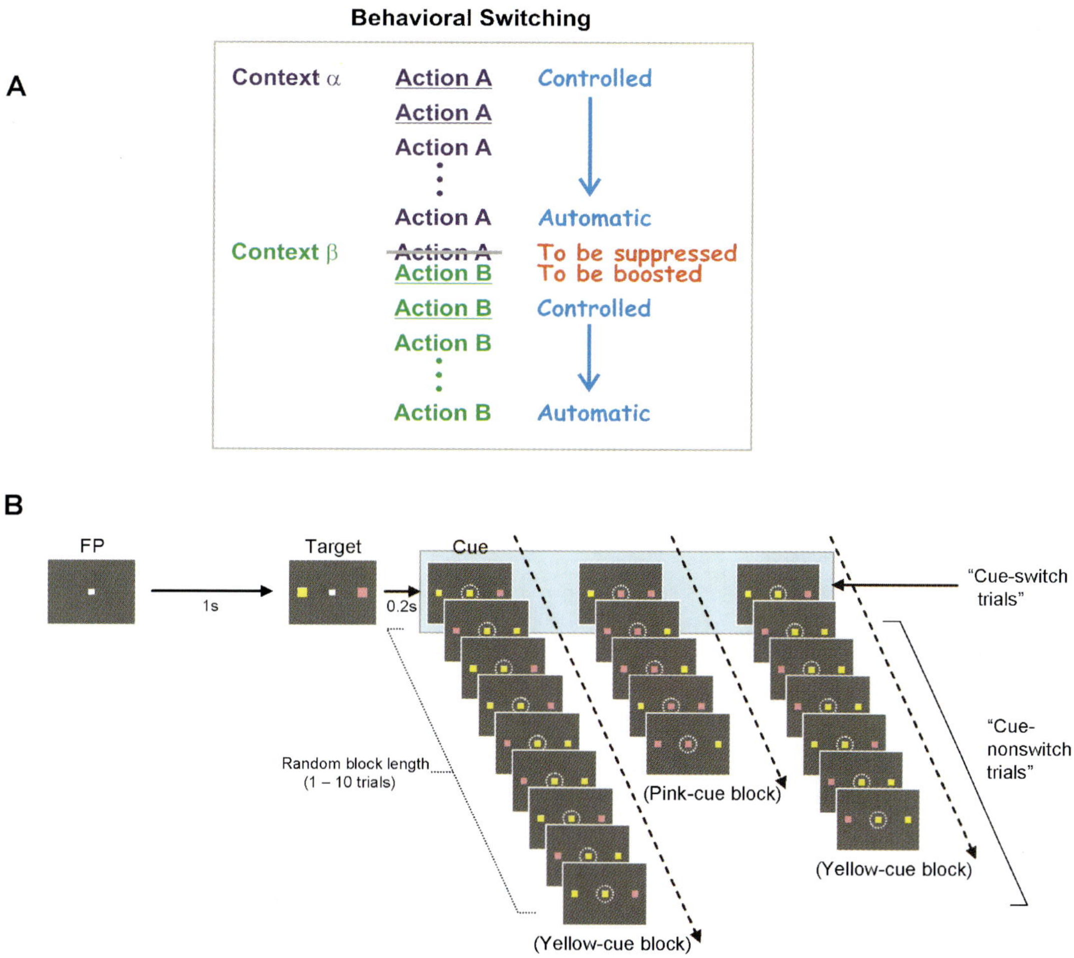

Plate 5.1.1. (A) Scheme of switching from automatic to controlled action. (B) Saccade overriding task — a behavioural task to examine switch-selective neuronal activity.

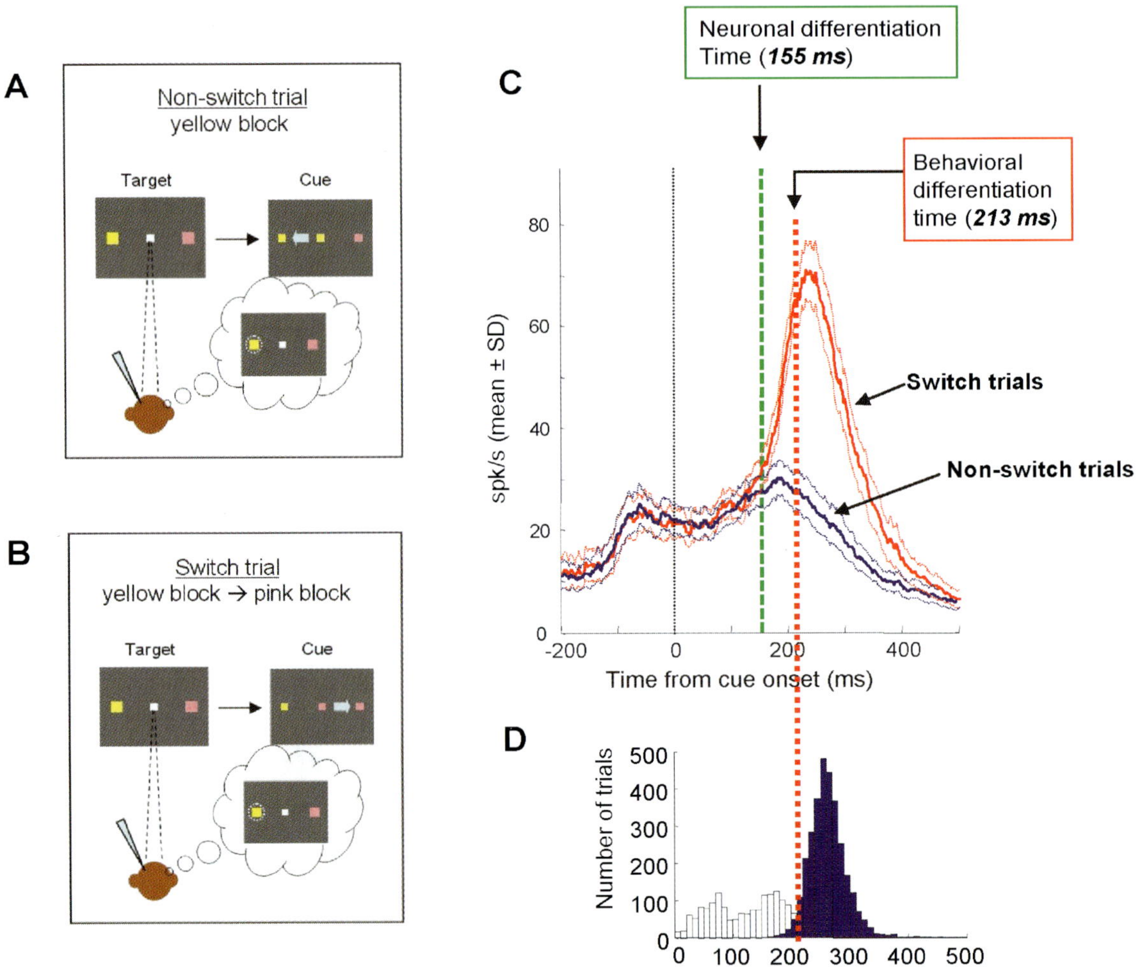

Plate 5.1.3. Switch-selective neuronal activity in the monkey pre-SMA. (A–B) In the saccade overriding task the monkey prepares for a habitual saccade to the stimulus whose colour is the same as the target colour in preceding trials (i.e., yellow). This preparation is valid in nonswitch trials (A), but not in switch trials (B). (C) The population activity of switch-selective pre-SMA neurons for nonswitch trials (blue) and switch trials (red). (D) The distribution of saccade reaction times in switch trials, shown separately for incorrect saccades (light blue) and correct saccades (dark blue).

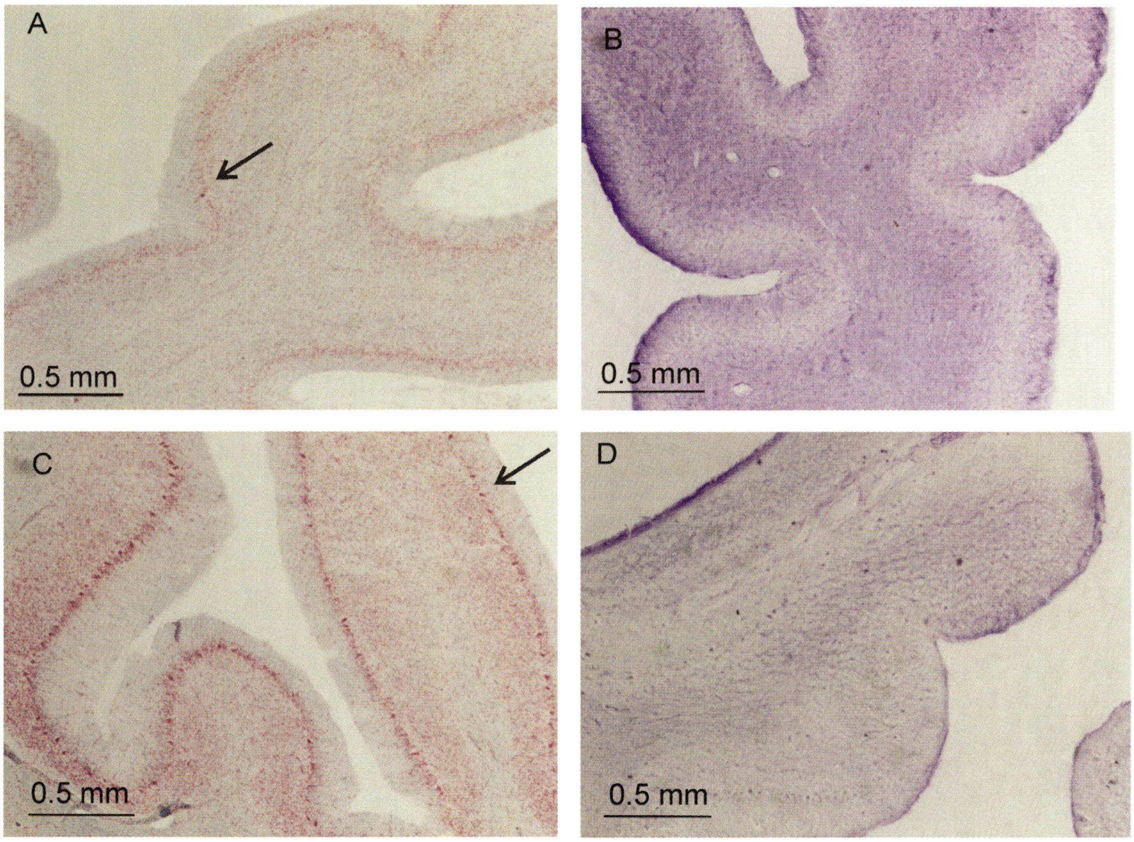

Plate 6.11.1. Cerebellar vermis photomicrographs. (A) Luxol stain through vermis lobules reveals a nearly complete loss of Purkinje cells (one visible at arrow) and atrophy of the granular cell layer. (B) Holzer stain through vermis lobules reveals a thick area of violet staining, consistent with granular cell layer and white matter gliosis. (C) Luxol stain through the nodulus with preservation of Purkinje cells (arrow) and granular cell layer. (D) Holzer stain through the nodulus appears normal without gliosis.

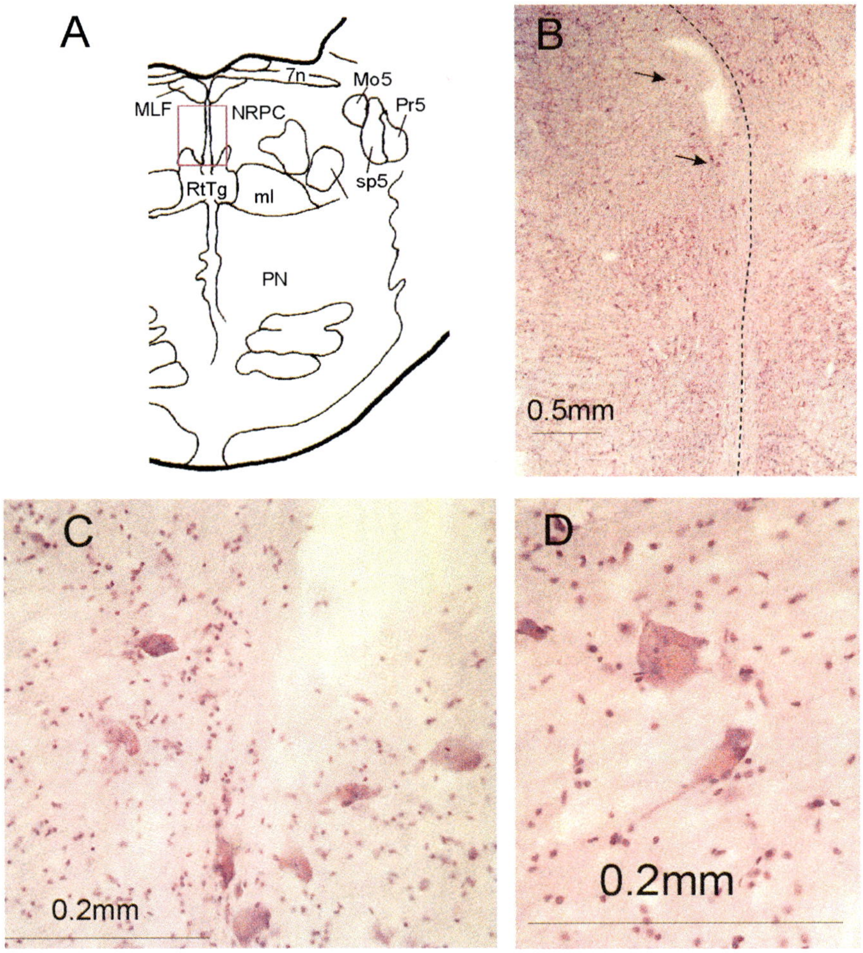

Plate 6.11.2. (A) Drawing of a transverse caudal pontine section demonstrating the region of the nucleus raphe interpositus (RIP) (box), containing OPNs shown in B–D. MLF, medial longitudinal fasciculus; 7*n*, seventh nerve; NRPC, nucleus reticularis pontis caudalis; ml, medial lemniscus; PN, pontine nuclei; Mo5, trigeminal motor nucleus; RtTg, reticulotegmental pontine nucleus. (B) Low-power photomicrograph of the Luxol-stained RIP (box in A). Hatched line is pontine midline, with OPNs just off the midline (black arrows). (C and D) High-powered photomicrographs of Luxol-stained OPNs with massive inclusions.